HUMAN GEOGRAPHY

SEVENTH EDITION

WILLIAM NORTON

OXFORD
UNIVERSITY PRESS

OXFORD

UNIVERSITY PRESS

8 Sampson Mews, Suite 204, Don Mills, Ontario, M3C 0H5
www.oupcanada.com

Oxford University Press is a department of the University of Oxford.
It furthers the University's objective of excellence in research, scholarship,
and education by publishing worldwide in

Oxford   New York

Auckland   Cape Town   Dar es Salaam   Hong Kong   Karachi
Kuala Lumpur   Madrid   Melbourne   Mexico City   Nairobi
New Delhi   Shanghai   Taipei   Toronto

With offices in
Argentina   Austria   Brazil   Chile   Czech Republic   France   Greece
Guatemala   Hungary   Italy   Japan   Poland   Portugal   Singapore
South Korea   Switzerland   Thailand   Turkey   Ukraine   Vietnam

Oxford is a trade mark of Oxford University Press
in the UK and in certain other countries

Published in Canada
by Oxford University Press

Library and Archives Canada Cataloguing in Publication

Norton, William, 1944-
Human geography / William Norton. – 7th ed.

Includes bibliographical references and index.
ISBN 978-0-19-543184-1

1. Human geography–Textbooks. I. Title.

GF41.N67 2010   304.2   C2009-907227-0

Cover image: Philippe Renault/Getty Images

Planet Friendly Publishing
✔ Made in the United States
✔ Printed on Recycled Paper
Text: 10%   Cover: 10%
Learn more: www.greenedition.org

GREEN EDITION

At Oxford University Press Canada we're committed to producing books in an
earth-friendly manner and to helping our customers make greener choices.
According to Environmental Defense's Paper Calculator, by using this innovative paper
instead of conventional papers, we achieved the following environmental benefits:

Trees Saved: 42 • Air Emissions Eliminated: 3,970 pounds
Water Saved: 19,120 gallons • Solid Waste Eliminated: 1,161 pounds

Mixed Sources
Product group from well-managed
forests, controlled sources and
recycled wood or fiber
www.fsc.org   Cert no. SW-COC-002985
©1996 Forest Stewardship Council

FSC

Printed and bound in the United States of America.

1 2 3 4 — 13 12 11 10

Contents

INTRODUCTION: THE ROADS AHEAD   1

1 WHAT IS HUMAN GEOGRAPHY?   17

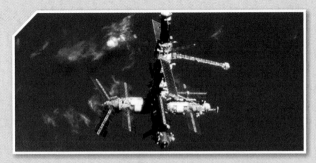

6 CULTURAL IDENTITIES AND LANDSCAPES 211

7 SOCIAL IDENTITIES AND LANDSCAPES 255

8 POLITICAL IDENTITIES AND LANDSCAPES 303

9 A GLOBALIZING WORLD 351

10 AGRICULTURAL LIVES AND LANDSCAPES 389

12 URBAN FORM AND GOVERNANCE 475

11 SETTLEMENT PATTERNS 437

13 LIVING AND WORKING IN CITIES 517

14 INDUSTRIAL LIVES AND LANDSCAPES 557

APPENDIX 1: GLOBAL PHYSICAL GEOGRAPHY 605

CONCLUSION: WHERE NEXT? 595

APPENDIX 2: THE EVOLUTION OF LIFE 618

Figures, Tables, and Boxes

● ● ● ● ● ●

FIGURES

TABLES

BOXES

Links to Other Chapters

As you read through *Human Geography*, seventh edition, you will find that many key themes and topics recur throughout the text. The following lists identify key concepts in each chapter and provide links to chapters where the topic is further discussed. The parenthetical comments specify aspects of the key concept that are expanded on in the other chapter(s).

CHAPTER 1

Because this chapter traces the roots and development of human geography over time, it has links with all later chapters.

- Philosophies, concepts, and techniques: Chapter 2
- Humboldt, Ritter, Vidal, and Sauer: Chapter 3 (human impacts on land); Chapter 6 (approaches to landscape study)
- The history of geography: Chapter 7 (changing types of society); Chapter 8 (European political expansion)

CHAPTER 2

- Philosophies, concepts, and techniques: Chapter 1 (history)
- Environmental determinism and possibilism: Chapter 3 (natural resources and environmental ethics); Chapter 6 (cultural landscapes); Chapter 10 (agricultural regions)
- Positivism: Chapters 10, 11, and 14 (spatial analysis in economic geography, and location theories of von Thünen, Christaller, and Weber, respectively)
- Humanism: Chapters 6 and, especially, 7 (landscapes); Chapter 13 (contemporary urban experience)
- Marxism: Chapter 4 (Marxist theory of population growth); Chapter 5 (world systems approach); Chapter 7 (unequal landscapes); Chapter 10 (Marxist economic geography); Chapter 13 (contemporary urban experience)
- Concepts of space, location, and distance: Chapters 9, 10, 11, 12, and 14 (interaction and location theories)

- Concept of place: Chapters 6, 7, and 13 (understandings of landscape)
- Concept of region: Chapters 4 and 5 (demographic maps); Chapter 6 (cultural regions); Chapter 10 (agricultural regions)
- Spatial scale: Tables and maps generally, especially in Chapters 4 and 5 (population)
- Concept of diffusion: Chapter 5 (migration); Chapter 9 (diffusion processes)
- Concept of perception: Chapter 5 (migration); Chapters 6 and 7 (group identities); Chapter 12 (urban areas)
- Concept of development: Chapter 5 (more and less developed worlds); Chapter 14 (uneven spatial development)
- Concept of discourse: Chapter 7 (feminism and postmodernism)
- Globalization: Chapter 3 (ecosystems and global impacts); Chapter 4 (population growth, fertility decline); Chapter 5 (refugees, disease, more and less developed worlds); Chapter 6 (cultural globalization); Chapter 7 (popular culture); Chapter 8 (political globalization); Chapter 9 (economic globalization); Chapter 10 (agricultural restructuring); Chapter 11 (global cities); Chapter 14 (industrial restructuring)
- Map projections: World maps (all chapters)
- GIS and remote sensing: Data collection and analysis (all chapters)
- Qualitative methods: Chapters 7 and 13
- Quantitative methods: Chapters 9, 10, 11, 12, and 14 (spatial analysis)

CHAPTER 3

- Globalization: Chapter 2 (concepts); Chapter 4 (population growth, fertility decline); Chapter 5 (refugees, disease, more and less developed worlds); Chapter 6 (cultural globalization); Chapter 7 (popular culture); Chapter 8 (political globalization); Chapter 9 (economic globalization); Chapter 10 (agriculture and the world economy); Chapter 11 (global cities); Chapter 14 (industrial restructuring)

- The interrelatedness of all things: All chapters
- Relations between humans and land: Chapter 1 (Humboldt, Ritter, Vidal, and Sauer); Chapter 2 (environmental determinism; possibilism)
- Energy use and technology: Chapter 14 (especially the Industrial Revolution and current circumstances)
- More developed and less developed regions: Chapter 5
- Agricultural revolution: Chapter 6 (beginnings of civilization); Chapter 10 (agriculture)
- Natural resources and human values: Chapter 2 (possibilism)
- Human impacts: Chapters 9, 10, 11, 12, 14 (economic geography); Chapter 10 (agriculture); Chapter 14 (industrial activity)
- The catastrophist–cornucopian debate: Chapters 4, 5, 8

CHAPTER 4

- Population growth—especially the exponential growth that has taken place over the last 250 years: Chapter 5 (population distribution and density, migration)
- Demographic measures (CBR, LE, etc.): Chapter 5 (inequalities)
- Human origins: Appendix 2
- Human impacts on earth: Chapter 3
- Inequalities between the more and less developed worlds: Chapter 5
- Changes in social and cultural landscapes: Chapter 6; Chapter 7
- The political world today: Chapter 8
- Agricultural change: Chapter 10
- Urbanization: Chapter 11
- Industrial change: Chapter 14
- Transportation and trade: Chapter 9
- The catastrophist–cornucopian debate: Chapter 3 (environmental futures); Chapter 5 (refugees, the less developed world); Chapter7 (landscapes, power relations)
- Explaining population growth: Chapter 2 (philosophical approaches); Chapter 7 (feminism)
- Globalization: Chapter 2 (concepts); Chapter 3 (ecosystems and global impacts); Chapter 5 (refugees, disease, more and less developed worlds); Chapter 6 (cultural globalization); Chapter 7 (popular culture); Chapter 8 (political globalization); Chapter 9 (economic globalization); Chapter 10 (agriculture and the world economy); Chapter 11 (global cities); Chapter 14 (industrial restructuring)

CHAPTER 5

- Globalization: Chapter 2 (concepts); Chapter 3 (ecosystems and global impacts); Chapter 4 (population growth, fertility decline); Chapter 6 (cultural globalization); Chapter 7 (popular culture); Chapter 8 (political globalization); Chapter 9 (economic globalization); Chapter 10 (agriculture and the world economy); Chapter 11 (world cities); Chapter 14 (industrial restructuring)
- Population distribution and density: Chapter 2

(environmental determinism and possibilism); Chapter 4 (population growth); Appendix 1 (physical geography, especially Fig. A1.8: global environment regions)
- Explanations of migration: Chapter 2 (positivism and humanism)
- Forced migration, immigration to Canada, restrictive immigration policies: Chapter 8 (types of society; landscapes and power relations; ethnicity; racism)
- Free migration within countries: Chapter 11 (rural settlements in transition; rural–urban fringe); Chapter 12 (urban spread)
- Illegal migration: Chapter 9 (economic globalization)
- Refugees: Chapter 8 (political geography, especially Box 8.6)
- The less developed world: Chapter 3 (energy and technology, earth's vital signs); Chapter 4 (fertility and mortality); Chapter 8 (colonialism); Chapter 10 (agriculture); Chapter 13 (settlement); Chapter 14 (industry)

CHAPTER 6

- Landscape: Chapter 1 (Humboldt, Ritter, Vidal, and Sauer)
- Globalization: Chapter 2 (concepts); Chapter 3 (ecosystems and global impacts); Chapter 4 (population growth, fertility decline); Chapter 5 (refugees, disease, more and less developed worlds); Chapter 7 (popular culture); Chapter 8 (political globalization); Chapter 9 (economic globalization); Chapter 10 (agriculture and the world economy); Chapter 11 (global cities); Chapter 14 (industrial restructuring)
- World cultural divisions: Chapters 7 and 8
- 'Culture' and 'society' as terms in human geography: Chapter 1 (disciplinary evolution, especially Sauer and Vidal)
- Social scale: Chapter 2 (concept of scale)
- Beginnings of civilization: Chapter 2 (Marxism); Chapter 3 (energy sources); Chapter 7 (races and racism); Chapter 10 (origins of agriculture)
- World regions: Chapter 2 (concept of region); Chapter 3 (environmental regions); Chapter 4 (demographic regions); Chapter 5 (more and less developed regions); Chapter 8 (political and cultural regions); Chapter 10 (agricultural regions); Chapter 11 (urban regions); Chapter 14 (industrial regions)
- Cultural adaptation: Chapter 2 (environmental determinism and possibilism); Chapter 3 (natural resources and human impacts in general)
- Language and identity; religion and identity: Chapter 7 (ethnicity); Chapter 8 (nationalism)
- Language and landscape; religion and landscape: Chapter 2 (humanism); Chapter 7 (symbolic landscape)
- Women and religion: Chapters 4, 5, 7, and 9 (gender)
- Religion and land use: Chapter 10 (agricultural land use)

CHAPTER 7

- Globalization: Chapter 2 (concepts); Chapter 3 (ecosystems and global impacts); Chapter 4 (population growth, fertility decline); Chapter 5 (refugees, disease, more and less developed worlds); Chapter 6

(cultural globalization); Chapter 8 (political globalization); Chapter 9 (economic globalization); Chapter 10 (agriculture and the world economy); Chapter 11 (global cities); Chapter 14 (industrial restructuring)
- Rethinking culture: Chapter 6 (culture and society)
- Capitalism and socialism: Chapter 1 (disciplinary history); Chapter 2 (Marxism); Chapter 5 (world systems analysis)
- Humanism and Marxism: Chapter 2
- Modernism: Chapter 3 (environmental ethics); Chapter 5 (global inequalities)
- Feminism: Chapter 4; Chapter 5 (gender); Chapter 10 (feminist approaches to economic geography)
- Postmodernism: Chapter 10 (postmodern approaches to economic geography); Chapter 13 (postmodern urban theory)
- Concept of place: Chapter 2 (humanism)
- Vernacular regions: Chapter 2 (concepts); Chapter 6 (regions, including Mormon landscape)
- Ethnicity: Chapter 6 (language and religion)
- Gender and sexuality: Chapter 5 (global inequalities, especially indexes of human development); Chapter 8 (politics of protest)
- Folk culture: Chapter 6 (landscape)
- Popular culture: Chapter 13 (intra-urban issues); Chapter 14 (tourism)
- Ecotourism: Chapter 3 (sustainable development); Chapter 5 (less developed world)
- Group engineering: Chapter 6 (divided world)

CHAPTER 8

- Globalization: Chapter 2 (concepts); Chapter 3 (ecosystems and global impacts); Chapter 4 (population growth, fertility decline); Chapter 5 (refugees, disease, more and less developed worlds); Chapter 6 (cultural globalization); Chapter 7 (popular culture); Chapter 9 (economic globalization); Chapter 10 (agriculture and the world economy); Chapter 11 (global cities); Chapter 14 (industrial restructuring)
- State creation, nation-states, nationalism, state stability: Chapter 6 (language and religion); Chapter 7 (ethnicity)
- Group formation and group identity: Chapters 6 and 7
- Colonialism, dependency theory, socialism in less developed states: Chapter 1 (history of geography); Chapter 5 (the less developed world, world systems analysis)
- European colonial activity: Chapter 1 (European overseas movement)
- Core and periphery: Chapter 14 (uneven development)
- State groupings (NAFTA, EU, ASEAN); globalization: Chapter 9 (trade)
- Nation–state discordance in Africa: Chapter 5 (refugees)
- Politics of protest: Chapter 3 (Box 3.14); Chapter 7 (contested landscapes)
- Possible future conflicts: Chapters 3 and 5 (future trends in population, food, and environment, the role of bad government); Chapter 6 (religious identities)

CHAPTER 9

The study of distance is geography par excellence; hence this chapter links with all other chapters in this book.
- Globalization: Chapter 2 (concepts); Chapter 3 (ecosystems and global impacts); Chapter 4 (population growth, fertility decline); Chapter 5 (refugees, disease, more and less developed worlds); Chapter 6 (cultural globalization); Chapter 7 (popular culture); Chapter 8 (political globalization); Chapter 10 (agriculture and the world economy); Chapter 11 (global cities)
- Five concepts of distance and space: Chapter 2 (concepts); Chapters 6 and 7 (cognitive and social distance in relation to cultural regions and landscapes); Chapters 10, 11, 14 (time and economic distance in relation to land-use theories)
- Diffusion research in cultural geography: Chapter 6 (creation of cultural landscapes)
- 'Neighbourhood effect': Chapter 10 (von Thünen theory)
- Diffusion and political economy: Chapter 10 (political ecology and agriculture)
- Regional integration and globalization: Chapter 8 (groupings of states); Chapters 10, 11, 12, 14 (economic restructuring)

CHAPTER 10

- Globalization: Chapter 2 (concepts); Chapter 3 (ecosystems and global impacts); Chapter 4 (population growth, fertility decline); Chapter 5 (refugees, disease, more and less developed worlds); Chapter 6 (cultural globalization); Chapter 7 (popular culture); Chapter 8 (political globalization); Chapter 9 (economic globalization); Chapter 11 (global cities); Chapter 14 (industrial restructuring)
- Cultural and political considerations: Chapter 6 (religion); Chapter 7 (ethnicity); Chapter 8 (role of the state)
- Location and interaction theories; the economic operator concept; von Thünen theory: Chapter 2 (positivism, concepts of space, location, and distance); Chapter 9 (gravity and potential models, regional science)
- Correlation and regression analysis: Chapter 2 (quantitative techniques)
- Origins of agriculture: Chapter 2 (environmental determinism, possibilism); Chapter 3 (energy and technology); Chapter 4 (population growth through time); Chapter 6 (beginnings of civilization)
- Second agricultural revolution: Chapter 2 (Marxist terminology); Chapters 3 and 14 (Industrial Revolution); Chapter 4 (Malthusian concepts)
- Nine agricultural regions: Chapter 2 (environmental determinism, possibilism); Appendix 1, especially Figure A1.8 (global environments)
- Plantation agriculture; green revolution; agriculture in the less developed world: Chapter 4 (fertility transition); Chapter 5 (migration [especially Box 5.3], population and food, world systems analysis)
- Industrialization of agriculture; global trade in food: Chapter 9 (globalization); Chapter 14 (restructuring)

- Political ecology of agriculture: Chapter 2 (Marxism); Chapter 3 (environmental ethics); Chapters 5 and 7 (gender); Chapter 7 (structuration theory)

CHAPTER 11

- Globalization: Chapter 2 (concepts); Chapter 3 (ecosystems and global impacts); Chapter 4 (population growth, fertility decline); Chapter 5 (refugees, disease, more and less developed worlds); Chapter 6 (cultural globalization); Chapter 7 (popular culture); Chapter 8 (political globalization); Chapter 9 (economic globalization); Chapter 10 (agriculture and the world economy); Chapter 14 (industrial restructuring)
- Urbanization: Chapter 4 (population growth); Chapter 5 (more and less developed worlds)
- Rural settlement: Chapter 10 (second agricultural revolution, world agricultural regions)
- Canadian prairie settlement: Chapter 6 (language); Chapter 7 (ethnicity)
- Rural settlement location theories; central place theory: Chapter 2 (positivism, concepts of space, location, and distance); Chapter 9 (distance); Chapter 10 (agricultural location theory); Chapter 14 (industrial location theory)
- Rural depopulation: Chapter 5 (free migration); Chapter 10 (sectoral shift from agricultural to urban economies)
- World cities: Chapter 8 (political globalization); Chapter 9 (economic globalization)

CHAPTER 12

- Globalization: Chapter 2 (concepts); Chapter 6 (cultural globalization); Chapter 7 (popular culture); Chapter 8 (political globalization); Chapter 9 (economic globalization); Chapter 10 (agriculture and the world economy); Chapter 14 (industrial restructuring)
- Intra-urban theories: Chapter 2 (positivism, concepts of space, location, and distance); Chapter 5 (migration, less developed world); Chapter 7 (postmodernism); Chapter 9 (distance); Chapter 10 (agricultural location theory); Chapter 14 (industrial location theory)
- City governance and planning: Chapter 8 (the role of the state)
- Urban transportation: Chapter 9 (concepts of distance and space, transportation)
- Urban expansion: Chapter 5 (migration); Chapter 8 (political globalization); Chapter 9 (economic globalization); Chapter 10 (agriculture and the world economy); Chapter 14 (industrial restructuring)

CHAPTER 13

- Globalization: Chapter 2 (concepts); Chapter 6 (cultural globalization); Chapter 7 (popular culture); Chapter 8 (political globalization); Chapter 9 (economic globalization); Chapter 10 (agriculture and the world economy); Chapter 13 (industrial restructuring)
- Housing: Chapter 6 (cultural identity, language, religion); Chapter 7 (cultural identity, ethnicity)
- Poverty in the city: Chapter 2 (Marxism); Chapter 7 (feminism, crime, health)
- Retailing and industry; post-industrial trends: Chapter 9 (economic globalization); Chapter 14 (industrial change)
- Cities in the less developed world: Chapter 3 (human impacts); Chapter 4 (fertility and mortality, history of population growth, explaining population growth); Chapter 5 (less developed world, world systems analysis); Chapter 8 (colonial activity and the colonial experience)

CHAPTER 14

- Globalization: Chapter 2 (concepts); Chapter 3 (ecosystems and global impacts); Chapter 4 (population growth, fertility decline); Chapter 5 (refugees, disease, more and less developed worlds); Chapter 6 (cultural globalization); Chapter 7 (popular culture); Chapter 8 (political globalization); Chapter 9 (economic globalization); Chapter 10 (agriculture and the world economy); Chapter 11 (global cities)
- Industrial location theory: Chapter 2 (positivism); Chapter 10 (agricultural location); Chapter 11 (settlement location)
- Industrial Revolution: All chapters
- Energy: Chapter 3 (changing energy sources)
- Differences in industrial geography of more and less developed worlds: Chapter 5 (inequalities); Chapter 10 (differences in agriculture); Chapter 13 (differences in settlement)
- Industry in Japan; NICs; industrial geography of China: Chapters 9, 10, and 11 (multinationals)
- Industrial restructuring (especially in the service industry): Chapter 11 (industry and services in cities, global cities); Chapter 13 (cities as centres of production and consumption)
- Industry and society: Chapter 2 (Marxism); Chapter 7 (Marxism, class, gender)
- Environmental considerations in industry: Chapter 3 (environmental issues)
- Uneven development: Chapter 9 (trade and trade blocs); Chapter 11 (core and periphery)

Preface

The need for an education in geography—knowing where things are, why they are where they are, and why this knowledge matters—has always been paramount in all societies. It is no different today. Both geographic knowledge and an appreciation of the value of a geographic perspective are essential to help individuals and groups make sense of the changing worlds in which we live.

This book attempts to capture both the spirit and the practical value of our contemporary human geography. Like the discipline itself, it encompasses an extraordinarily broad range of subject matter and no single approach or methodology dominates. Loosely organized around three recurring themes—relations between humans and land, regional studies, and spatial analysis—this text emphasizes how human geography has developed in response to society's needs, and how the discipline continues to change accordingly. It also stresses the links between human geography and other disciplines, not only in order to clarify the various philosophies behind different types of human geographic work, but also to encourage students to apply their human geographic knowledge and understanding in other academic contexts.

Seventh Edition: Special Features

In response to the changing facts of human geography and to changing understandings of those facts, this edition has been changed from the sixth edition in four substantive ways. First, there is much new content, both in some new sections and in new material added to existing discussions. Second, as in previous editions, a renewed effort has been made to provide information for students in easily readable ways: new figures, tables, and boxes have been added; all of the maps have been redrawn; and the listing of helpful websites at the end of each chapter has been enhanced considerably. Third, much of the detailed factual content in this seventh edition is new, especially relating to topics such as environmental change, population, political developments, and economic change. Finally, the entire text has been revised for clarity and readability, and a number of errors have been corrected.

Among the most significant additions and changes are the following:

- **Introduction**, **Chapter 1**, and **Chapter 2**: These chapters have been updated as appropriate and there is new content, especially in Boxes I.1, I.2, 1.3, 1.6, and 1.11, and in the Chapter 1 discussions of early British geography and the idea of physical geography as cause.

- **Chapter 3**: There is a new section on polluted landscapes, which includes a new figure. A new box on the politicization of climate science introduces students to the important idea that some opinions about scientific matters are related more to ideological standpoint than to careful appraisal of evidence. Also, significant new content has been added on renewable energy sources and on human impacts on land, air, and climate.

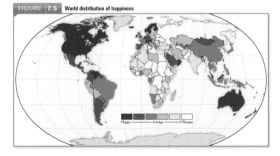

- **Chapter 4**: Substantial new content on fertility, mortality, and natural increase helps clarify important details of population growth. A significantly enhanced discussion of government population policies provides new insights.

- **Chapter 5**: New material is provided on the links between government and development, including a new figure. A new box on the food crisis of 2007–8 raises important questions about links between politics, environmental change, and global economic health. There are expanded and updated accounts of refugees in the Horn of Africa and of Ethiopia as a less developed country. New material on forced labour, refugees, the 2008 Myanmar cyclone, and the selectivity of disease adds much to the previous content. There is also some discussion of the onset of a global recession in 2008, as this affected economic growth in the less developed world.

- **Chapter 6:** The account of religion has been revised and expanded so that this section now provides a fuller and more informative account of the world's major religions, the links between religion and conflict, religion as a factor in human identity, and religious landscapes. A box on disappearing peoples has been moved to this chapter, and some new content on language has been added.

- **Chapter 7**: A new section on the geography of happiness includes a new figure. The major section on tourism and recreation has been moved here from Chapter 14, as it ties in well with the account of popular culture. New material on regional identity and sense of place, gender and human rights, and aspects of folk and popular culture—especially music and sport—is added.

- **Chapter 8**: A significant rewriting and reordering of material on geopolitics and the stability of states facilitates comprehension of this content. As well, African conflicts, European separatist movements, conflicts in the former USSR, and the geography of nuclear war are discussed in greater detail. New content on civil wars and drug production and movement includes much current material. A revised discussion of the global spread of democracy includes a new table and figure.

- **Chapter 9**: The discussion of modes of transport includes a new figure, and new discussions of containerization and transport in China have been added. There is significant updating of the economic globalization content.

- **Chapter 10**: A new figure shows areas of agricultural domestication and early diffusion. A substantial restructuring of material on technological changes in world agriculture includes new content on organic farming. A new section, including a new table and figure, discusses no-till strategies, and new material has been added on consumption and place.

- **Chapter 11**: Global urban data have been updated as appropriate. In addition, the chapter includes new content on counterurbanization and an enhanced discussion of world cities with particular reference to China.

- **Chapter 12**: New content has been added on Canadian cities, planned cities, and suburban spread, and a new box discusses innovative solutions to urban congestion.

- **Chapter 13**: Discussions on inequalities in cities and on retailing have been extended, and a significantly enhanced discussion of urban slums in the less developed world provides additional insights into this aspect of urban growth. There are additional examples of city environments in the less developed world, including Chinese cities.

- **Chapter 14**: The account of world industrial geography includes new content on the origins of the Industrial Revolution and on industry in China. New sections on the 'new economic geography' and on development issues in the less developed world provide additional insights into these topics. The material on tourism and recreation has been moved to Chapter 7.

- **Conclusion**: Some material has been moved into earlier chapters with the result that the Conclusion is now shorter and more direct.

- **Appendices 1 and 2**: These remain largely unchanged. Updating of material on human evolution reflects new understandings.

---

416   Human Geography

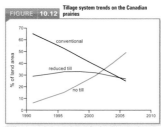

FIGURE 10.12 Tillage system trends on the Canadian prairies

SOURCE: Agriculture and Agri-Food Canada. 2008. Flexibility of No Till and Reduced Till Systems Ensures Success in the Long Term 2008. http://www4.agr.gc.ca/AAFC-AAC/display-afficher.do?id=1219778199236&lang=eng. Reproduced with the permission of the Minister of Public Works and Government Services Canada, 2009.

prairies under the three different tillage systems for each of the last four agricultural census years.

Huggins and Reganold (2008: 77) conclude that no-till, although 'not a cure-all', is a sustainable strategy that is able to 'deliver a host of benefits that are increasingly desirable in a world facing population growth, environmental degradation, rising energy costs and climate change.'

World Agriculture Today: Types and Regions

The noted American geographer Whittlesey identified nine major regional types of agriculture, and Figure 10.13 shows the global distribution of these nine types as elaborated upon by Grigg (1974); it includes a twofold division of one of the types and identifies those areas with little or no agriculture.

Even a cursory review of Figure 10.13 highlights the relationships between agriculture and climate (Figure A1.7), between agriculture and generalized global environments (Figure A1.8), and between agriculture and population distribution and density (Figure 5.1). The figure also suggests the complex links between areas of production and areas of consumption.

PRIMITIVE SUBSISTENCE AGRICULTURE

One of the earliest agricultural systems, 'primitive subsistence' or 'shifting' agriculture, is today practised almost exclusively in tropical areas. It involves selecting a location, removing vegetation, and sowing crops on the cleared land. Land preparation is minimal, and little care is given to the crops. Typically, agricultural implements are limited and livestock are not normally part of the system, although fowl and pigs may be present. After a few years, the land is abandoned and a new location is sought. In this farming system, land is not owned by an individual or family but rather by some larger social unit such as the village or tribe. Shifting agriculture has traditionally been for subsistence purposes, that is, for local consumption. Today, however, in addition to their subsistence crops many such cultivators also produce a different crop for sale so that they have at least some market orientation.

Superficially, shifting agriculture appears to be wasteful and indicative of a low technology. Yet, evidence suggests that it is a highly appropriate farming method. Not only is it an effective way of maintaining soil fertility in humid tropical areas, and well suited to circumstances of low population density and ample land, but it provides adequate returns for minimal capital and labour inputs. Shifting agriculture, then, can be explained by reference to a number of variables beyond economic considerations, including environment and population numbers.

WET RICE FARMING

Most of the rural population of East Asia is supported by wet rice farming, an intensive type of agriculture that requires only a small portion of the total land area—farms are small, perhaps only 1–2 ha (2.5–5 acres) and subdivided into fields—but large amounts of human labour. There are many versions of wet rice farming, but in all instances the crop is submerged under slowly moving water for much of the growing period. Wet rice farming is restricted to specific local environments, most notably flat land adjacent to rivers. Low walls delineate the fields and hold the water. Because this technology minimizes soil depletion, continuous cropping is often possible, producing multiple harvests each year. Cultivation techniques changed little until the green revolution of the 1960s, when some

---

CHAPTER 13 | Living and Working in Cities   541

FIGURE 13.6 The incidence of urban slums

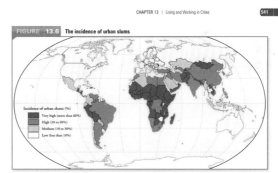

Incidence of urban slums (%)
- Very high (more than 60%)
- High (30 to 60%)
- Medium (10 to 30%)
- Low (less than 10%)

SOURCE: Updated from United Nations Human Settlements Program, *The State of the World's Cities, 2004/2005 Globalization and Urban Culture* (Sterling, Va: Earthscan, 2004), 106.

population—were slum dwellers, with the number increasing rapidly unless significant action is taken, possibly reaching 1.4 billion by 2020 and 3.5 billion by 2050. Annual slum growth rates in Africa are estimated to be 4.5 per cent, in Asia 2.5 per cent, and in Latin America 1.3 per cent.

- In Africa, about 70 per cent of the urban population live in slums. The highest incidence is in Sierra Leone, where slum dwellers comprise 96 per cent of the urban population; the figure for the Central African Republic is 92 per cent, and for Nigeria 79 per cent.

Children scavenging at the municipal dump of La Chureca in Managua, Nicaragua, in November 2006.
AP photo/Cristobal Herrera

Features

Learning Tools

In addition to comprehensive chapter openers and summaries, the text is packed with learning tools. 'Links to Other Chapters' underline connections between topics that may not be apparent at first glance; 'Further Explorations' at the end of each chapter present annotated references for further study; and 'On the Web' provides useful links to online resources that expand on topics covered in the text.

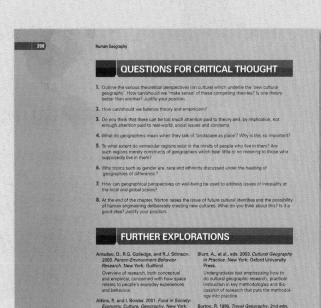

CHAPTER 3 SUMMARY

USE OR ABUSE?
Today there is considerable evidence that we are damaging our home—the earth—and increasing recognition of how fragile that home is.

SCALE
The value of a global perspective in any discussion of human use of the earth is clear: everything is related to everything else.

THE INCREASING HUMAN IMPACT
Two principal variables—sheer numbers of people and increasing use of technology and energy—explain why human activities today have such great impact on the earth.

ECOSYSTEMS
Combining systems logic and ecological principles produces the concept of the ecosystem: any self-sustaining collection of living organisms and their environment is an ecosystem. A whole series of ecosystems together make up the global ecosystem—the ecosphere or biosphere—which is the home of all life on earth. The ecosphere includes air, water, land, and all life.

ECOSYSTEM SIMPLIFICATION
When humans change an ecosystem, the result is usually a simplification. We cause changes largely because of our numbers, our technology, and our energy use. Today we number in excess of 6.5 billion, and our technological ability—that is, our ability to convert energy into forms useful to us—is constantly increasing. As human culture (especially technology) changes, so does the value attributed to resources. Different interest groups evaluate resources according to different criteria and hence have differing views. Stock resources are finite in quantity, whereas renewable resources are relatively unlimited in quantity.
Our simplification of ecosystems is now sufficiently evident that a new environmental ethic is needed—one involving co-operation with nature, not domination over it.

HUMAN IMPACTS
The presence of humans on the earth changes the planet. Some of the changes brought about by humans are inevitable because they are necessary to human survival. Other changes, however, are the result of inappropriate actions motivated by human greed and misunderstanding. Over time, humans' ability to affect ecosystems at all levels has increased. Although it may be inappropriate to focus on individual parts of an ecosystem—which is by definition an inseparable whole—it is convenient to consider the impact of human activities on specific parts of ecosystems such as vegetation, animals, land and soil, water, and climate. In each case, we have caused many changes.
Two important issues today are tropical rain forest depletion and desertification; in both cases, human activity is at least partially explained by the problems experienced by poor people in less developed countries. Impacts on animal life include those associated with domestication and species extinction. Extensive changes to land surfaces result from a host of human activities such as resource extraction and the creation of urban industrial complexes. Soil is especially susceptible to abuses involving salinization, increased laterite content caused by removal of vegetation, and erosion. Water, an essential ingredient for life, is typically used in ways that result in shortages and contamination, and the oceans continue to be polluted.
The ecosystem concept of interrelatedness is especially useful with respect to climate. We are having a significant effect on the atmosphere by increasing the quantity of greenhouse gases and damaging the ozone layer. Climate change is undeniable and humans are a cause, but we also need to understand more about climatic change that is not caused by humans.

THE HEALTH OF THE PLANET
Although debates continue concerning the consequences of human activity for the health of the planet, most agree that human-induced ills are many and serious. Probably the best solution is prevention—an argument advanced by many environmentalist groups.

SUSTAINABLE DEVELOPMENT
Sustainable development serves our present needs, but does not compromise the ability of later generations to meet their needs.

QUESTIONS FOR CRITICAL THOUGHT

1. Outline the various theoretical perspectives (on culture) which underlie the 'new cultural geography'. How can/should we 'make sense' of these competing theories? Is one theory better than another? Justify your position.

2. How can/should we balance theory and empiricism?

3. Do you think that there can be too much attention paid to theory and, by implication, not enough attention paid to real-world, social issues and concerns.

4. What do geographers mean when they talk of 'landscape as place'? Why is this so important?

5. To what extent do vernacular regions exist in the minds of people who live in them? Are such regions merely constructs of geographers which bear little or no meaning to those who supposedly live in them?

6. Why topics such as gender are, race and ethnicity discussed under the heading of 'geographies of difference'?

7. How can geographical perspectives on well-being be used to address issues of inequality at the local and global scales?

8. At the end of the chapter, Norton raises the issue of future cultural identities and the possibility of human engineering deliberately creating new cultures. What do you think about this? Is it a good idea? Justify your position.

FURTHER EXPLORATIONS

Amadeo, D., R.G. Golledge, and R.J. Stimson. 2009. *Person-Environment-Behavior Research*. New York: Guilford.

Overview of research, both conceptual and empirical, concerned with how space relates to people's everyday experiences and behaviour.

Atkins, P., and I. Bowler. 2001. *Food in Society: Economy, Culture, Geography*. New York: Oxford University Press.

Covers a wide range of topics, including food consumption, regional differences in food preferences, and links between food and gender.

Bale, J. 2003. *Sports Geography*, 2nd edn. New York: Routledge.

Material from a disparate literature; probably includes something to interest every geography student, regardless of philosophical persuasion or conceptual preference.

Blunt, A., et al., eds. 2003. *Cultural Geography in Practice*. New York: Oxford University Press.

Undergraduate text emphasizing how to do cultural geographic research; practical instruction in key methodologies and discussion of research that puts the methodology into practice.

Burton, R. 1995. *Travel Geography*, 2nd edn. London: Pitman.

A factually detailed overview of the world's geographical resource base for tourism and of the spatial patterns of world tourist activity.

Cartier, C., and A.A. Lew, eds. 2005. *Seductions of Place: Geographical Perspectives on Globalization and Touristed Landscapes*. New York: Routledge.

A collection of studies that employ recent innovations associated with the new cultural geography.

New Aids to Student Learning

New questions for review and critical thinking and a running glossary have been added to the existing pedagogical tools to establish better understanding of core concepts and encourage discussion and further research on key issues.

138   Human Geography

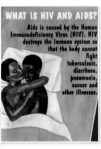

Much of the effort to combat HIV/AIDS focuses on education. Here are two posters from Africa. The one at the right, from Namibia, was produced by the local Catholic church and highlights the links between AIDS and other illnesses. The poster above, from Lesotho, stresses the need for people to behave responsibly and inform themselves.
Das Fotoarchiv

estimate is that a 15-year-old boy in Botswana today has an 80 per cent chance of dying of AIDS. Writing about one clinic in Malawi, Toolis (2000: 29) stated:

> If you have no money, then there is no choice; you must queue in the morning heat of the Boma's outpatients clinic and hope to be admitted. Demand outstrips supply. There are 130 beds, and every morning another 250 would-be patients appear at the gates. All but the sickest are turned away, as well as the medically hopeless, who are sent back home to

die. There is not enough staff, not enough drugs, not enough needles, not enough anything.

It seems clear that poverty is the biggest single factor contributing to the spread of AIDS. Poverty means that infections remain untreated, greatly increasing the risk of further transmission. Poverty also keeps children out of school, increasing the likelihood that the sale of sex will be a primary means of supporting themselves. But of course AIDS only serves to exacerbate poverty. The virus is ravaging African agricultural productivity; in parts of Kenya, for example, the amount of cultivated land has declined by 68 per cent, while in Rwanda there has been a 60–80 per cent decrease in farm labour. Yet in many countries the disease is still seen as so shameful that it cannot even be named; in Mozambique, for example, death certificates for AIDS patients often bear the words 'cause unknown'. As a result of these attitudes, the frank discussion needed to raise public awareness is still lacking in many cases.

Russia, China, and India

Most accounts of AIDS today focus on sub-Saharan Africa, but the disease may soon have major impacts elsewhere. According to Eberstadt (2002: 24), 'it is quite possible that the center of the global HIV/AIDS crisis, in

TABLE 4.3	Countries with Highest Levels of Adult (age 15–49) HIV/AIDS Prevalence, 2008
Country	Adults with HIV/AIDS as % of Total Population
Swaziland	26.1
Botswana	23.9
Lesotho	23.2
South Africa	18.1
Zimbabwe	15.3
Namibia	15.3
Zambia	15.2
Mozambique	12.5
Malawi	11.9

SOURCE: Population Reference Bureau, *2008 World Population Data Sheet* (Washington: Population Reference Bureau, 2008).

Richly Illustrated

With newly redesigned visuals, the seventh edition has an increased emphasis on providing information for students in easily readable ways. New figures, tables, and redrawn maps are included throughout the text to enhance student understanding of the material.

National and Global Perspectives

A carefully selected blend of up-to-date Canadian and international examples are used throughout the text. Students will gain a global perspective and greater depth of understanding through the carefully balanced material including such boxes as 'Exurbanization in Southwestern Ontario', 'Psychogeography: The Sense of Oklahomaness', and 'The "Green Revolution" in India'.

450   Human Geography

Urban sprawl encroaches on a farm property south of Calgary.
CP photo/Larry MacDougal

exurbanization The movement of households from urban areas to locations outside the urban area but within the commuting field.

work in these urban centres. Examples of such growth in the 2001–6 period include Sylvan Lake near Red Deer (36.1 per cent growth) and Strathmore near Calgary (34.2 per cent growth). Not surprisingly, rural and small-town

population decline was most evident in areas at greater distances from urban centres.

In Europe, which never had extensive areas of agricultural land, creeping urbanization is leading inexorably to the elimination of what might be thought of as the traditional countryside. With urban spread, the related development of transportation networks and nodes, and a declining agricultural sector, it increasingly seems that the urban never really ends and the rural never really begins. A report by the Campaign to Protect Rural England (2005) suggests that the English countryside will have all but disappeared by 2035.

The rural–urban fringe

Transition zones between rural and urban areas often have multiple land uses and varied social and demographic characteristics. Indeed, the absence of clear boundaries between rural and urban areas is a distinctive feature of contemporary land use and lifestyle. One way to look at this situation is to focus on the centrifugal forces that encourage urban land uses to expand into the fringe belt—a process that has come to be known as deconcentration, in contrast to the concentration more traditionally associated with urban development. The term **urban sprawl** is often used to describe the deconcentration that involves low-density expansion of urban land uses into surrounding

Box 11.3   Exurbanization in Southwestern Ontario

In many areas, deconcentration is taking the form of residential movement to the rural–urban fringe. Such movement is not simply a matter of the extension of existing suburbs as it does not involve a steady outward expansion of urban areas but rather scattered and often relatively isolated pockets of settlement. Does it perhaps represent a form of counter-urbanization, or rural repopulation? The fragmented landscapes of exurbanization are most evident in North America, especially in the western United States where many of these new places are experiencing dramatic growth. This case study looks at southwestern Ontario.

For Woodstock in southwestern Ontario, Davies and Yeates (1991) show that both perspectives are relevant. Focusing on *exurbanization*—the movement of households from urban areas to locations outside the urban area but within the commuting field—they distinguish between exurbanite households located in dispersed rural settlements

(54 per cent of the total in the area around Woodstock) and exurbanite households located in villages (46 per cent of the total in the area around Woodstock). As shown in Table 11.9, 'the exurban-rural group is part of the new urban-to-rural migration pattern subsumed within the idea of counter-urbanization, while the exurban village group is part of the long-standing trend of suburbanization related to a search for more housing space and privacy at lower prices' (ibid., 186). This conclusion is based on the relatively high valuation that the exurban-village group places on lower taxes and lower house prices and on the relatively high valuation that the exurban-rural group places on the amenity attractions of rural areas. Both groups place a high value on privacy, larger houses, more land, the attractive rural landscape, a better quality of life, and less crime.

Another way to think about these trends is in terms of the push and pull factors introduced in the discussion of migration in Chapter 5.

Instructor and Student Supplements to Accompany the Text

The seventh edition of *Human Geography* is supported by a wide range of supplementary items for students and instructors alike, all designed to enhance and complete the learning and teaching experience.

For Students

Online Resources—Available at **<www.oupcanada.com/Human7e>**.
An updated and expanded student study guide of review material—including an expanded section of links to human geography websites, new self-grading practice quizzes, short answer exercises, and in-depth chapter summaries and objectives—is designed to reinforce student understanding of the material and provide directions for further research.

LAB MANUAL (9780195436228)

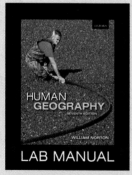

The lab manual contains a wide variety of classroom-tested field exercises related to each chapter. Students will apply what they have learned in the text to exciting research situations.

For Instructors

The following instructors' resources are available online to qualifying adopters. Please contact your OUP Canada sales representative for more information.

- **Instructor's manual**. A comprehensive resource package featuring lecture outlines, teaching objectives, media resources, small group and project ideas, and review questions.
- **PowerPoint slides**. Dozens of images, including maps, charts, and diagrams, digitized for classroom use.
- **Test Generator.** A comprehensive bank provides hundreds of multiple choice, true/false, and short answer questions.
- **CBC DVD**. A wide range of videos from the CBC archives designed to support the text and enhance the classroom experience. This resource is available to qualifying adopters. Please contact your OUP Canada sales and editorial representative for more information.
- **Image Bank**. A dynamic lecture resource containing hundreds of full-colour figures, photographs, and tables that makes the classroom presentation of material more engaging and relevant for students.

Acknowledgements

The author is pleased to acknowledge two colleagues for their significant contributions to this new edition: Doug Fast has rethought and redrawn all of the maps; Barry Kaye has continued to draw my attention to numerous new books and relevant news items. I am most grateful for their ongoing interest and involvement in this work. In addition, many other geographers and many of my students have commented on earlier editions and most of their suggestions are incorporated in this edition.

As always, staff at Oxford University Press have supported this book in every way possible. I am particularly grateful for the enthusiastic support shown by Katherine Skene, Phyllis Wilson, Peter Chambers, Euan White, and Sophia Fortier, vice president and director of the Higher Education Division. Richard Tallman edited the manuscript and, as with the previous edition, has improved the readability of the book both through his attention to detail and his insightful understanding of the subject matter. Finally, many thanks to my wife, Pauline, for her unceasing support of my academic activities.

Oxford University Press and the author would also like to thank the many reviewers whose comments have proved invaluable over the years. In addition to those who provided anonymous feedback on this seventh edition, the author and the publisher would like to thank the following reviewers whose thoughtful comments and suggestions have helped to shape this text:

Sean Doherty
Wilfrid Laurier University

Matthew Evenden
University of British Columbia

Walt Peace
McMaster University

Scott Bell
University of Alberta

Michael Fox
Mount Allison

Michael Mercier
McMaster University

Raj Navaratnam
Red Deer College

Marilyn Lewry
University of Regina

Edgar L. Jackson
University of Alberta

Introduction

THE ROADS AHEAD

As a young man, my fondest dream was to become a geographer. However, while working in the customs office I thought deeply about the matter and concluded that it was far too difficult a subject. With some reluctance, I then turned to physics as a substitute.

—Albert Einstein

You know that this book is an introduction to human geography—but what does that term mean? Note down in a few key words your idea of what human geography is. It should be interesting to compare your initial perception with your understanding after you have read this introductory chapter.

Consider the word 'geography'. Its roots are Greek: *geo* means 'the world', *graphei* means 'to write'. Literally, then, what geographers do is write about the world. The sheer breadth of this task has demanded that it be divided into two relatively distinct disciplines. *Physical geography* is concerned with the physical world, for example, climates and landforms, and *human geography* is concerned with the human world, for example, agricultural activities and settlement patterns.

So human geographers write about the human world. But what questions do they ask? How do they approach their work? What methods do they employ? Not surprisingly, human geographers ask various questions and have diverse approaches and methods from which to choose, and the later chapters of this book reflect this diversity. Nevertheless, three themes are central to any study of the human world: relations between humans and land, regionalization, and spatial analysis.

Three Recurring Themes

Keeping these three themes in mind from the start will help you to follow and understand the complex connections among the many aspects of human geography. The intellectual origins and evolution of these themes are outlined in Chapter 1, and a philosophical justification for each theme is introduced in Chapter 2.

HUMANS AND LAND

The human world is not in any sense preordained. It is not the result of any single cause such as climate, physiography, religion, or culture. Rather, it is the ever-changing product of the activities of human beings, as individuals and as group members, working within human and institutional frameworks to modify pre-existing physical conditions. Thus human geographers often focus on the evolution of the

Corrientes Avenue in Buenos Aires. The thick haze of smoke is the result of massive brush fires set by cattle ranchers and soybean farmers.
Daniel Garcia/AFP/Getty Images

human world with reference to people, their cultures, and physical environments. We usually describe the human world as a **landscape**. There are two closely related aspects to 'landscape' in this sense:

1. First, landscape is what is there as a result of human modifications of physical geography. This aspect of landscape includes crops, buildings, lines of communication, and other visible, material features.

2. Second, landscape has significant symbolic content—in other words, meaning. It is hard to view a church, a statue, or a tall office building without appreciating that it is something more than a visible, material human addition to the physical geography of a place: such features are expressions in landscape of the cultures that produced them.

As human geographers, we are interested in landscape both for what it is and for what it means to live in it: we interpret landscape as the outcome of particular relationships between humans and land. For some geographers, both physical and human, the study of humans and land together is geography. This view of geography, relating human and physical variables and identifying links, is sometimes called ecological analysis.

REGIONAL STUDIES

To facilitate their task of writing about the world, human geographers often divide large areas into smaller areas that exhibit a degree of unity—that share one or more features in common. These smaller areas are **regions**. To regionalize is to classify on the basis of one or more variables. The fact that we are able to regionalize tells us that human landscapes make sense; they are not random assemblages of features. Groups of people occupying particular areas over a period of time create regions: human landscapes that reflect their occupancy and that differ from other landscapes. Much contemporary regional study focuses on this social organization of space, on the impact of creating regions on social and economic life, on interactions between regions, and on regional boundaries as barriers.

Building on a long tradition, contemporary human geography considers regions at a wide range of scales—from the local (a suburban neighbourhood, for example) all the way through to the global (the world as a single region). In acknowledging the relevance of different spatial scales of analysis, human geography reveals the importance of place in all aspects of our lives.

SPATIAL ANALYSIS

Understanding the human world requires that we explain **location**: why things are where they are. Typically, a geographer using a spatial analysis approach tackles this question through theory construction, models, and hypothesis testing, using quantitative methods. For example, we may find that towns in an area are spaced at relatively equal distances apart, or that industrial plants are located close to one another. The primary goal of spatial analysis is to explain such locational regularities. Thus we have already identified two types of spatial organization: the creation of regions, and the identification of distinct patterns of locations. A secondary goal is often that of identifying alternative locational patterns that might be more efficient or more equitable.

Central to this spatial analysis is the realization that all things are related—and for geographers, to say that things are related is usually to say that they interact. Perhaps the best examples of interaction in the contemporary world have to do with what is often described as **globalization**—a complex set of processes with economic, political, and cultural dimensions.

A COMMON THREAD

Our three recurring themes reflect three separate but overlapping traditions (to be outlined in Chapter 1). One common thread among them, however, is the fact that the human world is always changing. Thus human geography, regardless of the specific focus, typically incorporates a time dimension. Change can be a response to either internal or external factors. Some changes are rapid, some gradual. Landscapes change in content and meaning; regions gain or lose their distinctive features; and locations adjust to changing circumstances.

One compelling example of changing geographies can be found in the tragic events of 11 September 2001, when a series of concerted terrorist attacks destroyed the twin towers of the World Trade Center in New York City and part of the Pentagon in Washington,

---

landscape
A major concern of geographic study; the characteristics of a particular area, especially as created through human activity.

location
A term that refers to a specific part of the earth's surface; an area where something is situated.

globalization
A complex combination of economic, political, and cultural changes that have long been evident but that have accelerated markedly since about 1980, bringing about a seemingly ever-increasing connectedness of both people and places.

region
A part of the earth's surface that displays internal homogeneity and is relatively distinct from surrounding areas according to some criterion or criteria; regions are intellectual creations.

DC, killing more than 3,000 people. The consequences of those attacks, some immediate, some longer-term and still uncertain, have served to reshape both local and global geographies. The skyline of New York City will never be the same again. Nor will the pattern of global politics, since the United States responded to the attacks by invading first Afghanistan, where the ruling Taliban had supported terrorist groups, and then Iraq. Indeed, by 2006 Saudi Arabia had solidified plans to construct a double-fence of steel and barbed wire along its 900-km (560-mile) border with Iraq, just as politicians in the United States, in reaction to fears of lax security following the terrorist attacks, developed plans to build a fence along parts of the US–Mexico border to shut out illegal immigrants. Many commentators responded to the attacks by suggesting that global politics would henceforth be characterized by a 'clash of civilizations', and there was widespread acknowledgement that fighting terrorism would require in-depth understanding of local geographies, both physical and human.

In identifying our three principal themes, we explicitly acknowledge the legitimacy of multiple approaches to our subject matter: the human world. It is also important to note how the application of these approaches changes over time. Such change is equally legitimate—and necessary—in any dynamic discipline.

Why 'Human' Geography?

Although many geographers have striven to make their field a truly integrated one, combining physical and human components as in the humans-and-land theme described above, human geography and physical geography are typically acknowledged to be relatively distinct disciplines. Contemporary human geographers do not deny the relevance of physical geography to their work, but they do not insist that their studies always include a physical component. Because human geography studies human beings, it has close ties with social sciences such as history, economics, anthropology, sociology, psychology, and political studies. One of the particular strengths of human geography is that it considers the human world in multivariate terms, incorporating physical

and human factors as necessary. Where appropriate, then, this book introduces aspects of physical geography. For example, to understand world population distributions we obviously need to know about global climates. However, we also need to know about human perceptions of those climates. Thus human geography is related to but separate from physical geography in much the same way as it is related to but separate from the other disciplines that study human beings.

The central subject matter of human geography is *human behaviour* as it affects the earth's surface. Expressed in this way, the subject matter of human geography is very similar to that of the various other social sciences, all of which focus on human behaviour in some specific context.

DEFINING HUMAN GEOGRAPHY

Providing a meaningful one-sentence definition of any academic discipline is a real challenge. In the case of human geography, however, the American geographer Charles Gritzner has suggested a useful definition in the form of three closely related questions: 'What is where, why there, and why care?' (Gritzner, 2002). Every exercise in human geography, regardless of which theme it highlights, begins with the spatial question: 'where?' Then, once the basic environmental, regional, and spatial facts are known, the geographer focuses on understanding, or explaining, why things are where they are. Finally, the third part of the question—'why care?'—draws attention to the pragmatic nature of human geography: geographic facts matter because they reflect and affect human life. This is true of any geographic fact, whether it is something as seemingly mundane as the distance between your home and the nearest convenience store or something as far-reaching or serious as a drought or a civil war that causes peasant farmers in Ethiopia to lose all their crops.

The Goal of Human Geography

Writing about the human world to increase our understanding of it has been the goal of human geography at least since the time of the ancient Greeks. Two twentieth-century statements of this continuing goal help to explain why it remains so important:

The function of geography is to train future citizens to imagine accurately the conditions of the great world stage and so help them to think sanely about political and social problems in the world around. (Fairgrieve, 1926: 18)

Geography is the only subject that asks you to look at the world and try to make sense of it. The field never stops being exciting, because that's what geography is all about—trying to make sense of the world. (Lewis, 2002: 4)

Although written some 80 years apart, these two statements express essentially the same view. Human geography is a practical and socially relevant discipline that has a great deal to teach us about the world we live in and how we live in the world. The goal of this particular book is to provide a basis for comprehending the human world as it is today and as it has evolved. If it succeeds, it will have made a contribution to the more general goal of advancing a just global society. Students of human geography are in an enviable position to accomplish this more general goal. A holistic discipline, it is not restricted to any narrowly defined subject or single theme that might limit our ability to appreciate what we might call the interrelatedness of things. Traditional links with physical geography clearly play a key role, but human geographers do not hesitate to draw on material and ideas from many other disciplines. This book explicitly acknowledges the controlled diversity of human geography.

The Human Geographer at Work

What do human geographers actually do? What questions do they ask and what problems do they strive to solve? These questions raise others. What is the nature of the world we live in? How do we live in the world? The following four vignettes offer some preliminary insights into the concerns and activities of human geographers that will be further expanded in later chapters. The first two vignettes focus on the ways in which our world is divided; the second two look at some of the ongoing transformations of global life.

A WORLD DIVIDED 1: POLITICAL DIVISIONS

Figure I.1 is a simple map showing how the land surface of the earth is divided into countries—one example of the diversity of human life. What do these divisions tell us? Have they always existed? Are the divisions changing today—possibly even between the time when I am writing and the time when you read these words? Is the total number of countries increasing or decreasing? What do these political divisions reflect—population distribution, cultures, development? Is it possible to aggregate countries? Are there meaningful groupings? Do basic aggregations such as North versus South, or East versus West, have any value? These are the sorts of questions that human geographers ask.

Figure I.2 is similar to Figure I.1, except that it depicts the political divisions that existed in 1938. To say that the two maps are different seems elementary, but it is a crucial point. The human world does indeed change—substantially. The major changes evident when we compare the two figures reflect the decolonization of Africa and Asia, the political consequences of World War II, and the post-1989 political transformation of Europe and the former USSR (see Box I.1). The most general consequence is a significant increase in the number of countries, from about 70 in 1938 to more than 190 in 2010. There is little reason to believe that a political world map in 50 years' time will closely resemble that of today. The contemporary political world is characterized by two tendencies in particular: first, a tendency for regions to begin a process of integration, as in the case of the European Union, and, second, a tendency for portions of some countries to seek a separate political identity, as in the case of Quebec within Canada.

Human geographers are interested in these issues for at least three general reasons. First, the political partitioning of the world is an example of one of our recurring themes: regionalization, or the organization of space. Second, political units usually play a major role in determining such matters as the numbers of births and deaths and the availability of paid employment. Third, many political divisions reflect differences in ethnicity and/or culture (for instance, in language or religion). Human geographers are able to identify relationships between culture and political divisions, often

FIGURE I.1 World political divisions, 2010

FIGURE I.1 World political divisions, 2010

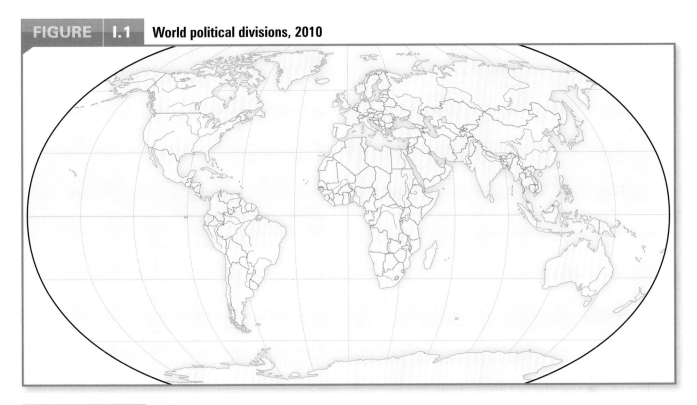

FIGURE I.2 World political divisions, 1938

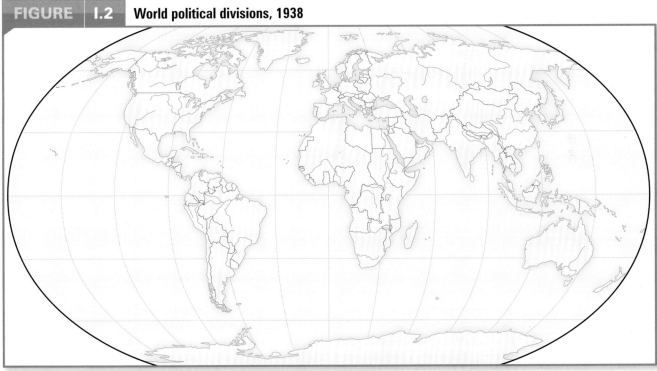

with the help of maps. These and related issues are addressed specifically in Chapter 8.

A WORLD DIVIDED 2: ECONOMIC DIVISIONS

The world is divided not only into political units but also into larger areas to which labels such as 'more developed' and 'less developed' are often applied. Our human world is full of diversity: languages, religions, ethnic identities, and standards of living all vary from place to place. There are differences among individuals and also more general differences between groups of individuals. Similarly, landscapes

Pedestrians cross a street in front of The Bay department store in downtown Vancouver, BC.

Darryl Dyck/The Canadian Press

Farm labourer with his family in front of their house in rural Rivas department west of Lake Nicaragua.

Dennis Cox/Alamy/GetStock

Box I.1   Germany Reborn

The new Germany was born on 3 October 1990 as a result of the unification of West Germany and East Germany. That this unification was achieved peacefully is remarkable, given the political tensions that had plagued Europe since 1945. A united Germany was one product of the sweeping changes that occurred throughout Eastern Europe beginning in 1989. Germany first appeared as a country on the map of Europe in 1871, following the Franco-Prussian War. The specific territory changed in accord with a series of military activities (most notably, of course, the two world wars), and in 1945 Germany was occupied by the four Allied powers and divided into four zones: American, British, French, and Soviet. In 1949 the three western zones linked to create West Germany. The Soviets chose not to co-operate and instead created East Germany. Gradually, during the 1950s, links between the two diminished and an 'Iron Curtain' was put in place separating not only East and West Germany but also Eastern (communist) and Western (capitalist) Europe. The most notorious indicator of this division was the Berlin Wall, initially set up in 1961, which divided the then international city of Berlin. Berlin, although located in East Germany, was a microcosm of Germany, with American, British, French, and Soviet sectors. In 1989, however, the wall (literal and figurative) dividing Germany began to crumble.

The changes in Eastern Europe began in Poland and quickly spread, removing Communist governments in favour of democratic political systems. Even today, political observers find it difficult to explain why such remarkable changes occurred: after all, these changes were neither predictable nor inevitable. Popular mass movements resulted in democracies replacing repressive Communist governments. For Germany, the remarkable outcome was a decision, agreed to in February 1990 at a meeting in Ottawa, to unify Germany. Technically, East Germany disappeared some eight months later, absorbed into West Germany, and the united country is now known, once again, simply as Germany.

Germany is again a major European power with a population of 82.2 million (as of 2008) and a new capital in Berlin, but numerous problems remain to be faced. There is increasing concern about the exceptionally low fertility rate, which, in 2008 at 1.3 children per woman, is well below the replacement level of 2.1 children per woman. Factors identified as contributing to low fertility include secularization (i.e., a decrease in religious practice and belief) and economic uncertainty. In Germany today over 30 per cent of women choose not to have children and many others have just one child. Economic problems in the former East Germany include the need to make industry more competitive and to modernize agriculture. In the former West Germany, economic recession and inexpensive imports have given rise to an economic and social crisis in the major industrial region of the Ruhr. Major environmental problems, including high levels of pollution and related industrial hazards, are apparent in the east. It is suspected, for example, that large areas of forest have died because of sulphur dioxide pollution, and that groundwater is substantially polluted. There are serious social problems, especially in the urban areas, resulting from the differences between the former states and the immigration of labour from elsewhere. The collapse of communism in the east caused massive unemployment, which has prompted a tightening of the rules for admitting refugees. Right-wing nationalists exploit social discontent and direct racist violence against foreign workers, especially Turks and Jews. For example, in 1992, 80 Jewish cemeteries were desecrated, roughly the same number as in the five-year period from 1926 to 1931 during the rise of the xenophobic and racist German Nazis, and in 2006 there were 1,046 violent racist crimes reported, and probably many others unreported. Despite all these problems, however, the new Germany is clearly playing a key role in the Europe of the twenty-first century.

vary. Each particular location is unique, of course, different from every other—but it is often more useful to note the more general differences that exist between regions. The photographs on page 6 suggest a basic distinction in our contemporary world. In the first we see middle-class people in a prosperous area of a modern Western urban centre; we can probably assume that most of them are employed, own or rent comfortable homes, and have surplus ('disposable') income for purposes such as recreation. In the second image we see a family in a rural area of what is often called the less developed world: small farmers with little more than what they need for subsistence. Pictures like these suggest that the human world we are studying is in reality at least two worlds.

It is important to realize that major differences exist not only between major world regions but also between other spatial entities, such as regions inside countries (Box I.2). Even within a single city, the quality of life can differ enormously between the outer suburbs and the inner-city skid row. For more on this theme, see especially Chapters 5, 7, and 13.

A WORLD TRANSFORMED 1: THE IMPACT OF MODERNIZATION

Table I.1 provides basic sample information on population totals, births and deaths, and life expectancy for various years from 1700 to 2008. Population totals have increased, the birth rate has declined, the death rate has declined, and life expectancy has increased. In

Box I.2   Less Developed Canada?

Although Canada is a part of the developed world, as measured by various cultural and economic criteria, the benefits are not distributed equally among all Canadians. Throughout Canada, Aboriginal people live different lives from most other Canadians. Aboriginal communities are characterized by more youthful populations, shorter life expectancies, and higher birth rates that result in a high rate of natural increase (about double the national average and comparable to rates in many less developed countries). More generally, Aboriginal people in Canada are disadvantaged; they are more likely than other Canadians to be unemployed, to rely on social assistance, to live in poor-quality housing, to have limited access to medical services, and to have little formal education.

One of the more notorious settings was Davis Inlet, a government-created Innu community on the Labrador coast of about 680 people that lacked even running water and had deep-seated social problems. In late 2002 and early 2003, the community moved—at a cost to the federal government of some $152 million—15 kilometres across the Labrador Sea to the new village of Natuashish, in the hope that a new place would allow for a fresh start.

The situation on reserves in the north of many provinces is also deplorable. A survey of housing on northern Manitoba reserves estimated that 1,800 (29 per cent) of the 7,200 units were in need of major repair and that an additional 1,400 units were required. More than 50 per cent of reserve residents in Saskatchewan are dependent on welfare, and on some isolated reserves the rate is close to 90 per cent. Further, many communities lack piped water and sewage systems—a situation that has clear implications for health. For residents of these reserves, the rate of tuberculosis is

eight times the provincial average. Frustration, hopelessness, depression, and apathy are common, with high rates of violent death, family breakdown, and substance abuse. Another tragic circumstance came to light in late 2005 when many residents were evacuated from the Kashechewan reserve on the shore of James Bay in northern Ontario because a polluted water supply caused serious health problems. More generally, a 2007 newspaper article noted that the new territory of Nunavut is 'racked by violence, with rates of homicide, assault, robbery, rape and suicide stunningly above the national average' (Harding, 2007). That such conditions should persist in relatively close proximity to some of the most affluent communities in the world is a scandal; yet most Canadians are still able to ignore it because Aboriginal people are typically marginalized, both spatially and socially.

Explaining this state of affairs is far from easy. Writing with general reference to the Canadian North, Bone (1992: 197) noted that 'the notion of "ethnically blocked mobility" arises; that is, Native Canadians may value educational and occupational achievements less than other Canadians because of cultural differences such as the sharing ethic and because of perceived or experienced discrimination.' Other important reasons have to do with the history of relations between indigenous people and European Canadians and the nature of capitalism as a social and economic system.

Canada is not alone in this respect. In Australia, the death rate for Aboriginal children is three times higher than the rate for other Australians, and life expectancy is about 20 years shorter. As in Canada, Aboriginal people often live in overcrowded housing, have unsafe drinking water, poor sanitary conditions, and higher rates of diabetes, heart disease, cancer, and violence.

| Table I.1 | Selected Demographic Data |

(a) World Population Totals (millions)		(b) Birth Rates and Death Rates (per 1,000), Finland			(c) Life Expectancy at Birth (number of years), France	
Year	Total	Year	Birth Rate	Death Rate	Year	Life Expectancy
1700	680	1755	45.3	28.6	1750	27
1800	954	1805	38.4	24.7	1825	41
1900	1,600	1909	31.0	17.7	1905	49
2008	6,705	2008	11.0	9.0	2008	81

NOTES:

1. Finland and France are selected because both countries have better than average records.
2. Birth rate refers to the total number of live births in one year for every 1,000 people living.
3. Death rate refers to the total number of deaths in one year for every 1,000 people living.

sum, these demographic changes reflect a process that we might label 'modernization'. Birth rates decline primarily because of changing social and economic conditions; death rates decline and life expectancy increases primarily because of technological advances related to sanitation and health. Population totals increase because of differences between birth and death rates. Clearly, however, this process of modernization has not had the same effects in all areas of the world.

Modernization involves much more than these demographic changes; cultural and economic aspects of landscape and life have also been transformed over the past three centuries, and human geographers employ their various

United Nations headquarters, New York. Founded in 1945 with an initial membership of 51 countries, the UN now has 192 members. Its goals are to maintain international peace and security; develop friendly relations among nations; co-operate on solving international economic, social, cultural, and humanitarian problems and promoting respect for human rights and fundamental freedoms; and help to harmonize members' efforts to meet those ends. The United Nations family of organizations is made up of the UN Secretariat, programs and funds such as the UN Development Program (UNDP) and UN Children's Fund (UNICEF), and various specialized agencies. Individual programs, funds, and agencies have their own governing bodies and budgets, and set their own standards and guidelines. Together they provide technical help and other practical assistance in virtually all areas of economic and social endeavour.

Visual&Written SL/Alamy/GetStock

perspectives—those suggested in our three recurring themes—to describe and explain these changes. Questions of population growth and change are discussed in Chapter 4.

One aspect of modernization evident today in advanced economies, which is causing changes both in landscape and in way of life, is the rise of post-industrial society. Among the changes associated with this process are an increase in the importance of information technologies, an economic transformation from a goods-producing to a service economy, the rise of professional and technical workers as a new middle class, and the related decline of the working class. It is possible that these ongoing changes will have consequences for our social and economic identities that will be just as dramatic as those of the Industrial Revolution. In North American and European cities, for example, traditional industries are declining, the professional workforce is growing, and the process of decentralization is accelerating. Such post-industrial cities include large numbers of people who are disadvantaged, unemployed, or the 'working poor' (many of the latter are part-time workers).

There is a new geography of employment, a new geography of housing, and a new geography of neighbourhoods. Here again, the human geographer uses her or his particular approach (any one of our three recurring themes) to identify these ongoing trends, the reasons for them, and their character, and to analyze their impacts on landscapes and ways of life. These and related topics are discussed in detail in Chapters 11–14.

A WORLD TRANSFORMED 2: GLOBALIZATION

Our fourth vignette addresses what many observers consider to be the most important change—or, more accurately, set of changes—evident in the early twenty-first century. Put simply, our long history of creating distinctive lives and landscapes in essentially separate parts of the world is ending. More and more, we are seeing groups of people and the places they occupy changing in accord with larger global (as opposed to regional or local) forces. These global forces are something more than the movement between places that began in the fifteenth century as Europeans moved overseas, and they are affecting all aspects of people and place. There is no single reason for these changes, but human geographers

Kensington Market, Toronto. In the Kensington neighbourhood, shops open out onto the streets, forming an old-fashioned market. Local markets offer a pleasant alternative to the more common urban experience of indoor malls and major grocery chains. Kensington's appeal is enhanced by its ethnic diversity. In earlier years many Jewish and Italian immigrants settled here, but eventually they moved on to more affluent neighbourhoods, and today the dominant cultures are Portuguese and Caribbean. Close by the market are two other distinctive areas: the city's original Chinatown and the College Street strip of upscale cafés and clothing stores.

Dick Hemingway Photographs

are always quick to point out that a key factor is humans' increasing ability to overcome what we call the friction of distance. Today we can move our goods and ourselves from one place to another much more rapidly than ever before, while ideas and capital can travel anywhere in the world almost instantaneously. We will consider some detailed implications of this point throughout this book, but especially in Chapter 9.

Culturally, as we will see in Chapters 3–7, a particular set of attitudes, beliefs, and behaviours is spreading that reflects a more ecocentric environmental ethic, increasing insistence on gender equality, and a better understanding of the linguistic, religious, and ethnic bases for differences between groups of people. More generally, we are seeing the extension of some American characteristics to other parts of the world—what has been described as 'Coca colonization'. Politically, as discussed in Chapter 8, many countries are working together for strategic reasons—the United Nations is the prime example of such a grouping. Economically, as discussed in Chapters 9–14, the connections between different parts of the world are constantly increasing in terms of product movement and capital investment; corporate headquarters in major cities now control economic activities throughout the world.

But this description is incomplete, and therefore misleading. It is tempting to jump on the globalization bandwagon and begin interpreting all human geographic change in those terms. But even a brief reflection on our personal lives or a cursory glance at newspaper headlines will make it clear that the local still matters to us. Think of your own life, of the places you frequent and the people who matter to you. For most of us, our home and local community, our family, friends, and neighbours, give shape and meaning to our lives. Distance has not become irrelevant, and the local has not been submerged under a tidal wave of globalization. Today, many groups of people want to see their specific culture recognized and their particular region attain some autonomous status; in the Canadian context alone, we can point to several separatist movements, the political aspirations of First Nations, and the tribal urge sometimes evident in the behaviours of some cultural groups. Certainly the forces associated with globalization are powerful, and their impacts are undoubted, but they are not the only forces at work today.

Indeed, 'globalization' is a highly contested term and a much debated topic. One attempt to make some sense of the many and diverse viewpoints evident today identifies three theses concerning what globalization is (and is not) and how it is (or is not) transforming the world. The *hyperglobalist* thesis views globalization as a new age, while *skeptics* question the very existence of globalization. In between these two contrasting viewpoints is the *transformationalist* thesis. Table I.2 summarizes these three important theses, providing useful background information for several of the debates introduced in later chapters.

About This Book

The diversity of human geography means that we have much subject matter and many methods to introduce. Chapters 1 and 2 address two basic issues.

First, in Chapter 1 we investigate the nature of human geography and its history as an endeavour of research and exploration. Our

Table I.2	Three Theses about Globalization		
	Hyperglobalist	**Skeptic**	**Transformationalist**
What is happening?	The global era.	Increased regionalism.	Unprecedented interconnectedness.
Central features	Global civilization based on global capitalism and governance	Core-led regionalism makes globe less interconnected than in late nineteenth century.	'Thick globalization'. High intensity, extensity, and velocity of globalization.
Driving processes	Technology, capitalism, and human ingenuity.	Nation-states and the market.	'Modern' forces in unison.
Patterns of differentiation	Collapsing of welfare differentials over time as market equalizes.	Core–periphery structure reinforced, leading to greater global inequality.	New networks of inclusion/exclusion that are more complex than old patterns.
Conceptualization of globalization	Borderless world and perfect markets.	Regionalization, internationalization, and imperfect markets.	Time–space compression that increases interaction between places.
Implications for the nation-state	Eroded or made irrelevant.	Strengthened and made more relevant.	Transformed governance patterns and new state imperatives.
Historical path	Global civilization based on new transnational elite and cross-class groups.	Neo-imperialism and civilizational clashes through actions of regional blocs and neo-liberal agenda.	Indeterminate; depends on construction and action of nation-states and civil society.
Core position	Triumph of capitalism and the market over nation-states.	Powerful states create globalization agenda to perpetuate their dominant position.	Transformation of governance at all scales and new networks of power.
Moral message	Pro-globalization. Globalization is real, and is a moral good and force for progress.	Mostly anti-globalization. Globalization is a discourse propagated by powerful interests.	Anti-globalization. Globalization is real and requires regulation to optimize it.

SOURCE: Adapted from: D. Held, A.G. McGrew, D. Goldblatt, and J. Perraton, *Global Transformations: Politics, Economics and Culture* (Cambridge: Polity Press, 1999),10; W.E. Murray, *Geographies of Globalization* (New York: Routledge, 2006), 353–4.

discipline has a long and distinguished history, involving many different groups of people and many noteworthy individuals working in different places at different times. The close ties between human and physical geography are evident. By the time you reach the end of this chapter, the origins of our three recurring themes should be clear.

Second, in Chapter 2, we look at the many ways in which human geography is studied today. This subject naturally includes many technical matters. Before turning to them, however, we will look at the philosophies behind each of our three recurring themes. Many of the ideas introduced in Chapter 2 may seem too abstract on your first reading, but as you read further you will find that many of them are central to the issues discussed in later chapters.

Chapter 3 offers an overview of humans' global impact on and adjustment to the physical environment. Surveying humans' occupation of the earth through time, we consider the use and abuse of resources and related environmental issues, which are of crucial importance to contemporary life. The theme of the relationship between humans and land is especially evident in this chapter.

The next two chapters are closely related: both deal with population issues, specifically questions of growth through time, both globally and in major world regions. Chapter 4 describes and suggests explanations for the exponential growth in global population that has occurred especially since about 1650, and identifies some of the consequences of that growth. Understanding population growth is essential if we are to understand many of the other issues discussed in this book. The use of population data as indicators of regional inequalities is discussed in Chapter 5, which provides a fuller account of the idea, introduced in Chapter 3, that the countries of the world can be divided into two general groups, less developed and more developed. All three of our recurring themes come into play in these two chapters.

Much human geography is best studied on a group scale, employing the concepts of culture and society as they relate to human landscapes. Chapter 6 focuses on the evolution and regionalization of landscapes and includes discussions of such cultural factors as language and religion. Chapter 7 introduces some current theoretical concepts in order to clarify the relevance to landscape of social factors such as ethnicity, class, gender, and sexuality. Like Chapter 5, Chapter 7 encourages us to think about various examples of inequality in the human experience from place to place. Two of our three recurring themes, humans and land and regional studies, are central to these discussions.

Political groupings of people in space are the subject of Chapter 8, which looks at the reasons behind the current partitioning of the world into territories. Because division into political units is the most fundamental of the ways in which humans divide the world, this chapter includes discussions of the significance of these divisions for our way of life. Perhaps the most compelling aspect of this chapter is its discussion of the considerable evidence indicating that many of the tensions in our contemporary world—tensions reported regularly in the news media—result when politics and culture disagree on the way space should be organized. Here again, as in the two preceding chapters, two themes—humans and land and regional studies—are crucial.

One feature of the contemporary world that will become obvious in Chapters 3 through 8 is the increasing necessity for human geographers to discuss topics at a global (as opposed to a regional or national) scale. To point out this trend is one way of saying that our world is becoming more and more integrated and interconnected. Located at a pivotal point in this textbook, Chapter 9, 'A Globalizing World', builds on much of the content of Chapters 3 through 8 and anticipates much of the content of Chapters 10 through 14. Discussions of the means by which humans overcome distances through transportation and trade lead logically to consideration of globalization processes. At the same time the opportunity is taken to reinforce some of the fundamental human geographic concepts introduced in Chapter 2.

The next five chapters include a good deal of spatial analysis—our third recurring theme— while also reflecting our other two themes. Chapter 10, on agriculture, asks why specific agricultural activities are located where they are; describes how the various major agricultural regions around the world have evolved; discusses contemporary agricultural landscapes in both the less and the more developed worlds, using a political economy perspective and with explicit reference to restructuring processes; and considers issues related to food consumption.

Chapter 11, the first of three chapters on settlement, discusses the reasons behind settlement patterns, both rural and urban; the differences between rural and urban experiences; the origins and growth of urban centres; and world urbanization and the growth of world cities. Chapter 12 looks at the location of activities inside cities; at how cities are governed and planned; at how people move around cities; and at the spread of cities into previously rural areas. Chapter 13 considers cities as homes, with accounts of housing and an emphasis on the formation and meaning of neighbourhoods; examines reasons for poverty and deprivation inside cities; describes cities as centres of production and consumption; and also looks at city life in the less developed world. Chapter 14, on industrial activity, looks at the reasons behind industrial location decisions; the period known as the Industrial Revolution; contemporary industrial circumstances in a global context; the complex processes of industrial restructuring (possibly related to a new post-industrial phase); and the general tendency for countries to exhibit uneven industrial development. This last discussion is closely linked to the earlier discussions of inequalities in Chapters 5 and 7. Finally, the Conclusion highlights some of the changing ways that human geographers study our changing world.

For a summary of the structure of this book, indicating the chapters of most relevance for the various subdisciplines of human geography, see Table I.3.

As you are beginning to realize, human geographers seek to understand both the world we live in and how we live in the world. To achieve these aims, this book incorporates not only a wide variety of facts about humans and their world, but a number of different approaches to those facts. Our understanding of human geography is enhanced by acknowledging the links between human geography and physical geography, as well as other social science disciplines. Where appropriate, these links are highlighted.

About Being a Human Geographer

Human geography is worth studying not only for the importance of the subject matter but also for the career training that it offers. Many graduating students find that a background in human geography, especially in conjunction with physical geography, leads to diverse employment opportunities in such areas as education, business, and government. Geography students are employed as cartographers, geographic information specialists, demographers, land-use and environmental consultants, social service advisers, and industrial and transportation planners, to mention only a few possibilities.

Three distinctive features of an education in geography make graduates particularly valuable to employers:

1. subject matter, physical and human, that provides the geographer with a specific perspective and knowledge base;
2. technical experience with data collection and analysis; and
3. recognition that any attempt to understand our contemporary world must take into account the crucial roles played by space and place.

Table I.3	Subdisciplines of Human Geography and Text Coverage
Subdiscipline	**Chapters**
Geography of natural resources	3
Geography of energy	3 and 14
Population geography	4 and 5
Cultural geography	6 and 7
Social geography	7, 13, and 14
Political geography	8
Economic geography	9, 10, 11, 12, and 14
Transportation geography	9
Geography of trade	9
Agricultural geography	10
Settlement geography	11, 12, and 13
Rural geography	10 and 11
Urban geography	11, 12, and 13
Industrial geography	14
Geography of tourism	7
Regional development and planning	14

SUMMARY

GEOGRAPHY

Literally, writing about the world.

THREE RECURRING THEMES

1. Humans and land: the human world, a landscape, is continuously changing because of human actions within institutional and physical frameworks.
2. Regional studies: describing the earth and dividing the whole into parts—regions—in which one or more variables exhibit some uniformity.
3. Spatial analysis: explaining the locations of geographic phenomena using abstract arguments and quantitative procedures.

Because the human world is constantly changing, human geography typically includes a time dimension.

HUMAN GEOGRAPHY/PHYSICAL GEOGRAPHY

Related disciplines, both traditionally and for logical reasons. Best introduced separately at the introductory level because physical geography is a physical science, while human geography is a human science.

OUR SUBJECT MATTER

Human behaviour as it affects the earth's surface.

GOAL OF THE BOOK

To provide a basis for comprehending the human world as it is today and as it has evolved.

THIS BOOK IS ABOUT

The world we live in and how we live in the world.

ABOUT BEING A HUMAN GEOGRAPHER

As a human geographer, you will have an understanding of our world and a sound training for a variety of careers.

QUESTIONS FOR CRITICAL THOUGHT

1. What distinguishes human geography from other disciplines in the social sciences?

2. What are the similarities and differences between a human geography of the world in the early twentieth century compared to today?

3. Why does the geographical perspective matter?

FURTHER EXPLORATIONS

Cloke, P., P. Crang, and M. Goodwin, eds. 2005. *Introducing Human Geographies*, 2nd edn. London: Hodder Arnold.

A collection of valuable chapters on a range of human geographic topics. A strong British focus.

Demko, G.J., J. Agel, and E. Boe. 1992. *Why in the World: Adventures in Geography*. Toronto: Anchor Books.

Aimed at the popular market, this entertaining volume asks and answers a multitude of questions to demonstrate the practical relevance of geography.

Douglas, I., R. Huggett, and C. Perkins, eds. 2006. *Companion Encyclopedia of Geography: From the Local to the Global*. London: Routledge.

A revised edition of the volume noted immediately below with place as the unifying principle.

Douglas, I., R. Huggett, and M. Robinson, eds. 1996. *Companion Encyclopedia of Geography: The Environment and Humankind*. London: Routledge.

A comprehensive volume centred on the idea that geography reflects the interdependence of humans and environment.

Hanson, S.E., ed. 1997. *Ten Geographic Ideas That Changed the World*. New Brunswick, NJ: Rutgers University Press.

A cleverly conceived book written in a style accessible to introductory-level students and containing statements by leading practitioners of the discipline; a good read.

Kitchin, R., and N. Thrift. 2009. *International Encyclopedia of Human Geography*, 12 vols. New York: Elsevier.

This is the authoritative source of information about the concepts and practice of human geography, with over 1,000 entries. Available in print and on-line.

Phillips, M., ed. 2005. *Contested Worlds: An Introduction to Human Geography*. Burlington, Vt: Ashgate.

A well-conceived volume that covers some of the important debates about the discipline of human geography and about the topics studied by human geographers.

Rogers, A., H. Viles, and A. Goudie, eds. 2003. *The Student's Companion to Geography*, 2nd edn. Oxford: Blackwell.

A clear and comprehensive student guide to both physical and human geography comprising over 50 entries on such diverse topics as the history of geography, the art of interviewing, world libraries, and careers for geographers.

Warf, B., ed. 2006. *Encyclopedia of Human Geography*. London: Sage.

Useful for students, although with a marked American emphasis.

ON THE WEB

GEOGRAPHY: FACTS, NEWS, DISCUSSION

Central Intelligence Agency: The World Factbook

www.cia.gov/library/publications/the-world-factbook/

Major source of basic factual information, with maps and country data. Updated every two weeks.

Geography in the News

www.geographyinthenews.rgs.org/

Up-to-date source for all things geographical. Maintained by the Royal Geographical Society with the Institute of British Geographers. A good source of quality information.

Global Issues

www.globalissues.org/

A broad-ranging website with material on all major global issues.

About Geography

www.geography.about.com/science/geography/?once=true&

A site designed to attract a wide audience; addresses basic popular geographic questions.

Geography Videos

geovideos.fliggo.com/

GeoTube is a collection of videos from YouTube and other on-line sites.

GEOGRAPHY IN CANADA

The Atlas of Canada

atlas.nrcan.gc.ca/site/index.html

An excellent resource for learning about the geography of Canada, both physical and human.

Statistics Canada

www.statcan.gc.ca/start-debut-eng.html

The aim of Statistics Canada is to provide statistics about all aspects of the country—population, resources, economy, society, and culture. Home to population and agricultural census data.

Resources for Teaching and Learning about the Geography of Canada

www.canadainfolink.ca/geog.htm

Numerous useful resources for teachers and students of Canadian geography. Includes much basic factual information.

Canadian Heritage

pch.gc.ca/index-eng.cfm

Canadian Heritage is responsible for national policies and programs that promote Canadian content and foster cultural participation, active citizenship, and participation in Canada's civic life.

GEOGRAPHY IN THE UNITED STATES

United States Census Bureau

www.census.gov/

Wealth of statistical data on many topics of interest to human geographers, including population, employment, housing, and poverty.

Library of Congress: Map Collections

lcweb2.loc.gov/ammem/gmdhtml/gmdhome.html

Information about various maps and atlases.

United States Geological Survey

geography.usgs.gov/

Although mostly of interest to physical scientists, this site includes information on natural resources and environment.

WHAT IS HUMAN GEOGRAPHY?

This chapter provides an overview of the origins and evolution of human geography, describing advances in geographic knowledge through time and noting how this knowledge gradually was organized to form a new academic discipline. The chapter concludes with a look at the changing character of contemporary geography, the links between physical and human geography, and the current status of human geography as a social science.

The approach taken here is chronological: from the major contributions of the Greek, Chinese, and Islamic civilizations, to the period of European overseas exploration, which began in the fifteenth century and led to the development of new techniques of mapping and improved descriptions of the world. Beginning with the attempts by seventeenth- and eighteenth-century writers such as Varenius and Kant to organize geographic knowledge, the formal discipline of geography emerged in European and North American universities during the late nineteenth and early twentieth centuries, inspired by the ambitious writings of the great nineteenth-century geographers such as Humboldt and Ritter.

Monument to the Discoveries in Lisbon, Portugal celebrates and honours the Portuguese who took part in the Age of Discovery of the 15th and 16th centuries. Henry the Navigator stands at the tip of the statue followed by other famous explorers, scientists, missionaries, and cartographers.
lillisphotography/iStock

To understand contemporary human geography, it is essential to understand how it has changed through time. Indeed, there would be cause for great concern if such change did not occur. *Geography, like most other academic disciplines, functions to serve society. As society and societal requirements change, so does geography.* Geographers have always had a consistent purpose: to describe and explain the world. Only the manner in which they approach this task changes. For example, for many years geographers were principally involved in discovering, describing, and explaining an increasingly better-known world. As unknown areas became known, geographers worked feverishly either to fit new facts into established knowledge or to propose radically new knowledge bases. Their most important tool was the map. More recently, since the nineteenth century, geographers have reoriented their activities. Once basic global descriptions were in place, the emphasis shifted to developing better explanations and clearer understandings of geographic facts.

Full of drama and intriguing individuals, the history of geography is a fascinating subject in its own right. At the same time, it provides the background that is essential if we are to fully understand contemporary geography (Box 1.1).

Preclassical Geography

It seems likely that the earliest geographic descriptions took the form of maps—a simple but effective means of communicating spatial information. No doubt maps have been created, temporarily at least, by all human groups. Roughly sketched with a stick in sand or soil, or carefully scratched into rock or wood, maps could be used to show the location of water, game, or a hostile group. To the extent that geography is about maps, humans have always been geographers. Lack of hard evidence need not prevent us from recognizing the centrality of geography to our human existence. Our ancestors could not function without maps any more than we can today.

The world's first civilization—for a discussion of this often-contentious term, see Chapter 6—emerged in Mesopotamia (now southern Iraq) some time after 4000 BCE. Surviving Mesopotamian maps, drawn on clay tablets, typically show local areas, reflecting limited knowledge of areas beyond the immediate environment. Geographic knowledge probably was similarly limited for all the civilizations before that of classical Greece: namely, the civilizations in the Nile Valley, the Indus Valley, China, Crete, the Greek mainland (Minoan and Mycenaean), and southeast Mexico (Olmec).

Classical Geography

With the emergence of classical Greece, shortly after 1000 BCE, geographic understanding—and hence maps—for the first time began to cover more than small local areas. The reason was that the Greeks were the first civilization to become geographically mobile and to establish colonies. Greek scholars initiated two major geographic traditions.

The first tradition is *literary*, involving written descriptions of the known world. Many Greek scholars contributed to this tradition, although relatively few of them were geographers. Herodotus (484–*c*. 425 BCE), for example, is best known as the first great historian, but he was also an accomplished geographer. His descriptions of lands and peoples—based

The GA.SUR or Nuzi map. This early map was engraved on a clay tablet *c*. 2200 BCE and discovered in 1930–1 by archaeologists digging at an ancient site in what is now northeastern Iraq. The original tablet has since been lost. Measuring just 7.6 × 6.5 cm, the map was oriented with the east at the top and showed an area bounded by two ranges of hills and bisected by a stream. It seems to have been intended to indicate the location of a parcel of land to the west of the stream, perhaps for purposes of defining ownership or assessing taxes.

Courtesy of the Semitic Museum, Harvard University.

Box 1.1   Telling a Story of Geography

This chapter aims to tell the story of geography as a tradition of scholarly inquiry, as one way of looking at and seeking understanding of the world in which we live. This sounds straightforward enough at first blush, but 'the deceptively simple word "geography" embraces a deeply contested intellectual project of great antiquity and extraordinary complexity' (Heffernan, 2003: 3). There are two closely related challenges faced by anyone who tries to write the story of geography.

First, there is more than one story of geography. In any recounting of the past, some facts are included and others are excluded. The conventional telling of the story of geography involves a focus on the achievements of influential scholars, on ever-more sophisticated key ideas, on advances in factual knowledge, on influential books, and on technical advances in mapping.

One of the other stories that might be told is that of changing geographical ideas, both true and false. This way of thinking about the past of geography was advocated some years ago and labelled *geosophy* (Wright, 1926, 1947). Following this argument, the important subject matter to be included in the story of geography is what people believed at the time. After all, what people believe is a principal cause of their behaviour and therefore of the changing world. Wright favoured a story about geographical ideas rather than the more usual story of geographical realities.

Second, there is more than one way of telling the chosen story of geography. Inevitably, facts are interpreted and presented in accord with a particular way of thinking. In the English-speaking world the conventional version of the story of geography emphasizes achievement and progress from humble beginnings to the successes of today. Histories are written this way, at least partly, because we judge the past from the perspective of the present and we make the assumption that progress is an integral part of change over time. Indeed, Nisbet (1980: 4) claimed that 'no single idea has been more important than, perhaps as important as, the idea of progress in Western civilization for nearly three thousand years.'

A different version of the story of geography might focus on a theme other than economic 'progress'. Both the choice of facts to be included and the interpretation of these facts reflect the prevailing **discourse** as this validates a particular version of a story and a particular way of telling that story. In other words, who is telling the story and whose story is it? For example, was the period of European 'exploration' and 'discovery' not also a time of 'invasion' and 'conquest'? It is always important while reading to wonder, why *this* story, and why *this version* of the story. With these cautionary comments in mind, this chapter now tells its version of the story of geography.

---

on observations made during extensive travels to such areas as Egypt, Ukraine, and Italy—make it clear that he saw geography as a necessary background to history and vice versa. Possible relationships between latitude, climate, and population density were noted by Aristotle (384–322 BCE), who also speculated about the ideal locations for cities and the conflicts between rich and poor groups.

Eratosthenes (*c.* 273–*c.* 192 BCE)—often considered the father of geography because he coined the word—contributed to the literary tradition by writing a book describing the known world, though unfortunately no copy of this work has survived. He also mapped the known world at the time: the Mediterranean region and adjacent areas in Europe, Africa, and Asia (Figure 1.1). We know about the contributions of Eratosthenes and others because the literary tradition they began was summarized by Strabo (64 BCE–20 CE) in a multi-volume work that has survived, entitled *Geographia*. An encyclopedic description of the entire world known to the Greeks, it consists of two

introductory books, eight books on Europe, six on Asia, and one on Africa. Much of what we know today about Greek geographers before Strabo comes from his detailed summaries of their work.

The second tradition is *mathematical*. Thales (*c.* 625–*c.* 547 BCE), one of its originators, took a scientific view of the world and successfully predicted an eclipse of the sun in 585 BCE. By the fifth century BCE the Greeks knew that the earth was a sphere, and in the second century BCE Eratosthenes calculated the circumference of the earth. Shortly after that, Hipparchus devised a grid system of imaginary lines on the surface of the earth based on the poles and the equator—these, of course, were the lines of **longitude** and **latitude**, and they made mapping considerably more accurate. Determining the exact position of any location, however, remained difficult. Latitude was relatively easy to calculate by observing the angle of the sun's shadow using an early version of a sundial, but longitude continued to require estimation because there was no way

discourse
A system of ideas or knowledge that serves as the context through which new facts and ideas are understood.

longitude
Angular distance on the surface of the earth, measured in degrees, minutes, and seconds, east and west of the prime meridian (the line of 0° longitude that runs through Greenwich, England); lines of constant longitude are called meridians.

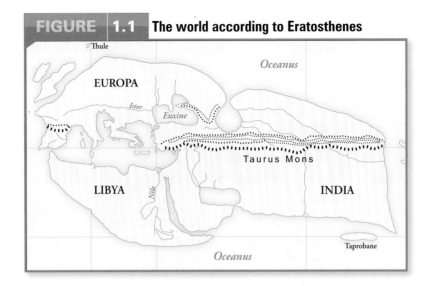

FIGURE 1.1 The world according to Eratosthenes

to measure time precisely, especially at sea. In addition, Hipparchus was the first person to tackle a problem that remains with us today: how to map the curved surface of the earth on a flat surface.

Much of this mathematical tradition was summarized by the Alexandrian Ptolemy (fl. second century CE) in his eight-volume *Guide to Geography*. Using the grid system devised by Hipparchus, Ptolemy produced the first index of places, or gazetteer, of the world with coordinates provided for about 8,000 locations. Although his gazetteer is sometimes fanciful (for instance, he believed that a great continent must exist in the southern ocean; see Box 1.4) and includes numerous errors because latitude was not carefully measured and longitude was necessarily estimated, Ptolemy also produced a world map (Figure 1.2) that includes a grid system and that in its details is generally

a clear improvement over Eratosthenes's map (Figure 1.1).

Together, the written works of Strabo and Ptolemy and the world map of Ptolemy provide a useful indication of the achievements of Greek civilization in geography, for they make it clear that geography emerged and evolved in response to larger societal requirements. The civilization of ancient Rome added little to geographic knowledge, despite the expansion of the Roman Empire. Apart from some geographic guidebooks published in response to exploratory and expansionist needs, there was little development of either the literary or the mathematical tradition begun by the Greeks. Although Roman civilization continued until the fifth century CE, classical geography effectively ended with Ptolemy, and it seems that Roman leaders continued to rely on Ptolemy's work.

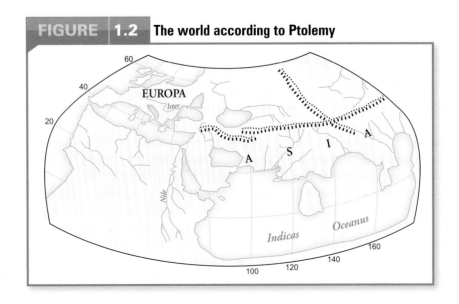

FIGURE 1.2 The world according to Ptolemy

Thus, the classical geography of the Greeks is the true beginning of our contemporary discipline. Geography in the classical world was concerned with the locations and interconnections of places on the surface of the earth. In addition to the word 'geography', the Greeks introduced 'cosmography', 'chorography', and 'topography'. **Cosmography** refers to the universe, heavens and earth. **Chorography** refers to places smaller than the earth, such as countries, while **topography** refers to local areas within countries. The Greeks' introduction of these terms indicates that they recognized the importance of differences in scale. As we have seen, the Greeks also produced relatively accurate maps using a sophisticated grid system. What is perhaps most significant is that the mapping procedures devised by Ptolemy have persisted to the present. Both the basic style and the language of the maps that we use today originated with Ptolemy.

The Fifth to Fifteenth Centuries: Geography in Europe, China, and the Islamic World

The period from the fifth to the fifteenth centuries was one of only sporadic and limited geographic work in Europe. Elsewhere, however, especially in China and the Islamic world, geography flowered. Again we find a close relationship between expanding societies and a thirst for geographic knowledge.

THE EUROPEAN DECLINE

The word 'geography' did not enter the English language until the sixteenth century. Medieval Europeans knew little beyond their immediate environment. Over the centuries, much of the knowledge gained by the Greeks was lost, and the only place where any kind of geographic work continued was in the monasteries. The general assumption, inside and outside the monasteries, was that God had designed the earth for humans (this doctrine is called **teleology**). In effect, during the Middle Ages geography as such no longer existed.

The clearest evidence of decline can be seen in maps of the period. The ancient Greek maps had been drawn by scholars with expertise in astronomy, geometry, and mathematics. The medieval European map-makers, by contrast,

were more interested in symbolism than scientific facts. Stylizing geographic reality in order to arrive at a predetermined structure, they produced maps that are less detailed and accurate than those produced 1,500 years earlier by the Greeks. The best examples are the 'T-O' maps produced between the twelfth and the fifteenth centuries. Consisting of a T drawn within an O (Figure 1.3), they show the world as a circle divided by a T-shaped body of water. East is at the top of the map; above the T is Asia; below left is Europe, and below right is Africa. The cross of the T is the Danube–Nile axis, the perpendicular part is the Mediterranean, and the map is centred on Jerusalem. In these maps depicting scriptural dogma—what Christians were expected to believe—symbols triumphed over facts.

Another type of medieval map divided the world into climatic zones, largely hypothetical, on either side of the equator. Still others were lavishly decorated: the Ebstorf *Mappemonde* (*c.* 1284) had as background a picture of the crucifixion, while the Hereford map (*c.* 1300) was really an encyclopedia. Perhaps the only medieval maps that served a practical purpose were the ones known as Portolano (meaning 'handy') charts (Figure 1.4). Dating from about 1300, these maps depicted a series of radiating lines. The lines did not serve to locate positions on the map, nor did the maps use any projection. Nevertheless, they often succeeded in locating coastlines accurately. Although there are no lines of latitude and longitude, a network of overlapping lines radiates from several centres on the map. Usually there are 8 or 16 such lines from each centre and these correspond to the points of the compass. Sailors were able to lay out compass courses using these lines.

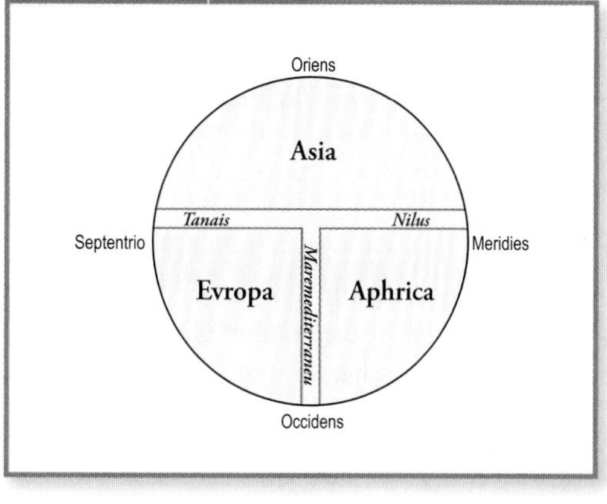

cosmography
The science that maps and describes the entire universe, both heavens and earth; this Greek term is rarely used today.

chorography
A Greek term revived by nineteenth-century German geographers who used it to refer to regional descriptions of large areas.

topography
A Greek term, revived by nineteenth-century German geographers to refer to regional descriptions of local areas.

teleology
The doctrine that everything in the world has been designed by God; also refers to the study of purposiveness in the world and to a recurring theme in history, such as progress or class conflict.

FIGURE 1.3 An example of a T-O map

FIGURE 1.4   An example of a Portolano chart

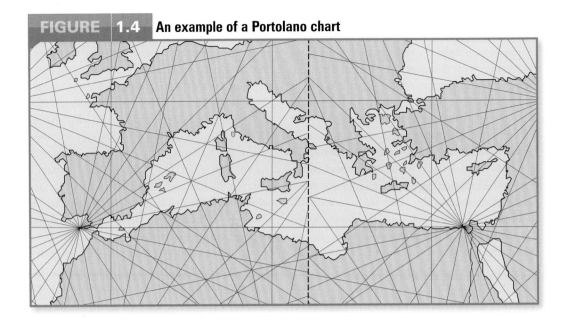

Medieval Europeans made very few contributions to geographic knowledge. Although Norsemen sailed to Greenland and North America, and Christian Europeans embarked on a series of crusades and military invasions to the Holy Land, the results, as far as geographic knowledge was concerned, were minimal. The most significant exploratory journey was that of Marco Polo (1254–1323), a Venetian who visited China and wrote a description of the places he visited. He was unable to add to Greek knowledge because, not being a scholar, he was largely unaware of it. The distinction is not always an easy one to grasp, but Marco Polo was an explorer, not a geographer.

GEOGRAPHY IN CHINA

During the period after the Greeks and before the fifteenth century, the major geographic advances took place in China and the Islamic world. In China a great civilization—clearly the ancestor of contemporary China—developed before 2000 BCE. The longest-lasting civilization in the world, China inevitably made important contributions to geographic knowledge. Writings describing the known world of the Chinese date back to at least the fifth century BCE. Later, the Chinese also explored and described areas beyond their borders; in 128 BCE, for example, Chang Chi'en discovered the Mediterranean region, described his travels, and initiated a trade route. Other Chinese geographers reached India, central Asia, Rome, and Paris. Indeed, Chinese travellers reached Europe before Marco Polo reached China (Box 1.2).

There is one important respect in which early Chinese geography differed from the European equivalent. It is a difference of geographic perspective—a different way of looking at the world. Traditionally, Chinese culture has viewed the individual as *a part of* nature, whereas the Greeks and subsequent European culture have typically viewed the individual as *apart from* nature. This distinction reflects the differing attitudes underlying Confucianism (which dates from about the sixth century BCE) and Christianity. Given their view of humans and land as one, it was natural for Chinese descriptive geographers to integrate human and physical description.

Maps were central to geography in China, as elsewhere, and there is evidence that a grid system was in use during the Han dynasty (third century BCE to third century CE). It appears that the first Chinese map-makers were civil servants who drew and revised maps in the service of the state. Their maps were symbolic statements, asserting the state's ownership of some territory.

GEOGRAPHY IN THE ISLAMIC WORLD

The second contribution to geography outside Europe came from the Islamic world. As the religion of Islam—founded in the seventh century CE by the prophet Muhammad (d. 632)—spread, it served as a unifying force, bringing together previously disparate tribes. Consequently, at the time when Europe was immersed in the Dark Ages, civilization

Box 1.2   Did China 'Discover the World' in 1421?

China was a great naval power in the early fifteenth century—so great that Chinese navigators may have reached North America 70 years before Columbus, circumnavigated the globe 100 years before Magellan, and reached Australia 350 years before Cook. We know that Emperor Zhu Di sponsored several voyages during his reign (1403–24) for the purposes of exploration, mapping, and collecting tribute from other peoples. We also know that in 1421 the great admiral Zheng set sail with many ships across the Indian Ocean to the east coast of Africa. However, a recent book—*1421: The Year China Discovered the World* (Menzies, 2002)—suggests that the expedition continued around the southern capes of Africa and South America and across the Pacific before arriving back in China in 1423. With the emperor's death the following year, the country entered a long period of isolation, and presumably this was why the story of the voyage remained unknown in Europe.

Menzies's theory—based in part on the existence of European maps from as early as 1428 depicting regions that Europeans themselves had not yet seen—is fascinating, although cartographers are not likely to rewrite their histories without some additional evidence. A 1763 Chinese map, discovered and unveiled in 2006, purports to be the needed additional evidence but has not resolved this debate. The map shows the Americas and Australia and is claimed to be a copy of a 1419 map, a claim that is not generally accepted. A Canadian writer, Chiasson (2006), supports the claims made by Menzies, suggesting that the Chinese established a colony on Cape Breton Island following the discovery of gold.

flowered in Arabia. As Islamic conquests spread beyond the Arab region, the geographic knowledge base expanded to include North Africa, the Iberian peninsula, and India.

By the ninth century Islamic geographers were recalculating the circumference of the earth. From then until the fifteenth century, they and their successors produced a wealth of geographic writings and maps based on earlier Greek work as well as Islamic travels. Among the most notable contributions were those of al-Idrisi (1099–1180), whose book on world geography corrected many of Ptolemy's errors. Perhaps the best-known traveller was ibn-Batata (1304–c. 1368), who journeyed extensively in Europe, Asia, and Africa. A third major addition to geography came from ibn-Khaldun (1332–1406), a historian who wrote extensively about the relations between humans and the environment. Maps produced by Islamic geographers, including al-Idrisi, centred on Arabia.

An important eleventh-century Arabic atlas, previously unknown to modern geographers, was discovered in a private collection in 2002, and is now housed in a library in Oxford, England. Widely regarded as a missing link in the history of cartography, this two-volume, 96-page manuscript includes 17 maps, two of them depicting the world as it was known at the beginning of the second millennium. The fact that some of the maps show travel routes suggests that, unlike earlier Greek and later European maps, they were intended not to represent actual landscapes but to serve a practical purpose as memory aids for travellers.

Chinese and Islamic geographies prior to the fifteenth century were roughly comparable to Greek geography. Two traditions,

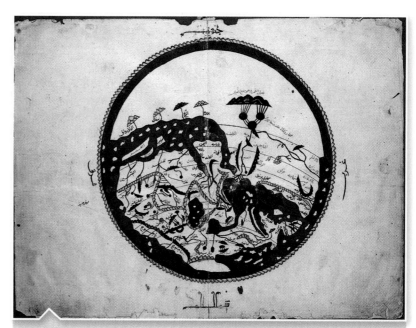

Al-Idrisi's circular world map: a copy (dated 1456) of the twelfth-century original. As early as the ninth century Muslim traders had travelled east to China by land and by sea; south along the Indian Ocean coast of Africa; north into Russia; and west as far as the Atlantic.

Al-Idrisi combined the knowledge acquired through these journeys with that available from the works of Greek and Persian geographers to produce the geographic description of which this map was a part. Depicting Europe and Africa as well as Asia, it is oriented with south at the top.

Bodleian Library, Oxford Pococke Manuscript 375, fols. 3c-4r. © Bodleian Library, University of Oxford.

mathematical and literary, developed, and map-making was central to most geographic work. In all cases, the geographers' work reflected the knowledge and needs of particular societies. As Box 1.3 suggests, maps in particular are perhaps best understood as one component of the gradual development of human knowledge about the world and ourselves.

The Age of European Overseas Movement

By 1400, then, geographic knowledge had grown considerably, but the 'known world' on which it was based was still limited to a small portion of the earth. In the fifteenth century European scholars began to recognize this fact, and the impact on geography was dramatic and significant. All three components of the geography discussed so far—mathematical, literary, and cartographic—underwent rapid change. Between the fifteenth and seventeenth centuries, Europeans embarked on a period of unprecedented exploratory activity that happened to coincide with a decline in Chinese and Islamic explorations.

Many factors, among them the desire to spread Christianity and to establish trade routes, contributed to the surge in European exploration, but there were two interesting additional factors. Printing technology, which was first applied to maps by the Chinese in 1155, was first used for map production in Europe in 1472. In 1410, editions of the work of Ptolemy were printed; editions including maps (redrawn using new projections) appeared in 1477. Printing allowed information, including that contained on maps, to diffuse rapidly. A copy of Ptolemy's Atlas that was printed in 1477 sold at a 2006 auction for almost US$4 million.

The second additional factor, which reflected the period's thirst for knowledge, was the establishment of what might be called centres of geographic analysis. The first, initiated by Prince Henry of Portugal in 1418, focused on key geographic questions such as the size of the earth and the suitability of tropical environments for habitation. In addition, techniques of navigation were taught there, and Prince Henry set in motion a series of explorations along the western coast of Africa. Interestingly, the maps produced following these voyages did not always reflect new discoveries. It seems that a 1459 map by Fra Mauro deliberately concealed new information in order to maintain secrecy.

EXPLORATION

Exploration is not, of course, geography, but exploration furnished new facts and provided the basis for new maps and new descriptive geographies. The relationship between geography and exploration is symbiotic, with maps and books encouraging exploration and exploration in turn generating new maps and books. The major explorations were led by Bartolomeu Dias around southern Africa (1486–7); by Christopher Columbus to North America (1492–1504); by Vasco da Gama to India (1497–9); and by Ferdinand Magellan, who reached Asia by sailing west (1519–22). The last in this line of great explorers was James Cook, who made three voyages into the Pacific (1769–80). It was Cook who corrected one of Ptolemy's greatest errors, revealing that the supposed southern continent did not exist, at least not as it had been envisioned (see Box 1.4). By 1780 the basic outlines of the world map were established, adding considerably to Europeans' knowledge of the world—and thus to the geographer's task.

Exploratory activities became easier as several basic hurdles were overcome. Most important was the absence, recognized over a thousand years earlier by the Greeks, of an accurate method of determining location. As we have seen, establishing latitude was no problem, but an instrument for establishing longitude at sea was not available until 1761 and was not used on a major voyage until Cook's second voyage, in 1772–5.

MAPPING

During the early phase of European overseas movement science in general changed from being a practice controlled by the church to one concerned with the acquisition of knowledge. In response to the demands of sea travellers, maps in this period returned to the model developed by the Greeks—a model in which facts triumphed over imagination. Unlike medieval maps, with their mixture of fantasy and dogma, the typical map from the fifteenth century onward was functional. Maps showing the grid of latitude and longitude replaced T-O maps centred on Jerusalem.

Box 1.3 Evaluating Our Place in the World

Maps are useful indicators not only of what we know about the world, but also of how we see ourselves in the world. It might be said that while we create maps, maps in turn recreate us. T-O maps, for instance, served to reinforce the Christian world view, while the ability to map an area suggests power—perhaps even ownership rights—over it. Maps can be, and often are, used to convey messages and are thus powerful symbolic representations. Nevertheless, most early maps were explicitly practical, whether they were used by mariners for navigation or by state officials as aids to tax collection.

The long history of map-making might be seen as one of slow intellectual growth and gradual spatial expansion. We begin with knowledge of our own bodies before moving on to our immediate surroundings, gradually progressing to a partially known world, and finally to detailed knowledge of the world. Throughout this process, the unknown is usually seen as dangerous, and often it is only with great reluctance that we allow newly discovered fact to triumph over familiar fiction. Herodotus did not accept Phoenician accounts of a circumnavigation of Africa about 600 BCE, and when Pytheas sailed out of the Mediterranean to England and Iceland in about 320 BCE he was branded a liar by Strabo because he was describing places presumed not to exist. In both of these cases, evidence contradicting existing maps was rejected because it did not fit the story that was accepted and understood.

Even today, maps are both enabling and constraining. They enable us to see what our own vision does not permit, but the fact that they necessarily simplify complex realities constrains our understanding. Indeed, the proliferation and increasing use of Internet maps, such as those provided by Google, and satellite navigation technologies have been blamed (by Britain's most senior cartographer in 2008) for wiping the rich history and geography of Britain off the map. This is because such maps are designed only to provide directional information, and not to reflect major landscape features. For more on the idea that maps are both enabling and constraining, see Chapter 2, especially Boxes 2.7 (on map projections) and 2.8 (on maps that are intentionally misleading).

One of the pioneers in these endeavours was the Polish scientist Nicolaus Copernicus (1473–1543), who produced maps of Poland that were used to settle a variety of boundary disputes.

The demand for geographic knowledge, and hence for maps, coincided with the rise of printing to encourage new developments. Gerardus Mercator (1512–94) was undoubtedly the most influential of the new map-makers. He tackled the crucial problem of projection: how to represent a sphere on a flat surface. His answer was the famous 1569 Mercator projection, still used extensively today. This projection, which showed the earth as a flat rectangle with a grid of latitude and longitude lines, was enormously useful to sea travellers because a straight line on the map was a course of constant compass bearing. By the early seventeenth century, Mercator's map had replaced all earlier charts used at sea, including the Portolanos.

Another map-maker, Abraham Ortelius (1527–98), produced the first modern atlas in 1570; such was the demand for geographic knowledge that it ran into 41 editions by 1612. From Mercator onward, map-making has steadily improved and consistently reflected new geographic knowledge.

Mercator's map of the North Pole, 1595.

Library and Archives Canada.

GEOGRAPHIC DESCRIPTION

The awakening of Europe and the burst of exploratory activity had an impact not only on map-making but also on geographic description. European scholars had to make sense of the 'new world'. Once again, society turned to geography to provide answers to a multitude of questions concerning the shape of the earth, the location of places, physical processes, and human lifestyles. Just as earlier Greek, Chinese, and Islamic geographers had described the worlds known to them, so the European geographers of the sixteenth century onward needed to describe the world they knew—one that was constantly expanding and changing. Now geographers faced an enormous task: writing about all aspects of the entire world.

The resulting flurry of geographic writing was of mixed quality. Some works of fiction were regarded as factual. Imaginative works by an author of uncertain identity, Sir John Mandeville, were reprinted three times in 1530 alone. Other authors perpetuated Ptolemy's error concerning the 'great southern continent'; as late as 1767 a Scottish geographer, Sir Alexander Dalrymple (1737–1808), wrote that almost all of the unknown areas between the equator and 50°S were land (Box 1.4).

Other geographers, however, largely reflected available knowledge and excluded content that was essentially surmise. Their works, along with developments in map-making, provide a clear indication of the progress of geography and geographic understanding. Among them was Peter Apian (1495–1552), a map-maker and writer who in 1524 published a book that divided the earth into five zones (one torrid, two temperate, and two frigid), provided notes on each continent, and listed major towns. Sebastian Münster (1488–1552), a contemporary of Apian, produced *Cosmography* in 1544, the first major work following the initial burst of European expansion activities that included descriptions of the major regions of the earth. The book is a careful compilation of existing knowledge, but is relatively weak for areas other than Europe. Nevertheless, it

---

Box 1.4   The Southern Continent

The Greeks were the first to suggest that a large land mass should exist in the southern hemisphere, to balance the large known areas of land in the North. Along with so many other provocative ideas, this notion was lost during the medieval period; the medieval belief in a flat world did not require any such symmetry. The first knowledge of the basic distribution of land and sea in the southern hemisphere was gained with the beginnings of European overseas exploration in the fifteenth century. Early voyages often were prompted by incorrect information and sometimes they returned with fiction as well as fact. For example, in the early sixteenth century Gonneville claimed to have discovered a tropical paradise in the South, inhabited by an easygoing, contented people. All records of this voyage were lost, but the claim clearly influenced French exploration for the next 200 years. Subsequent French and other explorers sought to rediscover this idyllic 'Gonneville land'.

When Magellan, on his voyage of 1518, discovered the Straits of Magellan, separating the island of Tierra del Fuego from the mainland of South America, it was thought that the island was the edge of the huge southern continent. Cartographers such as Orontius, in 1531, responded with maps that included the continent. By 1578, however, Drake had shown that Tierra del Fuego was not a large continent. Exploration then turned east with the discovery of the islands of South Georgia by de la Roche in 1675. The French

continued to search for Gonneville's lost land. When, in 1772, Kerguelen reported that he had discovered a place promising all the resources the mother country desired—wood, minerals, diamonds, and rubies—he called his discovery South France, and many thought he had finally located the long-sought Gonneville land. Find Kerguelen Island in your atlas and you will see just how wrong Kerguelen was: the island named after him is in an isolated, far from tropical location in the south Indian Ocean. Clearly Kerguelen discovered what he wanted to discover, just as Gonneville had almost 200 years earlier.

The distribution of land and sea in the southern hemisphere became clearer in the late eighteenth century, even as Dalrymple was unequivocally asserting that almost all of the area between the equator and 50°S was land. In 1773 Cook spent much of a long voyage searching for the southern continent, sailing farther south than any previous explorer; but instead of land, all he discovered was pack ice. Although, on his 1775 voyage, Cook discovered numerous small islands, all were inhospitable environments. The coastline of Antarctica was finally mapped during the nineteenth century.

This intriguing example is not necessarily atypical. Once a false idea is established, considerable effort may be needed before the true facts are known. False ideas can and do influence human behaviour.

remained the standard geographic work for at least 100 years and was printed in 46 editions and six languages.

The period from the early fifteenth century to the mid-seventeenth was one of enormous growth in geographic fact and fiction alike. Geography itself continued in the tradition begun by the Greeks. The process of discovering and making sense of ourselves and our world meant questioning old ideas as well as exploring. For geography, the result was a dramatic increase in factual content. Map-makers and writers strove to keep pace with these questionings and explorations and to encourage more. By the mid-seventeenth century, geography's main concerns still were mapping and world description.

Geography Rethought

VARENIUS

Contemporary geography continues to be concerned with map-making and description, but it also addresses many other questions. The first indication of these new questions appeared in 1650 with the publication of *Geographia Generalis* by Bernhardus Varenius (1622–50), which remained the standard geographic text for at least a century; the last English-language edition appeared in 1765. What made this work so distinctive and important was that in it Varenius provided an explicit definition of geography:

> Geography is that part of mixed Mathematics, which explains the State of the Earth, and of its Parts, depending on Quantity, viz. its Figure, Place, Magnitude, and Motion, with the Celestial Appearances, etc. By some it is taken in too limited a Sense, for a bare description of the several Countries; and by others too extensively, who along with such a Description would have their Political Constitution.

Clearly, Varenius was interested in establishing an appropriate place for geography in the system of knowledge as a whole. He went on to clarify the definition above:

> We call that *Universal Geography* which considers the whole Earth in general, and explains its Properties without regard to particular Countries: But *Special* or *Particular Geography*, describes the Constitution and Situation of each single country by itself which is twofold, viz. *Chorographical*, which describes Countries of a considerable Extent; or *Topographical*, which gives a View of some place or small Tract of the Earth.

It is worth noting that the universal/special distinction noted by Varenius was an important intellectual issue for all branches of knowledge at the time. Universal geography, more typically called *general geography*, was seen as crucial. It was not adequate merely to provide detailed descriptions of particular places; it was also necessary to develop an understanding of laws that applied to all places. In geography as in any other subject, this is a basic distinction. Varenius saw the two as mutually supportive. Interestingly, as we shall see, twentieth-century geographers sometimes failed to appreciate the need for both types of study, arguing for one at the expense of the other.

Varenius confirmed geography as the study of the earth's surface, both physical and human. Detailed description—special geography—is essential, as is the development of generalizations—general geography. There was little that was really new here; the Greeks, for example, had contributed to both of these areas. But the explicit recognition of two different approaches—a recognition prompted by the remarkable increases in knowledge and the prevailing intellectual trends at the time—did mark a new phase in geography's evolution.

BROADENING VISTAS: 1650–1800

The one and a half centuries, 1650 to 1800, following the publication of Varenius's major work witnessed major advances in a wide range of geographic issues. Exploratory activity continued apace, as did map-making. In addition, geographic questions were asked that, while often prompted by exploration, had little to do with the acquisition of new facts and the mapping of newly discovered places. Questions were asked about such fundamental issues as the role of the physical environment as a cause of the growth of civilization, the unity of the human race, and the relationship between population density and productivity.

These broad questions generated a variety of responses. Most of the individuals who tackled them were not geographers as such—in fact, knowledge had not yet been divided into the disciplines that we know today. Nevertheless, the views of J. Bodin (*c.* 1528–96) and C. Montesquieu (1689–1755) on the role of the physical environment, of G.-L. Buffon (1707–88) on the unity of the human race, and of T.R. Malthus (1760–1834) on population and food supply all were eminently geographic. Each of these issues will be considered later in this textbook. For now, the point to note is simply that all of them were raised before 1800.

Other geographic writing reflected the current trends. Thus several important geographies were published after 1650 that stressed encyclopedic description. Most such studies concentrated on describing political units. Both A.F. Büsching (1724–93), in a six-volume description of Europe published in 1792, and J.C. Gatterer (1727–99), who saw geography as earth description and description of humans using the earth, focused almost entirely on regional description.

IMMANUEL KANT

Immanuel Kant lived in Königsberg, Germany, from his birth in 1724 until his death in 1804. For most of his adult life he taught at the University of Königsberg, becoming a professor of logic and metaphysics in 1770. Through his writings, Kant has had a substantial influence over subsequent philosophy, but he also was interested in the natural sciences. His early publications were in astronomy and geophysics, and he lectured in physical geography (broadly interpreted to include much human geography) for 40 years beginning in 1756. He has been described as 'the outstanding example in western thought of a professional philosopher concerned with geography' (May, 1970: 3).

To introduce his lectures on physical geography, Kant emphasized that the subject involved the description or classification of facts in their spatial context. As with Varenius, this focus was not particularly novel. However, it seems that Kant has been interpreted by some later geographers, especially Alfred Hettner, as an explicit advocate of geography as a regional study. This interpretation is based on Kant's argument that geography is description according to space and that history is description according to time. In *Physische Geografie*, published in 1802, Kant asserted that geography and history together comprise all knowledge.

Although, again, there was nothing especially new about these ideas, it is possible that Kant contributed heavily to the notion of geography as a descriptive regional science. In an important sense, his view of geography as special, not general, represents a departure from Varenius. At the same time, there is little doubt that Kant was very much in accord with prevailing late eighteenth-century thought.

UNIVERSAL GEOGRAPHY: 1800–1874

The first three-quarters of the nineteenth century saw the decline of the geography associated with exploratory activity and centred on mapping and description. This period was one of radical change in the way knowledge was organized, in which the academic disciplines that we know today formally emerged at universities. The first geography departments were established in 1874, at several Prussian universities. Simply put, geography could not continue to be primarily mapping and description—the late nineteenth century required a much more academic discipline—and it was necessary that mid-nineteenth-century scholars lay the groundwork for the formal discipline. The years 1800–74 witnessed the culmination of all that we have discussed so far and the beginnings of much that has happened since.

The first major work of this period was published by a Danish geographer, Conrad Malte-Brun (1775–1826), between 1810 and 1829. Following in the established tradition, this all-embracing study includes both general and special geography as defined by Varenius. Mathematical, physical, and political principles were discussed along with physical phenomena, including animals and plants, and human matters, including race, language, beliefs, and law. Describing all areas of the known world, Malte-Brun succeeded in producing a complete geography.

HUMBOLDT AND RITTER

However, geography in the first half of the nineteenth century was dominated by two German scholars, Alexander von Humboldt and Carl Ritter (Boxes 1.5 and 1.6). Humboldt's greatest work was *Cosmos*, a five-volume study published between 1845 and 1862; the title is the Greek term for an orderly universe (as opposed to chaos). Like earlier geographers, Humboldt wanted to describe the universe, but that was not all. As he wrote, 'my true purpose is to investigate the interaction of all the

Box 1.5 | Humboldt

Alexander von Humboldt (1769–1859) was born in Berlin into the landowning aristocracy. Before attending university, he had the opportunity to mix with a diverse group of liberal intellectuals, and this experience, coupled with an inquiring mind, meant that he was interested in a wide range of academic and related topics. Following brief spells at two other universities, he commenced studies at the university in Göttingen and there met George Foster, who had recently returned from a world voyage with Cook. Apparently it was this encounter that started Humboldt along the geographic path. His studies continued at Freiburg, and when he inherited a substantial income, he decided to use it to support travel and related publication.

From 1796 until his death, Humboldt travelled extensively and published prolifically. His major travels were to Central and South America from 1799 to 1804. His American travels, which culminated in a visit to Washington and a series of meetings with Thomas Jefferson, resulted in a 30-volume work that was published between 1805 and 1834 and in many languages.

The writing of *Cosmos*—one of the most important and successful pieces of scientific writing ever accomplished—occupied Humboldt from the late 1820s until his death. In it he strove to investigate the interaction of all natural and human forces, drawing on all the sciences to explain everything from the Milky Way galaxy to microscopic organisms.

In many areas of thought he was well ahead of his time; he insisted, for example, that all races had a common origin and that no one race was inherently superior to any other.

Although he made no single great discovery, Humboldt was one of the most admired men of the nineteenth century and might justifiably be regarded as 'the last man who knew everything' (Johnson, 2005: 3). For several years he was the toast of Europe with crowds flocking to his talks. Another great scientist, Darwin, was inspired by Humboldt's writings and even sought out Humboldt's opinions as he prepared his own monumental work on evolution. Today, as when he lived, Humboldt is acknowledged as a great geographer. His conception of geography as an integrating science and his insistence on seeing humans as a part of nature, not separate from it, remain key ideas for many people. Indeed, perhaps more so today than at any other time since his death, Humboldt's insistence on looking at the earth as a whole is a message that cannot be ignored.

Humboldt
Metropolitan
Toronto Reference
Library.

forces of nature.' For Humboldt, humans were part of nature—a major shift of emphasis in the European world. In addition to conventional regional descriptions, Humboldt strove to offer a complete account of the way all things are related. General concepts were carefully blended with precise observation.

Ritter, too, was concerned with relationships and argued for coherence in describing the way things are located on the earth's surface. Like Humboldt, he expressed interest in moving from description alone to description and laws. This, of course, was exactly where Varenius had directed the attention of scholars. These interests are paramount in *Die Erdkunde*, an only partially complete world geography comprising 19 volumes published between 1817 and 1859.

All three of our recurring themes are evident in the work of Humboldt and Ritter. The study of humans and land is central to their conception of geography; a focus on regions and the concerns of spatial analysis are anticipated

in their interest in the formulation of concepts as general statements that aid in the understanding of specific facts. Together, Humboldt and Ritter represent a fitting conclusion and an appropriate beginning. Although much of the world was still unknown to Europeans—Africa was largely a mystery, as was Asia—these two men produced two great works of geography. Both recognized the need for complete geographic descriptions and yet neither could truly fulfill that aim. By the year of their deaths, 1859, there was simply too much geography. A complete, encyclopedic description of the world was now beyond the reach of a single scholar. In this sense, they conclude an era.

Yet in another sense they represent a beginning. Humboldt and Ritter were the first geographers to pay full attention to concept formulation, that is, to the derivation of general statements from the detailed factual information available. Their interest in the relations between things and their inclusion of humans as part of nature established the basis for much

Box 1.6   Ritter

Carl Ritter (1779–1859) was born 10 years after Humboldt and died the same year as Humboldt. Although he never became as well known in the larger scientific world, Ritter proved to have a greater influence on German geography.

Ritter had the good fortune to be selected at age five for involvement in a new system of education that encouraged understanding rather than conventional rote learning. One of his teachers was a geographer who emphasized that humans were part of nature, not some separate entity—a viewpoint that would become central to Ritter's own work. He was able to fund his studies at universities, including Halle and Göttingen, by working as a private tutor. Becoming professor of history at Frankfurt in 1819, in 1820 he obtained the post he would hold for the rest of his life: the first chair of geography in Germany at the University of Berlin. As early as 1811, Ritter published a two-volume textbook on Europe; the first volume of his most important work, *Die Erdkunde*, was published in 1817. At the time of his death *Die Erdkunde* had reached 19 volumes, most of which focused on Asia; he had not even begun to discuss Europe in this series.

For Ritter, geography was clearly an empirical and descriptive science reflecting the unity of humans and nature. Ritter had first met Humboldt in 1807 and the two established an important relationship. There was a clear agreement regarding the importance of the unity of all geographic facts. Unlike Humboldt, Ritter did not undertake any of the exploratory activity that led to the discovery of new facts, but he was an inspiring lecturer who influenced many subsequent geographers. In both his lectures and his writings, Ritter employed a conventional regional focus, one that continued to dominate in late nineteenth-century German geography.

Ritter

The home of the Royal Geographical Society in London. The Royal is the largest geographical society in Europe and one of the largest in the world. In the nineteenth century, such societies helped geographers to organize their field of study and promoted recognition of their work. Most also encouraged exploration and implicitly—if not explicitly—supported colonial activity. In 2009 the Royal hosted a spirited, even contentious, debate regarding the role the Society needed to play in the contemporary world. Some members argued for continued funding of major exploratory activity, while others argued for more and less expensive academic projects—the latter viewpoint was successful, heralding a symbolic end to the geography-as-exploration era (Royal Geographical Society).

CCL-wikipedia

subsequent geography. Thus, they represent both an end to the long road begun by the Greeks and a beginning to a geography combining regional description and concept formulation.

MAP-MAKING, EXPLORATION, AND GEOGRAPHICAL SOCIETIES

Despite considerable new thinking about geography, there was also a steady continuing interest in some of the traditional geographic concerns. Map-making was considered so important that governments began to assume responsibility for the task (you will recall that in China civil servants were responsible for map-making, probably as early as the third century CE) and in England the Ordnance Survey was founded in 1791. Large-scale topographic maps, showing small areas in considerable detail, became possible with the development of exact survey techniques in eighteenth-century France. Atlases also became much more sophisticated. Exploration continued and was greatly assisted by the founding of geographical societies in Paris (1821), Berlin (1828), London (1830), St Petersburg (1845), and New York (1851). The ever-growing interest in overseas areas encouraged prosperous individuals as well as governments to support these organizations.

Institutionalization: 1874–1903

The year 1874 marks the formal beginning of geography as an institutionalized academic discipline. In 1903, the first North American university department of geography was established. This 30-year period was one of dramatic developments in geography.

First, once geography became legitimized in institutions of higher learning, it was necessary that university geographers clearly define their discipline and distinguish it from the various other disciplines emerging at the same time. Not surprisingly, not all geographers immediately agreed on the discipline's content and methods. Second, the mid- to late nineteenth century was a period of major advances in various areas of knowledge that had direct effects on geography, when a mechanistic view of science, emphasizing cause and effect, prevailed. Third, the publication in 1859 of *On the Origin of Species* by Charles Darwin (1809–82) had repercussions throughout the scientific world. Fourth, Europe now had a 400-year history of expansion and movement into regions clearly different from the home area. Finally, the rapid growth of physical science encouraged the development of topics that were traditionally geographic in a number of other disciplines, notably geology.

At the time when geography was institutionalized, then, the intellectual environment was complex and changing. Not only did geographers have to consider the appropriate weight to give the many aspects of established geographic traditions—mapping, description, general and special geography, regions, human–land relations—but now they also had to take into account ongoing developments in the physical sciences, the social sciences, and mechanistic science, as well as the hugely controversial issues raised by Darwinian thought and the moral questions associated with expansion overseas. In brief, from 1874 onward, geographers had a great deal to accommodate. How did they respond?

GERMANY

In 1874 the Prussian government established departments of geography in all Prussian universities. Why? Possibly because the time was ripe; possibly because the value of geography was especially evident following the Franco-Prussian War of 1870–1; possibly because of a belief that geographic knowledge would

Box 1.7 | Ratzel

Following university studies in Heidelberg and Jena, Friedrich Ratzel (1844–1904) fought in the Franco-Prussian War (1870–1), and then, after another brief spell at the university in Munich, he travelled in Europe and then in North and Central America in 1874–5. The American travels in particular impressed upon him the importance of humans as makers of landscapes and inspired him to pursue a career in the new field of geography.

He taught geography in Munich from 1875 until 1886 and then succeeded Richthofen as professor of geography at Leipzig, where he remained until his death in 1904. The first volume of Ratzel's *Anthropogeographie*, published in 1882, was specifically designed to be a study of the effects of physical landscapes on history. The second volume, published in 1891, focused attention on relations between humans and land, with a primary emphasis on the role of humans. These two volumes together represent a major contribution to human geography. The first volume proved to be especially influential in the United States. Ratzel also made major contributions in political geography, regarding states as spatial organisms.

In many respects, several of Ratzel's ideas were used somewhat indiscriminately by later geographers. Volume 1 of *Anthropogeographie* was seen as a powerful argument for viewing human landscape as secondary to—even caused by—physical landscapes, and the political geographic ideas that Ratzel expressed were used by Nazi geographers in the 1930s to justify spatial aggression. Nevertheless, Ratzel remains an important figure, largely because he was the first influential geographer to explain clearly the concept of human landscapes as ever-changing additions to original physical landscapes—an idea integral to the thinking of later German, French, and American geographers.

Ratzel

Metropolitan Toronto Reference Library.

chorology
A Greek term revived by nineteenth-century German geographers as a synonym for 'regional geography'.

géographie Vidalienne
French school of geography initiated by Paul Vidal de la Blache at the end of the nineteenth century and still influential today, focusing on the study of human-made (cultural) landscapes.

facilitate political expansion. Whatever the reasons, it became the responsibility of these new departments to clarify what geography entailed. Several leading scholars attempted this task.

For Ferdinand von Richthofen (1833–1905), geography was—as it had been earlier—the science of the earth's surface. Research centred on field studies and observation. Richthofen maintained the distinction between special and general, first noted by Varenius, and further argued that the two could be combined to form a chorological (regional) approach. The subject matter of this regional geography (or **chorology**) included human activities, but only in relation to the physical environment.

Friedrich Ratzel focused on human geography, or what he termed 'anthropogeography' (Box 1.7). His major work was published in two volumes (1882 and 1891). Volume 1 had as an overriding theme the influence of physical geography on humans, while the second volume focused on humans using the earth. Interestingly, the first volume had the greatest immediate impact, possibly because its inherently simple cause-and-effect logic was in

accord with mechanistic views and helped to assert the importance of physical geography. It is appropriate to regard Ratzel as a founder of human geography, since he probably was the first to focus on human-made landscape.

Alfred Hettner (1859–1941) was the most influential follower of Richthofen. In his methodological and descriptive work, geography was unequivocally regarded as the chorological science of the earth's surface, a view that had been clearly enunciated first by Kant. A persuasive advocate of regional geography, Hettner was highly influential in the United States. However, his work excluded studies of human–land relations, as well as studies involving time.

Thus, German geography following 1874 consisted of three quite different interpretations. First, there was geography as chorology (Richthofen and Hettner); second, there was geography as the influence of physical geography on humans (Ratzel, volume 1); and, third, there was geography as the study of the human landscape (Ratzel, volume 2). These differences would continue well into the twentieth century.

Box 1.8   Vidal

Paul Vidal de la Blache (1845–1918) studied history and geography in Paris, and then taught successively at Nancy, Paris, and the Sorbonne. Vidal laid down ideas that, once accepted and amplified by others, established a dominant French geographic tradition, *géographie Vidalienne*, or *la tradition Vidalienne*. This tradition was not, of course, established in an intellectual vacuum, as Vidal had been introduced to the great works of Humboldt and Ritter during his schooling and also travelled often to Germany, meeting with such geographers as Richthofen and Ratzel.

The first clear methodological statement by Vidal was in his Sorbonne inaugural address in 1899; other articles appeared in the journal he founded in 1891, *Annales de Géographie*. Vidal rejected the idea that the physical landscape determined the human. Rather, he believed that geography should focus on the reciprocal relationship between humans and land. Central to his blueprint for geography were the ideas of *genre de vie* ('way of life', or culture), *milieu* (local area), and the regional concept of *pays* (natural region). Vidal's conception of geography involves chorological analysis, but with a distinctive humans-and-land focus that was not clearly

formulated by either Richthofen or Hettner. Similarities between Vidal and the Ratzel of *Anthropogeographie*, volume 2 are evident, and also between Vidal and the German geographer Otto Schlüter (1872–1952).

In France, Vidal's followers included such geographers as Jean Brunhes (1869–1930) and Albert Demangeon (1872–1940). The absence of any alternative French conception of geography did not mean, however, that Vidal's view was unchallenged. There was an ongoing debate between Vidal and the great French sociologist Émile Durkheim (1858–1917). Simply put, Durkheim was critical of Vidal and geography in general for emphasizing physical landscapes at the expense of an adequate emphasis on society, but Vidal, like Durkheim and his positivist sociology, founded a distinct school of geographic thought.

Vidal
Metropolitan Toronto Reference Library.

FRANCE

Geography in France followed a route independent of German developments, although it reflected many of the same ideas expounded by Ratzel in the second volume of *Anthropogeographie*. One influential scholar, Pierre Le Play (1806–82), was not a geographer but a sociologist; he focused on human–land relations and the impact of technology on social groupings. Elisée Réclus (1830–1905), along with Le Play, paved the way for later French geography. This geographer and anarchist, who was barred from France and imprisoned at various times, published a descriptive systematic geography of the world and a 19-volume universal geography. His work was very much in the Ritter tradition. Interestingly, another well-known anarchist, Peter Kropotkin, a former Russian prince, was also a leading geographer in the late nineteenth century.

The most important French geographer, Paul Vidal de la Blache (Box 1.8), followed both Ratzel and Le Play and succeeded in introducing a well-articulated geographic method. His overriding concerns were the relations between humans and land, the evolution of human landscapes, and the description of distinctive local regions. For Vidal, geography should consider both physical geographic impacts on humans and human modification of physical geography. The parallels with Humboldt, Ritter, and Ratzel (volume 2) are clear. Vidal and his many followers continue to exert enormous influence on human geography.

BRITAIN

As was the case in Germany, where Ritter held a Chair in Geography beginning in 1820, long before the first departments of geography were established, a Professorship of Geography was established at University College London in 1833. This post was held until 1836 by Alexander Maconochie, who subsequently turned his attention to questions of penal reform (Ward, 1960). In 1887 geography lectureships were established at Oxford and Cambridge universities with funding provided by the Royal Geographical Society, but it was not until 1900 that the first British geography department was established at Oxford, again largely as a result of efforts by the Royal Geographical Society. The dominant influence at first was Halford J. Mackinder

(1861–1947). Interestingly, he determined to become the first European to climb Mount Kenya, in Africa, to ensure that his geographic views would be favourably received; to be respected as a geographer, he believed, one had to be a successful explorer! For Mackinder, geography and history were closely related: a global geographic perspective was essential, and, following Richthofen, physical geography was a prerequisite for human studies. In Britain, then, the traditionally close association between geography and mapping, description, and exploration continued: the emphasis was on physical rather than on human geography.

UNITED STATES

The establishment of the first North American department of geography, at the University of Chicago in 1903, came at a time when American geography was influenced largely by German scholars. Physical geography (physiography) was dominant, possibly because of the powerful personality of William Morris Davis (1850–1934), a geologist who promulgated the views (imported from Germany) that physical geography influenced human landscapes and that geography was essentially a regional science. Ellen Churchill Semple (1863–1932) followed the early Ratzel to become a leading proponent of the 'physical influences' school of thought, while many American geographers quickly adopted a regional focus. At the time there was no evidence of a distinctive school analyzing the relations between humans and physical geography. Neither the general legacies of Humboldt and Ritter nor the views of Vidal were in evidence, although George Perkins Marsh (1801–82), an American geographer and congressman, had earlier focused on the same issues.

GEOGRAPHY IN 1903

Once again we have reached an important turning point. By 1903 geography was a full-fledged academic discipline in many European countries and in the US. Elsewhere geography was institutionalized somewhat later, largely as a consequence of academic contacts with these pioneering countries. In Canada, for example, a partial department of geography was created at the University of British Columbia in 1923, 12 years before the first full department, at Toronto, was established in 1935. Today, Canada has more than 40 geography departments. An interesting recent trend in Canada and some other countries, notably Australia, involves the explicit linking of geography and environmental programs in a single university department.

In 1903 the general subject matter was clear—there had been no real change since Greek times—and a number of different approaches were advocated. Many people studied physical geography in its own right; some focused on physical geography as the cause of human landscapes; others concentrated on regional description; still others centred their attention on humans and nature combined. These general threads continued through much of the last century. Geography also continued its association with mapping and, to a lesser extent, exploration.

Prelude to the Present: 1903–1970

The period from 1903 to 1970 was characterized by several different approaches to geographic subject matter. As in the past, academic geography continued to emphasize three principal areas of study—physical geography as cause, humans and land, and regional studies—to which was added a fourth, relatively novel, approach: spatial analysis. The physical-geography-as-cause approach proved relatively short-lived; the other three, of course, are our recurring themes.

PHYSICAL GEOGRAPHY AS CAUSE

Much of the attraction of the emphasis on physical geography as cause (most clearly stated in Ratzel's volume 1) lay in its relative simplicity. Arguing that human landscapes and cultures resulted primarily from physical geography reduced the need to think about economics, politics, societies, and so forth. Among the well-known geographers who favoured this approach were Semple, Ellsworth Huntington (1876–1947), and Griffith Taylor (Box 1.9). Nevertheless, the principal significance of the physical-geography-as-cause perspective lies in the fact that it simply was taken for granted as a self-evident truth. Much twentieth-century geography that did not explicitly take this perspective—also known as **environmental determinism**—did so implicitly.

The greatest problem with this approach is the tendency to use it in isolation. There are important and often intimate relations between physical and human geography, but these are never so straightforward as to allow us to assume that the former is the cause and the latter the effect. Examples of these intimate relations include much evidence that movements of early humans and technological changes such as the development of agriculture often were linked to changes in climate (noted in Chapters 3 and 10). It is clear that the global distribution of humans corresponds closely to global environmental regions (discussed in Chapter 5). Of particular concern today are suggestions that human-induced climate change might provoke new conflicts as groups compete for dwindling resources. Indeed, historical evidence suggests that past climate changes were a factor in conflicts in parts of Southeast Asia: specifically, a period of drought in the fifteenth century contributed to the decline of Angkor (in present-day Cambodia) and invasions by other groups, while a second period of drought in the eighteenth century contributed to Burmese invasions into Siam (now Thailand).

Fortunately for geography, environmental determinism as an explicit identification of physical cause and human effect, although popular in the early twentieth century and occasionally taken to extremes, never became the focus of a formal school of geography and is now discredited.

HUMANS AND LAND

The view of geography as the study of all things physical and human, specifically as the study of relationships between physical and human facts, has a long tradition. Humboldt, Ritter, and Ratzel (volume 2) all took this view, and their specific legacy is evident in much subsequent geography. However, such a vast subject required a more solid foundation than the physical-geography-as-cause school could provide. In particular, it needed a better definition of its subject matter and a formal methodology. These were provided by three scholars.

Vidal, as we have seen, developed a distinctive French school, *géographie Vidalienne*, beginning about 1899. Otto Schlüter (1872–1952) founded a German school of **Landschaftskunde**—'landscape science', or 'landscape geography'—in 1906, which provided a clear definition of subject matter. The third scholar was Carl Sauer, an American who, in 1925, effectively introduced the various European ideas to North America and elaborated on

environmental determinism
The view that human activities are controlled by the physical environment.

Landschaftskunde
A German term, introduced in the late nineteenth century, best translated as 'landscape science'; refers to geography as the study of the landscapes of particular regions.

Box 1.9 | Taylor

Griffith Taylor (1880–1963), born in Britain, was responsible for introducing geography as a university discipline to Australia and Canada. A member of the Scott expedition to the Antarctic, 1910–12, Taylor had been trained as a geologist, and this background led him into geography. He first taught geology at the University of Melbourne and was appointed the first professor of geography at the University of Sydney in 1920.

Taylor was a dedicated and effective teacher with strong opinions on the influence of physical geography on human activities. For Taylor, physical geography determined the basic direction of human landscape change, but the economic factors of labour and capital determined the rate of that change, a view he labelled 'stop and go' determinism. His dim view of the settlement potential of the arid interior of Australia—diametrically opposed to the propaganda of the Australian government—brought him into considerable disrepute. At least partly as a result of this controversy, in 1928 he moved to Chicago, where he taught until 1935. In that year he moved to Toronto, where he was responsible for the founding of the first full Canadian department of geography and remained until his retirement in 1951.

Always a stimulating writer, Taylor produced some 20 books and 200 articles. In Canada, as in Australia, he was involved in discussions about settlement of environmentally difficult areas, in this case the Canadian North. When Taylor returned to Australia at the age of 70, he was treated as a national hero, his earlier views having been vindicated. At the time of his death, Taylor was being considered in Australia for a knighthood.

---

them (Box 1.10). The **landscape school** that grew out of his work focuses on the transformation of the physical geographic landscape by human cultural groups over time.

Together, the ideas of Vidal, Schlüter, and Sauer provide us with a clearly defined approach to our subject matter. Landscape geography explicitly rejects any suggestion of physical-geography-as-cause by rejecting environmental determinism in favour of **possibilism**: the view that human activity is determined not by physical environments, but rather by choices that humans make.

REGIONAL STUDIES

Regional geography, or chorology, proved to be the most popular focus during the first half of the nineteenth century. Pioneering work by German geographers such as Richthofen and Hettner (in his later arguments) was carried over into English-language geography. In Britain a view emerged that the ultimate task of geography was to delimit regions. In America this view was attractive for physical geographers in particular, and W.M. Davis, for example, produced a map of regions as early as 1899. In 1905 the British geographer A.J. Herbertson (1865–1915) proposed an outline of the world's natural regions. These developments culminated with the 1939 publication of *The Nature of Geography* by the American Richard Hartshorne. This substantial contribution to geographic scholarship argued forcefully for geography as the study of regions—what Hartshorne often called **areal differentiation**—a view that was very much in accord with prevailing American opinion and that continued to dominate geography until the mid-1950s.

Geography in the North American world by 1953 was thus characterized by two related but different emphases: (1) analysis of the relationship between humans and land and (2) regional studies. Both were the products of long geographic traditions, and both have continued to change. In 1953, however, a new focus was forcefully introduced.

SPATIAL ANALYSIS

A paper by F.W. Schaefer, published in 1953 (see Box. 2.3), is usually seen as the beginning of an approach called spatial analysis. This approach, as we saw in the Introduction, focuses on explaining the location of geographic facts. Schaefer argued that geographers should move away from simple description in regional studies to a more explanatory framework based on scientific methods such as the construction of theory and the use of quantitative methods. Here again we can recognize the special/general distinction identified by Varenius. Schaefer objected to what he saw as the overly special focus of regional geography. Regardless of the specific merits of Schaefer's arguments, his views struck home to many. What followed might be called a revolution, in which the emphasis shifted to quantitative studies. The decade of the 1960s was characterized by

---

landscape school American school of geography initiated by Carl Sauer in the 1920s and still influential today; an alternative to environmental determinism, focusing on human-made (cultural) landscapes.

areal differentiation From Hartshorne, a synonym for 'regional geography'.

possibilism The view that the environment does not determine either human history or present conditions; rather, humans pursue a course of action that they select from among a number of possibilities.

Box 1.10   Sauer

For students of English-language human geography, Carl Sauer (1889–1975) is a major figure. Born in Missouri, Sauer obtained a doctorate from the University of Chicago and taught at the University of Michigan from 1915 to 1923, but he is most closely associated with the department of geography at the University of California at Berkeley, where he taught from 1923 to 1957. Proponents of Sauer's conception of geography are often labelled the 'Berkeley school' or 'landscape school'.

Sauer articulated that conception in his opening lecture upon appointment to the university at Berkeley. Entitled 'The Morphology of Landscape', the address presented a highly original version of earlier European ideas. Sauer saw geography as a chorological discipline, but one largely concerned with the manner in which humans modify physical landscapes. Humboldt, Hettner, Schlüter, and the core French school of Vidal all are evident influences. Sauer explicitly opposed the popular view that physical geography was a cause of human landscapes. He employed the term 'landscape' as the object of geographic study in preference to 'region', which he felt was too closely associated with overly detailed descriptions. Although Sauer's view of geography was largely derivative of others, it proved to be highly influential and has contributed significantly to the development of historical and cultural emphases.

Sauer succeeded in establishing a landscape geography that continued from the 1920s to the present with relatively little change until the 1980s. This was no mean achievement, and it reflects both the quality of his methodological and research writings and the legitimacy of his basic arguments. In addition to launching the study of landscape geography, Sauer produced highly original studies of early America and the origins of agriculture. Indeed, although Sauer is often, and rightly, applauded for his pioneering methodological statements, his substantive field studies are a major contribution to human geography and indicate great sensitivity to cultural diversity and to the intimate relationships between groups and the landscapes they occupy and modify.

---

phenomenal growth in detailed analyses based on the proposal and testing of hypotheses by means of quantitative procedures.

By 1970, regional geography was receding in popularity and spatial analysis had found a niche alongside somewhat modified versions of both the landscape and regional approaches.

Contemporary Geography

Since 1970, geography has seen a number of important revisions and additions to traditional interests and approaches. Thus, geography today, while clearly the product of the earlier developments discussed above, also incorporates some relatively new components. One is the welcome addition of a wide range of new (for the human geographer) philosophies that are prompting major shifts in the traditional approaches (human–land relations and regional studies). Another is the ever-increasing impact of technological advances in data collection and analysis.

Seven major trends in contemporary human geography, considered at greater length in Chapter 2, can be discerned:

1. an increasing separation of the physical and human components of geography;
2. a revitalized landscape approach;
3. a revitalized regional geography;
4. an ongoing interest in spatial analysis;
5. recognition of the need for a global perspective;
6. an increasing concern with applied matters;
7. an increasing emphasis on technical content.

PHYSICAL AND HUMAN GEOGRAPHY

Our discussion so far has not made any formal attempt to distinguish between physical geography and human geography. As we have seen, geography from the Greeks onward has had both physical and human components. Prior to Humboldt and Ritter, however, the clear tendency was to treat them separately. Descriptions of the earth typically dealt first with physical and second with human matters. Humboldt and Ritter, by contrast, focused closely on integrating the two. Subsequently, some geographers saw physical geography as paramount (the environmental determinist school), some saw the human landscape as the relevant subject matter (the landscape school), and some saw the two as

Lunenburg, Nova Scotia, from the harbour. Geographers often suggest that landscapes can speak to us, communicating information about the people who live there, what they do, and the values they hold. Established by the British in 1753 as their second outpost in Nova Scotia (the former French colony of Acadia), Lunenburg was first settled by Protestants specifically recruited from Germany, Switzerland, and France. Farming, fishing, shipbuilding, and ocean-based commerce provided the foundations of a vibrant economy, and today tourism is a major source of income. This photograph reflects elements of the town's economic history and its character. For example, there is no evidence of the ongoing physical change—whether construction or demolition—so typical of most urban areas. Rather, the presence of many carefully maintained old buildings suggests permanence and stability.

Gerald Hallowell photo.

separate (the regional school). It was not unusual for geographers to assert a unity between physical and human geography, but there was not much evidence of such unity. Very few geographers deny the relevance of physical to human geography, but equally few see the one as necessary to the other. Today, therefore, we tend to teach and research the two separately, an approach that applies even in most of the newly formed departments that include both geographers and environmental scholars.

The separation of physical from human geography and the recognition of human geography as a separate discipline are thus relatively recent. Of course, some regional accounts and some specific issues require consideration of both dimensions. Chapter 3 of this book is a clear instance of the value of a geography that is both physical and human. And a basic understanding of global physical geography is obviously relevant to much work in human geography (see Appendix 1). Overall, though, the two are no longer seen as so closely related that we need to discuss them as one. Contemporary human geography is a social science, but one with special and valuable ties to physical sciences, especially physical geography.

CONTEMPORARY LANDSCAPE GEOGRAPHY

At least since the writings of Vidal, Schlüter, and Sauer, landscape geography has maintained a consistent place in human geography alongside the regional approach. Also known as social geography (in Britain) and cultural geography (in North America), landscape geography typically involves studies of human ways of life, cultural regions and related landscapes, and relationships between human and physical landscapes. But the landscape approach has recently been enhanced by the inclusion of new conceptual concerns. Since about 1970, ideas associated with humanism, Marxism, and feminism, along with a generally increased awareness of advances in other social science disciplines, especially those relating to postmodernism, have enriched landscape studies.

Today landscape geography is concerned both with visible features (such as fields, fences, and buildings) and with symbolic features (such as meaning and values). Thus, many contemporary landscape geographers focus on the human experience of being in landscape; here landscapes are regarded as relevant not simply because of what they are but also because of what we think they are.

Further, there is explicit acknowledgement that landscapes, like regions, both reflect and affect cultural, social, political, and economic processes.

Over time, studies in human–land relations have become increasingly subtle and profound, and they are now closely tied to conceptual developments in disciplines such as psychology, anthropology, and sociology.

CONTEMPORARY REGIONAL GEOGRAPHY

The rise of spatial analysis came largely at the expense of the regional geography ('areal differentiation') articulated by Hartshorne. Since 1970, however, regional geography, like landscape geography, has resurfaced in a somewhat different guise and has once again become a central perspective. Most human geographers accept that the earlier regional geography is no longer appropriate; there is a clear need to move beyond regional classification and description. Yet, the regional approach is obviously a valuable and distinctive geographic device. One geographer described regional geography as 'the highest form of the geographer's art' (Hart, 1982: 1).

Today regional geography emphasizes the understanding and description of a particular region and what it means for different people to live there. This emphasis reflects at least three general concerns, each of which is tied to a particular set of conceptual constructs. First, there is a concern with regions as settings or locales for human activity. Second, there is a concern with uneven economic and social development between regions, including a focus on the changing division of labour. Third, there is a concern with the ways in which regions reflect the characteristics of the occupying society and in turn affect that society. The underlying concepts for each of these three regional geographic concerns are humanism and Marxism; both of these are introduced in Chapter 2. Looking at regional geography in this light, we can see it as increasingly satisfying the requirement that human geography serve society by addressing a range of economic and social problems.

The new directions in regional geography and landscape geography are sufficiently similar that these two approaches now share several common interests.

CONTEMPORARY SPATIAL ANALYSIS

The spatial analytic approach that first became a prime interest of human geographers in the mid-1950s was very influential until about 1970. Since that time it has continued to be an integral part of human geography, but it is now generally regarded as only one among several widely accepted approaches. A central concern is that the theoretical constructs it uses to explain locations are somewhat limited, and hence that it tends to emphasize generalizations at the expense of specifics. The topics studied by spatial analysts typically are economic (for example, the locations of industrial plants), as opposed to political, cultural, or social. This tendency reflects the influence of various economic location theories.

There is a clear distinction between the spatial analytic approach and the contemporary regional and landscape approaches. Indeed, some geographers believe that spatial analysis is overly concerned with spatial issues, perhaps to the point of seeing space as a cause of human landscapes, and as a result ignores the full range of human variables as causes of landscapes. Contemporary regional and landscape geographers explicitly argue that space is important only when it is analyzed in terms of the use that humans make of it. This is an important philosophical question that will be discussed in the next chapter.

GLOBAL ISSUES

Geography has always been a global discipline. Much of the work of the early Greek, Chinese, and Islamic geographers, for example, was concerned with describing the limits of the known world and identifying links between places. European geography, as it developed from about 1450 onward, focused on understanding where places were, mapping routes, and describing what was previously unknown. Much contemporary human geography continues in these traditions.

Nevertheless, there is something behind our current concern with global issues. The fact is that people and places in the contemporary world increasingly are connected and interconnected. Today we are interested in the global movements of people, products, ideas, and capital. We are interested in how the many and diverse peoples and places scattered throughout the world are increasingly associated with one another. Human impacts on the earth, population growth, the spread of diseases (such as AIDS, SARS, and swine flu), international migration and refugee problems, food shortages, vanishing languages, the spread of democracy, agricultural change, urban growth, and industrial restructuring are just a few of the major issues in the world today that require thinking on the global scale. More than ever before, human geographers are finding it necessary to look at the big picture—the world—to understand the detail at the local level.

APPLIED GEOGRAPHY

Because geography is an academic discipline that serves society, it has always had an applied component in which geographic skills are used to solve problems. Geography as exploration is a prime example of such applied work. Since about 1930, geographers have played a major role in land-use studies. For example, the studies of the British Land Utilization Survey directed by L.D. Stamp in the 1930s proved enormously valuable during World War II, and American geographers made important contributions in both land-use and military matters.

| Box 1.11 | **Why Learn about the History of Geography?** |

Geography matters, but why does the history of geography matter? The simple answer is that 'what's past is prologue' (Shakespeare, *The Tempest*, Act 2, Scene 1). Necessarily, then, understanding the history of geography enhances our understanding of geography today and even enables us to think about what might happen in the future. This point is demonstrated through reference to two examples.

First, the story told in this chapter shows that for many years a principal concern of geography was that of acquiring factual information. Facts were required to enable geographers to accomplish the basic task of description, including mapping. Strabo's *Geographia*, Ptolemy's *Guide to Geography*, Varenius's *Geographia Generalis*, and many more books were attempts to reflect the then current geographic knowledge base. As each new book was written, the body of geographic knowledge became more comprehensive and better established, with earlier knowledge verified or corrected.

It is no different today. All geographic research relies on a body of factual knowledge, and much geographic writing reports on new factual knowledge or on reinterpretations of established knowledge. Much of our new knowledge is acquired by different means from those employed in the past, for example by using satellites, but the basic goal is no different. Also, much new knowledge may refer to recently defined geographic problems, such as homeless living, but again the basic goal is no different. Geographic research, like any other research, requires careful observation as a first step.

Second, looking at the history of geography also clarifies our understanding of how geographers interpret facts. For most of the past 3,000 years, new facts resulted from people travelling to lands new to them. These facts were then understood in the context of the existing knowledge base and recorded accordingly. Consider, for example, the maps shown in Figures 1.2, 1.3, and 1.4. Each map reflects not only the geographic knowledge base but also the larger understanding that the creators of the map had of their world. Consider also the knowledge that Europeans acquired about other lands and peoples as they moved overseas. How did they accommodate this new knowledge? The answer is that the new facts were understood in the context of the prevailing discourse, or what can be more simply called the larger social and intellectual context. Specifically, as the process of European overseas expansion proceeded, new lands and people were reinterpreted, especially through colonial and racist lenses.

Today, still, geographic facts are interpreted and recorded in ways that reflect prevailing discourses. For example, in Chapter 3 we consider human impacts on the earth, in Chapter 4 we consider whether or not there are too many of us living on earth, and in Chapter 5 we ask why the world is such an unequal place. In all three of these cases, answers are provided that are necessarily framed in the context of our larger intellectual framework, that is, the prevailing discourse. But our study of the history of geography has taught us that these discourses are always changing—overt colonialism and racism no longer play the important role they once did—so in this textbook we are careful always to acknowledge that our explanations are necessarily somewhat uncertain and that alternative understandings of facts may be appropriate.

The story told in this chapter has made it clear that all knowledge is contextual and thus needs to be understood as it relates to other knowledge. This is central to understanding how we, as humans, interact with and shape the natural environment, and it is an important lesson to learn.

A security guard wearing a surgical mask and gloves as protection against H1N1, Hong Kong, May 2009.

Ulana Switucha/Alamy/GetStock

Contemporary human geography responds to the wide range of social and environmental issues that confront us with regular contributions on such topics as peace, energy supply and use, food availability, population control, and social inequalities. These issues are covered in this book, for human geographers recognize that geography has responsibilities to society and the world. Much recent geography of this type reflects the influence of other disciplines (for example, in the use of Marxist analysis). There are also close links between applied geography and both spatial analysis and technical advances.

TECHNICAL ADVANCES

Human geography today is greatly aided by a variety of technical advances in areas such as navigation-assisted exploration. Among the technical advances that have facilitated data acquisition are aerial photography and both infrared and satellite imagery. Similarly, advances in computer technologies and geographic information systems have facilitated mapping and data analysis. These important geographic techniques will be discussed in detail in the next chapter.

Postscript

As befits geographers, we have travelled a long way in this chapter (Box 1.11). From the earliest map-makers to the present, we have seen the evolution of geography as an academic discipline. We know that human geography has a rich heritage and exciting contemporary developments. But we have not yet travelled far enough. The next chapter presents fuller discussions of the various contemporary developments that have received only brief mention so far.

Human geography today is a responsible social science with the basic aim of advancing knowledge and serving society. In this respect it is no different from Greek, Chinese, Islamic, or later European geographies. Our subject matter is clear and our methods various. Our key goal—to provide a basis for comprehending the human world as it is today and as it has evolved—is of crucial importance to contemporary world society. As a brochure produced by the Association of American Geographers expresses it, 'Without geography you're nowhere.'

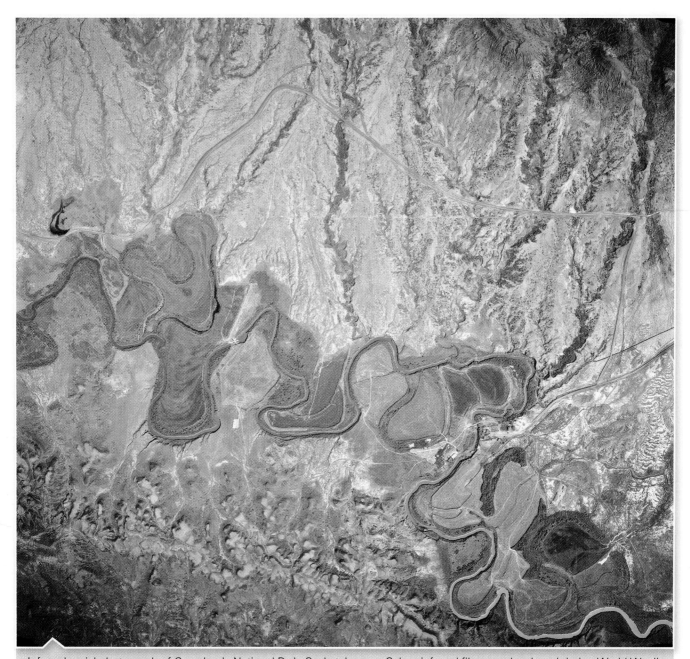

Infrared aerial photograph of Grasslands National Park, Saskatchewan. Colour infrared film was developed during World War II to detect painted targets camouflaged to look like vegetation. In this image healthy grass and tree leaves appear red.

Infrared aerial photograph of Grasslands National Park, Saskatchewan. Saskatchewan Grasslands 2006, www.nrcan.gc.ca. Reproduced with the permission of the Minister of Public Works and Government Services Canada, 2009.

CHAPTER 1 SUMMARY

A RAISON D'ÊTRE FOR GEOGRAPHY

To serve society.

THE STORY OF GEOGRAPHY

There is more than one story about geography and more than one way of telling the chosen story. In common with other intellectual projects, geography is deeply contested.

PRECLASSICAL GEOGRAPHY

The earliest geographic descriptions were maps. The earliest surviving maps, from ancient Mesopotamia, showed only local areas.

CLASSICAL GEOGRAPHY

Because the Greeks were geographically mobile, they began a descriptive literary tradition that included the mapping of the entire world as they knew it. Major figures in this tradition include Herodotus, Eratosthenes, and Strabo. The Greeks also initiated a mathematical tradition; Eratosthenes measured the circumference of the earth and Hipparchus devised a grid system for maps. Ptolemy summarized the mathematical tradition. Roman civilization added little to geographic knowledge.

THE FIFTH TO FIFTEENTH CENTURIES

From the fifth century to the fifteenth century, only sporadic and limited geographic work was accomplished in Europe. Christian dogma determined the contents of T-O maps. Only Portolano charts represented a development in mapping. Thus, geographic study largely disappeared. In China from c. 2000 BCE onward, and in the Islamic world from the seventh century CE on, geography flourished; significant progress was made in exploration, mapping, description, and mathematical work.

A EUROPEAN RESURGENCE

Beginning in the fifteenth century, Europe expanded overseas; cartography was transformed with new map projections, especially that of Mercator; geographic description reappeared with major works by Apian and Münster. By the mid-seventeenth century, geographers had added significantly to our knowledge of ourselves and our world; geography itself continued to be concerned with mapping and describing the earth.

RETHINKING GEOGRAPHY

Varenius was the first scholar to formally define geography and distinguish between general and special geography (in 1650). Kant contributed further to the understanding of geography by identifying it as the science that describes or classifies things in terms of area (c. 1765). Between roughly 1650 and 1800, geography broadened to include discussion of such issues as physical-geography-as-cause, the growth of civilizations, the unity of the human race, and population density and productivity relationships. Descriptive geography and mapping continued.

UNIVERSAL GEOGRAPHY

The mid-nineteenth century was the high point for geography as mapping and description. By this time considerable geographic knowledge had been accumulated, and specialist sciences were emerging. Major nineteenth-century geographies include those by Malte-Brun, Humboldt, and Ritter. The latter two made novel contributions by discussing human–land relations in addition to providing comprehensive descriptions, and they considered geography to require both general and special studies.

INSTITUTIONALIZATION

Geography was firmly established as a university discipline in Prussia in 1874. Other countries rapidly followed suit. Programmatic statements about geography were plentiful: Richthofen and Hettner formulated geography as chorology; Ratzel introduced anthropogeography and Vidal followed his lead; Mackinder focused on a world view; and Davis pioneered physiography. The first North American department of geography was established at the University of Chicago in 1903.

PRELUDE TO THE PRESENT

Four principal emphases were evident between 1903 and 1970. Physical-geography-as-cause flowered initially. Sophisticated human and land views developed under Vidal, Schlüter, and Sauer and have persisted to the present. Regional geography dominated North American work until the 1950s. Spatial analysis appeared in the 1950s and flourished in the 1960s, but is now regarded as just one of several valid approaches.

HUMAN GEOGRAPHY TODAY

Contemporary human geography has close ties to the long and illustrious history of geography and is affected by several relatively new influences. Today human geography is easily distinguished from physical geography, includes revitalized regional and landscape approaches, takes an ongoing interest in spatial analysis, is increasingly concerned with applied issues in the global context, and has a continually expanding technical component.

CONCLUSION

Knowledge of geography is essential to the education of all humans. Its fundamental goal is still to serve the society of which it is a part.

QUESTIONS FOR CRITICAL THOUGHT

1. What are the major developments in the evolution of human geography as an academic discipline since the time of classical Greece?

2. Why are maps so important to the study of geography? Can we do geography without them?

3. How does the perspective of geography differ from that of other academic disciplines?

4. Why is it important to study the history of one's discipline? What do we gain by knowing the 'history of geography'?

FURTHER EXPLORATIONS

Benko, G., and U. Strohmayer, eds. 2004. *Human Geography for the 21st Century*. New York: Oxford University Press.

Informative discursive essays on several of the principal human geographic subdisciplines.

Buisseret, D., ed. 2007. *The Oxford Companion to World Exploration*, 2 vols. Oxford: Oxford University Press.

Comprehensive and detailed work describes all aspects of world exploration from Egyptian civilization onward and includes biographies of major figures, as well as photographs, maps, and annotated primary materials.

Cloke, P., P. Crang, and M. Goodwin, eds. 2004. *Envisioning Human Geographies*. New York: Oxford University Press.

A challenging and stimulating text comprising a series of essays on current approaches to human geography.

Crane, N. 2003. *Mercator: The Man Who Mapped the Planet*. New York: Henry Holt.

A good read. Both entertaining and informative.

Dickinson, R.E. 1969. *The Makers of Modern Geography*. Boston: Routledge & Kegan Paul.

A traditional history of geography that emphasizes German and French scholars and contains much detailed information.

Dueck, D., H. Lindsay, and S. Pothecary, eds. 2005. *Strabo's Cultural Geography: The Making of a Kolossourgia*. New York: Cambridge University Press.

Detailed scholarly study of Strabo's *Geographia*.

Edson, E. 1997. *Mapping Time and Space: How Medieval Mapmakers Viewed Their World*. London: The British Library.

For students especially interested in the history of map-making, this is an excellent and often entertaining survey of the years 500 to 1500; stresses the meaning and purpose of maps following the logic that the hand of the map-maker is responding to a particular spatial and temporal context.

Fernandez-Armesto, F. 2006. *Pathfinders: A Global History of Exploration*. Oxford: Oxford University Press.

This wonderful book begins by noting that history has two big stories to tell; the first is a very long story about how human cultures diverged and developed differently, while the second is a much shorter story about how they came back together as a result of European overseas movement. The book tells the second story.

Hartshorne, R. 1939. *The Nature of Geography: A Critical Survey of Current Thought in the Light of the Past*. Lancaster, Penn.: Association of American Geographers.

The leading methodological statement arguing for regional geography; worth a look, but difficult to read in depth.

Harvey, D.W. 1969. *Explanation in Geography*. London: Arnold.

A comprehensive and pioneering statement about geography as a science that, like the Hartshorne book, is not an easy read for the beginning geographer, but rewards even a cursory inspection.

Holt-Jensen, A. 1988. *Geography: Its History and Concepts*, 2nd edn. Totowa, NJ: Barnes and Noble.

A stimulating evaluation of geography as a discipline that is especially readable for the new geographer.

Howgego, R.J. 2008. *Encyclopedia of Exploration: 1850-1940, Continental Exploration*. Sydney, Australia: Hordern House.

Exceptionally detailed, informative, and entertaining encyclopedia.

Johnston, R.J., and J.D. Sidaway. 2004. *Geography and Geographers: Anglo-American Human Geography Since 1945*, 6th edn. London: Arnold.

An overview of English-language geography since 1945; thoughtful debate of the role of geography in society.

Livingstone, D.N. 1992. *The Geographical Tradition*. Oxford: Blackwell.

A fascinating account of the last 500 years of geography; comprises a series of discussions of intellectual history, the first of which is entitled 'Should the History of Geography be X-Rated?'

Martin, G.J. 2005. *All Possible Worlds: A History of Geographical Ideas*, 4th edn. New York: Oxford University Press.

The best comprehensive history of geography, with detailed discussions of geographic developments and early geographers.

Minshull, R. 1967. *Regional Geography: Theory and Practice*. London: Hutchinson.

A clear account of the regional approach, focusing on current issues in method and practice.

Morrill, R.L. 1984. 'The Responsibility of Geography', *Annals, Association of American Geography* 74: 1–8.

A thoughtful assessment of the obligations of geographers, including those to society and humanity.

Sachs, A. 2006. *The Humboldt Current: Nineteenth-Century Exploration and the Roots of American Environmentalism*. New York: Viking.

A book that discusses the many insights offered by and influence of Humboldt, with a focus on the activities of four American explorers and environmentalists (Clarence King, J.N. Reynolds, George Wallace Melville, and John Muir).

Sauer, C.O. 1925. 'The Morphology of Landscape', *University of California Publications in Geography* 2: 19–53.

The landmark article introducing the landscape concept to North American geography.

Semple, E. 1911. *Influences of Geographic Environment*. New York: Henry Holt.

A classic example of the physical-geography-as-cause interpretation of human geography; very readable today.

Smart, L. 2004. *Maps That Made History: The Influential, the Eccentric and the Sublime*. Toronto: Dundurn Group.

A scholarly and entertaining discussion of 25 maps, ranging from Ptolemy's representation of the world to a map of the London Underground, that highlights what maps say and what they do not say.

Taafe, E.J. 1974. 'The Spatial View in Context', *Annals, Association of American Geographers* 64: 1–16.

A brief and effective summary of major geographic approaches prior to the 1970s.

ON THE WEB

PROFESSIONAL GEOGRAPHY ASSOCIATIONS

Canadian Association of Geographers

🔍 www.cag-acg.ca/en

Useful information on geography in Canada; intended primarily for professional geographers and those whose employment is closely related to their geographic education.

Association of American Geographers

🔍 www.aag.org/

The most important American site for professional geographers; a valuable resource that offers information on many topics, including careers for geographers. See also the websites for professional geography associations in Australia (www.iag.org.au/home/), New Zealand (www.nzgs.co.nz/), and South Africa (www.ssag.co.za/).

National Council for Geographic Education

🔍 www.ncge.org/i4a/pages/index.cfm?pageid=1

American organization that focuses primarily on the development and dissemination of teaching materials.

The Royal Geographical Society with the Institute of British Geographers

🔍 www.rgs.org/HomePage.htm

A UK-based but globally focused site with useful information on studying geography and potential careers.

The International Geographical Union

🔍 www.igu-net.org/

International organization that includes data on university geography departments globally and has some useful links.

POPULAR GEOGRAPHICAL PUBLICATIONS

Canadian Geographic

🔍 www.cangeo.ca/

The website for *Canadian Geographic* magazine; an informative source that deals with a wide range of geographic issues for Canada.

National Geographic

🔍 www.nationalgeographic.com/

Undoubtedly the best-known geography magazine. Very informative, well-written, inspiring photography. Popular yet often of scholarly significance.

Geographical Review and Focus

🔍 www.amergeog.org/

The American Geographical Society publishes *Geographical Review* and also the very readable and student-friendly *Focus*.

Geographical

🔍 www.geographical.co.uk/Home/index.html

Aimed primarily at a student audience, this informative publication includes a wide variety of articles and news, with an emphasis on the UK.

STUDYING HUMAN GEOGRAPHY

The broad subject matter of human geography has remained relatively constant over several thousand years and in various cultural settings. Approaches to that subject matter have changed, however, in response to the specific needs of time and place. Today human geography comprises a wide—some might say bewildering—array of questions, concepts, and techniques of analysis.

In order to make sense of such a complex subject, this chapter is divided into three parts: philosophical, conceptual, and analytical. First, we consider the philosophical component: why geographers ask the questions they do. The answer may seem self-evident, but it is not. Different geographers ask different questions, and we need to know why. Second, we consider the conceptual element: what concepts human geographers use to assist their research activities. Once again, we find several core concepts that are central to almost any piece of geographic research, and several other concepts that tend to be associated with particular types of questions. Third, we look at the many analytical techniques that are available to help human geographers answer the questions they ask. In fact, the three parts of this chapter are intimately related.

The space station Mir over New Zealand
World Perspectives/Getty Images

Charles Darwin and the title page from his landmark work, *On the Origin of Species*, 1859.

Why is philosophy important to the beginning geographer? Our discussion of the discipline's evolution made it clear that the specific content of human geography is essentially the product of three distinct eras: a pre-institutional phase, before the first university geography departments were established in 1874; the period when the discipline was institutionalized, and geographers needed to articulate a specific content for it; and more recent developments. We can fully appreciate our content—the way we analyze and conceptualize problems—only if we first understand our philosophies. It is our philosophical perspectives that explain our specific content, concepts, and techniques of analysis.

In human geography, as in other disciplines, facts, concepts, and techniques are logically interrelated. But it is not enough simply to accept that these are related: we need to know *why*. This means we must understand the philosophical viewpoints that serve as the 'glue'. Much (though not all) geographic work is guided by a particular philosophical viewpoint. Before we embark on such work, we need to learn about our philosophical options—to understand why we ask the questions we do, why we conceptualize as we do, and why we use particular techniques. Philosophy is at the heart not only of this chapter but also of much of this book. We begin with an overview of our philosophical options.

Philosophical Options

First, it may be helpful to realize that we have already started discussing such matters, if only indirectly, in the Introduction and Chapter 1. To clarify this point, let us return to the concept of 'physical geography as cause', or environmental determinism. This term, as we have seen, implies that the physical environment is the principal determinant of all human matters, including human landscapes. It refers to what Ratzel (volume 1) called anthropogeography and was subsequently used by many others, notably Semple, Huntington, and Taylor. But we have already seen that there are other views. Possibilism implies that the physical environment does not determine human matters, but simply offers various possibilities. This is the cultural landscape view variously introduced by Vidal, Schlüter, and Sauer. Immediately, we find that what appeared to be merely two alternative views about human geography is in fact the tip of a massive philosophical iceberg: determinism versus free will.

Determinism is a philosophical concept postulating that all events, including human actions, are predetermined. Free will, on the other hand, postulates that humans are able to act in accordance with their will. Environmental determinism is a specific version of the larger concept of determinism, while possibilism is a specific variant of the larger concept of free will—a variant that does not necessarily require acceptance of free will itself. It is not difficult to understand why environmental determinism proved attractive to geography:

1. According to the prevailing late nineteenth-century view, it was the job of science to explain phenomena in terms of cause and effect. Thus, a geography centred on environmental determinism was a geography with scientific credibility. Specifically, its emphasis on cause and effect linked it philosophically to Darwin's landmark work, *On the Origin of Species*, published in 1859.

2. A long history of environmental determinist thought began with the Greeks and included such philosophers as Bodin and Montesquieu. We might argue that Ratzel, in 1882, was merely introducing to geography a long-established and appropriate concept.

3. When geography was in its institutional infancy, any approach that explicitly asserted the importance of geography to human affairs would enhance the new discipline's status.

4. The apparent logic of the environmental determinist argument was attractive in itself. As Taylor once wrote, 'as young people we were thrilled with the idea that there was a pattern anywhere, so we were enthusiasts for determinism' (cited in Spate, 1952: 425).

Instead of engaging in any substantial debate about determinism versus free will, those geographers influenced by Ratzel's volume 1 simply accepted the logic of determinism. It was not until the 1950s that essentially philosophical cases for environmental determinism and possibilism were made (Box 2.1).

Possibilism emerged as a component of the newly institutionalized discipline of geography at much the same time as environmental determinism, for three reasons:

1. A substantial literature, centred on humans and land, already presumed that humans could make decisions not explicitly caused by the physical environment. (Humboldt and Ritter were its two most notable authors.)
2. Geographers could point to many instances in which different human landscapes were evident in essentially similar physical landscapes.
3. Possibilism corresponded to the historically popular view that every event is the result of individual human decision–making.

This brief discussion of environmental determinism and possibilism reminds us that almost every geographic question we ask has philosophical underpinnings. The only aspects of geography that are essentially independent of philosophy are exploration and mapping. Our concern now is to identify the major philosophies that play a part, implicitly or explicitly, in contemporary geography: empiricism, positivism, humanism, and Marxism.

EMPIRICISM

Much human geography, even today, may appear to be aphilosophical. Certainly most human geographers before the 1950s ignored philosophical issues; they simply conducted

Box 2.1   Environmental Determinism

'In the early part of this century, a major stumbling block to meaningful objective research in American geography was the assumption that the physical environment largely determined the cultural landscape' (Dohrs and Sommers, 1967: 121).

There is no denying the long scholarly pedigree of this view in Western thought. Both Plato and Aristotle regarded Greece as having an ideal climate for government. Montesquieu contended that areas of high winds and frequent storms were particularly conducive to human progress. As the French philosopher Cousin put it:

Yes, gentlemen, give me the map of a country, its configuration, its climate, its waters, its winds and all its physical geography; give me its natural productions, its flora, its zoology, and I pledge myself to tell you, a priori, what the man of this country will be, and what part this country will play in history, not by accident but of necessity, not at one epoch but at all epochs. (quoted in Febvre, 1925: 10)

Environmental determinism was not only in accord with the prevailing views of science in the late nineteenth century, but was seen by some as offering a distinct and necessary basis for the new discipline of geography.

Nevertheless, in retrospect it still is surprising that so many geographers accepted environmental determinism in a relatively uncritical manner. Davis, for instance, considered any statement to be geographic if it linked physical factors to some human response, while Semple (1911: 1) wrote, 'Man is a product of the earth's surface.' Most significantly, until the 1950s environmental determinism was implicitly accepted in much geography, including much of the regional geography that was the dominant approach at the time.

Taken out of context, many of the apparently extreme statements of this view, such as those quoted from Cousin and Semple above, are easily misinterpreted. However, careful reading suggests that such dramatic statements are not substantiated in the larger body of their work—see, for example, G.R. Lewthwaite (1966: 9). Indeed, Huntington, a human geographer usually regarded as an archetypal determinist, wrote, 'Physical environment never compels man to do anything: the compulsion lies in his own nature. But the environment does say that some courses of conduct are permissible and others impossible' (Huntington, 1927: vi).

Debates about the relative merits of determinism and alternative interpretations became common in the 1950s. Environmental determinists posited that it was essential to talk in terms of cause and effect, and therefore it was necessary to be a determinist. The basic counter to this assertion was that environmental determinist statements were incapable of being tested; they were merely working assumptions, useful only for suggesting research directions. According to this logic, the fundamental error of much environmental determinism is that it simply assumes a certain cause and effect to be true, when in actuality there is no basis for any such assumption.

Many other social scientists regarded geography with some dismay because of what they perceived as an over-emphasis on the role of physical factors in discussions of human behaviour. As noted in Chapter 1, Durkheim even objected to the inclusion of physical factors in the French school of geography initiated by Vidal, a school that itself opposed environmental determinism.

Regardless of where one stands on the relative importance of physical and human factors, it is always an oversimplification to treat physical geography as a cause of human behaviour.

empiricism
A philosophy of science based on the belief that all knowledge results from experience and, therefore, gives priority to factual observations over theoretical statements.

positivism
A philosophy that contends that science is able to deal only with empirical questions (those with factual content), that scientific observations are repeatable, and that science progresses through the construction of theories and derivation of laws.

scientific method
The various steps taken in a science to obtain knowledge; a phrase most commonly associated with a positivist philosophy.

theory
In positivist philosophy, an interconnected set of statements, often called assumptions or axioms, that deductively generates testable hypotheses.

hypothesis
In positivist philosophy, a general statement deduced from theory but not yet verified.

research that was considered appropriate in the light of the historical development of the discipline. Such work, particularly in regional and cultural studies, made no claim to have a philosophical base; human geographers simply were doing what human geographers did. Nevertheless, it can be argued that there was an implicit philosophy in such regional and cultural work. This is the philosophy of **empiricism**, which contends that we know through experience, and that we experience only those things that actually exist. Empiricism typically sees knowledge acquisition as an ongoing process of verifying and, as necessary, correcting factual statements. In practical terms, an empiricist approach allows the facts to speak for themselves. By definition, empiricism rejects any philosophy that purports to be an all-embracing system.

Empiricism, in turn, is rejected by most other philosophies. It is, however, a fundamental assumption of positivism.

POSITIVISM

For some human geographers **positivism** is a very attractive philosophy because it is rigorous and formal. In principle a clear and straightforward philosophy for human geography, positivism makes the following arguments:

1. Human geography needs to be objective; the personal beliefs of the geographer should not influence research activity. Do you agree with this principle? If so, do you believe that it is possible to research human geography without being affected by your personal beliefs? According to humanism

and Marxism (outlined in the two following sections) objectivity is not only undesirable but in fact not possible.

2. Human geography can be studied in much the same way as any other science. For the positivist, there is really no such thing as a separate geographic method; all sciences rely on the same method. Specifically, positivism first found favour in the physical sciences, and its applications in human geography reflect the belief that humans and physical objects can be treated in a similar fashion. Once again, humanism and Marxism reject this assumption, believing that it dehumanizes human geography.

3. The specific method that positivism sees as appropriate for all sciences, physical and human, is known as the **scientific method** (Figure 2.1): reflecting the empiricist character of the philosophy, research begins with facts; a **theory** is derived from those facts, together with any available laws or appropriate assumptions; a **hypothesis** (or a set of hypotheses) is derived from the theory; and that hypothesis becomes a **law** when verified by the real world of facts. Thus, the scientific method consists of the study of facts, the construction of theory, the derivation of hypotheses, and the related recognition of laws (Box 2.2). For the positivist, any science rests on the twin pillars of facts and theory, and a disciplinary focus on one at the expense of the other is wrong.

Positivist philosophy has roots in the work of the French social philosopher Auguste Comte (1798–1857) and was introduced into

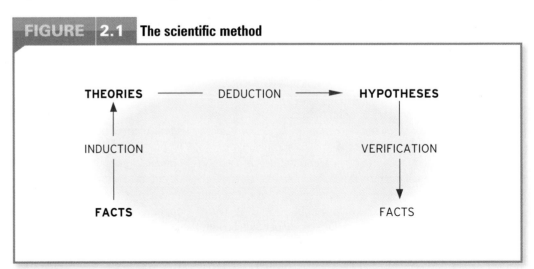

FIGURE   2.1   The scientific method

THEORIES ——— DEDUCTION ——▶ HYPOTHESES

INDUCTION   VERIFICATION

FACTS   FACTS

Science stands on the twin legs of facts and theory; theoretical work depends on and stimulates factual work.

SOURCE: Adapted from J.G. Kemeny, *A Philosopher Looks at Science* (Englewood Cliffs, NJ: Prentice-Hall, 1959), 81.

human geography relatively late (1953). The principal impetus behind Comte's initial argument for positivism was to distinguish between science on the one hand and metaphysics and religion on the other. In human geography, positivism was closely associated with quantitative methods and theory development during the 1960s. Positivism is an integral part of what we described earlier as the spatial analysis approach. Its introduction to geography was controversial because it directly challenged the regional approach dominant at the time. Box 2.3 offers a brief sketch of that controversy.

HUMANISM

Compared to positivism, **humanism** is a much more loosely structured set of ideas. It comes in several significantly different versions. Contemporary humanistic geographers argue for intellectual origins in earlier geographic work, such as that of Vidal and Sauer; in fact, however, this earlier work was essentially aphilosophical, whereas the contemporary

work is typically rooted in some philosophical ground.

Humanistic geography developed from about 1970 onward, initially in strong opposition to positivism. It focuses on humans as individual decision-makers, on the way humans perceive their world, and emphasizes subjectivity in general. Thus, there are substantial differences between the two philosophies. Positivists argue that the world exists as an objective reality, independent of the mind that perceives it, whereas humanists believe it is impossible to separate the two—and, hence, that research is inevitably influenced by the particular qualities of the individual researcher. Positivists seek to generalize, whereas humanists focus on the individual and specific. Positivists base their research on theories, test hypotheses, and develop laws; humanists have no such package of standard, reproducible methods. Rather, humanistic research is essentially personal, the product of the individual geographer, involving intuition and interpretation.

law
In positivist philosophy, a hypothesis that has been proven correct and is taken to be universally true; once formulated, laws can be used to construct theories.

humanism
A philosophy centred on such aspects of human life as value, quality, meaning, significance, and spirituality.

Box 2.2 | Positivistic Human Geography

An example of a positivistic approach to research in human geography will demonstrate the value of working within this philosophical framework. This research, analyzing the retailing and wholesaling activities in urban centres, proceeds through five conventional stages (Yeates, 1968: 99–107).

1. *Statement of problem:* The purpose of the study is to determine the relationships between the numbers of retailing and wholesaling establishments and the size of urban centres in a part of southern Ontario. This problem is suggested by earlier work, both factual and theoretical, suggesting that urban land uses and activities may be explained in terms of variables such as size.
2. *Hypotheses:* Three specific hypotheses are identified on the basis of available human geographic theory (the theory in question is known as central place theory and is discussed in Chapter 11). These are: (1) the number of retail establishments and the population size of an urban centre are directly related; (2) the number of retail and service functions in an urban centre increases with the size of that centre, but at a decreasing rate; (3) the number of wholesale establishments and the population size of an urban centre are directly related.
3. *Data collection:* Population, retail, and wholesale data are collected for all urban centres of 4,000 or more population

(1966). In addition to fieldwork, the data sources used include city directories, telephone directories, and censuses. Note that the use of a minimum population of 4,000 is an example of an operational definition (a way of describing how an individual object can be identified as belonging to a general set).

4. *Data analysis:* This is the stage in the research when information is subjected to appropriate analysis to formally test the stated hypotheses (the verification stage). In this study, the data are graphed (for example, the number of establishments against the population of urban centres), and a statistical procedure known as correlation analysis is used to calculate the extent of the relationship between the two sets of data.
5. *Stating conclusions:* In standard positivistic research, the formal testing of a hypothesis results in a statement concerning the validity of the hypothesis. In this study, all three of the stated hypotheses are confirmed by the statistical analysis; such analysis increases our confidence in the theory that generated the hypotheses and also provides insights into the specific landscape investigated.

As this example shows, it is no exaggeration to describe positivism as a rigorous and formal philosophy.

pragmatism
A humanistic philosophy that focuses on the construction of meaning through the practical activities of humans.

phenomenology
A humanistic philosophy based on the ways in which humans experience everyday life and imbue activities with meaning.

Not surprisingly, positivistic and humanistic geographers usually study different types of research problems. Humanists tend to focus on humans and land, landscapes, and regions, whereas positivists tend to focus on spatial analysis. These differences will become apparent in later chapters.

Discussions of humanism easily become complicated because there are several humanistic philosophies, among them pragmatism and phenomenology.

The philosophical approach known as **pragmatism** developed in the United States at the end of the nineteenth century and was especially influential before 1945. Pragmatism centres on the idea that human actions are structured by our subjective interpretations of the world we live in. Moreover, those interpretations have a marked practical—pragmatic—content. The central argument is that every human action taken in the world is based on human perceptions and observation of the apparent conditions for and practical consequences of previous actions—in short, on experience. Thus, pragmatism is closely associated with empiricism. There is little evidence to date of explicit uses of this philosophy in human geography, although it is basic to some of our qualitative research techniques.

Unlike some other humanistic philosophies, pragmatism has a marked focus on the group or society (as opposed to the individual). One influential offshoot of pragmatism sees individual behaviour as comprehensible only by reference to the behaviour of the appropriate larger groups to which the individual belongs. This offshoot is known as interactionism.

Box 2.3   The Schaefer–Hartshorne Debate

Because overviews of the evolution of an academic discipline, particularly with respect to philosophical changes, are supposed to be dispassionate, rarely can they do justice either to the individuals involved or to the events in which they participate. On this occasion, however, we will depart from tradition.

By the mid-1950s, geography seemed ripe for change. The combination of regional geography and environmental determinism was coming to be seen as unproductive, while the landscape school led by Sauer appealed only to a minority, albeit a substantial one. The transition to spatial analysis was not easy. One important element in that transition was the Schaefer–Hartshorne debate (see Martin, 1989). In September 1952, Fred Schaefer submitted his manuscript 'Exceptionalism in Geography: A Methodological Examination' to a major geographical periodical, the *Annals of the Association of American Geographers*. In the winter of 1952–3, Schaefer suffered a heart attack. On 6 June 1953, at the age of 48, he died of a second attack, and his paper (Schaefer, 1953) was published shortly thereafter. In it he refuted the claim that geography was necessarily a regional discipline and advocated the adoption of a positivist approach and related methods. Specifically, Schaefer objected to the argument he saw as implicit in Hartshorne's work, to the effect that geography studied unique places or regions and was therefore unlike other disciplines.

The article, the greatest achievement of Schaefer's relatively short career, faced strong opposition when it was submitted to the journal in the fall of 1952, and two of the three peer reviewers expressed serious reservations. These criticisms, which further drained his energies, included that Schaefer used emotional language without defending the validity of his words. Schaefer wrote a lengthy response to the editor but chose not to make any revisions. Despite his refusal to make revisions in light of the peer review process, the article was accepted for publication in May 1953, shortly before Schaefer died.

Hartshorne first saw the article in October 1953. He corresponded with the *Annals* editor several times in that month, complaining that Schaefer's article was 'a palpable fraud, consisting of falsehoods, distortions and obvious omissions' (quoted in Martin, 1989: 73), and later published a series of substantial rebuttals (Hartshorne, 1955, 1959).

One geographer (Bunge, 1968) wrote a biographical sketch of Schaefer's life that was later condensed (Bunge, 1979). As a socialist in Germany during Hitler's rise to power, Schaefer was 'immediately subjected to terrorism. . . . He was under constant police surveillance' (1979: 128). He became a political refugee in England and subsequently moved to Iowa City, where he joined the University of Iowa geography department. Bunge notes the harassment that Schaefer faced in the early 1950s, when a US Senate committee headed by Senator Joseph McCarthy was attempting to find and punish suspected Communists in the US. 'Hartshorne, in a commanding position in geography, turned his fury against Schaefer. With sheer brilliance of argument as his only weapon, Schaefer fought to be heard. Fire engine sirens vividly reminded him of Hitler's terrorism. . . . McCarthyism repulsed and sickened Schaefer.' At the time of his death, Schaefer's 'article—his "reason for being"—had not yet been published' (Bunge, 1968: 20–1).

The philosophy of **phenomenology** has many variations, all of them based on the idea that knowledge is subjective. To understand human behaviour, therefore, it is necessary to reconstruct the worlds of individuals. A central component of phenomenology is the idea that researchers need to demonstrate **verstehen**, or sympathetic understanding of the issue being researched. The distinction with respect to positivism is clear. Phenomenology seeks an empathetic understanding of the lived worlds of individual human subjects, whereas positivism seeks objective causal explanations of human behaviour, without reference to individual human differences. One especially thoughtful geographer has written a series of books and articles that are phenomenological in focus (Box 2.4).

Several other humanistic philosophies, such as **existentialism** and **idealism**, have been advocated by geographers, but they have not yet proved to be highly influential. Our brief discussion of humanism has identified two general issues. First, the basic distinction between positivism and humanism is one of objectivism versus subjectivism. Positivism contends that the study of human phenomena can be objective; humanism says that it cannot. This argument is an acknowledged, long-established, and unresolved issue in the social sciences. Three questions are relevant here:

1. Is there an interaction between the researcher and the subject researched that invalidates the information collected?
2. Does the researcher have a personal background that effectively influences his or her choice of problem, methods, and interpretation of results?
3. If we view humans objectively, does this mean we see them as objects? If so, is this approach dehumanizing?

Answering 'no' to these three questions means that you have positivistic tendencies; 'yes' means that you have humanistic tendencies. If your answer is 'don't know', that is perfectly understandable at this stage in your studies.

verstehen
A research method, associated primarily with phenomenology, in which the researcher adopts the perspective of the individual or group under investigation; a German term, best translated as 'empathetic understanding'.

existentialism
A philosophy that sees humans as responsible for making their own natures; stresses personal freedom, decision-making, and commitment in a world without absolute values outside of individuals' personal preferences.

Box 2.4 | Humanistic Human Geography

What questions does humanistic geography ask? What types of research issues does it address? What kinds of answers does it look for? Although the 'humanistic' name covers many different approaches, one way to think about these questions is to consider the work of a leading practitioner, Yi-Fu Tuan.

Tuan was born in China in 1930 and educated at London and Oxford universities in England and at Berkeley in California. An initial interest in physical geography soon expanded into a broad focus on questions of humans and nature. Tuan, the leading humanistic geographer during the 1970s and 1980s, was the first to make explicit reference to philosophy, although he never claimed to subscribe to any single approach: 'My point of departure is a simple one, namely, that the quality of human experience in an environment (physical and human) is given by people's capacity—mediated through culture—to feel, think and act' (Tuan, 1984: ix).

With that aim in mind, Tuan was critical of much geography for ignoring individuals and the relationships they have with one another. For Tuan, there was a need to be comprehensive, like a novelist: 'comprehensiveness—this attempt at complex, realistic description—is itself of high intellectual value; places and people do exist, and we need to see them as they are even if the effort to do so requires the sacrifice of logical rigour and coherence' (Tuan, 1983: 72).

Three examples of Tuan's work, all well received within and beyond our discipline, illustrate the broad range of his interests: the concept of landscapes of fear, which proposes that fear is a component part of some environments and that, in turn, it helps create environments (Tuan, 1979); the idea of gardens as an example of our attempts to dominate environments (Tuan, 1984); and his discussion of the links between territory and self (Tuan, 1982). In the latter, Tuan contends that the presence of segmented spaces—for instance, the many discrete spaces in any urban area—leads to segmented societies. Humans retreat into segmented social worlds in response to the difficulties of coping in a complex society. There is a clear distinction between a small settlement in the less developed world, which has few human-constructed barriers and few spaces assigned for exclusive activities, and a modern city with numerous physical and social barriers. The human experiences in these two environments are very different, with significant implications for individual and group behaviour.

The differences between this humanistic geography and the positivistic variety are striking. The two approaches tackle different problems; they raise different questions; and they reach different answers.

idealism
A humanistic philosophy according to which human actions can be understood only by reference to the thought behind them.

Marxism
The body of social and political theory developed by Karl Marx, in which mode of production is the key to understanding society and class struggle is the key to historical change.

capitalism
A social and economic system for the production of goods and services based on private enterprise.

historical materialism
An approach associated with Marxism that explains social change by reference to historical changes in social and material relations.

forces of production
The raw materials, tools, and workers that actually produce goods.

relations of production
The ways in which the production process is organized, specifically the relationships of ownership and control.

The second issue that is becoming apparent is that of social scale. Do we as human geographers study individuals or groups of people? Traditionally, we have focused on groups, but you will have noted that the classic humanistic philosophies place some emphasis on individuals. For many human geographers, this emphasis on individuals is inappropriate, and much humanistic work in geography has been done on a group scale. Typically, groups are defined by culture, religion, language, ethnicity, or class.

MARXISM

Whether or not one espouses a class-based materialist analysis of society, the ideas of Karl Marx (1818–83) are foundational to an understanding of the human–land interface and must be considered. Unfortunately, Marx and his collaborator, Friedrich Engels, are difficult writers to summarize. Nowhere are his ideas presented in a clear, unequivocal fashion, and any interpretation of **Marxism** is bound to give rise to disagreement. In broad terms, Marx was a political, social, economic, and philosophical theorist who worked to construct a body of social theory that would explain how society actually worked. He was also a revolutionary who aimed to facilitate a change in the economic and political structure of society, from **capitalism** to communism.

Before reading about Marx's ideas, it is important to recognize the enormous influence they have had on recent world history. Before the collapse of the former USSR in 1991 and related changes in Eastern Europe, approximately one-third of the world's population lived in societies that traced their origins to Marx's ideas (Box 2.5).

What is it about those ideas that led so many revolutionary political groups to try to adopt them for national governance—and at the same time attracted so much academic elaboration by so many different schools of thought? The simple answer is that Marx has something relevant to say, both pragmatically and intellectually.

The Marxist perspective is often described as **historical materialism**. This term refers to Marx's concern with the material basis of society and his effort to understand society and social change by referring to historical changes in social relations. Marx summarized the character of society as follows. First, there are **forces of production**: the raw materials, implements, and workers that actually produce

goods. Second, there are **relations of production**: the economic structures of a society, that is, the ways in which the production process is organized. The most important relations are those of ownership and control. In a capitalist society, for example, those relations are such that workers are able to sell their labour on the open market. Thus, although workers—the labour force—produce goods, the relations of production determine how the production process is organized.

Together, these forces and relations make up what is called the **mode of production**. This concept is the key to understanding the composition of society. Examples of modes of production include slavery, feudalism, capitalism, and socialism. It is helpful to think of the mode of production as a stage in a process.

Marx used these ideas to sketch a history of economic change that traces the transitions from one mode of production to another—for example, from feudalism to capitalism to socialism. Marx was especially critical of the capitalism that flourished in his lifetime because of his belief that one class (owners) exploited the other (workers) in order to maximize its own profits. Exploitation itself was not new; slavery and serfdom both were obvious forms of exploitation. What Marx recognized was that under capitalism the same kind of exploitation continued, except it was disguised. For this fundamental reason—the exploitation of workers by owners—Marx called for a socialist revolution in the form of a class struggle: a struggle to overturn one mode of production and replace it with another mode of production under the control of a different class—the workers themselves.

Two more terms are important to our human geographic understanding of Marxism. **Infrastructure (base)** is simply another word for the relations of production; **superstructure** is the level of appearances, that is, the legal and political system, the social consciousness, and, for our purposes, the larger human geographic world. These terms are important to us because they are connected: the infrastructure is seen as the cause of the superstructure. Thus a primary concern of Marxist human geography is to understand the crucial role played by the infrastructure—in this case the economic structure of capitalist society—in determining the superstructure of the human geographic world. The human geographer who works within Marxism thus analyzes superstructures, at any location at any time, as they derive from the relevant infrastructures (essentially sets of economic processes).

Marxism differs profoundly from both positivism and humanism in that it sees human behaviour as being constrained by economic processes. Marx held human societies alone responsible for the conditions of life within them. In other words, he believed that social institutions were created by humans—and that when those institutions no longer served a society's needs, they could and should be changed.

Thus, as Box 2.5 suggests, Marxism has a significant revolutionary dimension. Working within Marxism usually implies that one is striving both to understand the human world and to change it. Box 2.6 summarizes one example of the Marxist concern with the global spread of capitalist culture. The long-term aim of such work is to facilitate a social and economic transformation from capitalism to socialism. In the shorter term, Marxism is an attractive philosophy for many human geographers who feel strongly that their work should focus explicitly on social and environmental ills and contribute to solutions to those ills. Other philosophies, such as positivism, are seen as merely describing the superstructure—which they implicitly accept as appropriate. It is worth noting that Marxist geography is not the only kind of geography that strives to contribute solutions to environmental and social ills; however, it is fair to say that that focus is most explicit in Marxist work.

The ideas of Marx live on in many different interpretations and elaborations. His writings and those of his followers are essential reading

---

mode of production
The organized social relations through which a human society organizes productive activity.

infrastructure (base)
The economic structure of a society, especially as it gives rise to political, legal, and social systems.

superstructure
The political, legal, and social systems of a society.

Box 2.5   Marxist Thought and Practice

To understand the broad outlines of Marxist thought, it may help to look at how that thought has been applied from the mid-nineteenth century to the present and how it has been interpreted and modified by others. Three overlapping stages can be identified in the history of Marxism:

1. During the first period, from the publication of the *Communist Manifesto* in 1848 to roughly 1917, followers of Marx focused on four classic concerns: (1) explaining the means by which capitalists exploited workers; (2) advocating a change from private ownership to state control of property and industry; (3) denouncing religion as the 'opiate of the people'; and (4) preparing for the inevitable revolution that, when it came, would eliminate the class system.

2. In the second period Marx's ideas were put into practice in some countries, beginning with Russia. Following the Russian Revolution of 1917, in which Communist (or Bolshevik) forces overthrew the czar, communism expanded into Eastern Europe, parts of the less developed world, and, by 1949, China. Until the collapse of Communist systems in the former USSR and Eastern Europe in the early 1990s, about one-third of the population of the world lived in a political and social system inspired by Marxism. It is important to point out that in none of these cases was the system that Marx described fully implemented. In the former USSR (Union of Soviet Socialist Republics) and those countries influenced by it, it was the government—controlled by a Communist

political party—that controlled economic activity. Contrary to Marx's own expectations, in the USSR workers had minimal influence on economic policies. In fact, the system in place has sometimes been described as 'state capitalism' rather than 'communism'. The Chinese variant of communism was closer to Marx's ideas until the 1976 death of Mao Zedong; since then the government, though still Communist, has made substantial moves towards a capitalist economy (some of the important agricultural and industrial changes since 1976 are addressed in Chapters 10 and 14).

3. The third stage, which might be called Western academic Marxism, began in Europe in the 1920s but became influential in North America only in the 1960s. This academic Marxism developed in opposition to the Soviet version and includes numerous variants. Two major schools of thought in this Western tradition, which you may encounter if you read further in this area, are the Frankfurt school of critical theory, which flourished in the 1960s, and the structural Marxism developed by Althusser.

By 1970, when 'Marxist' ideas were first introduced to human geography, the term itself could refer to a multitude of related but often quite different ideas.

Karl Marx

Manchester cotton mills, c. 1940. Similar landscapes developed in many parts of northern England during the Industrial Revolution. Marx's collaborator, Friedrich Engels, wrote of Manchester a century before this photograph was taken: 'If any one wishes to see in how little space a human being can move, how little air—and such air!—he can breathe, how little of civilisation he may share and yet live, it is only necessary to travel hither' (*The Condition of the Working-Class in England in 1844* [London, 1892, p. 53]). In 'dark satanic mills' like these, workers manufactured cotton goods, spinning the yarn and weaving it on huge looms. Some factories were attached to orphanages that offered children board and lodging in return for their labour. Small children—employed to clear out cotton fluff from the looms—were sometimes crushed and killed by the fast-moving machines. In the nineteenth century, mill owners who used child labour were often praised for their philanthropy in keeping the children off the streets and providing them with food and shelter. Today some mills remain as museum pieces, cleaned up and Disneyfied for tourist consumption.

Getty Images

idiographic
Concerned with the unique and particular.

nomothetic
Concerned with the universal and the general.

for any social scientist, especially for those who are critical of our current societies.

PHILOSOPHY AND HUMAN GEOGRAPHY

It was to aphilosophical human geography, or empiricism, that positivism objected in the mid-1950s. By about 1970 positivism, in turn, was facing criticism from humanism and Marxism—philosophies that also object to certain aspects of one another. Not surprisingly, human geographers continue to welcome further additions to an already broad range of options. Among the current developments in social thought that you will encounter later in this book are feminism and postmodernism.

First, however, one more important distinction needs to be drawn: between idiographic and nomothetic methods. An **idiographic** approach is concerned with individual phenomena; traditional empiricist regional geography—Hartshorne's 'areal differentiation'—was idiographic in focus, as is much humanistic geography. A **nomothetic** approach, on the other hand, formulates generalizations or laws. Research in a positivist tradition is nomothetic, as is much Marxist human geography.

The philosophical diversity of contemporary human geography reflects the diversity of our subject matter, namely human behaviour in a spatial context. It also means that we draw on a wide variety of concepts and methods. Some aspects of the human geography presented in this book are clearly aphilosophical or empiricist, some are positivist, some are humanist, some are Marxist, and some reflect the influence of other aspects of current social thought.

Human Geographic Concepts

Like any other academic discipline, human geography involves two basic endeavours. First, is the need to establish facts. A vital starting point, certainly, is to know where places are and to be aware of their fundamental characteristics. This is what we might call *geographic literacy*. Second, is the need to understand and explain the facts—to know why the facts are the way they are. This is what we might call *geographic knowledge*. Understanding and explaining require that we ask intelligent questions; this in turn requires the conscious adoption of an appropriate philosophical stance.

But much more is involved in the acquisition of geographic knowledge. Philosophy guides us, but it does not necessarily provide all the tools we need. Our task now is to identify those tools, first in the form of concepts and then, in the next section, in the form of techniques. Some of these concepts and techniques are relevant regardless of philosophical bent, while others are philosophy-specific. Where appropriate in this account, concepts are linked to the parent philosophy. Concepts and techniques that are not tied to particular philosophies focus on factual matters, such as determining where things are on the surface of the earth; those tied to a philosophy focus on understanding or explaining why things are where they are. A useful way to think about concepts is to understand them as specifically human geographic vocabulary or, indeed, as part of the prevailing discourse of human geography as this term was introduced in Chapter 1.

SPACE

This account of concepts begins with a trio of closely related terms: space, location, and place. We have already had occasion to use the word **space** several times in this book. In fact, it is not uncommon to describe human geography as a spatial discipline, one that deals primarily with space. (Note that the term is

space
Areal extent; a term used in both absolute (objective) and relative (perceptual) forms.

Box 2.6   Marxist Human Geography

Much contemporary human geography employs—implicitly or explicitly—a Marxist philosophical framework. The following discussion of the global spread of a capitalist culture may help to clarify what doing Marxist research really means (Peet, 1989).

What is culture? How is it formed? Why has capitalist culture been able to spread across the globe and supplant previously existing cultures? A Marxist response to these questions begins with the idea of the consciousness of a culture—that is, the general way that humans think about things. Put simply, different human experiences in different regions of the world result in differences in regional consciousness and culture. Before the introduction of capitalism as a new economic and cultural system (a new mode of production), the world comprised a variety of regional cultures, each with its own distinct kind of consciousness and its own specific means and modes of production. One way to explain these regional variations is to recognize that human culture was closely related to physical geography (this is not an environmental determinist assertion but a simple recognition that humans with limited technology are especially closely related to their environment).

To understand cultural change as a result of the rise of capitalism, a Marxist acknowledges that human history is a story of domination by forces beyond human control. Initially, as we have seen, the dominating force was nature—physical geography—but over time capitalism came to replace nature as the dominating force.

Like nature, capitalism has the ability to cause human wealth or poverty, success or failure. It is a powerful and all-pervading economic and cultural system that can overwhelm and replace earlier regionally based economies and cultures. Capitalist culture is now a world culture, while regional non-capitalist cultures in the less developed world either have been or are being dramatically transformed. In explicitly Marxist terminology, this process can be seen as reflecting change in the mode of production.

A human geographer with a Marxist perspective regards these developments as undesirable because the regional cultures that are disappearing had adapted to deal with a wide range of environmental problems, while the new culture, with its new ways of life and thought, is unsuited to handling those issues.

spatial separatism (or spatial fetishism)
A phrase critical of spatial analysis for treating space or distance as a cause without reference to humans.

location
A term that refers to a specific part of the earth's surface; an area where something is situated.

site
The location of a geographic fact with reference to the immediate local environment.

situation
The location of a geographic fact with reference to the broad spatial system of which it is a part.

place
Location; in humanistic geography, 'place' has acquired a particular meaning as a context for human action that is rich in human significance and meaning.

sense of place
The deep attachments that humans have to specific locations such as home and also to particularly distinctive locations.

not being used in the context of 'outer space', but rather in the context of the surface of the earth.) To understand the geographic meaning of this term, it is helpful to distinguish between absolute and relative space. *Absolute space* is objective: it exists in the areal relations among phenomena on the earth's surface. This conception of space is at the heart of map-making, chorology, and spatial analysis, and it is central to the ideas of Kant, who saw the geographer as ordering phenomena in space. *Relative space*, by contrast, is perceptual: it is socially produced and, therefore, unlike absolute space, is subject to continuous change.

Some human geographers, notably those with a primarily humanistic perspective, are critical of spatial analysis for what they see as its **spatial separatism (or spatial fetishism)**: they believe that a human geography based on spatial analysis focuses on space alone as an explanation of human behaviour. Such critics argue that space itself is devoid of content, and becomes important only in the context of human life.

LOCATION

Regardless of philosophical persuasion, every human geographer begins with one central question: where? **Location** is the basic concept; it refers to a particular position within space, usually (but not necessarily) a position on the earth's surface. As we have just noted with regard to space, there are two ways to describe location. Absolute location identifies position by reference to an arbitrary mathematical grid system such as latitude and longitude; for example, Winnipeg, Canada, is located at 49° 53′N latitude, 97° 09′W longitude. Such mathematically precise statements often are essential. Sometimes, though, they may not be as meaningful as a statement of relative location: the location of one place relative to the location of one or more other places. Absolute locations are unchanging, whereas some relative locations do change. The location of Winnipeg, relative to lines of communication such as roads and air routes, changes over time.

There are other means by which the location of a geographic fact can be identified. A location can be described simply by reference to its place name or toponym: for instance (with increasing degrees of exactness), Canada, Manitoba, Winnipeg, Portage Avenue.

Geographers also distinguish between site and situation. **Site** refers to the local

characteristics of a location, whereas **situation** refers to a location relative to other locations. Figures 2.2 and 2.3, respectively, illustrate the site and the situation of Winnipeg.

PLACE

The term **place** has developed a special meaning in human geography. It refers not only to a location, but also and more specifically to the values that we associate with that location. We all recognize that some locations are in some way distinctive to us or to others. A place is a location that has a particular identity. Our home, our place of worship, and the shops we frequent are locations with particular identities. Clearly, for one person or another, many locations on the surface of the earth qualify as places. To distinguish place from concepts such as space and location, we might note that 'place' is not about where we live but rather about how we live where we live.

The term **sense of place** was popularized in the 1970s by humanistic geographers with philosophical roots in phenomenology. It refers to the attachments that we have to locations with personal significance, such as our home, and also to especially memorable or distinctive locations. It is possible to be aware that a particular location evokes a sense of place without necessarily visiting that location; Mecca, Jerusalem, Niagara Falls, and Disney World are all examples.

A closely related concept in sociology, now also used by humanistic and other geographers, is that of **sacred space**. This term refers to landscapes that are particularly esteemed by an individual or a group, usually for a religious reason but possibly also for some political or other comparable reason. For example, members of the Church of Jesus Christ of Latter-day Saints (commonly known as Mormons), a religious group based in Utah, regard a number of spaces as sacred. All their temples are sacred; they are open only to Mormons and are considered to be locations where communication between earth and heaven occurs. Another sacred Mormon space is a location near Palmyra in New York state, where they believe their founder received God's message (Jackson and Henrie, 1983). By contrast, mundane space is occupied by humans but has no special quality. Humanistic geographers use the related concept of **placelessness** to identify landscapes that are relatively homogeneous and standardized; examples include many tourist landscapes, urban commercial strips, and sub-urbs (Relph, 1976). There is an interesting parallel here with the Marxist perspectives outlined in Box 2.6: places may be thought of as especially characteristic of pre-capitalist cultures, whereas placelessness is more evident in the industrial and post-industrial world.

Tuan, the humanistic geographer whose work is discussed in Box 2.4, introduced the concept of **topophilia**, literally 'love of place'. This term refers to the positive feelings that link humans to particular environments. By contrast, the less frequently used term **topophobia** refers to dislike of a landscape that may prompt feelings of anxiety, fear, or suffering.

These concepts add valuable refinement not only to the more general concept of location, but also to any human geographic work that aspires to understand people and their relationships with land. Indeed, we can think of places as emotional anchors for much human activity.

REGION

Region is one of the most useful and yet at the same time one of the most confusing of geographic concepts:

> So much geography is written on a regional basis that the idea of the region and the regional method is as familiar and as accepted as is Mercator's map in an atlas. Yet as with so many other familiar ideas which we use every day and take for granted, the concept of the region floats away when one tries to grasp it, and disappears when one looks directly at it and tries to focus. (Minshull, 1967: 13)

Chapter 1 identified regional geography as a traditional focus. In the simplest sense, geographers have long recognized that geographic accounts cannot encompass the earth as a whole; hence the logical tendency to subdivide. In similar fashion, historians have subdivided the span of human history into periods. By the early twentieth century, however, regional geography had come to a formal definition of region as 'a device for selecting and studying areal groupings of the complex phenomena found on the earth. Any segment or portion of the earth surface is a region if it is homogeneous in terms of such an areal grouping' (Whittlesey, 1954: 30). Such a definition can accommodate many types of region, both physical and human.

FIGURE 2.2 The site of Winnipeg

Dividing a large area into regions, or **regionalization**, is a process of classification in which each specific location is assigned to a region. Human geographers recognize various types of regions, most notably the **formal** (or uniform) **region**, an area with one or more traits in common, and the **functional** (or nodal) **region**, an area with locations related either to each other or to a specific location. Thus, an area of German-speaking people qualifies as a formal region, while the sales distribution area of a city newspaper qualifies as a functional region. Delimiting formal regions is notoriously troublesome. Consider,

FIGURE 2.3 The situation of Winnipeg within North America

sacred space A landscape particularly esteemed by an individual or a group, usually (but not necessarily) for religious reasons.

placelessness Homogeneous and standardized landscapes that lack local variety and character.

topophilia The affective ties that humans have with particular places; literally, love of place.

topophobia The feelings of dislike, anxiety, fear, or suffering associated with a particular landscape.

regionalization A special kind of classification in which locations on the earth's surface are assigned to various regions, which must be contiguous spatial units.

formal region A region identified as such because of the presence of some particular characteristic(s).

functional region
A region that comprises a series of linked locations.

'Honeymoon capital of the world': Many North Americans would easily recognize Niagara Falls from this label alone. One of the most readily identifiable places in the world, Niagara Falls is a distinctive physical landscape with a distinctive cultural image. The reason for the association with honeymoons is debated, but there is no doubt that the Falls has traditionally represented a 'trip of a lifetime' destination.

imac/Alamy/GetStock

vernacular region
A region identified on the basis of the perceptions held by people inside and outside the region.

for example, how you might delimit a wheat-growing area in the Canadian prairies. Would you delimit it according to the percentage of wheat farmers per township? the percentage of acreage under wheat? the percentage of farm income derived from wheat? These are only some of the possibilities. Clearly, the choice of measure involves considerable subjectivity. Another problem inherent in region delimitation is the implication that regions are geographically meaningful: in fact, a prairie wheat-growing region may or may not be a significant portion of geographic space.

In the 1960s, with the rise of spatial analysis, the traditional concept of formal regions declined and the concept of functional regions increased in popularity. Then, by about 1970, humanistic geographers began arguing for a revitalized regional geography involving **vernacular regions**: regions perceived to exist by people either within or outside them. The 'Bible belt' of the American South and Midwest is one example. This region is readily

identifiable to most Americans as an area with a particularly strong commitment to various conservative Protestant religions. For many of those living in the region, the name is a source of pride; Oklahoma City declares itself to be the 'buckle on the Bible belt'.

The concept of region is fundamental. We have identified four separate applications:

1. Regionalization is a valuable simplifying device, comparable to periodization for historians; the exercise of classifying in itself may be a valuable aid to the understanding of landscapes.
2. Delimitation of formal regions was central to the chorological approach that dominated during much of the first half of the twentieth century.
3. Delimitation of functional regions was an important aspect of spatial analysis.
4. Many contemporary geographers see vernacular regions as crucial to our understanding of human landscapes.

DISTANCE

Since our first four concepts—space, location, place, and region—all refer to portions of the earth's surface, logically we are also interested in the **distance** between locations, between places, and between or within regions. Not surprisingly, because distance is quantifiably measurable, spatial analysts tend to be particularly interested in it. Distances may be measured in many ways, but most measuring systems typically involve some standard unit such as the kilometre. As discussed in Chapter 9, human geographers also measure distances in terms of time and cost, or—rather more difficult to measure—in terms of cognitive or social variables.

The **distribution** of geographic facts—such as churches or towns—can often be explained by reference to the distance between them and other geographic facts. Geographers often talk about distributions, patterns, or forms in reference to the mapped appearance of spatial facts (see Figure 2.4).

It is often the case that such maps reveal some sort of order. Specialty retail stores (jewellers and shoe stores, for example) tend to locate near one another. Similarly, financial institutions prefer to be close to one another. It is useful to characterize such distributions as resulting from a clustering process. Urban centres, on the other hand,

typically develop at some distance from one another, since one of the principal reasons for a particular location is to serve surrounding rural populations. Urban centres also locate at least partly in response to competition. Other geographic facts locate apart because they involve the provision of services without involving competition; hospitals, community centres, and recreation areas are examples.

Much of what has been said about distance so far is evident in what W. Tobler (1970) referred to as the first law of geography: *everything is related to everything else, but near things are more related than distant things*. This is the notion of **distance decay**, graphically depicted in Figure 2.5 and sometimes referred to as the effect or **friction of distance**; typically both time and cost are involved in overcoming distance. This concept lies at the heart of much spatial analysis (for example, see Box 2.2) and is subject to criticism from humanistic geographers who consider it to be an example of spatial separatism, in that distance is seen as having causal powers.

The distance concept is related to several other ideas. **Accessibility** refers to the relative ease with which a given location can be reached from other locations and therefore indicates the relative opportunities for contact and interaction; it is a key concept in the

distance
The spatial dimension of separation; a fundamental concept in spatial analysis.

distribution
The pattern of geographic facts (for example, people) within an area.

distance decay
The declining intensity of any pattern or process with increasing distance from a given location.

friction of distance
A measure of the restraining effect of distance on human movement.

accessibility
A variable quality of a location, expressing the ease with which it may be reached from other locations.

FIGURE 2.4 **Clustered, random, and uniform point patterns**

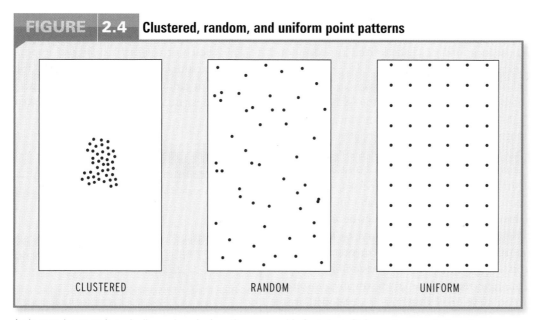

CLUSTERED   RANDOM   UNIFORM

A clustered pattern is typically produced when the geographic facts benefit from close proximity—for example, specialty retail outlets such as jewellery stores. A random pattern may be the result of a random process; may reflect the influence of more than one process; or may represent a temporary stage during changing circumstances. A uniform, or dispersed, pattern is produced when geographic facts benefit from separation because they compete for surrounding space—for example, urban centres in a region.

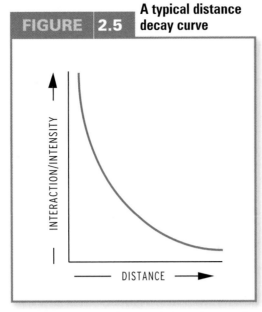

FIGURE 2.5 — A typical distance decay curve

The specific slope is a function of the particular variable being plotted against distance. The slope usually becomes less steep over time, reflecting decreasing distance friction as a result of constantly improving technology.

interaction
The relationship or linkage between locations.

agglomeration
The spatial grouping of humans or human activities to minimize the distances between them.

deglomeration
The spatial separation of humans or human activities so as to maximize the distances between them.

scale
The resolution levels used in any human geographic research; most characteristically refers to the size of the area studied, but also to the time period covered and the number of people investigated.

agricultural, settlement, and industrial location theories that we will encounter later in this text. **Interaction** refers to the act of movement, trading, or any other form of communication between locations. **Agglomeration** describes situations in which locations (usually of activities related to production or consumption) are in close proximity to one another. Conversely, **deglomeration** refers to situations in which those locations are characterized by separation from one another.

Distance is conventionally measured as absolute distance using some standard unit. Distance measured in economic, temporal, cognitive, or social terms is relative. Such distances can be, and frequently are, quite different from corresponding absolute distances. Minimum-cost distances between two locations often involve greater absolute distances than do straight-line distances because of the lower cost of water compared to land transport, or because of physical or political barriers. Distance may also be measured in terms of the time required: times are usually shorter in more developed than in less developed areas; they also tend to decrease with increasing levels of technology. Cognitive and social distances vary according to individual and group circumstances; such differences often reflect differences in knowledge of the areas between the two locations.

SCALE

One of the first decisions made in any piece of geographic research relates to the selection of appropriate scales—spatial, temporal, and social. It is possible to study a large area or a small area, a long period of time or a particular moment in time, a great many people or a single individual. The choice of **scale** is usually determined by the questions posed. But it is important to recognize that different scales can generate different answers.

Geographers use the concept of *spatial scale* in two distinct ways: first, according to a technical meaning associated with the use of maps, in which case, scale is the ratio of distance on the map to distance on the ground. In this sense we might describe the world maps in this book as being at a 'small scale', the maps of regions as being at 'intermediate scales', and maps of local areas as being at a 'large scale'. Note that maps of large areas are at a small scale and maps of small areas are at a large scale.

Second, if we ask whether the locations in a given set are clustered together (agglomerated) or dispersed (deglomerated), the answer will vary with the area selected (Figures 2.6 and 2.7). Spatial scale also needs to be carefully identified whenever we make statements about density. In Canada, for example, a single statistic masks enormous variations; although the average population density for Canada as a whole is 3.25 persons per km^2, for an urban centre the density may be 10,000 per km^2.

The choice of *temporal scale* is also important. If questions are asked concerning the evolution of landscape, then a temporal perspective is essential; but if questions are asked

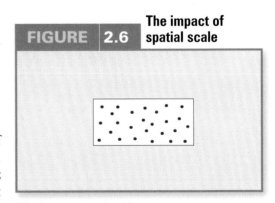

FIGURE 2.6 — The impact of spatial scale

Whether we describe this point pattern as clustered or dispersed depends on the area within which it is contained. Using the inner boundary, the pattern is dispersed; using the outer boundary, it is clustered.

concerning the manner in which a given area functions, then a temporal perspective may be quite irrelevant. Historical and cultural geography usually emphasize time, while chorological work and spatial analysis tend to emphasize the present. Contemporary human geography uses whatever temporal scale is most appropriate.

The relevance of *social scale* has emerged as a major concern with the rise of humanistic perspectives in human geography. Although the focus traditionally has been on groups, generally delimited by reference to culture, some humanistic geographers today favour an individual scale of analysis. Selecting the scale most appropriate to the particular question posed, and that will therefore lead most directly to the correct answer, is not simple. Scales must be selected with care, and properly justified, if we are to be confident of our results.

DIFFUSION

Landscapes, regions, and locations—our three recurring themes from Chapter 1—all are subject to change. **Diffusion**—the spread of a phenomenon over space and growth through time—is one way change occurs. The migration of people, the movement of ideas (for example, the spread of a religion), and the expansion of land use (for instance, wheat cultivation) are examples of diffusion. Diffusion-centred research has long been central to cultural geography because of the need to understand landscape evolution. Probably the greatest impetus for such research was the work of a pioneering Swedish geographer, T. Hagerstrand, who developed a series of diffusion-related concepts in 1953; his work was beginning to become known in the English-speaking world by the early 1960s, but was not translated *in toto* until 1967. Hagerstrand and the work that he inspired, which was largely positivistic in character, introduced three important ideas.

First is the *neighbourhood effect*. This term describes situations where diffusion is distance-biased, that is, where a phenomenon spreads first to individuals or groups nearest its place of origin. Other situations, however, involve a *hierarchical effect*. In this case, the phenomenon first diffuses to large centres, then to centres of decreasing size. One geographer noted that the 1832 cholera epidemic in North America diffused in a neighbourhood fashion, whereas the 1867 epidemic diffused hierarchically. Why the difference? North America in 1832 had a

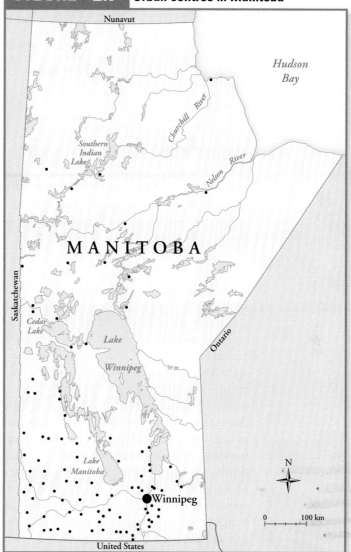

FIGURE 2.7 **Urban centres in Manitoba**

This real-world pattern encompasses two relatively distinct regions: a sparsely settled centre and north and, by comparison, a densely settled south.

limited number of urban centres and a limited communication network, whereas North America in 1867 had many urban centres and a well-developed communication network. A third contribution of Hagerstrand's focused on the well-established fact that most diffusion situations proceed slowly at first, then very rapidly, ending with a final slow stage to produce an *S-shaped curve* (Figure 2.8). The diffusion concept is perhaps best described as a process that prompts changes in landscapes, regions, and locations.

PERCEPTION

In 1850, Humboldt noted that, 'in order to comprehend nature in all its vast sublimity,

diffusion
The spread of any phenomenon over space and its growth through time.

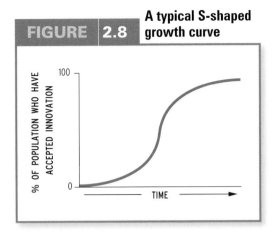

FIGURE | 2.8   **A typical S-shaped growth curve**

Adoption of the new idea begins slowly, then proceeds rapidly, only to slow down as the adoption rate reaches 100 per cent.

it would be necessary to present it under a twofold aspect, first objectively, as an actual phenomenon, and next subjectively as it is reflected in the feelings of mankind' (quoted in Saarinen, 1974: 255–6). Despite this pioneering statement, geographers paid relatively little attention to subjective matters, especially the perceived environment, until the late 1960s. We now recognize that all humans relate not to some real physical or social environment but rather to their **perception** of that environment—a perception that varies with knowledge and is closely related to cultural and social considerations. Humanistic geographers in particular discuss the mental images of places and other people, and seek to describe and understand the mental maps that we carry in our heads. These images and maps are important for at least six reasons:

1. Mental images of other places and other people are always changing. We all live in a shrinking world. We are becoming increasingly aware that when we pollute a specific location, we are polluting the world.
2. Research into mental maps demonstrates that humans have varying perceptions of environments. These perceptions help explain population movements. In the less developed world, the most obvious migrations are from rural to urban areas; in the US, there has been a continuing movement to the south, especially since 1950.
3. The mental maps of particular individuals are of great importance—for example, those of people who make decisions about where

to locate industrial plants. The location of a plant that will employ perhaps 1,000 workers will often reflect the biases and values of a single person or small group. Today the traditional industrial areas of the developed world are rarely favoured by such decision-makers. Instead, many prefer to locate close to recreational amenities and/or large cities. In the US the South and West generally are favoured over the North and East; and in England the London region is favoured over the Midlands and North. In Canada there is an additional factor, as many locational decisions reflect the biases and values of Americans, not Canadians. The mental maps of decision-makers are especially important because so many industries today are footloose—not tied to specific raw materials or to power sources.

4. Serious problems can arise when people in positions of power have distorted mental maps. We may find it humorous when an American president mistakes one country for another, but the consequences could be disastrous. Research in the 1980s showed that the mental maps of the typical American military officer completely confused the sizes of areas and relative populations, and tended to see only one threatening country (the former USSR).
5. Mental maps do change. In Brazil, the movement of the capital from Rio de Janeiro to Brasilia completely changed the image of the interior from one of a worthless wasteland to one of a region with huge potential. In Canada, the image of the North changed with the beginning of oil exploration, and the image of Quebec changes in response to changes in the political fortunes of the separatist movement.
6. Mental maps of relatively unknown areas are especially subject to error. Figure 2.9 is an interesting composite of the image of North America held by Europeans in the mid-eighteenth century.

Clearly, we acquire our mental maps from a variety of sources, but one is of particular relevance to us: human geography teaches us about the world, about where things are located, why they are there, and what they really are. By the time you have completed this course, you may notice a significant improvement in your own mental maps and images!

FIGURE 2.9 Images of North America in 1763

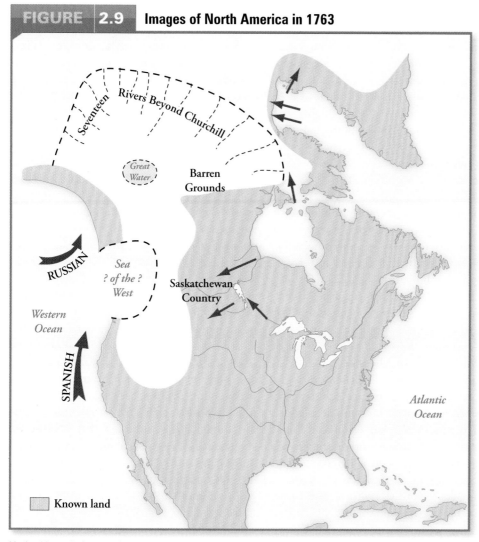

Limited knowledge resulted in many inappropriate decisions. As reality became better understood, spatial behaviour became increasingly appropriate.

SOURCE: R.I. Ruggles, 'The West of Canada: Imagination and Reality', *Canadian Geographer* 15 (1971): 237. Published by Blackwell Publishing Ltd.

DEVELOPMENT

In their studies of the landscapes created by humans, human geographers recognize that any one area changes through time and that different areas have different landscapes. Such conditions are interpreted in terms of **development**. The general meaning of this term involves measures of economic growth, social welfare, and modernization. It is possible to define, for any given area at any given time, the level or state of development. Following this logic, certain areas are qualified as more developed and others as less developed. (This distinction will be pursued in some detail in Chapter 5.)

Such basic identifications can be of real value, but we need to use caution in applying them. It is important that human geographers highlight spatial disparities in economic well-being—but it is also important that we interpret variations with reference to cultural and social considerations. If, for example, we use income levels as a measure of development, then we must not neglect to recognize that income level may reflect both economic success and cultural values.

Contemporary human geographers analyze development while remaining fully aware of the risks of oversimplification. A Marxist might view underdevelopment as a consequence of the rapid diffusion of the capitalist economic and social system, arguing that areas brought into the expanding capitalist system become dependent. Another Marxist perspective might

development
A term that should be handled with caution because it has often been used in an ethnocentric fashion; typically understood to refer to a process of becoming larger, more mature, and better organized; often measured by economic criteria.

suggest that a capitalist system tends to create depressed areas within any given country, thus prompting uneven development. Such variations are evident throughout what we generally label the more developed world. In Canada, for example, an economically advantaged core area, the St Lawrence lowlands of Quebec and Ontario, is surrounded by a number of relatively disadvantaged peripheries, such as the Maritime area and the North.

DISCOURSE

The root meaning of the word 'discourse' is 'speech'. In the social sciences, however, the term also refers to a way of communicating, in speech or in writing, that serves to identify the person communicating as a member of a particular group. (Note that language, dialect, or accent can serve the same function.) For example, the technical vocabulary introduced above—'space', location', and so on—is part of the discourse of human geography, and serves to identify those who use that vocabulary as members of the group of human geographers.

However, as implied, but not detailed, in Box 1.1, 'discourse' also has a more profound meaning, derived from the work of the French social theorist Michel Foucault. Foucauldian theory was introduced into the literature of human geography in the 1980s, as one aspect of contemporary social theory. According to Foucault, the history of ideas is a history of changing discourses in which (a) there is a fundamental connection between power and knowledge, and (b) truth is not absolute but relative, dependent on the power relations within the societies that construct it. This more complex meaning of 'discourse' is briefly pursued in Chapter 7, in connection with feminism and postmodernism. The latter are bodies of social theory that challenge established discourses because those discourses are seen as products of people in positions of academic power who are able to define the truth in their terms, to the exclusion of the concerns of other, usually marginalized, groups (see Box 1.1).

GLOBALIZATION

Our final concept, **globalization**, integrates— some might even say replaces—several of the concepts introduced above: namely space, location, place, region, and distance. Earlier discussions of globalization suggested that our complex and varied human worlds are coming, however unevenly, to look more and more like a single world.

Accordingly, we now identify globalization as an overriding metaconcept, providing human geographers with a body of ideas that may facilitate analysis of environmental, cultural, political, and economic topics. At the same time, as detailed in Table I.2, we acknowledge the many uncertainties concerning what globalization is and how it is affecting the world. The term 'globalization' came into widespread use only in the 1980s. In general, it refers to the idea that the world is becoming increasingly homogenized—economically, politically, and culturally. Globalization is both a result and a cause of an ever-increasing connectedness of places and peoples as economic, political, and cultural institutions, flows, and networks all combine to bring previously separated peoples and places together. Among the most obvious components of globalization are advances in communications technologies and the increasing dominance of **transnationals** (corporations whose activities are not confined to a single state). Among the most obvious consequences is the fact that, because we are generally able to overcome distances much more easily today than we were in the past, distance no longer plays the critical role it once did in promoting the development of separate human geographic worlds.

Globalization increases both the quantities of goods, information, and people that move across national boundaries and the speed with which they do so. Most readers of this book are accustomed to having significant world happenings—sports competitions, concerts, political events—beamed directly to their living rooms, to receiving information and goods from faraway places, and to seeing local decisions affected by larger global circumstances. For many who regularly travel to other countries, the sense of place-to-place difference is vanishing as a few dominant brands become ever more ubiquitous—in the area of information (CNN, the BBC) no less than that of consumer goods (McDonald's, Coca Cola, Benetton, Nike). Many of us have also had direct experience of financial homogenization: today we can purchase goods or local currency almost anywhere in the world using only one major credit card. Virtually all of these developments have come about since 1980.

Some observers go so far as to predict the emergence of a global village, a global culture.

---

transnational
A large business organization that operates in two or more countries.

---

globalization
A complex combination of economic, political, and cultural changes that have long been evident but that have accelerated markedly since about 1980, bringing about a seemingly ever-increasing connectedness of both people and places.

Although such a development may still be far in the future, we do appear to be entering a new era involving the creation of global landscapes, in which places separated by great distances are nevertheless very similar in character. Think of shopping malls, major hotels, fast-food restaurants, and airports. In these instances, similar settings may be patronized by people with similar lifestyles and similar aspirations. At the centre of all this increasing interaction and homogenization is the capitalist mode of production. We will see evidence of the trend towards globalization throughout this book, yet, interestingly, the 2008–9 global economic recession threatened economic globalization as the decline in trade prompted national governments to introduce various measures of protectionism.

CONCEPTS: A SUMMARY

This discussion of human geographic concepts has introduced the terms that are central to understanding how human behaviour affects the earth's surface—space, location, place, region, distance, scale, diffusion, perception, development, discourse, and globalization—along with various concepts that can be conveniently subsumed under those headings. In most cases, as we have seen, a concept can be understood and used in multiple ways, depending on philosophical approach; think especially about region, distance, scale, and development. All the terms introduced here will reappear in subsequent accounts of the substantive work of human geographers; all the concepts will be put into practice.

Techniques of Analysis

Now that we have a basic understanding of why human geographers ask particular research questions, what their philosophical options are, and the concepts they use to structure their research activity, we are ready to look at the techniques they use to collect, display, and analyze data.

The first three techniques discussed are cartography, computer-assisted cartography, and geographic information systems. Each of these closely related techniques is inherently geographic, having been largely developed within geography, and each involves inputting, storing, analyzing, and outputting spatial data. A group of techniques that relate primarily to the collection of data—remote sensing—is discussed

next. Finally, we consider both qualitative and quantitative methods of collecting and analyzing data.

Although many of these techniques are philosophically neutral, this section is not entirely philosophy-free. Data collection by fieldwork clearly includes some procedures that are positivist and others that are humanist. We will return to the question of underlying philosophies—and what they mean for the 'human' in 'human geography'—in the conclusion of this chapter.

CARTOGRAPHY

Chapter 1 highlighted the centrality of maps to the geographic enterprise. The science of map-making is known as **cartography**. Until the 1960s, cartography was essentially limited to map production, following data collection by surveyors and preceding analysis by geographers. There was much emphasis on manual skills. The primary purpose of such maps was to communicate information. Quite simply, maps are an efficient means of portraying and communicating spatial data. Today, however, cartography is less dependent on manual skills and is closely integrated with analysis.

As communication tools, maps describe the location of geographic facts. As analytical tools, they can be used to clarify questions and suggest research directions. In the production of maps, cartographers need to decide on questions that can significantly affect map appearance and quality: scale, type, and projection.

As we have already noted, map scale relates to the area covered and the detail presented. Large-scale maps portray small areas in considerable detail, while small-scale maps portray large areas with little detail. Scale is always indicated on a map, whether as a fraction, a ratio, a written statement, or a graphic scale. Canada has a National Topographic System, with maps at a scale of 1:250,000 covering approximately 15,539 km^2 (6,000 square miles) and maps at a scale of 1:50,000 covering 1,036 km^2 (400 square miles) in greater detail. On a 1:50,000 map, farm buildings are located and contours (lines of equal elevation) are shown at 7.5-m (25-ft) intervals. Figures 2.10 and 2.11 provide examples of each type.

The type of map constructed depends on the information being presented. Dot maps are useful for data such as towns, wheat farming, cemeteries, incidence of diseases, and so forth. Typically, each dot represents one

cartography
The conception, production, dissemination, and study of maps.

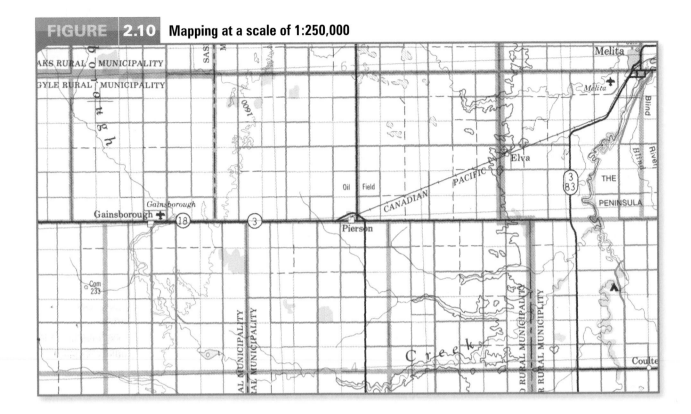

FIGURE 2.10   Mapping at a scale of 1:250,000

choropleth map
A thematic map using colour (or shading) to indicate density of a particular phenomenon in a given area.

isopleth map
A map using lines to connect locations of equal data value.

projection
Any procedure employed to represent positions of all or a part of the earth's spherical (three-dimensional) surface onto a flat (two-dimensional) surface.

occurrence of the phenomenon being mapped (Figure 2.12). A **choropleth map** displays data using tonal shadings that are proportional to the density of the phenomena in each of the defined areal units (Figure 2.13). Choropleth maps sacrifice detail for improved appearance. An **isopleth map** consists of a series of lines—isopleths or isolines—that link points having the same value (Figure 2.14). Examples include contour maps, isochrome maps showing lines of equal time, and isotim maps showing lines of equal transport cost.

Small-scale maps (of large areas such as the earth) raise a fundamental question: how can we best represent a nearly spherical earth on a flat surface (a process of conversion known as a **projection**)? No satisfactory answer to this question has yet been found, although hundreds of projections have been developed (Box 2.7).

Map users need to understand the cartographic design process, including the significance of the chosen scale, the types of symbols, and the projection, in order to interpret a map correctly (Box 2.8).

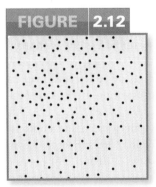

FIGURE 2.12

Schematic representation of a dot map

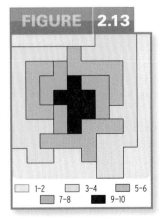

FIGURE 2.13

	1–2		3–4		5–6
	7–8		9–10		

Schematic representation of a choropleth map

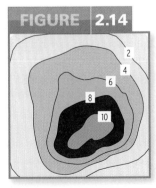

FIGURE 2.14

2
4
6
8
10

Schematic representation of an isopleth map

FIGURE 2.11 Mapping at a scale of 1:50,000

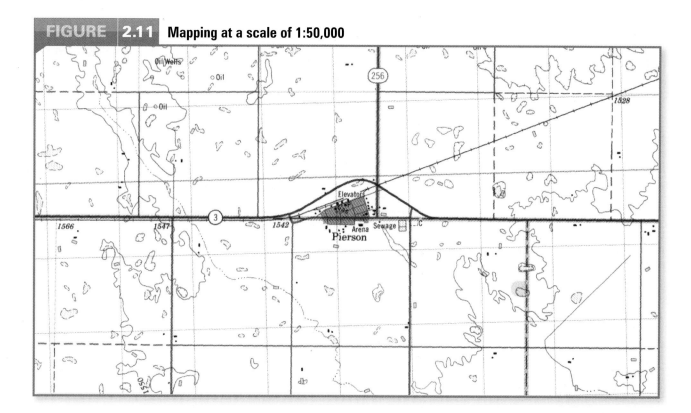

COMPUTER-ASSISTED CARTOGRAPHY

Computer-assisted cartography, sometimes called digital mapping, is discussed separately from traditional cartography because it represents much more than just another evolution in production techniques, although technical advances have significantly reduced the need for manual skills.

Computers allow us to amend maps by incorporating new and revised data; they enable us to produce various versions of the mapped data to create the best version; through the use of mapping packages, they diminish the need for artistic skills and allow for desktop map creation, although this still requires considerable design skills as decisions are made about colouring, shading, labelling, and other aspects of map creation. Computer-assisted cartography has introduced maps and map analysis into a wide range of new arenas: for example, businesses can now use cartography to realign sales and service territories. Computer-generated maps facilitate decision-making and are becoming important in both academic and applied geography.

GEOGRAPHIC INFORMATION SYSTEMS

A **geographic information system** (GIS) is a computer-based tool that combines several functions: storage, display, analysis, and mapping of spatially referenced data. A GIS includes processing hardware, specialized peripheral hardware, and software. The typical *processing hardware* is a personal computer or workstation, although mainframe computers may be used for especially large applications. *Peripheral hardware* may be used for data input—for example, digitizers and scanners—while printers or plotters produce copies of the output. *Software* production is now a major industry, and numerous products are available for the GIS user; examples include IDRISI, a university-produced package designed primarily for pedagogic purposes, and ARC/INFO, a package developed by the private sector that is widely used by governments, industries, and universities.

The origins of contemporary GIS can be traced to the first developments in computer-assisted cartography and to the Canada GIS of the early 1960s. These developments centred on computer methods of map overlay and area measurement—tasks previously accomplished by hand. Since the early 1980s, however, there has been an explosion of GIS activity related to the increasing need for GIS and the increasing availability of personal computers.

The roots of GIS are clearly in cartography, and maps are both its principal input and its

geographic information system (GIS) A computer-based tool that combines the storage, display, analysis, and mapping of spatially referenced data.

Box 2.7 Map Projections

'Not only is it easy to lie with maps, it's essential' (Monmonier, 1991: 1). Monmonier is referring to the fact that even the best-intentioned cartographers are forced to distort reality because they must always use a map projection: a systematic two-dimensional transformation of the three dimensions of a sphere. This issue is most important when the map is of a large area (small scale).

A projection attempts to retain one or more of the following characteristics of a sphere: shape, size, direction, and distance. The correct projection to use is the one that best serves the objectives of the map. The main decision to be made concerns the choice between a conformal projection, which depicts shape correctly, and an equal area projection, which maintains the relative areas of all parts of the earth.

Figures 2.15, 2.16, and 2.17 show three important projections. The Mercator projection, introduced in 1569, greatly exaggerates the size of high-latitude land masses (Figure 2.15). Note that Alaska and Brazil appear to be of similar size; yet in reality Brazil is some six times larger. The Mercator is a conformal projection. The great advantage of the Mercator projection is that it can be easily used for navigational purposes; one has only to draw a straight line between two locations on the projection and that line gives the necessary compass direction. It has, however, been criticized for depicting Europe (and North America) as larger than they are, thus giving these areas disproportionate prominence. Certainly, it is not well-suited as a general world map today, but it is important to realize that the projection was designed by Mercator to solve a technical problem in navigation.

The Mollweide projection, introduced in 1805, depicts all regions in correct relative size, but compresses high-latitude regions (Figure 2.16). Mollweide is an equal-area projection.

The third projection, shown in Figure 2.17, was introduced in 1963 by the geographer Robinson and was adopted in 1988 by the National Geographic Society (although it was, in turn, replaced by the similar Winkel-Tripel projection in 1998). The Robinson projection has distortions both at the equator and at the poles and is accurate only at latitudes 38°N and 38°S. Because this projection offers a good compromise between geographic accuracy and aesthetic—meaning, it 'looks right'—it is the one used in this text. *Even so, whatever projection a cartographer uses, getting one thing right will always mean getting something else wrong.*

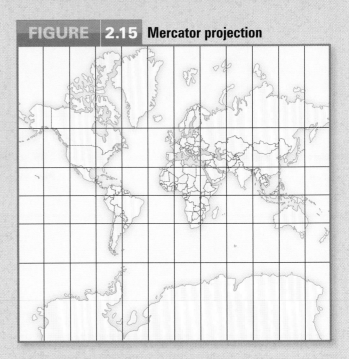

FIGURE 2.15 Mercator projection

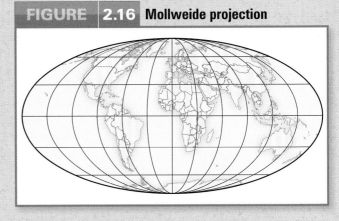

FIGURE 2.16 Mollweide projection

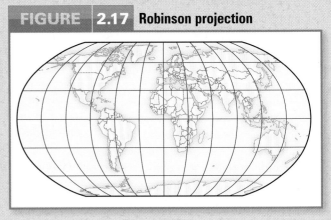

FIGURE 2.17 Robinson projection

principal output. But computers are able to handle only characters and numbers, not spatial objects such as points, lines, and areas. Hence

GISs are distinguished according to the methods they use to translate spatial data into computer form. There are two principal methods

Box 2.8 | The Power Of Maps

The authoritative appearance of modern maps belies their inherent biases. To use maps intelligently, the viewer must understand their subjective limitations' (Wood, 1993: 85). While Box 2.7 outlines the unavoidable errors in maps that result from the need to employ a projection, this discussion centres on how maps reflect the assumptions of their creators and how they may be deliberately employed to convey specific misleading messages.

Recall the T-O maps introduced in Chapter 1; these European maps clearly reflect medieval Christian values, with Jerusalem placed at the centre of the world. Other maps from that place and time show terra incognita to the south and make extensive use of decorative pictures. Nineteenth-century European maps reflect the political and economic concerns of their creators: colonial possessions, for example, are prominently displayed. These maps, building on Ptolemaic traditions, established the conventions that most of us take for granted today: north at the top, 0° longitude running through Greenwich, England, and the map centred on either North America or Western Europe. These conventions are totally arbitrary—and yet their influence is enormous, because in effect they tell us that certain countries are world centres and others are outliers.

Other maps may convey a biased message either deliberately or because the map-maker has some ulterior motive. As Monmonier (1991) suggests in his book *How to Lie with Maps*,

there are many ways to deceive. Political propaganda maps represent an extreme example. Less extreme but still significant are maps that simply ignore important features: for example, maps of the Love Canal area in Niagara Falls, New York (Figure 2.18), that simply fail to include extremely important information. The message here is that 'a single map is but one of an indefinitely large number of maps that might be produced for the same situation or from the same data' (Monmonier, 1991: 2).

FIGURE 2.18 | Two topographic maps of the Love Canal area, Niagara Falls, New York

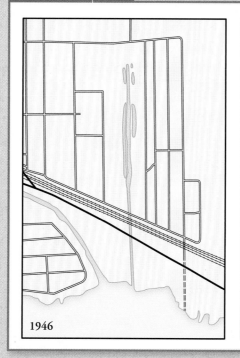
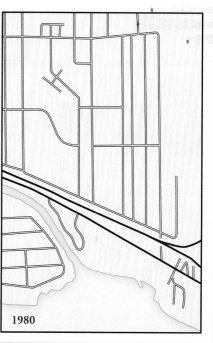

1946   1980

The 1946 map shows the canal (the long vertical feature in the centre of the map), but does not show the use made of it, beginning in 1942, as a dump for chemical waste. The 1980 map does not show the filled-in canal, nor does it indicate that dumping continued until 1954. In 1978 the New York health commissioner declared a state of emergency and relocated 239 families.

SOURCE: M. Monmonier, *How to Lie with Maps* (Chicago: University of Chicago Press, 1993), 121–2. 1946: US Army Map Service, 1946, Tonawanda West, NY, 7.5-minute quadrangle map. 1980: US Geological Survey, 1980, Tonawanda West, NY, 7.5-minute quadrangle map.

of translation. The **vector** approach describes spatial data as a series of discrete objects: points are described according to distance along two axes; lines are described by the shortest distance between two points; and areas are described by sets of lines. The **raster** method represents the area mapped as a series of small rectangular cells known as pixels (picture elements); points, lines, and areas are approximated by sets of

pixels, and the computer maintains a record of which pixels are on or off. Box 2.9 summarizes the sorts of activities involved in using a GIS.

What is the value of a GIS? What applications are possible? The short answer is that GISs have numerous and varied applications in any context that may be concerned with spatial data: biologists analyze the effect on wildlife of changing land-use patterns; geologists search

vector
A method used in GIS to represent spatial data; describes the data as a collection of points, lines, and areas, and describes the location of each of these.

raster
A method used in GIS to represent spatial data; divides the area into numerous small cells and pixels, and describes the content of each cell.

remote sensing
A variety of techniques used for acquiring and recording data from points that are not in contact with the phenomena of interest.

for mineral deposits; market analysts determine trade areas; and defence analysts select sites for military installations. We will encounter several applications of GIS in later chapters. The common factor is that the data involved are spatial. After handling spatial data for about 3,000 years, geographers now have important new capacities. In brief, a GIS achieves a whole new range of mapping and analytical capabilities—additional ways of handling spatial data.

REMOTE SENSING

No map can be produced without data. GISs and analytical methods in general also require data. Where do these data come from? Early map-makers obtained their data from explorers and travellers. The contemporary geographer collects different data in different ways.

One particularly important group of collection methods focuses on gathering information about objects from a distance. The term **remote sensing** describes the process of obtaining data using both photographic and non-photographic sensor systems. In fact, we all possess remote sensors in the form of our eyes, and it has been one of humanity's ongoing aims to improve their ability to acquire information. Improving our eyes themselves (using telescopes), improving our field of vision by gaining altitude (climbing hills or using balloons and aircraft), and improving our recording of what is seen (using cameras) are advances in remote sensing. Today, most applications of remote sensing rely on electromagnetic radiation to transfer data from the object of interest to the sensor. Using satellites as far as 36,000 km (22,370 miles) from the earth, we are now able to collect information both about the earth as a whole and about what is invisible to our eyes. Electromagnetic radiation occurs naturally at a variety of wavelengths, and there are specific sensing technologies for

Box 2.9 | Using a GIS

A GIS can be used to perform as many as eight different functions (Raper and Green, 1989):

1. *Data capture:* Vector data are recorded by manual tracing or electronic digitizing of the spatial data. Raster data are recorded by a scanner that uses a light sensor to record the spatial data as 'on' pixels.
2. *Editing:* Once captured, the data are edited to correct errors: for example, to relocate misplaced points in vector data or to remove stray pixels in raster data.
3. *Structuring:* This involves storing the data in a form suitable for rapid spatial retrieval; a number of technical procedures are available to achieve this goal.
4. *Restructuring:* Changing the level of map detail or converting between vector and raster data types.
5. *Manipulation:* One of the greatest advantages of using a GIS is the ease with which spatial data can be manipulated and revisions made.
6. *Search or retrieval:* Another major benefit of working with a GIS is the ease of identifying and retrieving information if necessary.
7. *Analysis:* The ability to analyze, not merely represent, spatial data is perhaps the most important activity that a GIS is able to perform, and is the clearest distinction between GIS and computer-assisted cartography. The quality of GIS software is often judged primarily in terms of this activity.
8. *Integration:* One output possible from a GIS is the integration of two or more maps of an area to form a single map; in the simplest terms, a map overlay procedure is used to create a new composite map.

Several of these activities are technical and are both prompting and responding to further technical advances. Not surprisingly, given the difficulty of understanding the internal workings of a GIS, the mapped output must always be interpreted and used with care.

There are some important differences between traditional maps and a GIS. In fact, some are so significant that they are changing the nature of geographic information and the role that this information plays in society. Whereas a traditional map is a single product, a GIS can be 'customized' by individual users, who may change data, update details, and superimpose multiple scales. Further, a GIS challenges one of the most basic of geographic conventions, namely, the idea that regions are separated by lines. Of course, we know that this is not the case, but the very act of drawing maps with neat boundaries effectively reifies both boundaries and the regions they circumscribe—it misleads us into thinking of them as fixed realities. A GIS is much more capable than a conventional map of depicting continuous spatial change, and thus requires users to think about geographic worlds without precise boundaries. This is an example of how a new technology can generate new ideas.

A debate is ongoing concerning what some human geographers see as the philosophical limitations of GIS, which has close associations with positivism but no connection at all to any of the post-positivist philosophies such as humanism and Marxism. Today, as critical analyses of the social implications of GIS technology begin to appear, there are encouraging signs that GIS may yet play a bigger role in the broader practice of human geography.

the principal spectral regions (visible, near-infrared, infrared, and microwave).

The conventional camera was the principal sensor used until the introduction of earth orbital satellites in the 1960s. Aerial photography is still used for numerous routine applications, especially in the visible and near-infrared spectral regions. The near-infrared spectral region has proved particularly useful for acquiring environmental data. Today, the emphasis is on satellite imagery, especially since the launching by the United States of Landsat (initially called ERTS—Earth Resources Technology Satellite) in 1972 as the first unmanned satellite with the express purpose of providing detailed geophysical data. Satellite scanners numerically record radiation and transmit numbers to a receiving station; these numbers are used to computer-generate pixel-based images. Satellite images are not photographs.

The principal advantages of satellite remote sensing include, first, repeated coverage of an area facilitates analysis of land-use change. The most recent Landsat satellite covers most of the earth's surface every 16 days; these data are homogeneous and comprehensive. Second, because the data collected are in digital format, rapid data transmission and image manipulation are possible. Third, for many parts of the globe, these are the only useful data available. Finally, remote sensing allows the collection of entirely new sets of data: satellite data first alerted us to the changing patterns of atmospheric ozone in high-latitude areas.

For the human geographer, remote sensing is especially valuable in aiding understanding of human use of the earth; much of the discussion in Chapter 3 relies on remotely sensed data. Remote sensing is less useful if we are concerned with underlying economic, cultural, or political processes.

Most satellites have been launched by either the US or the European Space Agency. Canada launched a satellite, Radarsat, in 1996. The Earth Observing System platforms are scheduled to be placed in orbit from 1998 to 2016. A recent substantial achievement—indeed, a breakthrough in the science of remote sensing—was the remarkably detailed mapping of the earth's surface in 2000 by a manned NASA space shuttle. This mission involved a partnership among the military, intelligence-gathering, and environmental communities, and resulted in a topographic map of the earth's land mass between 60°N and 56°S that is about 30 times

When geographers collect data in the field, they need to specify the geographic locations at which they take measurements. These locations are mapped to facilitate identification in a corresponding remote sensing image. Today, precise field locations can be established using a global positioning system (GPS). Originally developed by the US military, a GPS involves a group of satellites circling the earth in precisely known orbits and transmitting signals that are recorded by ground-based, usually hand-held, receivers like the one shown. The US GPS consists of 24 satellites. There is also a Russian GPS, and a European system is in development.

Hugh Threlfall/Alamy/GetStock

as precise as the best maps available before the mission.

There is another means, besides remote sensing, by which some data can be collected. In the early 1990s geographers began to make use of another new digital geographic technology, the global positioning system (GPS). A GPS is an instrument (either hand-held or installed in a personal computer) that uses signals emitted by satellites to calculate location and elevation. Along with remotely sensed data, data from a GPS can be integrated into a GIS.

QUALITATIVE METHODS

Human geographers collect and analyze data using a broad range of **qualitative methods**—a term widely used in other social sciences that refers to research with a focus on the attitudes, behaviour, and personal observation of human

qualitative methods
A set of tools used to collect and analyze data in order to subjectively understand the phenomena being studied; the methods include passive observation, participation, and active intervention.

Box 2.10   Some Qualitative Resources

The term 'qualitative resources' is used to refer to both methods and data sources. Holland et al. (1991) have identified nine of the many such resources available. Here is their list, accompanied by some comments that point to the breadth of humanistic research today:

1. *Talking and listening:* The typical quantitative interview uses a carefully structured questionnaire to ensure that multiple respondents answer a specific set of questions. By contrast, a qualitative interview seeks to gain insight into the human worlds of individuals. It is particularly important to talk with and to listen to respondents when the human world under investigation is different from the interviewer's own. In planning such research, you should always try to take into account the respondents' likely perspectives and be ready to change your questions, depending on what they say.

2. *A feminist perspective:* Feminist research focuses on the social significance of gender and especially the unequal distribution of power between women and men. A feminist perspective will influence the questions posed as well as the techniques used to carry out research.

3. *Literature:* Why do humans change landscapes as we do? Some geographers have turned to literature to answer this question. Creative writing often reflects cultural attitudes.

4. *Cartoons:* Although, like works of literature, cartoons are usually the work of an individual, they often reflect a broad social consensus on political, environmental, or economic issues.

5. *Photographs:* Human geographers frequently turn to photographs for information about the past. Photos can be invaluable not only as records of historical landscapes but as evidence of cultural attitudes. Choice of subject matter (including what the photographer omits or ignores) and manner of presentation can be extremely revealing.

6. *Art:* Many works of art have been interpreted as reflecting the distinctive character of a particular time and place.

7. *Popular music:* Music is often closely associated with place—just think of blues and the American South. Today much popular music is a direct reflection of inner-city urban experiences, highlighting social and environmental ills.

8. *Buildings:* Buildings are among the most important features of human landscapes, especially in urban areas. Different styles and functions reflect larger geographies, past and present.

9. *Monuments, memorials, and cemeteries:* Careful analysis of these landscape features can tell us a lot about the cultural identities of their creators.

These are only a few examples of the many qualitative possibilities open to the skilled researcher.

"Damn! Somebody's pinched our spot!"

Cartoon by Chic
© Punch Ltd. www.punch.co.uk

subjects. Qualitative methods are a part of **ethnography**, a general approach that requires researcher involvement in the subject studied.

Much **fieldwork**—a traditional term for the methods that geographers use to obtain primary data (for example, observation and interviews with informants)—is qualitative in character. Observation of landscapes was central to much early regional and cultural geography, but such observation decreased in importance around 1960 as geographers began to focus more on secondary data. Recently, however, new types of fieldwork have appeared in response to humanistic concerns, and human geographers now use a range of qualitative methods for collecting and analyzing data (Box 2.10). Early fieldwork was not philosophically motivated, although it was implicitly empiricist because it assumed that reality was present in appearance. By contrast, contemporary qualitative field-work is by nature humanistic—a response to the humanistic requirement that human geography strive to understand the nature of the social world.

For the humanist, qualitative methods that involve a researcher's observation of and involvement in everyday life are central to understanding humans and human landscapes. **Participant observation**, a standard method in anthropology and sociology, is now a popular geographic approach. The principal advantage of this method is its explicit recognition that people and their lives do matter.

To conduct research using qualitative methods requires considerable skill. A subjective procedure such as participant observation does not provide any means for the researcher to objectively control the relationship between observer and observed. You will recall that this was one of the key issues in our discussion of the differences between humanism and positivism. It may be the case that the researcher, who is often of a higher social status, is ethnocentric (**ethnocentrism** is the presumption that one's own culture is normal and natural and that other cultures are inferior); or the researcher may identify with the group being observed and become their advocate. Contemporary geographers and other social researchers pursuing field research, however, seek to bring a 'reflexivity' to their fieldwork, which includes an awareness of their own real or potential biases, of how their presumed status and gender may affect the data they collect from human

subjects, and of how their simple presence inevitably will alter the dynamic of that which they seek to observe and understand. Other disadvantages of qualitative procedures include the risk that the researcher will begin with a biased or otherwise inappropriate idea about the data to be collected, or that the subjects of the study may not be sufficiently representative to provide an accurate picture. You will have an opportunity to assess the quality of some of this work, especially in Chapters 7 and 13.

QUANTITATIVE METHODS

Some fieldwork is explicitly quantitative in character, notably the use of a **questionnaire** to survey people. Like the traditional qualitative fieldwork procedure of observation, a questionnaire is part of an empiricist research activity. Unlike qualitative fieldwork, however, a questionnaire asks all individuals the same questions in the same way. The value of questionnaire results depends on the response rate achieved and the way potential respondents are selected, namely the **sampling** method. Proper sampling methods, based on statistical sampling theory, allow the results of a sample to be treated as representative of the population, within certain error limits. The most common technique used for selecting respondents is random sampling.

During the 1960s, **quantitative methods** in general developed extensively in association with the spatial analysis school and the general acceptance of a positivist philosophy. The principal methods used were statistical, and the purposes were to describe data (by calculating a mean and a standard deviation, for example) and to test hypotheses generated by theory (using correlation tests, e.g., as in Box 2.2, point 4).

The spatial analysis school recognized early that models could play a much greater role in analyzing data. A **model** is an idealized, simplified representation of the real world that highlights its key properties and eliminates incidental information. Many of the earliest spatial models employed by human geographers were based on generalizations about the relationships between the distribution of geographic facts and distance.

Geographers use quantitative techniques for a wide variety of purposes, but especially for analyzing relationships between spatial patterns and for classifying data. Describing

ethnography
The study and description of social groups based on researcher involvement and first-hand observation in the field; a qualitative rather than quantitative approach.

fieldwork
A means of data collection; includes both qualitative (for example, observation) and quantitative (for example, questionnaire) methods.

participant observation
A qualitative method in which the researcher is directly involved with the subjects in question.

ethnocentrism
A form of prejudice or stereotyping that presumes that one's own culture is normal and natural and that all other cultures are inferior.

questionnaire
A structured and ordered set of questions designed to collect unambiguous and unbiased data.

sampling
The selection of a subset from a defined population of individuals to acquire data representative of that larger population.

quantitative methods
A set of tools used to collect and analyze data to achieve a statistical description and scientific explanation of the phenomena being studied; the methods include sampling, models, and statistical testing.

model
An idealized and structured representation of the real world.

relationships is fundamental in producing explanations and is usually broached by proposing a functional relationship such that one variable is dependent on one or more other variables. The relationship specified is, ideally, derived from appropriate theory in accordance with the scientific method outlined earlier. Classifying imposes order on data, and a number of techniques facilitate that activity.

A Concluding Comment

We conclude this chapter with a provocative, if somewhat crude, distinction.

- Positivism and the quantitative procedures associated with it have a tendency to exclude the individual human element from research, preferring to focus on aggregate data; critics consider such work to be dehumanized human geography. We might call it a geography of people.
- Humanism and the qualitative procedures associated with it, on the other hand, emphasize the integration of researcher and researched, thus generating a geography for, and possibly even with, people.

We do not, however, need to debate the relative merits of differing research strategies; we simply need to acknowledge that contemporary human geography is an exciting and diverse discipline incorporating a variety of useful philosophies, concepts, and techniques that complement each other and combine to offer an enviable range of procedures.

Aerial photograph of the Ottawa region.
SOURCE: © 2006. Produced under licence from Her Majesty the Queen in Right of Canada, with permission of National Resources Canada.

CHAPTER 2 SUMMARY

THE IMPORTANCE OF PHILOSOPHY

The subject matter of human geography, the questions asked, and the aids used in obtaining answers are best understood by reference to various underlying philosophies. Most human geography is explicitly or implicitly guided by one particular philosophical viewpoint.

PHILOSOPHICAL OPTIONS

Contemporary human geography draws on four principal philosophies. *Empiricism* gives priority to facts over generalizations; traditional regional geography, although presumed to be aphilosophical, was empiricist in character. *Positivism* argues for use of the scientific method, as in the physical sciences, and purports to be objective. The approach known as spatial analysis follows a positivistic logic. *Humanism* is a set of subjective philosophies, including pragmatism and phenomenology, with a focus on humans. Early cultural geography had some humanistic content, but humanist philosophies did not begin to play a major role in human geography until about 1970. *Marxism* is a loosely structured philosophy that aims to understand and change the human world; work inspired by Marxist thought tends to centre on social and environmental problems.

HUMAN GEOGRAPHIC CONCEPTS

A great many concepts are used in human geography. We constantly refer to space (indeed, even thinking of geography as a spatial discipline), locations (where a geographic fact is present), place (the quality of a given location), region (a grouping of similar locations or places), and distance (the interval, however measured, between locations or places). Geographic research is conducted at specific spatial, temporal, and social scales. Explaining and understanding landscape change is often enhanced by considering both diffusion—the spread and growth of geographic facts—and perception—how landscapes are viewed by individuals and groups. Analyses of landscapes frequently distinguish different degrees of development. The technical vocabulary of human geography—for example, conceptual terms such as 'place' and 'location'—is an example of discourse. Most notably, human geographers identify globalization as an overriding metaconcept, providing a body of ideas that may facilitate analysis of environmental, cultural, political, and economic topics.

TECHNIQUES OF ANALYSIS

Geographic data are collected and handled by a variety of techniques. Maps are used to store, display, and analyze data, and can be produced manually or by computer. Geographic information systems are computer-based tools that analyze and map data, and have a wide range of applications. An important method of collecting data involves remote sensing techniques, such as aerial photography and satellite imagery. Fieldwork, both observation and interviewing, is an essentially qualitative procedure for the collection and analysis of data and is favoured by humanists. The use of models and quantification is associated with positivistic spatial analysis and involves abstraction and empirical testing of ideas.

QUESTIONS FOR CRITICAL THOUGHT

1. Why does Norton state 'it is always an oversimplification to treat physical geography as a cause of human behaviour' in Box 2.1?

2. What are the major differences among the three major philosophies in contemporary geography (i.e., humanism, positivism, and empiricism)?

3. Why is it important to be familiar with the various philosophical underpinnings of contemporary geography?

4. Why are 'geographic literacy' and 'geographic knowledge' important tools or skills?

5. Why are maps arguably the most important tools of geographers? Could we study geography without maps?

6. In what ways are qualitative and quantitative methods in geography complementary?

FURTHER EXPLORATIONS

Bird, J.H. 1993. *The Changing Worlds of Geography: A Guide to Concepts and Methods*, 2nd edn. New York: Oxford University Press.

A comprehensive account of concepts and methods, many of which are discussed in great detail.

Campbell, J.B. 2008. *Introduction to Remote Sensing*, 4th edn. New York: Guilford.

Text introducing students to methods and approaches and with discussion of many and varied applications.

Flowerdew, R., and D. Martin, eds. 2005. *Methods in Geography: A Guide for Students Doing a Research Project*. Toronto: Pearson.

A useful resource book with focus on how to prepare for a research project, how to collect appropriate data, how to analyze data, and how to produce the final research report.

Goodchild, M.F. 1988. 'Geographic Information Systems', *Progress in Human Geography* 12: 560–6.

A brief history and assessment of geographic information systems that helps to place this major growth area in an appropriate perspective; avoids technical terminology.

Gould, P., and R. White. 1986. *Mental Maps*, 2nd edn. Boston: Allen and Unwin.

A highly readable and well-researched discussion of the topic; the inspiration for this original research is one of the stories told in Eyles (1988).

Harvey, F. 2008. *A Primer of GIS: Fundamental Geographic and Cartographic Concepts*. New York: Guilford.

Focuses on the key concepts and skills needed to understand and apply a GIS. Numerous practical examples and exercises facilitate understanding.

Honderich, T. 2005. *The Oxford Companion to Philosophy*, new edn. New York: Oxford University Press.

An excellent guide to specific philosophical ideas and terms that will be useful for students wishing to appreciate the larger intellectual context for some of the philosophies introduced in this chapter.

Johnston, R.J. 1986. *Philosophy and Human Geography: An Introduction to Contemporary Approaches*, 2nd edn. London: Arnold.

Although rather dated, this is a clear, readable statement of our various philosophies; employs a tripartite division into positivism, humanism, and Marxism.

———, D. Gregory, and D.M. Smith, eds. 2000. *The Dictionary of Human Geography*, 4th edn. Oxford: Blackwell.

A remarkably comprehensive, detailed, and clear dictionary with major entries often treated in an almost encyclopedic fashion; highly recommended.

Limb, M., and C. Dwyer. 2001. *Qualitative Methodologies for Geographers: Issues and Debates*. New York: Oxford University Press.

An innovative text that covers many methodologies as they are used in practical research contexts; covers research design, interviewing, group discussions, participant observation and ethnography, interpretative strategies, and writing.

Monmonier, M. 2004. *Rhumb Lines and Map Wars: A Social History of the Mercator Projection*. Chicago: University of Chicago Press.

A scholarly and entertaining book that explains both the history and geometry of map projections with particular attention paid to the often misunderstood Mercator projection.

Northey, M., and D.B. Knight. 2005. *Making Sense: A Student's Guide to Research and Writing*, updated 2nd edn. Toronto: Oxford University Press.

A valuable guide for students that provides insights and guidelines concerning all aspects of the research and report-writing process from a student perspective.

Obermeyer, N.J., and J.K. Pinto. 2008. *Managing Geographic Information Systems*. New York: Guilford.

Thorough account of GIS systems and their many uses.

Peet, R. 1998. *Modern Geographical Thought*. Oxford: Blackwell.

An overview of human geographic thought from about 1970 to the present; also useful for the conceptual content of Chapter 7.

Philipps, M. 2005. 'Philosophical Arguments in Human Geography', in Philipps, ed., *Contested Worlds: An Introduction to Human Geography*. Burlington, Vt: Ashgate, 13–85.

Informative overview of philosophies in human geography highlighting the important role played in research.

Relph, E.C. 1976. *Place and Placelessness*. London: Pion.

A good example of writing from a phenomenological perspective.

Robinson, A.H., and B.B. Petchenik. 1976. *The Nature of Maps: Essays towards Understanding Maps and Mapping*. Chicago: University of Chicago Press.

A highly readable book full of useful ideas. Robinson devised the projection used for world maps in this text.

Stoddard, R.H. 1982. *Field Techniques and Research Methods in Geography*. Dubuque, Iowa: Kendall Hunt.

A comprehensive book covering such topics as data collection and analysis; provides clear guidelines concerning how to conduct positivist research.

Watson, M.K. 1978. 'The Scale Problem in Human Geography', *Geografiska Annaler* 60B: 36–47.

This is a valuable read for all geographers.

ON THE WEB

PHILOSOPHY FOR GEOGRAPHERS

Internet Encyclopedia of Philosophy

🔍 www.utm.edu/research/iep/

Includes details on all of the philosophical terms introduced in this chapter.

Dictionary of Philosophical Terms and Names

🔍 www.philosophypages.com/dy/

Includes links to electronic texts and detailed discussions of particular topics.

MAPPING THE WORLD

Worldmapper

🔍 www.worldmapper.org/index.html

A quite wonderful and thought-provoking collection of maps on a wide range of topics. An innovative technique is employed: the maps are equal-area cartograms, also known as density-equalizing maps. The cartogram re-sizes each territory according to the variable being mapped.

Map projections

🔍 www.colorado.edu/geography/gcraft/notes/mapproj/mapproj_f.html

Detailed information on projections with clear definitions and numerous examples.

REMOTE SENSING AND GIS

Canada Centre for Remote Sensing

🔍 ccrs.nrcan.gc.ca/index_e.php

A Natural Resources Canada website, with a wide range of information on remote-sensing technologies and applications; includes sample images.

Canadian Institute of Geomatics

🔍 www.cig-acsg.ca/page.asp?intNodeID=7

This scientific and technical association is a geospatial knowledge network; includes information on careers.

NASA Earth Observatory

🔍 www.esri.com

Informative site with numerous global images and information.

Environmental Systems Research Institute

🔍 www.esri.com

The home page of ESRI, the creator of two of the most widely used GIS software packages, ArcView and ArcInfo; includes useful general information on GIS technologies.

A FRAGILE HOME

Most human impacts on the earth are the result of efforts to improve our well-being. Yet, these efforts sometimes have the opposite effect. To understand their impacts, we need to keep in mind five facts:

1) Everything in nature is related.

2) Besides physiological needs, humans have many culturally based wants.

3) Human impacts increase with greater levels of technology, increased use of various energy sources, and larger populations.

4) Different cultures have different attitudes towards the environment; also, attitudes within a single culture change over time.

5) A shift appears to be occurring in Western culture, from an anthropocentric (human-centred) world view to an ecocentric (environment-centred) one.

Following a discussion centred on these five points, this chapter outlines specific human impacts on the environment—globally, regionally, and locally. Finally, a discussion of the current state of the earth raises a provocative question that will resurface in various forms in later chapters: *Is a global disaster looming, or will human ingenuity and technology be able to solve our problems?* We conclude with a look at the progress currently being made towards a sustainable world.

South Korean soldiers clean up the seashore covered with crude oil spilled from a tanker off the coast in 2007. The clean up of over 10,500 tons of crude oil spilled into the West Sea was hampered by the lack of available equipment.
Chung Sung-Jun/Getty Images

For most of the time that humans and their ancestors have been using the earth, their impacts on the planet have been slight, for two reasons: technologies were limited and populations were small. Technologies such as stone tools and the use of fire and a total human population of about 4 million as recently as 12,000 years ago meant that the few environmental changes brought about by humans were temporary and restricted to a local scale. Today, with a substantial array of agricultural and industrial technologies and a population of 6.8 billion (2009), the situation is vastly different. Our interactions with and impacts on the environment are now numerous, relatively permanent, and often on a global scale.

A Global Perspective

Our topic is the impact of human groups on the global ecosystem. *A global perspective is employed because of the fact that everything is related to everything else; one cannot change one aspect of nature without directly or indirectly affecting other aspects.* In principle, then, human activity in any one area has the potential to affect all other areas; in practice the evidence that it does so is now overwhelming. One way to approach these issues is to consider the question of how a civilization survives. Smil (1987: 1) answered, 'It survives by harnessing enough energy and providing enough food without imperilling the provision of irreplaceable environmental services. Everything else is secondary.'

In exploring how harnessing energy and producing food have changed our environment, and whether the changed environment threatens human survival, many human geographers find that the most valuable concepts are those of systems and ecology.

SYSTEMS, ECOLOGY, AND ECOSYSTEMS

In principle, the concept of the **system** is simple and widely applicable. It is surprising, therefore, that it has not been more widely used in studies of the relationship between humans and land. Systems can be defined usefully as sets of interrelated parts. The attraction of the systems concept is its ability to describe a wide range of phenomena and offer a simplified description of what is usually a complex reality. Descriptions of systems typically focus on distinctions between the parts of the system and the relationships between the parts. Open systems interact with elements outside the system, necessitating the study of inputs and outputs of energy and matter. Closed systems—lacking inputs and outputs—are less common. Finally, relationships between the parts of a system, or between a system and some external elements, often are described as feedbacks. Positive feedback reinforces some change that is occurring; negative feedback counters some change.

The roots of the term **ecology** are Greek: 'eco' comes from *oikos*, 'house' or 'place to live', and the common suffix '-logy', meaning 'study of', comes from *logos*, 'word' or 'reason'. Thus ecology is the study of organisms in their homes. The concept of an ecosystem, integrating systems and ecology, was formally developed by an English botanist, A.G. Tansley, in 1935.

Ecosystems can be identified on a wide range of scales. The global ecosystem (often called the ecosphere or biosphere) is the home of all life on earth. It is a thin (about 14-km/8.5-mile) shell of air, water, and soil. Ecosystems can be identified within the larger ecosphere: any self-sustaining collection of living organisms and their environment is an ecosystem. Thus the ecosystem concept refers to distinct groupings of things and the relationships between these things.

On the global scale, the three basic parts of the ecosystem are the atmosphere, the hydrosphere, and the lithosphere: air, water, and land. This global ecosystem has one key input: the sun. The sun is the source of the energy that warms the earth, providing energy for photosynthesis and powering the water cycle to provide fresh water. In order to be self-sustaining, our global ecosystem depends on the cycling of matter and the flow of energy. Figure 3.1 outlines the cycling of critical chemicals and the one-way flow of energy. Matter must be cycled; the law of conservation of matter states that matter cannot be created or destroyed, only changed in form. Energy flows through the system because of the second law of thermodynamics: that energy quality cannot be recycled. We call the circular pathways of chemicals biogeochemical cycles; 'bio' refers to life, 'geo' to earth. Our global ecosystem,

ecology
The study of relationships between organisms and their environments.

ecosystem
An ecological system; comprises a set of interacting and interdependent organisms and their physical, chemical, and biological environment.

system
A set of interrelated components or objects linked together to form a unified whole.

FIGURE 3.1 **Chemical cycling and energy flows**

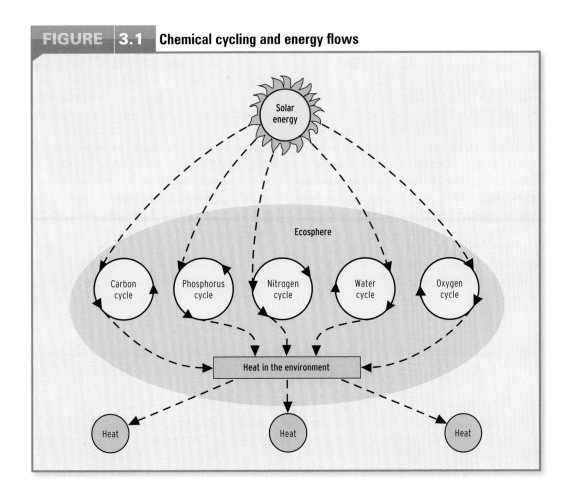

Humans as Simplifiers of Ecosystems

Any human change to an ecosystem usually is a simplification, and a simplified ecosystem usually is vulnerable (Box 3.1). Ecologists agree that, if the global ecosystem is to survive, systems modified by humans must be balanced by systems that have not been modified by humans. For example, imagine a cultivated lowland ecosystem. Such a system depends on an upland forest system nearby that releases water and minerals to it. If humans remove the forest system, the cultivated area will suffer. As the basic principle of ecology states, all things are related. Today, one of the most urgent issues facing us is the need to change the way we live inside our ecosystems—to change our

indeed every ecosystem, is dynamic. As we are about to find, one of the most important—and least understood—causes of ecosystem change is the human population.

long-standing habits of domination for new ones of co-operation. Nothing illustrates this domination better than our use of energy and our high valuation of technology.

ENERGY AND TECHNOLOGY

Humans have basic physiological needs for food and drink; we also have a host of culturally based wants that appear to have no upper limit. Needs and wants are satisfied largely as a result of humans' using their energy to harness other forms of energy. **Energy** is the capacity to do work. The more successfully we can utilize other energy, the more easily we can fulfill our needs and wants and the more profoundly we can affect the environment. We become aware of other energy sources and acquire the ability to use those sources through the development of new technology.

All forms of **technology** represent ways of converting energy into useful forms. An important technological advance was the human use of fire to convert inedible plants into an edible form for human use. In domesticating

energy
The capacity of a physical system for doing work.

technology
The ability to convert energy into forms useful to humans.

domestication
The process of making plants and/or animals more useful to humans through selective breeding.

agricultural revolution
The slow transition, beginning about 12,000 years ago, from foraging to food production through plant and animal domestication.

plants, humans gained control over a natural energy converter: the plants that convert solar energy into organic material via photosynthesis. Similarly, in domesticating animals, humans took control over another natural converter: the animals that change one form of chemical energy (inedible plants, usually) into another form usable by humans (such as animal protein). In this sense, the **domestication** of plants and animals, sometimes known as the **agricultural revolution**, is an example of how humans have used technology to tap new energy sources.

Between the agricultural revolution (that began approximately 12,000 years ago) and the eighteenth century, many other technological changes permitted increasing human control of energy sources: new plants and animals were

domesticated and new tools and techniques were invented. In addition, three new energy converters were developed to utilize the energy in water and wind: the water mill, the windmill, and the sailing craft. But with the **Industrial Revolution** of the eighteenth century, most particularly the invention of the steam engine, humans began using inanimate converters on a large scale to tap into new energy sources: coal in the second half of the eighteenth century, oil and electricity in the second half of the nineteenth century, and nuclear power in the middle of the twentieth century.

This brief account raises two key points. First, energy resources such as coal and oil are not literally 'new': they are 'fossil fuels', formed from ancient organic matter, and they

Box 3.1   Lessons from Easter Island

One of the most remote inhabited places on earth, Easter Island, in the Pacific Ocean, is 3,747 km (2,328 miles) off the coast of South America and 2,250 km (1,400 miles) southeast of Pitcairn Island, the nearest inhabitable land. When Europeans first visited Easter Island on Easter Sunday, 1722, they encountered a population of about 3,000 living in a state of warfare, with minimal food available from limited resources and a treeless landscape (Ponting, 1991: 1–7). They also encountered clear evidence of a once-flourishing earlier culture: between 800 and 1,000 huge stone statues, or moai. Because visiting Europeans believed that these statues could not have been carved, transported, and erected by the impoverished local population, it was long assumed that Easter Island must have been visited by a more culturally and technologically advanced group.

There is a simpler but more disturbing answer to this 'mystery' of Easter Island (Diamond, 2005). Radiocarbon dating of evidence suggests that Easter Island was settled by the seventh century, and possibly as early as the fifth century; the first settlers were likely Polynesians, not South Americans. They arrived at an island with few species of plants and animals and limited opportunities for fishing, but with considerable areas of woodland. Their diet consisted of sweet potatoes (an easy crop to cultivate) and chicken. The considerable amount of free time available allowed the Easter Islanders to engage in elaborate rituals and construct the moai. Agricultural activities, cooking food, cremating the dead, and building canoes required the removal of some trees, but most of the deforestation was carried out for the purpose of moving statues. It is likely that statue construction involved competition between different groups. By about 1550, the population peaked at roughly 7,000. Deforestation,

which would, over time, have led to soil erosion, reduced crop yields, and caused a shortage of building materials for both homes and boats, probably was complete by 1600. Without wood for canoes, Easter Islanders were unable even to build canoes to catch porpoises (their principal source of protein) or to escape the island. The Easter Islanders impoverished by the consequences of deforestation were those people the Europeans encountered.

It appears that what began the collapse of the Easter Island culture and economy was total deforestation. What is the significance of this lesson? Bahn and Flenley offer this suggestion:

> We consider that Easter Island was a microcosm which provides a model for the whole planet. Like the Earth, Easter Island was an isolated system. The people there believed that they were the only survivors on Earth, all other land having sunk beneath the sea. They carried out for us the experiment of permitting unrestricted population growth, profligate use of resources, destruction of the environment and boundless confidence in their religion to take care of the future. The result was an ecological disaster leading to a population crash. (Bahn and Flenley, 1992: 212–13)

Diamond (2005) makes the point that the collapse of many ancient cultures, as well as more recent examples such as Rwanda and Darfur, can be explained in terms of two circumstances—internal social and political factors that affect the use of resources, and external climatic variability that affects the availability of resources. The relative importance of these two varies, but always both are present.

are not renewable. Second, we are using these resources in ways that are often harmful to our ecosystem. Environmental degradation resulting from **pollution** is now commonplace in industrial societies. According to Smil (1989: 10), 'Environmental pollution, previously a matter of regional impact, started to affect more extensive areas around major cities and conurbations and downwind from concentrations of power plants as well as the waters of large lakes, long stretches of streams and coastlines, and many estuaries and bays.' Much pollution can be attributed to twentieth-century developments such as the thermal generation of electricity and the mass use of automobiles, plastics, nitrogenous fertilizers, and pesticides.

Energy use is spatially variable. China, with 1.3 billion people, is now the world's largest user of energy, having overtaken the United States. Although Europeans and Japanese use less energy than we do in North America, countries in the **more developed world** generally use more energy per capita than do countries in the **less developed world**. Energy sources are also spatially variable. Globally, oil accounts for 35 per cent of energy used, coal for 25 per cent, natural gas for 21 per cent, biomass and waste for 10 per cent, nuclear power for 6 per cent, hydroelectricity for 2 per cent, and other renewables for 1 per cent. As this list suggests, in many parts of the less developed world, the primary energy sources are not oil, coal, and gas but the traditional **biomass** sources of wood, crop waste, and animal dung. And energy use is increasing. It has been estimated that in one year humans use an amount of fossil fuel that took one million years to produce. Clearly we cannot continue indefinitely in this fashion, because we are draining the supply of non-renewable energy sources. We must realize that human survival depends on our retaining access to appropriate energy sources—and we must act on that realization.

NATURAL RESOURCES AND HUMAN VALUES

Humans continually evaluate physical environments. As human culture (especially technology) changes, so do those evaluations. Thus, something becomes a 'resource' only if humans perceive it as useful in some way: technological, political, economic, or social. Different groups do not necessarily agree on what is and is not a resource. Some people may see an area of wetland in prairie Canada as a valuable scientific or recreational landscape; others will see it as potential farm or building land. Different interest groups use different evaluation criteria, and hence may hold radically different views.

Traditionally, geographers have divided resources into two types. **Stock resources** include all minerals and land and are essentially fixed, as they take a very long time, by human standards, to form. **Renewable resources**, such as air and water, are continually forming. This simple distinction is generally useful, but it does blur some key issues. Numerous resources lie somewhere between the two extremes of stock and renewable, and continuing availability depends on how we manage those resources. Obvious examples include game populations for hunting economies and fish populations for much of the contemporary world. In both cases, there may be a perceived need for conservation, but also some compelling reasons to continue depleting the resources. In Canada, the east coast fishery has been especially susceptible to controversies over such issues.

RENEWABLE ENERGY SOURCES

Perhaps the biggest technological challenge of the early twenty-first century is to decrease our reliance on non-renewable fossil fuels (coal, oil, natural gas) as sources of energy and find ways of generating more energy from renewable sources (Chapter 14 includes an overview of current global production and consumption). There are many reasons to favour increasing use of renewable energy (Goodall, 2008). Environmental reasons include the impact on the atmosphere of burning fossil fuels and the global warming that is a consequence (discussed later in this chapter). Political reasons include the pressure exerted on governments by voters who are increasingly aware of environmental issues; the need for national economies to be self-sufficient in energy; and the need to diversify in order to reduce dependence on particular suppliers of energy sources. Economic reasons centre on the fact that coal, oil, and natural gas will become more expensive as supplies become scarcer. Many of these arguments first came to the fore as a result of the energy crisis of the early 1970s, although interest waned when energy costs fell again in the 1980s. Today, these concerns are, once again, front and centre, especially as

Industrial Revolution
The process that converted a fundamentally rural society into an industrial society beginning in England *c*. 1750; primarily a technological revolution associated with new energy sources.

pollution
The release into the environment of substances that degrade land, air, or water.

more developed world
(According to a United Nations classification) Europe, North America, Australia, Japan, New Zealand, and the former USSR.

less developed world
All countries not classified as 'more developed' (see 'more developed world').

biomass
The mass of biological material present in an area, including both living and dead plant material.

stock resources
Minerals and land that take a long time to form and hence, from a human perspective, are fixed in supply.

a result of the dramatic increases in oil prices that occurred in 2007–8. Indeed, such concerns are being acknowledged even by those uncomfortable with environmental arguments, as evidenced by the generally well-received book authored by the conservative pro-globalization American scholar, Thomas Friedman (Friedman, 2008). In *Hot, Flat, and Crowded*, Friedman joins those calling for a transformation of energy systems, specifically in the United States, involving a move away from fossil fuels, less use of electricity, and requirements that power companies buy energy from cleaner sources. Such a transformation would involve federal government intervention, including a new tax regime.

Renewable energy sources are tapped in only a few areas, on a relatively local scale. The main potential sources are the sun, the wind, biomass (e.g., wood and grasses), geothermal energy (the natural heat found inside the earth), and water (rivers, waves, and tides). Nuclear power is also considered renewable, since uranium reserves are expected to last about 1,000 years if an efficient process is used.

To date, the principal technology for generating renewable energy has involved the damming of rivers. Many large dams have been constructed in parts of the less developed world,

often with the financial support of the World Bank. Unfortunately, although such projects do make effective use of a renewable energy source, they can cause serious damage to the environment. The massive Three Gorges project in China was refused World Bank funding, partly because of the anticipated human and environmental impacts. Nevertheless, the raising of the world's third largest river, the Yangtze, began and the dam, the largest in the world, opened in 2006. Between 1 and 2 million people have been displaced, with evidence of further misery through the brutal crushing of protests and of officials pocketing funds intended for resettlement projects. Many ancient fortresses, temples, and tombs were submerged under water; it is also feared that the reservoir may become a giant cesspool filled with sediment washed down from the deforested mountains that surround it. As early as 2007, these fears appeared to be becoming reality.

Another Chinese megaproject, the principal purpose of which is to divert water from the south to the north, is subject to similar criticisms. Other major projects include a dam being constructed in Laos that is expected to help lift much of the population out of poverty, and a series of five dams planned for construction on the Salween River along the border between Myanmar (Burma) and Thailand.

Nuclear power per se does not pollute, but the residue from nuclear power plants is extremely polluting and long-lasting and nuclear power has not been established as a source of safe and inexpensive energy. Indeed, cost may be a very real issue here. As just one example, a power station being built on the island of Olkiluoto in western Finland was scheduled to open in 2008 but is delayed until 2012, at least partly because the cost has doubled. In the United States, power companies tend not to favour nuclear precisely because of concerns of escalating costs. There are more than 400 nuclear reactors in the world, most of which use the basic nuclear fission process, although future reactors are likely to use the more efficient 'fast breed' process. The likelihood is that there will be increased use of nuclear power, at least partly in response to concerns about the global warming caused by fossil fuel burning. Still, several serious nuclear accidents have occurred. As discussed in Box 3.2, the most devastating accident was the 1986 Chernobyl explosion in northern Ukraine (then part of the USSR).

Ships emerge from the five-stage lock at the Three Gorges dam on the Yangtze River. More than two kilometres wide and 185 metres high, the dam has created a reservoir extending nearly 650 km into the country's interior. Ocean-going vessels will have access to agricultural and manufactured products from a vast region, while the dam's hydropower turbines will generate as much electricity as 18 nuclear power plants. The dam is the largest construction project in China since the Great Wall.

CP/AP photo/Greg Baker

Solar power has been widely discussed, and, although it remains expensive, new technologies are being developed that will reduce costs. Spanish and German companies are installing large-scale solar power plants in North Africa, a development that requires power transmission over long distances. There is also great potential for the use of solar panels in private homes as well as businesses. The prospects for solar power production at reasonable costs seem positive.

Several countries in the more developed world are now making effective use of wind power and, as with solar power, the prospects for this becoming a significant source are improving. Denmark is a leader in wind farm technology, and the numbers of such farms in Canada, the United States, and in various European countries are increasing.

Recent years have witnessed increasing use of crops to produce biofuels as an alternative to gasoline (a key factor in the problem of rising food prices noted in Chapter 5). But biofuels might be a problem rather than part of the solution to climate change. A 2009 report commissioned by the International Council for Science concluded that farming biofuel crops such as corn and canola releases enough nitrous oxide (N_2O), a potent greenhouse gas, that the benefits of reduced emissions of CO_2 resulting from replacing gasoline are negated. Finally, several countries have experimented with wave and tidal power, but these have not established themselves as viable sources; biomass sources, such as wood, remain important in the less developed world, and geothermal energy is a minor contributor at the present time.

How the energy picture will change in the twenty-first century is difficult to predict, at least partly because the amounts of energy available from renewable sources can be affected by weather conditions. In the United States, for example, consumption of hydroelectric power from renewable sources fell in 2001 because of widespread drought conditions.

Environmental Ethics

WESTERN ENVIRONMENTAL CONCERN BEFORE 1900

Use or abuse? It is not always easy to distinguish the two; they are, after all, relative terms. Yet there is considerable evidence to suggest that we are currently causing damage—perhaps

irreparable damage—to our environment. Concerns about the consequences of human activities were raised by the ancient Greeks; Plato himself noted the detrimental effects of agricultural activities on soil. Despite such early observations, this general question received relatively little attention in the Western world until the eighteenth century. Before that time, geographers were most interested in the earth as a home for humans made by God (teleology) and the land as a cause of human activity (environmental determinism). Significantly, the Europeans' general failure to appreciate the potential dangers of certain human activities

In September 2003 ENMAX Corporation and TransAlta Corporation announced the official opening of the $100-million McBride Lake Wind Farm in southern Alberta. With 114 turbines, this facility will generate approximately 235,000 megawatts of electricity a year—enough energy to power more than 32,500 homes. Similar projects are underway in many parts of the world, but local responses are not always favourable. In the Lake District of northwest England, for example, opponents object that turbines are an eyesore in a landscape widely considered to be idyllic.

Courtesy of TransAlta Energy Corporation

Box 3.2   The Chernobyl Nuclear Accident

On 26 April 1986, one of four nuclear reactors at the Chernobyl power station exploded following a failed experiment to determine whether the cooling system could operate effectively if the auxiliary electrical supply failed. The accident was especially devastating because the Soviet-built reactor was not housed in a reinforced concrete shell, as was usually the practice in most countries. The building itself was severely damaged; more importantly, large quantities of radioactive debris, at least 100 times more than the atom bombs dropped on Nagasaki and Hiroshima in 1945, were released into the atmosphere. Figure 3.2 maps the spread of the radioactive fallout as measured one week after the accident.

Most fallout was deposited in the immediate vicinity of Chernobyl in Ukraine and in nearby areas of Russia and Belarus, but fallout affected almost all countries in the northern hemisphere. The specific details shown in Figure 3.2 reflect wind direction and rainfall.

Determining the human consequences of accidents such as this is very difficult. About 350,000 people have been resettled away from the worst affected areas, but another 5.5 million remain. For several years it was estimated that as many as 7,000 people died and up to 3.5 million suffered from diseases related to the release of radioactive material. More recently, a United Nations report estimated the number of deaths that had resulted as of 2005 at less than 4,000. However, a 2006 report from Greenpeace claimed that in the coming years about 100,000 will die of cancer, especially thyroid cancer, caused by the accident.

The UN declared the emergency phase ended in 2007 and urged a move to a recovery phase, helping communities to begin reversing the domino effect of poor health, poverty, and fear. One small positive sign is the revival of wildlife in the worst affected area; indeed, birds have successfully nested in the reactor building.

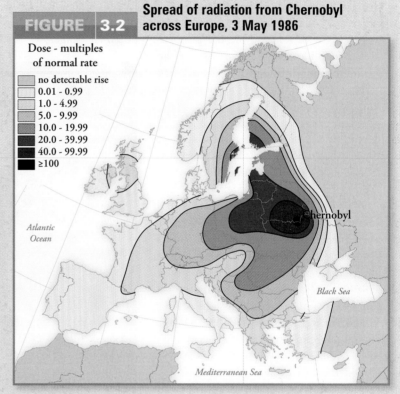

FIGURE 3.2   **Spread of radiation from Chernobyl across Europe, 3 May 1986**

Dose - multiples of normal rate

- no detectable rise
- 0.01 - 0.99
- 1.0 - 4.99
- 5.0 - 9.99
- 10.0 - 19.99
- 20.0 - 39.99
- 40.0 - 99.99
- ≥100

Atlantic Ocean

Chernobyl

Black Sea

Mediterranean Sea

SOURCE: Adapted from <news.bbc.co.uk/2/shared/spl/hi/guides/456900/456957/html/nn3page1.stm>.

appears to have been unique: many other cultures have recognized the necessity of protecting natural 'resources'. The eighteenth-century origins of Western environmental concern reflected the overseas movement of Europeans, particularly their colonization of tropical areas that, in Europe, had long been regarded as pristine Utopias. It soon became obvious that European activity in those areas was environmentally destructive.

The general question of human impact on the land first received scholarly attention from Buffon (1707–88) in discussions of the contrasts between settled and unsettled areas and of the human domestication of plants and animals. Buffon believed that humans inhabit the earth

in order to transform it. Malthus (1760–1834) established the terms of the current debate by focusing attention on the relationship between available resources and numbers of people. On a more specific level, Humboldt, during his travels in South America, explicitly identified lowered water levels in lakes as human impacts, and explained that they were caused by deforestation for the purpose of agricultural activities.

Probably the earliest systematic work on human impacts was done by G.P. Marsh (1801–82), an American geographer and congressman. His book *Man and Nature, or Physical Geography as Modified by Human Action* (1864, with revised editions in 1874 and 1885) was intended:

to indicate the character and, approximately, the extent of the changes produced by human action in the physical conditions of the globe we inhabit; to point out the dangers of imprudence and the necessity of caution in all operations which, on a large scale, interfere with the spontaneous arrangements of the organic and of the inorganic worlds; to suggest the possibility and the importance of the restoration of disturbed harmonies and the material improvement of wasted and exhausted regions; and, incidentally, to illustrate the doctrine that man is, in both kind and degree, a power of a higher order than any of the other forms of animated life, which, like him, are nourished at the table of bounteous nature. (Marsh, 1965 [1864]: 3)

This powerful statement seems more reminiscent of the 1960s than the 1860s. The late nineteenth-century Western world, heavily involved in colonial expansion, became concerned about environmental change only when that change had negative impacts on human economic interests. 'If a single lesson can be drawn from the early history of conservation, it is that states will act to prevent environmental degradation only when their economic interests are shown to be directly threatened. Philosophical ideas, science, indigenous knowledge and people and species are, unfortunately, not enough to precipitate such decisions' (Grove, 1992: 47).

THE CURRENT DEBATE: ORIGINS

The first sign of a real shift in our appreciation of human impacts on ecosystems came in the 1960s with the publication of Rachel Carson's *Silent Spring* (1962), which revealed the dangers associated with indiscriminate use of pesticides. A few years earlier, a seminal academic work, totalling 1,194 pages and entitled *Man's Role in Changing the Face of the Earth* by Thomas et al. (1956), had come to many of the same conclusions. The 1960s saw increasing pressure from advocates of wilderness preservation and new scientific evidence about worsening air pollution. Two popular explanations for the environmental 'crisis' pointed to, first, the Judeo-Christian belief that humans had been placed on earth to subjugate nature and, second, the failings of capitalism. The first of these explanations is too simplistic

and ignores the complexity of Christian attitudes. The second explanation is one aspect of an ideological approach that stresses the links between different parts of the world: 'Clearly there are problems, many of them—and all of them intertwined in the operations of a capitalist world economy, which is hell-bent on annihilating space and place. Those problems are severe now at a global scale, and life-destroying in some places' (Johnston and Taylor, 1986: 9). Today, our awareness of environmental impacts outside the capitalist world economy, in Eastern Europe and the former Soviet Union prior to the major political changes that began in 1989, should lead us to question such assertions, although the basic idea of interrelatedness is sound.

THE CURRENT DEBATE: POLITICAL OVERTONES

Environmental issues entered the political arena in the early 1970s with the creation in the United States of the Environmental Protection Agency, while the first major international meeting, the UN Conference on Human Environment, was held in Stockholm in 1972. By 1980 the Western world was becoming increasingly aware of environmental and related food supply problems in the less developed world, with food shortages in India and droughts in the Sahel region of Africa. A series of disasters and discoveries during the 1980s ensured that the environment was always in the news. These included the 1984 leak of methyl isocyanate from a pesticide plant in Bhopal, India, that killed perhaps as many as 10,000 and disabled up to 20,000; the 1986 nuclear disaster in Chernobyl, Ukraine; the 1985 discovery of a seasonal ozone hole over Antarctica; recognition of both the rapidity and the consequences of tropical rain forest removal, especially in Brazil; and, more generally, an increasing concern for numerous local environmental problems.

By the late 1980s the environment was on the national agenda of many countries, as well as the international political agenda. At the national level, green political parties first appeared in West Germany in 1979 and were present in most countries in the more developed world by 1990. Further, many countries have some form of green plan; an encyclopedic survey of the Canadian environment is available as a part of Canada's Green Plan (Supply and Services Canada, 1991).

International agreement is the best way to solve those environmental problems that transcend national boundaries, some of which have global impacts. There have been many calls for the creation of international institutions and policies, most notably by the Brundtland Commission (World Commission on Environment and Development, 1987). Not surprisingly, although most governments agree on the need for international policies, many are unwilling to sacrifice their sovereignty; before they can reach agreements, countries need to work together to resolve their conflicting goals and priorities. Major international developments include the 1987 Montreal Protocol aimed at the reduction and eventual elimination of chlorofluorocarbons (CFCs), which are one cause of global warming; the 1992 UN Conference on Environment and Development (the Rio Earth Summit); the 1997 Kyoto Protocol designed to reduce emissions of greenhouse gases; and the 2002 UN World Sustainable Development Summit held in Johannesburg. Both of the UN-sponsored meetings were attended by thousands of delegates in addition to various world leaders.

THE CURRENT DEBATE: THREE CONTENTIOUS ISSUES

Before we address the impacts that humans are currently having on the environment, three issues should be identified. The first concerns relationships between the environment and the economy. It is usually argued that market forces are unlikely to solve environmental problems as market-based decisions rarely consider environmental factors as equal to or above the need to make a profit, even at the national level. A detailed analysis of economic decision-makers in Canada showed that only 6 per cent gave significant consideration to the environment. Such findings suggest that the integration of economic and environmental concerns will be a major challenge in the future (Gale, 1992). But the relationships between economic growth and environment are far from simple.

Box 3.3   The Tragedy of the Commons or Collective Responsibility?

Imagine that you and a group of friends are dining at a fine restaurant with an unspoken agreement to divide the check evenly. What do you order? Do you choose the modest chicken entrée or the pricey lamb chops? The house wine or the Cabernet Sauvignon 1983? If you are extravagant you could enjoy a superlative dinner at a bargain price. But if everyone in the party reasons as you do, the group will end up with a hefty bill to pay. And why should others settle for pasta primavera when someone is having grilled pheasant at their expense? (Glance and Huberman, 1994: 76)

This accurately depicts the clash between individual and collective attitudes that, in the context of human use of the environment, was so forcefully put forward by Hardin (1968) as follows:

- a group of graziers use an area of common land;
- they continually add to their herds so long as the marginal return from the additional animal is positive, even though the common resource is being depleted and the average return per animal is falling;
- indeed, individual graziers are obliged to add to their herds because the average return per animal is falling;
- clearly, efficient use of the common resource requires restricted herd sizes;

- but individuals will not reduce herd sizes on the common land unless all other group members similarly reduce their herd numbers;
- hence the metaphor 'the tragedy of the commons'.

Individual rational behaviour does not result in a collectively prudent outcome when individuals have access to common resources.

Both examples prompt the same question: How do we ensure that individuals act for the common good rather than for personal gain? With reference to the environment, three solutions have been proposed (Johnston, 1992):

1. Resources may be privatized, with the private owners implementing strategies for environmental preservation not available to group owners.
2. The group owners may be able to devise an agreement about the use of the common resource that they are able to implement themselves. According to some recent work in social theory, the success of local and regional recycling programs depends on such group co-operation (Glance and Huberman, 1994: 80).
3. The common resource may be subject to some external control. However difficult it may be to implement, this third proposed solution appears most necessary for ensuring the reduction of deleterious human impacts on the environment.

Increasing evidence suggests that economic growth leads to a reduction of environmental problems, as long as growth is accompanied by good governance. The richer a country is, and the better it is governed, the more it invests in environmental protection such as cleaning water supplies, reducing pollution, and improving sanitation.

Second, environmental problems are increasingly affecting relationships between countries—not only because of the international implications of many human impacts, but also because environmentalists in, for example, North America are anxious to impose their standards on other countries. Payment is one solution: the 1987 Montreal Protocol included a fund to assist those countries most likely to suffer economically as a result of the agreement. International disapproval is another possible solution: Britain eventually agreed to cease dumping sewage sludge in the North Sea because of the political costs of the dumping. Finally, trade policies are among the few weapons available to one national government to persuade another national government to amend its environmental behaviour.

The third issue, closely related to the second, concerns the behaviour of individuals as group members (Box 3.3). The ecophilosopher Arne Naess argued that humans need to develop a new **ecocentric** world view recognizing how we are all connected; that we need to work with and not against nature; and that a central goal of human activity is the preservation of ecosystems. 'Deep ecology', introduced by Naess in 1979, is a term sometimes used to describe this viewpoint (see Katz et al., 2000), and there are close conceptual links to the Gaia concept as described in Box 3.4. These ideas have informed many of the ideas of contemporary green movements. Table 3.1 provides a comparison of deep and shallow views of ecology. Both represent improvements over some traditional attitudes towards land—for instance, the **anthropocentric** notions that humans are the source of all value and that land exists for human use, as well as the misconception that energy and other resources are unlimited.

To the extent that deep ecology calls for significant changes to contemporary lifestyles, it finds parallels in the concepts of sustainability and sustainable development. Before we examine these important concepts in detail, however, we need to take a closer look at various specific human impacts on the earth, so

Table 3.1	Shallow and Deep Ecology Compared
Shallow Ecology (Spaceship earth)	**Deep Ecology (Sustainable earth)**
Views humans as separate from nature	Views humans as part of nature
Emphasizes the right of humans to live (anthropocentrism)	Emphasizes the idea that every life form has in principle a right to live; recognizes that we have to kill to eat, but that we have no right to destroy other living things without sufficient reason based on ecological understanding
Concerned with human feelings (anthropocentrism)	Concerned with the feelings of all living things; deep ecologists feel sad when another human or a cat or dog feels sad and grieve when trees and landscapes are destroyed
Concerned with the wise management of resources for human use (anthropocentrism)	Concerned about resources for all living species
Concerned with stabilizing the population, especially in less developed countries	Concerned not only with stabilizing the human population worldwide, but also with reducing the size of the human population to a sustainable minimum without revolution or dictatorship
Either accepts by default or positively endorses the ideology of continued economic growth	Replaces this ideology with that of ecological sustainability and preservation of biological and cultural diversity
Bases decisions on cost-benefit analysis	Bases decisions on ethical intuitions about how the natural world really works
Bases decisions on short-term planning and goals	Bases decisions on long-range planning and goals and on ecological intuition when all facts are not available
Tries to work within existing political, social, economic, and ethical systems	Questions these systems and looks for better systems based on the way the natural world works

SOURCE: G.T. Miller Jr, *Living in the Environment: Principles, Connections, and Solutions*, 4th edn (Belmont, Calif.: Wadsworth, 1985), 456. Reprinted with permission of Brooks/Cole, a division of Thomson Learning: <www.thomsonrights.com>. Fax 800 730-2215.

that we can understand why our current social and economic systems must change if the environment is to be safeguarded for future generations.

Human Impacts

Human history is a history of impacts on the natural environment; we must have an impact on land, air, and water to survive. Only recently have we realized that some of our impacts actually threaten our continued survival. Here are five fundamental driving forces.

1. Small, often insignificant, changes to the environment can have major impacts if they are repeated often enough. Arable and pastoral activities can lead, over time, to major environmental problems.

ecocentric Emphasizing the value of all parts of an ecosystem rather than, for example, placing humans at the centre, as in an anthropocentric emphasis.

anthropocentric Regarding humans as the central fact of the world; stressing the centrality of humans to the detriment of the rest of the world.

2. Technological changes related to demands for energy continually change the environment.
3. The lifestyles promoted by technological changes also work to change the environment.
4. Increasing human populations are a threat to the environment.
5. Increasing connections between different regions of the globe mean that human activities that used to have merely local or regional consequences are now more likely to be global in their impact. The most obvious example is global warming.

Our ability to collect and analyze information about current human impacts is greatly enhanced by technologies such as remote sensing and GIS (introduced in Chapter 2). Different wavebands of the electromagnetic spectrum can be used to gather information on particular land covers, such as areas under crop, pasture land, forest, and wetlands. It is possible to measure such properties of ecosystems as soil surface moisture and temperature, to map soils, and to 'sense' biogeochemical cycles. Much of the detailed information in this discussion of human impacts is derived from remote sensing by satellite and analyzed in a GIS.

IMPACTS ON ECOSYSTEMS IN GENERAL

Hunter-gatherers affect ecosystems only in specific areas and on a short-term basis. Their environmental impacts are limited, both spatially and temporally, by the small numbers of people involved, low levels of technology, and some deliberate conservation strategies. It seems probable that most ecosystems recover from hunter-gatherer activities, as they do from sustainable development of resources. Natural energy flows are not greatly altered except in cases of regular use of fire or animal overkill, and the latter has typically transpired only in recent times with increased population numbers and the impact of a capitalist mode of production. Examples include the Canadian prairies when the flourishing buffalo robe trade

Box 3.4   Introducing Gaia

'Gaia' is the Greek name for the goddess of the earth. Today the term is used to denote a self-regulating system, with all components of the ecosphere—chemical, physical, and biological systems—in a stable balance that keeps the planet habitable. This remarkable concept was first introduced in a book by James Lovelock (1979). In brief, Gaia is seen as a self-regulating entity that keeps the environment relatively constant and comfortable for life; earth and all life on it have evolved together as one. Specifically, the Gaia hypothesis supposes 'that the atmosphere, the ocean, the climate, and the crust of the Earth are regulated at a state comfortable for life because of the behaviour of living organisms' (Lovelock, 1988: 19). According to Gaia, the earth itself is 'alive'. This revolutionary idea contradicts the standard view in which life and the earth evolved separately and life had to adapt to conditions on earth. Initially rejected by most scientists, the Gaia concept was popularized mostly by environmentalists.

The Gaia concept obliges us to assume a global perspective. It is the planet as a whole that matters, not any particular species—including humans. If we humans continue to change the earth, then we may precipitate our replacement. If, on the other hand, we act as part of a living organism, then perhaps we will remain on earth for a long time.

The Gaia concept has helped to explain some crucial scientific issues. For example, the surface temperature of the earth has remained relatively constant since life first appeared, even though the heat from the sun has increased by about 25 per cent. How has this occurred? The level of CO_2 in the atmosphere has decreased over this period, reducing the natural greenhouse effect and allowing surface temperatures to remain relatively constant. Life, specifically photosynthesis, is the cause of the long-term CO_2 decline. Without life, CO_2 would probably increase once again. (These comments should help to place our present greenhouse crisis, discussed later in this chapter, in perspective: by adding CO_2 and other greenhouse gases to the atmosphere, we are countering a 3.5-billion-year trend towards CO_2 reduction.) A second example that supports the Gaia concept concerns the salinity of sea water. The fact this has barely changed over the long term is puzzling, given that salt is continually being added to the sea via rivers. How is salt being removed from the seas? The Gaia response is that various life forms in the seas promote the segregation of sea water in lagoons, allowing the sun to evaporate the water and remove the salt.

Both of these examples illustrate the meaning of Gaia as a self-regulating entity, keeping the environment fit for life. The existence of Gaia is not a proven fact; indeed, many scholars consider the hypothesis too general to be tested (Kirchner, 1989). Nevertheless, it is both scientifically stimulating (in 20 years, Gaia has gained considerable scientific legitimacy as evidenced especially by Lovelock's summary of recent developments in the prestigious science journal *Nature*) and valuable for the global perspective it represents.

of the nineteenth century led to the extirpation of the plains bison, and the Newfoundland and Labrador cod fishery when factory trawlers in the last decades of the twentieth century decimated groundfish stocks and seabeds.

Agriculturalists, both cultivators and pastoralists, affect far more extensive areas over longer time periods. Cultivators change normal energy flows, often permanently, and often consciously direct new flows. Pastoralists have less effect on energy flows, as domesticated animals may merely replace previously wild populations. Again, we find that agricultural impacts are most evident with the rise of commercial agriculture to feed large non-local populations, as in the case of Amazon deforestation.

Industrialists have affected virtually the entire surface of the earth. Energy flows have substantially increased with the use of fossil fuels and nuclear-based power. Today, however, oil supplies are often restricted not only by the limited amounts available, but by political circumstances. Ecosystem stability or instability is now very much a product of human decision-making rather than natural processes. We have now 'advanced' sufficiently to cause global instabilities, primarily through our depletion of the ozone layer and the increasing concentration of atmospheric carbon dioxide.

IMPACTS ON VEGETATION

When we consider human impacts on vegetation, animals, land and soil, water, and climate, it is appropriate to discuss vegetation first, since modification of plant cover results in changing soils, climates, geomorphic processes, and water: 'Indeed, the nature of whole landscapes has been transformed by man-induced vegetation change' (Goudie, 1981: 25). Figure 3.3 summarizes some of the consequences of vegetation change. Table 3.2 summarizes deforestation through time by major world regions, showing that most clearing has occurred since 1850 and that Europe, Asia, and the former USSR were the first regions to experience any considerable impact. It is only relatively recently that the tropical and subtropical areas of Central and South America have been subjected to significant deforestation.

Remote sensing is a versatile and effective tool for monitoring forestry operations. In Canada, Landsat and other imagery allow forest managers to collect data on forest inventory, depletion, and regeneration in areas as small as 2 ha (5 acres) and with boundary accuracy within 25 m (27 feet). Such data are invaluable in the accurate mapping of, for example, clear-cut areas.

| FIGURE | 3.3 | **Some consequences of human-induced vegetation change** |

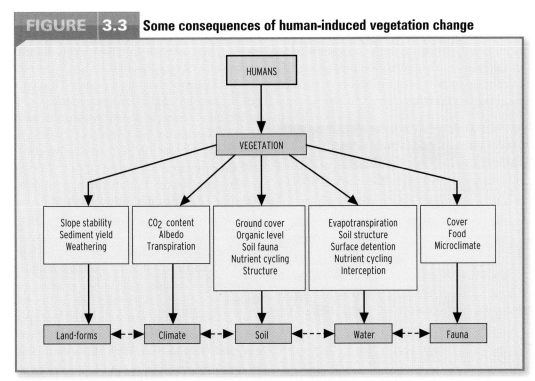

SOURCE: Adapted from A. Goudie, *The Human Impact: Man's Role in Environmental Change* (Oxford: Blackwell, 1981), 25.

Box 3.5   The Threatening Forest

Europe was once heavily wooded, but by the eighteenth century it was a largely agricultural region with few remnants of the once dominant forest. For eighteenth-century and later Europeans, cleared land represented progress and the triumph of technology.

When Europeans moved to temperate areas overseas, they confronted a very different environment. Eastern North America in particular was densely wooded. Aboriginal populations typically cleared only small areas and then moved elsewhere, allowing the forest to regrow. Among the nineteenth-century British in Ontario, the prevailing attitude towards the forest was antagonistic: 'settlers stripped the trees from their land as quickly as possible, shrinking only from burning them as they stood. They attacked the forest with a savagery greater than that justified by the need to clear the land for cultivation, for the forest smothered, threatened and oppressed them' (Kelly, 1974: 67).

Indeed, for the majority of settlers, the forest was a symbol of nature's domination over humans. Deforestation in Ontario proceeded apace as humans established their dominance over the land. By the 1860s, settlers were becoming aware of the disadvantages of deforestation, such as lack of shelter belts, fuel, and building materials. But by then the damage was done.

Fire

For at least 1.5 million years, humans used fire deliberately to modify the environment. Initially, vegetation removal probably resulted in increased animal numbers and greater mobility for human hunters. Fire also offered security and a social setting at night, and encouraged movement to colder areas. For later agriculturalists, fire was a key method of clearing land for agriculture and improving grazing areas; fire continues to serve these and similar functions today. Indeed, deforestation by fire or other means has been prompted largely by the need to clear land for agricultural activities, both pastoral and arable. In Europe, large-scale deforestation continued from about the tenth century onward, and in temperate areas of European overseas expansion it was carried out largely in the nineteenth century. Box 3.5 describes the European attitude towards forests in one part of the 'New World'. Together, deliberate burning and the natural fires that are much more common—it has been estimated that lightning strikes some 100,000 times each day (Tuan, 1971: 12)—have drastically modified vegetation cover.

Fire has played a major role in creating some vegetation systems—savannas, mid-latitude grasslands, and Mediterranean scrub lands are prime examples—and has probably affected all vegetation systems except tropical rain forests. Areas significantly affected by fire typically possess considerable species variety.

Plant domestication

Domestication is a process whereby a plant is modified in order to fulfill a specific human desire; once domesticated, the plant is permanently different from the original. This process is ongoing and is an important part of agricultural research today. Associated with plant domestication has been the labelling of plants that are not domesticated as weeds—and their removal. Once again, such human activity contributes to ecosystem simplification. Early domesticates included wheat, barley, oats (Southwest Asia), sorghum, millet (West Africa), rice (Southeast Asia), yams (tropical areas), potato (Andes), and manioc and sweet potato (lowland South America). Other domesticates include such pulses as peas and beans and such trees/vines as peach and grape.

It is thought that many fruit and nut species, including apples, apricots, plums, cherries, and walnuts, were first domesticated in an extensive forested area in the mountainous landscape of Central Asia—Almaty, the former

TABLE 3.2	Global Deforestation: Estimated Areas Cleared (000s km²)				
Region	Pre–1650	1650–1749	1750–1849	1850–1978	Total
North America	6	80	380	641	1,107
Central America	15	30	40	200	285
Latin America	15	100	170	637	922
Oceania	4	5	6	362	377
Former USSR	56	155	260	575	1,046
Europe	190	60	166	81	497
Asia	807	196	601	1,220	2,824
Africa	161	52	29	469	711
Total	1,254	678	1,652	4,185	7,769

SOURCE: Adapted from M. Williams, 'Forests', in B.L. Turner et al., eds, *The Earth as Transformed by Human Action: Global and Regional Changes in the Biosphere over the Past 300 Years* (Cambridge: Cambridge University Press, 1990), 180.

Stretching from Prince George in central British Columbia south to central Idaho, the world's largest remaining temperate rain forest includes at least fifteen tree species—western red cedar, western hemlock, mountain hemlock, ponderosa pine, Douglas fir, western larch, lodgepole pine, western white pine, subalpine fir, western yew, trembling aspen, paper birch, and three species of spruce—and is a priceless habitat for a wide range of life forms.

In the background of this photo is a hillside that has been 'clear-cut'. Such indiscriminate logging destroys both the trees themselves and the biologically diverse ecosystem that they support. It can also have drastic effects on streams like this one when logging debris accumulates after flooding.

Philip Dearden

capital of Kazakhstan, means 'Father of Apples'. In recent years this area has been subject to extensive deforestation because of overgrazing and other human activities, and there is concern that the loss of original wild species in this biological Eden might threaten the future of these foods, especially given the uncertainties of climate change.

Tropical rain forest removal

Human removal of vegetation—particularly of tropical rain forest—and desertification continue to be of major concern. Without human activity, forests would cover most of the earth's land surface. Large-scale deforestation accompanied the rise of the Chinese, Mediterranean, and Western European civilizations, as well as the nineteenth-century expansion of settlement in North America and Russia. Today, deforestation is concentrated in the tropical areas of the world. Viewed in this historical perspective, the current removal of rain forest may not seem excessive, but there are important differences between temperate and tropical deforestation. Tropical forests typically grow on much poorer soils that are unable to sustain the permanent agriculture now practised in temperate areas. Also, tropical rain forests now play a major role in the health of our global ecosystem—a fact that has been significantly acknowledged only in recent years. Figure 3.4 shows the past and present distribution of tropical rain forests. These rain forests cover about 8.6 million km^2 (3.3 million square miles).

The current rate of depletion and the loss to date are both open to dispute. Fortunately, remote sensing using Landsat, together with other satellite data, permits some objective assessment of rates of clearance. These data show that annual clearance rates in Brazil and elsewhere were several times greater than the early 1980s estimates made by the UN Food and Agriculture Organization (Repetto, 1990). On the other hand, the World Bank's 1989 estimate of a 12 per cent loss of Brazilian rain forest was shown to be about twice the actual

FIGURE 3.4   Past and present location of tropical rain forests

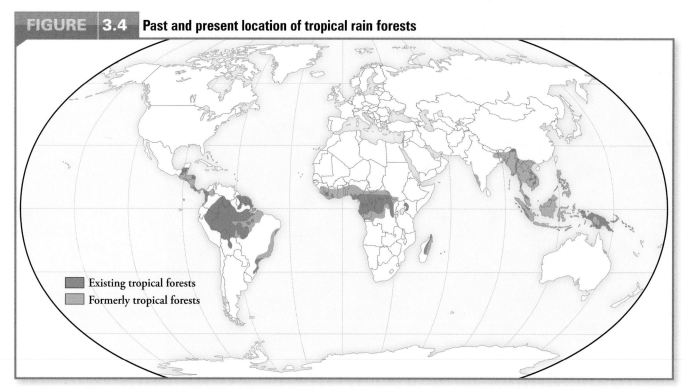

Existing tropical forests
Formerly tropical forests

SOURCE: W.E. Murray, *Geographies of Globalization* (New York: Routledge, 2006), 328. Copyright© 2006 Routledge. Reproduced with permission of Taylor & Francis Books UK.

loss. Although there are still uncertainties about the details of rain forest removal—the Food and Agriculture Organization acknowledged in 2008 that it is difficult to demonstrate convincingly that deforestation is occurring—estimates derived from satellite imagery in 2005 suggest annual rates of about 15,500 km² (6,000 square miles) due to selective cutting and a similar reduction due to clear-cutting.

Regardless of the details of tropical rain forest removal, two key questions are: Why are the rain forests being removed, and what are the ecological impacts?

As shown in Figure 3.4, rain forests are located primarily in less developed areas of the world, such as Bolivia, Brazil, Colombia, Venezuela, Gabon, DR of Congo, Indonesia, and Malaysia, but the more developed areas of the world are a leading cause of deforestation because of their enormous appetite for tropical timber and the inexpensive beef produced in areas cleared of tropical timber. There is a relationship between high commodity prices and deforestation, with about a six-month lag between the two. In addition, poor people in the less developed world use cleared land for some subsistence farming. Often such farming is possible only for a few years because

of cultivation techniques that rapidly deplete the soil of key nutrients. Cattle ranching also quickly becomes less profitable, because rain forest soil supports grazing for only a few short years. Nevertheless, as already noted, from the perspective of the countries experiencing rain forest clearance, such activity can be seen as a means of reducing population pressure elsewhere and as generally equivalent to the resource and settlement frontier of, say, North America during the past 200 years and of Europe during the past 1,000 years.

Rain forest removal has two principal ecological consequences. First, it is a major cause of species extinction because the rain forests are home to at least 50 per cent of all species (the total has been estimated at 30 million). From a strictly utilitarian viewpoint, many tropical forest species are (or may prove to be) important to humans as foods, medicines, sources of fibres, and petroleum substitutes.

The second principal consequence of rain forest removal involves global warming. Carbon is stored in trees, and when burning occurs the carbon is transferred to the atmosphere as carbon dioxide. In addition, soil is a source of carbon dioxide, methane, and nitrous oxide, all of which are released into the atmosphere as a

Box 3.6   Defeating Desertification

Led by 2004 Nobel Peace Prize winner Wangari Maathai, the Kenyan Green Belt Movement had its origins in a 1974 Nairobi tree-planting scheme that focused on the value of working on a community level, especially with women (Agnew, 1990). The objectives of the movement are many and varied, as befits any effort to solve so difficult a problem as desertification.

The central objective is to reclaim land lost to desert and to guarantee future fertility. Specific objectives include conserving water, increasing agricultural yields, limiting soil erosion, and increasing wood supplies. The ideas of working on the local level and involving women are crucial.

Reclaiming land is important to the local people, and their direct involvement makes the exercise much more meaningful to them, while women (the principal wood-gatherers in Kenya and in most of the Sahel) are increasingly aware of future needs. Planting crops around trees improves yields—an important and direct consequence in a subsistence economy.

Tree-planting to combat desertification is far from a panacea, but it is a positive development. Other parts of the world are adopting the strategies of the Kenyan movement, particularly its community focus. The parallels with the Grameen Bank (Box 5.12) are intriguing.

---

result of forest removal and farming. Each of these gases contributes to what we now call the greenhouse effect—a topic that we will consider shortly.

Desertification

Desertification is land deterioration caused by climatic change and/or by human activities in semi-arid and arid areas: 'It is the process of change in these ecosystems that can be measured by reduced productivity of desirable plants, alterations in the biomass and the diversity of the micro and macro fauna and flora, accelerated soil deterioration, and increased hazards for human occupancy' (Dregne, 1977: 324). The significance of desertification lies not simply in the clearing of vegetation but also in the consequences of clearing, which include soil erosion by wind and water and possible alterations of the water cycle. A 2007 UN report estimated that 2 billion people live in drylands susceptible to desertification in sub-Saharan Africa and in Central Asia, and that as many as 50 million are in danger of being driven from their homes by about 2017.

Deserts are natural phenomena, but desertification is the expansion of desert areas. The human causes are complex, but typically involve removal of vegetation as a result of overgrazing, fuel-gathering, intensive cultivation, waterlogging and salinization of irrigated lands. The reasons behind these activities are usually population pressure and/or poor land management. Climate change likely also plays a role, but it is much more difficult to measure this impact. Fortunately, the technology

to combat desertification is available; unfortunately, the will to use it is rare (Box 3.6). A major international effort to combat desertification, the 1977 UN Plan of Action, was a failure. There were two reasons for this failure: (1) technical solutions were applied to areas where the key causes were economic, social, and political, and these underlying causes were not addressed; and (2) local populations were not involved in the search for solutions. A potentially important development is discussion concerning an international convention to combat desertification as first proposed at the 1992 Earth Summit in Rio de Janeiro.

Estimates of the spatial extent of desertification in 1992, by the UN Environment Program, arrived at a figure of 3.5 million ha (8.6 million acres). Since remote-sensing imagery and local area surveys have not confirmed this figure, considerable confusion surrounds the actual spatial extent. The most publicized area experiencing desertification is the sub-Saharan Sahel zone of West Africa, an area first brought to world attention following the 1968–73 drought. Here, desertification is caused by population pressure, inappropriate human activity, human conflict, and periods of drought. Elsewhere, desertification similarly has multiple causes and no simple solution. Many of the human causes of desertification may be related to the pressures placed on local people by the introduction of capitalist imperatives into traditional farming systems.

As is the case with the tropical rain forests, areas subject to desertification are home to poor people who often are unable to adopt

desertification
The process by which an area of land becomes a desert, typically involves the impoverishment of an ecosystem because of climate change and/or human impact.

Desertification outside Benguela City, Angola, 1991. Agriculture in the area was disrupted by war, and the soil, left untended, became desert. The traces of former crop furrows are still visible.
© CIDA photo: Bruce Paton

appropriate remedies and lack the necessary political influence. A proper solution requires that the dryland ecosystems be treated as a whole—land management is needed. Population pressures must be reduced, land needs to be more equitably distributed, greater security of land tenure is required, and global warming must be combatted.

IMPACTS ON ANIMALS

Animal domestication serves many purposes, providing foods such as meat and milk (cows, pigs, sheep, goats), as well as draft animals (horses, donkeys, camels, oxen), and pets (dogs, cats). Once domesticated, animals have often been moved from place to place, both deliberately and accidentally. Some deliberate introductions, such as that of the European rabbit to Australia, have had drastic ecological consequences. Whalers and sealers were probably the first to introduce rabbits into the Australian region, but the key arrival was in 1859, when a few pairs were introduced into southeast Australia to provide so-called sport for sheep-station owners. Following this introduction, rabbits spread rapidly across the non-tropical parts of the continent, prompting a series of 'unrelenting, devastating' (Powell, 1976: 117) rabbit plagues. Rabbits consume vegetation

needed by sheep and remain a problem today despite the introduction of the disease myxomatosis and rabbit-proof fences (Figure 3.5). The European rabbit in Australia has probably caused more damage than any other introduced animal anywhere in the world—but it is only one of many unwanted guests (Box 3.7).

As human numbers and levels of technology have increased, so also have animal extinctions. It is possible that humans have caused species extinction since 200,000 BCE, although the evidence is uncertain. What is certain is that hunting populations have caused extinctions. After Europeans arrived in New Zealand, all moa species became extinct, and there is much evidence to suggest that animal extinctions in North America coincided with human arrival. The 1859 publication of Darwin's *On the Origin of Species* helped to place the extinction of plant and animal species in context and was followed by protectionist legislation in several of the British colonial areas; an early example was the 1860 law in Tasmania to protect indigenous birds.

Today our role in animal extinction is clearer. When natural animal habitats are removed, as in the case of tropical rain forest, extinction follows. This is one component of a major threat to biodiversity.

Biodiversity loss

E.O. Wilson, the best-known Darwinian thinker today, believes that the earth is entering a new evolutionary era involving the greatest mass extinction since the end of the Mesozoic era, 65 million years ago. Humans are the cause of the current extinction phase, in which species around the world are dying off as humans remove or destabilize their environments. This loss of biodiversity is irreversible by any reasonable human standards of time, and future generations are certain to live in a world that is biologically impoverished. The current rate of biodiversity loss is difficult to measure, with suggestions ranging from 100 to 10,000 times normal rates as recorded in the fossil record. A 2006 report compiled by the World Conservation Union listed 16,000 species of animals, birds, fish, and plants as under serious threat of becoming extinct—among the animals included were the polar bear and hippopotamus—while a 2008 report by the World Wildlife Fund (WWF) and related organizations concluded that biodiversity had plummeted by one-third between 1970 and 2005.

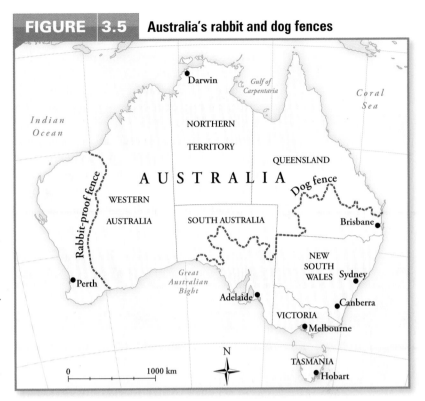

FIGURE 3.5 Australia's rabbit and dog fences

Fences built to protect Australian agriculture from rabbits and dingoes (wild dogs). Stretching 1,833 km (1,139 miles), the rabbit fence was built in the 1890s in an attempt to confine the introduced species to the central desert, but it was too late; rabbits had already entered western Australia. The dingo fence in eastern Australia stretches 5,321 km (3,307 miles) and was built to protect sheep; dingoes are thought to have been introduced into Australia by Asian seafarers more than 3,500 years ago.

The Carolina parakeet was one of the many species that humans have hunted to extinction. The last flock, numbering 30 birds, was sighted in Florida in 1920. From John James Audubon's *Birds of America*.

Plate XXVI: Toronto Public Library (TRL), Special Collections, Archives and Digital Collections Centre

Most commentators explain biodiversity loss by reference to population growth and increasing consumption of energy and resources. With more and more people around the world aspiring to the lifestyle of the affluent minority, the human ecological footprint is outgrowing the resources needed to support it. Wilson extends this argument in a more controversial direction to suggest that the impulse towards environmental destruction is innate, hard-wired into us—in short, that clearing forests and killing animals is instinctual for humans. Furthermore, he says, we are unable to see things in the long term. Whether or not Wilson is correct, the evidence of biodiversity loss is compelling; however, as much of the content of this chapter makes clear, the significance of this loss is still hotly debated. Indeed, there is debate not only about the rate of loss but also uncertainty about specific species. An example is the ivory-billed woodpecker, assumed to be extinct because of deforestation but reportedly sighted in Arkansas in 2004.

IMPACTS ON LAND, SOIL, AND AIR

Because we humans live on the land and use soil extensively, we are geomorphic agents that change land and affect that thin and vulnerable resource, soil. The earlier account of desertification, an impact both on vegetation and soil, is a clear example. Also, industrial activities add substances to the atmosphere that can cause harm to people and to environments; these include carbon monoxide, nitrogen oxides, sulphur oxides, hydrocarbons, and particulate matter (solid and liquid).

Land

Most human activities actually create landforms. Excavation of resources such as common rocks (limestone, chalk, sand, gravel), clays (stoneware clay, china clay), minerals (dolomite, quartz, asbestos, alum), precious metals (gold, silver), and fossil fuels (oil, coal, peat) can have major impacts: changing ecosystems, lowering land surfaces, flooding, building waste heaps, creating toxic wastes, and leaving scenic scars. These are not the only human activities that make us geomorphic agents: when we modify river channels, for example, sand dunes are affected; other effects include coastal erosion and coastal deposition. There are many causes and consequences of human impact on land.

Degradation and loss of arable land are occurring throughout the world as a result of population increases, industrialization, and

Box 3.7   Unwanted Guests

Beginning in 1788, the European colonization of Australia ended a long period of isolation and introduced numerous animals and plants. Many of these species, whether introduced deliberately or accidentally, proliferated in their new environment, where the natural controls on their numbers (such as predators and diseases) were absent. Inevitably, the presence of introduced species reduces the populations of native animals that evolved for millions of years in isolation.

Rats and mice that arrived in the holds of ships multiplied rapidly after being accidentally introduced. Sheep and cattle were brought over to satisfy British economic needs, while animals brought over to satisfy the 'sporting' habits of British settlers included deer, fox, and, of course, rabbits. The interior deserts prompted the introduction of pack animals such as burros, asses, and camels, and soon there were more camels in Australia than in all of Arabia. Many varieties of introduced livestock and pets have reverted to a feral (wild after previously being domesticated) existence, including pigs, horses, camels, water buffalo, and even house cats.

Feral water buffalo damage forest ecosystems by destroying trees and eroding soil while feral pigs damage sugar cane and banana crops, threaten wildlife, and spread disease—there are an estimated 23 million feral pigs, more than the number of Australian people. More generally, because all feral species require a supply of that most precious resource in the interior, water, they reduce the water available for native species. Many of the native animals of Australia have disappeared because of the presence of the introduced species. Of the smaller marsupials, 17 have become extinct and another 29 are considered endangered.

A recent introduction causing considerable damage is the cane toad, first brought to Queensland from Hawaii in the 1930s to combat the cane grub, which was damaging the sugar cane crop. Unfortunately, cane toads do not eat only cane grubs—they eat almost anything. They are also capable of much more rapid reproduction than native toad species, with females producing up to 40,000 eggs a year; they compete effectively with the native species; and they have no natural predators. Cane toads have now moved far beyond the sugar cane area, advancing about 50 km (31 miles) per year, and there are no signs of an end to their movement. In response, authorities announced in 2005 that they are offering a reward to anyone who comes up with an effective way to limit further advance. One initiative undertaken in 2009 was the declaration of a 'Toad Day Out', with everyone in Queensland asked to capture as many toads as possible and hand them to the authorities alive and unharmed, after which they were humanely killed.

Damage to farmland in Australia by feral pigs digging into the soil.

National Geographic Image Collection/Alamy

improper agricultural practices. Smil (1993: 67) reported that the average annual loss of farmland in China between 1957 and 1980 was a mammoth 1 million ha (2.5 million acres). About 40 per cent of the world's agricultural land is seriously degraded, with Latin America the most impoverished at 75 per cent.

One example of a current human impact on land is in Canada where exploitation of the Alberta oil sands began in earnest in the 1990s, prompted by technological advances that reduced costs, resulting in significant environmental damage. The oil sands are a mixture of sand, water, and heavy crude oil, and the oil is difficult and expensive to extract. After boreal forest is cleared and peat bog removed, what remains is oil-saturated sand. Extracting the oil requires large amounts of natural gas and water. The water, drawn from the Athabasca River and consequently affecting water levels and the health of communities downstream, cannot be reused and ends up in tailings ponds (a generic term for mining by-products) with other waste materials. In April 2008, about 500 ducks and geese returning north landed on what would have looked to them like a lake, but they died on contact with an oil slick on top of the tailings pond. This incident was a public relations disaster for the Alberta government and a major embarrassment for the company involved, Syncrude, a consortium that includes Petro-Canada. In 2009, Syncrude was charged by Environment Canada under the 1994 Migratory Birds Convention Act.

Soil

By its very nature, soil is especially susceptible to abuse, and humans use and abuse soils extensively. Agricultural activities have the greatest impact on soil, not only through erosion but also through chemical changes. Humans increase the salinity (salt content) of soil largely through irrigation—and yet increased salinity has a negative effect on plant growth. Humans increase the laterite content of soil (laterite is an iron- or aluminum-rich duricrust naturally present in tropical soils) by removing vegetation—and yet laterite is essentially hostile to agriculture.

Soil erosion is associated with deforestation and agriculture. Forests protect soil from runoff and roots bind soil. Probably the best-known example of human-induced soil erosion is the 'Dust Bowl' of the 1930s in the North American mid-latitude grasslands. Among the causes of this phenomenon were a series of low-rainfall years, overgrazing, and inappropriate cultivation procedures associated with wheat farming. The 'black blizzards' that were the result led many people to leave the prairies.

Air

Although air pollution is a problem almost everywhere, the pollution related especially to smoke-producing industries in more developed countries is significantly reduced today, but by no means eliminated, as a result of changes to industrial processes and careful management and control. Many large cities throughout the world suffer the problem of chemical smog (smoke + fog) largely as a result of automobile emissions: Los Angeles is a particularly notorious example.

An important atmospheric concern is the **ozone layer**. Ozone (O_3) is a form of oxygen that occurs naturally in the cool upper atmosphere. It serves as a protective sunscreen for the earth by absorbing the ultraviolet solar radiation that can cause skin cancer, cataracts, and weakening of the human immune system; ultraviolet radiation is also damaging to vegetation. Ozone depletion was first recognized in 1985 by scientists with the British Antarctic Survey. An ozone hole over Antarctica appears to be about one-half the size of Canada. The principal culprits are chlorofluorocarbons (CFCs) and some greenhouse gases (Box 3.8). As these rise into the atmosphere, chemical reactions occur and ozone is destroyed. However, a UN study concluded that depletion of the ozone layer had peaked by 2005, possibly

ozone layer
Layer in the atmosphere 16–40 km (10–25 miles) above the earth that absorbs dangerous ultraviolet solar radiation; ozone is a gas composed of molecules consisting of three atoms of oxygen (O_3).

Mining trucks carry loads of oil-laden sand at the Albian Sands oils sands project at Fort McMurray, Alberta.

Jeff McIntosh/The Canadian Press

Box 3.8   Chlorofluorocarbons and the Atmosphere

Chlorofluorocarbons (CFCs) exemplify the devastating environmental impact of many efforts to satisfy human demands through the use of technology. CFCs were not even synthesized until the late 1920s; at that time they represented a remarkable advance. They are ideal as coolants because they vaporize at low temperatures and also serve well as insulators. Most important, they are easy and therefore inexpensive to produce: hence their widespread use since World War II as coolants in refrigerators, as propellant gases in spray cans, and as ingredients in a wide range of plastic foams.

Recognition of the impact that CFCs have on the atmosphere, especially on ozone, prompted 24 countries to gather in Montreal in 1987 and agree to reduce CFC production by 35 per cent by 1999. Most environmental experts argued that this target was inadequate. Subsequently, a 1990 London agreement set the goal of eliminating CFC production by the year 2000. Neither of these goals was achieved. Because CFCs have a long atmospheric lifetime, peak levels will not be reached until roughly a decade after production ceases.

What we really need, of course, is a substitute for CFCs. So far, however, any possible substitutes are more expensive to produce, and world economies, whether capitalist or socialist, are not keen to sacrifice short-term economic benefits for long-term planetary stability.

as a consequence of the banning of CFCs (the Montreal Protocol and later agreements).

IMPACTS ON WATER

Water is an essential ingredient of all life. We know this and yet we choose to ignore it. Rather than carefully safeguarding water quantity and quality, we cause shortages and continually contaminate it. The two issues of scarcity and contamination dominate our consideration of the human impact on water.

The global water cycle

How much water is available? Figure 3.6 outlines the global water cycle, identifying three principal paths—precipitation, evaporation, and vapour transport. Total annual global precipitation is estimated at 496,000 km³ (119,000 cubic miles), most of which (385,000 km³/92,000 cubic miles) falls over oceans and cannot be easily used. Water returns to the atmosphere via evaporation from the oceans (425,000 km³/102,000 cubic miles) and from inland waters and land as well as by transpiration from plants (71,000 km³/17,000 cubic miles combined). In addition, some of the precipitation that falls on land is transported to oceans via surface runoff or groundwater flow (41,000 km³/9,800 cubic miles), and some water evaporated from the oceans is

FIGURE   3.6   The global water cycle

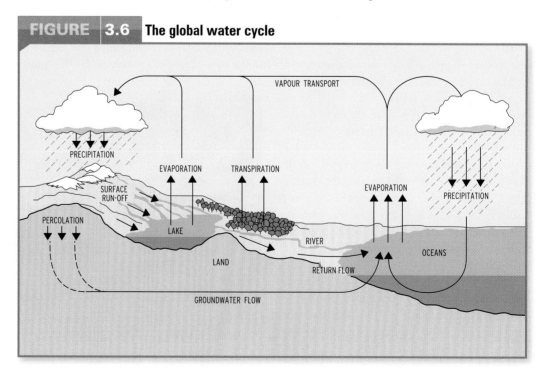

transported by atmospheric currents and subsequently falls as precipitation over land (again, some 41,000 km³/9,800 cubic miles). In principle, 41,000 km³ (9,800 cubic miles) of water are available each year globally, but this figure is reduced by 27,000 km³ (6,500 cubic miles) lost as flood runoff to the oceans and by another 5,000 km³ (1,200 cubic miles) flowing into the oceans in unpopulated areas. Perhaps 9,000 km³ (2,200 cubic miles) are readily available for human use.

This total of 9,000 km³ (2,200 cubic miles) is possibly enough water for 20 billion people. However, some states have a plentiful amount and others have an inadequate amount. Where water is abundant, it is treated as though it were virtually free; where it is scarce, it is a precious resource. Thus the average citizen of the US annually consumes 70 times more water—through combined household, industrial, and agricultural uses—than does the average citizen of Ghana.

Using water

Agriculture consumes 73 per cent of global supplies of water, often highly inefficiently. In second place is industry, which consumes perhaps 10 per cent of global supplies. The bulk of what is left goes to supply basic human needs. As each of these uses increases in any given area, whether through agricultural and industrial expansion or population increases, water quantity may become a problem. Shortages are common in many areas, some continuous and some periodic. Bahrain, for example, has virtually no fresh water and relies on the desalinization of sea water. Groundwater depletion is common in the US, China, and India, and water levels have fallen in both Lake Baikal (southern Siberia, the world's deepest lake) and the Aral Sea (Kazakhstan–Uzbekistan) (Box 3.9). As of 2009 Australia had experienced a 10-year drought that has devastated agricultural landscapes, while periods of drought in Brazil and South Africa have meant there is

---

Box 3.9   A Tale of Three Water Bodies

Until the 1960s, the Aral Sea, on the border between the central Asian states of Kazakhstan and Uzbekistan (both former parts of the Soviet Union), was the fourth largest inland water body in the world. By the early 1990s, however, remote-sensing imagery showed that the area covered by the sea had been reduced by more than 40 per cent. The explanation was that, since the 1950s, much of the water that formerly flowed into the sea had been diverted to irrigate cotton. Although data for the late 1990s show some signs of recovery—water levels have risen as a result of dam construction and reductions in water use—the Aral Sea remains vulnerable to a range of human activities. Tragically, because cotton requires large quantities of defoliants, pesticides, and fertilizers, the Aral Sea has also become dangerously polluted, and in the local area there are high rates of infectious disease, cancer, miscarriage, and fetal abnormalities. These problems are not likely to be solved any time soon.

The Rhine River, one of the most important international waterways in the world, is 1,320 km (820 miles) in length and has a catchment area of 185,000 km² (71,428 square miles) populated by about 50 million people in six European countries. The river receives pollutants from several heavily industrialized areas in addition to domestic sewage and shipping discharges. By the late 1980s, as a result of these human activities, the quality of Rhine water was so compromised that treatment was necessary before the water could be used for drinking or crop irrigation in the Netherlands, and

local ecosystems were changed, destroying much animal and plant life. However, considerable success in combatting these problems has been achieved through the activities of the International Commission for the Protection of the Rhine against Pollution, established in 1950. Today the Rhine is in relatively good health; one indicator was the return of salmon in 2000.

When Europeans first reached Lake Erie, the water was clear and supported a large fish population. By the 1960s, however, Erie was one of the most seriously polluted large water bodies in the world. With some 13 million people and much heavy industry in its watershed, over time the lake was polluted by domestic sewage, household detergents, agricultural wastes, and industrial wastes. Clear water became green, fish perished, and algae growth was enhanced by the addition of phosphates and nitrates. These problems were acknowledged and the Canadian and American governments, urged to take action by newly formed environmental groups, initiated a comprehensive program to solve these problems in the 1970s. Known as the Great Lakes Water Quality Agreement, this program has been remarkably successful, and today Lake Erie has been rehabilitated.

These three examples demonstrate that, given effective management strategies, seriously polluted water bodies can recover; however, such strategies are more likely to be employed in the more developed than in the less developed world.

insufficient water to generate hydroelectricity. Consider also that many of the great rivers of the world, some of which flow through major grain-growing regions, no longer reach the sea—these include the Huang He River in China, Murray-Darling in Australia, Indus in India and Pakistan, Rio Grande in the United States and Mexico, and Colorado in the United States.

But do these regional problems mean that there is, or soon will be, a global water shortage? The simple, but misleading, answer to this question is no. This is because humans use only about 9 per cent of the water that flows through the global water cycle. But, of course, it is not only humans and their activities that use water—so does all other life on earth. Further, human use of water is increasing significantly because of increasing population numbers, improved living standards, and climate change. For example, most of the addition to the world population is taking place in the cities of less developed countries and urban dwellers use more water than rural dwellers. Also, improved living standards typically involve a shift in diet from vegetarian to meat-eating, and growing 1 kg of wheat uses about 1,000 litres of water while producing 1 kg of beef requires 15,000 litres.

Polluting water

The second major issue, after water quantity, is water quality. As water passes through the cycle described in Figure 3.6, it is polluted in two ways. Organic waste (from humans, animals, and plants) is biodegradable, but can still cause major problems in the form of oxygen depletion in rivers and lakes and in the form of water contamination, causing such diseases as typhoid and cholera. Much industrial waste is not easily degraded (paper, glass, and concrete are exceptions), and such wastes are a major cause of deteriorating water quality. These pollutants enter water via pipes from industrial plants, diffuse sources (runoff water containing pesticides and fertilizers), and the atmosphere (**acid rain**).

Both inland waters and oceans are suffering the consequences of pollution. In addition to the three examples detailed in Box 3.9, recent satellite evidence suggests that Lake Chad (north-central Africa) is disappearing because of drought and irrigation: it has shrivelled from 23,000 km² (14,300 square miles) to only 900 km² (350 square miles) since about 1960. An especially notorious example of pollution is the toxic chemical waste in waters at Love Canal in the US (see Figure 2.18), while an explosion at a petrochemical plant near Harbin, China, in 2005 released toxic pollutants into a major river resulting in water being shut off to almost 4 million residents for five days.

Acid rain is a difficult issue because the pollution itself may occur far away from its source. Acid rain is one general consequence of urban and industrial activities that release large quantities of sulphur and nitrogen oxides into the atmosphere. The effects of acid rain are not entirely clear, but there is no doubt about its negative impacts on some aquatic ecosystems. One of the most severe problems involved in reducing water pollution is the difficulty of securing the necessary international co-operation. The Rhine Action Plan involves only four countries; efforts to combat acid rain, especially in Europe, require co-operation among many more countries.

Nowhere is the need for international agreement more evident than in combatting ocean pollution. Many states exploit oceans, but no state is prepared to assume responsibility for the effects of human activities. Estimates suggest that only 3 per cent of the world's oceans—remote icy waters near the poles—remain undamaged, with overfishing and climate change the main causes of damage. More than 50 per cent of the world's population live close to the sea, and most of the world's ocean fish are taken from coastal waters. Water quality in the oceans, especially coastal zones, is seriously endangered, and many ocean ecosystems have been damaged. News reports often now refer to 'dead zones', usually areas close to land that are starved of oxygen because fertilizer flows down rivers into the sea with a resultant loss of fish and underwater vegetation. Remote sensing enables objective assessments of water pollution; images of the Mediterranean Sea show a marked contrast between the northern shore, heavily polluted by water from major European rivers and coastal towns, and the southern shore. Remote sensing permits effective monitoring of oil spills, as in the *Exxon Valdez* incident off the Alaskan coast in 1989. Once again, we do not know enough about the consequences of our activities, but we do know that restoring the quality of ocean water is likely to be much more difficult than is the case with surface inland water.

acid rain
The deposition on the earth's surface of sulphuric and nitric acids formed in the atmosphere as a result of fossil fuel and biomass burning; causes significant damage to vegetation, lakes, wildlife, and built environments.

POLLUTED LANDSCAPES

Pollution of land, soil, air, and water is most severe in cities in poor less developed countries, where it can be 'like living under a death sentence' as it is a 'major source of death, illness and long-term environmental change' (Blacksmith Institute, 2006: 3). Tragically, any discussion of badly polluted landscapes has no shortage of examples from which to choose, and nine are described here. These nine, along with Chernobyl, comprise the top 10 of the world's worst polluted places as described by the Blacksmith Institute (2006), an environmental charity based in the United States (Figure 3.7).

1. *Dzerzhinsk, Russia.* About 300,000 people are potentially affected. The principal pollutants are chemicals and toxic by-products from the manufacturing of chemical weapons during the Cold War period. About 25 per cent of the local population continue to work in factories that produce toxic chemicals. Toxic waste has polluted aquifers that are still used for the local water supply. Average life expectancy is in the mid-40s, and death rates exceed birth rates by 2.6 times.

2. *Linfen, Shanxi Province, China.* About 200,000 people are potentially affected. This is the heart of China's coal-mining region, with many mines operated illegally and ignoring all regulations. Linfen is probably the most

Chernobyl, Ukraine.

Ivan Nesterov/Alamy/GetStock

polluted city in China in terms of air quality, with many cases of bronchitis, pneumonia, lung cancer, and other health problems. Water supplies for drinking and other domestic use are limited because of the amount used in coal mining.

3. *Kabwe, Zambia.* About 250,000 people are potentially affected. Located about 150 km north of the capital, Lusaka, Kabwe is the second largest city in Zambia and one of

FIGURE 3.7 The 10 most polluted places on earth

SOURCE: Map drawn using information detailed in Blacksmith Institute, *The World's Most Polluted Places: The Top Ten*. New York: Blacksmith Institute, 2006

six principal towns located in the industrial Copperbelt region. Unregulated mining and smelting activities have taken place since the early twentieth century, leading to high concentrations of lead in local soil and water. Most at risk are children who play in the soil and men who scavenge the area for metal.

4. *Norilsk, Siberia, Russia.* About 134,000 people are potentially affected. This city was founded in 1935 as a slave labour camp and is the location of the world's largest heavy metals smelting complex. Respiratory and other diseases are common, with people living close to the industrial plants most at risk. Children and pregnant women are especially vulnerable. Since 2001 this city has effectively been closed to outsiders, and consequently the details of health problems are difficult to determine.

5. *Haina, Dominican Republic.* About 85,000 people are potentially affected by an area of lead contamination from a now closed automobile battery recycling smelter. There are exceptionally high levels of lead in blood and soil, with children's health most affected. Although Haina is especially polluted, there are many other examples of cities in the less developed world that have similar problems as a result of the presence of battery recycling plants.

6. *La Oroya, Peru.* About 35,000 people are potentially affected. Toxic emissions from a poly-metallic smelter run by a US company cause unacceptably high levels of lead in children's blood. Estimates suggest that 99 per cent of local children are affected. High lead levels in blood are known to affect mental development. Local vegetation has been destroyed by acid rain caused by emissions of sulphur dioxide.

7. *Ranipet, India.* About 3,500,000 people are potentially affected. About 160 km from Chennai, Ranipet is the location of a factory manufacturing chemicals used locally in the leather tanning process. Solid waste is stacked in an open area. Soil and groundwater are severely polluted, affecting health and agriculture. Many wells and public hand pumps can no longer be used.

8. *Rudnaya Pristan, Russia.* About 90,000 people are potentially affected. The residents of this lead-mining area in eastern Russia suffer in an area with badly polluted air, water, and soil. Analysis suggests that lead in children's blood is 8 to 20 times the acceptable level. The smelter is now closed.

9. *Mailuu-Suu, Kyrgyzstan.* About 23,000 people are potentially affected as of now, with millions more possibly to be affected. Home to a former Soviet uranium plant, this area is now the site of a massive amount of radioactive mining waste. Local soil and water are polluted and the population has exceptionally high levels of various cancers. Because this is an area of seismic activity, there is a danger that a much larger area might be affected, hence the possible impacts on millions.

IMPACTS ON CLIMATE

Nowhere is the concept of interrelations better demonstrated than in a consideration of human impacts on climate. Scientists and environmentalists feel that our most damaging impacts of all are on global climate. Unfortunately, as in our discussions of other human impacts, we are confronted with two general areas of uncertainty: the role played by humans (as opposed to natural physical factors) and the extent of any human-induced change. But note that while these two areas of uncertainty do prevail, *what is certain is that humans are causing climate change.*

Physical causes of climatic change

Not all climatic changes are caused by humans; in fact, such changes can occur for various reasons and on various time scales. It appears that some changes result from external variables, such as variation in the input of solar radiation. Yet it also seems that climate oscillates regardless of such factors; it has recently been determined, for example, that long-term changes are related to the changing distribution of land and sea, with the current distribution conducive to the emergence of glacial periods in North America and Eurasia. The most recent glacial period ended some 12,000 years ago. Other climatic changes occur at time scales of a few hundred to a few thousand years, and these changes are not well understood. During the past 12,000 years, our current interglacial period, there was a cool phase 10,000–11,000 years ago, a distinctly warm phase 6,000 years ago, and another warm period about 1,000 years ago. Historical evidence shows that, before the fourteenth century, England was warm enough for wine cultivation, and that Europe in general was unusually cold in the sixteenth and seventeenth centuries.

It is important to consider human impacts in the above perspective. We know that we are capable of, and are, changing climate, but we are not sure of precisely how we may affect future climates, because we do not know how our activities may interact with other, non-human variables. We know that, in the very long term, we are moving towards another glacial period, but we do not know what shorter-term changes may naturally occur prior to the onset of a new ice age.

The natural greenhouse effect

Temperatures on the earth's surface result from a balance between incoming solar radiation and loss of energy from earth to space. If the earth had no atmosphere, then the average surface temperature would be about −19°C (−3°F), but the presence of an atmosphere results in an actual average surface temperature of about 15°C (60°F). The atmosphere causes an increase in surface temperatures because it prevents about one-half of the outgoing radiation from reaching space; some is absorbed and some bounces back to earth. This natural 'greenhouse' effect is not related to human activity but is the result of the presence in the atmosphere of water vapour, carbon dioxide (CO_2), ozone (O_3), and other gases. These greenhouse gases are only a fraction of the atmosphere—nitrogen and oxygen make up 99.9 per cent of it (excluding the widely varying amounts of water vapour)—but their impact is nevertheless considerable. Today we are increasing the greenhouse effect by adding CO_2 and some other gases that perform a similar function, such as sulphur dioxide (SO_2), nitrous oxide (N_2O), methane (CH_4), and a variety of CFCs. How are we doing this?

Human-induced global warming

In the most general sense, the human additions to the natural greenhouse effect are the product of our increasing population numbers and advancing technology. More specifically, they are the result of burning fossil fuels, increased fertilizer use, increased animal husbandry, and deforestation. Until recently, much carbon was stored in the earth in the form of coal, oil, and natural gas. Burning these resources releases CO_2, water vapour, SO_2, and other gases that are then added to the atmosphere. Burning wood also adds CO_2 to the atmosphere. Further, soil contains large quantities of organic carbon in the form of humus, and agricultural activity speeds up the process by which this carbon adds CO_2 to the atmosphere. Estimates suggest

that the concentration of CO_2 in the atmosphere has increased from 260 ppm (parts per million) 200 years ago to 370 ppm today and might be 400 to 550 ppm in 2030. The current level is higher than at any time in the past 650,000 years. Agricultural activity also adds CH_4 (methane) and N_2O (nitrous oxide) to the atmosphere. CH_4 is increasing both as a result of paddy rice cultivation and the large number of flatulent farm animals. Recall from the earlier reference to biofuels that growing crops such as corn and canola releases N_2O.

Although most research rightly focuses on what is happening today, it is possible that deforestation resulting from the spread of agriculture through much of Europe and China about 5,000 years ago initiated warming comparable to that evident since the Industrial Revolution. Indeed, Ruddiman (2005) suggests that the spread of early agriculture prevented the onset of a period of much colder temperatures.

Each of the greenhouse gases noted so far (CO_2, N_2O, SO_2, CH_4) is naturally present in the atmosphere; human activities simply increase the natural concentrations of them. However, we are also adding new greenhouse gases, the most important being two of the CFCs—$CFCL_3$ and CF_2CL_2. These gases are added primarily by our use of aerosol sprays, refrigerants, and foams (see Box 3.8).

It is widely accepted that this human-induced global warming is raising the average

Because they are dependent on the Arctic sea ice for hunting and other important activities, the polar bear is severely affected by disrupted climate paterns in the Arctic.

Peter Van Wagner/iStock

temperature of the earth. The fourth report of the UN's authoritative Intergovernmental Panel on Climatic Change (IPCC), published in 2007, predicted a low-scenario increase of 1.8°C (with a likely range of 1.1 to 2.9°C) and a high-scenario increase of 4.0°C (with a likely range of 2.4 to 6.4°C) by 2100. Some scientists think the increase will be less dramatic because the panel overestimated the rate of economic growth in the less developed world, while others think the increase might be greater for several reasons, including that warmer soils exhale CO_2, accelerating the global warming process.

It is important to appreciate that many of the popular accounts of global warming, and some of the scientific ones, too, make little if any reference to the uncertainties involved in predicting either the magnitude or the consequences of the human-induced greenhouse effect. In particular, there is considerable uncertainty about the regional consequences of global warming, a reflection of the basic ecological fact that all things are related. One of the few specific predictions that scientists feel confident in making is that, as warming occurs, the earth's polar regions will be the most seriously affected. By the late 1990s, it was clear that such warming was already underway, and not all the evidence to this effect came from scientific sources. In northern Canada, Inuit elders and hunters reported that glaciers were receding and coastlines eroding, that the fall freeze was arriving later, and winters were becoming less severe. The combination of scientific data and traditional ecological knowledge makes for a compelling argument. Although at present the evidence of warming in lower latitudes is less clear, it is generally agreed that the next several decades will see a poleward retreat of cold areas, a corresponding expansion of forests and agricultural areas, changes in the distribution of arid areas, and, of course, rising sea levels as the ice caps melt (see Boxes 3.10 and 3.11 and Figure 3.8).

Responding to human-induced global warming

The principal response to the fact of global warming and its probable consequences has been an effort to implement policies that will reduce the emission of greenhouse gases. Most notably, the UN-sponsored Kyoto Protocol established goals that, if met, would slow the rate of global warming. This Protocol, a legally binding agreement to cut greenhouse gas emissions, was agreed in 1997 by about 150 countries but only came into force in early 2005 with most participating countries agreeing to reduce emissions by a specific percentage. It is notable, however, that neither the United States nor Australia has ratified the agreement, principally because it is seen as unfair that neither China nor India are required to cut emissions during the first phase of the agreement (through to 2012). This is significant especially because the United States and China are first and second respectively in the amount of greenhouse gases emitted.

How have the participating countries performed? Two countries, Germany and Britain, are generally judged to have responded positively with about 10 per cent reductions, but most others have failed to meet their targets. Canada, for example, agreed to cut emissions by 6 per cent from the 1990 level but, by 2006, emissions had increased by 21.3 per cent. Overall, emissions of greenhouse gases by the major industrial countries increased 2.3 per cent between 2000 and 2006, with the biggest increases in the countries of the former Soviet Union, China, and Canada. Discussions concerning the details of human-caused global warming and possible responses to that warming continue on a regular basis, but tend to be compromised by the perceived need of many national governments to focus on economic growth.

Progress on Kyoto was assessed during UN conferences on climate change held in Ottawa in 2005 and Poznan, Poland, in 2008. Most observers judged these meetings as limited successes, with progress made in four broad areas. First, many remaining details about Kyoto were finalized, including what is to happen if countries fail to meet targets. Second, talks to focus on further emission reductions after 2012 were planned. Third, all participants agreed to talk about a possible UN climate pact that will include countries that have not signed up to Kyoto, most importantly the United States. Fourth, participants agreed to promote carbon capture and sequestration technologies.

A separate international agreement was reached in mid-2005. Six countries—the US, China, India, South Korea, Japan, and Australia—agreed to reduce emissions, but this pact has received much criticism because it is non-binding. A feature of this agreement is the emphasis on tackling emissions through new

Box 3.10   Climate Change—Front and Centre

Should we worry about global warming? Yes, most definitely, respond scientists. Although the future is uncertain, and specific regional predictions are difficult to make, scientists are expressing real concerns about the possible consequences of human-induced global warming. In the first few years of the twenty-first century, global warming without doubt became *the* critical environmental issue. Few see any realistic likelihood that the major contributors to CO_2 emissions—the United States, the countries of the former Soviet Union, and the emerging industrial economies of China and India—will cut back these emissions in the foreseeable future, which means that, notwithstanding Kyoto and other agreements, there seems little hope of slowing the rate of change.

Consider that:

- The best way to detect global temperature change is to use satellites to measure ocean temperatures (because ocean temperatures are not subject to the same diurnal and seasonal fluctuations that air temperatures are). These data indicate that global temperatures have increased by about 0.74°C over the past 100 or so years.
- The six hottest years since effective record-keeping began in the 1890s were, in descending order, 2005, 1998, 2002, 2003, 2004, and 2007.
- The world's oceans are warming (about 0.5°C since the 1940s) from top to bottom, a fact that is best explained by human activity.
- The area covered by sea ice in the Arctic is shrinking each year with the annual rate of shrinkage estimated to be about 8 per cent, meaning that there might be no ice remaining by about 2050.
- The Greenland ice sheet, second in size only to the Antarctic ice sheet, appears to be crumbling with a dramatically increased rate of shrinking reported by recent satellite data.
- Most glaciers in the world are melting so rapidly that they are likely to disappear sometime during the next few hundred years. Glacier loss will be particularly significant for the large numbers of people in lower-lying areas, who rely on glacial meltwater for farming activities and domestic use.
- It is likely that the increased intensity of extreme weather conditions, such as hurricanes, is related to increased sea surface temperatures that destabilize the atmosphere.
- The Antarctic ice sheet, which contains 90 per cent of the world's ice, is melting more rapidly that previously

thought according to 2009 research; this ice shelf disintegration in the Antarctic likely results from melting from underneath and from surface melting caused by global warming. Satellite data indicate a 0.6°C warming in the past 50 years.
- Permafrost is melting rapidly in the Siberian tundra, a process that is releasing additional methane into the atmosphere.

These and other facts about the consequences of global warming are reported regularly in major scientific journals, including *Nature*, *Scientific American*, and *Science*.

There are real concerns about the negative impacts of global warming, especially because we have made settlement decisions and developed lifestyles with the prevailing climate as background. What are some likely consequences of global warming?

- Reports from the World Bank link much illness and death in the less developed world to infectious and respiratory diseases, including malaria and dengue fever, that are spreading as climate changes.
- A report from the Australian Medical Association expressed concern that increased incidence of disease might lead to conflict, population displacement, and more authoritarian governments.
- The impact on crop yields is unclear. On the one hand, it is well recognized that additional carbon dioxide in the atmosphere is likely to act as fertilizer, potentially increasing yields of all major cereal crops except corn and sorghum. On the other hand, longer droughts and more ground-level ozone may significantly reduce yields.
- The impact on drought conditions in Africa is also unclear. According to many scientists the region contributing least to global warming, Africa, will be the most adversely affected with increased incidence and length of droughts exacerbating current problems. But other research claims that rainfall in the semi-arid Sahel region will increase, not decrease.
- Some research suggests devastating impacts on biodiversity, with up to 25 per cent of all plant and animal species extinct by 2050 according to a 2004 UN report. The most vulnerable species may be those in extensive areas with little variation in altitude as species would have to move long distances to find suitable environments.
- The most significant possible negative consequence of global warming is sea level rise caused by ice melt (see Box 3.11).

technologies rather than through any reductions in economic growth. In 2006, the new Conservative government in Canada indicated that it was skeptical of Kyoto, favouring instead this alternative voluntary approach to emission reductions, a fact that highlights the close links between environmental action and political ideology.

Box 3.11   Melting Ice and Rising Sea Levels

Global warming is expected to affect sea levels. A rise of as much as 1.5 m (almost 5 feet) has been predicted for 2050 given current rates of melting of the Antarctic and Greenland ice sheets. If this occurs, the consequences for many currently populated areas may be catastrophic. Up to 15 per cent of Egypt's arable land would be at risk, and many coastal cities such as New York and London would be below sea level. The consequences for two of the most densely populated areas in the world—coastal Bangladesh and the Netherlands—will be disastrous unless we are able to adapt. Indeed, even before sea levels rise substantially, many coastal areas would be in danger because of storm surges.

The Netherlands has successfully adapted to a situation where much of the current area is already almost 4 m (13 feet) below sea level. In principle, the Netherlands model—construction of levees and dikes—can be followed. Venice is already pursuing a similar strategy, constructing a flexible seawall to protect the city against Adriatic storms. But what are the costs of such projects? One estimate places the cost at approximately US $300 billion to protect only major areas, not including coastal margins.

In theory, there is an alternative to adaptation: moving away from the threatened areas. But for many people in the less developed world, this is hardly an option, and for many in the more developed world, it is culturally and economically unthinkable at the present time.

On the other hand, predictions of a sea-level rise are at best uncertain. In 1985, one major United States scientific committee predicted a rise of 1 m (3 feet) by 2100, and subsequently, in 1989, amended the prediction to a rise of 0.3 m (1 foot). Even the higher of these two is less than the prediction referred to earlier. Why such uncertainty? The principal unknown factor is the effect of warming on the Antarctic ice sheet. Instead of increasing ice melt, it is possible that, through time, warmer weather could increase snowfall, which would in turn help build up the ice sheet. However, in 2009 at a climate change conference in Copenhagen, scientists warned that the rise in sea levels would be between 1–2

metres by 2100, putting many of the world's coastal areas at risk.

The problem is serious. We need to adapt soon to an unknown situation. Dikes and population movements may or may not be needed. If they are needed, we do not know the extent of the need. In this way global uncertainty becomes local political uncertainty.

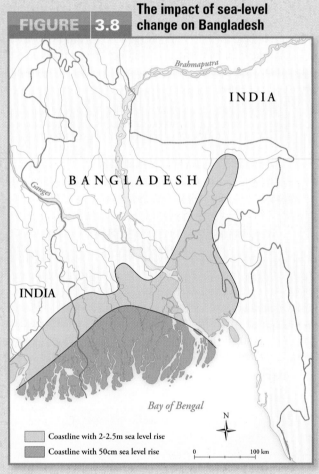

FIGURE 3.8   The impact of sea-level change on Bangladesh

Coastline with 2-2.5m sea level rise
Coastline with 50cm sea level rise

SOURCE: G.A. McKay and H. Hengeveld, 'The Changing Atmosphere', in C. Mungall and D.J. McLaren, eds, *Planet Under Stress* (Toronto: Oxford University Press, 1990), 66.

But are these agreements the best route to take? One other option is to do nothing on the grounds that there might be positive consequences of global warming. For example, the melting of Arctic ice might open the Northwest Passage, significantly reducing the distance for ships sailing between Europe and Asia compared to travelling through the Panama Canal. (Of course, such an eventuality would also bring to the fore acrimonious international conflict regarding Canadian sovereignty in the

Far North.) The melting of Arctic ice might also mean that drilling for oil in the region will be more economically feasible, a circumstance prompting further interest in issues of Arctic sovereignty, with at least five countries—Canada, the US, Russia, Denmark, and Norway—having legitimate sectoral claims to portions of the Arctic seabed—and many other countries, notably Britain and Japan, claiming an interest in the Arctic waters being 'international' for shipping purposes. Another possible advantage is that the growing season might be extended in some areas. But most observers think differently, with the general consensus being that the probable negative consequences of warming significantly outweigh any possible advantages (see Boxes 3.10 and 3.11).

The point of no return?

So, is it possible to arrive at any definitive conclusions? James Lovelock, originator of the Gaia concept—the idea that the earth possesses a planetary-scale control system that functions to keep it fit for life—has despaired. His depressing conclusion is that global warming caused by humans has already moved beyond the point of no return, meaning that the planetary control system is no longer able to function properly and earth will never be the same place again. In *The Revenge of Gaia*, Lovelock (2006) contends that during the current century average temperatures will rise about 8°C in temperate areas and 5°C in the

---

Box 3.12   The Politicization of Climate Science

Two facts about human impacts on climate are clear: humans are adding additional amounts of greenhouses gases to the atmosphere, and (ignoring any possible contrary impacts of natural change) an increase in global temperature is therefore inevitable.

But public understanding of these two incontrovertible facts is not helped by what is best described as the politicization of climate science. Most notably, individuals and organizations with a vested interest in the continued burning of fossil fuels (and they are legion in today's profit-oriented economy) have routinely questioned, even rejected, these two facts. Consider the comments made by Chrysler's chief economist in 2007 to the effect that any climate change is a far-off risk of uncertain magnitude. The publication of such comments suggests to many people that a legitimate scientific debate continues about whether humans are adding greenhouses gases and whether global warming is resulting. A 2009 conference held in New York by the Heartland Institute, a Chicago-based think-tank funded by Exxon Mobil until 2005, was billed as the largest-ever meeting of climate change deniers. A major cause of embarrassment for Europe was that among the speakers was the President of the Czech Republic, at a time when that country held the rotating presidency of the European Union.

In fact, there is no such scientific debate. As repeatedly noted in this chapter, human-caused climate change is a fact, and the scientific basis for denying this is non-existent (as demonstrated in the 2007 report of the IPCC). Still, elements of the media convey a very different impression, with many conservative publications regularly implying that scientific debate continues. The debate taking place is not scientific; rather, it is politically motivated, with some individuals and groups ideologically committed to continuing the current practices, to the achievements of technology, and thus to denial of human-caused climate change.

Scientists acknowledge it is impossible to predict in detail what will happen even in the short term, but the pattern of temperature increase is clear. Those who deny human-caused climate change take every opportunity to highlight the lack of specifics, suggesting that this reflects uncertainty about the overall trend, which of course it does not. Those who deny also focus on news items such as the 2008 research noting that there might be little additional warming before about 2020 because the earth is probably entering a short, about 10-year, cooling phase related to the natural cycle of ocean temperatures, but to interpret this possible natural effect as evidence that humans are not causing change is quite simply wrong.

Those on the ideological right, who support business, free markets, and short-term gain, also tend to focus on some of the regional uncertainties, implying that these indicate doubt about the global trends. For example, scientific evidence suggests that the higher precipitation and related melting of snow and ice in northern areas that is likely to accompany global warming will lead to increased river discharge into the Arctic and North Atlantic oceans. The resulting decrease in ocean salinity may disturb the balance of water flow (known as the thermohaline circulation) that involves a warm northward flow of surface water and a returning southward flow of deep water. A possible consequence of reduced thermohaline circulation might be to flip the North Atlantic back into a glacial mode and to substantially lower temperatures in Northwest Europe (Anderson, 2000). This scenario is simply one example of the many uncertainties involved in attempting to anticipate the regional consequences of global temperature changes—but it is not an argument against human-caused global warming.

It is perhaps inevitable that some pundits and think-tanks, especially those with vested interests in the oil industry, have been able to muddle the issue. What is needed is a clear public acknowledgement of increases in greenhouse gases and in global temperatures, accompanied by an understanding that details are often lacking. Responsible media can play a key role in encouraging this understanding through an explicit separation of scientific fact and political interpretation and preference.

tropics. These predicted temperature increases are so high because Lovelock argues that the consequences of humans damaging the control system are non-linear and likely to accelerate. According to Lovelock we have abandoned our responsibility for keeping the global eco-system fit for life.

James Hansen, a leading US researcher on human-caused climate change, arrived at a somewhat more sanguine but no less alarmist conclusion: 'The Earth's climate is nearing, but has not passed, a tipping point beyond which it will be impossible to avoid climate change with far-ranging undesirable consequences. These include not only the loss of the Arctic as we know it, with all that implies for wild-life and indigenous peoples, but losses on a much vaster scale due to rising seas' (quoted in McKibben, 2006: 18).

Explaining why we have allowed global warming to proceed as far as it has is not easy. Boxes 3.1 and 3.3 provide a context for under-standing this important matter. More generally, making the needed changes is complicated by six general factors: uncertainty about what is really happening; uncertainty about the bal-ance of negative and positive consequences;

uncertainty about technological solutions; vested economic and political interests; cultural barriers to change; and lack of awareness (Box 3.12).

Urban climates

Any discussion of human impacts on climate must also note urban climates. Not surprisingly, any urban development affects local climate. Built-up areas store heat during the day and release it at night; they also generate artificial heat, creating an 'urban heat island'. Built-up areas also affect cloud formation and precipi-tation, but these impacts are more difficult to determine and measure.

THE ROLE PLAYED BY GLOBALIZATION

Many human impacts on the environment are exacerbated by current trends towards glo-balization, especially the activities of trans-nationals (see Chapter 2). It seems clear that transnationals are among the principal actors in many environmental issues. For example, fast-food corporations have been blamed for the destruction of tropical rain forests because they use meat from animals raised in areas that have been cleared specifically to create pasture. Similarly, oil companies that explore for and produce oil in such environmentally sensitive areas as the Amazon basin, the Canadian Arctic, and the mangrove swamps of the Niger delta are particularly condemned by environmen-talists, as are the trawlers of transnational fish-processing companies that have been a major factor in the collapse of local fisheries from New Zealand to Newfoundland.

Amnesty International suggests that the recently completed oil and gas pipeline from the Caspian Sea port of Baku in Azerbaijan to the Mediterranean Turkish port of Ceyhan has infringed on the human rights of thousands of people and will cause massive environ-mental damage. The route of the pipeline—constructed by a BP-led consortium—was designed to avoid passing through Russian or Iranian territory. As a result, according to Amnesty International, 30,000 villagers have been forced to give up their land rights. In addition, Amnesty International claimed that the health and safety precautions in place are inadequate and that local protestors faced state oppression. BP, for its part, argued that the project followed the highest international stan-dards, that appropriate compensation has been

"Hey, let's not underestimate science! I bet they're working on a replacement to the ozone layer right now."

Cartoon by J.B. Handelsman, from *Punch*, 14 Sept. 1990.

© Punch Ltd. www.punch.co.uk

paid, and that there will not be any environmental damage.

Transnational companies have sometimes been charged with actively perpetuating local social injustices. Talisman Oil, the largest independent Canadian oil and gas producer, has operations in Canada, the North Sea, and Indonesia. It is currently conducting explorations in Algeria and Trinidad, and between 1998 and 2003 was also active in Sudan. Its Sudanese operations were severely criticized by social justice and church groups in Canada because of claims that Talisman actively supported the government of Sudan in an ongoing civil war. That government, headed by the National Islamic Front, has consistently been identified as violating basic human rights. The oil fields in question were located in southern Sudan—traditionally a rebel stronghold—and there was evidence that the government forcibly removed people to make the area safe for Talisman. However, the company denied complicity in any such social injustices and in 1999, in response to a request from the Canadian government, formally adopted the International Code of Ethics for Canadian Business, developed in 1997 by a group of Canadian companies with transnational operations. It was at least partly because of the difficult political and social environment in Sudan that Talisman sold all its interests there in 2003.

Notwithstanding these comments, it is important to appreciate the many uncertainties about globalization. Using the three globalization theses found in Table I.2—hyperglobalist, skeptical, and transformationalist—Table 3.3 highlights varying understandings of the links between globalization and environmental issues.

Earth's Vital Signs

As the preceding account demonstrates, our current impacts on ecosystems, from the global to the local, are greater than ever before, and are increasing as a general result of the growth of population and technology (Figure 3.9). Further, international co-operation will certainly be required to address many of the problems that human activities have caused (see Box 3.3); as already noted, the first such international agreement was the 1987 Montreal Protocol limiting the production of CFCs. Debate continues, however, concerning the present condition of the environment and the

Table 3.3	Globalization Theses and Environmental Issues
Hyperglobalist: The Global Era	**Skeptical: Increased Regionalism**	**Transformationalist: Unprecedented Interconnectedness**
Sustainability solved by the market. Market will price the environment efficiently and technology will solve scarcity.	Sustainability threatened by global capitalism. Spread of modernization and consumerism pushes environment to limits.	Sustainability attainable through political action. Environmental problems and concerns globalize. Regulation from above and below required.

SOURCE: Adapted from W.E. Murray, *Geographies of Globalization* (New York: Routledge, 2006), 353–4.

probable future scenario. As Smil (1993: 35) put it, 'Confident diagnoses of the state of our environment remain elusive', and this observation seems as valid today as it was in the early 1990s.

APOCALYPSE NOW, DEFERRED, OR NEVER?

In fact, there are many different opinions about the impact of human activities on the environment. At one extreme are **catastrophists**, who view the current situation and future prospects in totally negative terms—a view articulated by Kaplan (1994, 1996). At the other extreme are **cornucopians**, who believe that the gravity of current problems has been greatly exaggerated and that human ingenuity and technology will overcome the moderate problems that do exist (Simon and Kahn, 1984). This is a broad and complex debate, which we will revisit in Chapters 4, 5, and 8.

A balanced view of such an ideologically and economically charged topic is not easy. Certainly the environment is sustaining considerable damage. Yet, as Smil (1993: 5) argues, since the 1960s the environment has been the subject of many ill-informed commentaries and predictions: 'Once the interest in environmental degradation began, the Western media, so diligent in search of catastrophic happenings, and scientists whose gratification is so often achieved by feeding on fashionable topics, kept the attention alive with an influx of new bad news.' As our discussion of human impacts has demonstrated, it is difficult to separate the many very real problems from the numerous exaggerated claims (Box 3.13).

RESPONDING TO UNCERTAINTY

In view of the many uncertainties that surround environmental questions, Smil writes:

catastrophists Those who argue that population increases and continuing environmental deterioration are leading to a nightmarish future of food shortages, disease, and conflict.

cornucopians Those who argue that advances in science and technology will continue to create resources sufficient to support the growing world population.

FIGURE 3.9 Global distribution of some major environmental problems

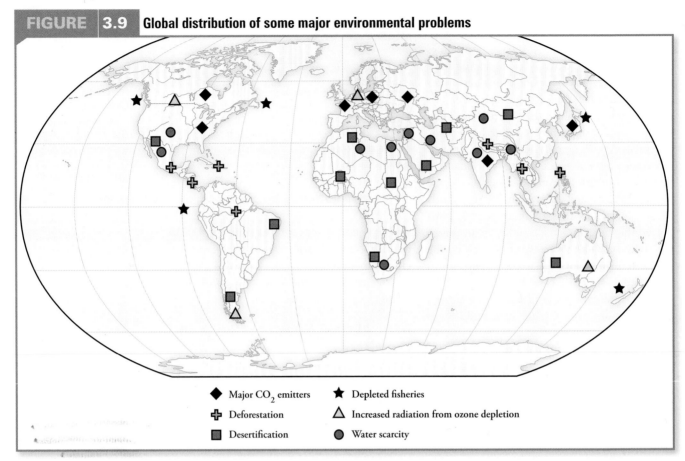

Major CO$_2$ emitters ◆ ★ **Depleted fisheries**
✚ **Deforestation**   △ **Increased radiation from ozone depletion**
■ **Desertification**   ● **Water scarcity**

This map shows the general regional occurrence of six current environmental stresses. Symbols are not intended to identify specific locations, or countries, but rather to indicate general regions affected by particular kinds of stress.

SOURCE: Adapted from *Current History* 95, 604 (Nov. 1996).

The task is not to find a middle ground: the dispute has become too ideological, and the extreme positions are too unforgiving to offer a meaningful compromise. The practical challenge is twofold. First, to identify and to separate the fundamental long-term risks to the integrity of the biosphere from less important, readily manageable concerns. The second task is to separate effective solutions to such problems from unrealistic paeans to the power of human inventiveness. (Smil, 1993: 36)

The first challenge is continually being addressed by geographers and other environmental scientists. Given our current understanding of global environmental problems, there are three possible general responses to the second challenge:

1. We can attempt to develop new technologies to counter our deleterious impacts; for example, some scientists have suggested that dust might be deliberately spread in the upper atmosphere to reflect sunlight, replacing the depleting ozone layer. Overall, though, it seems inappropriate to rely too much on such technological solutions. Despite massive evidence that we are changing the earth's climate on a variety of scales—local to global—we still know too little to change it deliberately in ways we desire. The results of attempts at rain-making, hurricane modification, and fog dispersal have been mixed. But there are perhaps some simple changes we might make that will have positive consequences. Two examples are noted. First, eating less meat will reduce methane emissions from animals and was advocated by the UN climate chief in 2008. Second, transforming dark urban surfaces into white will increase the reflection of sunlight and reduce warming. California recognized the value of this strategy and in 2005 required commercial premises with flat roofs to repaint those roofs white.

Adaptation
The process by which humans adjust to a particular set of circumstances; changes in behaviour that reduce conflict with the environment.

Box 3.13 The Skeptical Environmentalist

An impressive body of scientific research indicates that the environment is in bad shape. For example, a 2002 study by the WWF warned that the earth is being plundered at a rate that outstrips the capacity to support life, and that unless drastic action is taken, humans will have to colonize other planets by the year 2050. Indeed, this is exactly what noted physicist Stephen Hawking proposed in 2006. Another example: a 2008 Living Planet report argued that humans are using 30 per cent more resources than the earth can replenish each year, meaning that we will need two planets like earth by 2030. Nevertheless, at several points in this chapter we have had occasion to note uncertainties concerning both the facts of environmental change and the consequences of such change. Interestingly, comparable uncertainties will be evident in the two chapters that follow concerning the related matters of population growth and food supply, and also in Chapter 9, concerning the consequences of economic globalization.

Perhaps nothing better demonstrates these uncertainties than the sharply divergent responses sparked by the 2001 book *The Skeptical Environmentalist: Measuring the Real State of the World*, by Bjorn Lomborg. Some view Lomborg's massive compendium of facts as confirmation that the prophets of environmental doom and gloom are either exaggerating or just plain wrong, while others view his work as preposterous; one critic expressed his opinion on the matter by throwing a pie at the author. In short, conservative and business-oriented commentators hailed Lomborg as the voice of sanity, while environmentalists and most scientists rejected his views in no uncertain terms. Thus the book was reviewed favourably in *The Economist,* the *New York Times*, and the *Washington Post*, but reviewed critically in *Nature*, *Science*, and *Scientific American*.

Bjorn Lomborg is a Danish statistician who wrote *The Skeptical Environmentalist* following a failed attempt to disprove the late Julian Simon's claim that the state of the environment was not getting worse, but rather was improving all the time. Unable to find any flaws in either Simon's data or his logic, Lomborg conducted his own analyses and then wrote his book endorsing Simon's claims. According to Lomborg, we are not running out of energy or natural resources; the world's species are not disappearing at an alarming rate; air and water supplies are becoming less polluted; and warnings of a possible 5°C temperature rise this century are unrealistic. Lomborg even goes so far as to suggest that climate change may offer positive benefits, including improvements to human health and increases in crop yields.

Lomborg's critics, including the WWF, maintain that he is wrong on all these counts; that he mischaracterizes the environmental movement and commits precisely those sins for which he attacks environmentalists, namely exaggeration, gross generalization, presentation of false choices, selective use of data, and factual errors. For his part, Lomborg suggests that bad news attracts more attention than good news and that scientists are more likely to receive funding for research into identifiable problem areas.

Whatever the rights and wrongs of Lomborg's claims, and he has restated, updated, and clarified these on a regular basis, especially in the 2007 publication, *Cool It: The Skeptical Environmentalist's Guide to Global Warming*, debating them will help us to reach the best possible understanding both of what is happening now and of what may happen in the future.

2. We can acknowledge that environmental impacts are inevitable and emphasize our need to adapt to such changes. In the case of possible climatic change, **adaptation** might involve new population movement, water-supply systems, and coastal defences.

3. The third response, probably the most popular and most logical, involves the conservation of resources and the prevention of harmful impacts. Further, even if we are uncertain about the details of change, the most logical way to approach the environmental crisis is to acknowledge the likelihood of significant negative consequences and to take action now that is designed to mitigate the anticipated consequences. **Conservation** refers generally to any form of environmental protection. Prevention involves limiting the increase of greenhouse gases, reducing the use of certain materials, and reusing and recycling. **Recycling** often makes sense economically as well as environmentally.

Prevention is central to many current moves to protect environments. It can be argued that economic systems, including capitalism, do not properly reward the efficient use of resources, and that the physical environment has been seen as irrelevant in the final economic accounting. Only recently have we begun to understand that price and value are not equal. But how do we assign a monetary value to a

recycling
The reuse of material and energy resources.

conservation
A general term referring to any form of environmental protection, including preservation.

wilderness landscape, for example? One answer, though imprecise, is to assert that certain ecosystems or landscapes are sufficiently distinct as to merit protection or preservation. The largest protected ecosystem today is probably that of Antarctica. Other protected landscapes include wilderness regions such as Canada's national and provincial parks. (The first areas to be preserved as national parks were Yosemite [1864] and Yellowstone [1872] in the US.)

To improve the health of the earthly patient of which we are all a part, solutions are needed at all spatial and social scales. Above all, perhaps, what we require is education about and understanding of the need for reducing harmful human impacts.

Sustainability and Sustainable Development

To conclude this chapter, we return to the key concepts of sustainability and sustainable development that were briefly introduced earlier. Clearly, we are transforming the earth in ways that we do not intend. And, equally clearly, we need to manage the earth along appropriate pathways. Management requires us to understand what kind of earth we want, to reach consensus, and to find an appropriate balance between the values of economic development and conservation. Although these aims will be difficult to achieve, if only because people in different areas live in very different circumstances and have very different value systems, it is imperative that a balance be established: the relationship between humanity and the land is such that environment and economics must both be central concerns.

The term 'sustainability' was introduced in the late 1970s to refer to the idea that our current way of life, based on ever-increasing consumption of resources, could not continue indefinitely and that we would have to find a more sustainable way of life. It was difficult to argue against the desirability of sustainability, but there was considerable disagreement as to what changes were needed to achieve the desired state. The debate on this theme advanced significantly with the introduction of the concept of **sustainable development**, loosely defined as development that accounts for social, economic, and environmental concerns. This term

sustainable development
A term popularized by the 1987 report of the World Commission on Environment and Development, referring to economic development that sustains the natural environment for future generations.

was introduced by the influential 1987 report of the World Commission on Environment and Development, titled *Our Common Future*.

The term was defined in *Our Common Future* as follows: 'Sustainable development is development that meets the needs of the present without compromising the ability of future generations to meet their own needs.' A more elaborate definition was put forward in 1989: 'Sustainability is the nascent doctrine that economic growth and development must take place, and be maintained over time, within the limits set by ecology in the broadest sense—by the interrelations of human beings and their works, the biosphere and the physical and chemical laws that govern it' (Ruckelshaus, 1989: 167). This second definition is reminiscent of the deep ecology perspective summarized in Table 3.1. The concept of sustainability can also be explained by reference to the systems concept we looked at earlier. The earth can be regarded as a closed system in that, although energy enters and leaves the system, matter only circulates within it. This type of system can reach a state of dynamic equilibrium—one that involves optimal energy flow and matter cycling in such a way that the system does not collapse. In a sense, sustainable development would represent a similar state of dynamic equilibrium.

Once the term was introduced, the idea of sustainable development became a key focus at the UN Conference on Environment and Development held in Rio de Janeiro in 1992. The essence of sustainable development is the attempt to blend into one process what are often seen as opposites, namely sustainability and development, and for some critics the phrase is an oxymoron—a contradiction in terms. After all, to limit damage to environments, it might be necessary to limit growth. This contradiction is most easily resolved, in principle if not in practice, by suggesting that rich countries should strive for sustainability by limiting their own growth, while poor countries, which have not reached an adequate level of economic development and should have a chance to do so, should strive for sustainable development.

How do we move towards a sustainable world where environmental changes are in accord with sound ecological principles? Any such move clearly requires a significant—and deliberate—shift towards a new attitude. Four principles are essential to that new attitude:

Box 3.14   Clayoquot Sound: The Case for Sustainability

Clayoquot (pronounced Klak-wat) Sound, on the west coast of Vancouver Island in British Columbia, is one of the largest (about 3,500 km² or 1,351 square miles) areas of coastal temperate rain forest in the world. It is a distinctive ecosystem, with the highest biomass (weight of organic matter to land area) of any forest type. It also contains at least 4,500 known plant and animal species, and possibly several thousand other insects and micro-organisms. Several of the species, such as the sea otter, are on the Canadian endangered species list. Aboriginal groups, including the Nuu-chah-nulth (meaning 'all along the mountains'), have lived in the area for more than 8,000 years without causing environmental degradation, but in recent decades the area has become attractive to lumber companies, and clear-cut logging is posing real threats to the ecosystem.

There are, then, several conflicting interests at stake. Conservationists, including groups such as the Sierra Club, argue for preservation of the ecosystem; Aboriginal people ask that their values and lifestyles, which are intimately related to the forest, be recognized and respected; and lumber companies seek to make profits in a province that has 20 per cent of its economy in the forestry sector. In 1984 the first open confrontation occurred when local Aboriginal and other residents set up a blockade to prevent logging. The provincial government established a Wilderness Committee in 1986, then three mediating forums in 1986, 1989, and 1992. Meanwhile, in 1988, a second confrontation began in another location.

Discussions at the various forums failed to arrive at an agreement satisfactory to all interest groups, and in 1993 the provincial government announced the Clayoquot Sound Land Use Decision, permitting logging in two-thirds of the area and protecting one-third. A massive civil disobedience movement followed, with about 900 people arrested by the end of 1993. In 1994 the forum set up in 1992 (the Commission on Resources and the Environment) proposed arrangements for the area that would result in the loss of about 900 logging jobs. This prompted a mass demonstration by loggers.

In 1995 the provincial government endorsed a report of the Clayoquot Sound Scientific Panel, another advisory forum it had established following the 1993 protests; the report made a series of recommendations that included significant reductions in the rate of cutting but did not call for permanent protection. The recommendations represented a substantial change from previous reports, as they explicitly acknowledged the importance of maintaining the rain forest ecosystem. This change of attitude appears to be continuing. In 1996, at their annual meeting, members of the World Conservation Union, including the BC government, endorsed a proposal supporting the designation of Clayoquot Sound as a UN Biosphere Reserve. Further, in 1999 the provincial government reached agreements with MacMillan Bloedel

concerning compensation for the company's loss of cutting rights following the creation of new parks; and, in a move welcomed by environmental groups, the company announced new forest management policies involving the phasing out of clear-cutting, increased conservation of old growth, and independent validation of their forest practices. In 1999 MacMillan Bloedel was acquired by Weyerhaeuser to create the largest North American forest products company. Currently, the annual total yield from the limited logging in Clayoquot Sound ranges between zero and 100,000 cubic metres—a massive reduction from the late 1980s high of almost 1 million cubic metres per year—and Aboriginal people share in the logging and in the profit, a small but perhaps not insignificant instance of 'sustainable livelihoods'.

This example highlights many important aspects of current human impacts on land. Above all, it suggests that governments find it hard to appreciate the arguments for a sustainable economy—not surprising, given that governments need political support from both industry and trade unions and also benefit financially from economic activities (in this case from the sale of logging licences). But it also indicates that public education campaigns and the intense public pressure can have an impact on government attitudes and policies. In 2006 the BC government announced that about one-third of the Great Bear Rainforest, which stretches along the Pacific coast between Vancouver Island and Alaska, is to be preserved. The remaining two-thirds will see some logging using sustainable practices. Environmental groups, including Greenpeace, continue to lobby for less logging.

In the summer of 1993 more than 800 people were arrested for blockading a logging road into Clayoquot Sound. This non-violent civil disobedience campaign was designed to attract media coverage and pressure the provincial and federal governments into putting a stop to clear-cut logging in the area.

Philip Dearden

1. We need to recognize that humans are a part of nature. To destroy nature is to destroy ourselves.
2. We need to account for environmental costs in all our economic activities.
3. We need to understand that all humans deserve to achieve acceptable living standards. A world with poor people cannot be a peaceful world.
4. We need to be aware that even apparently small local impacts can have global consequences. This is one of the basic themes of ecology—'think globally, act locally'—and it becomes ever more obvious as globalization processes unfold.

There are, of course, vast differences between acknowledging the need for a new attitude, seeing it adopted globally, and, finally, putting it into practice. The dilemma facing humans today was expressed forcefully some years ago: 'To continue with the trial and error procedures of the past means to risk irreparable damage to [our] habitat. To accept responsibility for full environmental control means to anticipate change and decide in which direction to go' (Wilkinson, 1963: 32). Since these words were written, many strides have been made in the right direction. Clean air acts, environmental impact assessments, and environment ministries are now standard in many countries. Although it would be an exaggeration to assert that solutions are in sight, or even that they are being sought in all cases (Box 3.14), environmental issues are increasingly recognized as relevant at all levels, from individuals to governments. But this is not equally the case for all countries. Thus it is the clear responsibility of the more developed world to demonstrate environmental concern by example:

in creating the consciousness of advanced sustainability, we shall have to redefine our concepts of political and economic feasibility. These concepts are, after all, simply human constructs; they were different in the past, and they will surely change in the future. But the earth is real, and we are obliged by the fact of our utter dependence on it to listen more closely than we have to its messages. (Ruckelshaus, 1989: 174)

Postscript

At the beginning of this chapter it was noted that everything is related to everything else, that one cannot change one aspect of nature without directly or indirectly affecting other aspects. This fact points to the difficulty of predicting the future with any degree of certainty: 'The principal reason why even the cleverest and the most elaborate explorative scenarios are ultimately so disappointing is that they may get some components of future realities approximately right, but they will inevitably miss other ingredients whose dynamic interaction will create profoundly altered outcomes' (Smil, 2005: 202).

What we can reasonably state is that, through a combination of our large and growing population, our advancing technologies, and our improving living standards we are changing the global ecosystem in many and varied ways. Some of these ways are fairly easy to predict, at least in general terms, while others are likely unanticipated. Certainly, our global future is uncertain, as evidenced by the multitude of often contradictory reports that include results and forecasts that, sometimes, flow from preconceived outcomes. Because the authors of some research assume never-ending progress while other authors assume inevitable downfall, the best message is to read and think about environmental issues both thoughtfully and critically.

CHAPTER 3 SUMMARY

USE OR ABUSE?

Today there is considerable evidence that we are damaging our home—the earth—and increasing recognition of how fragile that home is.

SCALE

The value of a global perspective in any discussion of human use of the earth is clear: everything is related to everything else.

THE INCREASING HUMAN IMPACT

Two principal variables—sheer numbers of people and increasing use of technology and energy—explain why human activities today have such great impact on the earth.

ECOSYSTEMS

Combining systems logic and ecological principles produces the concept of the ecosystem: any self-sustaining collection of living organisms and their environment is an ecosystem. A whole series of ecosystems together make up the global ecosystem—the ecosphere or biosphere—which is the home of all life on earth. The ecosphere includes air, water, land, and all life.

ECOSYSTEM SIMPLIFICATION

When humans change an ecosystem, the result is usually a simplification. We cause changes largely because of our numbers, our technology, and our energy use. Today we number in excess of 6.5 billion, and our technological ability—that is, our ability to convert energy into forms useful to us—is constantly increasing. As human culture (especially technology) changes, so does the value attributed to resources. Different interest groups evaluate resources according to different criteria and hence have differing views. Stock resources are finite in quantity, whereas renewable resources are relatively unlimited in quantity.

Our simplification of ecosystems is now sufficiently evident that a new environmental ethic is needed—one involving co-operation with nature, not domination over it.

HUMAN IMPACTS

The presence of humans on the earth changes the planet. Some of the changes brought about by humans are inevitable because they are necessary to human survival. Other changes, however, are the result of inappropriate actions motivated by human greed and misunderstanding. Over time, humans' ability to affect ecosystems at all levels has increased. Although it may be inappropriate to focus on individual parts of an ecosystem—which is by definition an inseparable whole—it is convenient to consider the impact of human activities on specific parts of ecosystems such as vegetation, animals, land and soil, water, and climate. In each case, we have caused many changes.

Two important issues today are tropical rain forest depletion and desertification; in both cases, human activity is at least partially explained by the problems experienced by poor people in less developed countries. Impacts on animal life include those associated with domestication and species extinction. Extensive changes to land surfaces result from a host of human activities such as resource extraction and the creation of urban industrial complexes. Soil is especially susceptible to abuses involving salinization, increased laterite content caused by removal of vegetation, and erosion. Water, an essential ingredient for life, is typically used in ways that result in shortages and contamination, and the oceans continue to be polluted.

The ecosystem concept of interrelatedness is especially useful with respect to climate. We are having a significant effect on the atmosphere by increasing the quantity of greenhouse gases and damaging the ozone layer. Climate change is undeniable and humans are a cause, but we also need to understand more about climatic change that is not caused by humans.

THE HEALTH OF THE PLANET

Although debates continue concerning the consequences of human activity for the health of the planet, most agree that human-induced ills are many and serious. Probably the best solution is prevention—an argument advanced by many environmentalist groups.

SUSTAINABLE DEVELOPMENT

Sustainable development serves our present needs, but does not compromise the ability of later generations to meet their needs.

QUESTIONS FOR CRITICAL THOUGHT

1. In what ways is the earth 'fragile'? Do you agree with this description of the earth? Why or why not?

2. How and why have human impacts on the earth changed during the Holocene?

3. In his discussion of the 'three contentious issues' regarding the current debate about the environment, Norton states that 'market forces are unlikely to solve environmental problems'. Do you agree? Why or why not? What other factors (apart from market forces) might play a key role in our attempt to deal with environmental issues?

4. Are humans part of nature or are they separate from nature?

5. Of the various global environmental issues discussed in this chapter (tropical rain forest destruction, desertification, biodiversity loss, pollution, and climate change) which one do you think is most serious? Why? What do you think should be done about this particular issue?

6. How serious is climate change? Have we already reached the 'point of no return' as James Lovelock claims in *The Revenge of Gaia*?

7. Are you a catastrophist or a cornucopian? How would you explain your position?

FURTHER EXPLORATIONS

Blackbourn, D. 2006. *The Conquest of Nature: Water, Landscape and the Making of Modern Germany*. London: Jonathan Cape.

Detailed study of the last 250 years of German environmental history with emphasis on how humans address the natural world.

Cipolla, C.M. 1974. *The Economic History of World Population*, 6th edn. Harmondsworth: Penguin.

A survey that focuses on the importance of energy sources in understanding economic change.

Cutter, S.L. 1994. 'Environmental Issues: Green Rage, Social Change and the New Environmentalism', *Progress in Human Geography* 18: 217–26.

Insightful discussion focusing on the usefulness of a variety of new environmental philosophies.

Flannery, T. 2006. *The Weather Makers: How We Are Changing the Climate and What It Means for Life on Earth*. New York: Atlantic Monthly Press.

A very readable account of human impacts on climate that argues passionately for the need to do something now, not later.

Giddens, A. 2009. *The Politics of Climate Change*. Malden, Mass.: Polity Press.

A wide-ranging book by a renowned sociologist arguing that the scale of the problem and the enormous challenges faced in reconciling green views with economic growth can be resolved in a democratic framework.

Goodall, C. 2008. *Ten Technologies to Save the Planet*. London: Profile Books.

This important contribution to issues of human impacts challenges much of thinking concerning green technologies. For example, Goodall shows that solar power is not too expensive, wind power is not too unreliable, and marine energy is not a dead-end.

Goudie, A. 2006. *The Human Impact on the Natural Environment*, 6th edn. Oxford: Blackwell.

An excellent general textbook by a well-known geographer.

Hughes, J.D. 2001. *An Environmental History of the World: Humankind's Changing Role in the Community of Life*. New York: Routledge.

A history of the world organized around the concept of ecology that treats humans as part of the environment and assumes that population is outstripping resources; includes a series of interesting and informative case studies.

Johnson, C. 1991. *The Green Dictionary: Key Words, Ideas and Relationships for the Future*. London: McDonald Optima.

A useful source for definitions, and sometimes discussions, of a wide variety of terms relevant to human impacts on the environment.

Johnson, D.L., ed. 1977. 'The Human Face of Desertification', *Economic Geography* 53: 317–432.

A theme issue that contains 19 articles, many of which offer specific examples of desertification in a wide range of spatial, cultural, and technological contexts.

Kemp, D.D. 1998. *The Environment Dictionary*. New York: Routledge.

A remarkably comprehensive dictionary, written by a geographer.

Krech, S., III, J.R. McNeill, and C. Merchant. 2004. *Encyclopedia of World Environmental History*, 3 vols. New York: Routledge.

Concise overviews of relevant topics, places, and people that stress the need for a global perspective on environmental issues.

Lovelock, J. 2009. *The Vanishing Face of Gaia*. London: Allen Lane.

In his ninetieth year, Lovelock continues to write passionately about the Gaia concept, asserting that global warming is irreversible.

Mackay, A., R. Battarbee, J. Birks, and F. Oldfield, eds. 2003. *Global Change in the Holocene*. New York: Oxford University Press.

Informative text that addresses natural global change over the past 12,000 years.

Matthews, J.A. et al., eds. 2003. *The Encyclopaedic Dictionary of Environmental Change*. London: Arnold.

A comprehensive survey of both natural and human-induced environmental change; an invaluable source of facts and definitions.

Park, C. 2001. *The Environment*, 2nd edn. New York: Routledge.

A comprehensive text on the environment and human interactions with it; covers all the topics addressed in this chapter and includes numerous detailed regional examples.

Phillips, D.E., A. Wild, and D.S. Jenkinson. 1990. 'The Soil's Contribution to Global Warming', *Geographical Magazine* 62, 4: 36–8.

A brief popular piece that succinctly summarizes the important links between soils and the greenhouse effect.

Porter, G., and J.W. Brown. 1991. *Global Environmental Politics*. Boulder, Colo.: Westview Press.

A good discussion of the links between environmental issues and other aspects of public policy; includes evaluations of how well national governments and international organizations are responding to environmental challenges.

Roberts, N., ed. 1993. *The Changing Global Environment*. Oxford: Blackwell.

Includes details of global environmental change, with emphasis on climate, oceans, and water; also very useful for Chapter 4.

Smil, V. 2002. *The Earth's Biosphere: Evolution, Dynamics, and Change*. Cambridge, Mass.: MIT Press.

A readable, thoughtful overview of the biosphere emphasizing the gaps in our understanding that make it impossible to diagnose accurately the current state of the world.

Williams, M. 2003. *Deforesting the Earth: From Prehistory to Global Crisis*. Chicago: University of Chicago Press.

With a focus on the close links between agricultural aspirations and deforestation, this book provides a valuable global overview making it clear that logging and clearing have been integral to the development of society almost everywhere.

Worldwatch Institute. 2010. *State of the World, 2009: Into a Warming World*. Washington: Worldwatch Institute.

An important annual publication on the progress being made, or not made, towards a sustainable world. This volume focuses on global warming.

Worster, D. 2008. *A Passion for Nature: The Life of John Muir*. New York: Oxford University Press.

Worster, an environmental historian, has written a readable and sensitive biography of one of the most important figures in the early American conservation movement.

ON THE WEB

REPORTING ON HUMAN IMPACTS ON ENVIRONMENT

Resources

www.wri.org/

The home page of the World Resources Institute; includes information on the full range of environmental issues, including climate change, biodiversity, ecosystem changes, and resource use and abuse, as well as regional analyses. Also useful for Chapter 10.

Environmental Knowledge for Change

www.grida.no/

Probably the most comprehensive, informative, and reliable website on global environmental issues; published by the UN Environmental Program (UNEP). Numerous maps and graphics.

United Nations Systems Wide Earthwatch

earthwatch.unep.net/

Earthwatch is a broad UN initiative aimed at co-ordinating, harmonizing, and encouraging environmental observation activities among all UN agencies. Covers all major issues and world regions.

The Canadian Environment

www.ec.gc.ca/

Environment Canada offers a comprehensive overview of the Canadian environment and of environmental issues.

HUMAN IMPACTS

Climate Change

www.epa.gov/climatechange/
ec.europa.eu/environment/climat/
home_en.htm

Both the US Environmental Protection Agency and the European Union provide detailed and factually sound climate change information that is easily understood.

Biodiversity

www.arkive.org/

This website, launched in 2003, is intended to serve as a 'digital ark' for the planet's flora and fauna, bringing together pictures, sounds, and information.

Eden Project

www.edenproject.com/

Owned by the Eden Trust, an educational charity, the Eden Project located in southwest England reflects global diversity. The focus is on educational projects.

Deforestation

www.wrm.org.uy/deforestation/index.html

The World Rainforest Movement website includes details on deforestation around the world.

Desertification

www.unccd.int/

This UN website provides news and details about desertification with a focus on prevention.

Pollution

www.pollutionprobe.org/

Pollution Probe is a Canadian charitable environmental organization concerned with disseminating information about and proposing solutions to pollution issues.

Earthlights

antwrp.gsfc.nasa.gov/apod/image/0011/
earthlights_dmsp_big.jpg

Striking view of the world at night.

TOWARDS A SUSTAINABLE WORLD

Vision for a Sustainable World

www.worldwatch.org/

The Worldwatch Institute offers a regularly updated survey of human impacts on environment with the fundamental goal of facilitating the creation of an environmentally sustainable society.

Investing in Our Planet

www.gefweb.org/default.aspx

Established in 1991 as a World Bank initiative but now a separate facility, the Global Environment Facility is a partnership of 178 countries addressing environmental issues and supporting national sustainable development.

UN Division for Sustainable Development

www.un.org/esa/dsd/index.shtml

This UN agency is tasked with providing leadership on sustainable development initiatives.

Holding Corporations Accountable

www.corpwatch.org/

A site established to be a self-proclaimed 'watchdog on the web' that provides critical evaluations of the environmental impacts that may result from the activities of some leading corporations.

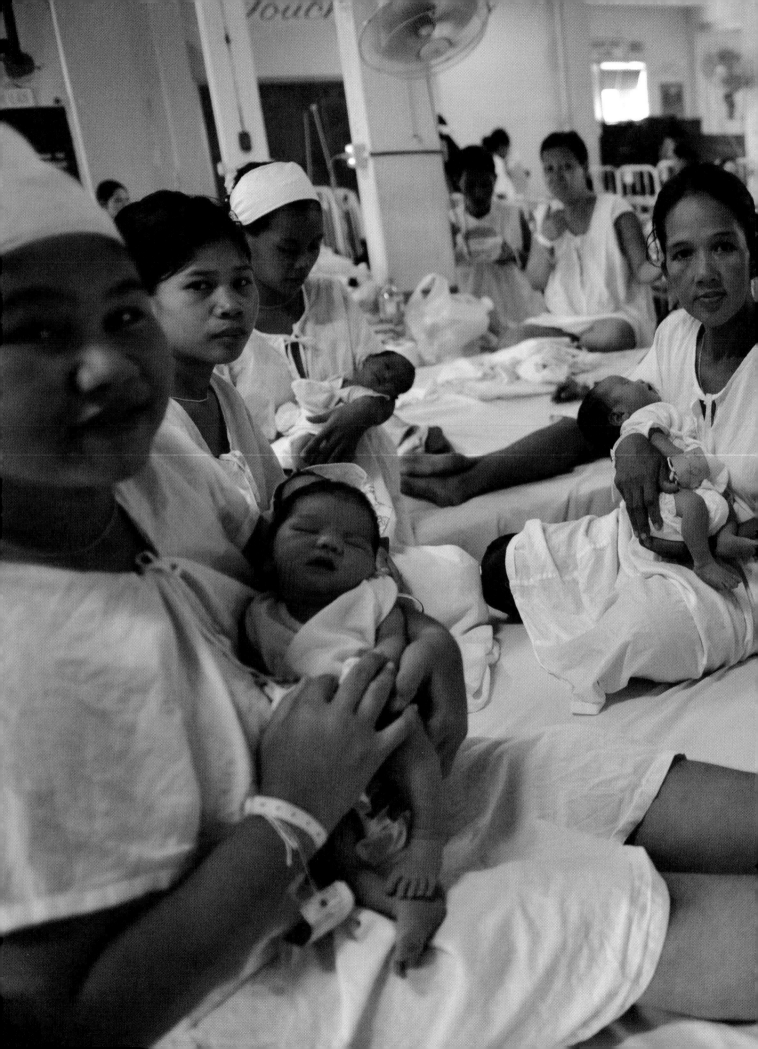

A CROWDED HOME?

Together, Chapters 4 and 5 present an overview of the subdiscipline of population geography. This chapter focuses on the fundamentals of human population growth through time, at both global and regional scales. The two factors on which all changes in population size depend are fertility and mortality. This chapter provides a brief introduction to the various measures used to track fertility and mortality rates; some of the factors that influence those rates; and the temporal and spatial variations that occur in both cases. We look at recent evidence indicating that fertility is declining not only in the more developed parts of the world but in the less developed parts as well—even though populations there are still growing much more rapidly than those in the more developed world.

This fundamental difference between the two worlds is a direct reflection of the many economic, political, social, and other human inequalities that will be discussed in detail in Chapter 5. We conclude this chapter with a look at six theories that have been proposed to explain the history of population growth and to predict possible scenarios for the future.

Mothers holding their newly-born babies in Manilla. The Philippine population now stands at around 90 million, with an annual growth rate of 2.04 percent, one of the highest in Asia and above the government's target of 1.9 percent.

Ted Aljibe/AFP/Getty Images

At the time of writing in 2009, the world population is estimated at 6.8 billion. By the time you read these words, that figure will be higher; the United Nations estimates 8.0 billion by the year 2025, 9.4 billion by 2050, and a stable population of about 10 billion by 2200. More significant even than this numerical increase, however, is the fact that the increases are occurring in less developed parts of the world that are currently least capable of supporting increased numbers, whereas population in the more developed world is expected to remain essentially unchanged through to 2050. Hence an ever-present theme in this chapter and the one following is the uncertainty that such increases in the less developed world imply for human well-being. A number of human geographers and other scholars have suggested links between population numbers and such problems as famine, disease, and, as discussed in the previous chapter, environmental deterioration. Others, however, are more optimistic about our ability to cope with increasing numbers, arguing that past increases in population have been accompanied by improvements in human well-being through technological change.

Much of this chapter draws on measures and procedures developed in **demography**: the science that studies the size and makeup of populations (according to such variables as age and sex), the processes that influence the composition of populations (notably fertility and mortality), and the links between populations and the larger human environments of which they are a part.

Fertility

At the global level, all changes in population size can be understood by reference to two factors: fertility and mortality. Thus:

$$P_1 = P_0 + B - D$$

where

P_1 = population at time 1
P_0 = population at time 0 (before time 1)
B = number of births between times 0 and 1
D = number of deaths between times 0 and 1

Fertility and mortality rates vary significantly according to time and location, and both rates are affected by many different variables.

At the sub-global level a third factor becomes relevant: migration into and out of that particular subdivision of the world. Thus:

$$P_1 = P_0 + B - D + I - E$$

where

I = number of immigrants to area between times 0 and 1
E = number of emigrants from area between times 0 and 1

The effects of migration are discussed in Chapter 5.

FERTILITY MEASURES

The simplest and most common measure of **fertility** is the *crude birth rate* (CBR): the total number of live births in a given period (usually one year) for every 1,000 people already living. Thus:

$$\text{CBR} = \frac{\text{number of live births in one year}}{\text{mid-year total population}} \times 1,000$$

The CBR for the world in 2008 was 21. Historically, measures of CBR have typically ranged from a minimum of about 10 (the theoretical minimum is, of course, 0) to a maximum of 55, which is a rough estimate of the biological maximum. Clearly, the CBR is a very useful statistic, but it may be somewhat misleading because births are related to the total population and not to that subset of the population that is able to conceive (it is called 'crude' for this reason). To more accurately reflect underlying fertility patterns by reference to the biological concept of female **fecundity**—the ability of a woman to conceive—two other, more sophisticated measures are used: general fertility rate and total fertility rate. Both relate births as directly as possible to that segment of the population responsible for them, excluding those unable to conceive, namely males, and those unlikely to conceive, namely very young girls and women over the age of 50.

First, the *general fertility rate* (GFR) refers to the actual number of live births per 1,000 women in the fecund age range: those years in which a woman has the ability to conceive, typically defined as ages 15 to 49 (sometimes 15 to 44). It is calculated as follows:

fertility
Generally, all aspects of human reproduction that lead to live births; also used specifically to refer to the actual number of live births produced by a woman.

demography
The study of human populations.

fecundity
A biological term; the ability of a woman or man to produce a live child; refers to potential rather than actual number of live births.

FIGURE 4.1 **World distribution of crude birth rates, 2008**

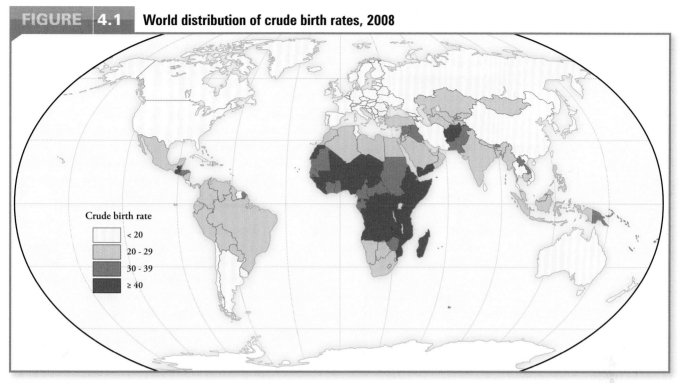

SOURCE: Population Reference Bureau, *2008 World Population Data Sheet* (Washington: Population Reference Bureau, 2008).

$$GFR \quad \frac{\text{number of live births in}}{\text{mid-year number of females}} \times 1,000$$
$$\text{aged 15–49 years}$$

Second, and more useful, the *total fertility rate* (TFR) is the average number of children a woman will have, assuming she has children at the prevailing age-specific rates, as she passes through the fecund years. This is an age-specific measure of fertility that is useful because child-bearing during the fecund years varies considerably with age. It is calculated as follows:

$$TFR = 5 \sum_{A=1}^{7} \frac{\text{number of births to women in age group A in a given period}}{\text{mid-year number of females in age group A}}$$

where 'A' refers to the seven five-year age groups of 15–19, 20–4, 25–9, 30–4, 35–9, 40–4, and 45–9. The 5 preceding the summation sign is necessary because each age group covers five years. The world TFR for 2008 was 2.6, meaning that the average woman has 2.6 children during her fecund years.

Although there are various other measures of fertility, with different measures used in different circumstances and with different aims in mind, we will concentrate on two measures, the CBR and the TFR. The CBR is a factual measure reflecting what has actually happened in a given time period—x number of children per 1,000 members of the population were born. The TFR, on the other hand, reflects an assumption that as a woman reaches a particular age, she will have the same number of children as do the women of that age today; hence it is a predictive measure.

The impact of a particular CBR or TFR on population totals is related to mortality. Generally, a TFR of between 2.1 and 2.5 is considered **replacement-level fertility**; that is, it maintains a stable population. The lower figure, 2.1, applies to areas with relatively low levels of mortality and the higher figure, 2.5, to areas with relatively high levels of mortality.

FACTORS AFFECTING FERTILITY

Fertility, the reproductive behaviour of a population, is affected by biological, economic, and cultural factors. It is not difficult to identify many of these factors; it is difficult, however, to assess their specific effects. But perhaps the most surprising fact to emerge as the discussion of population unfolds throughout this chapter is that human beings might

replacement-level fertility
The level of fertility at which a couple has only enough children to replace themselves.

be the only species that reproduces less when they are well fed.

Biological factors

Age is the key biological factor in fecundity. Fecundity begins at about age 15 for females, reaches a maximum in the late twenties, and terminates in the late forties. The pattern for males is less clear, with fecundity commencing at about age 15, peaking at about age 20, and then declining, but without a clear termination age. Some females and males are sterile—incapable of reproduction; data are not readily available, but it is probably not unrealistic to say that in 10 per cent of all married couples in the more developed world there is one sterile partner.

Reproductive behaviour also is affected by nutritional well-being; populations in ill health are very likely to have impaired fertility. Thus, periods of famine reduce population growth by lessening fertility (as well as by increasing mortality). Box 4.1 reviews the situation in contemporary tropical Africa, a region with generally high birth rates but some particular areas of low fertility.

A third biological factor affecting fecundity also is related to diet. Women with low levels of body fat tend to be less fecund than others. Members of nomadic pastoral societies are particularly likely to have low levels of body fat because they eat a low-starch diet, and they are characterized by low fertility.

Economic factors

Another group of variables affecting fertility are principally economic. Indeed, until recently, fertility changes in more developed countries were essentially one part of the process of economic development. With increasing industrialization and urbanization, fertility declines. This economic argument suggests

Box 4.1   Fertility in Tropical Africa

Fertility is high throughout tropical Africa—in several countries the total fertility rate (TFR) is between 6 and 7—but the national statistics mask some remarkable variations within countries. For some areas, TFRs are as low as 2 to 5. Two geographers working at Syracuse University identified a belt of low rates from southwestern Sudan, through the Central African Republic, and into the Democratic Republic of Congo. Other low-fertility areas include parts of Cameroon and Gabon, the Lake Victoria area, isolated locations in the savanna zone of West Africa, parts of Namibia and Botswana inhabited by San peoples, parts of Ethiopia inhabited by nomadic pastoralists, and parts of the East African coast (Doenges and Newman, 1989). In each of these cases, there appears to be a close correlation between a low TFR and a particular ethnic group.

Various explanations have been offered for the fact that fertility rates in some areas are impaired (abnormally low for the region). Four will be noted here. First, there are cultural variations in the length of time a baby is breast-fed; breast-feeding limits ovulation and often extends the period of infertility following birth from about two months to 18 months. The second cause of localized areas of low fertility is the impact of diseases such as gonorrhea and syphilis, which can cause sterility and also tend to reduce the likelihood of sexual intercourse. Poor nutrition is the third cause: fertility declines following famine and is consistently low in areas experiencing chronic undernutrition. The fourth cause relates to marriage. Although almost all women marry, and the age at first marriage is usually in the mid-teens, there are exceptions.

For example, among the Rendille of northern Kenya, cultural practices result in one-third of the women not marrying until their mid-thirties; and among nomadic pastoralists, periods of prolonged spousal separation are not uncommon.

Doenges and Newman (1989: 111) concluded their study of tropical African fertility as follows: 'If in the future impaired fertility ceases to be a significant concern for African populations, deliberate birth control will have a better chance of becoming an accepted social norm.' In this respect, impaired fertility acts very much like high infant mortality: both are important limitations on social well-being that require resolution before the state of regulated fertility can be reached.

A Rendille family.

Thomas Cockrem/ Alamy/GetStock

that traditional societies, in which the family is a total production and consumption unit, are strongly *pro-natalist*: that is, they favour large families. By contrast, modern societies emphasize small families and individual independence; in economic terms, whereas children once were valued for their contributions to the household, today they represent an expense. The economic argument is that the decision to have children is essentially a cost-benefit decision. In the traditional (often, extended-family) setting, children are valuable both as productive agents and as sources of security for their parents in old age—hence large families. Neither of these factors is important in modern societies.

This economic argument, which implies that any reductions in fertility are essentially caused by economic changes, is central to the demographic transition theory—one of the six explanations of population growth through time to be discussed later in this chapter.

Cultural factors

A host of complex and interrelated cultural factors also affect fertility; in fact it is now often suggested that the reasons behind current reductions in fertility are primarily cultural rather than economic. This suggestion is central to the fertility transition theory, another of the six models of population growth through time to be discussed later.

Most cultural groups recognize *marriage* as the most appropriate setting for reproduction. There are several measures of **nuptiality**, the simplest of which is the nuptiality rate:

$$\text{nuptiality rate} = \frac{\text{number of marriages in one year}}{\text{mid-year total population}} \times 1,000$$

The age at which females marry is important because it may reduce the number of effective fecund years. A female marrying at age 25, for example, has 'lost' 10 fecund years. Until recently, perhaps the most obvious instance of such loss was Ireland. In the 1940s, the average age at marriage for Irish women was 28 and for Irish men 33; in addition, high proportions of the population lost all their fecund years as a result of not marrying—as many as 32 per cent of women and 34 per cent of men. Late marriage and non-marriage are usually explained by reference to social organization and economic aspirations. Interestingly, since about

1971 the Irish people have tended to marry earlier and more universally. Other cultures actively encourage early marriage. Some Latin American countries, such as Panama, have legal marriage ages as low as 12 for females and 14 for males. Overall, delayed marriage and celibacy are uncommon throughout Asia, Africa, and Latin America. Some of the recent success in reducing fertility in China can be explained by the government requirement that marriage be delayed until age 25 for females and 28 for males.

Another cultural factor that affects fertility is *contraceptive use*. Attempts to reduce fertility within marriage have a long history. The civilizations of Egypt, Greece, and Rome all used contraceptive techniques. Today contraceptives are used around the world, but especially in more developed countries. The less developed countries typically have lower rates of contraceptive use, although most evidence shows that the rates are increasing. Data for 2008 contraceptive use are shown in Table 4.1. Two statistics are included for each area: the percentage of married women practising contraception and the percentage of married women using modern contraceptive methods such as the pill, IUD, and sterilization. It is noteworthy that for many countries, and for many of the larger regions of the world, data are not available. The practice of contraception is closely related to government attitudes and to religion. Today most of the world's people live in countries that actively encourage limits to fertility; yet as recently as 1960 only India and Pakistan had active programs to reduce fertility. The complex issue of government policies designed to impact on fertility is discussed later in this chapter.

Another important cultural factor affecting fertility is *abortion*: the deliberate termination of an unwanted pregnancy. It is estimated that for every three births in the world, at least one

nuptiality
The extent to which a population marries.

Table 4.1	Contraceptive Use by Region, 2008	
	% Married Women Using Contraception	
	All Methods	**Modern Methods[1]**
World	62	55
More developed world	69	58
Less developed world	61	55
Less developed world (excluding China)	51	43

[1]'Modern' methods include the pill, IUD, condom, and sterilization.

SOURCE: Population Reference Bureau, *2008 World Population Data Sheet* (Washington: Population Reference Bureau, 2008).

A family-planning fieldworker talks about contraceptives with village women in Bangladesh.
© CIDA photo: Roland Pirker

pregnancy is deliberately terminated. Abortion is an even more complex moral question than contraception. Although it is a long-standing practice, it is still subject to widespread condemnation, usually on religious and moral grounds. Only about half of the estimated 42 million abortions performed annually are legal.

In many countries, governments actively try either to promote abortion or to discourage it, for their own pragmatic reasons. China has a liberal abortion law that reflects the need for population control; indeed, abortion is obligatory for women whose pregnancies violate government population policies. Sweden has a liberal abortion law on the grounds that abortion is a human right. In other countries, however, especially those where the dominant faith is either Catholicism or Islam, access to abortion is either denied or limited to instances in which the woman's life is threatened—about 40 per cent of women live in countries with very restrictive abortion laws. Romania has the highest rate of abortions with about 75 per cent of pregnancies terminated, while in Russia about 66 per cent are terminated. In some other countries, such as Canada, Australia, the United Kingdom, and the United States, abortion is one of the most hotly debated biological, ethical, social, and political issues. A 2005 UN-sponsored conference on equality for women highlighted the abortion debate, and it is notable that UN documents do not refer to abortion as a right because of the opposition of some member countries.

As noted, attitudes towards contraception and abortion often reflect religious beliefs, and it is frequently assumed that religion is a major influence on fertility in general. For example, accounts of fertility in Pakistan and India often note how much higher the fertility rate is in Pakistan (4.1 in 2008) than in India (2.8 in 2008) and attribute the disparity to religious differences between Hindus and Muslims. But this simple cause-and-effect relationship is clearly flawed, as a more detailed regional analysis shows. For example, the Indian state most successful in reducing fertility—Kerala—is second only to Kashmir in the proportion of its residents who are Muslim. Indeed, throughout India, fertility rates among Muslims usually are closer to those of neighbouring Hindu groups than they are to those of Muslims in Pakistan. Similarly, the fertility rate of 2.7 in Muslim Bangladesh is similar to the Indian rate. So what does explain these differences in fertility rates? As Box 4.2 suggests, and as the account of the fertility transition later in this chapter makes clear, regional differences in fertility reflect differences not in religion but rather in the empowerment of women and hence in gender inequality.

VARIATIONS IN FERTILITY

Spatial variations in fertility today correspond closely to spatial variations in levels of economic development, a fact that offers some support for the importance of the economic factors outlined earlier. It appears that, especially since the onset of the Industrial Revolution in the mid-eighteenth century, modernization and economic development have prompted lower levels of fertility. Thus, in 2008, at the broad regional level, the more developed world had a CBR of 12 and a TFR of 1.6, while the less developed world had a CBR of 23 and a TFR of 2.8. More specifically, in the more developed world, Italy was among the group of countries with low fertility, with a CBR of 9 and a TFR of 1.3, while in the less developed world, Liberia had a CBR of 50 and a TFR of 6.8. Figure 4.1 maps CBR by country. In general, variations of this type reflect economic conditions.

Current evidence, however, indicates that although fertility rates in the less developed world still are high, they are decreasing, and that, as already suggested, the reasons are more cultural than economic. As recently as 1990 the less developed world had a CBR of 31 and a TFR of 4.0—considerably higher than the

Box 4.2 Declining Fertility in the Less Developed World

The less developed world is currently experiencing a decline in fertility that began in the 1970s and appears to have little to do with economic factors and everything to do with cultural factors (Robey et al., 1993). Birth rates are falling rapidly without any prior improvement in economic or living conditions.

The magnitude of this decline is evident in Thailand, which had a 1975 TFR of 4.6 and a 2008 TFR of 1.6; similar reductions have occurred in many other countries in Asia and Latin America, including Indonesia, Turkey, Colombia, and Morocco. The fact that several sub-Saharan African countries—Kenya, Botswana, Zimbabwe, and Nigeria—also are experiencing reduced fertility is especially significant, as this area has always appeared to be resistant to any such trend.

The explanation for these declines appears to be related to the availability and acceptance of new contraceptive technologies, the success of family-planning programs, and the educational power of the mass media. Traditional methods of contraception, such as periodic abstinence and withdrawal, are being replaced by modern methods. Female sterilization is the favoured method of contraception in Asia and Latin America, but is less popular in Africa and the Middle East.

Although this decline in fertility in the less developed world apparently began in the 1970s, the reasons behind it were not apparent until recently because data on such matters were not readily available. The data now available—derived largely from 44 surveys of more than 300,000 women carried out during the 1990s—are substantial and well-documented.

These fertility declines are in addition to the case of China, where government policies, often compulsory, have resulted in a below-replacement fertility level of 1.6 only about 30 years after the introduction of the policies.

One detailed analysis of Botswana (VanderPost, 1992) showed that, geographically, the fertility decline there is proceeding in an uneven manner; as is so often the case, national statistics disguise local details. Indeed, although in some areas of Botswana fertility is decreasing, in other areas it is actually increasing. The difference is essentially urban/rural, with lower TFRs in urban areas.

The importance of fertility declines in the less developed world is difficult to exaggerate, both practically and in terms of our understanding of fertility change. Because family planning is now practised throughout much of the less developed world (typically it appears without any substantial prior improvement in economic circumstances), many earlier accounts of anticipated population growth are being revised. The conventional wisdom in the 1970s was well expressed by Demeny (1974: 105): countries in the less developed world 'will continue their rapid growth for the rest of the century. Control will eventually come through development or catastrophe.' It now appears that rapid growth is slowing down, but not for either of these two reasons: instead, cultural changes mean that people are becoming more willing to employ modern contraceptive methods in order to have smaller families. Fertility decline in the less developed world is discussed in the section on the fertility transition later in this chapter.

2008 rates of 2.3 and 2.8. There is evidence, also, of declining fertility in several countries in the more developed world that may be prompted by cultural factors (Box 4.3).

As suggested in the case of Botswana noted above, fertility also varies significantly within any given country. There is, for example, usually a clear distinction between urban areas with relatively low fertility and rural areas with relatively high fertility. This distinction can be found in all countries, regardless of development levels. Similarly, fertility within a country is higher for those with low incomes and for those with limited education.

Mortality

MORTALITY MEASURES

Like fertility, **mortality** may be measured in a variety of ways. The simplest, which is equivalent to the CBR, is the *crude death rate* (CDR):

the total number of deaths in a given period (usually one year) for every 1,000 people living. Thus:

$$CDR = \frac{\text{number of deaths in one year}}{\text{mid-year total population}} \times 1{,}000$$

The CDR for the world in 2008 was 8. Measures of CDR have typically ranged from a minimum of 5 to a maximum of 50.

The CDR does not take into account the fact that the probability of dying is closely related to age (it is called 'crude' for this reason). Usually, death rates are highest for the very young and the very old, producing a characteristic J-shaped curve (Figure 4.2). Other mortality measures take into account the age structure of the population. The most useful of these for our purposes is the *infant mortality rate* (IMR): the number of deaths of infants under 1 year old per 1,000 live births in a given year. Thus:

mortality
Deaths as a component of population change.

Box 4.3   Declining Fertility in the More Developed World

'The last European will die on 6 August 2960' (Nelson, 2006: 24). This prediction assumes that current demographic trends continue and is the outcome of a European world where sex is separated from child-bearing and people are too busy to have large families. Indeed, the lowest fertility rates in the world today are found in several Central and Eastern European countries; the TFR is as low as 1.2 in Slovakia, Bosnia-Herzegovina, and Slovenia. The extraordinarily low figures for much of Europe reflect both a desire to postpone starting families and a desire for smaller families. For many Europeans, the ideal family now includes only one child; a 1989 survey in France showed that 19 per cent favoured a one-child family, compared to only 3 per cent in 1979. Such preferences may be prompted by uncertain economic circumstances and the growth of employment opportunities for women (Hall, 1993; King, 1993).

The figure for Germany, 1.3, reflects an unprecedented trend in the former East Germany, which appears to have come as close to a temporary suspension of child-bearing as any large population in the human experience. Eastern Germans have virtually stopped having children. The explanation is not an increase in abortions, which have also fallen abruptly. One possible explanation is the trauma associated with the transition from communism to capitalism and, specifically, concern about employment opportunities for future generations in a region with high levels of unemployment. It may be appropriate, then, to interpret the low fertility rate in Germany—actually a form of demographic disorder—as one temporary outcome of the transition from the 'old' to the 'new' political order. This suggestion appears to have some merit, as fertility also declined between 1989 and 1993 by 20 per cent in Poland, 25 per cent in Bulgaria, 30 per cent in Romania and Estonia, and 35 per cent in Russia (*The Economist*, 1993: 54).

Whether or not the low fertility evident in many European countries will continue at or close to the current low values remains to be seen. It may be that if economies boom and employment opportunities increase, then fertility will rise.

As of 2008 there was little indication that such a rise was occurring, and the current rates typically imply net losses in population, since the replacement level in such countries is a TFR of about 2.1.

It is often argued that population decline goes hand in hand with economic decline (although the precise cause-and-effect relationship is debated). Indeed, in Russia, there is much talk about a population and economic catastrophe as fertility declines. In response to concerns about economic decline, several European countries are actively looking at ways to increase population, for example, through improved child-care programs. Both Germany and Poland have placed fertility high on the political agenda; Sweden and Norway in particular have a range of policies intended to help couples to want to have children, notably generous maternity and paternity leave arrangements, heavily subsidized daycare, and flexible work schedules; and France has the most extensive state-funded child-care system in Europe. Several Asian countries are also concerned: both Japan and Hong Kong are actively promoting fertility.

Although Canada does not belong to the group of lowest-fertility countries, both CBR and TFR statistics have been declining or stable in recent years. Most European countries have slightly lower birth and fertility rates than Canada, but in the United States they are slightly higher (CBR is 14 and TFR is 2.1). Table 4.2 summarizes the current demographic situation in Canada.

Table 4.2	Population Data, Canada, 2008
Total population	**33.3 million**
CBR (crude birth rate)	11
CDR (crude death rate)	7
RNI (rate of natural increase)	0.3%
IMR (infant mortality rate)	5.4
TFR (total fertility rate)	1.6

SOURCE: Population Reference Bureau, *2008 World Population Data Sheet* (Washington: Population Reference Bureau, 2008).

$$IMR \frac{\text{number of infant deaths under one year old}}{\text{number of births in that year}} \times 1{,}000$$

The world IMR for 2008 was 49, although figures for individual countries ranged from as low as 1.3 for Iceland to as high as 163 for Afghanistan. The IMR is sensitive to cultural and economic conditions, declining with improved medical and health services and better nutrition; reductions in the IMR usually precede overall mortality decline. In 2007 UNICEF reported that, globally, fewer children under the age of five were dying, with a drop from 13 million in 1990 to fewer than 10 million in 2006. The principal reasons for this fall were measles vaccinations, mosquito nets, and increased rates of breast-feeding.

Although not a mortality measure, another useful statistic that reflects mortality is *life expectancy* (LE), the average number of years to be lived from birth. World life expectancy in 2008 was 68, although for Japan and some European countries the LE was in the low 80s, while for several African countries it was in the 40s.

FIGURE 4.2 Death rates and age

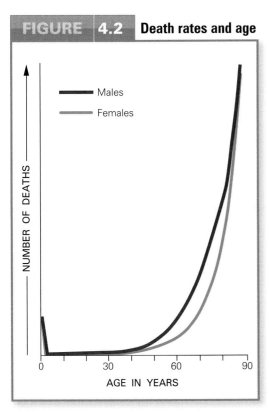

Generalized J-shaped curve of death rates and age.

FACTORS AFFECTING MORTALITY

Humans are mortal. Whereas it is possible in principle for the human population to attain a CBR of 0 for an extended period, the same cannot be said for the CDR. Unlike fertility, which is affected by many biological, cultural, and economic variables, mortality is relatively easy to explain—despite the fact that the World Health Organization has recognized some 850 specific causes of death! As the CDR and LE figures cited above suggest, mortality generally reflects socio-economic status: high LE statistics are associated with high-quality living and working conditions, good nutrition, good sanitation, and widely available medical services—and vice versa.

VARIATIONS IN MORTALITY

The study of death statistics tells a great deal about how people lived their lives because many people die not as a result of aging but rather as a result of environmental conditions and/or lifestyle. The world pattern for CDR shows much less variation than does that for CBR (Figure 4.3). This is a reflection of the

FIGURE 4.3 World distribution of crude death rates, 2008

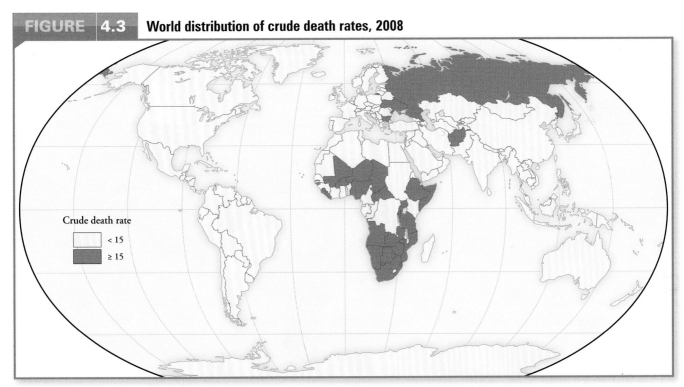

As this map shows, the world pattern of death rates varies relatively little. Most countries have rates of 15 or less with the striking exception in most sub-Saharan African countries of high rates. Rates of 15 or above also prevail in Afghanistan (21), Russia (15), and Ukraine (16). Generally speaking, rates lower than 15 are acceptable. It is notable that in some countries rates are increasing because of the increasing percentage of elderly people in a national population; the greater the proportion of elderly people, the higher the death rate. For example, both Norway and the United Kingdom have a rate of 9, whereas Malaysia and the Philippines have a rate of 5: these different statistics do not reflect differences in health and the quality of life, but rather different age structures—all four countries have 'low' death rates.

SOURCE: Population Reference Bureau, *2008 World Population Data Sheet* (Washington: Population Reference Bureau, 2008).

general availability of at least minimal health-care facilities throughout the world. Figure 4.4 maps LE by country. In this case, major variations are evident: as already noted, LE figures are more sensitive to availability of food and health-care facilities than are death rates. The map of LE is a fairly good approximation of the health status of populations. Low LE statistics are found in tropical countries in Africa and in South and Southeast Asia. High LE statistics are particularly common in Europe and North America.

Mortality measures also vary markedly within countries. In countries such as the United Kingdom, Canada, the United States, and Australia, certain groups have higher CDRs and IMRs and lower LEs than the population as a whole. These differences reflect what we might call the social inequality of death. Two examples suffice. In the United Kingdom, 2007 data suggest that, while the national average is 79 years, the poorest men in Glasgow have a life expectancy of 54 years. In the US, the national IMR in 1993 was 8.6, but Washington, DC, in which blacks comprise a high percentage of the population, reported a figure of 21.1, while for both South Dakota (with a large Native American population) and Alabama (with a large black population) the figure was 13.3. Black babies in the US are almost twice as likely to die in their first year as white babies—a state of affairs that has remained essentially unchanged for the last 50 years.

Box 4.4 provides two examples of LE and IMR data that clearly reflect environmental and health problems at least partly attributable to inefficient political systems.

TWENTY-FIRST-CENTURY NIGHTMARE?

Although mortality rates have declined significantly in the less developed world since about 1900, they remain painfully high in some countries. One reason for high mortality is the all-too-frequent famine circumstances discussed more fully in the account of the less developed world in the following chapter. The intent here is to discuss the increasing mortality caused by AIDS (acquired immune deficiency syndrome). AIDS itself was first identified in 1981, and HIV (the human immunodeficiency virus) was recognized as its cause in 1984. The immune system of an individual infected by HIV weakens over time, leaving the body less and less able to combat infection. The majority of HIV-infected people develop AIDS within a few years. Today, in the more developed areas of the world, drugs make AIDS relatively manageable, but the worst-affected regions are in

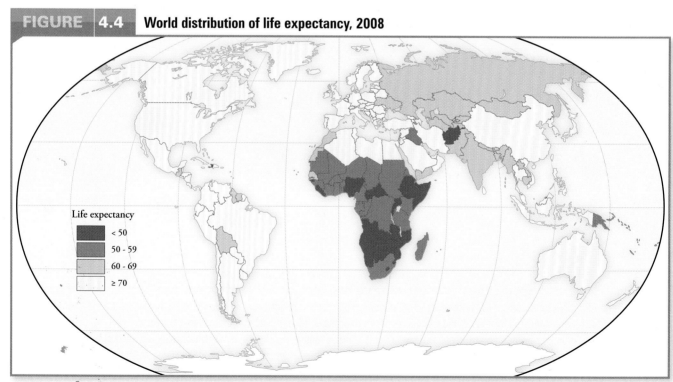

FIGURE 4.4 **World distribution of life expectancy, 2008**

Life expectancy
- < 50
- 50 - 59
- 60 - 69
- ≥ 70

SOURCE: Population Reference Bureau, *2008 World Population Data Sheet* (Washington: Population Reference Bureau, 2008).

Box 4.4 Problems in Central Asia and Russia

Measures of infant mortality and life expectancy are especially useful indicators of quality of life. In both cases, statistics reflect living conditions with respect to the quality of the diet available, the health-care system, public sanitation, and disease control. Two examples that indicate serious problems are outlined here, both in the former USSR.

In one area on the shores of the Aral Sea in Central Asia, the IMR is estimated to be as high as 111 (see Box 3.9). This high rate reflects a number of basic health problems—a contaminated water supply, an abominable sewage system, and unsanitary hospitals. These problems are exacerbated in the summer as a consequence of increased temperatures and related disease problems. These basic health problems, in turn, reflect a host of social problems resulting from years of neglect by political leaders.

A remarkable example of an unexpected decline in LE is that of Russia. In 1987, the LE was 67 for Russian males and 74 for Russian females, but since then there have been significant decreases. The 2008 data show that LE has fallen to 60 for males and, less dramatically, to 73 for females. These figures are below those of all major industrialized countries; comparable Canadian data for 2008 are 78 for males and 83 for females. The explanation appears to involve a combination of issues. The transition from communism to capitalism may have created a new class of wealthy entrepreneurs, but it has also resulted in economic problems for other Russians. For many, diet is worsening because of high prices for meats, fruits, and vegetables; the industrial policies of the previous Communist regime, which led to pollution of the air and drinking water, contributed to the high IMR noted earlier and are undoubtedly affecting the health of many Russians; there is an epidemic of alcohol abuse; and the high rate of abortions may have left many females unable to bear children.

---

less developed areas—notably sub-Saharan Africa—where drugs are largely unavailable, and it is not unusual for a person developing AIDS to die just six months after infection.

AIDS is the greatest threat to human health in the world, not because of the number of people dying from AIDS but because the virus is able to spread through a population with astonishing rapidity. Consider, for example, that in 1990 the affected population in South Africa was 1 per cent, but by 2006 it was about 22 per cent. As of 2008, it was estimated that some 33 million people globally were infected with HIV. More than 20 million have died of AIDS since the early 1980s. Although the overall picture remains grim, a 2008 UN report suggested that the incidence of new infections peaked in the late 1990s and that the global percentage of people living with AIDS has stabilized—obviously at an unacceptably high level—since about 2000.

Given these numbers, and the fact that AIDS has now spread to all parts of the world, the disease is appropriately described as a **pandemic**. Although it is especially prevalent in sub-Saharan Africa and among the poorest members of individual populations, AIDS honours no social or geographical boundaries. It is therefore very difficult to predict its spatial spread. Because AIDS is most often transmitted sexually, the most vulnerable population is the 15–49 age group—the very group that ought to be most highly productive, and that is most likely to have dependants, both young and old.

Sub-Saharan Africa

The area of the world most devastated by AIDS is sub-Saharan Africa. In 1991 a UNICEF-sponsored study forecast an AIDS crisis in several African countries, predicting that 5.5 million children on the continent would lose their mothers to the disease in the 1990s. By the end of the 1990s it was clear that this projection was a tragic underestimate; in fact, approximately 10 million African children had lost their mothers to AIDS since 1991.

Since the early 1980s, some national death rates have doubled; rates of infant mortality have increased sharply; and life expectancy has been reduced by as much as 23 years. Table 4.3 provides data on the 10 countries with the highest percentages of adults living with HIV/AIDS. These rates are devastatingly high. In sub-Saharan Africa, HIV/AIDS is most common in two age groups: infants and adults aged between 20 and 40 years. Women are at greater risk of contracting AIDS than men, by a ratio of about 1.5:1 (Daniel, 2000: 47), and often infect their infants, either before birth or after, through breast-feeding. No amount of data can do justice to the human suffering caused by AIDS. Nevertheless, a particularly poignant

pandemic
A term used to designate diseases with very wide distribution (a whole country, or even the world); 'epidemic' diseases have more limited distribution.

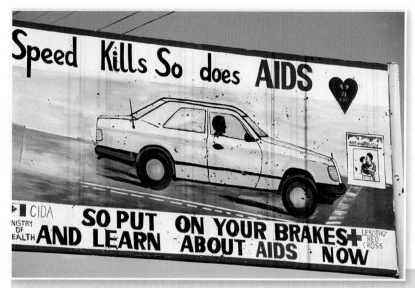

Much of the effort to combat HIV/AIDS focuses on education. Here are two posters from Africa. The one at the right, from Namibia, was produced by the local Catholic church and highlights the links between AIDS and other illnesses. The poster above, from Lesotho, stresses the need for people to behave responsibly and inform themselves.

Das Fotoarchiv

estimate is that a 15–year-old boy in Botswana today has an 80 per cent chance of dying of AIDS. Writing about one clinic in Malawi, Toolis (2000: 29) stated:

> If you have no money, then there is no choice; you must queue in the morning heat of the Boma's outpatients clinic and hope to be admitted. Demand outstrips supply. There are 130 beds, and every morning another 250 would-be patients appear at the gates. All but the sickest are turned away, as well as the medically hopeless, who are sent back home to

die. There is not enough staff, not enough drugs, not enough needles, not enough anything.

It seems clear that poverty is the biggest single factor contributing to the spread of AIDS. Poverty means that infections remain untreated, greatly increasing the risk of further transmission. Poverty also keeps children out of school, increasing the likelihood that the sale of sex will be a primary means of supporting themselves. But of course AIDS only serves to exacerbate poverty. The virus is ravaging African agricultural productivity; in parts of Kenya, for example, the amount of cultivated land has declined by 68 per cent, while in Rwanda there has been a 60–80 per cent decrease in farm labour. Yet in many countries the disease is still seen as so shameful that it cannot even be named; in Mozambique, for example, death certificates for AIDS patients often bear the words 'cause unknown'. As a result of these attitudes, the frank discussion needed to raise public awareness is still lacking in many cases.

Russia, China, and India

Most accounts of AIDS today focus on sub-Saharan Africa, but the disease may soon have major impacts elsewhere. According to Eberstadt (2002: 24), 'it is quite possible that the center of the global HIV/AIDS crisis, in

TABLE 4.3	Countries with Highest Levels of Adult (age 15–49) HIV/AIDS Prevalence, 2008
Country	Adults with HIV/AIDS as % of Total Population
Swaziland	26.1
Botswana	23.9
Lesotho	23.2
South Africa	18.1
Zimbabwe	15.3
Namibia	15.3
Zambia	15.2
Mozambique	12.5
Malawi	11.9

SOURCE: Population Reference Bureau, *2008 World Population Data Sheet* (Washington: Population Reference Bureau, 2008).

terms of absolute numbers, will shift from Africa to Eurasia over the coming generation.' According to a 2008 UN report, in most regions outside sub-Saharan Africa those most affected are injecting drug users, men who have sex with men, and sex workers.

As is so often the case with HIV/AIDS, in Russia, India, and China the data are unreliable. Nevertheless, it appears that the infection rate in Russia may be three times that in the United States, in part because of the social and economic changes that have been underway since the fall of communism. Together, increasing poverty, increasing social freedom, and (especially) increasing illegal drug use are making Russians particularly vulnerable to infection.

In the case of China, a 2002 UN report suggested that the number of people with HIV was between 800,000 and 1.5 million, but 2008 reports placed the number higher, suggesting perhaps 10 million cases by 2010. One secondary cause of AIDS transmission in China is the sale of blood by impoverished peasants to illegal brokers (*The Economist*, 2002: 75).

Finally, India may have about 3 million people infected with HIV. Most of the current epidemic there is concentrated in major cities and is associated with heterosexual behaviour. Because there is little open discussion of AIDS, the risks of infection are not widely recognized.

In all three countries it is possible that the disease soon will spread well beyond impoverished and at-risk groups, largely because governments are unwilling to take the necessary public health measures. Eberstadt (2002: 34–8) painted a dire picture of the spread and growth of AIDS, with an 'intermediate epidemic' in all three countries resulting in 259 million new HIV cases and 155 million deaths by 2025. Such an epidemic would represent a humanitarian crisis of staggering proportions, with significant impacts both on population growth and on national economies. UN data for 2008, however, do not suggest that this rapid increase is occurring.

The future of AIDS

So many factors can influence the course of the battle against AIDS that it is impossible to predict the future with any certainty. However, some positive signs are emerging. Overall, the fact that impacts are being kept at a relatively low level in more developed areas suggests that negative consequences can be contained. Some of the differences between the United States

On the eve of world AIDS Day (1 December) 2002, this student at a government high school in Gauhati, India, pinned a red ribbon—the symbol of AIDS awareness—on a classmate. Along with education, peer support is an important part of the fight against AIDS.
CP/AP photo/Anupam Nath

and Canada may be instructive. The infection rate in Canada is 0.3 per cent, half of that in the US—a significant difference attributed to Canada's having more open sex-education programs, easier access to condoms, better and more widely available needle exchange programs, and free treatment provided by its universal health-care system. Elsewhere, countries such as Thailand, the Philippines, and Brazil have been fighting the disease with much success since the 1990s through prevention programs designed to educate people about such matters as condom use.

Perhaps most notably, there are positive signs in parts of sub-Saharan Africa. Attitudes towards AIDS are changing and governments are actively supporting safer sex practices. The greatest success story appears to be Uganda, where the adult infection rate has dropped from 30 per cent in 1992 to about 5 per cent in 2008. Perhaps surprisingly, this reduction is attributed not to increased use of condoms, but to a government campaign stressing abstinence and fidelity, although reports in 2005 suggested that advocating abstinence is no longer working and that Uganda needs to provide free condoms as the infection rate is rising once again. In many other countries, including South Africa, people are increasingly well educated concerning the causes of AIDS, and there are clear signs that attitudes towards sexual behaviour are changing. In Botswana, some mining companies that employ large numbers of adult males are providing hospital facilities for those already infected and are working to prevent AIDS not only through education but also by

making free condoms readily available. Perhaps the greatest success has been achieved in Brazil, where the strategy involved is the widespread distribution of free condoms.

A controversy erupted in 2009 following remarks by Pope Benedict XVI prior to a visit to Africa. Stating that condoms did not help solve the problem but rather exacerbated the effect of HIV/AIDS, the Pope advocated instead the traditional church teachings of abstinence and fidelity. In a response that was unprecedented in its forcefulness and condemnation, the prestigious medical journal, *Lancet*, wrote that the Pope was publicly distorting well-established scientific facts in order to promote Catholic doctrine.

Another important factor in the world-wide fight against AIDS is the response of the more developed world, and here, too, the signs are increasingly positive. A 2001 UN summit agreed to set up a global fund to fight AIDS, along with tuberculosis and malaria. Despite initial funding problems, there is now reason to believe that the errors of the 1990s, when the more developed world quite literally looked away, are now being corrected. In 2003 the United States committed up to $15 billion over five years to combat AIDS in Africa and the Caribbean, although much of this funding is targeted to programs advocating abstinence. It is far too early to be confident of success, but changing attitudes in both less and more developed countries may well help to limit the spread of AIDS in the coming years.

It is also important to place mortality related to AIDS in context. For example, the annual global death toll from AIDS in the early years of the twenty-first century is estimated at about 3.0 million, while the annual death toll

from diarrhea and tuberculosis combined is about 3.4 million (World Health Organization, 2003). The significance of this circumstance is not only the larger number of deaths, but rather the fact that both diarrhea and tuberculosis are preventable. As discussed more fully in the following chapter, the possible twenty-first-century nightmare does involve much more than AIDS.

Natural Increase

The *rate of natural increase* (RNI) is determined by subtracting the CDR from the CBR; thus it measures the rate (usually annual) of population growth. In 2008 the world CDR was 8 and the CBR was 21, producing an RNI of 12 per 1,000; this figure is typically expressed as a percentage of total population—hence 1.2 per cent (note that it is not 13 per 1,000 because the CBR and CDR are being rounded to whole numbers). This RNI has been relatively constant in recent years. Because RNI data take into account only mortality and fertility, not migration, they generally do not reflect the true population growth of any area smaller than the earth. Not surprisingly, the highest RNI figures are for those countries in the less developed world.

In 2008 the world's population increased by about 82 million people, with 139 million births and 57 million deaths. The size of the annual increase is declining each year—or, to put it another way, *world population is still increasing, but at a decreasing rate*. The reduction in annual increases began about 1990, when the annual increase reached 87 million—and the reason behind it is that world fertility, most easily measured by the TFR is declining. This decrease in the size of the annual increase is very significant.

Although, as suggested in Boxes 4.2 and 4.3, fertility is declining in parts of both the less and the more developed worlds, the number of females of reproductive age continues to rise. Thus the total world population continues to grow rapidly because of **population momentum**.

REGIONAL VARIATIONS

Of course, to use world numbers in this way is to ignore important regional differences. Tables 4.4 and 4.5 show RNI data for two groups of countries, while Figure 4.5 maps RNI data for individual countries.

---

population momentum
The tendency for population growth to continue beyond the time that replacement-level fertility has been reached because of the relatively high number of people in the child-bearing years.

TABLE 4.4	Countries with Highest Rates of Natural Increase (3.0 and above), 2008	
Country	**Natural Increase (%)**	**Population (millions)**
Mali	3.3	12.7
Malawi	3.2	13.6
Niger	3.1	14.7
Uganda	3.1	29.2
DR of Congo	3.1	66.5
Congo	3.1	3.8
Guinea-Bissau	3.1	1.7
Liberia	3.1	3.9

SOURCE: Population Reference Bureau, *2008 World Population Data Sheet* (Washington: Population Reference Bureau, 2008).

RNI data included in Tables 4.4 and 4.5 and mapped in Figure 4.5 provide a useful commentary on the present world population. Low RNI are essentially restricted to Europe, especially some former Communist countries. High RNI are concentrated in tropical African countries (countries that have temporarily inflated RNI statistics because of refugee in-movement are excluded from Table 4.4).

In addition to considering RNI data, it is important to recognize that large populations grow faster than small ones even when they have lower fertility rates—simply because the base population is so much larger. At present, China and India are home to almost 38 per cent of the world's population; hence even a small increase in fertility in one of these countries will mean a significant increase in the total world population. (Such a development is more likely in India, which appears to have ended a period of fertility decline and is again growing rapidly.)

Table 4.6, showing projected data for selected world regions/countries, highlights the fact that much of the growth predicted between 2008 and 2050 will occur in African countries and India. Indeed, of the anticipated world growth of 2,647 million, Africa will contribute 965 million and India 606 million, together accounting for 59 per cent of

the world's total growth. Most of the rest of the growth will be in other high-population Asian countries. More generally, almost 99 per cent of the growth projected by 2050 will take place in the less developed world. These regional projections also have implications for the larger arena of political power and economic growth. For example, the population of Africa is projected to double by 2050, whereas

TABLE 4.5	Countries with Lowest Rates of Natural Increase (less than 0.0), 2008	
Country	Natural Increase (%)	Population (millions)
Ukraine	−0.6	46.2
Bulgaria	−0.5	7.6
Hungary	−0.4	10.0
Serbia	−0.4	7.4
Latvia	−0.4	2.3
Lithuania	−0.4	3.4
Belarus	−0.3	9.7
Russia	−0.3	141.9
Croatia	−0.3	4.4
Germany	−0.2	82.4
Romania	−0.2	21.5
Moldova	−0.1	4.1
Estonia	−0.1	1.3

SOURCE: Population Reference Bureau, *2008 World Population Data Sheet* (Washington: Population Reference Bureau, 2008).

FIGURE 4.5 World distribution of rates of natural increase, 2008

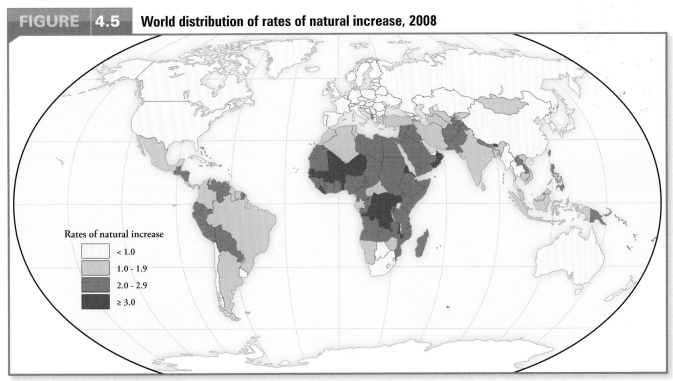

SOURCE: Population Reference Bureau, *2008 World Population Data Sheet* (Washington: Population Reference Bureau, 2008).

that of Europe will decline. Precisely how such changes will play out politically and economically remains to be seen.

DOUBLING TIME

Related to the rate of natural increase is the useful concept of **doubling time**: the number of years needed to double the size of a population, assuming a constant RNI. For example, at the current 1.2 per cent growth rate, the world population will double in approximately 58 years. Relatively minor variations in RNI can significantly affect the doubling time. An RNI of 0 (where CDR and CBR are equal) results in a stable population with an infinite doubling time; on the other hand, an RNI of 3.5 per cent produces a doubling time of a mere 20 years. (Note: 70 divided by the RNI provides a good approximation of doubling time.)

> **doubling time**
> The number of years required for the population of an area to double its present size, given the current rate of population growth.

Government Policies

Thus far we have paid only incidental attention to governments' impact on population. Most governments attempt to influence various aspects of population, such as total numbers and spatial distribution. Governments can also attempt, directly or indirectly, to control deaths, births, and migrations.

All policies to control deaths have the same objective: to reduce mortality. Such policies, usually adopted for both economic and humanitarian reasons, include measures to provide medical care and safe working conditions. Despite the near universality of such policies, many governments actively raise mortality levels at specific times—for instance, in times of war. Further, many governments do not ensure that all members of their population have equal access to the same quality of health care. This is true throughout the world, regardless of levels of economic development.

Many governments have actively sought to influence fertility, and the policies in place may be a major factor affecting fertility at the national scale. However, unlike policies to reduce mortality, those related to fertility have varying objectives. While many governments choose not to establish any formal policies, either because of indifference to the issue or because public opinion is divided, others are either actively *pro-natalist* or actively *anti-natalist*. Of course, government policies are always subject to change.

PRO-NATAL POLICIES

Pro-natalist policies typically are in place in countries dominated by a Catholic or Islamic theology (e.g., Italy, Iran), in countries where the politically dominant ethnic group is in danger of being numerically overtaken by an ethnic minority (e.g., Israel), and in countries where a larger population is perceived as necessary for economic or strategic reasons (see Box 4.5).

Two Asian countries, both of which belong in the Asian group of newly industrialized countries (see Chapter 14), began to actively encourage fertility increases in the 1980s (Dwyer, 1987). Both Singapore and Malaysia had succeeded in lowering their TFRs by following an anti-natalist line, but in 1984 the governments of these countries decided that new policies were required to reverse the trend and increase fertility again. In Singapore this reversal was related to the economic difficulties perceived to be the result of reducing the TFR to 1.6 by 1984; there was a perceived need to provide a larger market for domestic industrial production and a larger workforce. The Malaysian case was similar in that the motivation was economic, although the TFR remained above replacement level in 1984 at 3.9. The extent to which such pro-natalist policies are successful is debatable; the 2008 TFR for Singapore was only 1.4 and for Malaysia 2.6 (both decreases since 1984).

In addition to such direct attempts to increase fertility, many countries indirectly support increased fertility through 'baby bonuses', establishment of daycare centres, and state-supported low-cost, high-quality housing. Russia is greatly concerned about low fertility—in 2006 President Vladimir Putin stated that the situation is critical. In 2008 the

Table 4.6	Projected Population Growth, 2008–2050		
Region or Country	Population, 2008 (millions)	Projected Population, 2050 (millions)	Amount of Projected Increase (millions)
World	6,705	9,352	2,647
Africa	967	1,932	965
Europe	736	685	−51
US and Canada	338	480	142
China	1,325	1,437	112
India	1,149	1,755	606

SOURCE: Calculated from Population Reference Bureau, *2008 World Population Data Sheet* (Washington: Population Reference Bureau, 2008).

Box 4.5 Fertility in Romania, 1966–1989

In the October 1986 issue of the German magazine *Der Spiegel*, the president of Romania was quoted as saying that 'The fetus is the socialist property of the whole society. Giving birth is a patriotic duty, determining the fate of our country. Those who refuse to have children are deserters, escaping the laws of national continuity.'

Alarmed by a crude birth rate of 16 in 1966, the Communist government in Romania ended all legal access to abortion and set a goal of increasing the national population to between 24 and 25 million by 1980 (a 30 per cent increase). Contraceptives were banned, and all Romanian employed women under 45 had to take a monthly gynecological examination. Unmarried people over 15 and married couples without children or a valid medical reason were assessed an additional 30 per cent income tax. The TFR increased from 1.9 in 1966 to 3.6 in 1968, but since that time the rate has steadily dropped, and in 2008 it stood at 1.3. There is reason to believe that many women had illegal abortions despite the establishment of a special unit within the State Security Police whose job it was to combat abortion. Interestingly, all the strategies devised to increase the fertility rate were negative and coercive: they punished people for not having children rather than rewarding them for having children with such positive inducements as maternity leave or family income supplements.

The government's principal reasons for attempting to increase fertility were probably to strengthen national security (more people equals more strength) and improve productivity (by alleviating labour shortages).

In general, it seems reasonable to say that Romania before 1989 had the most stringent fertility policy in the world. Among other previously Communist countries, the former USSR encouraged higher fertility, but at the same time probably had the highest national legal abortion rate in the world (about one abortion for every birth).

Directly or indirectly, most states have an influence on fertility. What types of policies are evident in Canada? Does the Canadian government encourage or discourage births? What role do churches play?

president of Turkmenistan announced financial incentives for women who have more than eight children. Japan also has introduced financial incentives with the aim of encouraging a baby boom. The concern in Japan is explicitly economic, as it is feared that an aging population supported by fewer working people would keep the country in a state of permanent recession. In addition, daycare facilities are being expanded and employers are making their companies more family-friendly. Canada has similar policies and practices in place, with subsidized daycare and with many employers offering generous maternity and paternity leaves. One effective tactic, at least in the short term, was employed in Georgia by the head of the Georgian Orthodox Church. Government concern about low fertility prompted Patriarch Ilia II to issue a promise in 2007 that he would personally baptize any baby born to parents who already had two children. In a country where about 80 per cent of the population are adherents of the Georgian Orthodox faith and the Patriarch is held in great esteem, the immediate result was a 20 per cent increase in births in 2008.

It seems likely that more and more countries will adopt pro-natalist policies if their fertility declines to the point where not only the economic but the political and cultural consequences come to be seen as unacceptable—for instance, if it is feared that a smaller population will mean a loss of national identity.

ANTI-NATAL POLICIES

Today, however, the most common policies relating to births are anti-natalist. Since about 1960, many of the less developed countries have initiated policies designed to reduce fertility. The argument behind such policies is that overpopulation is a real danger and that **carrying capacity** has been (or will soon be) exceeded. Regardless of whether this threat is real, there is a common assumption that certain areas are too densely populated and that the best solution is reduced fertility. As we noted earlier, birth rates are indeed falling in many of the less developed countries as a result of changes in attitudes, government programs, and such general social advances as improvements in literacy and rural development.

India was the first country to intervene actively to reduce fertility. Beginning in 1952, the Indian government introduced a series of programs designed to encourage contraceptive use and sterilization. The first program offered financial incentives; others have been coercive or educational. Fertility rates have declined,

carrying capacity
The maximum population that can be supported by a given set of resources.

but with a 2008 RNI of 1.6 per cent, mean–ing a doubling time of 44 years, it is clear that the various programs have not achieved the desired result.

China, on the other hand, has made great strides towards reducing fertility. The 2008 RNI was 0.5 per cent and the doubling time 140 years. Much of this success can be attributed to the programs introduced, mostly in the late 1970s, by China's Communist government. Families were restricted to having one child, and marriage was prohibited until the age of

Box 4.6   Population in China

Data from the 1990 census and more recent estimates provide a relatively sound factual description of the Chinese demographic situation. As of 2008, China continued to enjoy a declining CBR, one that is low by Asian standards (12), a CDR that is among the lowest in the world (7), an RNI of 0.5, and an LE that is approaching the average for the more developed world (73).

A useful way to depict a critical period in the recent history of Chinese population growth is with a population pyramid (Figure 4.6). The indentations in the 1990 pyramid show several changes in fertility and mortality during the 1960–90 period. The 1960s were a period of generally high fertility and low mortality, reflected by the bulge for the 10-year cohort aged 18–27 in 1990; these were people born in the high-fertility period, 1963–72. In 1964 more than 40 per cent of the population was under age 15. The high fertility of 1972 was followed by dramatic decreases, beginning in 1973, that resulted in a narrowing of the pyramid, and the 1990 percentage under 15 was reduced to 28.

The broadening at the base of the pyramid reflects the fact that fertility is highest for women in their late twenties; if the number of births to married women of this age remains constant, the CBR will remain high. As Jowett (1993: 405) noted, 'China is currently faced with the "echo" effect from demographic developments that occurred 20–25 years ago. The large-small-large-small cohorts of population aged eight, six, three and zero are a condensed form of the fluctuations from the previous generation aged 32, 29, 27, and 23.'

Recent data for China highlight three developments. First, it seems that if current fertility trends continue, China's population will begin to decline about the year 2042. This astounding prospect prompted authorities to informally relax the one-child policy, and it is now common in rural China for couples to have two children, especially if the first child is a girl. The policy continues to be applied in some parts of the country by local officials. Indeed, several recent news items have reported very serious violations of human rights with forced abortions and forced sterilizations designed to reduce births. As of 2008, the uneven application of the policy is leading to concerns of increasing fertility and a possible population rebound.

The second significant development concerns an increasingly skewed sex structure because of a marked disparity in the sex ratio of children born in recent years. The 2000 census data show a ratio of 118 males to 100 females. These numbers suggest that increasing numbers of female fetuses have been aborted. Female infanticide is practised, but is likely much less evident than selective abortion. There is also evidence of a higher infant mortality rate for girls than for boys because of neglect or maltreatment. All of this happens because parents prefer sons over daughters since a daughter is responsible for her birth family only until she marries, whereas a son remains responsible after his marriage. The other country with a similar pattern of female abortion is India, where many parents still have to offer dowries in order to find husbands for their daughters.

The third significant development concerns an increasingly skewed age structure with the rapid aging of the population. It is estimated that the rate of aging of China's population will be the fastest in the world. Assuming current trends continue, the ratio of people of working age to retirees will fall from about 6:1 in the early twenty-first century to about 2:1 by 2040. Considering that China is on the way to becoming an economic powerhouse, one way to express what this aging means is to say that China will get old before it gets rich. Certainly, the huge elderly non-working population is expected to place great stress on the national economy. China is approaching what can be broadly described as a '4–2–1' scenario, that is, four grandparents, two parents, and one child.

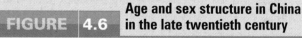

FIGURE 4.6   Age and sex structure in China in the late twentieth century

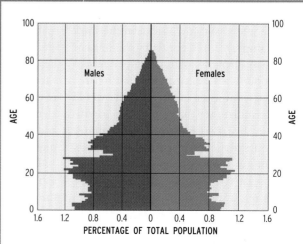

SOURCE: J. Jowett, 'China's Population: 1,133,709,738 and Still Counting', *Geography* 78 (1993): 405. Reprinted by permission of the Geographical Association, www.geography.org.uk

28 for men and 25 for women. Contraception, abortion, and sterilization were free. There were financial incentives for families with only one child and penalties for those with more than one. As noted in Box 4.6, this policy has been applied unevenly in recent years.

It is interesting, if a little perplexing, to note that China's strict anti-natalist policy did not reflect the conventional Marxist position. As noted in the discussion of population models later in this chapter, according to classical Marxism, human problems such as apparent limits to resources and apparent overpopulation are caused by inappropriate modes of production and social organization; therefore the solution to population issues is not to reduce fertility but to correct the economic and social conditions at the root of those problems. This does not mean that Marxism is pro-natalist—simply that it sees social change as the first priority; fertility is only secondary. This view does not preclude the possibility that an anti-natalist policy may be a step in the right direction.

Of course, governments may also attempt to influence migration, especially between countries. Most countries formally restrict immigration, and until about 1989 some countries, especially Communist countries in Eastern Europe, limited emigration.

THE BEST POLICY?

Is it possible that the best government population policy is neither pro-natal nor anti-natal, but rather no policy at all? The logic behind having no policy is simply that the inhabitants of a country ought to be allowed to determine for themselves how many children they have. Anti-natal policies especially have destroyed lives, through forced sterilization, for example, and at root are derived from imperialist ideas of supremacy and the presumed ignorance of others.

But perhaps the most compelling reason not to have an anti-natal policy is that such policies may not be needed. The last few decades have made it clear that with education and improved lifestyles, women choose to have fewer children and do not need to be told to have or, worse, coerced into having fewer children. Even in the case of China, it can be argued that fertility was already declining prior to the introduction of the one-child policy and that improved educational opportunities for women, combined with the ongoing process of urbanization and industrialization, would

Liufu village, Anhui Province, China. In China there is an abnormally high ratio of young males as a result of the one-child policy.

Homer W Sykes/Alamy/GetStock

have been sufficient to lead to the desired low fertility. As discussed later in this chapter, education and the related emancipation of women is the key factor contributing to declines in fertility in the less developed world.

Viewed retrospectively, many anti-natal policies were put in place by government officials, or by presumed experts from more developed countries, on the arrogant grounds that people do not know what is best for themselves—it is perhaps a fine line between arrogance, assuming what is best for others, and inhumanity, requiring that people behave in ways deemed appropriate by others.

The Composition of a Population

AGE AND SEX STRUCTURE

Fertility and mortality vary significantly with age. Inevitably, then, the growth of a population is affected by the age composition of the population. Age composition is dynamic. As we have seen in Box 4.6, it is usual to represent age and sex compositions with a **population pyramid**. Three general patterns can be suggested (Figure 4.7). First, if a population is rapidly expanding, a high proportion of the total population will be in the younger age groups. Because fertility is high, each successive age group is larger than the one before it. Second, if a population is relatively stable, then each age group, barring the older groups that are losing

> **population pyramid**
> A diagrammatic representation of the age and sex composition of a population. By convention, the younger ages are at the bottom, males are on the left, and females on the right.

numbers, is similarly sized. Third, if a population is declining, then the younger groups will be smaller than the older groups.

The three generalizations in Figure 4.7 are useful, but specific age and sex composition pyramids are more revealing. Usually, population pyramids distinguish males and females and divide them into five-year categories. Each bar in the pyramid indicates the percentage of the total population that a particular group (such as 30–34-year-old females) makes up. Figure 4.8 presents the population of Brazil for the period 1975 to 2025 in a series of three pyramids

Discussions of the number of males and females in a population refer to the **sex ratio** (technically, a masculinity ratio as it relates the number of males to the number of females). Sex ratio data for individual countries frequently are estimated and, in particular, the number of males is usually under-enumerated in more developed countries while females are under-enumerated in less developed countries (a result, perhaps, of who is at home during the survey and of various political, cultural, and social factors that can affect the accuracy of enumeration, not to mention that enumerators can hardly know where everyone is). Globally, it is generally accepted that there are about 101 males for each 100 females. Under normal circumstances, in most parts of the world about 104–8 boys are born for each 100 females. The major exceptions are those countries, especially India, Taiwan, South Korea, and China, where there is a preference for sons and the number of girls is reduced through abortion and female infanticide, as discussed in Box 4.6.

The surplus of males at birth is reduced through time as male mortality rates are generally higher and male life expectancy generally lower. In most countries the number of males in the population of a country will be overtaken by the number of females by middle age, and the elderly population usually is

sex ratio
The number of males per 100 females in a population.

population aging
A process in which the proportion of elderly people in a population increases and the proportion of younger people decreases, resulting in increased median age of the population.

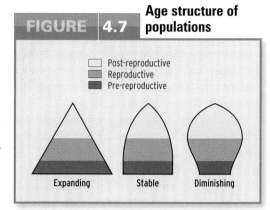

FIGURE 4.7 Age structure of populations

Post-reproductive
Reproductive
Pre-reproductive

Expanding   Stable   Diminishing

Expanding populations have a high percentage in the pre-reproductive age group. Stable populations have relatively equal pre-reproductive and reproductive age groups. Diminishing populations have a low percentage in the pre-reproductive age group.

predominantly female. In short, sex ratios vary with age.

GLOBAL POPULATION AGING

The year 2000 was a watershed as the first year in which people under 14 were outnumbered by people over 60. The process by which older individuals come to make up a proportionally larger share of the total population is known as **population aging**, and it is occurring in most countries around the world, albeit at different rates.

The data are compelling. In 1900, less than 1 per cent of the world's population was over age 65; by 2008 this figure had increased to about 7 per cent, and the estimate for 2050 is 20 per cent. Furthermore, global aging is occurring at an increasing rate: the older population is growing faster than the total population in most parts of the world, and the difference between the two growth rates is increasing. Table 4.7 provides data on the global aging trend, while Figure 4.9 shows the expected impact of aging on the Canadian population.

Causes of population aging

Population aging is the result of ongoing changes in two areas: fertility and mortality (United Nations, 2002).

1. Fertility is declining in both more developed and less developed countries. This is the primary reason for population aging: the fewer the young people in a population, the larger the proportion of elderly people. Globally, the TFR has declined by about one-half in the past 50 years. As fewer babies

Table 4.7	Global Aging, 1950–2050					
Region	**Median Age (years)**			**% Aged 60 or Older**		
	1950	1999	2050	1950	1999	2050
World	23.5	26.4	37.8	8.1	9.9	22.1
More developed world	28.6	37.2	45.6	11.7	19.3	32.5
Less developed world	21.3	24.2	36.7	6.4	7.6	20.6

SOURCE: Population Division of the Department of Economic and Social Affairs of the United Nations Secretariat, *The World at Six Billion*, 1999. The United Nations is the author of the original material.

are born, the base of the population pyramid narrows, producing a relative increase in the older population.

2. At the same time, mortality is declining—and hence life expectancy is increasing (particularly for women) everywhere as well. Globally, life expectancy has increased by about 20 per cent in the past 50 years, with females expected to live 70 years and males 67 years. Japan had the highest life expectancy in 2008, with the female rate at 86 years and the male rate at 79 years. Life expectancy is increasing because of ongoing improvements in health care and living conditions.

As a result of these changes, the median age of the world's population is expected to increase from 27 years in 2008 to about 38 years in 2050. Among specific countries, Italy and Japan have the highest median age today (40 years), but it is estimated that by 2050, 10 countries will have a median age above 50, with Spain heading the list at 55 years and Italy and Austria at 54 years.

Although population aging is indeed a global phenomenon, there are significant regional differences, most obviously related to differences between the more and less developed worlds in their patterns of fertility decline and increasing life expectancy. In the case of fertility, the decline started later in the less developed world but is proceeding more rapidly. In the case of life expectancy, the

increases are lower throughout the less developed world, and in some places they are plummeting because of AIDS.

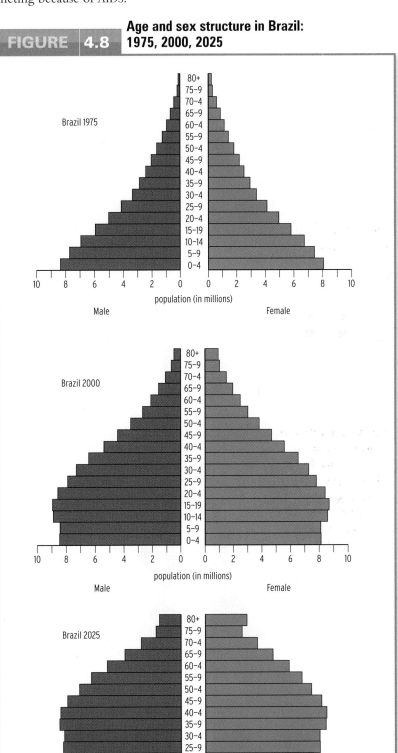

FIGURE 4.8 Age and sex structure in Brazil: 1975, 2000, 2025

These three age–sex pyramids show the changes and predicted changes in Brazil's population over a 50-year period. In 1975, the wide base and gently sloping sides indicate high rates both of births and deaths. In 1975 much of the population was under 20, ensuring that births would continue to be high in the future. The 2000 pyramid shows a slight reduction in the number of births and a decrease in deaths: this is essentially the second stage in the demographic transition. Throughout this period, the population increases rapidly because birth rates are high and death rates are dropping. The 2000 pyramid indicates that Brazil's population growth has not yet reached its maximum acceleration—a fact that does not bode well for a nation already experiencing difficulties supporting its current population. The predicted 2025 pyramid shows a significantly changed situation with the narrowing base reflecting a reduction in births and evident signs of population aging.

SOURCE: US Census Bureau, at: www.census.gov/ipc/www/idbpyr.html

Consequences of population aging

One commentator has suggested that the greying of the population in the less developed world may do more to reshape our collective future than even the proliferation of chemical and other weapons, terrorism, global warming, and ethnic tribalism (Peterson, 1999: 42). The reasoning behind this prediction has to do with changes in the ratio between what might be described as the dependent elderly population and those of working age. Estimates suggest that this ratio will double in much of the more developed world and triple in much of the less developed world.

Perhaps most notably, an aging population will place increasing stress on retirement, pension, and related social benefits, necessitating radical changes in social security programs. Further, global aging will lead to quite different patterns of disease and disability; for example, degenerative diseases associated with aging, such as cancer, heart problems, and arthritis, will become increasingly common. The need for adjustments to national health-care programs is apparent.

At the same time, national economies will face enormous strain as the numbers of workers available to support the growing

FIGURE 4.9
Age and sex structure in Canada: 1861, 1921, 1981, 2036

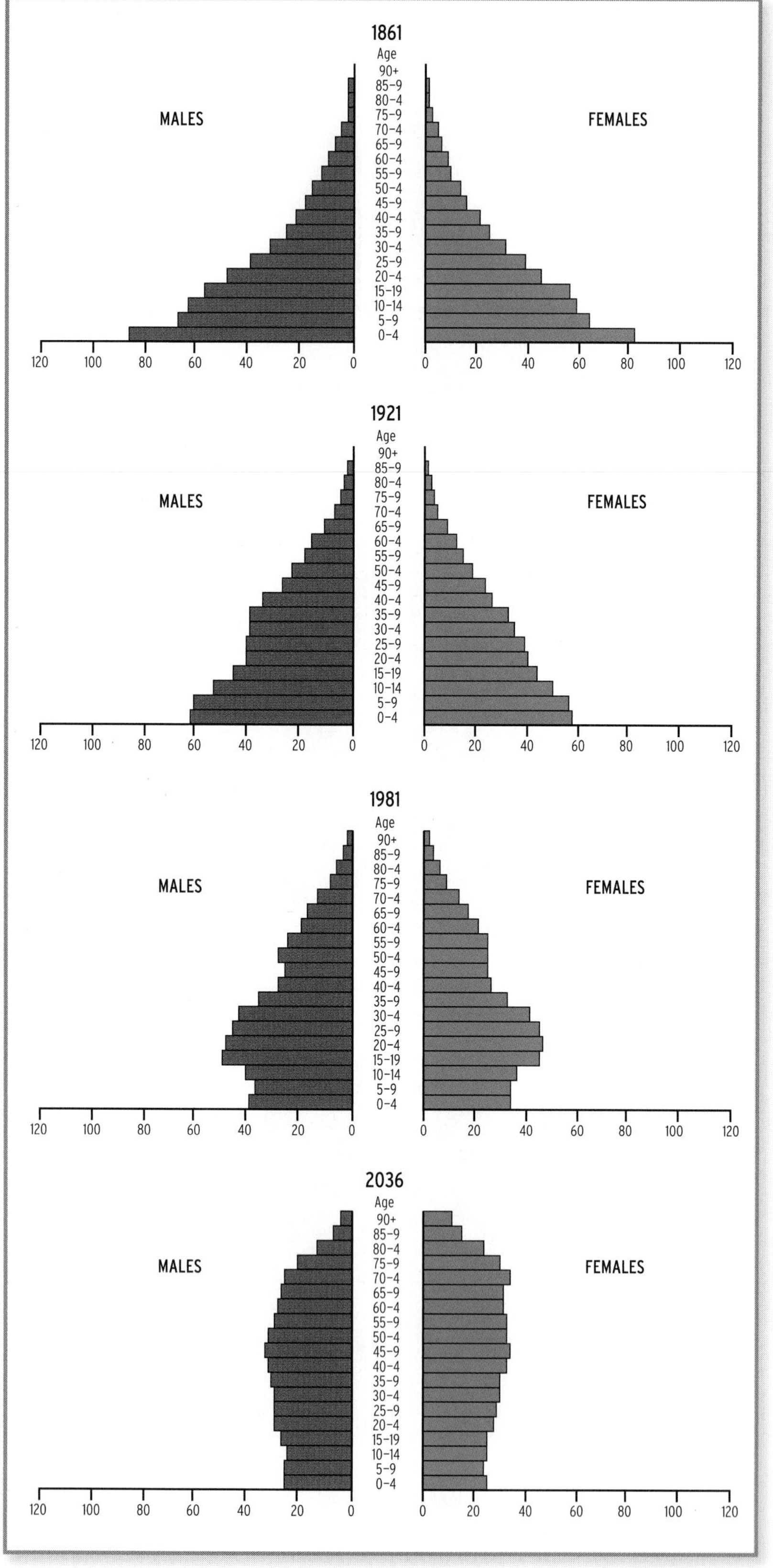

This series of four population pyramids provides considerable information about the population composition of Canada through time. The numbers along the base of each pyramid indicate the number of people in each age and sex group per 1,000 of the total population. In 1861 immigration and high fertility combined to produce a pyramid that shows cohorts steadily decreasing in size with age. The 1921 pyramid shows the effects of reduced fertility in a narrowing of the base, but the effects of immigration still are apparent. In 1981 the Canadian population is characterized by extremely low birth and death rates; the age–sex structure exhibits a 'beehive' shape. The pyramid bulge between the ages of 15 and 35 represents the post-war baby boom, which accelerated population growth. The birth rate has long since returned to its normal low; without the baby boom, the pyramid would have had almost vertical sides. A continuing low birth rate and an aging population are reflected in the projection for 2036; this age–sex structure may well represent the future for most countries in the more developed world.

NOTE: As Figures 4.6, 4.8, and 4.9 demonstrate, population pyramids may be presented in various styles.

SOURCE: *Report on the Demographic Situation in Canada*, 1992, Catalogue no. 91–209, pp. 132–3 November 1992. Reproduced with the permission of the Minister of Public Works and Government Services Canada, 2009.

population of non-working elderly gradually decline. It is possible that a decrease in the number of dependent young will free up some resources for the dependent elderly; however, the amount of public spending required to support an elderly person is two to three times that required to support a young person. Countries with a strong tradition of children supporting elderly parents, such as China, will face particular stress as increasing life expectancy makes this more and more difficult (see Box 4.6).

Inevitably, the problems of aging will be exacerbated in those parts of the world that already lack financial and other resources. Many babies born today in less developed countries can expect to live into their seventies. In the two very large population countries of China and India the number of people aged over 80 is expected to increase about 5 times by 2050. The fact that the more developed world became rich before aging, whereas the less developed world is aging first, before it has a chance to get rich, is a very significant difference between the two. The implications of population aging for poorer countries are considered sufficiently serious that a 2002 UN conference vowed to defend the rights of the old through improvements in education, employment, pension guarantees, housing, health care, and human rights.

History of Population Growth

Like all animal populations, early human populations increased in some periods and declined in others. The principal constraints on growth were climate and the (related) availability of food. Unlike other animals, humans gradually increased their freedom from such constraints as a consequence of the development of culture. Cultural adaptation has enabled humans to increase in numbers and—so far—to avert extinction. For most of our time on earth, however, our numbers have increased very slowly.

Among the cultural changes that permitted human numbers to increase before about 12,000 years ago were the evolution of speech, which facilitated co-operation in the search for food; the introduction of monogamy, which increased the chances that children would survive; and the use of fire and of clothing, which together made it possible to move into cooler

A pair of elderly Chinese women beg on the busy Nanjing Road shopping area in Shanghai. Many argue that China is woefully unprepared for the task of caring for a rapidly aging population, a challenge that will hit with full force within just a few years.

Mark Ralston/AFP/Getty Images

areas. Nevertheless, the cumulative effect of these advances was not great; 12,000 years ago, the human population totalled perhaps 4 million. The rapid growth since that time has not been regular. Rather, there have been several relatively brief growth periods and longer intervals of slow growth. Each growth period can be explained by a major cultural advance.

The first such advance, beginning about 12,000 years ago, was the so-called agricultural revolution: the gradual process, over thousands of years, by which humans domesticated animals and plants. As agriculture spread, the first human economic activities, hunting and gathering, became marginal. This 'revolution' took place in various centres, beginning in the Tigris and Euphrates valleys of present-day Iraq. Some 9,000 years ago the first region of high population density appeared, stretching from Greece to Iran and including Egypt. By 4000 BCE agriculture and related population centres had developed on the Mediterranean coast and in several European locations, as well as in Mexico, Peru, China, and India. Around the beginning of the Common Era new centres arose throughout Europe and Japan, and the total world population reached an estimated 250 million—a dramatic increase in the 10,000 years since the beginnings of agriculture.

Before the cultural innovation of agriculture, population numbers had changed little, increasing or decreasing depending on cultural advances and physical constraints. Birth and death rates were high, about 35 to 55 per 1,000, and life expectancy was short, about 35 years. Following the introduction of agriculture, birth rates remained high but death rates fluctuated; as the epidemics listed in Table 4.8 suggest, humans still had relatively little control over the environment. This state of affairs continued until the seventeenth century. The 250 million at the beginning of the Common Era had reached 500 million by 1650, but in the absence of any major cultural advance, the pace of growth remained slow.

From about 1650 onward, however, the world population increased rapidly in response to improved agricultural production and the beginning of a demographic shift to cities, especially after the onset of the second major cultural advance, the Industrial Revolution, which originated in England around 1750. The Industrial Revolution initiated a rapid growth and diffusion of technologies, and industry replaced agriculture as the dominant productive sector. The agricultural revolution had involved more effective use of solar energy in plant growth. This second revolution involved the large-scale exploitation of new sources of energy—coal, oil, and electricity. The result was a rapid reduction in death rates, with a delayed but equally significant drop in birth rates. World population increased to 680 million in 1700, 954 million in 1800, and 1.6 billion in 1900.

THE CURRENT SITUATION

The growth initially spurred by industrialization has largely ceased in the more developed world, but it continues in the less developed world. As a result, the total world population continues to increase substantially. Table 4.9 places the recent and expected growth in context by noting the number of years taken to add each additional one billion people, and Figure 4.10 displays these data graphically. What is clear from these data is that the world population is no longer growing at an increasing rate—indeed, the largest annual increments to the world population occurred in 1990, when about 87 million people were added. As noted in the account of natural increase, by 2008 this annual increment had fallen to about 82 million. The principal reason for this reduction is the current fertility decline in the less developed world; additional details of, and reasons for, this decline are discussed later in this section.

To summarize the current situation: the world population growth rate—the rate of natural increase—has fallen from a late 1960s high of 2.04 per cent per year to the current 1.2 per cent per year. Because of this decline, the world population will grow less rapidly in the current century than it did in the twentieth century, although it will continue to increase substantially because of population momentum. Specifically, there will be continued rapid growth until about 2050, followed by much slower growth after that date and, perhaps, a stable population by 2200.

Of course, future increases in population will not be uniformly distributed. Because of spatial variations in fertility, especially between the more and less developed worlds, the percentages of the world population in the various major regions are expected to change dramatically by the mid-twenty-first century, with the most notable changes being an increase in Africa and a decrease in Europe. These changes in distribution were noted earlier in this chapter and are detailed at the beginning of the next.

This brief account explains the basis of population growth. Numbers have increased in response to cultural, specifically technological, advances. The dramatic increases since about 1650 reflect the fact that the death rate

Table 4.8	Major Epidemics, 1500–1700
Date	**Event**
1517	Smallpox in Mexico, death of one-third of population, followed by other diseases in Mexico and other American countries
1522–34	Series of epidemics in Western Europe resulted in significant decrease of population
1563	Plague in Europe
1580–98	Epidemics in France
1582–3	Epidemics and famines in southern Sudan
1588	Epidemics and famines throughout China
1618–48	German population declined from 15 million to 10 million because of epidemics and war
1628–31	Disastrous plague epidemics in Germany, France, and northern Italy
1641–4	Epidemics, famines, and conflicts reduced Chinese population by 13 million
1648–57	Plague in Spain; 1 million died
1650–85	Plague and conflict in Britain; 1 million died
1651–3	Plague and conflict in France; 1 million died
1690–4	Food shortages and epidemics throughout Europe

began to fall before the birth rate did. In what is now the more developed world, populations thus grew rapidly after about 1650; elsewhere numbers increased rapidly in the last century, especially since the 1940s.

POPULATION PROJECTIONS

Predicting population growth is hazardous. In the early 1920s, Pearl predicted a stable population of 2.6 billion persons by 2100. A second example is the 1945 prediction by an eminent American demographer, Frank Rotestein, of 3 billion by 2000. These are just two of many examples. But despite some unimpressive precedents, there is good reason to suggest that we are in a position now to make better forecasts. The principal reason is that both fertility and mortality rates now lie within narrower ranges than was previously the case. In the more developed world, fertility and mortality rates are relatively low. In the less developed world, however, both rates remain relatively high.

Current United Nations 'medium' projections suggest a world population of 9.4 billion for 2050. This projection assumes that the mortality transition will be complete at that time, such that the CDR is approximately equal throughout the world. A look further ahead might see the fertility transition in the less developed world completed by the year 2100, with a relatively stable world population of about 10 billion by 2200.

What these population projections do not take into account is the possibility that there may be **limits to growth** (this phrase became popular after it was used as the title of a report published by the Club of Rome [Meadows et al., 1972]). Many environmentalists and ecologists argue cogently that there are definite limits to the growth both of populations and of economies. The earth is finite, and many resources are not renewable. The classic argument in this genre is, of course, that of Malthus (see below), but the 1972 Club of Rome report broadened the debate to include natural resources and environmental impacts. Judging by trends in the early 1970s, the authors of the report argued that world population was likely to exceed world carrying capacity, leading to a population and economic collapse. A well-publicized earlier work, *The Population Bomb* (Ehrlich, 1968), similarly anticipated widespread famine, raging pandemics, and possibly nuclear war by about 2000 as a result of worldwide competition for scarce resources.

Table 4.9	Adding the Billions: Actual and Projected	
Billions	Year Reached	Years Taken
1	1804	
2	1927	123
3	1960	33
4	1974	14
5	1987	13
6	1999	12
7	2012	13
8	2028	15
9	2054	26
10	2183	129

SOURCE: Updated from Population Division of the Department of Economic and Social Affairs of the United Nations Secretariat, *The World at Six Billion*, 1999. The United Nations is the author of the original material.

As we saw in Chapter 3, this thesis is articulated by catastrophists, but there is another, more positive view, articulated by cornucopians. According to this second view, technological advances will make new resources available as old resources are depleted: 'there are limits to growth only if science and technology cease to advance, but there is no reason why such advances should cease. So long as technological development continues, the earth is not really finite, for technology creates resources' (Ridker and Cecelski, 1979: 3–4).

Indeed, some believe people themselves to be the ultimate resource. Japan, for example, has few resources apart from people, and that resource is slowly decreasing (see Box 4.7). According to some neo-liberal commentators, a slow-down in population growth will lead to a slow-down in economic growth—and therefore is to be avoided. The ideological debate between the left-leaning catastrophists and the right-leaning cornucopians is far from over. Perhaps the best we can do at present is to recognize the wide range of views on the subject, seek to set aside opposing ideologies, and weigh the demographic and scientific evidence with extreme care.

limits to growth The argument that both world population and world economy may collapse because available world resources are inadequate.

Explaining Population Growth

Making sound predictions about future population numbers is not easy; nor is it easy to explain past population growth. In this section six different theories and models are described and evaluated.

THE S-SHAPED CURVE MODEL

All species, including humans, have a great capacity to reproduce, which is regulated by constraints such as space, food supplies, disease, and social strife. Our first explanation proposes a simple and natural statement of population growth known as the S-shaped curve (see Figure 2.8).

The S-curve is produced under carefully controlled experimental conditions. The growth process begins slowly, then increases rapidly (exponentially), and finally levels out at some ceiling. It seems probable that such growth curves never actually occur in nature. It may not be unusual for growth to be first slow, then rapid, but it does appear to be unusual for any population (plant or animal) to remain steady at some ceiling level. A more characteristic final stage would involve a series of oscillations. Various scientists neverthe-less predicted that world human population growth would correspond to the S-shaped curve. We have already referred to Pearl's pre-diction in the 1920s of a stable population of 2.6 billion by 2100; clearly, using the curve as a predictive tool is risky. Simply put, the curve model does not take into account the variety of cultural and economic factors that affect human populations. As seen in regard to fertility, stable populations result not from some natural law but from human decisions to control birth rates knowingly and will-ingly. On the other hand, it is still possible,

as suggested in Figure 4.10, that the history of human population growth will reflect an S-shaped curve by 2200.

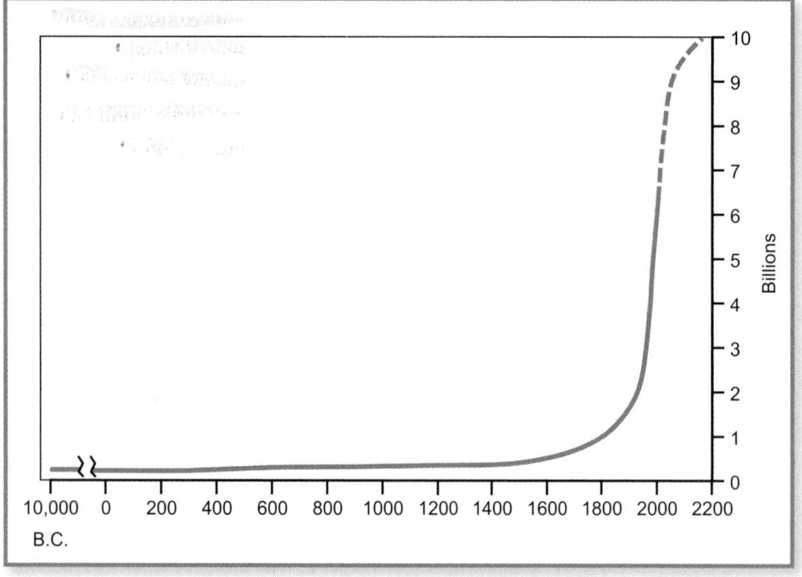

MALTHUSIAN THEORY

Our second model, unlike the S-shaped curve, is rooted in real-world facts. Malthusian theory was first presented in 1798 and remains rel-evant today. Thomas Robert Malthus (1766–1834) was born in England at a time of great technological and other changes associated with industrialization and rapid population growth. He opposed the prevailing school of economic thought, mercantilism, which was explicitly pro-natalist; for mercantilism, more births meant more wealth, since a large labour force was needed both in England to increase national productivity and overseas to increase English strength in the colonies.

Malthus expressed his views in a book titled *An Essay on the Principle of Population*, first published in 1798. Basic Malthusian logic is straightforward. It can be usefully presented in terms of two axioms and a hypothesis:

Axiom 1: Food is necessary for human exis-tence; further, food production increases at an arithmetic rate, i.e., 1, 2, 3, 4, 5 . . .

Axiom 2: Passion between the sexes is neces-sary and will continue; further, population increases at a geometric rate, i.e., 1, 2, 4, 8, 16 . . .

Given these two axioms, which Malthus regarded as reasonable assumptions, the follow-ing hypothesis is deduced:

Hypothesis: Population growth will always create stress on the means of subsistence.

According to Malthus, this conclusion—the inevitable result of the different growth rates of food supplies and population—applied to all living things, plant or animal. As rational beings (that is, with culture), humans in principle had the capacity to anticipate the consequences and therefore avoid them by deliberately reducing fertility through adoption of what Malthus called *preventive checks*: means such as moral restraint and delayed marriage. However, he believed that in practice humans were inca-pable of voluntarily adopting such checks: they would do so only under the pressure of extreme circumstances such as war, pestilence,

FIGURE 4.10 World population growth

and famine—what he called *positive checks*. Hence, the human future would be one of famine, vice, and misery.

Malthus's central concern was imbalance between population and food. His arguments are intriguing, if not especially prophetic. In considering his theory, we must first note that there is no particular justification for the specific rates of increase proposed in the two axioms. Indeed, Malthus himself acknowledged in later editions of his work that the rates of population and food increase could not be definitely specified. Malthus failed to anticipate that food supplies could be increased not only by increasing the supply of land (which he correctly saw as finite) but also by improving fertilizers, crop strains, and so forth. Nor could he have predicted that contraception would become normal and accepted. He was aware of contraception such as existed at the time, of course, but saw it as an immoral preventive check (homosexuality was another), unlike the moral preventive checks of delayed marriage and self-restraint, of which he approved.

Events have proven Malthus incorrect in his predictions, and his theory lost favour in the mid-nineteenth century as birth-rate reductions and emigration eased population pressures in Europe. Today, what might be called neo-Malthusian theory is often purported to be relevant in the less developed world, especially in several sub-Saharan African countries. It is generally accepted that China has avoided Malthusian circumstances because its political philosophy actively encourages state intervention in matters of everyday life and hence it has been able to implement a rigorous population policy, although as noted in the earlier account of government policies it can be argued than even without such a policy there might have been a fertility reduction.

Box 4.7   The Prospect of Underpopulation

It is possible that the population explosion will prove to be a short-term phenomenon, a period of perhaps 250 years in which life expectancy increased dramatically as a result of improvements in health and living conditions. As our discussion of population projections notes, it seems likely that the explosion is now over in the sense that the world population, although continuing to increase, is doing so at a decreasing rate. Some commentators are even beginning to raise the spectre of underpopulation. But we should not be misled by suggestions of this kind. The shrinking populations those commentators worry about are mostly in the more developed world. Meanwhile, serious problems continue in many parts of the less developed world, largely as a result of imbalances between population and resources. Table 4.10 identifies the countries expected to experience population losses of 20 per cent or more between 2008 and 2050. Note that the inclusion of Swaziland in this table is exceptional, reflecting the high death rates associated with AIDS rather than a very low fertility rate.

It is important to recognize that the percentages in Table 4.10 are susceptible to even slight changes in fertility rates; therefore the predictions for 2050 are especially uncertain. Consider, for example, what will happen to the population of Japan if the current fertility rate trends continue. According to the Japanese government, the country's total population—128 million in 2008—will fall to only 500 people by the end of the current millennium, and to a single person by the year 3500. This is an exercise in fantasy, of course, as was the earlier comment about the death of the last European. But the percentages provide a hint of what the future may hold in the more developed world as fertility rates stabilize below the replacement level of about 2.1.

TABLE 4.10	Anticipated Largest Population Declines, 2008–2050
Country	**Percentage Population Decline 2008–2050**
Bulgaria	−35
Swaziland	−33
Georgia	−28
Ukraine	−28
Japan	−25
Moldova	−23
Russia	−22
Serbia	−21
Belarus	−20
Romania	−20
Bosnia-Herzegovina	−20

SOURCE: Population Reference Bureau, *2008 World Population Data Sheet* (Washington: Population Reference Bureau, 2008.

MARXIST THEORY

One of the earliest and most powerful critics of Malthusian theory was Marx, who objected to the rigorous axioms and hypothesis used by Malthus and believed that population growth must be considered in relation to the prevailing mode of production in a given society. For Marx, Malthus represented a bourgeois viewpoint in which the primary aim was to maintain existing social inequalities. Malthus saw population growth as the primary cause of poverty, whereas Marx saw the capitalist system as the primary cause. The difference is fundamental: for Marx the problem was not a population problem at all, but a resource distribution problem caused by capitalism.

A related distinction can be made between the concepts of overpopulation and surplus population. These terms are not synonyms. Malthus was concerned with overpopulation: he believed that a society could be said to be overpopulated when food and other necessities of human life were in such short supply that life-threatening circumstances arose. For Marx, by contrast, the key concept was surplus population: surplus—unemployed—workers represented a 'reserve' labour force. According to Marx, capitalism depends on the existence of those 'surplus' workers to keep wages low and profits high; surplus population is, therefore, an inevitable consequence of capitalism. Both concepts are valuable to geographers.

BOSERUP THEORY

A fourth theory is available in the writings of Ester Boserup on historical changes in agriculture in subsistence societies (Boserup, 1965). According to this argument, subsistence farmers select farming systems that permit them to maximize their leisure time, and they will change these systems only if population increases and it becomes necessary to increase the food supply accordingly.

Boserup argued that in subsistence societies, population growth requires that farming be intensified, first by reducing the amount of land left fallow and then employing multi-cropping, so that the supply of food is increased to feed the additional population. Thus, though population increase prompts an increase in gross food output, food output per capita decreases. Further, a series of negative changes ensues because most of the agricultural areas with high and increasing populations are already areas of poverty and often limited agricultural technology.

The contrast with Malthus is clear. Malthus saw population as dependent on food supply; Boserup reversed this relationship by proposing population as the independent variable. Evidence suggests that this theory is applicable in a subsistence context, especially throughout much of Africa, but not in the more developed world, where technology plays a much larger role and agricultural populations are declining. For many geographers, the principal attraction of these ideas is the questioning of Malthusian claims.

THE DEMOGRAPHIC TRANSITION

So far in our discussion of theories about population growth, we have considered one natural law (the S-shaped curve) and three influential writers (Malthus, Marx, and Boserup). Each has something to offer; each is at least partially flawed. Clearly, population growth is not easily explained. Our fifth explanation is in a somewhat different category.

The **demographic transition** model is a descriptive generalization; it simply describes changing levels of fertility and mortality, and hence of natural increase, over time in the contemporary, more developed world. Because it is based on known facts rather than specific axioms or general assumptions, it has a major advantage: although simplified, it is clearly more realistic than most of the other explanations we have looked at. Inevitably, it also has a major disadvantage: because it is descriptive, it does not offer a bold, provocative perspective or hypothesis. Figure 4.11a presents the demographic transition in conventional graphic form; Figure 4.11b shows that the transition can be graphed as an S-shaped curve.

The first stage is characterized by a high CBR and a high CDR—the two rates are approximately equal. The CDR fluctuates in response to war and disease, but the principal reason for the high death rates is the lack of clean drinking water and of an effective sewage system. Throughout this stage infant mortality is high and life expectancy low. The high birth rates are characteristic of a situation where there is no incentive to reduce birth. This stage involves a low-income agricultural economy. Population growth is limited, with doubling times as long as 5,000 years. This stage applied to all human populations around the world until about the mid-eighteenth century.

---

demographic transition The historical shift of birth and death rates from high to low levels in a population. Mortality declines before fertility resulting in substantial population increase during the transition phase.

a The demographic transition model. This model describes population change over time. It is based on the observed changes, or transitions, in birth and death rates in industrialized societies over time, with its focus on about the past 250 years.

b Below the diagram of the model, the corresponding growth of population is shown, highlighting the fact that the transition in births and deaths can be graphed as an S-shaped curve.

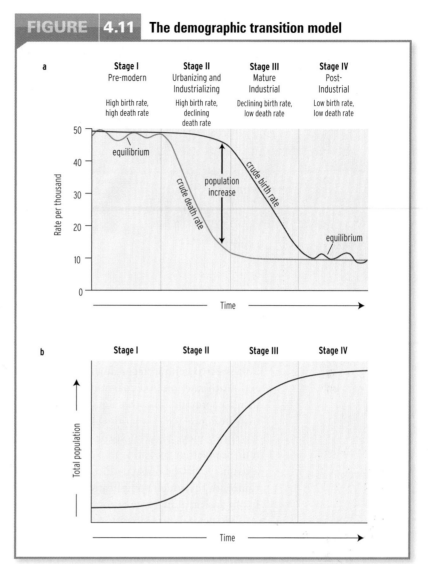

FIGURE 4.11 The demographic transition model

In the second stage, CDR is reduced dramatically as a result of the onset of urbanization and industrialization and related medical and health advances. This reduction is not accompanied by a parallel reduction in CBR; hence the RNI is high and population growth is rapid. This second stage occurred first in Northern and Western Europe and soon spread east and south throughout most of Europe and then to other industrializing countries. Within the larger context of urbanization and industrialization, there are several specific reasons for the decline in deaths. First, improvements in agricultural technologies, especially crop rotation and selective breeding, preceded the Industrial Revolution, as did the introduction of new crops from the Americas, especially the potato and corn. Second, as the Industrial Revolution progressed, city planners became increasingly concerned to improve public health through more effective sewage systems and improved living conditions generally (a topic discussed more fully in Chapter 13). Perhaps the most notable demographic feature of this second stage is a decline in infant mortality with the result that populations become more youthful. The specific details of CDR decline in any given country are debated and there are suggestions that some of the population increase in this stage was related to a rising CBR, in turn related to earlier marriage. The length of Stage 2 varied from place to place but it usually lasted several decades.

The principal feature of the third stage is a declining CBR, the result of voluntary decisions to reduce family size—decisions facilitated by advances in contraceptive techniques. The evidence suggests that voluntary birth control is related to declining death rates, as parents realize that fewer births are needed, and, more generally, to increased standards of living and, critically, to the empowerment of women as evidenced by improvements in

female literacy and in employment opportunities. The RNI falls during this third stage as the CBR approaches the already low CDR. The fact that this stage occurs contradicts the Malthusian claim that death rates are the key to explaining changes in population. The first clear signs of the onset of this third stage were in the late nineteenth century, again in Northern and Western Europe.

In the fourth (and final?) stage, the CBR and CDR are once again, as in the first stage, approximately equal—but now, instead of being wastefully high, both rates are low, as is the RNI. The process of population momentum ensures that population continues to increase for some years following the onset of Stage 4. Key demographic features of this stage include fertility rates below replacement level and an aging population.

Growth rates are similar in the first and fourth stages, but the ways in which the rates

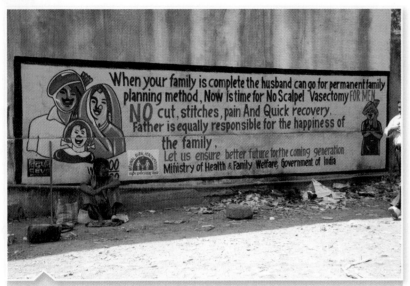

A government family planning poster in the city of Panaji, Goa, India.

Stuart Forster/ Alamy/Get Stock

understand this change, we need to consider the current circumstances of the areas being affected. Although the current fertility declines match the pattern described by the demographic transition model, they are occurring more rapidly than they did in the more developed world and under very different social and economic circumstances. In fact, the currently available data for the less developed world appear to conform more closely to what is called the fertility transition model (Robey et al., 1993).

During the period since 1970, the most powerful influence on fertility in the less developed world has been improvement in the education and lives of women and the related extent to which modern contraceptive methods are employed. Education leads to family planning, hence, fertility drops. In the less developed world (not including China), about 51 per cent of married females now practice family planning, and about 85 per cent of these employ modern as opposed to traditional methods of contraception. Information about family planning is widespread because of the influence of the mass media.

According to the fertility transition model, large families have fallen out of favour because of the obvious problems—such as pressure on agricultural land and poor quality of urban life—associated with rapid and substantial population increases. The evidence is clear: females—who are generally experiencing a rise in social status—favour later marriage, smaller families, and more time between births.

The principal reason fertility is declining so rapidly in the less developed world is that women are better-educated. Also, effective contraception is available to meet the demand as soon as the necessary cultural impetus is in place; this was not the case when the more developed world was first culturally predisposed to smaller families, a time when abstinence, withdrawal, and illegal abortion were the only techniques available. The fertility decline occurring today may be best described as a 'reproductive revolution' because it is rapid and substantial.

Economic development does help to create a climate conducive to reductions in fertility, but the key point of the fertility transition argument is that these reductions are caused by a new cultural attitude related to educational advances—a desire for smaller families and the willingness to employ modern contraceptive

are generated are very different. The transition involves two central stages during which birth and death rates are unequal and a high rate of increase prevails. The three population pyramids in Figure 4.8 exhibit the characteristics of a country moving through the transition. The four population pyramids in Figure 4.9 show a country that has passed through the transition, with the added complications of the baby boom and an aging population.

The demographic transition model accurately describes in a simplified form the experience of the more developed world. Most of that world reached the fourth stage during the first half of the twentieth century—but what about the less developed world? Throughout much of Asia, Africa, and Latin America, the situation resembles the third stage of the model. Can we assume that it is only a matter of time before economic development guarantees passage through the third stage and entry into the stable fourth stage? To put the question another way: Is the demographic transition a valuable predictive tool? The answer appears to be 'yes', but for rather different reasons. As we have already seen in Box 4.2, compelling evidence suggests that the fertility decline evident in the third and fourth stages of the demographic transition is indeed occurring in the less developed world.

THE FERTILITY TRANSITION

Since about 1970 fertility has declined throughout much of the less developed world and is continuing to decline. To help methods—and by the ready availability of these methods.

EVALUATING THE AVAILABLE EXPLANATIONS

How do the six theories and models above aid our understanding of population growth through time? In general, human geographers are far from satisfied with their explanatory and predictive powers. Moreover, as we saw in Chapter 2, the varied philosophies of individual human geographers give rise to a wide range of interpretations.

An *empiricist* philosophy focuses on facts, not theory. Remember that much traditional regional and cultural geography was—at least implicitly—empiricist. Much the same can be said of population geography; its focus is on facts, to the general exclusion of interpretation, understanding, or explanation. Both the demographic and fertility transition models are empiricist descriptions based on available data, although both also suggest explanations for the trends shown by the data. It is now reasonably clear that the demographic transition model, which centres on changing economic circumstances, provides a good account of the post-1650 experience of what is now the more developed world and is a useful predictive tool for the less developed world, albeit for different specific reasons. The fertility transition model suggests that the changes currently occurring in the less developed world can be explained by improved education of women and changing cultural preferences.

A *positivist* philosophy focuses on theory construction, related hypothesis and law formulation, and statistical analysis. The S-shaped curve model, Malthusian theory, and the Boserup thesis all have positivistic overtones in that they explicitly relate cause and effect. In the S-shaped curve, population size is related to time and the availability of space and food. In Malthusian theory, population growth is similarly explained, while the Boserup thesis sees population as cause and food supplies as effect. All three contributions are useful, but they are far from universally valid. The S-shaped curve is of limited use, Malthusian theory has clear limitations, and the Boserup argument applies only to subsistence societies. A problem with the use of a positivist philosophy is that it requires data to test hypotheses statistically—data that are rarely available at the full range of spatial and temporal scales in question.

A slum neighbourhood in Malabo, the capital of Equatorial Guinea. Despite an oil boom that by 2003 had attracted more than $3 billion in direct investment from US businesses, the country's ordinary people continue to be among the poorest on earth, and human rights groups have described the regime of Teodor Obiang as one of the most repressive in the world today.

Christine Nesbitt/AP/The Canadian Press

A *humanistic* perspective has rarely been employed in analyses of population growth. Yet there are some obvious applications, given the humanistic emphasis on individual experience. What, for example, was the reaction of Romanian women to the strict pro-natalist policies of their country prior to 1989? Or how did Chinese peasant couples respond to the one-child policy? The fertility transition model seems to be one case where a humanist approach might well be useful.

A *Marxist* philosophy applied to population issues sees population growth as an outcome of the society's particular mode of production. Fertility behaviour, for example, is seen as the product of society rather than of free decision-making. It seems that the value of Marxist theory is essentially restricted to the period of early industrial capitalism.

ALTERNATIVE EXPLANATIONS?

In these discussions we rarely have chosen to subdivide human populations on the basis of age, ethnicity, class, religion, gender, or other relevant variables. The decision to treat humans as a cohesive whole reflects our favoured scale of analysis, which is large: either the world as a whole or large areas within it. (Material focusing on population subgroups will be introduced in later chapters.) Nevertheless, the human experience is not unitary: in fact, it is closely tied to a number of variables. As this chapter makes clear, one important variable that merits discussion is gender.

Feminist geography has developed as an area of interest in human geography since the 1970s. Like Marxist geography, it is concerned with criticizing and changing established procedures. Specifically, feminist geography addresses the question of gender inequality in our human experience. The aim is not only to recognize different gender geographies but, much more fundamentally, to develop gender-specific theory.

How might feminist geography contribute to our discussions of population? Answers are many and varied. We will briefly consider gender differences in the more developed world as an example. The principal change currently occurring involves women's role: from wife and mother to wife, mother, and worker outside the home. In the more developed world, the typical family structure remains the conjugal variety.

The family has responsibility for reproduction, and the changing role of women in society is linked clearly to fertility. In all the more developed countries, fertility rates are low and women's participation in the labour force is increasing. Thus, women and men alike now have links with a world outside the family—a fact that should, in principle, foster greater commonality of experience between men and women. Whether this actually is happening is debatable.

A brief review of philosophies and related attempts at explaining changes in fertility and mortality cannot do justice to them. Still, it does appear that current explanations, though useful, are far from adequate. What would a more complete explanation need to include?

Regardless of philosophical affiliation, new explanations need to explain reality at a variety of spatial, social, and temporal scales. As an example, let us consider the reality of the demographic transition. How are we to explain changing death and birth rates? An appropriate theory needs to incorporate economic circumstances (such as industrialization) and cultural issues (such as desirable family size). But we also need to consider the behaviour of individual males and females and the motivations for their behaviour. We also need to recognize that the transition was not the same in every region. In France, for example, the decline in fertility was underway by the 1830s, but in England it did not become evident until the 1890s. In reality, significant spatial variations occur in the timing and details of what we presented above as a uniform transition. Detailed analysis of such variations might facilitate the construction of a more adequate theory by identifying specific causal variables. A full theory to explain the transition, and indeed many other aspects of population change through time, is still lacking.

CHAPTER 4 SUMMARY

HOW MANY?

A population of 6.8 billion in 2009, increasing to about 9.4 billion by 2050 and possibly stabilizing at 10 billion by 2200. Almost all future population increases will be in the countries of the less developed world.

FERTILITY

There are various measures of fertility; these include the crude birth rate, the general fertility rate, and the total fertility rate. In 2008 the world birth rate was 21 (21 live births per 1,000 members of the population), while the total fertility rate was 2.6 (the average woman has 2.6 children). A total fertility rate of about 2.1 to 2.5 is sufficient to maintain a stable population. Fertility is affected by many variables: age and related fecundity, nutrition, level of industrialization, age at marriage, celibacy, governmental policies, contraceptive use, abortion, and empowerment of women. Spatial variations in fertility are closely related to level of development; the total fertility rate in the more developed world is 1.6; in the less developed world it is 2.8. At present there is evidence of declining fertility in much of the less developed world (related to advances in women's education and related widespread acceptance of family planning) and in much of the more developed world (possibly related to uncertain economic prospects).

MORTALITY

Three mortality measures are the crude death rate, the infant mortality rate, and life expectancy. In 2008 the death rate was 8 and life expectancy was 68 years. Major causes of death are old age, disease, famine, and war. Around the world, death rates show much less variation than birth rates; sub-Saharan Africa is the last region of high mortality. Data on infant mortality and life expectancy are good indicators of health. Many countries exhibit significant internal variations in mortality patterns.

THE AIDS PANDEMIC

Increased mortality and reduced life expectancy as a result of AIDS are a major humanitarian and economic concern in much of sub-Saharan Africa, and there are fears that incidence of the disease will increase significantly in some other countries, including China, India, and Russia. AIDS usually is seen as the greatest threat to human health in the world. As of 2008, it was estimated that approximately 39 million people around the world were infected with HIV and that some 20 million people had died of AIDS since the early 1980s.

NATURAL INCREASE

In 2008 the rate of natural increase was 1.2 per cent. If maintained, which seems highly unlikely, this rate will result in a doubling of world population in about 58 years. Natural increase is affected by the age composition of a population.

POPULATION AGING

The age structure of the world's population is changing significantly as the proportions of elderly people increase relative to other age groups. This change is the result of declines in fertility and increases in life expectancy. The year 2000 was a watershed: the first year in which people under 14 were outnumbered by people over 60. The social and economic implications of this shift are considerable.

GOVERNMENT INTERVENTION

Most governments intervene directly and/or indirectly to influence growth. Policies aimed at limiting unnecessary early death are normal and designed to reduce mortality. Birth control policies can be laissez-faire, pro-natalist, or anti-natalist. A pro-natalist approach may be related to a dominant religion or economic and strategic motives. Anti-natalist policies, of varying degrees of success, are common in many of the less developed countries. The two most populous countries, China and India, have employed different approaches; China has been highly successful, India less so. Many anti-natalist policies are highly intrusive.

PREDICTING GROWTH

Past forecasts have often erred seriously. As of 2008, population is estimated to continue to increase rapidly for about another 50 years and then increase more slowly. There is much uncertainty concerning the relative merits of two arguments. The limits-to-growth, or catastrophist, thesis sees definite limits to population and economic growth because the earth is finite. The cornucopian thesis sees technology as continually making new resources available and hence enabling the earth to accommodate increasing numbers of people.

WORLD POPULATION GROWTH

The world's population increases in response to cultural change. An agricultural revolution began about 10,000 BCE and prompted increases from about 4 million to 250 million by the beginning of the Common Era. A second major change, the Industrial Revolution, made it possible for the world's population to increase from 500 million in 1650 to 1.6 billion in 1900.

EXPLAINING GROWTH

The S-shaped curve is a useful biological analogy, and may be applicable in the long term. Malthus, Marx, and Boserup each contributed usefully to our understanding of population growth—Malthus saw population as limited by food supplies; Marx saw population as a response to a particular social and economic structure; Boserup saw population increases as prompting food supply increases—but none is of general applicability. Malthus is the theorist discussed most frequently, and a contemporary version of his theory is known as neo-Malthusianism. The demographic transition model is a summary of birth and death rates over the long period of human history in what is now the developed world; for the less developed world an accurate description is provided by the fertility transition model. There is a clear need for more and better theory—whether positivist, humanist, or Marxist.

QUESTIONS FOR CRITICAL THOUGHT

1. How do culture and development affect fertility and mortality in the more developed and less developed parts of the world?

2. By virtually all population indicators, Africa faces more population-related challenges than any other continent. What can be done about high rates of natural increase, high mortality rates, low life expectancy, and other related issues in Africa?

3. To what extent can/should a government take measures to control the population of a country through: (1) placing limits on how many children an individual can have, and (2) encouraging (or limiting) immigration?

4. What are the social, economic, and political impacts of an aging population in countries of the more developed world?

5. Most experts agree that the rate of world population increase is slowing down and that the predictions made in the 1970s and 1980s for the twenty-first century were overestimates. Does this mean that world population is less important as a global concern now than it was three or four decades ago? How significant will population issues be in the future?

6. How are population trends (births and deaths) tied to levels of economic development as described in the demographic transition model? What are the alternative explanations to this model?

FURTHER EXPLORATIONS

Barrett, H., and A. Browne. 2005. 'Global Crises? Issues in Population and Environment', in M. Phillipps, ed., *Contested Worlds: An Introduction to Human Geography*. Burlington, Vt: Ashgate, 127–51.

An investigation of the complex links between population and environment.

Cipolla, C.M. 1974. *The Economic History of World Population*, 6th edn. Harmondsworth: Penguin.

A stimulating and provocative overview of world population growth emphasizing our use of energy sources; dated but still interesting.

Clarke, J.I. 2000. *The Human Dichotomy: The Changing Number of Males and Females*. New York: Pergamon.

A detailed account of this intriguing and often neglected aspect of population geography.

Connelly, M. 2008. *Fatal Misconception: The Struggle to Control World Population*. Cambridge, Mass.: Harvard University Press.

A challenging, insightful book that critiques recent population policies in many countries, painting an unflattering portrait of population planners who think they know how many children people ought to have. The book argues that population policies are not needed.

Epstein, H. 2007. *The Invisible Cure: Africa, the West, and the Fight against AIDS*. New York: Farrar, Straus and Giroux.

A compelling account arguing, among other things, that the spread of AIDS is not a result of Africans having more sexual partners than do people elsewhere, as often claimed, but rather a result of a preference, among men and women, for more than one sexual partner at the same time.

Magnus, G. 2008. *The Age of Aging: How Demographics Are Changing the Global Economy and Our World*. New York: Wiley.

A good, easy-to-read account of the greying of the world population.

Newbold, K.N. 2006. *Six Billion Plus: Population Issues in the Twenty-first Century*. New York: Rowman & Littlefield.

An excellent human geography book, fully updated from the earlier 2002 edition, that covers the key population issues in the context of larger global concerns.

Newman, J.L., and G.E. Matzke. 1984. *Population: Patterns, Dynamics and Prospects*. Englewood Cliffs, NJ: Prentice-Hall.

This population geography textbook can be used by students with minimal previous study of the topic.

Omran, A.R. 1992. *Family Planning in the Legacy of Islam*. London: Routledge.

An examination of the Islamic view of marriage, family formation, and child-rearing.

Scientific American. 1974. *The Human Population*. San Francisco: Freeman.

A collection of articles all originally published in *Scientific American*. The data are somewhat dated, but most articles remain of value, especially those dealing with the genetics of human populations.

ON THE WEB

POPULATION DATA

Population Reference Bureau

www.prb.org/

The most fundamental and reliable of many websites dealing with population matters. The Population Reference Bureau publishes an annual *Population Data Sheet* that provides basic demographic data for most countries in the world; these data are posted on the website. This site also includes commentaries on specific issues.

Daily Population Updates

www.census.gov/cgi-bin/ipc/popclockw

This US Census Bureau website provides the estimated total world population on a daily basis.

Canadian Population Data

www.statcan.ca/start.html

Statistics Canada maintains this comprehensive site, which includes recent census data.

United States Population Data

www.census.gov/

Home page of the United States Census Bureau.

HEALTH AND DISEASE

World Health Issues

who.int/en//

Home page of the World Health Organization. Especially useful are the many news items and the annual reports.

Center for Disease Control and Prevention

www.cdc.gov/

Comprehensive source for data and information on disease, including HIV/AIDS.

Global Issues

www.globalissues.org/article/218/diseases-ignored-global-killers

Information on diseases worldwide; one part of a comprehensive global issues website.

POPULATION AGING

Global Information

www.popcouncil.org/socsci/aging.html

The Population Council website covers many important population-related topics, including aging.

Canadian Information

www.mta.ca/about_canada/aging/index.htm

Useful discussion covering many aspects of aging in the Canadian context.

POPULATION AND DEVELOPMENT

Problems and Issues

www.unfpa.org/

Home page for the United Nations Population Fund; covers a wide range of topics and regularly updated.

AN UNEQUAL HOME

In Chapter 4 we saw that current population growth is a significant—and possibly temporary—deviation from the rates that prevailed during most of human history. In this chapter we examine several general questions about population patterns:

- Where do we live—and why?
- Why have so many people, past and present, moved from one location to another?
- Why are so many people today effectively without a home country?
- Why is the world divided between a stable population in the more developed world and a rapidly increasing population in the less developed world?
- What are the consequences of this division?

These questions are related to one another, and they all relate to one crucial question: Why is the world divided into more and less developed areas? To answer this complex question we must consider various factors: quantity and quality of food supplies in different parts of the world; spatial distribution of debt; regional selectivity of disasters; measures employed to distinguish levels of development; and, of profound significance, political and economic relationships between the world's countries.

Indonesia slum dweller Budi cradles his son Indra while his wife Ida sits behind under a bridge in Jakarta on July 7, 2009, as they spend their day waiting to collect floating plastic scraps from the river below. The bridge is a shelter to six families numbering about 50 people. Jakarta is a city of up to 12 million people, where nearly half the population live in slums jammed between shopping malls and luxury homes and some under bridges.
Romeo Gacad/AFP/Getty Images

Distribution and Density

Determining population distributions and densities—where people are located and in what numbers—is one of the human geographer's central concerns. Establishing these basic facts is not as easy as you might think. There are three related problems. First, the required data are not available for all countries—let alone all regions within countries—and even the data that are available may not always be reliable. Second, the available data have been collected for various purposes, and hence may reflect different divisions from the ones we are interested in (see Box 5.1). Third, since distribution and density statistics are closely related to spatial scale (recall Figure 2.6), they can often suggest an inaccurate picture of reality.

MEASURING DENSITY

Distribution refers to the spatial arrangement of geographic facts; **density** refers to the frequency of occurrence of geographic facts within a specified area. Conventionally, population density is arithmetic density: the total number of people in a unit area. Maps of population often combine two characteristics: distribution and density. Figure 5.1, showing population density and distribution for the world, is a good example.

Many data sources use a single statistic to identify the population density of a country, but that practice can be seriously misleading. Canada is a prime example: because its population is not evenly distributed—in some areas the density is very high, in others it is very low—a single density statistic for the entire country would be meaningless. We mentioned this problem in Chapter 2, in reference to spatial scale. For Canada the 2008 population

density was 3.2 people per square kilometre—but this single statistic is merely an average produced by combining large areas that are virtually unpopulated and relatively small areas that are densely populated. In most cases, the local scale is the only one where density measures are appropriate; the smaller the area, the less likely it is to include significant spatial variations.

More refined density measures relate population to some other measure, such as cultivated or cultivable land; for example, **physiological density** is the relation between population and that portion of the land area deemed suitable for agriculture.

There is one exception to the general rule about density measures and scale. Simple measures employed at a regional scale can be useful for showing the broad outlines of world population patterns. In Table 5.1, for instance, which presents 2008 population data by continent, the use of a simple measure highlights the disparity between Asia (60.4 per cent of the world's population today) and Europe (only 11.0 per cent today). Similarly, as we noted in Chapter 4, the projections for 2050 point to some quite dramatic changes at this spatial scale: a significant increase in the percentage of world population located in Africa; a slight increase in North America; a significant decrease in Europe; and slight decreases in the Latin America/Caribbean region and Asia.

Table 5.2 identifies the 10 countries with the largest total populations in 2008. China ranks first, followed by India; these two countries have much larger populations than any other country. Four other Asian countries—Indonesia, Pakistan, Bangladesh, and Japan—also are in the top 10. Table 5.2 also includes projections for 2050. Although China and India continue to dominate, India has

physiological density
Population per unit of cultivable land.

density
A measure of the number of geographic facts (for example, people) per unit area.

Table 5.1	World Population Distribution by Major Area (percentage): Current and Projected	
Region	2008	2050
Africa	14.4	20.7
Asia	60.4	58.0
Europe	11.0	7.3
Latin America and Caribbean (including Mexico)	8.6	8.3
North America (Canada and US only)	5.0	5.1
Oceania	0.5	0.5

SOURCE: Calculated from Population Reference Bureau, *2008 World Population Data Sheet* (Washington: Population Reference Bureau, 2008).

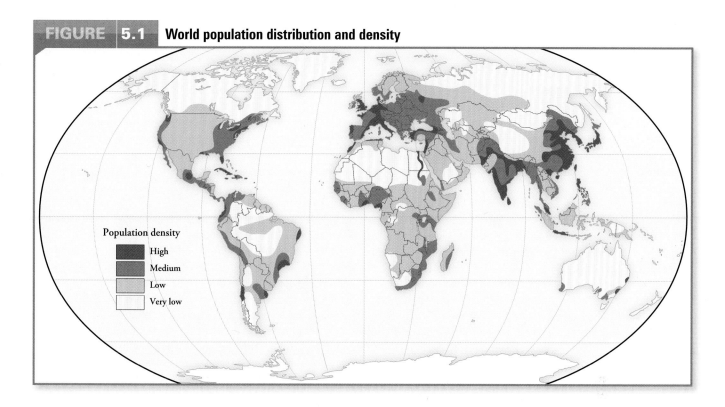

FIGURE 5.1 **World population distribution and density**

Population density
- High
- Medium
- Low
- Very low

overtaken China as the world's most populous country. African countries are also more prominent by 2050, with Nigeria moving up from ninth to seventh place and the DR of Congo, which today has a population of only 67 million, occupying ninth place. Two countries in the top 10 for 2008 have slipped off the list by 2050: Russia and Japan are expected to lose population between 2008 and 2050 because of low fertility. In 2008 only the 10 countries listed in Table 5.2 and Mexico (108 million) had populations of more than 100 million, but by 2050 there are expected to be eight more (DR of Congo, Philippines, Vietnam, Turkey, Iran, Egypt, Ethiopia, and Uganda).

Table 5.3 shows the number of people per square kilometre in each of the top 10 countries; perhaps surprisingly, the countries with the largest populations are not those with the highest densities. Of the 10 most populous countries, Bangladesh is by far the most densely populated, followed by India and Japan.

MAPPING WORLD POPULATION

Although the world map of population distribution and density (Figure 5.1) is valuable, it presents a static picture of a dynamic situation, providing no indication of how the distribution developed or of how it might change in the future. It shows three areas of population concentration: East Asia, the Indian

subcontinent, and Europe. Both of the Asian areas are long-established population centres, locations of early civilizations and early participation in the agricultural revolution. Today, both include areas of high rural population density, especially in the coastal, lowland, and river valley locations, and large urban centres. In Europe densities are high, but lower than in the high-density areas of Asia, and these are related primarily to urbanization. In all three

Table 5.2	The 10 Most Populous Countries: Current and Projected		
2008		**2050**	
Country	Population (millions)	Country	Population (millions)
China	1,325	India	1,755
India	1,149	China	1,437
United States	305	United States	438
Indonesia	240	Indonesia	343
Brazil	195	Pakistan	295
Pakistan	173	Nigeria	282
Nigeria	148	Brazil	260
Bangladesh	147	Bangladesh	215
Russia	142	DR of Congo	190
Japan	128	Philippines	150

SOURCE: Population Reference Bureau, 2008 *World Population Data Sheet* (Washington: Population Reference Bureau, 2008).

Box 5.1   Counting People in Nigeria

The principal sources of information about the demographic characteristics of a country are its censuses. The enumeration of populations has a long history, but the first modern censuses were conducted in some European countries in the eighteenth century. Relatively reliable data are available for most European countries from at least the early nineteenth century. The united Province of Canada conducted its first comprehensive **census** in 1851. Almost all countries have conducted censuses in recent decades. In order to be useful, censuses need to be comprehensive (that is, to include all members of the population) and taken at regular intervals (in many countries, at least once every 10 years).

In 1991 Nigeria conducted its first official census since 1963 (Porter, 1992). The long delay between counts reflected both the cost involved (at least Cdn$150 million) and, more important, two related political issues. First, because the spatial distribution of Nigerian government spending on services and amenities was linked to population numbers, some areas inflated their returns in the 1963 census, which as a result contained many errors; 10 years later, the 1973 census was actually declared invalid because the results, both total and regional, clearly were false. Second, since Nigeria gained independence from Britain in 1960, different ethnic groups have used population numbers as arguments in the ongoing debate over which group will exercise political control; the

principal conflict is between a primarily Muslim north and a primarily Christian south.

During the early 1990s, the Population Reference Bureau, based in the United States, estimated the total population of Nigeria at 119 million, a figure based on the recorded 1963 total and assumptions about subsequent natural increase rates. The 1991 census therefore provided a major surprise when it reported a total population of only 88.5 million. Based on this census information, the 2008 estimate, as detailed in Table 5.2, is a total population of 148 million. Although the possibility of significant errors in the 1991 census cannot be discounted, there are sound reasons to be reasonably confident of the quality of the data.

The most recent census was conducted in 2006, and it proved to be a very difficult event. Once again, funding and political representation were at stake, although sensitive questions about ethnicity and religion were not included. About one million enumerators moved from house to house throughout the country counting residents over a period of several days, with the police preventing people from moving around to make sure that they were counted once and only once.

This example offers an important lesson: all data have to be carefully assessed for their reliability, and any uses made of data need to acknowledge possible limitations.

---

areas, population densities are clearly related to land productivity.

Other scattered areas of high density are northeastern North America, around large cities in Latin America, the Nile Valley, and parts of West Africa. Perhaps the most compelling impression conveyed by Figure 5.1, however, is that a large proportion of the earth's surface is only sparsely populated.

EXPLAINING THE MAP OF WORLD POPULATION

Physical variables

There is a simple—but potentially misleading—correlation between Figure 5.1 and basic physical geography. Certainly humans have favoured some physical environments and not others. Even a cursory comparison of Figure A1.8 (our map of global environments) and Figure 5.1 suggests that three particular environments—monsoon, Mediterranean, and temperate forest areas—are associated with high population densities, while three others—desert, tundra, and polar areas—are associated with very low densities. Once again, though, it is important to remember that physical geography does not cause human geography; the correlation between the two simply reflects the fact that humans have recognized the relative

Table 5.3	Population Densities of the 10 Most Populous Countries, 2008
Country	**Population Density per km²**
Bangladesh	1,023
India	350
Japan	338
Pakistan	217
Nigeria	160
China	139
Indonesia	126
US	32
Brazil	23
Russia	8

SOURCE: Population Reference Bureau, *2008 World Population Data Sheet* (Washington: Population Reference Bureau, 2008).

attractiveness of certain areas, especially with respect to productivity.

At the global scale, four physical variables are relevant: temperature, availability of water, relief (the physical contours of the land), and soil quality. High-density areas typically have temperatures that permit an agricultural growing season of at least five to six months a year. Water is essential, whether it comes as precipitation or in the form of irrigation water. Optimum temperature levels and precipitation amounts are not easy to specify, since they vary with technology, but it is clear that extremely high temperature and precipitation together, as in tropical rain forests, are not associated with dense population.

Cultural variables

A second set of factors related to global population densities has to do with cultural organization. In many of the high-density areas shown in Figure 5.1, a form of state organization was established relatively early; China, India, Southern Europe, Egypt, Mexico, and Peru all were centres of early civilizations. A strong state organization facilitates the concentration of population; conversely, the collapse of such organization (usually with the onset of war) works against continued concentration. However, not all areas of high density today experienced early state development: in Western Europe and northeastern North America, high densities developed much later, in conjunction with the Industrial Revolution and the urbanization associated with it.

Together, the distribution and density of world population today represent the dynamic outcome of a long historical process. Is it possible that the pattern is stabilizing today? We know that certain areas are continuing to experience high rates of natural increase and are therefore likely to experience density increases. We also know that population continues to move from one location to another; indeed, some areas have grown largely as a consequence of migration.

Migration

Humans have always moved from one location to another—early pre-agricultural humans moved out of Africa to populate all major land masses of the world except Antarctica. Such movements expanded the resource base available, facilitated overall population increases,

and stimulated cultural change by requiring ongoing adaptations to new environmental circumstances. Migration has continued to be important.

For our purposes, migration may be defined as a particular kind of mobility that involves a spatial movement of residence. Thus we do not regard the journey to work, or the trip to the store, or the seasonal movements of some pastoralists and agriculturalists as migration. Focusing on those movements that involve a change of residence, we are particularly concerned with the distance moved, the time spent in the new location, the political boundaries crossed, the geographic character of the two areas involved, the causes of the migration, the numbers of persons involved, the cultural and economic characteristics of those moving, and some consequences of movement.

WHY PEOPLE MIGRATE

To ask why people migrate is not to suggest that a single reason is behind the multitude of migrations that have taken place in human history. We identify five approaches to this topic.

Push–pull logic

The first approach might best be described as a useful generalization—people move from one location to another because they consider the new location to be more favourable, in some crucial respect, than the old location. This is the key idea of inequality from place to place. How much more favourable the new location needs to be is a matter of individual judgement.

In mid-nineteenth-century Ireland, for example, there was virtually unanimous agreement that some overseas location such as the United States, Canada, or Australia was preferable to famine-stricken Ireland. In this case, the advantages of a new location appeared to be so overwhelming that the decision to stay behind may have been just as difficult as the decision to leave, if not more so. In other cases, though, only a small percentage of the population will decide to migrate, suggesting that the perceived difference between old and new locations is not so substantial. If we rank each area (however defined) of the world on an attractiveness scale from, say, 1 to 10—1 being the least attractive and 10 being the most attractive—who will want to move where? The answer is that those living in low-ranked areas will wish to move to high-ranked areas; thus there will be large-scale movements from areas

census
The periodic collection and compilation of demographic and other data relating to all individuals in a given country at a particular time.

Eastern European immigrants to North America on board ship, c. 1910–14.

Library and Archives Canada/ Page Toles/PA-131596

ranked low to areas ranked high. This scenario is a simple way of saying that migration decisions can be conceptualized as involving a push and a pull. Being located in an unattractive area is a push; being aware of an attractive alternative area is a pull.

The internal migration or population shift over the past several decades in the United States, from the Rust Belt states of the Middle West and Northeast to the Sun Belt states of the South and Southwest, is a good example. The warmer climate of the South, especially with the widespread advent of air conditioning, was attractive to many people, just as

the heavy manufacturing jobs in the North began to disappear in a post-industrial economic environment. Of note, a generation or two earlier, internal migration in the United States moved in the other direction, a result of both economic and social factors, as blacks from the American South and whites from the Appalachian region migrated to the manufacturing centres of the North, such as the Detroit area.

Table 5.4 summarizes some common push and pull factors. Typical push–pull factors can be sorted into four categories: economic, political, cultural, and environmental. A simple *economic* explanation is that migration is a consequence of differences in wages, with people moving from low- to high-wage areas. Relatively low wages are a push; relatively high wages are a pull. Another economic explanation involves the relative availability of agricultural land. From perhaps the sixteenth century to the twentieth century, land was less readily obtainable in Europe than in temperate areas overseas. A third economic explanation, and perhaps the most fundamental, applies in situations where life itself is threatened because of inadequate food supplies; in this case, any

Table 5.4	Some Push and Pull Factors
Push Factors	**Pull Factors**
Localized recession because of declining regional income	Superior career prospects and increased regional income
Cultural or political oppression or discrimination	Improved personal growth opportunities
Limited personal, family, career prospects	Preferable environment (climate, housing, medical care, schools)
Disaster, such as floods, earthquakes, wars	Other family members or friends

SOURCE: After D.J. Bogue, *Principles of Demography* (New York: Wiley, 1969), 753–4.

viable agricultural alternative area will be more attractive.

Another respect in which areas can be seen as unequal has to do with *politics*. After 1945, the desire of many Eastern Europeans to migrate reflected their assessments of the relative merits of Communist and democratic political orders. In some extreme cases, people have felt obliged to seek refuge in a country other than their own. In recent years, the political environments in Afghanistan, Ethiopia, the former Yugoslavia, and Rwanda have prompted massive refugee movements. This topic is so important to human geographers that it will be addressed separately in the following section.

Culturally, migrants are most likely to move to areas that they perceive as similar in terms of language and religion. Much of the movement from former colonies to the former colonial power relates to these considerations and also, of course, to the relative ease of such movement in terms of the legal requirements of entry. Migration to a new location is also more likely when friends or relatives have migrated to that location previously.

Relative attractiveness can also be measured *environmentally*. Migration may be induced by flooding and desertification, for example, and the historical experience indicates that most migrations have been towards the temperate climatic areas.

As the preceding account suggests, push factors at one place depend on pull factors at other places. Understanding migration requires both sets of factors to be considered, as well as the relationship between these factors. The balance between push and pull factors is complex. For example, high unemployment might be a push factor and low unemployment in another place a pull factor. But the difference in employment rates needs to be considered in the context of the economic cost of migration and the cultural upheaval associated with leaving family and friends.

As obvious as push–pull reasoning may seem, it does have a serious limitation: it assumes that people invariably behave according to the logic of a theory. Specifically, it is unable to explain why some people decide to stay in an unfavourable area when a more favourable alternative is available. In other words, the concept of push–pull factors assumes that the decision to migrate is essentially beyond the control of the individual.

Laws of migration

The logic behind the push–pull concept is place inequality. Although this logic is valuable as a general explanation, applicable in many different circumstances, clearly it is not especially useful to assert that people move from *a* to *b* because *b* is in some way preferable to *a*.

A classic attempt to formulate specific 'laws' of migration was made in the late nineteenth century by E.G. Ravenstein (1876, 1885, 1889). Based on analyses of population movements in Britain, Ravenstein's laws are not laws in a formal, positivistic sense, but generalizations with varying degrees of applicability. The 11 generalizations that Ravenstein developed (Box 5.2) still are among the most valuable concepts we have for understanding migration. However, like push–pull logic generally, they fail to take into account individual differences.

The mobility transition

A third effort to explain why people migrate was made by Zelinsky (1971). This theory is labelled the mobility transition. The term is derived from the concept of demographic transition (see Chapter 4). Zelinsky proposed five phases of temporal changes in migration:

1. the pre-modern traditional society;
2. the early transitional society;
3. the late transitional society;
4. the advanced society;
5. a future, super-advanced society.

Each of these five societies has particular migration characteristics. In Phase 1 there is minimal residential migration and only limited human mobility. Zelinsky sees this phase as temporally parallel to the first stage of the demographic transition (high birth rates, high and fluctuating death rates). In Phase 2, numerically significant migration begins in the form of rural-to-urban and overseas movements. This phase is temporally parallel to the second stage of the demographic transition (continuing high birth rates, rapidly falling death rates) and includes the mass movements associated with industrialization and European overseas expansion. In Phase 3, rural-to-urban mobility declines somewhat but remains numerically significant, while overseas migration is drastically reduced. This phase is temporally parallel to the third stage of the demographic transition (declining birth rates, low death rates). In Phase 4 residential mobility continues apace;

Box 5.2   The Ravenstein Laws

More than a century ago, E.G. Ravenstein wrote three articles on migration that have been highly influential in much subsequent research on the subject. On the basis of information contained in the British censuses of 1871 and 1881, Ravenstein identified 11 'laws'—or, more correctly, generalizations. In the following list, each generalization is followed by a brief comment.

1. The majority of migrants travel only a short distance. The idea of distance friction, noted in Chapter 2, is one of the most fundamental geographic concepts; for more on the 'tyranny of distance', see Chapter 9, especially Box 9.2.
2. Migration proceeds step by step. Thus, a migrant from Europe to Canada might go first to a port city such as Montreal and then to rural Quebec.
3. Migrants moving long distances generally head for one of the great centres of commerce or industry. This reflects the fact that large centres are usually better known than small ones to people from far away.
4. Each current of migration produces a compensating countercurrent. Any such countercurrent is usually relatively small.
5. The natives of towns are less migratory than those of rural areas. This law reflects the frequency of rural-to-urban migration.
6. Females are more migratory than males within their country of birth, but males more frequently venture beyond. Females often move within a country in order to marry. International migrants usually are young males.

7. Most migrants are adults. Families rarely migrate out of their country of birth.
8. Large towns grow more by migration than by natural increase. Remember that Ravenstein was writing at a time of dramatic industrial and urban growth.
9. Migration increases in volume as industries and commerce develop and transport improves. Such developments make urban centres more attractive and reduce distance friction.
10. The usual direction of migration is from agricultural areas to centres of industry and commerce. This is still the most common direction in the early twenty-first century, as evidenced by the ongoing rural depopulation in the Canadian prairies.
11. The major causes of migration are economic. This law, too, reflects Ravenstein's time: the present text includes several examples of migrations undertaken for social or political reasons.

Although these 11 statements have been modified by subsequent research, they have not been disproven. Clearly, they are somewhat time-specific, but all of them have some validity in different places and at different times (Grigg, 1977). Probably Ravenstein's principal limitations are his neglect of various forms of forced migration and failure to account for the current exodus from some large cities in the more developed world (see Chapter 12).

What is your impression of these laws? Do they make sense to you, both intuitively and with reference to the ideas introduced in this chapter? Or are they too general?

---

rural-to-urban movement lessens but continues; urban-to-urban movement is significant; and international migration increases, with both unskilled and skilled workers moving from the less developed to the more developed world. This phase is temporally parallel to the fourth stage of the demographic transition (low birth rates, low death rates). Finally, in Phase 5—Zelinsky's attempt to predict future trends—most migration is between urban centres. There is no equivalent stage in the demographic transition model.

Zelinsky's mobility transition theory is a useful summary of temporal changes in mobility, but it is not without its critics. The mobility transition can be regarded as an example of **developmentalism**—as a 'geography of ladders' in which 'the world is viewed as a series of hearth areas out of which modernization diffuses, so that the Third World's future can be

explicitly read from the First World's past and present in idealized maps and graphs of developmental social change' (Taylor, 1989: 310). An alternative way of looking at the mobility history of a particular country would be to see it as reflecting global processes as well as those taking place within the country. The same issue arose, somewhat less formally, in our discussion of the demographic transition; we will pursue it in a more general context later in this chapter.

A behavioural explanation

Another way of approaching the question of why people migrate is from a more humanistic perspective. None of the three approaches outlined so far has focused primarily on the people themselves. Push–pull logic, the 11 laws of Ravenstein, and the mobility transition model all have positivistic overtones: they

developmentalism Analysis of cultural and economic change that treats each country or region of the world separately in an evolutionary manner; assumes that all areas are autonomous and proceed through the same series of stages.

can be interpreted as implying that each individual human responds in an identical fashion to various external factors. Of course, any human geographer knows that this is not so, but many of us have been willing to sacrifice some reality for the sake of simplicity. Our fourth attempt to explain why people migrate shifts attention to the people themselves and away from the forces presumed to be affecting their decisions.

For convenience, we will label this approach the behavioural view of migration, since it centres on the behaviour of individuals rather than on aggregate, usually large-group, behaviour. Interestingly, behavioural approaches usually employ a version of push–pull logic, but typically do so at the level of the individual.

Place utility is a measure of the extent to which an individual is satisfied with particular locations. Typically, the place utility that people attribute to their current place of residence is much better grounded in fact than the place utility they attribute to other locations. Place utility is thus an individually focused version of push–pull logic. This concept was introduced first by Wolpert (1965), who argued that it was necessary to research an individual's **spatial preferences**.

Clearly, such preferences are based on perceptions, not objective facts. All people have what we call 'mental maps' (see Chapter 2): mental images of various places that can contribute to migration decisions. To determine what these maps are, geographers typically use questionnaires asking individuals for their perceptions of different areas. The pioneering work in this area was done by Gould and White (1986), who gathered the data on group perceptions (spatial preferences) mapped in Figure 5.2. In principle, we can use maps such as these to predict migration behaviour.

Moorings

The studies in migration that we have looked at are not conceptually sophisticated by the standards of much contemporary human geography; indeed, 'migration research is in danger of being left behind by recent developments in social theory' (Halfacree and Boyle, 1993: 337). Perhaps the greatest limitation, particularly in push–pull logic, the laws of migration, and the mobility transition theory, is the tendency to base ideas on relatively inflexible assumptions about human behaviour. Even though

the choice to migrate is typically a major life decision, determined to a large extent by individual personality and the culture of the group to which the individual belongs, much migration research focuses exclusively on the material aspects of the decision.

One way of advancing studies in this area is to acknowledge that migration reflects a personal decision made within a larger political and economic framework—for example, by drawing on theories of human motivation as developed in social psychology and putting greater emphasis on the cultural influences on migration. An example is our fifth and final attempt to explain why people migrate. This is the idea of 'moorings'—issues through which individuals give meaning to their lives—suggested by Moon (1995). Table 5.5 identifies some typical 'moorings'.

The moorings approach centres on the idea that individuals' perception of their current location, and hence the likelihood of their either remaining there or migrating to another location, depends on the value they place on their various moorings. It seems probable that future efforts to explain why people migrate will increasingly focus on conceptual formulations that incorporate both the personal and the cultural aspects of the migration decision.

place utility
A measure of the satisfaction an individual derives from a location relative to his or her goals.

spatial preferences
Individual (sometimes group) evaluation of the relative attractiveness of different locations.

Table 5.5	Some Typical Moorings
Life-Course Issues	
household/family structure	
career opportunities	
household income	
educational opportunities	
caregiving responsibilities	
Cultural Issues	
household wealth	
employment structure	
social networks	
cultural affiliations	
ethnicity	
class structure	
socio-economic ideologies	
Spatial Issues	
climate features	
access to social contacts	
access to cultural icons	
proximity to places of recreation interest	

SOURCE: B. Moon, 'Paradigms in Migration Research: Exploring "Moorings" as a Schema', *Progress in Human Geography* 19 (1995): 515. Copyright © Sage Publications Ltd., 1995. Reprinted by permission.

FIGURE 5.2 Mental maps

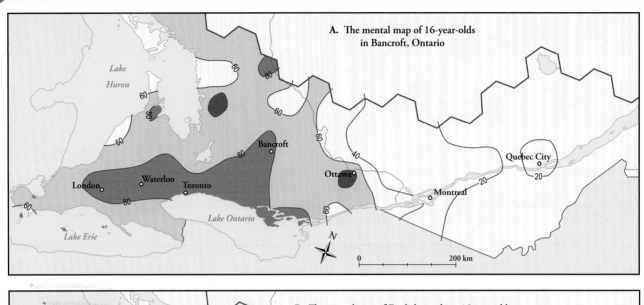

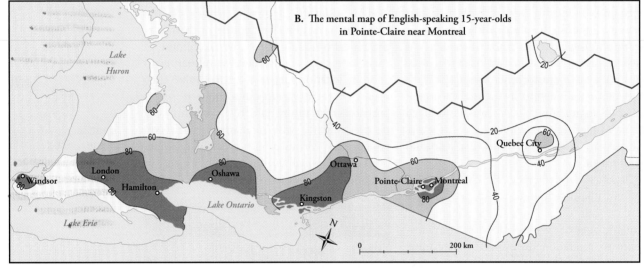

Mental maps; numbered 'isolines' represent spatial preferences on a scale from 1 to 100. These particular examples demonstrate that, in some cases, cultural considerations are more important than distance. Map A represents the mental map of 16-year-olds in English-speaking Bancroft, Ontario. Note the preference for larger urban areas and, especially, the sharp drop in ratings across the border with Quebec. Map B represents the mental map for English-speaking 15-year-olds in Pointe-Claire, near Montreal, in the predominantly French-speaking province of Quebec. Instead of favouring their local area, these young anglophones favour various areas in English-speaking Ontario.

SOURCE: Adapted from P. Gould and R. White, *Mental Maps*, 2nd edn (Boston: Allen and Unwin, 1986), 77–9.

The selectivity of migration

Migration is a selective process. Objectively speaking and as our discussion of place utility emphasized, areas in themselves are neither unattractive nor attractive. Relative attractiveness is a matter of subjective, individual perception: what is attractive or unattractive to one person may not be so to another, and migration decisions are, in the final analysis, made by individuals. We are not all equally affected by the general factors prompting migration. Who moves and who stays?

Among the factors that appear to influence individual decisions about migration are age (most migrants are older adolescents or young adults); marital status (today, as in the past, most of the people migrating from the less developed world are single adults); gender (males typically are more migratory, but there are many exceptions to this generalization); occupation (higher-skilled workers are most likely to move); and education (migrants have higher levels of education than non-migrants). In regard to migration across national borders, all of these factors are determined to some degree by the immigration criteria of the receiving country. In the most general sense, however, there is a useful relationship between **life cycle** and the likelihood of individual migration.

In addition, there is often a substantial difference between what people would like to do and what they are able to do. People who want to migrate may be unable to leave their home for political reasons, and people who want to move to a specific new area may not be able to do so because of the immigration policies of that area. Similarly, potential migrants need to consider the economic and personal costs; some may be unable to pay for the move; some may not be able to move because of health, age, or family circumstances (Lee, 1966).

TYPES OF MIGRATION

One of the most useful attempts to classify migration was made by the sociologist Petersen (1958), who identified four classes of migration:

1. *Primitive* migration, associated with pre-industrial peoples and caused by some ecological necessity, specifically an inability to cope with the natural circumstances of the current location.
2. *Forced* migration, in which people have little or no alternative but to move, usually as a result of political circumstances or because of social pressures exerted by other groups.
3. *Free* migration, in which people decide to move or stay on the basis of place utility as they seek to improve their lives.
4. *Mass* migration is a specific form of free migration that involves a great many people making a specific migration decision at about the same time. Distinguishing free and mass migrations is useful as it highlights the relevance of social scale; free migrations are decisions made by individuals and families in response essentially to their understandings of push and pull factors. Mass migrations involve circumstances where migration is a collective decision.

In addition to these four, it is useful to add *illegal* migration, which can take two forms: illegal exit occurs when a country prohibits out-movement; illegal entry occurs when people enter a country without official approval.

Figure 5.3 provides an overview of the principal forced and free/mass migrations that have taken place at the global scale since about 1500.

Primitive migration

Primitive migration is really a specific instance of adaptation to environment, in which people

life cycle
The process of change experienced by individuals over their lifespans; often divided into stages (such as childhood, adolescence, adulthood, old age), each of which is associated with particular forms of behaviour.

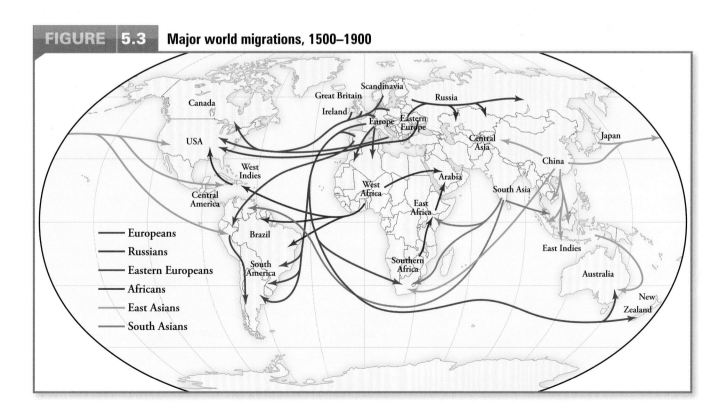

FIGURE 5.3 Major world migrations, 1500–1900

- Europeans
- Russians
- Eastern Europeans
- Africans
- East Asians
- South Asians

respond to an unfavourable environment by leaving it for a more favourable one. Pre-industrial societies tend to make such adaptation decisions on a group rather than on an individual basis. Thus, hunting and gathering groups might migrate regularly as the resources of an area are depleted or as game animals move on; some agricultural groups might move as soil loses fertility. Another instance of primitive migration occurs when populations increase in size so that additional land is needed. In pre-industrial societies, primitive migration was a normal part of the human search for appropriate environments—the search that was responsible for the human occupation of most of the earth. As we noted in Box 3.1, by the fifth century CE, even Easter Island, one of the most isolated locations in the world, was settled.

Forced migration

Forced migration has a long history. **Slavery** was a significant institution in early civilizations such as those of Greece and Rome. It appears that slaves forced to migrate from areas occupied by the Romans made up the largest portion of Rome's population at its peak. The Europeans who colonized the Caribbean and the warm coasts of North and South America—areas not conducive to large-scale European migration—also relied on slave labour. Perhaps as many as 11 million Africans were moved as slaves between 1451 and 1870.

A second example of forced migration can be seen in the late nineteenth century, when workers from China, Java, and India were shipped to the new European-controlled plantations of Malaysia, Sumatra (now Indonesia), Burma (now Myanmar), Ceylon (now Sri Lanka), and Fiji. These workers were supposed to be engaged voluntarily, on the basis of contracts, but in fact force often was used (Box 5.3). A third and quite differently motivated example was the post-1938 movement of Jewish populations in areas controlled by Nazi Germany. In each of these instances, movement was literally forced on people.

A variant of forced migration is the situation in which the migrant has some voice, however small, in the decision-making process. Examples of such *impelled* migration would

slavery
Labour that is controlled through compulsion and is not remunerated; in Marxist terminology, one particular mode of production.

Box 5.3   A New System of Slavery: Indian Indentured Labour in Mauritius

From a broad global and historical perspective, forced (or unfree) labour has been normal: free labour, in which the labourer has the right to choose an employer, did not become usual in Europe until the late eighteenth century. Indeed, in parts of Europe, slavery persisted into the nineteenth century; serfdom (an essential social relationship in feudalism that involved the legal subjection of peasants to a lord) was not abolished in Poland until 1800 and in Russia until 1861. Forms of forced labour also were central to the evolution of the European colonial world. To produce tropical products and precious metals overseas, European powers relied first on slavery. Following the abolition of slavery by the British in 1834, the French in 1848, and the Dutch in 1863, these powers turned to indentured (that is, contracted) labour. The migration of indentured labourers became a key component in the global system by which Europeans combined cheap labour and abundant land to make big profits in their colonial areas.

The majority of indentured labourers—perhaps as many as 1.5 million—came from India between the years 1830 and 1916 (Tinker, 1974). The British were able to establish and maintain this system as a result of their penetration into the Indian economy and society, a penetration that brought commercialization of agriculture, payment of rents in cash rather than kind, a decline in traditional crafts, and discriminatory

taxation. Together, these changes had a severe impact on the lower agricultural classes and served to make the prospect of emigration attractive. In the course of the nineteenth century more than 500,000 indentured labourers left India for the island of Mauritius (a British colony in the Indian Ocean), where most worked on sugar cane plantations.

In 1837 British legislation laid down the rules to be followed in organizing the emigration of indentured labourers. In principle, the labourers contracted to work for a certain period (normally five years) in exchange for their passage. In reality, however, forced banishment, kidnapping, and deception were common parts of the emigration process. Once overseas, indentured workers were not slaves, but their freedom was severely constrained. Movement outside the plantations was restricted, and the opportunities available after contracts expired were limited by measures such as taxes and vagrancy laws. Overwork, low wages, poor food, illness, low-quality housing, and inadequate medical and educational facilities were usual. Typically, any attempt at resistance on the part of workers was met by stringent labour legislation. Although one of the conditions of indenture was to offer all workers transport back to India after 10 years, this responsibility was neglected in Mauritius after 1851. Very few indentured labourers ever returned home, and their descendants are the majority population in Mauritius today.

include the many cases in which people have chosen to flee oppressive political regimes, war zones, and areas of famine. The dividing line between forced and impelled is not clear, nor is the line between impelled and free. Most contemporary refugee movements would qualify as impelled. Most forced and impelled migrations are related to the actions of others who directly influence the migration decision.

Tragically, forced labour, often involving forced migration, remains a real problem today. The UN judges the traffic in human beings to be the third most lucrative in the world today, after drugs and arms. The International Labour Organization suggested a cautious minimum of 12.3 million people working as forced labourers, effectively slaves, about one-sixth of whom are victims of human trafficking. Young children and women are especially vulnerable, with many forced to work as soldiers or in the sex trade. The largest numbers originate from some of the poorest less developed countries. In a few cases—notably Myanmar and Sudan, but also China—the state is directly involved in this forced labour. It is noteworthy that most developed countries are indirectly involved as many companies rely on forced labour, although this is often not clear because of the complex webs of subcontracting and supply chains.

Free migration

In free migration the person has the choice either to stay or to move. Much free migration today takes place within rather than between countries. In the United States, the South and West, as noted above, are attractive destinations for a combination of reasons having to do with climate and job opportunities. Movement to destinations perceived as attractive is also evident in England, where the southeast is the most attractive area, and in Canada, where Alberta is especially attractive. For Canada, during the intercensal period 2001–6, there was an overall decline in movement. Americans also are much less likely to move today than in the recent past—in 2007 about 10 per cent moved whereas in the 1960s about 20 per cent moved. In addition to these regional movements within countries, people continue to move from rural to urban areas and from central to suburban zones within urban areas. Because these movements tend to be selective, reflecting the migrants' stage in the life cycle, gender, and ethnic background, they often lead to significant changes in the structure and composition of local populations. Most migration within countries, regional or more local, continues to be primarily economic in motivation, being related to employment, income potential, and the housing market.

Free migration between countries continues today, but now all of the more developed countries are popular destinations, including the European states that so many earlier migrants left behind. This situation confirms our observation that individuals' migration decisions often reflect their assessments of relative place utility. However, about 40 per cent of international migration takes place between poor or middle-income countries. Of course, there are significant restrictions on free migration, with the principal reason a country restricts immigration being economic, specifically concerns in rich countries about a flood of migrants taking jobs or collecting welfare benefits.

An interesting example of restricting anticipated labour migration arose in May 2004 when 10 new countries, mostly relatively poor Eastern European countries, joined the European Union. Because of concerns about a flood of immigrants, 12 of the 15 member countries of the European Union chose not to permit immigration from the 10 new member countries in 2004. Only three countries, the United Kingdom, Ireland, and Sweden, elected not to impose restrictions on labour movement from the new members. As a result, these three countries received many immigrants from Eastern Europe after 2004, a direct consequence of what is really unprecedented: a few relatively rich countries opening their doors to immigrants from a few relatively poor countries.

As this example suggests, many decisions to migrate cannot be carried out because most developed countries have implemented restrictive immigration policies. Most notoriously, in the early twentieth century, the immigration policies of countries such as the United States, Canada, Australia, New Zealand, and South Africa were explicitly racist (see Box 5.4).

Mass migration

Mass migrations are free migrations prompted by push–pull factors that are widely experienced and thus these migrations involve large numbers of people. Historically, mass migrations included movement from densely settled

Box 5.4   Restrictive Immigration Policies

Discriminatory immigration policies were introduced in the nineteenth and early twentieth centuries typically in those areas experiencing large-scale immigration. Although in some cases certain European groups also were targeted, in general such policies were specifically intended to limit immigration from Asia.

As we saw in Box 5.3, the termination of a slave trade prompted the introduction of an indentured labour system, initially involving Indian labourers on the island of Mauritius but soon involving others as well in the Caribbean, East Africa, South Africa, and some Pacific Islands, notably Fiji. The success of this system, as measured by the plantation owners, resulted in its expansion to include Chinese indentured labourers, most of whom went to the Caribbean and South America.

In the mid-nineteenth century, Chinese also moved to North America and Australia as free migrants, prompted in both cases by the discovery of gold. Many others worked in transcontinental railroad construction, first in the United States in the mid-1860s and then in Canada in the 1880s. But their presence was quickly opposed by local populations motivated by some combination of racist (see Appendix 2) and economic fears.

In countries that were part of the British Empire, the agreement of the British government was required before restrictive immigration policies could be implemented. Had this not been the case, it is probable that in some colonies restrictive policies would have been in place much earlier. In Australia, restrictions were imposed following the immigration of Chinese labourers and gold field workers. The first legislative action, in the state of Victoria in 1855, was followed by similar legislation in other states. But the most explicit restriction was introduced in the South African province of Natal in 1897. This was a requirement that all new immigrants be proficient in a European language. Although, in principle, such a restriction is 'colour-blind', in practice it is not. New Zealand introduced a restrictive immigration policy in 1881 but replaced it with a literacy requirement in 1899. In Australia, the in-movement of all coloured peoples was similarly limited: the requirement that all immigrants be literate in a European language marked the beginning of the 'White Australia Policy'.

In Canada, similar tactics were used in British Columbia, where Chinese immigrants had at first been welcomed as cheap labour; discrimination began in 1885 with the introduction of a head tax, and in 1923 Chinese immigration was virtually prohibited by an Act that was not repealed until 1947. Immigration from India began about 1900 but was largely restricted in 1908 by the requirement that immigrants arrive by a 'continuous voyage'—this at a time when there were no direct voyages between India and Canada. Japanese immigration, however, was not severely curtailed during the period of the Anglo–Japanese alliance (1902–22).

In the United States the Chinese Exclusion Act was in place from 1882 until 1943 and restrictions were imposed on most other Asian groups after 1917. These policies were greatly expanded in 1924 to include some European groups, especially from Eastern Europe, because of widespread concerns about political radicalism as well as the various nationalisms and economic dislocations of the period following the Great War. The 1924 Act included an intentionally discriminatory system and set quotas on the numbers of immigrants from specified groups. One term used to identify the outlook that prompted such policies is **nativism**: the protection of the interests of the native-born population against those of 'foreign' minorities. It was not until after World War II that these restrictions gradually were relaxed.

By the time of World War I, then, many of the countries of European overseas settlement had in place restrictive immigration policies directed specifically at Asians, policies that continued for several more decades. Of the motivations behind such policies, racist attitudes are the most obvious; British immigrants were typically regarded as the most desirable. In addition, however, there was fear of the economic competition that Asian immigrants—as an abundant source of cheap labour—would represent.

---

nativism
Intense opposition to an internal minority on the grounds that the minority is foreign.

countries to less densely settled ones, as in the period between 1800 and 1914, when some 70 million people migrated from Europe to temperate areas such as the United States, Canada, Australia, New Zealand, South Africa, and Argentina. As we saw in our discussion of the mobility transition, this migration was closely related to the demographic and technological changes that began in Europe after about 1750.

Thus, emigration relieved Europe of at least some of the population pressures that came with the second stage of the demographic transition. In 1800, people of European origin totalled 210 million; by 1900, the total was 560 million—a 166 per cent increase—and one in every three people in the world was of European origin. Most European countries participated in the nineteenth-century wave of migration. Irish, English, Scottish, Germans, Italians, Scandinavians, Austro-Hungarians, Poles, and Russians all moved overseas in large numbers; at the same time, many other Russians moved east to the Caucasus and Siberia. This historically brief period of movement had massive and wide-ranging effects in areas both of origin and of destination, redistributing a large number of people and bringing into contact many previously separate groups.

Indeed, free and mass migration lies at the root of some of the most difficult political issues in the world today, especially those involving the territorial claims of various minority groups: in Quebec, for example, non-francophone immigrants were blamed for the failure of the 'yes' side in the 1995 referendum on separation. Box 5.5 considers some of these issues in the context of the Canadian immigration experience.

Box 5.5 Migration and Ethnic Diversity in Canada

The population of Canada has changed dramatically since the beginning of the twentieth century, both in total numbers and in ethnic composition. In 1901 the country had 5.3 million people, most of British or French origin; in 2008 Canadians numbered 33.3 million, and made up one of the most ethnically diverse populations in the world. Four relatively distinct waves of migration can be identified during the twentieth century. The years 1901 to 1914 brought dramatic growth; many of these immigrants came from the new source areas of Eastern and Southeastern Europe and headed for the prairies. For some years during this period, the annual population growth rate was almost 3 per cent, but these rates slowed with the onset of World War I and remained low, at about 1.4 per cent, until the end of World War II.

The next major increase in immigration, between 1951 and 1961, joined with high fertility to produce an average annual growth rate of 2.7 per cent. In this period, during which the Canadian government was anxious to fill labour shortages in both the agricultural and industrial sectors, it nevertheless continued to show a strong preference for British and other European settlers. Potential immigrants from Africa, the Caribbean, and Asia were subject to quotas (see Box 5.4 for the larger context of such policies).

Since 1961, both immigration and fertility in Canada have declined, resulting in annual growth rates of about 1.3 per cent. In 1962, Canada's immigration rules dropped all references to race and nationality, and in 1967 a points system was introduced that reflected economic needs and gave particular weight to education, employment qualifications, English- or French-language competence, and family reunification. Thus, policies in this most recent phase have been relatively liberal, permitting significant diversification in the ethnic mix of immigrants. Immigrants from Europe and the United States accounted for about 95 per cent of the total in 1960, but make up only about 20 per cent today. Application of the points system has resulted in an increase in the number of qualified professionals, many from Asia and the Caribbean. More generally, there have been huge increases in immigrant numbers from less developed countries. A new Immigration Act in 1978 had three principal objectives: to facilitate family reunification, to encourage regional economic growth, and to fulfill moral obligations to refugees and persecuted people.

Although subsequent legislation has modified details, especially concerning refugee claimants and illegal immigration, the overall thrust has remained the same.

Canada's liberal immigration policies over the last four decades have benefited the country in many ways. Yet increasing ethnic diversity has made the establishment of a coherent cultural identity an elusive goal. Moreover, some Canadians' reactions against that diversity make it clear that we still have a long way to go if we are to free ourselves of racist attitudes.

New Canadians swear an oath of Canadian citizenship during a citizenship ceremony for 93 new Canadians in Edmonton, Alberta.

CP Photo/Jason Scott

Illegal migration

The term 'illegal migration' covers a wide variety of situations. Although it is obviously not possible to determine the exact number of people involved, it is certainly significant. The most important category of illegal immigrants consists of those who consciously violate immigration laws, but other immigrants become 'illegal' through no fault of their own, as a result of policy changes or the complexities of maintaining legal residency. One example of a policy change that created illegal immigrants involved several millions of West Africans who had legally immigrated to Nigeria during the 1970s and early 1980s; their status was changed after a new government changed the immigration regulations. Other migrants, however, deliberately violate immigration laws. The simple explanation sees such illegal movement as the result of desperate push factors (overpopulation, political turmoil, economic crises) combined with irresistible pull factors (high wages, plentiful job opportunities). This explanation is accurate in many cases, but it is not adequate. Indeed, much illegal movement takes place between more developed countries; for example, a 1993 analysis concluded that Italians made up the largest group of illegal immigrants in New York City. Most illegal immigrants are young, clustered in urban areas, and involved in industries such as construction and hospitality. The 2005 illegal alien population in the US was estimated at 11 million.

One of the best-known cases of illegal immigration is that of Mexicans moving into the US; every night, the border sees many attempts at illegal immigration. But there are numerous other examples. Approximately 10,000 Chinese illegal immigrants enter the US each year, and a significant Chinese illegal movement finds its way into Canada. Australia is experiencing overstays of legal temporary admissions; the total number of overstays recorded by 1992 was 81,500—a significant number when compared with the legal immigration in that year of 35,100. The countries of the European Union may have as many as 3 million illegal immigrants from Africa, Eastern Europe, and Asia. Major movements of illegal immigrants occur within Asia, where the favoured destinations are Japan and Singapore. Some of this movement occurs between more developed countries (for example, from South Korea to Japan). Before 1997, when it was returned to China, Hong Kong received many illegal immigrants from the mainland, while Singapore, despite very tight policies, has received many illegal immigrants from Malaysia.

These examples are only the best known. Although the motives for illegal immigration in many cases appear obvious, efforts at explaining other cases are complicated by the absence of precise data. What we can say with some certainty is that international migration, legal and illegal, reflects globalization processes, specifically a growing interdependence among the world's countries and a web of international relations that is becoming ever more complicated.

Refugees

REFUGEE MOVEMENTS

The first major movements of refugees after World War II took place in response to changing political circumstances. Immediately after 1945, about 15 million Germans relocated. The partition of India in 1947, which created the Muslim state of Pakistan, caused the movement of about 16 million people—8 million Muslims fled India for the new state, while 8 million Hindus and Sikhs fled Pakistan for India. A rather different migration took place prior to the construction of the Berlin Wall in 1961, when some 3.5 million moved from what was then Communist East Germany to democratic West Germany. Even after the wall was in place, about 300,000 succeeded in fleeing west before the country was reunited in 1989.

From about 1960 to the mid-1970s, the annual total number of refugees in the world was relatively low: between 2 and 3 million (Figure 5.4). However, in 1973 the end of the Vietnam War gave rise to one of the first large transcontinental movements of refugees; some 2 million people fled Vietnam, necessitating a major international relief effort. Because of its involvement in the war, the US received about half of these refugees. The flow of refugees from Vietnam continued until the early 1990s. More generally, the 1980s saw a huge increase in numbers of refugees for various political, economic, and environmental reasons. However, total refugee numbers began to fall in the twenty-first century. These reductions result from some successes in returning refugees to home countries as conditions improve. Figure 5.5 adds to this global picture, showing both the number of refugees and also the total number of persons of concern to the United Nations High Commission for Refugees (UNHCR) for the period 1997 to 2007. In 2005, refugees constituted 40 per cent of the total number of persons of concern. (It is important to note that the 2006 and 2007 data for persons of concern show significant increases. However, these increases are the result of changes in statistical calculation methodology such that the data cannot be meaningfully compared to earlier years.)

THE PROBLEM TODAY

It is generally agreed that refugees are people forced to migrate, usually for political reasons. There is, however, no agreement on how to determine the legitimacy of refugee claims; there are considerable logistical difficulties in counting refugees, and governments may have a vested interest in providing incorrect information on numbers. Accordingly, estimates of total numbers of refugees vary.

The most reliable source of information on refugee numbers today is the UNHCR. A few definitions are important at the outset. According to the 1951 Convention Relating to the Status of Refugees, a refugee is a person who, 'owing to a well-founded fear of being persecuted for reasons of race, religion, nationality, membership in a particular social group, or political opinion, is outside the country of his nationality, and is unable to, or unwilling to avail himself of the protection of that country.' Although most countries in the world offer protection to their citizens, some do not—either because they choose not to or because they are unable to do so—and these countries prompt refugee movement. Today, most refugee movements are prompted by civil wars and forms of ethnic conflict. A country that receives refugees is known as a 'country of asylum'.

In addition to refugees, the UNHCR includes three other classes of persons under the heading 'persons of concern to UNHCR':

1. Asylum seekers are people who have left their home country and applied for refugee status in some other country, usually in the more developed world.
2. Returnees are refugees who are in the process of returning home (the desired goal for most refugees); in recent years the UNHCR has been involved in major repatriations in many places, including Afghanistan, Iraq,

In August 1999 members of the Canadian Forces, Canadian Coast Guard, and RCMP rescued 190 Chinese migrants from a rusting unmarked ship off the coast of Vancouver Island. Suspected of intending to enter Canada illegally, they were taken first to Gold River, BC, where this photo shows them wrapped in blankets, waiting on the dock, and eventually to Esquimalt for processing.

CP photo/Chuck Stoody

Myanmar, Cambodia, and several African countries.

3. Internally displaced persons (IDPs) are people who flee their homes but remain within their home country; unlike refugees, they do not cross an international boundary. A recent example, among many, is the Tamil population of Sri Lanka, many of whom fled their homeland in the north of that country as civil war escalated.

FIGURE 5.4 Refugee numbers, 1960–2007

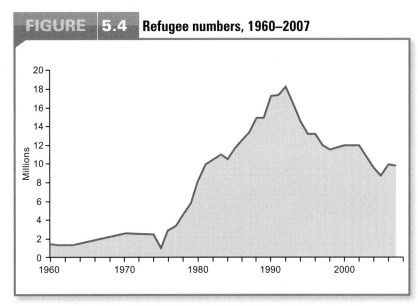

SOURCE: Updated from United Nations High Commissioner for Refugees, *The State of the World's Refugees: The Challenge of Protection* (London: Penguin, 1993).

FIGURE 5.5 — Refugees and total number of persons of concern to UNHCR, 1993–2007

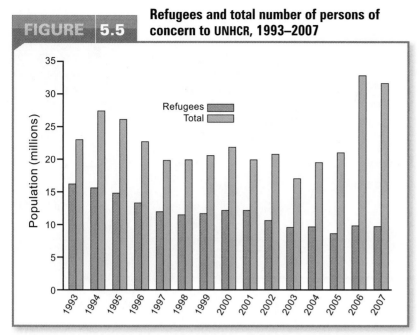

SOURCE: UNHCR.

As noted in Figure 5.5, for 2005 the total number of persons of concern to UNHCR globally was about 20.8 million: one in every 313 of the world's people at that time. The principal problem areas for both refugees and IDPs were Africa, Asia, and Europe. Most asylum seekers were located in countries in Europe and North America, and most of the returnees in Africa and West Asia. But even a figure of 20.8 million seriously underestimates the problem. According to UNHCR estimates, there are probably as many as 50 million people worldwide who have been forced to leave their homes—the difference between that figure and the official figure of 20.8 million reflects many distressing situations for which

reasonably detailed data are simply not available, as well as a variety of situations that are more difficult to categorize. There are, for example, people in refugee-like circumstances in Bangladesh (formerly East Pakistan), where about 258,000 Urdu-speaking Muslims, the Biharis, have been stranded since the partition of 1947. The Biharis are now in a difficult situation in Bangladesh because of their past support for the former West Pakistan (now Pakistan). Similarly, about 800,000 Iranians in Turkey are not recognized as refugees.

Table 5.6 provides more detailed data on the 10 largest refugee groups by country of origin. Iraq is the home country of most refugees—more than 2 million in 2007—followed by Afghanistan and four African countries, Sudan, Somalia, Burundi, and DR of Congo. Figure 5.6, which maps major source countries of refugees in 2007, highlights the key problem areas (East and Central Africa, West Asia, and Colombia in South America). Figure 5.7, which indicates IDPs protected/assisted by UNHCR in 2007, confirms these problem areas. Numerically less notable but nevertheless locally significant refugee or related problems are evident in several other countries.

It is often suggested that the vast majority of refugees and other persons of concern to UNHCR are women and children, but this is not the case. Globally, of the total of 20.8 million persons of concern, about 48 per cent are female and about 12 per cent are under the age of five. These percentages do not differ radically from standard gender and age distributions. It seems that when a population is

TABLE 5.6	Main Origins of Refugees, 2007
Country of Origin	**Total**
Iraq	2,279,245
Afghanistan	1,909,911
Sudan	523,032
Somalia	455,356
Burundi	375,715
DR of Congo	370,386
Palestine	335,219
Vietnam	327,776
Turkey	221,939
Eritrea	208,743

SOURCE: UNHCR.

TABLE 5.7	Main Countries of Asylum, 2007
Country of Asylum	**Total**
Pakistan	1,044,462
Iran	968,370
Germany	605,405
Jordan	500,229
Tanzania	485,295
China	301,027
United Kingdom	300,853
Chad	286,743
United States	280,841
Uganda	272,006

SOURCE: UNHCR.

displaced en masse, its demographic structure remains relatively balanced.

Most refugees move to an adjacent country, and this can result in some very complicated regional patterns when the same country is both a country of asylum for some refugees and the home from which others have fled. Such situations often arise in the context of ethnic conflict and civil war, and those affected are usually in serious need of aid and support. Today, examples include the East African countries of Sudan, Somalia, Ethiopia, and Djibouti; the Central African countries of Burundi, Rwanda, Uganda, and

FIGURE 5.6 Major source countries of refugees, end of 2007

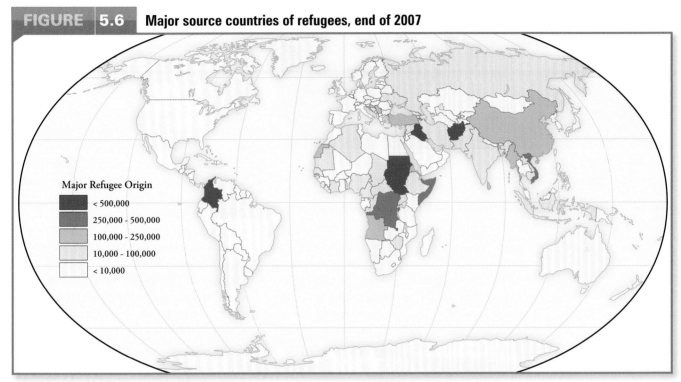

FIGURE 5.7 IDPs protected/assisted by UNHCR, end of 2007

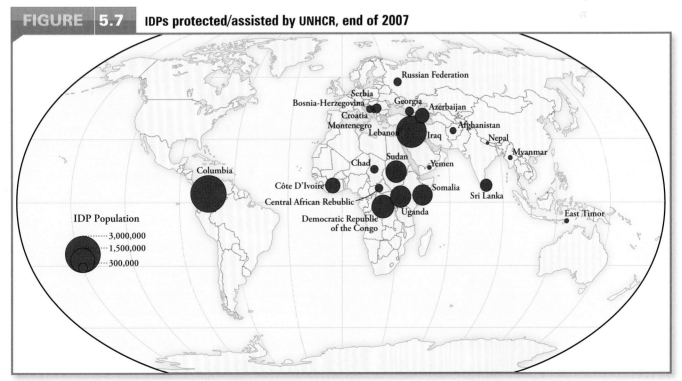

Hanifa Rehmat Khan, a Persian-speaking Afghan refugee from Kabul, actively takes part in the fight for the rights of women and children in New Nasir Bagh Refugee Camp in Pakistan, protesting against restrictions and lack of facilities. She lost her husband during the Afghan–Russian war in Afghanistan.

The Canadian Press

DR of Congo; and the West Asian countries of Afghanistan, Iran, and Iraq. Some other refugees are able to move to countries in the more developed world. Table 5.7 lists the principal countries of asylum. This table confirms that, although most refugees are in countries adjoining their homelands, many others have obtained legal status in the more developed world, where most countries receive locally significant numbers of refugees.

Refugee and related problems are greatest, however, in the less developed world, where few asylum countries have the infrastructure to cope with the additional pressures that refugees bring. In the more developed world, the collapse of communism in several European countries in the late 1980s, the breakup of the former USSR in 1991, and the conflict in the former Yugoslavia after 1991 all gave rise to mass refugee movements. Table 5.8 lists 37 refugee situations that the UNHCR has designated as protracted. These are circumstances in which refugees are in a seemingly permanent exile. Although their lives may not be in danger, basic human rights are lacking and poverty is endemic. Most of these protracted situations (23 of the 37) are in Africa and result from political impasses.

SOLUTIONS?

Different refugees have different reasons for moving, and there are no simple ways to resolve their many different cases. The UNHCR has traditionally proposed three solutions: voluntary repatriation, local settlement, and resettlement. More recently, the focus has turned to attacking the underlying causes of refugee problems. But all attempts at solutions are problematic.

The UNHCR tends to favour *voluntary repatriation*, but this is not possible for most refugees, since in most cases the circumstances that caused them to leave have not changed. However, some success is being achieved in the Horn of Africa region, as noted in Box 5.6.

Local settlement is difficult in areas that are poor and lack resources. Whatever the reason behind a refugee movement—political, environmental, or otherwise—nearby areas often face similar problems and hence are rarely able to offer solutions. People in areas that are already poor and subject to environmental problems, such as drought in parts of eastern and southern Africa, will be hard-pressed to provide refugees with the food, water, and shelter they need.

Resettlement in some other country is an option for only a few. No country is legally obliged to accept refugees for resettlement, and only about 20 countries do so on a regular basis.

Attacking the root causes of refugee problems is a mammoth challenge. Not only are those causes often a complex mixture of political, economic, and environmental issues for which no simple solutions are available, but refugee problems are greatest in parts of the less developed world that already face huge challenges. Certainly, a key first step in solving the many refugee and related problems is working towards national political stability.

The Less Developed World

Many of the topics introduced in earlier chapters are directly relevant to the emergence of what we are calling the less developed world. As we saw in Chapter 1, the growth and institutionalization of geography in the nineteenth century was largely spurred by the perceived importance of the discipline as a source of information about the parts of the world 'discovered' by Europeans during the period of overseas exploration.

Chapter 3 pointed out that a primary distinction between more and less developed countries concerns energy: both the quantities used and the sources exploited. Chapter 4 painted a vivid picture of global differences in such basic demographic measures as birth rates, death rates, rates of natural increase, life expectancy, and infant mortality. Also, the concept of 'race' (see Chapter 7) played an important part in justifying Europeans' colonization and exploitation of those parts of the world inhabited by 'inferior races'. These and related ideas inform the present section. They will reappear in later discussions of the political world, agriculture, settlement, industry, and, most generally, globalization processes.

WHAT IS THE LESS DEVELOPED WORLD?

The term 'Third World' was first used in the early 1950s, in the context of suggestions that former colonial territories might follow a different economic route from either the capitalist 'First' or the socialist 'Second' World. By 1960, 'Third World' was being used to designate a group of African, Asian, and Latin American countries that in 1969 would be described by Prime Minister Lee Kuan Yew of Singapore as 'poor, strife-ridden, [and] chaotic' (see Figure 5.9). Like all groupings based on broad generalizations, the 'Third World' contained many more variations than it did similarities. Nevertheless, there is some value in the classification, and other terms have been used in much the same way. For example, the Brandt Report, *North–South: A Programme for Survival* (Brandt, 1980), explicitly distinguished between 'North' and 'South', 'rich' and 'poor', as shown in Figure 5.13. Other common terms are 'developed' and 'underdeveloped', or 'developed' and 'developing'.

The principal advantage of the terms 'more developed' and 'less developed' is that they are frequently used in reports by organizations such as the United Nations, the World Bank, and the Population Reference Bureau, not to mention that the geopolitical bipolarity of the Cold War era and the 'First' and 'Second' Worlds effectively ceased to exist with the demise of the Soviet bloc and, thus, the term 'Third World' is anachronistic today. The more developed world comprises all of Europe and North America, plus Australia, New Zealand, and Japan. All other countries in the world are classed as less developed.

Table 5.8	Major Protracted Refugee Situations		
Region/Country of Asylum	**Origin**	**Total**	**% Assisted by UNHCR**
Burundi	DR of Congo	41,000	32
Central African Republic	Sudan	36,000	100
Chad	Sudan	110,000	50
DR of Congo	Angola	120,000	36
DR of Congo	Sudan	45,000	24
Rwanda	DR of Congo	35,000	100
Tanzania	Burundi	490,000	65
Tanzania	DR of Congo	150,000	100
Central Africa and Great Lakes		*1,027,000*	*67*
Djibouti	Somalia	25,000	100
Ethiopia	Sudan	95,000	100
Kenya	Somalia	150,000	100
Kenya	Sudan	63,000	100
Sudan	Eritrea	110,000	66
Uganda	Sudan	200,000	90
East and Horn of Africa		*643,000*	*93*
Zambia	Angola	160,000	45
Zambia	DR of Congo	58,000	93
Southern Africa		*218,000*	*59*
Cameroon	Chad	39,000	0
Cote d'Ivoire	Liberia	74,000	100
Ghana	Liberia	42,000	100
Guinea	Liberia	150,000	59
Guinea	Sierra Leone	25,000	60
West Africa		*330,000*	*67*
Algeria	Western Sahara	170,000	94
Egypt	Occupied Palestinian Territory	70,000	0
Iraq	Occupied Palestinian Territory	100,000	0
Iran	Afghanistan	830,000	100
Iran	Iraq	150,000	100
Pakistan	Afghanistan	1,120,000	100
Saudi Arabia	Occupied Palestinian Territory	240,000	0
Yemen	Somalia	59,000	100
C Asia, SW Asia, N Africa, Middle East		*2,739,000*	*85*
China	Vietnam	300,000	4
India	China	92,000	0
India	Sri Lanka	61,000	0
Nepal	Bhutan	100,000	100
Thailand	Myanmar	120,000	100
Asia and the Pacific		*673,000*	*34*
Armenia	Azerbaijan	240,000	
Serbia and Montenegro	Bosnia and Herzegovina	100,000	
Serbia and Montenegro	Croatia	190,000	
Europe		*530,000*	
Total		6,160,000	73

SOURCE: UNHCR.

Box 5.6   Refugees in the Horn of Africa

The Horn of Africa (Figure 5.8) comprises four countries: Somalia, Ethiopia, Eritrea, and Djibouti. Today, along with neighbouring Sudan and Kenya, the Horn is a land of refugees. Indeed, as detailed in Table 5.8, this region is home to several protracted refugee problems. The reasons behind this situation are a tragic mix of human and physical geographic factors. From 1962 until 1993 Eritrea was a province of Ethiopia but there were ongoing tensions between the two. For example, when, for several years in the 1980s, Eritrea and the neighbouring Ethiopian province of Tigray experienced severe droughts, the Ethiopian government refused to allow international aid efforts access to these areas. Following a nearly unanimous vote to secede in 1993, Eritrea became a separate state—and Ethiopia became a landlocked nation. Border disputes simmered after 1993 and about 80,000 people died during a 1998–2000 war that was the largest conventional conflict in Africa since World War II. UN peacekeeping forces withdrew from the border zone in 2008 and the dispute continues.

Politics is another factor in the Somalian conflict. Even though it is one of Africa's poorest countries, Somalia has placed economic and social development second to the political ideal of integrating all Somali people into one nation. Since independence in 1960, this has involved conflict with Ethiopia (in the Ogaden) and, especially, with Kenya. A full-scale war between Somalia and Ethiopia in 1977–8 coincided with drought. Since 1991, however, Somalia has not had an effective central government, being plagued by clan warfare and lawlessness. It is a country in crisis. A new government was established in 2004 but continued clan conflict and the rise of Islamist militias in the south of the country are preventing reconciliation. News reports in 2009 suggested that the world's largest refugee camp was at the Kenyan border, with about 300,000 people in a facility intended for 45,000. The ongoing absence of an effective national government has resulted also in Somali pirates becoming a major threat to international shipping off the Somali coast. Further, the continuing political instability has prevented any effective

response to several severe famines, which have exacerbated an already tragic scenario.

Neither of these political conflicts is merely regional. Both are closely related to European colonial policies. France, Britain, and especially Italy had a presence in the region during the colonial era, and, subsequently, the United States and the USSR (more recently Russia), because of its geographically strategic location, have been involved in the Horn of Africa. Despite the region's continuing political instability the UNHCR has achieved much success in repatriating refugees.

FIGURE 5.8   The Horn of Africa

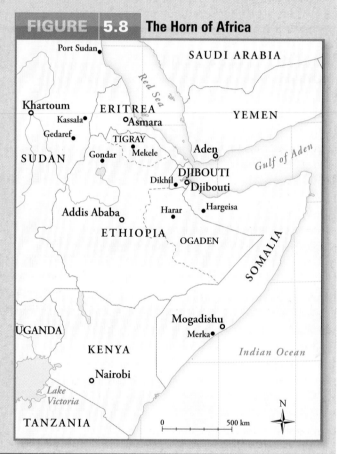

Thus, the less developed world is essentially the Third World as delimited by Dickenson et al. (1996) or the South as delimited by the Brandt Report (1980). This simple twofold division is, of course, unsatisfactory in many respects, particularly because it disguises significant differences among the countries classified as less developed. Indeed, some scholars identify a least developed world (mostly comprising countries in sub-Saharan Africa) as one means of acknowledging some of these differences.

In general, countries in the less developed world have relatively high levels of mortality and fertility and relatively low levels of literacy and industrialization; in addition, they are often beset by political problems stemming from ethnic or other rivalries. The basic demographic data are often unreliable; not only do the poorest countries have limited capital to conduct censuses, but low literacy levels may affect the quality of the data collected, and—as Box 5.1 pointed out—data may even be falsified for political reasons. Boxes 5.7, 5.8, and

FIGURE 5.9   The Third World

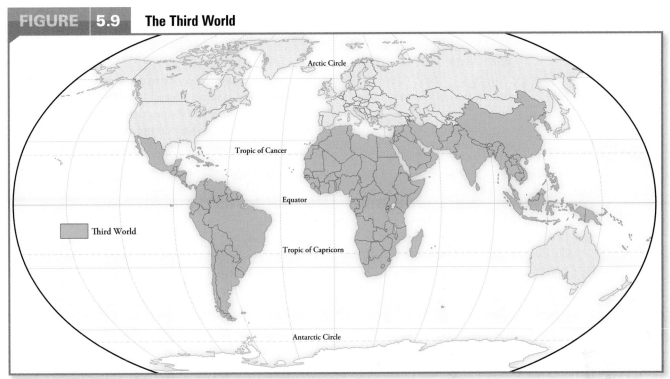

SOURCE: *A Geography of the Third World*, C.G. Clarke, J.P. Dickenson, W.T.S. Gould, S. Mather, R.M. Prothero, D.J. Siddle, C.T. Smith, E. Thomas-Hope. Copyright© 1996 Routledge. Reprinted by permission of Taylor & Francis Books, UK.

5.9 provide capsule commentaries on three case studies, one from each of Africa, Asia, and Latin America.

DEVELOPMENT: PROBLEMS OF DEFINING AND MEASURING

Traditionally, economic and social development have been measured by reference to **gross domestic product** (GDP) per capita or **gross national product** (GNP) per capita—now usually called **gross national income** (GNI) per capita—on the grounds that such macroeconomic indicators not only provide reliable data for comparing the economic performance of various countries but also serve as reliable surrogate measures of social development in the areas of health, education, and overall quality of life. Others, however, believe that such measures are inappropriate because they do not take into account either the spatial distribution of economic benefits or the real-life conditions that less developed countries face, such as population displacement, inadequate food supplies, and vulnerability to environmental extremes. It can be argued that for the less developed countries, GDP or GNI may indicate how the minority wealthy population are progressing, but tell us nothing about the poor majority, just as they tell us little about poor populations living in wealthier countries. The different opinions on measures of development reflect a lack of agreement on what 'development' itself means.

Some observers, for example, the Mexican activist Gustavo Esteva, argue that the entire concept of development is nothing more than a Eurocentric nonsense invented by the Western world and imposed on other places and cultures. For most people, the fundamental error in this way of thinking is that it appears to deny the need to eliminate poverty and disease.

One problem is that definitions of development frequently are ethnocentric. Thus, the standard view in the more developed world equates development with economic growth, the proliferation of wealth, and modernization.

Measuring development

The *World Development Report*, an annual publication of the World Bank, measures development on the basis of selected economic criteria, grouping countries into four categories—low income, lower-middle income, upper-middle income, and high income—according to GNI (formerly GNP) per capita. Figure 5.10 illustrates this classification of national economies. The problem with this way of ranking countries is that it reflects a developmentalist bias,

gross domestic product (GDP)
A monetary measure of the value at market prices of goods and services produced by a country over a given time period (usually one year); provides a better indication of domestic production than GNP.

gross national product (GNP) or gross national income (GNI)
A monetary measure of the value at market prices of goods and services produced by a country, plus net income from abroad, over a given period (usually one year).

suggesting that as countries become more technologically advanced they can—and should—increase their GNI. Moreover, as the World Bank itself has admitted, this measure 'does not, by itself, constitute or measure welfare or success in development. It does not distinguish between the aims and ultimate uses of a given product, nor does it say whether it merely offsets some natural or other obstacle, or harms or contributes to welfare' (World Bank, 1993: 306–7).

Measuring human development

An annual *Human Development Report* from the United Nations that first appeared in 1990 is intended to complement GNI measures of

Box 5.7   The Less Developed World: Ethiopia

As we saw in Box 5.6, Ethiopia is one of several countries in the Horn of Africa currently suffering as a result of a tragic combination of human and environmental factors.

Ethiopia was settled by Hamitic peoples of North African origin. Following an in-movement of Semitic peoples from southern Arabia in the first millennium BCE, a Semitic empire was founded at Axsum that became Christian in the fourth century CE. The rise of Islam displaced that empire southwards and established Islam as the dominant religion of the larger area; the empire was overthrown in the twelfth century. Ethiopia escaped European colonial rule, but its borders were determined by Europeans occupying the surrounding areas, and the Eritrea region was colonized by Italy from the late nineteenth century until 1945. Attempts by the central government in Addis Ababa to gain control over both the Eritreans and the Somali group in the southeast generated much conflict. In 1974–5, a revolution displaced the long-serving ruler, Haile Selassie, and established a socialist state.

Ethiopia is slightly larger than Ontario—1.2 million km² (463,400 square miles). Much of the country is tropical highlands, typically densely populated because such areas have good soils and are free of many diseases. In 2008 the population was estimated at 79.1 million; the CBR was 40, the CDR 15, and the RNI 2.5. Although it is a leading coffee producer, Ethiopia is one of the poorest states in Africa with many people reliant on overseas food aid. Population growth is highly uneven, and urban growth has actually declined since 1975 because of socialist land reform policies and the low quality of life in urban areas.

The majority of Ethiopians have little or no formal education, especially among the (predominant) rural population. The key social and economic unit is the family, in which women are subordinate, first to their fathers, then to their husbands, and then, if widowed, to adult sons. Health care varies substantially between urban and rural areas.

In addition, to the above, there is a particular regional problem. People in the Lower Omo River Valley already live in difficult circumstances, and their lives may become much worse. The Omo tribes are agricultural peoples who depend on the annual cycle of river flooding. Each year communities on the banks of the river move to higher ground prior to the flood, then return to plant their crops, mostly sorghum, in the replenished soil when the waters recede. But this resource and their cattle are often insufficient, and disputes between groups about water rights are common. Evidence of these disputes is that many of the males carry an automatic weapon, and the situation may be about to deteriorate further. The Ethiopian government is planning construction of a huge dam upstream that, the Omo people claim, will mean the end of the annual flooding that forms the basis for their livelihood. The government argues otherwise, claiming that it will be possible to regulate flooding as needed and that the Omo people will not be disadvantaged. Many scientists consider that the Omo concerns are legitimate.

Beggar woman, Lalibela, Ethiopia.
Louise Batalla Duran/Wandering Spirit Travel Images

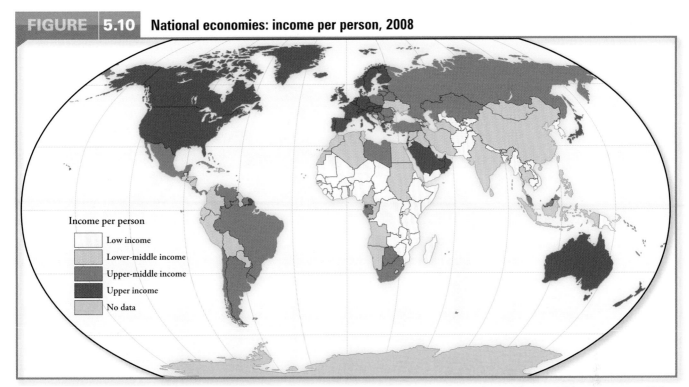

FIGURE 5.10 National economies: income per person, 2008

Income per person

Low income
Lower-middle income
Upper-middle income
Upper income
No data

Low-income countries have a GNI per capita of US$935 or less, lower-middle-income countries have from US$936 to US$3,705, upper-middle-income countries from US$3,706 to US$11,455, and upper-income countries US$11,456 and above.

development. This publication has three distinctive characteristics. First, the concept of development underlying it focuses on the satisfaction of basic needs, gender inequality, and environmental issues. Second, it uses a wide variety of data to construct a Human Development Index (HDI) based on three goals of development: life expectancy, education, and income. Third, it is explicitly concerned with how development affects the majority poor populations of the less developed world and recognizes a need to enlarge the range of individual choice. The HDI does not measure absolute levels of human development but ranks countries in relation to one another. The value of thinking in terms of the HDI rather than solely in economic terms was reinforced by some 2008 UNDP research, citing data from Angola and some parts of India, showing that economic growth does not necessarily translate into improvements in child mortality.

Figure 5.11 maps the HDI in three general categories, indicating the countries in each category. Table 5.9 presents data on 20 countries (out of the 179 countries for which data are available): the 10 with the highest HDI values and the 10 with the lowest HDI values using data from the 2007–8 report. Again, as

with GNI per capita data, African countries are the least developed. The HDI has a maximum value of 1.000 and a minimum value of 0.000; thus, Iceland, with a score of 0.968, has a shortfall in human development of a little over 3 per cent, whereas Sierra Leone, with a score of 0.329, has a shortfall of about 67 per cent. It is also important to note that, although a general relationship between economic prosperity and human development is evident, there is no direct link—some countries are more successful than others in translating economic success into better lives for people. For example, although Spain and Singapore have similar HDI levels, Spain's GNI per capita is only about half that of Singapore.

The HDI data in Table 5.9 paint a distressing picture of global inequalities. A positive side to the picture, however, is that most countries have managed over time to reduce their shortfall from the maximum value of 1.0, as demonstrated by Figure 5.12, which groups countries using a standard UNDP classification.

Another report, published by *Foreign Policy* magazine and the Fund for Peace think-tank, used 12 indicators to rank the vulnerability of states—a vulnerable state does not have effective control of its territory. This 2006

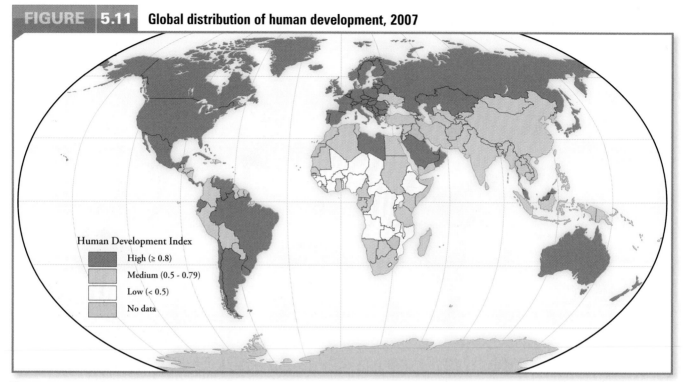

FIGURE 5.11   Global distribution of human development, 2007

There are 70 countries with a high level of human development, 85 with a medium level, and 21 with a low level. Data are not available for 16 member countries of the United Nations, most notably Afghanistan, Iraq, North Korea, Liberia, Somalia, and Serbia and Montenegro.

SOURCE: UNDP, *Human Development Report, 2007/2008. Fighting Climate Change: Human Solidarity in a Divided World* (New York: Macmillan, 2007), Table 1.

report identified Sudan as the most vulnerable state in the world, principally because of the humanitarian crisis in the Darfur region, followed by the DR of Congo, Côte d'Ivoire, Iraq, Zimbabwe, Chad, Somalia, and Haiti. Among the least vulnerable countries was Canada.

Table 5.9		Extremes of Human Development, 2007			
Top Ten			**Bottom Ten**		
Country	Rank	HDI Value	Country	Rank	HDI Value
Iceland	1	0.968	Chad	170	0.389
Norway	2	0.968	Guinea-Bissau	171	0.383
Canada	3	0.967	Burundi	172	0.382
Australia	4	0.965	Burkina Faso	173	0.372
Ireland	5	0.960	Niger	174	0.370
Netherlands	6	0.958	Mozambique	175	0.366
Sweden	7	0.958	Liberia	176	0.364
Japan	8	0.956	DR of Congo	177	0.361
Luxembourg	9	0.956	Central African Republic	178	0.352
Switzerland	10	0.955	Sierra Leone	179	0.329

SOURCE: UNDP, *Human Development Report, 2007/2008. Fighting Climate Change: Human Solidarity in a Divided World* (New York: Macmillan, 2007, Table 1).

WORLD SYSTEMS THEORY AND DEPENDENCY THEORY

Useful attempts to explain the presence of more and less developed worlds include discussions of agricultural potential at the continental scale (see the account of the shape of continents in Chapter 6) and modernization theory (see Chapter 14). A fundamentally flawed attempt is that which refers to presumed differences in human ability (see the discussion of the myth of race in Chapter 6).

Perhaps the single most important factor in explaining the plight of the countries in the less developed world is their relationship with more developed countries. Most of the less developed countries have a colonial history; even those (such as China, Thailand, Liberia, Saudi Arabia, Iran, and Afghanistan) that have not been colonies of European countries have been affected by Europe's world dominance between about 1400 and 1945. Why is this relationship so important?

First, on the world scale, **colonialism** has led to **dependence**. In the past, many former colonies became economically dependent on the more developed countries; more recently, aid

FIGURE 5.12 Improvements in human development, 1975–2006

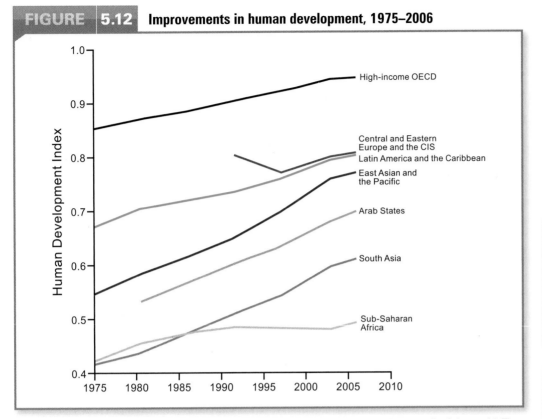

This graph highlights the significant improvements in human development throughout much of the world. The two exceptions to the prevailing trend are the former Soviet Union, which shows a marked drop following the collapse of communism but is now recovering, and sub-Saharan Africa, where improvements in development have stalled until recently, largely because of HIV/AIDS.

SOURCE: Updated from UNDP, *Human Development Report, 2005. International Cooperation at a Crossroads: Aid, Trade and Security in an Unequal World* (New York: Oxford University Press, 2005), 21. By permission of Oxford University Press, Inc.

intended to promote development has served to encourage increased dependence. Second, the indigenous cultures and social structures of former colonies have been largely relegated to secondary status, their place taken by European structures; thus, in the broadest sense, the less developed countries lack power, including the power to control and direct their own affairs.

An exciting contribution to human geography that addresses the issue of dependence is the **world systems theory** proposed by Wallerstein (1979) and related ideas of **dependency theory**. Describing the dynamic capitalist world economy from 1500 onward, world systems logic examines the roles that specific states play in the larger set of state interrelationships. It can be briefly summarized as follows. Capitalism emerged gradually from feudalism in the sixteenth century, consolidated up to 1750, and expanded to cover the world in the form of industrial capitalism by 1900; in 1917, however, the capitalist system

entered a long period of crisis that, some suggest, may eventually bring the world closer to a socialist system. Although the changes that have taken place, especially in Europe, since 1989 appear to make the prospect of further movement towards socialism less likely, this in no way detracts from Wallerstein's general observation.

The contemporary result of this historical process is a world divided into three principal zones: core, semi-periphery, and periphery (Figure 5.13). The *core* states benefit from the current situation, as they receive the surpluses produced elsewhere. The principal core states are Britain, France, the Netherlands, the US, Germany, and Japan—the countries where world business and financial matters are centred. The *semi-periphery* consists of states that are partially dependent on the core: for example, Argentina, Brazil, and South Africa. The *periphery* consists of those states that are dependent on the core and are effectively colonies;

colonialism
The policy of a state or people seeking to establish and maintain authority over another state or people.

dependence, dependency
In political contexts, a relationship in which one state or people is dependent on, and therefore dominated by, another state or people.

world systems theory
A body of ideas that suggests a division of the world into a core, semi-periphery, and periphery, stressing that the periphery is dependent on the core; has numerous implications for an understanding of the less developed world.

dependency theory
Centres on the relationship between dependence and under-development.

Box 5.8   The Less Developed World: Sri Lanka

Sri Lanka (Ceylon until 1972) is an island located in the Indian Ocean, south of India. Mostly low-lying, it has a tropical climate and a limited resource base. Minerals are in short supply, as are sources of power. The dominant economic activity is agriculture; about 36 per cent of the country is under cultivation.

In 2008 Sri Lanka had a population of 20.3 million, with a CBR of 19, a CDR of 7, and an RNI of 1.2. Although these figures are good by the standards of the less developed world, in 2008 Sri Lanka had a per capita gross national income of US$4,210—well below that of countries such as Singapore and Malaysia, with which its per capita GNI was on par in the 1960s.

Three factors account for Sri Lanka's continuing problems. First, although it has now effectively passed through the demographic transition (along with only a few other low-income countries such as China), in the past its population did increase rapidly.

Second, Sri Lanka has a long history of colonization: by the Portuguese from 1505 to 1655, the Dutch until 1796, and the British until independence was achieved in 1948. Problems of colonial dependency developed largely in the nineteenth century, when the ruling British established a system of plantation agriculture benefiting themselves rather than the local population. By 1945, tea plantations covered about 17 per cent of the cultivated area.

Third, Sri Lanka suffered a severe ethnic conflict between the Sinhalese (74 per cent of the population), who moved down from north India and conquered the island in the sixth century BCE, and the Tamils (18 per cent of the population), who arrived from south India in the eleventh century. Today the Tamils still have neither citizenship status nor voting rights, and in addition to seeking recognition as an indigenous people (in order to protect themselves from persecution by the Sinhalese majority) they were engaged in guerrilla activity from the 1970s until 2009. This conflict is discussed in a broader South Asian context in Chapter 8.

Contemporary Sri Lanka is a far cry from the tropical Indian Ocean island called 'Serendip' by the first Arab visitors and 'Paradise' by many later Europeans.

---

undernutrition
Diet inadequate to sustain normal activity.

malnutrition
A condition caused by a diet lacking some food necessary for health.

---

all the countries that we regard as less developed belong in this group.

Although this world system is dynamic, it is extremely difficult for a state to move out of peripheral status because the other states have vested interests in maintaining its dependency. Later, in Chapter 10, we will explore a more sophisticated conceptual basis for world systems theory, derived from Marxism.

The closely related logic of dependency theory, as developed by Frank (see, e.g., Chew and Denemark, 1996), stresses that, for some areas of the world to become developed, other areas have to become underdeveloped. This is because economic value is transferred in one direction only: from periphery to core. For example, in simple terms, most European countries benefited from extracting resources from their colonies but provided no substantial benefits in return. Thus the European colonizing countries grew economically, while the colonies lost potential.

POPULATION AND FOOD

Undernutrition and malnutrition

Until the nineteenth century, hunger and malnutrition were not uncommon in Europe. Today, however, large-scale food problems are for the most part limited to the less developed world. There is a world food problem, despite evidence that if the world's food production were equally divided among the world's population, nobody would go hungry.

A diet may be deficient in quantity, quality, or both. Requirements vary according to age, sex, weight, average daily activity, and climate, but an insufficient quantity of food (or calories) results in **undernutrition**. The best-known cases of acute undernutrition are the famines that attract media attention.

An adequate diet includes protein to facilitate growth and replace body tissue, as well as various vitamins. A diet deficient in quality results in **malnutrition**, usually a chronic condition. Among the health problems caused by undernutrition or malnutrition are kwashiorkor (too few calories), poor sight (inadequate vitamin A), poor bone formation (inadequate vitamin D), and beriberi (inadequate vitamin B_1). The most extreme consequence of undernutrition and malnutrition is death.

The extent of the problem

We all know that various parts of the world, especially in Africa, seem to be especially vulnerable to hunger and famine. What may not

Box 5.9   **The Less Developed World: Haiti**

The 'nightmare republic' of author Graham Greene, Haiti is a third example of a politically troubled country in the less developed world (Barberis, 1994). The basic demographic data for 2008 show a population of 9.1 million, a very high CBR (29), a high CDR (11), and a high RNI (1.8). The use of modern contraceptive techniques is low—although there are good reasons to believe that demand for them is high. There is a high population density of 328 per square kilometre.

Mostly mountainous with a tropical climate, Haiti has had a long history of political turmoil since gaining independence (the first Caribbean state to do so) from France in 1804 following a 12-year rebellion. After 30 years under the brutal rule of the Duvalier family, the first free election was held in 1990, but the victor was overthrown in 1992 and the country was again ruled by a military despot until democracy was re-established in 1994. However, the country has continued to be politically unstable.

Economic disparities are extreme, not only between the poor Creole-speaking black majority (95 per cent of the population) and the rich French-speaking minority (5 per cent), but also between the capital city of Port-au-Prince and the rural areas. It is estimated that the wealthiest 1 per cent of the population hold 44 per cent of the wealth. The rural population practise subsistence agriculture on soils that are generally poor, leading to erosion and declining soil fertility. Access to clean drinking water is difficult, hunger is widespread, and rates of infant mortality, tuberculosis, and HIV are all high. In 2009 Amnesty International initiated an on-line petition protesting government indifference to the high incidence of rape by gangs of armed men—rape was not a crime in Haiti until 2005. Also, a collapsed infrastructure and drug trafficking and related corruption prevail. Finally, bad government and lax building standards mean that Haiti is tragically unable to cope with natural disasters, such as tropical storms or the earthquake of January 2010 that killed many thousands and virtually destroyed Port-au-Prince.

Poverty and wealth in Haiti are starkly contrasted.

© CIDA photo/Hélène Tremblay; © CIDA photo/Benoit Aquin

be so well known is that world food supplies are increasing faster than world populations. A few examples:

- From 1960 to 2000, global food production more than kept up with global population increase.
- World cereal consumption doubled between 1970 and 2005.

- World meat consumption tripled between 1961 and 2005.
- The global fish catch grew more than six times between 1950 and 1997.

Clearly, however, these improvements at the world level do not translate into improvements at all of the regional levels, and as one extensive global research study recently warned, without

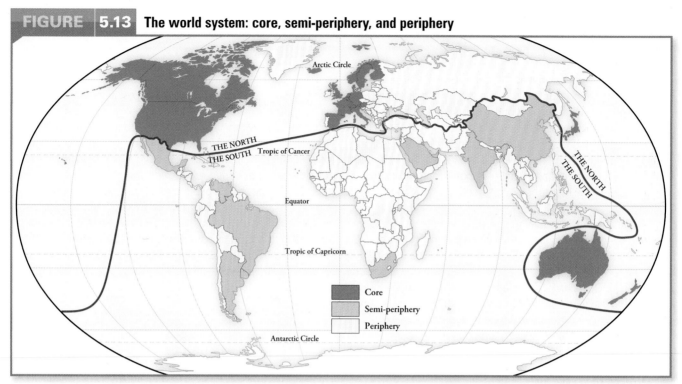

FIGURE 5.13   The world system: core, semi-periphery, and periphery

The line marks the 1980 Brandt Report division of North and South.

SOURCE: Adapted from P. Knox and J. Agnew, *The Geography of the World Economy*, 2nd edn (London: Arnold, 1994), 2.

major changes the global fish catch will dwindle to nothing by mid-century.

The World Bank estimates that over 1 billion people receive insufficient nourishment to support normal activity and work. It is important to recognize that without proper nutrition the body, including the brain, cannot develop properly. This means that malnutrition and undernutrition are both causes and consequences of poverty. Today, the highest levels of undernutrition are in such poor countries as Niger, Somalia, Mozambique, Sierra Leone, Bangladesh, and Bolivia.

Food aid

Food aid provided by the more developed world helps enormously, but has not proven to be a solution to famine. First, such aid tends to be directed to urban areas, even though the greatest need is usually in rural areas; indeed, much donated food goes to governments, which then sell it for profit. Second, food aid tends to depress food prices in the receiving country, thus reducing the incentive for the people to grow crops and increasing their dependence. Third, in many cases food aid is not effectively distributed, whether because of inadequate transport or because undemocratic governments in the receiving countries control the food supply and feed their armies before anyone else. In 2006, when suffering from a severe drought and food shortages, Eritrea expelled three charities working in the country in an apparent attempt to attract international attention to the long-running border disagreement with Ethiopia (see Box 5.6). Even more disturbing, a 2006 report from Save the Children stated that humanitarian workers and peacekeeping soldiers in Liberia were providing food aid in return for sex with young girls. Box 5.10 develops these comments about food aid in the larger context of combatting poverty.

Feeding the world

As we saw in Chapters 3 and 4, there is wide-ranging opinion on the number of humans the earth can support. The same is true of the world food problem. In a world that believed in consumption equity and nutritionally adequate diets for all of the world's people, there would be no problem in feeding many more than 10 billion people—although even the current 6.8 billion people could not be fed if the North American diet were taken for

Box 5.10 | How to Make Poverty History

The comments about food aid may also apply more generally to attempts to help countries emerge from poverty. There is, for example, much debate about the merits of the Make Poverty History campaign initiated in 2003 and strongly supported by the Irish rock singer, Bob Geldoff, other popular entertainers, and many politicians. A major focus of this campaign was collecting money from individuals and pressuring governments to provide more aid. Although such attempts to aid poor countries are laudable and well motivated, there is much debate as to whether they are effective. A country such as Malawi has received huge amounts of aid in recent decades but without any indication of meaningful improvement in the quality of life for the vast majority of the population. Raising this critical point is not to suggest that many countries do not require humanitarian aid, disaster relief, AIDS education, or drugs to combat disease. *They do.*

Rather, the criticism is that merely pouring in money does not achieve the desired end to poverty. Indeed, it can be argued that it only serves to hinder development at the local level by discouraging local people from taking their own initiatives. Also notable is that much global aid is not heading where it is most needed, to the poorest people in the poorest countries. A 2005 report by two major humanitarian agencies, Oxfam and ActionAid, accused Western governments of using aid to reward strategic allies and to support favoured economic development projects, and also suggested that only about one-fifth of aid money went to the countries most in need, with only half of that one-fifth being invested in health and education.

Perhaps, more importantly, aid funds are often sent to countries known to be governed by corrupt politicians. Botswana is an example of a country that has been governed well and has steadily improved in terms of quality of life; Malawi, Zimbabwe, and Zambia are contrary examples, not always well-governed and mired in poverty. In other countries, wealth achieved from oil or other natural resource revenues is not used to benefit the population but rather a select few. Indeed, Collier (2007) argues that natural resources lead to corrupt government. A recent notorious example involved the son of the president of Equatorial Guinea spending more than Cdn$2 million on luxury cars, in addition to owning

expensive homes in several countries around the world, all while the vast majority of the population of the country are poverty-stricken. More generally, about 40 per cent of Africa's private wealth is held overseas, much in Swiss bank accounts.

Of course, not being well-governed does not simply have to do with the occasional (or perhaps all too frequent) corrupt politician. A 2005 World Bank report, *Doing Business in 2005*, reported many of the regulatory and bureaucratic obstacles to conducting business and achieving prosperity in some less developed countries. Consider, for example, that it takes two days to incorporate a business in Canada, but 153 days in Mozambique, or that in Sierra Leone it costs 1,268 per cent of average annual income to register a company, whereas in Denmark it costs nothing. In Lagos, the commercial capital of Nigeria, recording a property sale involves 21 procedures and takes 274 days. Obtaining a licence to conduct business in Kenya is an opportunity available only to those with government connections. One consequence of these circumstances is that instead of creating wealth through private enterprise, many entrepreneurs resort to begging or criminal activity. These specific examples of inefficiency can be multiplied many times over and, together, such inefficiencies hinder economic and social development. Clearly, many of the solutions to poverty and related problems need to be addressed by the poor countries themselves. In many cases this is difficult because, tragically, poor countries are prone to civil war, especially if there are many young, uneducated men and several minority ethnic groups (Collier, 2007).

More positively, a 2006 report on the quality of government globally—measuring such things as free media, political stability, rule of law, and control of corruption—showed African countries with the greatest improvement, with particular successes in Kenya, Niger, Sierra Leone, Angola, and Rwanda. While Africa certainly faces numerous problems related to poverty, nutrition, health, and civil strife, it might be argued that the perception of Africa from outside is both partial and warped as media tend to cover only wars, disasters, and famines, and continually stress the need for aid and intervention. Thus, at least implicitly, these accounts suggest that Africa needs to be saved from Africans (Dowden, 2008).

---

the norm worldwide. The problem is complex. The number of undernourished people in the world is increasing, yet food is being produced today to feed everyone alive today. There seem to be good reasons to believe that current food shortages are not caused by inadequate supplies and that supplies will

be adequate in the future (Bongaarts, 1994; Smil, 1987). On the basis of a scientifically detailed analysis of the complete food cycle, Smil (2000: 315) arrived at a conclusion he described as 'encouragingly Malthusian'. In other words, the future may not be as bright as we might hope, but neither is it totally bleak. Thus, with reference to the competing catastrophist and cornucopian positions we might suggest that neither extreme is a realistic portrayal of our future.

Rather, like several other current problems (for instance, the problem of refugees), the food problem appears to be associated with specific areas and a multitude of complex interrelated causes—physical, economic, cultural, but primarily political.

EXPLAINING THE WORLD FOOD PROBLEM

Until recently, most explanations of the world food problem have focused on three factors:

1. *Overpopulation:* The root cause of the food problem is often said to be the sheer number of people in the world. Yet evidence suggests that in fact there are not too many people in the world. 'Overpopulation' is a relative term, and densely populated areas are not necessarily overpopulated. Table 5.10 confirms that countries with dense populations, such as the Netherlands and South Korea, are not overpopulated, while less densely settled countries, such as Ethiopia and Mexico, may be overpopulated. A compelling example is China, which experienced regular famines when it had a population of 0.5 billion but is now, with a population of 1.3 billion, essentially free of famine.

Customers become animated and compete to buy the next batch of fresh bread at this subsidized bakery in Cairo, Egypt. Egypt experienced a bread crisis in 2008 as the price of wheat on world markets increased dramatically.

Jason Larkin/Getty Images

2. *Inadequate distribution of available supplies:* Most countries have the transportation infrastructure to guarantee the intranational movement of food, but other factors prevent satisfactory distribution.

3. *Physical or human circumstances:* A 1984 drought in central and eastern Kenya is estimated to have caused food shortages for 80 per cent of the population. Other causes in specific cases include flooding and war. Circumstances such as these obviously aggravate existing problems, but they cannot be considered root causes.

Political and economic explanations

More recent theories have tended to focus on the political and economic aspects of the food problem. For example, the world systems perspective discussed earlier suggests that global political arrangements are making it increasingly difficult for many in the less developed world to grow their own food. Not only is the percentage of the population involved in agriculture in the less developed world declining (from more than 80 per cent in 1950 to less than 60 per cent today), but the vast majority of those remaining in agriculture are incapable of competing with the handful of commercial agriculturalists who are able to benefit from technological advances. Indeed, the vast majority have lost control over their own production because of larger global causes. For example, farmers in Kenya are actively encouraged to grow export crops such as tea and coffee instead of staple crops such as maize; peasant farmers in many countries are losing their freedom of choice because credit is increasingly controlled by large corporations; and many governments make it their policy to provide cheap food for urban populations at the expense of peasant farmers. The essential argument here is that the capitalist mode of production is affecting peasant production in the less developed world in such a way as to limit the production of staple foods, thus causing a food problem.

In the same way, global economic considerations make it increasingly difficult for people in the less developed world to purchase food. Throughout the world, food is a commodity; hence, production is related to profits. In other words, food is produced only for those who can afford to buy. In 1972, a year when famine was widespread in the Sahel region of Africa, farmers in the US were actually paid to take land out of production to increase world grain prices.

These arguments suggest that the cause of the world food problem is the peripheral areas' dependence on the core area. Yet any attempt to correct that problem (apart from clearly humanitarian action) is likely doomed to failure, because the more developed world is not about to initiate changes that would lessen its power and profits. If this argument is followed to its logical conclusion, the world food crisis can only get worse, regardless of technological change, because the cause of the problem lies in global political and economic patterns.

The fact is that feast and famine live side by side in our global village. There are rich and poor countries, and in the poor countries especially, there are rich and poor people. Increasingly, agriculture is a business, and it is in the nature of business that not all participants compete equally well. Food problems are not typically caused by overpopulation, and population density is a poor indicator of pressure on resources. As suggested in Table 5.10, if we

Table 5.10	Population Densities, Selected Countries, 2008
Country	**Population Density per km²**
South Korea	488
Netherlands	396
Ethiopia	72
Mexico	55

SOURCE: Population Reference Bureau, *2008 World Population Data Sheet* (Washington: Population Reference Bureau, 2008).

were to judge food scarcity by population density, we might assume that the people of South Korea and the Netherlands are going hungry, while the people of Ethiopia and Mexico are well-fed. This, of course, is not the case. We will not solve food problems merely by decreasing human fertility.

A concerted and co-operative international effort is needed to improve the quality of peasant farming and to reorient it towards the production of staple foods. But there is little indication that this will be achieved in the foreseeable future. Meanwhile,

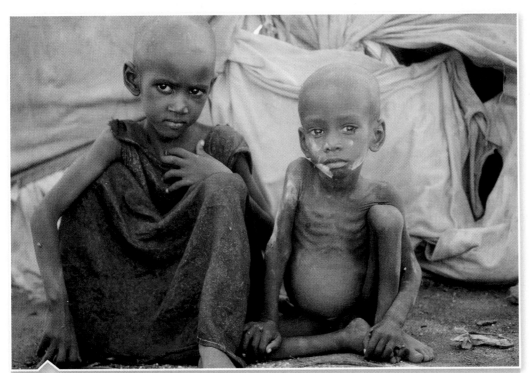

Severely malnourished children at a camp for refugees, returnees, and displaced persons in the Ogaden region of Ethiopia in 1991. Many factors contributed to the Ethiopian famine of the early 1990s, including drought in some regions; rains elsewhere that came at the wrong time, destroying the crops; returnees from the war in Somalia (many had gone to Somalia a few years before to escape the war in Ethiopia); and the escalation of the civil war in Ethiopia, during which the Mengistu government siphoned off government resources that could have helped famine victims.

© CIDA photo/Roger LeMoyne

undernutrition and malnutrition are not disappearing, and famines, typically prompted by events such as droughts and wars, will continue to occur.

What makes the world systems perspective particularly valuable is that it emphasizes the need to focus on structural causes—not just the fairly obvious immediate causes, such as drought and crop failure, that typically receive the attention of the media. But the basic world systems logic can be usefully applied at various spatial scales below that of the world as a whole. Indeed, the role of power, in the sense of the politics that govern access to food, is relevant at various scales.

The idea of entitlements

A case in point is the explanation of world food problems suggested by Young: 'Patterns of food distribution may be examined with reference to people's entitlements reflected in their ability to *command* food' (Young, 1996: 99). *Entitlements* are the factors and mechanisms that explain people's ability to acquire food in terms of their power. In addition to entitlements at the international scale (addressed by world systems theory), this argument identifies entitlements at three smaller scales: national, regional, and household.

At the *international* scale, as world systems logic suggests, among the historical legacies of colonialism is an emphasis on export production at the expense of local production and hence a vulnerability to global market changes. In other words, small changes in the price of a commodity can significantly affect entitlements. Especially in the 1980s, high levels of debt led the International Monetary Fund and the World Bank to require that countries in the less developed world adopt structural adjustment policies—typically involving even greater emphasis on export production—before additional loans would be issued or existing loans restructured. Contemporary globalizing trends are aggravating this situation as the demand for food continues to increase in parts of the world that already are poor and lack power.

At the *national* scale, governments may not be committed to ensuring that economic growth is accompanied by the elimination of food shortages. Because of the distribution of power, national governments may support urban activities and commercial agriculture at the expense of the rural peasant sector.

At the *regional* scale, governments may neglect areas inhabited by relatively powerless minority ethnic groups. Further, problems of various kinds are often regionally distinct: some areas may be more subject to conflict or environmental problems than other areas.

At the *household* scale, the most vulnerable family groupings are those that are poorest, that include many dependants, that are isolated, and that are powerless. Even within households, there are differences in ability to command food; females and the elderly are often the most vulnerable.

This multi-scale approach to understanding the world food problem stresses the inequalities that exist in people's ability to acquire food. Food shortages are placed in context using broad geographic and historical frameworks with emphasis on entitlements and the related ability of people to command food. We will introduce a conceptual basis for this political economy approach in Chapter 10. Those global areas, countries, regions within countries, households, and even individuals within households whose entitlements are most limited are the ones least likely to command adequate amounts of food. The basic equation is this: *lack of power equals lack of food.*

The role of bad government

In a similar vein, bad government is a cause of food problems and of lack of economic and social development more generally. Most notably, the Indian economist Amartya Sen argues that widespread hunger has nothing to do with food production and everything to do with poverty—which in turn is closely related to political governance. Sen points out that famine does not occur in democratic countries, because even in the poorest democracy famine would threaten the survival of the ruling government. Even a cursory review of current global famine areas confirms that famine seems to have six principal causes, with bad government front and centre. Of the other five causes, four are closely related to bad government: a prolonged period of underinvestment in rural areas; political instability related to conflict that causes refugee problems; HIV/AIDS and other diseases depriving families of productive members and damaging family structures; and continued population growth because of high birth rates. The final specific cause is bad weather. As of 2009, famine threatened much of sub-Saharan Africa, with many countries in need of food aid.

Links between quality of governance in a country and food production are emphasized in the 2008 *World Development Report*. Arguing that agriculture can play a key role in development, the *Report* notes the need for sound agricultural policies and investment. Unfortunately, many of the less developed countries perform poorly on a range of governance measures. Figure 5.14, employing a fourfold division of countries and six measures of quality of governance, shows a clear relationship between a poor governance score and emphasis on agriculture.

Box 5.11 extends the consideration of the role of bad government in an account of the food crisis that began in 2007. The larger question of the spread of democracy globally is considered in Chapter 8.

In recent years, many African countries—including Ghana, Niger, Mali, Gambia, Malawi, and Kenya—have shown enthusiasm for democratic institutions and increasing personal freedom. A 1999–2001 survey of more than 21,000 people in 12 African countries showed that more than 70 per cent of respondents favoured democracy over other forms of government. A former president of Botswana won a US$5 million prize to encourage good governance in Africa. Botswana has been one of the most politically stable countries in Africa since independence in 1966, never having had a coup and holding regular multi-party elections. Indeed, as of 2009, the repressive regime of Robert Mugabe in Zimbabwe, which has contributed to a food crisis affecting millions, is more the exception than the norm in sub-Saharan Africa.

Further support for the connection between famine and politics can be found in the historical record. Between 1875 and 1914, some 30–50 million people died in a series of famines in India, China, Brazil, and Ethiopia.

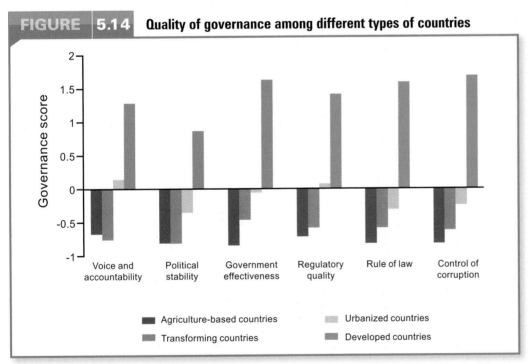

FIGURE 5.14 Quality of governance among different types of countries

Legend:
- Agriculture-based countries
- Transforming countries
- Urbanized countries
- Developed countries

The four country groupings are not mutually exclusive: agriculture-based countries are mostly in sub-Saharan Africa; transforming countries are mostly in South and East Asia, the Middle East, and North Africa; urbanized countries include most of Latin America and much of Europe and Central Asia; and the developed countries (a grouping that includes many of the urbanized countries) are those in the more developed world. The six dimensions of governance—voice and accountability, political stability and absence of violence, government effectiveness, regulatory quality, rule of law, and control of corruption—used to calculate the governance score are taken from a substantial World Bank research project (see Kaufmann et al., 2006). For the purposes of this figure, scores below 0 indicate poor governance and above 0 indicate better governance.

SOURCE: World Bank, *World Development Report: Agriculture for Development* (Washington: World Bank, 2008), 23, Figure 12, at: <siteresources. worldbank.org/INTWDR2008/Resources/2795087119211580172/WDROver2008-ENG.pdf>/12: Agriculture-based and transforming countries get low scores for governance, p. 23. Kaufmann, Kraay, and Mastruzzi 2006.

Box 5.11   A Food Crisis in the Early Twenty-First Century

In 2007–8 many basic foodstuffs trading on international commodity markets experienced dramatic price increases after about 30 years of stability; prices for wheat, corn, soya, rice, coffee, and meat all increased, in some cases more than doubling. While these increases affected consumers everywhere, the less developed world suffered most.

Attempts to explain this situation referred to the classical economic logic of an increased demand combined with a reduced supply. But several other factors played roles.

- Some explanations referred to the growing world population, especially to increased demand from the growing and wealthier populations of India and China.
- Other explanations centred on environmental pressure, for example, accelerating desertification that resulted in loss of agricultural land.
- Most observers agreed that increasing oil prices, which in turn raise the costs of agricultural production, were a key cause.
- There was general agreement that the shift to biofuels—related to concerns about global warming—shifted grains away from food to fuel. In 2007, for example, one-third of the United States corn harvest was directed to fuel. It is estimated that filling up an SUV fuel tank with ethanol uses enough corn to feed one person for a year.
- A variety of more locally specific causes, usually weather-related, included poor rice harvests in some Asian countries.

It can also be argued that bad government contributed to the crisis. In response to initial price rises and reduced surpluses, some governments ceased exports, thus increasing the likelihood of further price increases. More generally, especially in some African countries, poor governance may create barriers to distribution through inadequate transport infrastructure; may impose internal levies and tax technologies such as refrigeration and packaging and basic inputs such as fertilizer; may limit improvements in agriculture through landownership legislation, weak property rights, and a lack of the rule of law; may permit political elites to promote their interests at the expense of agricultural progress for the majority; and may operate marketing boards that oblige farmers to sell at below-market prices.

Knowing how best to respond to increases in the price of food is far from simple. For many, the solution is to repeat the successes of the 'green revolution' (discussed in Chapter 10) that saw massive production increases in much of the less developed world through new technologies of production. Although African countries benefited little from the 'green revolution', partly because of the varied topography and small farms, they might benefit from technologies tailored for local circumstances (Conway, 2008). Alternatively, the head of the UN Environment Program has argued for organic farming in Africa as the best way to increase yields, improve soils, and raise incomes. There is also a need to correct many of the government-related problems noted above, in particular through the provision of reliable and accessible input and output markets. Other observers argue that governments globally need to liberalize markets—eliminating export quotas and trade restrictions—not intervene in them.

Describing these famines, Davis (2001) recognizes the role played by meteorological conditions, but also points to the complex politics of colonialism and capitalism as essential causes.

THE WORLD DEBT PROBLEM

The long-term indebtedness of the less developed world—money owed to international lending agencies and commercial banks in the more developed world—increased from US$59.2 billion in 1970, to $445.3 billion in 1980, to $1,167.9 billion in 1990, and to more than $2,000 billion by the early twenty-first century. The 1970s saw high lending to less developed countries by the commercial banking sector, development agencies, and governments in the more developed world. These loans, intended for the establishment and support of economic and social programs, had long payback terms because it was generally expected that, in the long run, the less developed economies would boom and eventually provide good returns on the investments.

Unfortunately, the 1980s brought a recession, with rising interest rates, declining world trade, and steadily increasing debt. As a result, many countries are now so poor and owe so much that they need to borrow more money just to keep up the interest payments on their existing debt—a situation that first arose in 1982 in Mexico (Sowden, 1993). But the news is not all bad. First, agreement was reached in 2005 to reduce some debts, although the larger problem is far from resolved. Second, in 2006 Nigeria became the first African country to pay off its 'Paris Club' debt, of about US$12.4

billion, leaving a remaining debt to the World Bank and other lenders of about US$5 billion. This repayment means that Nigeria is able to invest significantly in health care and education and might also attract increased foreign investment.

The significance of debt to a country depends on the country's economy. The United States is the biggest net debtor in the world, but the effect of the debt on the economy is minimal because the US is a high-income country with a large volume of exports. Other countries such as South Korea, which borrowed heavily to finance industrialization, have been able to repay their loans because of their industrial and export success. In many less developed countries, however, the cost of servicing the foreign debt accounts for up to 30 per cent of all income from exports. For those countries, foreign debt is such a crushing burden that some, such as Colombia, have actually exported food to help repay it, even though their own populations are malnourished; not surprisingly, there is a concentration of severely indebted low-income countries in Africa.

Loans intended to help impoverished countries can have quite the opposite effect. Box 5.12 outlines one alternative approach, an innovative program in which loans are provided directly to poor individuals themselves

rather than to their country as a whole, and from within countries rather than from outside. It is also possible to rethink the whole question of world debt and argue that the debt owed by the less developed world is only one side of the coin. For example, a 1999 report published by Christian Aid, a leading British charity, asserted that the more developed world owes a huge debt to the less developed world because of the disproportionate environmental damage that the more developed world causes. Similarly, discussions at the 2001 World Conference Against Racism, Racial Discrimination, Xenophobia, and Related Forms of Intolerance, sponsored by the United Nations and held in Durban, South Africa, included arguments that rich countries should pay poor countries compensation for the abuses (such as slavery) inflicted on them in the past. The likelihood that rich countries would have heeded such a proposal was minimal; a few days later, the 9/11 terrorist attacks on the US shifted global priorities far away from poor country debt.

Recent data, however, suggest that the debt situation may be improving. More developed countries now are providing additional aid and debt relief while, more significantly, African countries have better-managed and growing economies. According to the World Bank, for the five years to 2008 the continent's 47 sub-Saharan

---

Box 5.12 | The Grameen Bank, Bangladesh

A remarkable transformation has occurred in parts of Bangladesh, a transformation brought about by a credit program that focuses exclusively on improving the status of landless and destitute people—especially women.

Bangladesh is one of the poorest countries in the world. It has a large population of 147.3 million, an exceptional population density of 1,023 per km[2], a CBR of 24, a CDR of 7, and an RNI of 1.7. The per capita gross national income is one of the lowest in the world at US$470.

The Grameen ('village') Bank program began in 1976, under the leadership of Muhammad Yunus, as a research project at the University of Chittagong to explore the possibility of providing credit for the most disadvantaged of the Bangladeshi people and of encouraging small-scale entrepreneurial activity. In 1983 the project became an independent bank, which by 1988 had 571 branches covering 17 per cent of the villages in Bangladesh. Essentially, the Grameen

program seeks to enhance the social and economic status of those most in need, especially women, by issuing them small loans (the first Grameen loan was $27 made to 43 women; today the institution's loans average $300). The bank requires that loan applicants first form groups of five prospective borrowers and meet regularly with bank officials. Two of the five then receive loans and the others become eligible once the first loans are repaid. The focus is on the collective responsibility of the group. As the recovery rate—a remarkable 98 per cent—shows, the Grameen Bank's innovative approach works.

This successful technique is now being replicated elsewhere in the world (see Mahmud, 1989; Todd, 1996), so that '[t]here are now 10,000 microfinance institutions in more than 100 countries, earning a modest but healthy profit and serving 100 million people' (*Toronto Star*, 2006). Yunus and the Grameen Bank were awarded the 2006 Nobel Peace Prize.

countries (including five small island countries and the larger island of Madagascar) were growing at a remarkable 5 per cent per annum.

THE SELECTIVITY OF DISASTERS

Human geographers have long used the term 'natural disasters' to refer to physical phenomena such as floods, earthquakes, and volcanic eruptions and their human consequences. In recent years, however, we have recognized that this label is inappropriate. Certainly such events are natural in the sense that they are part of larger physical processes—but not all such events become disasters. To understand why a natural event becomes a human disaster, we need to understand the larger cultural, political, and economic framework. In short, some parts of the world and some people are more vulnerable than others to natural events. Just as, on average, the rich live longer than the poor because they are better able to afford good nutrition and medical care, so some parts of the world are better able to protect themselves against natural events.

The United Nations designated the 1990s as a decade to focus on reducing the human disasters that so often accompany natural events, especially in the less developed world. Between the 1960s and 1980s, the number of major disasters increased fivefold, fatalities increased considerably, and regional disparities between the more and less developed worlds were widening.

As with food supplies and national debts, again the less developed world suffers the most. Adverse cultural, political, and economic conditions combine to place increasing numbers of people at serious risk in the event of any environmental extreme. 'The problems of staggering population/urban growth and crippling overseas debt that face many less developed countries have often become manifested in poor construction standards, poor planning and infrastructure, inadequate medical facilities and poor education—all of which exert a direct influence on vulnerability to natural hazards' (Degg, 1992: 201).

Box 5.13 provides details on a poor and densely populated area of the world that is regularly subject to devastating floods.

Earthquakes and tsunamis

If we were able to peel away the thin shell of air, water, and solid ground that is the outer layer of the earth, we would find that our planet is a furnace. Many of our major cities are a mere 35 km (22 miles) above this furnace, and in the deep ocean trenches the solid crust above the furnace is as little as 5 km (3 miles) thick. Not surprisingly, earthquakes and volcanic eruptions are regular occurrences. Between 200 and 300 earthquakes occur each year in Canada alone, while several major volcanic eruptions and earthquakes have made headlines recently (Mount Pinatubo in the Philippines and Mount Unzen in Japan, both in 1991, the Pakistan earthquake in 2005, and the China earthquake in 2008).

The impact of an earth movement on the Indian Ocean seabed was dramatically highlighted by the resulting devastating tsunami that hit parts of Asia and Africa in December 2004. The tsunami moved across the ocean with little loss of energy, resulting in much destruction when it reached land. Figure 5.15 shows the origin and movement of the tsunami and the areas affected. There were about 283,000 deaths and a further 1.1 million people displaced. An even greater tragedy was avoided as aid agencies moved in quickly to provide clean water in order to combat disease.

Contemporary scientists are continually attempting to improve our ability to predict earthquakes and volcanic eruptions. We know that most earthquakes occur at the margins of tectonic plates, when the plates move against each other; we also know some of the

People take shelter on the roof of a house on the outskirts of Dhaka, Bangladesh, in July 1998. Floods caused by torrential monsoon rains affected more than three million people and claimed at least 72 lives.

CP/AP photo/Pavel Rahman

Box 5.13 | Flooding in Bangladesh

Bangladesh is a low-lying country (mostly only 5–6 m/16–20 feet above sea level) that lies at the confluence of three large rivers, the Ganges, the Brahmaputra, and the Meghna. Floods are normal in this region; indeed, they are an essential part of everyday economic life, as they spread fertile soils over large areas. During the monsoon season, however, floods often have catastrophic consequences, especially if they coincide with tidal waves caused by cyclones in the Bay of Bengal (see Figure 3.8). Three factors contribute to flooding:

1. Deforestation in the inner catchment areas results in more runoff.
2. Dike and dam construction in the upstream areas reduces the storage capacity of the basin.
3. Coincidental high rainfall is a frequent occurrence in the catchment areas of all three rivers.

Regardless of specific causes, floods are difficult to control because of their magnitude and because the rivers often change channels.

The most extreme consequence of flooding, of course, is death; in a 1988 flood, over 2,000 died. Impelled migration is another serious consequence; also in 1988, over 45 million people were uprooted. Disasters occur regularly. In a normal year, over 18 per cent of Bangladesh is flooded, and even normal floods result in shifting of river courses and erosion of banks that result in population displacement. In an already poor country, such displacement aggravates landlessness and food availability. Most displaced persons move as short a distance as possible for cultural and family reasons. During these impelled migrations, the poorest suffer the most, and women-headed households are especially vulnerable. Some migrants move to towns, hoping to become more economically prosperous. Typically, however, such rural-to-urban migrants become further disadvantaged, clustering together in squatter settlements generally regarded as unwelcome additions to the established urban setting.

signs that can indicate a quake is likely in a particular area. To date, however, more specific predictions—predictions that might save lives—remain elusive.

Tropical cyclones

Tropical cyclones regularly kill people, but how many die may be far from natural. This obvious fact was starkly highlighted when Myanmar (Burma) was hit by a typhoon in May 2008 (typhoon is a regionally specific term for a tropical cyclone; the term used in the North Atlantic region, including the Gulf of Mexico, is hurricane). Myanmar is ruled by a secretive, isolated, non-democratic, military junta that chose not to respond effectively to the disaster. Some people died immediately, but the vast majority of deaths occurred in the following days as corpses spread disease and as victims were unable to obtain food and fresh water. Even six days after the cyclone struck the number of people affected was unclear. Other countries tried to send aid workers and aid but visas were refused and the military authorities insisted on handling, or in some cases rejected, all aid shipments. At a time when food was scarce for victims, Myanmar continued to export rice to other countries.

One month after the typhoon struck, the UN estimated that 2.4 million people were in need of food, shelter, or medical care. Even one year after the cyclone, the UN reported that several hundred thousand people were still in need of assistance, many continued to live in flimsy shelters, water supplies were contaminated, and reconstruction had barely begun. The inhumane government response might be explained by the fact that the government is more concerned with minimizing outside influences than it is with the welfare of its citizens. Put simply, the government chose not to save lives.

But disasters also can have devastating impacts in the more developed world, as evidenced by the hurricane that hit New Orleans in August 2005 (Figure 5.16). With a sustained wind speed of about 200 km/h, Hurricane Katrina passed directly through New Orleans, flooding much of the city and creating conditions that most people would have thought impossible in a major American city. As the city was evacuated it was clear that there was a close relationship between vulnerability and poverty, and it is possible to interpret the effects of this hurricane as a selective disaster impacting the poorest most severely.

FIGURE 5.15 Countries affected by the Asian tsunami, December 2004

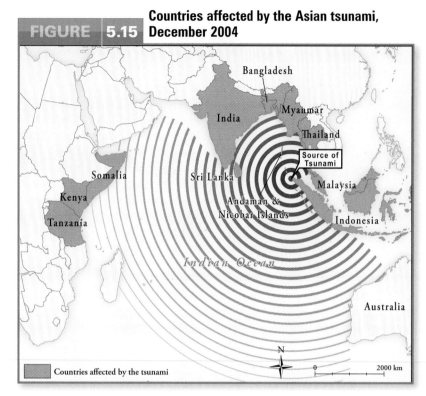

Countries affected by the tsunami

0   2000 km

Although much concern has been expressed about bird flu in Asia and elsewhere, about Marburg and Ebola viruses in Africa, and about swine flu throughout the world, the reality is that easily preventable diseases continue to kill large numbers of people. Malaria is a prime example. Figure 5.17 maps malaria in Africa, which has about 90 per cent of all cases, and large areas of South and Southeast Asia and of Latin America also are vulnerable to the disease. Note the extensive area of Africa where the disease is endemic, meaning constantly present in the population. In Africa, malaria accounts for an estimated US$12 billion in lost productivity each year. Mosquitoes carry malaria from person to person. On average, there are about 500 million cases of clinical infection and between 1.5 million and 3 million deaths from malaria each year. It merits stressing that this impact is comparable to AIDS, and yet malaria has never captured the public imagination in the same way as has AIDS. Most malaria deaths are in Africa and young children are most vulnerable. What makes these facts especially difficult to understand is that malaria can be prevented and can be treated. Of particular significance is that it is not difficult to predict a malaria outbreak, often several months in advance, because the disease is so closely related to local weather circumstances. Recent medical advances in the understanding of the disease, evidence from Kenya that the free distribution of mosquito nets dramatically reduces child deaths, and increased funding have prompted the UN-backed Global Malaria Action Plan to identify the goal of reducing malaria deaths to almost zero by 2015.

Other diseases that kill many poor people, typically several hundred thousand each year, include the following, most of which are preventable.

- Pneumonia and other lung diseases cause many deaths of young children.
- Polio, once believed to have been defeated, has returned to Africa and a major health campaign is underway to eradicate this disease permanently.
- Tuberculosis, a respiratory disease, might be the greatest killer in history with perhaps a billion deaths since about 1800. Poverty, poor nutrition, overcrowding, poor sanitation and hygiene, and HIV/AIDS all make tuberculosis infection more likely.

FIGURE 5.16 The path followed by Hurricane Katrina, August 2005

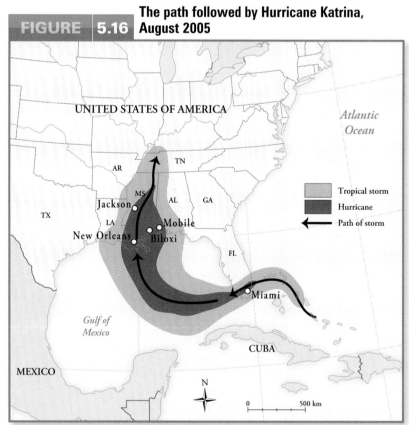

Tropical storm
Hurricane
Path of storm

0   500 km

THE SELECTIVITY OF DISEASES

In the previous chapter, the impact of HIV/AIDS was examined and links with poverty were noted. It is no different with several other diseases that threaten people in the less developed world today.

- A cholera epidemic that began in the summer of 2008 affected about 90,000 people in Zimbabwe; the epidemic peaked in March 2009.
- Whooping cough is still a major killer, with most cases in sub-Saharan Africa.
- Tetanus kills many other people, with most deaths in sub-Saharan Africa, South Asia, and Southeast Asia.
- Meningitis is an ongoing threat, with a 'meningitis belt' stretching across Africa from Senegal to Ethiopia.
- Syphilis is especially prevalent in Southeast Asia, sub-Saharan Africa, and Latin America.
- Lung cancer, heart disease, and other illnesses are closely linked to smoking, and it is estimated that tobacco products cause about 5.4 million deaths annually. Today, about 50 per cent of the world's 1.3 billion smokers live in China, India, and Indonesia.

Several of these diseases are being tackled and one recent success is the case of measles, with the number of deaths globally falling by half between 1999 and 2004. Vaccination is cited as the principal reason for the drop.

STRIVING FOR EQUALITY

The World Bank has identified eight 'millennium development goals':

1. *Eradicate extreme poverty and hunger.* Data for recent years show that GNI per capita in the less developed world is increasing and that the number of people living on less than US$1 per day is declining (Table 5.11). These improvements are most evident in Asia, especially China. Most regions are reducing the proportion of underweight children, and malnutrition rates in the less developed world have fallen from 46.5 per cent in 1970 to 27 per cent in 2000.
2. *Achieve universal primary education.* Education is essential both to the reduction of poverty and inequality and to the building of successful economies and democratic institutions. Data for 2000 suggested that 120 million primary-age children were not in school, mostly in South Asia and sub-Saharan Africa.
3. *Promote gender equality and empower women.* Here again, education is the key—yet in the less developed world, girls are even less likely than boys to attend school and more likely to drop out. Education allows people to improve their economic circumstances by helping them to find better jobs. As we saw in Chapter 4, empowering women is essential to reduce fertility rates.
4. *Reduce child mortality.* Deaths of infants and children have fallen dramatically in recent years, largely because of improved health

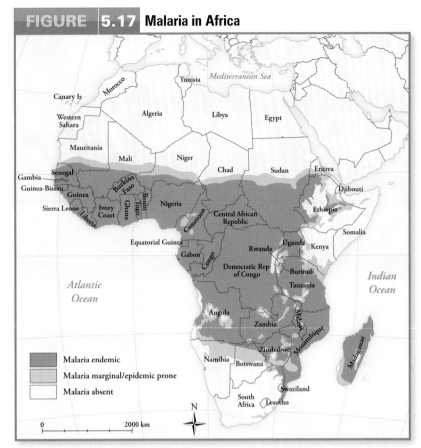

FIGURE 5.17 Malaria in Africa

- Malaria endemic
- Malaria marginal/epidemic prone
- Malaria absent

0   2000 km

Signs mark the location of various corn hybrids grown for use in ethanol production on a plot of farmland near Freeport, Illinois.

Scott Olson/Getty Images

Burmese family in the remains of their damaged house, in May 2008. It has been estimated that more than 100,000 people were killed by Cyclone Nargis and up to one million people were made homeless.

Getty Images

care. Still, the majority of children who die before the age of five do so because of disease and/or malnutrition. These deaths are preventable.

5. *Improve maternal health*. Complications related to pregnancy and childbirth are a leading cause of death among women of reproductive age in the less developed world, where a rate of one maternal death per 100 births is not unusual. The corresponding figure in the more developed world is about 1 in 100,000. Again, these deaths are preventable.

6. *Combat disease*. Many areas suffer from malaria and tuberculosis in addition to HIV/AIDS.

7. *Ensure environmental sustainability*. Wherever resources are depleted, the poorest suffer first and most severely. Sustainable practices are especially urgent with regard to fresh water, forests, and soils.

8. *Develop a 'global partnership for development'*. An open and non-discriminatory trading and financial system must take into account the special needs of the less developed world. It is crucially important to address the debt problem faced by many countries.

The global economic crisis that began in 2008 will inevitably impact negatively on progress towards these goals. The World Bank reported in 2009 that the economic downturn may even reverse some of the progress made in poverty reduction and also limit hoped-for reductions in child mortality. Following on the dramatic increases in food prices in 2007, declining global economic growth is having tragic consequences for those who bear no responsibility for the crisis, namely the poorest in the world.

It is now widely accepted that achieving these goals is a global responsibility. International aid is most effective when it goes to countries with good economic policies and sound governance. But prospects for achieving the eight millennial development goals are uncertain, and there is much disagreement about our global future with respect to the gap between more and less developed parts of the world. For some, there is limited prospect of achieving these goals, even partially, while others point to evidence of unprecedented progress against poverty. What does seem clear today is that the more developed world has become increasingly committed to reducing poverty and improving lives in the less developed world. Box 5.14 returns to this debate, introduced in Chapter 3, between catastrophists and cornucopians, this time in the context of development.

Table 5.11	Percentage of People Living on Less Than $1.00 per day, 1981–2001		
Region	1981	1990	2001
East Asia and Pacific	56.7	29.5	14.3
Europe and Central Asia	0.8	0.5	3.5
Latin America and Caribbean	10.1	11.6	9.9
Middle East and North Africa	5.1	2.3	2.4
South Asia	51.5	41.3	31.9
Sub-Saharan Africa	41.6	44.5	46.4
World	40.4	27.9	20.7

NOTE: These data suggesting that income poverty is declining in percentage terms need to be interpreted with care. Reducing income poverty is a result of both economic growth and the extent to which the very poor benefit from economic growth. The income poverty increase evident in Europe and Central Asia results from the economic difficulties following the collapse of communism, and in sub-Saharan Africa the increase is associated with the ravages of HIV/AIDS. The most notable declines in income poverty are in the East Asia and Pacific and the South Asia regions and result from economic growth in China and India, respectively. Indeed, the global decline from 40.4 per cent in 1981 to 20.7 per cent in 2001 is largely the result of poverty reduction in China.

SOURCE: UNDP, *Human Development Report, 2005. International Cooperation at a Crossroads: Aid, Trade and Security in an Unequal World* (New York: UNDP, 2005), 34. © The International Bank for Reconstruction and Development/The World Bank

Box 5.14 The Best or the Worst of Times?

Much of this chapter's discussion of the less developed world paints a grim picture. Yet there are also promising signs. What will our global future be? The answer is far from clear—and, as our discussion of world systems analysis demonstrates, it depends at least partly on ideological perspective. As we have seen, there is an ongoing debate between pessimists and optimists or, as they are often known, catastrophists and cornucopians. Similarly antithetical perspectives on the future of the world have existed for centuries. In the mid-1700s, just a few years before Malthus predicted a future of famine, vice, and misery, a Berlin vicar and statistician named Johann Peter Süssmilch claimed that the earth could feed 10 billion people.

The differences of opinion in the contemporary debate often seem almost equally extreme. Pessimists and optimists base their forecasts on very different kinds of assumptions. Pessimists such as Ehrlich and the Club of Rome, discussed in Chapter 4, usually assume an existing set of conditions, 'other things being equal'. Optimists, by contrast, argue that 'other things' are not equal; they maintain that circumstances change, and in particular that human ingenuity is always increasing, finding new ways to cope with problems (see Box 3.14).

One writer who has focused attention on specific problem regions in several African and Asian countries is Kaplan (1994, 1996). Reporting in a style that one reviewer described as 'travel writing from hell' (Ignatieff, 1996: 7), he details a nightmare scenario of population growth, ecological disaster, disease, refugee movements, ethnic conflict, civil war, weakening of government authority, empowerment of private armies, rural depopulation, collapsing social infrastructure, increasing criminal activity, and urban decay, and identifies West Africa as the region in the most severe trouble. Kaplan writes convincingly about the interrelatedness of problems such as population growth, environmental degradation, and ethnic conflict—all characteristic in many parts of the less developed world. The evidence that he finds, particularly concerning refugees and food supply, leads him to predict an immediate future of instability and chaos.

Others disagree with both the details of Kaplan's findings and his conclusions. Pointing to steady improvements in health, education, life expectancy, and nutrition, both globally and specifically in the less developed world, Gee (1994: D1) asserts that 'By almost every measure, life on earth is getting better.' It is certainly true that, by some measures, life is improving. Throughout much of the less developed world, fertility is declining, life expectancy is increasing, and adult literacy is increasing. One especially interesting development concerns Botswana, a country that has seen rapid growth in income per person over the past 35 years. Attempts to explain this surprising fact are varied, but it appears that Botswana has succeeded in aligning the interests of the elite with those of the masses, and has used aid money to improve the legal and court system and to fight disease. Elsewhere in Africa, Uganda is educating nearly all of its children of elementary school age.

Critics like Gee have compared Kaplan to such earlier doomsayers as Malthus and the Club of Rome, whose most dire predictions have not come to pass. But there are differences, and they may be significant. Unlike most earlier predictors of nightmare futures, Kaplan was looking at the relatively short term—perhaps two decades—and basing his judgements on specific circumstances in the mid-1990s. As of 2009, the overall picture looks brighter than a pessimist might have predicted. For example, the proportion of the world's people living in extreme poverty is falling (see Table 5.11). On the other hand, specific regional circumstances may well confirm the worst fears.

Some evidence shows improved standards of living and levels of education globally, but indications are that the very poorest countries are falling further behind. With reference to the three globalization theses noted in the Introduction (especially Table I.2), Table 5.12 introduces a further complication, namely the challenge of identifying the links between globalization and development. Clearly, the debate between cornucopianism and catastrophism remains unresolved.

Table 5.12	Globalization Theses and Development	
Hyperglobalist: The Global Era	**Skeptical: Increased Regionalism**	**Transformationalist: Unprecedented Interconnectedness**
Development attained by taking part in globalization. Poverty solved by market. Integration unavoidable.	Development threatened by global capitalist expansion. Marginalization increased by spread of market.	Globalization offers threats and opportunities for development. Progress contingent on careful management.

SOURCE: Adapted from W.E. Murray, *Geographies of Globalization* (New York: Routledge, 2006), 353–4.

CHAPTER 5 SUMMARY

DISTRIBUTION AND DENSITY

'Distribution' refers to the spatial arrangement of a phenomenon; 'density' refers to the frequency of occurrence of a phenomenon within a specified area. Population maps frequently combine the two characteristics. Asia has almost 61 per cent of the world's population; China is the most populous country, followed by India. A third area of population concentration is Europe. The Asiatic regions include some areas with very high rural population densities. Distribution and density of population are related to land productivity and cultural organization.

CAUSES OF HUMAN MOBILITY

Human populations have always been mobile. Migrations are often explained in terms of the relative attractiveness of different locations. This idea can be expressed by reference to simple push and pull logic, 'laws' such as Ravenstein's, or the concept of place utility. Another explanation for migration, related to the demographic transition, is called the mobility transition, and other explanations consider individual preferences and social contexts.

TYPES OF MIGRATION

'Primitive migration' is the process by which humans originally moved over the surface of the earth, gradually adapting to new environments. 'Forced migration' occurs when people have little or no choice but to move; slaves and refugees are examples of forced migrants. Refugees are prompted to move for political, social, and/or environmental reasons; major problem areas today include Ethiopia and neighbouring countries (Boxes 5.6 and 5.7), Israel and neighbouring countries, Iraq, and various Asiatic countries. 'Free migration' is the result of a decision made following evaluation of available locations. 'Mass migration' is a specific form of free migration that involves a great many people making a specific migration decision at about the same time. 'Illegal migration' is increasingly common in a world where many countries have immigration policies intended to ensure that only members of desired groups are admitted.

THE LESS DEVELOPED WORLD

'Less developed' countries may be identified according to various criteria: some, such as GDP or GNI, are solely economic; others reflect a number of measures of human development. All these countries have similar relationships with the more developed world. Many have been colonies and most are in a dependent situation today; virtually all have massive debt loads; most are vulnerable to environmental extremes such as droughts, floods, and earthquakes; and most are prone to disease. Sub-Saharan African is the location of many less developed countries.

WORLD SYSTEMS THEORY

Wallerstein's world systems approach posits a world divided into three zones: core, semi-periphery, and periphery. The world is organized in such a way that core countries benefit at the expense of peripheral countries. (All the countries in the less developed world are peripheral.) It has been argued that the world food problem can only get worse, regardless of technological change, because the causes of the problem lie in global political and economic patterns.

THE WORLD FOOD PROBLEM

Undernutrition is caused by insufficient food; malnutrition is caused by low-quality food lacking the necessary protein and vitamins. One explanation of the global food problem is overpopulation (too many people for the carrying capacity of the land); another is inadequate distribution of food. Some scholars have argued that the world food problem is a consequence of the imposition of a capitalist mode of production, which has made it impossible for increasing numbers of people in the less developed world either to grow or to purchase food. Much evidence also suggests that food shortages and related problems can often be traced to a lack of democratic institutions and to bad government more generally.

DISASTERS AND DISEASES

Disasters such as earthquakes are much more likely to have adverse impacts on people in the less developed world than on people in the more developed world because of a general inability to cope. The impact of diseases is also spatially unequal, with Africans, especially, suffering from several preventable diseases.

OUR GLOBAL FUTURE

There is much debate concerning how best to eliminate poverty in the less developed world, with some favouring huge inputs of aid money and others favouring helping poor countries to help themselves. There is also debate concerning what improvements in lives and livelihoods are taking place today.

QUESTIONS FOR CRITICAL THOUGHT

1. Describe the pattern of world population as shown in Figure 5.1. What factors account for this pattern?

2. The history of humanity can be viewed as an ongoing series of migrations spanning six million years. Why did our ancestors move throughout this history? Why do we continue to move? Does it make sense to think in terms of 'laws' of migration?

3. What role does government policy play as a factor affecting migration today?

4. How should governments respond to refugee situations? For example, who should be responsible for providing assistance for political and environmental refugees?

5. Why is development uneven (i.e., why is there a 'more developed world' and a 'less developed world')?

6. What, if anything, can/should be done by the more developed world to address issues such as malnutrition and debt in the less developed world? Justify your position.

7. From a moral point of view, is there an obligation on the part of the more developed world to send aid to the less developed world? Justify your position.

8. Under the heading 'Striving for Equality', Norton lists the World Bank's eight millennium goals. In your view, are these goals realistic? Attainable? Worth striving toward? Justify your position.

FURTHER EXPLORATIONS

Atkins, P., and I. Bowler. 2001. *Food in Society: Economy, Culture, Geography.* New York: Oxford University Press.

Covers a wide range of topics, including food supply and famine.

Blunt, A. 2007. 'Cultural Geographies of Migration: Mobility, Transnationality and Diaspora', *Progress in Human Geography* 31: 684–94.

A review of recent research on migration.

Brown, E. 2005. 'Unravelling the Web of Theory: Changing Geographical Perspectives on Development', in M. Philipps, ed., *Contested Worlds: An Introduction to Human Geography.* Burlington, Vt: Ashgate, 89–126.

Useful review of the idea of development as it has been employed by human geographers with a good overview of Marxist-derived approaches.

Castles, S., and M.J. Miller. 2009. *The Age of Migration: International Population Movements in the Modern World*, 4th edn. New York: Guilford.

Global overview of the causes and consequences of migration. Includes account of role of ethnic identity.

Champion, A., and A. Fielding, eds. 1992. *Migration Processes and Patterns*, vol. 1. New York: Belhaven Press.

A series of detailed analyses of contemporary migration trends; strong emphasis on links to cultural and economic change; focus on British examples.

Cohen, R. 1995. *The Cambridge Survey of World Migration.* Cambridge: Cambridge University Press.

Ninety-five contributions cover a multitude of migration topics, from the sixteenth century to the present; an invaluable source of information.

Desai, V., and R.B. Potter, eds. 2002. *The Companion to Development Studies.* New York: Oxford University Press.

Many short pieces on a wide range of topics; a useful source of basic factual information.

Dickenson, J., et al. 1996. *A Geography of the Third World*, 2nd edn. New York: Routledge.

An excellent overview that employs a developmentalist perspective to explore population and economic issues; relevant also for Chapters 10, 13, and 14.

Fukuyama, F., ed. 2009. *Falling Behind: Explaining the Development Gap between Latin America and the United States*. New York: Oxford University Press.

Rejecting geographic and climatic arguments, contributors favour arguments related to political culture to explain development differences between the US and Latin America.

Grigg, D.B. 1977. 'Ravenstein and the "Laws of Migration"', *Journal of Historical Geography* 3: 41–54.

An excellent summary and discussion of the topic, which helps our understanding of the contemporary relevance of Ravenstein and of the value of laws.

Guest, R. 2004. *The Shackled Continent: Power, Corruption and African Lives*. Washington: Smithsonian Institute Press.

Very readable account of Africa today, stressing that the principal problems are political.

Kaplan, R.D. 1996. *The Ends of the Earth: A Journey at the Dawn of the 21st Century*. New York: Random House.

Kaplan, a master of 'travel writing from hell', argues that our global future is fraught with difficulties and is being shaped at 'the ends of the earth'.

Keys, A., H. Masterman-Smith, and D. Cottle. 2006. 'The Political Economy of a Natural Disaster: The Boxing Day Tsunami, 2004', *Antipode* 38: 195–204.

A stimulating article highlighting that those who died were the poorest of the poor and that their deaths had no effect on global markets.

Loescher, G. 1993. *Beyond Charity: International Cooperation and the Global Refugee Crisis.* Toronto: Oxford University Press.

A powerful book that argues for a reform and strengthening of organizations such as UNHCR and for a concerted strategy by the industrial nations to tackle refugee problems effectively.

Manning, P. 2005. *Migration in World History.* New York: Routledge.

A thorough review of this substantial topic covering the entire span of human life on earth.

Meredith, M. 2005. *The Fate of Africa: From the Hopes of Freedom to the Heart of Despair—A History of Fifty Years of Independence*. New York: Public Affairs.

Overview of Africa, interpreting recent history in terms of a decline in the overall quality of life.

Moyo, D. 2009. *Dead Aid: Why Aid Is Not Working and How There Is Another Way for Africa*. London: Penguin.

Provocative book by a Zambian economist, arguing that Africa is treated in a patronizing way by the West, that aid has not brought meaningful benefits, and that parts of Africa have become aid-dependent. For Moyo, removal of obstacles to trade in agricultural products is necessary.

Ogata, S. 2005. *The Turbulent Decade: Confronting the Refugee Crises of the 1990s*. New York: Norton.

Ogata was the UN High Commissioner for Refugees from 1990 to 2000, and in this detailed study she explains why, in her view, there are no humanitarian solutions to humanitarian problems.

Peet, R., and E. Hartwick. 2009. *Theories of Development: Connections, Arguments, Alternatives*, 2nd edn. New York: Guilford.

A critical examination of theories of economic development in the context of current events and policy discussions.

Prothero, R.M. 2005. 'Malaria: Some Recent Developments', *Geography* 90: 180–5.

Informative article about this devastating disease.

Sheppard, E., P.W. Porter, D.R. Faust, and R. Nagar. 2009. *A World of Difference: Encountering and Contesting Development*, 2nd edn. New York: Guilford.

Detailed, comprehensive, and thoughtful overview of the causes of global inequality.

Storey, D. 2006. 'People on the Move', *Geography Review* 19, 5: 37–41.

Straightforward review of population movements with emphasis on the UK.

ON THE WEB

MIGRATION

Global Data and Information

www.migrationinformation.org/

A substantial source of data and discussion on a wide range of migration topics with a global focus.

US Migration

www.census.gov/population/www/
socdemo/migrate.html

Census and survey data and information on selected US topics.

Migration Research

www.uu.nl/uupublish/onderzoek/onderzoek
centra/ercomer/24638main.html

The European Research Centre on Migration and Ethnic Relations at the University of Utrecht offers some useful data and general information on research into migration topics, especially as these relate to ethnic identities.

REFUGEES

The UN Refugees Agency

www.unhcr.ch/

The United Nations High Commission on Refugees maintains this regularly updated site; provides a wealth of information and commentary.

US Committee for Refugees

www.refugees.org/article.
aspx?id=2196&area=About%20USCRI

An organization that aids refugees fleeing war and persecution and helps resettle families in the US.

Refugees in Canada

www.cic.gc.ca/english/refugees/index.asp

Citizenship and Immigration Canada website detailing refugee information.

DEVELOPING THE LESS DEVELOPED WORLD

UNDP

www.undp.org/

The home page of the United Nations Development Program; includes links to past issues of the annual *Human Development Report* and other useful publications.

World Food Program

www.wfp.org/

The home page for the United Nations World Food Program; a substantial resource on matters of hunger and famine in a global context.

World Bank

www.worldbank.org/

Another valuable web resource with comprehensive information on the policies and activities of the World Bank.

Human Development

hdr.undp.org/en/humandev/

UNDP website that provides information on the concept and measurement of human development.

Poverty

web.worldbank.org/WBSITE/EXTERNAL/TOPICS/
EXTPOVERTY/0,,menuPK:336998~pagePK:149018~
piPK:149093~theSitePK:336992,00.html

World Bank website focusing on global poverty.

GLOBAL JUSTICE

Human Rights

www.ohchr.org/EN/UDHR/Pages/
Introduction.aspx

One part of the UNHCR website that includes the authoritative documents prepared in 1948.

UNICEF

www.unicef.org/uwwide/

Created by the United Nations General Assembly in 1946 to assist children in Europe after World War II, UNICEF today helps children around the world get the care and stimulation they need in the early years of life; it encourages families to educate girls as well as boys; strives to reduce childhood death and illness; and works to protect children in the midst of war and natural disaster.

Amnesty International

www.amnesty.org/

Amnesty International is an NGO with a vision of a world in which every person enjoys all of the human rights enshrined in the Universal Declaration of Human Rights and other international human rights standards.

CULTURAL IDENTITIES AND LANDSCAPES

Both this chapter and Chapter 7 focus on culture, and both are closely related to the political geography that is the subject of Chapter 8. It may be helpful, therefore, to view the three chapters as an integrated unit. More specifically, Chapters 6 and 7 focus on three closely related subdisciplines of human geography—historical, cultural, and social geography—that share a concern with the behaviour of humans, both as individuals and as group members, and with the landscapes that humans create. The differences in focus between this chapter and Chapter 7—here, the geographic expression of culture in landscape; there, the spatial constitution of culture, its symbolic expression and social significance— are reflected in different approaches: whereas the emphasis here is strongly empirical, in the next chapter it will be much more theoretical and conceptual.

We begin with some provocative ideas about cultural identity in the contemporary world (a subject we will revisit in Chapters 7 and 8). Following a brief introduction to the terms 'culture' and 'society' as they are used by human geographers, we look at cultural evolution, the origins of 'civilization', cultural geographic regions—a good example of our second recurring theme, regional studies—and the making of cultural landscapes. We continue with discussions of the two cultural variables that are perhaps most important with respect to both human identity and the human creation of landscape: language and religion. The concluding material reviews the evidence for cultural globalization.

Two Tibetan girls in northwest China's Qinghai province with their lambs.

CP/AP photo/Tang Zhaoming

A World Divided by Culture?

Humans have 'brought into being mountains of hate, rivers of inflexible tradition, oceans of ignorance'. (James, 1964: 2)

In 1964 the eminent American geographer Preston James published an introductory geography textbook with an inspired title: *One World Divided*. We have already recognized some human divisions (demographic differences), and we have also identified, described, and attempted to explain the fundamental division of the world into two regions: more developed and less developed. Further, Appendix 1 describes some physical divisions (climate, relief, soil, vegetation, etc.). Our concern now is the human divisions that derive from differences in culture. In fact, the barriers created by human variables are often far more difficult to cross than any physical barrier. The divisions we noted in our accounts of the human population are only the tip of the iceberg.

The world we live in today is one of tremendous cultural diversity—but it is also one of increasing interaction between cultures. As a result, in many areas traditional groups are struggling to protect their established ways of life against 'foreign' influences; in others, the often uncontrolled passions that surround language, religion, and ethnicity are erupting in cultural and sometimes political conflicts. Ironically, our greatest human achievement—our culture—has been responsible not only for erecting barriers but for sparking conflict between peoples.

INTRODUCING CULTURE

Humans differ from all other forms of life in that they have developed not only biologically but culturally. Other forms of life, limited to biological adaptation, have become so highly specialized that most are restricted to particular physical environments; indeed, environmental change in many cases has resulted in species extinction. If, so far, humans have avoided such a fate, it is primarily because of our culture: our ability first to analyze and then to change the physical environments that we encounter. Unlike other animals, humans can form ideas out of experiences and then act on the basis of these ideas; we are capable not only of changing physical environments, but of changing them in directions suggested by experience. This is one of the meanings of the term 'culture'.

Why is it, then, that this remarkable human ability has tended to have the unfortunate consequences noted by James? A large part of the reason is that each cultural change, each new idea prompted by experience, brings with it new knowledge and responsibilities that, at least initially, do not fit easily into the cultural framework of existing attitudes and behaviours. The value that humans generally continue to place on tradition and the past means that our cultural frameworks rarely are conducive to the efficient use of change in culture. Once in place, differences in attitude and behaviour tend to be self-perpetuating and can be resolved only with the development of new, overarching values. Developing and establishing such new values is a challenging task, as it involves human engineering—literally changing ourselves in accord with these new values.

We can summarize this important introductory argument as follows:

1. Our world is divided, especially because of spatial variations in culture.
2. By culture, we mean the human ability to develop ideas from experiences and subsequently act on the basis of those ideas.
3. Once cultural attitudes and behaviours are in place, an inevitable tendency is for them to become the frame of reference—however inappropriate—within which all new developments in culture are placed and evaluated.
4. One task humans may choose to tackle today is to create new sets of values—in effect, to re-engineer ourselves.

We will return to this idea raised by James—that culture acts as a barrier between peoples at the world scale—in our discussions of language and religion in this chapter, our discussion of ethnicity in chapter 7, and again in the concluding section of Chapter 8, where we discuss possible geopolitical futures.

Humans in Groups

The terms 'culture' and 'society' are two of the most awkward employed in human geography, if not in all of social science. Both terms are associated primarily with disciplines other than human geography—culture with anthropology (and folklore), society with sociology—both of which study human behaviour at the group scale (the individual scale is the realm of psychology). Anthropology and sociology evolved as separate disciplines in the nineteenth century, the former focusing on non-Western and rural issues, the latter on Western and urban issues. As recently as 1958, however, eminent representatives of the two disciplines found it necessary to clarify what they meant by the terms 'culture' and 'society' (Box 6.1).

HUMAN GEOGRAPHY AND CULTURE

Following in the tradition pioneered by earlier geographers, especially Humboldt, Vidal treated humans as part of nature, not separate from it. As we saw in Chapter 1, he focused on the relations between humans and land, with emphasis on culture ('*genre de vie*'). A proponent of the possibilist view, he explicitly opposed the idea of determinism, whether environmental or social.

A view comparable to Vidal's did not develop in North America until the 1920s, when Sauer put forward a series of ideas that would become central to the 'landscape school' (see Box 1.10). These ideas, derived from possibilism, centred on the creation of cultural landscapes from earlier physical landscapes. Sauer emphasized that the object of geographic study was the landscape itself, not the culture that created it. The impact of the landscape school on North American human geography has survived largely intact from the 1920s to the present.

By now it will be apparent that **culture** is a complex concept in geography. Earlier, we suggested that 'culture' referred to our unique human ability to deliberately change physical environments in directions suggested by experience—a process that over time led to the development of distinct ways of life. But this is an extremely broad, general statement. To gain a more specific understanding, it is useful to note that culture can be divided into two categories, non-material and material (Huxley, 1966; Zelinsky, 1973: 72–4). *Non-material* culture has two components: (1) key attitudinal elements or values, such as language and religion, known as **mentifacts**, and (2) the norms involved in group formation, such as rules about family structure, known as **sociofacts**. *Material* culture comprises all the human-made physical objects and elements related to people's lives and livelihood, known as **artifacts**, and thus includes the human landscape.

Traditionally, then, human geographers have followed Vidal and Sauer (but especially the latter) in taking as their prime object of study the landscapes created by humans through their cultures. In our study of human landscapes, both mentifacts and sociofacts facilitate our understanding. Anthropologists study human cultures, whereas human geographers

culture
The way of life of the members of a society.

mentifacts
Those mental or non-physical elements of culture; the values held by members of a group.

sociofacts
Those elements of culture most directly concerned with interpersonal relations; the norms that people are expected to observe.

artifacts
The physical objects created by a culture for pleasure (art, toys), work (tools), living (vessels for cooking), or worship (crucifix).

Box 6.1   The Concepts of Culture and Society

The emerging disciplines of anthropology and sociology focused on 'culture' and 'society' respectively, but their definitions of these concepts do not differ in any significant way. The classic early definition of culture proposed in 1871 by E.B. Tylor—'that complex whole which includes knowledge, belief, art, morals, law, customs and any other capabilities and habits acquired by man as a member of society' (see Friedl and Pfeiffer, 1977: 288)—and developed by Franz Boas around the turn of the twentieth century, seems very close to the idea of society presented by the great early sociologists: Auguste Comte, Herbert Spencer, Émile Durkheim, and Max Weber. Thus the core concepts of late nineteenth- and early twentieth-century anthropology and sociology were not clearly distinguished. Rather, the distinction between the disciplines was operational: anthropologists applied those concepts largely to non-Western groups, while sociologists applied them to Western ones.

In 1958, distinguished representatives of both disciplines collaborated to clarify the terms, proposing that 'culture' be reserved for a narrower sense than it had been previously, to refer to the 'transmitted and created content and patterns of values, ideas and other symbolic-meaningful systems [that are] factors in the shaping of human behavior and the artifacts produced through behavior'. 'Society' or 'social system', on the other hand, would refer to the 'specifically relational system of interaction among individuals and collectivities' (Kroeber and Parsons, 1958: 583).

Although the distinction proposed by A.L. Kroeber and Talcott Parsons has been highly influential, today anthropologists and sociologists alike continue to use the two terms in various ways. Traditionally, human geographers have had their closest links with anthropology, and hence with the culture concept, but now they are turning to sociology and the society concept for new inspirations.

have traditionally paid attention to the landscape expressions of these cultures.

HUMAN GEOGRAPHY AND SOCIETY

Simply put, society is to sociology what culture is to anthropology: the central concept of the discipline. Like 'culture', 'society' may be defined in various ways, depending on theoretical perspective. According to the traditional view, the term **society** refers to a cluster of institutionalized ways of doing things. Whereas 'culture' refers to the way of life of the members of a society, 'society' refers to the system of interrelationships that connects those individuals as members of a culture (Giddens, 1991: 35). Thus 'society' refers to recurrent attitudes and behaviours among a particular group. Clearly, to the extent that 'society' includes non-material culture—languages, religions, and social structures are institutionalized ways of doing things—the term overlaps with 'culture'.

Let us turn to the concept of society as it is used in human geography. North American geographers who adhere to Sauer's landscape school have employed the term 'culture' rather than 'society' in their analyses of the human landscape; British and European geographers have done the reverse, using 'society' rather than 'culture'. In fact, there is no clear and absolute distinction between these two terms, either in social science generally or in human geography specifically. Not surprisingly, current research indicates that human geographers now recognize the need to integrate North American cultural and European social geography. Relatively nominal differences in emphasis cannot disguise a basic unity. Human geographers are able to use the two related concepts of culture and society to analyze landscape.

THE IMPORTANCE OF HUMAN SCALE

As human geographers, we are particularly aware of the importance of selecting appropriate spatial scales of analysis when conducting research. Our concern here is with a second choice of scale: human, or social, scale.

As students of human populations, we may in principle choose to work at any scale, from the individual to the world population, through the intermediate scales of nuclear families, extended families, friendship circles, voluntary associations, involuntary associations,

institutions, and nations. Selecting the appropriate scale is a function of the particular type of study being conducted. In order to improve understanding of our divided world, human geographers typically have focused on cultures. Thus, we study groups delimited on the basis of common operating rules (or what we might call common mentifacts or common sociofacts). For human geographers, an appropriate social scale allows delimitation of a group that is meaningful given the aims of the work. This point is more important than it may appear.

Philosophical emphasis and human scale

Choosing the most appropriate human scale is a thorny issue in all social science, including human geography. While those with a humanistic focus recognize the need to study the intentions and actions of people both as individuals and as members of groups, groups of people are the concern in most human geographic research. Those with a Marxist perspective focus on groups because they believe that individuals cannot be understood without reference to the appropriate larger cultural context and, more specifically, to the overarching social and economic mode of production. Most traditional cultural geography has favoured a group scale because it is best suited to the typical geographic interest in the world or regions of the world. Finally, most contemporary social theory favours the group scale, largely on the grounds that individual actions are determined by ideas and beliefs rooted in groups defined on the basis of interaction and communication.

The Evolution of Culture

EARLY HUMAN GROUPS

Our knowledge regarding the earliest cultural attainments of humans and the earliest organization of human groups is limited. Evidence of both language and tool-making dates from about 2.5 million years ago. Fire was first used perhaps 1.5 million years ago, prompting the identification of specific locations as home bases. Language appears to have expanded substantially about 400,000 years ago. Together, these developments facilitated the formation of social groups. Early human groups probably

In the margin image:

society
Refers to the interrelationships that connect individuals as members of a culture.

numbered between 10 and 30 (occasionally as many as 100) people linked by kinship. Groups of this size prevailed until the agricultural revolution.

No doubt many factors encouraged language, tool-making, and group formation, among them the requirements of basic subsistence, mating, child care, co-operative hunting, information sharing, and minimizing conflict. In early human groups—typically labelled 'bands'—there was probably little labour specialization (since economic activities were limited to the basic search for food) or variation in status. Age and gender were likely the two principal means by which individuals were differentiated, and any authority was probably exercised by heads of families, not separate political authorities. Similarly, religion did not have a separate existence; any beliefs or activities that we today might see as religious were simply a part of everyday life, centred on the search for food within a particular physical environment.

As a way of life, hunting and gathering was relatively undemanding, without either long-term food shortages or major outbreaks of disease. Nevertheless, it has been abandoned throughout the world with only a few major exceptions, such as the Kazakhs of Central Asia, and, to some degree, the Inuit of northern Canada. The cultural development that led to its abandonment was the agricultural revolution. Box 6.2 describes some current examples of relatively small groups of people who have maintained a traditional way of life but who are now close to disappearing.

Box 6.2   Disappearing Peoples

One effect of the collapse of spatial barriers is that the numbers of people who live in relatively traditional ways, detached from the larger global context, are rapidly decreasing. Tribal peoples use deceptively simple technologies to live in close harmony with their physical environment. All too often, however, they are in the way of what most of us consider 'progress', that is, materialism and higher standards of living based on consumerism and on a capitalist economic system. Such 'progress' has destroyed numerous tribal peoples perceived as backward and ignorant by the prejudiced, culturally arrogant advocates of these modern values. Although we are becoming increasingly aware of the negative consequences of such 'progress', little is being done to protect the few remaining tribal groups or towards encouraging others to rethink the values that continue to allow such extinctions to occur.

'Far from the eyes of the world, some sixty-four indigenous peoples living in voluntary isolation in Amazonian Ecuador, Peru, Brazil, and Bolivia—the Tagaeri, Huaorani, Taromenane, Corubo, Amamhuaca, Mascho, Kineri, Nanti, Nahua, and Kugapakori, among others—are condemned to gradual extinction' (UN Web Services Section, 2006). These tribes avoid contact with outsiders and little is known about them. As land developers penetrate into their traditional lands, threatening their fragile lifestyles, these groups retreat even further into the forest. Despite various attempts to protect them from outside influences, their numbers are declining with resultant loss of language and cultural identity. Indeed, news items about groups facing extinction are commonplace, with 2008 reports on the Kawesqar in Patagonia, the Dongria Kondh in eastern India, and the Penan in Borneo.

In Africa, the Batwa, members of a pygmy group that once roamed the forest area of southwest Uganda, are destitute. Although they had lived in the region for over 50,000 years as foragers and were generally recognized as belonging to the high forest area, in the nineteenth century they were displaced by other groups who began using much of the area for agriculture. Today the Batwa are socially subordinate to the newcomers, for whom they work, and are subject to widespread discrimination. Ironically, in 1991 the Batwa were forced out of some of their traditional territory when it was set aside as a refuge for the rare mountain gorilla. Only about 1,850 Batwa survive in Uganda, where they comprise about half of the Ugandan pygmy population and are mostly squatters or tenants on land owned by others and subsist largely by begging. Their culture is lost and their numbers are decreasing.

While the Batwa may be disappearing along with their culture, in other cases a people may simply abandon their traditional culture in the process of modernization. There are at least two ways of thinking about the disappearance of a traditional way of life. Most scholars in the field would probably regret the loss. Others, however, might argue that such regret suggests a preference for keeping primitive people primitive—in effect, reducing them to the equivalent of zoo animals.

The Iban people of Sarawak were once known as Borneo's most fearsome headhunters. Today they make up about a third of the population of Sarawak. The state's official tourism website describes them as a 'generous, hospitable and placid people' who have adopted 'a peaceful agrarian lifestyle'.

Andrew Leyerle

civilization
A culture with agriculture and cities, food and labour surpluses, labour specialization, social stratification, and state organization.

THE BEGINNINGS OF CIVILIZATION

The maintenance of a suitable environment . . . remains a critical precondition of human existence, [but] much more is needed for the development of civilizations. . . . The essence of these additional requirements is the appropriation of

energies, first merely by better management of human labour, later by extensively harnessing various renewable transformations of solar radiation . . . , finally by extracting fossil fuels and generating electricity. (Smil, 1987: 1)

In Chapter 3, we noted the importance of humans' use of energy sources. Here we will examine various theories about where, how, and why some human groups began the 'appropriation of energies' essential to the development of 'civilization'.

Prior to the agricultural revolution, as we have seen, the religious, political, and economic dimensions of life were so rudimentary that any variations in culture between bands were intimately related to the physical environment. It was only with the agricultural revolution that what we might now term **civilization** appeared: a particular type of culture that includes a relatively sophisticated economy, political system, and social structure. The first civilizations in the world evolved in areas where agriculture also evolved. For example, in Mesopotamia—the area between the Tigris and Euphrates rivers, which today is part of southern Iraq—irrigated agriculture, water management, and land reclamation were probably being practised as early as 5,000 years ago. The fact that this 'cradle of civilization', as it is often known, introduced regulations on water use indicates the complexity of its social organization, and the city of Babylon became a great administrative, cultural, and economic centre.

But what specific changes gave rise to the beginnings of civilization? There are many possible responses to this question. Racist explanations such as that proposed by Gobineau (discussed in the next chapter) were common in the nineteenth century, but have since been discredited. Explanations based on environmental determinism, such as Huntington's (1924), are no longer tenable because they are not supported by facts. The nineteenth-century anthropologist Lewis Henry Morgan proposed that all cultures evolve in the same basic way, through successive levels of material achievement (Table 6.1); but this explanation also is inadequate, because it has nothing to say about what brought about the necessary changes in material achievement and, needless to say, assumes a kind of societal evolution with Western civilization at the imagined apex.

Table 6.1	A Unilinear Evolutionary Model of Culture
Civilization	
	Alphabet and writing
	∧
Upper Barbarism	
	Iron tools
	∧
Middle Barbarism	
	Animal and plant domestication
	∧
Lower Barbarism	
	Pottery
	∧
Upper Savagery	
	Bow and arrow
	∧
Middle Savagery	
	Fish subsistence and fire
	∧
Lower Savagery	
	Fruit-and-nut subsistence

SOURCE: K.L. Feder and M.A. Park, *Human Antiquity*. Copyright © 1993 McGraw-Hill Companies, Inc.

As we examine civilization, then, we need to bear in mind its contested meaning, neatly reflected when Gandhi, asked what he thought of Western civilization, replied that he thought it would be a good idea.

Five suggested explanations for the development of civilizations, the first three of which presuppose an increasing population, merit more detailed comment.

The hydraulic hypothesis

Two theories centre on water. According to V.G. Childe (1951), who identified 10 characteristics of civilization (Table 6.2), the introduction of irrigation significantly increased agricultural productivity. More specifically, according to the German historian Wittfogel (1957), an expanding population would have required more agricultural land—but in a semi-arid area, the only way to achieve such expansion would have been through irrigation. In turn, irrigation would have required the construction of canals and aqueducts—which in turn would have required that the people work co-operatively, under the direction of some central authority. In short, a new form of cultural organization based on central control arose in response to the need for irrigation.

Coercion

Another theory supposes that the necessary social changes, particularly the imposition of central control, preceded agricultural development, as groups and their territories increased in size. This second suggested explanation proposes that the development of agriculture and civilization may have resulted from increases in population pressure in an area too small to support more people. In this case the population growth would have been accommodated not through intensification of agriculture (as in the hydraulic hypothesis) but through territorial expansion by means of force—the same kind of expansion that, over time, creates empires. According to Carneiro (1970: 734), 'only a coercive theory can account for the rise of the state.'

Marxism

A Marxist view sees civilization as the product of class differences, which developed as a consequence of the wealth accumulated by those individuals possessing domesticated animals. Once such class differences develop, class conflict follows, as it becomes necessary for those with wealth to defend themselves from those without wealth. At the same time, exchange begins between those with and those without animals, and as population increases, a class of specialists becomes involved in the exchange process. In this way people not directly involved in production begin to control economic life, adding a second dimension to the emerging class conflict. Such class conflict generates central control as a means for the more dominant class to maintain and even enhance its dominance. Civilization follows from these beginnings in the form of a culture consisting of rulers and ruled.

Climate challenges

It has long been popular to suggest that the rise of civilization was inextricably linked to a warming climate beginning about 12,000 years ago, but increasing evidence argues that subsequent climatic challenges, especially periods of low rainfall, might have shaped the rise and course of civilization in several different ways. Certainly, the last 10,000 years of human history, known as the Holocene period, has been a period of remarkable warmth overall, but it has also included periods of extreme drought and extreme cold.

According to Fagan (2004), it is possible that the early civilization in Mesopotamia was prompted by a period of reduced rainfall that began about 3800 BCE, causing many farmers to abandon their fields and seek work in a few favoured locations where irrigation was a practical possibility. Reduction in rainfall also obliged farmers in irrigated areas to innovate and intensify, for example, by using draft animals for the first time. This concentration of increasingly intensive grain production in a few advantaged areas may then have encouraged urban growth and other characteristics of civilization. This argument is difficult to prove and might be criticized for being reminiscent of earlier ideas that relate civilization to

Table 6.2	Childe's Characteristics of Civilization	
Primary	**Secondary**
City settlement	Monumental public works
Labour specialization	Long-distance trade
Concentration of surpluses	Standardized monumental artwork
Class structure	Writing
State organization	Arithmetic, geometry, astronomy

FIGURE 6.1   The shape of continents

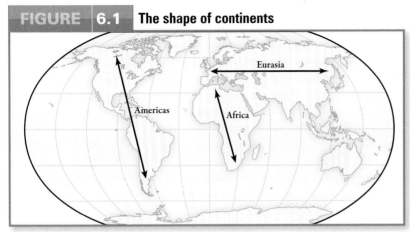

SOURCE: 'Major Axes of the Continents', from *Guns, Germs and Steel: The Fates of Human Societies* by Jared Diamond, published by Jonathan Cape. Copyright © 1997 by Jared Diamond. Reprinted by permission of The Random House Group Ltd., W.W. Norton & Company, Inc. and Jared Diamond.

environmental challenges, but it is certainly a thought-provoking idea.

The shape of continents

As with the previous account of climate change, some critics might understand this idea

FIGURE 6.2   Factors underlying the broadest patterns of history

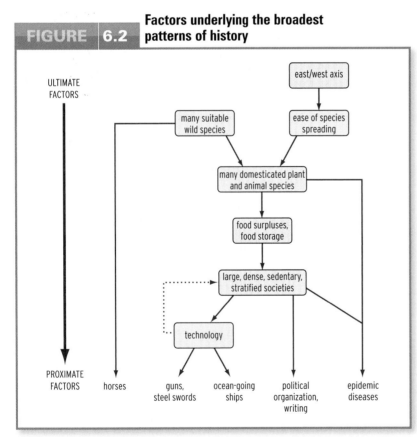

Diamond suggests a chain of causation that begins with the advantage of latitudinal extent and leads up to factors such as guns, diseases, and steel that eventually enabled Europeans to conquer other people.

SOURCE: *Guns, Germs and Steel: The Fates of Human Societies* by Jared Diamond, published by Jonathan Cape. Copyright © 1997 by Jared Diamond. Reprinted by permission of The Random House Group Ltd., W.W. Norton & Company, Inc. and Jared Diamond.

to be mere simplistic environmental determinism, but it is much more. Diamond (1997) contends that basic world geography is critical to an understanding of the rise and spread of agriculture and, thus, to the rise and spread of civilization. Figure 6.1 highlights the simple fact that the major axis of Eurasia is east–west while the major axes of the Americas and of Africa are north–south. Diamond proposes that these different continental shapes have been crucial as latitudinal extent allows agricultural technologies to spread great distances over areas of similar climate.

As we know, agricultural technologies are the precursor to the rise of civilizations. Building on this claim, it can be argued, as shown in Figure 6.2, that this simple geographic circumstance is the ultimate factor beginning a chain of causation that eventually leads to the ability of some societies to spread globally and dominate other societies. Recall that the issue of European overseas expansion was raised in the previous chapter and addressed in terms of world systems theory as this might explain changes since about 1450. It is possible to conceive of Diamond's ideas as aiding our understanding of why Europe (or it could have been China) was able to move overseas. As Figure 6.2 suggests, the latitudinal extent of Eurasia is the first link in the chain because it permitted agricultural technologies to spread and, combined with suitable plant and animal species, it led to a sophisticated agricultural region that, in turn, became what we are calling 'civilization'. The eventual outcome was a series of proximate factors—including the guns, germs, and steel that Diamond identifies in the title of his work—that enabled some people (Europeans) to move around the world and dominate most of those whom they contacted.

EARLY CIVILIZATIONS

Whatever the specific reasons motivating them, early civilizations represented major cultural advances and brought many changes. Although each one had its own distinctive character, they shared several features. Agriculture and urbanization are the two most obvious developments that have impacts on land, but civilization also involves changes in non-material culture: the rise of powerful elites, labour that is submissive to the elite, the disappearance of the egalitarian family-based group, the rise of armies and the technology of war, the growth of bureaucratic systems, opportunities for patronage for artists and scholars, and the institution of private landownership. In many cases religion was the basis of early power, often of a tyrannical system, but all civilizations became increasingly secular.

The great early civilizations are mapped in Figure 6.3; for a general time scale, see Table 6.3. The dating of Chinese civilization at about 3000 BCE reflects recent excavations of large cities in the Yangtze River valley. A second reflection of recent research is the inclusion of an Andean civilization in the table. This civilization was in the Norte Chico region of Peru. More generally, as the table makes clear, almost all the great early civilizations—sets of cultural advances—have been temporary; only the civilization of China has lasted to the present day. All others have collapsed, whether because of natural disasters (such as earthquakes or floods) or because of conflict (internal or external). Earlier cultural groupings might have been able to adapt and survive crises such as these, but the size, density, and organizational complexity of civilizations made them vulnerable.

Cultural Regions

The areas occupied by early civilizations are examples of **cultural regions**. In this section we will examine how the concept of cultural regions is used in contemporary human geography.

Traditionally, human geographers have been concerned not with humans themselves or human cultures, but with human impacts on landscape. Yet these impacts can vary considerably, depending on the specific characteristics of the human group in question—in other words, depending on culture. Because different cultures have emerged in different areas and because the earth is a diverse physical environment, there is a wide variety of human landscapes. In addition to describing and explaining these landscapes, human geographers have sorted them into cultural regions.

Delimiting cultural regions requires decisions on at least four basic points:

1. criteria for inclusion;
2. date or time period (since cultural regions change over time);
3. spatial scale;
4. boundary lines.

Irrigated terraces at the ancient Incan site of Pisac, Peru.

Alexey Stiop/Alamy/GetStcok

These four issues are interrelated, and they are not easy to resolve. As you read the following examples of geographers identifying and mapping regions, appreciate that the goal is that of enhancing understanding, not of arriving at some ideal scheme.

> **cultural regions** Areas in which there is a degree of homogeneity in cultural characteristics; areas with similar landscapes.

Table 6.3	Chronology of Some Early Civilizations
Egypt	3000–332 BCE
Minoan	3000–1450 BCE
China	3000 BCE–present
Andean	3000–1800 BCE
Indus	2500–1500 BCE
Mesopotamia	2350–700 BCE
Mycenaean	1580–1120 BCE
Olmec	1500–400 BCE
Greece	1100–150 BCE
Rome	750 BCE–375 CE
Monte Alban	200 BCE–800 CE
Moche and Nazca	200 BCE–700 CE
Teotihuacan	100–700 CE
Maya	300–1440 CE
Tiahuanaca	600–1000 CE
Toltec	900–1150 CE
Chimo and Inca	1100–1535 CE
Aztec	1200–1521 CE
Benin	1250–1700 CE

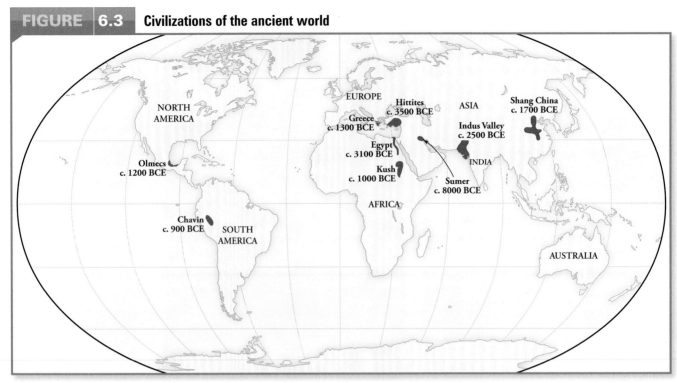

FIGURE 6.3   Civilizations of the ancient world

SOURCE: P.N. Stearns, M. Adas, S.B. Schwartz, and M.J. Gilbert, *World Civilizations: The Global Experience* (New York: Pearson, 2007), 3.

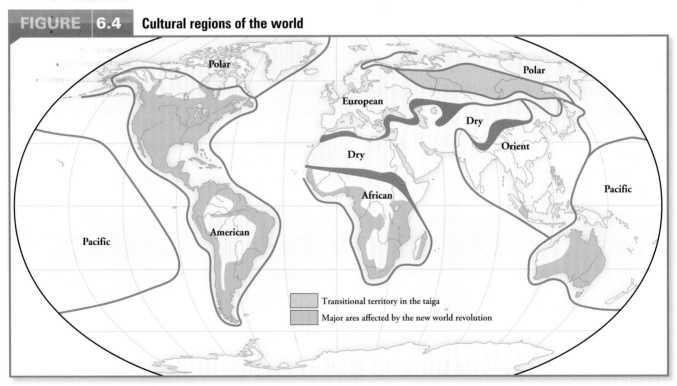

FIGURE 6.4   Cultural regions of the world

The varying width of the lines between regions is one means of acknowledging that boundaries may be zones of transition rather than lines that clearly divide regions from one another.

SOURCE: Adapted from R.J. Russell and F.B. Kniffen, *Culture Worlds* (New York: Macmillan, 1951).

WORLD REGIONS

On the world scale, many notable attempts have been made to delimit a meaningful set of cultural regions. The historian A.J. Toynbee, for example, identified a total of 26 civilizations as responses to environment, past and present: 16 'abortive' and 10 surviving (Toynbee, 1935–61). Three of the latter—the Polynesian,

Nomad, and Inuit civilizations—Toynbee described as 'arrested' because their overspecialized response to a difficult environment left them unable either to expand into different regions or to cope with environmental change. He identified the remaining seven living civilizations as Western Christendom, Orthodox Christendom, the Russian offshoot of Orthodox Christendom, Islamic culture, Hindu culture, Chinese culture, and the Japanese offshoot of Chinese culture; Toynbee considered most of Africa to be 'primitive'.

Although its limitations are clear, Toynbee's schema is instructive. His principal criterion obviously is religion, since he named most of his civilizations for their religions, but in a broader sense the most important feature of each is the manner in which it has responded to the environment—hence Toynbee's recognition of 'arrested' and 'abortive' civilizations.

The first geographers to attempt to delimit world regions were Russell and Kniffen (1951), who began by recognizing various cultural groups and then related them to areas so as to delimit seven cultural regions, each a product of a long evolution of human–land relations (Figure 6.4; the term 'new world revolution' in the legend refers to the diffusion of European ideas and technologies with European expansion). With the addition of one 'transitional' area, these regions effectively cover the world. Russell and Kniffen's regions are well justified, and their classification is a valuable contribution to our general understanding of the world. Nevertheless, it is important to emphasize that world regionalizations of this type do not represent the best use of the culture concept in geography. The central problem is spatial scale: the larger the area to be divided, the more likely it is that the regions identified will be either too numerous or too superficial to be helpful. Like regionalizations based on physical variables (see Appendix 1), cultural regionalizations are more appropriately seen as useful classifications than as insightful applications of geographic methods. The difficulties involved in delimiting Europe as one world cultural region are outlined in Box 6.3.

Cultural regions and the distinctive landscapes associated with them are more effectively considered on a more detailed scale. If, for example, we take North America as one example of a world region, we find that subdividing greatly enhances our appreciation of the relationship between culture and landscape.

NORTH AMERICAN REGIONS

North America is relatively easy to subdivide because the development of its contemporary cultural regions is recent enough that we are able to trace their origins. Figure 6.5 delimits 16 regions in North America on the basis of several criteria, such as physical and economic differences, not simply culture. Each of these regions is likely to be generally recognizable, and the authors of the map attempt to convey the 'feeling' of each region in their discussions. This regionalization is interesting because it focuses on broad regional themes rather than on specific criteria, and recognizes that the border between the United States and Canada is not a fundamentally geographic division. A thought-provoking alternative regionalization is included in the following chapter in the context of a discussion of vernacular regions.

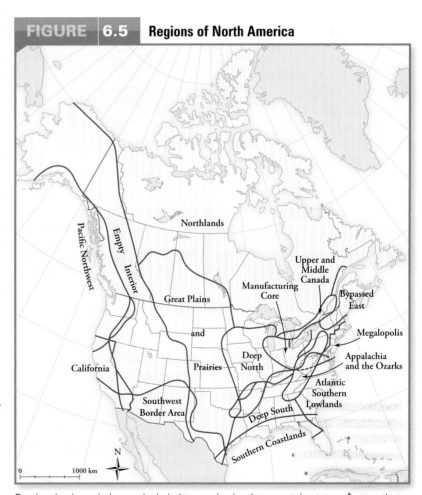

FIGURE 6.5 **Regions of North America**

Overlapping boundaries are included to emphasize the uncertain status of areas that might be seen as belonging to more than one region. This is another way of acknowledging that many of the boundaries between regions are zones of transition rather than firm divisions (see also Figure 6.4).

SOURCE: Adapted from S.S. Birdsall and J.W. Florin, *Regional Landscapes of the United States and Canada*, 2nd edn (New York: John Wiley & Sons, 1981), 18. Reprinted by permission of John Wiley & Sons, Inc.

Box 6.3   Europe as a Cultural Region

Although clearly an area of considerable diversity, Europe generally is seen as a single world region—'a *culture* that occupies a *culture area*' (Jordan, 1988: 6). Before the sixteenth century it was widely believed that Europe was a continent physically separated from Asia, and even after this error had been corrected, the image of a separate continent was so powerful that a new divide was needed: thus the idea became generally accepted that the Caucasus, the Urals, and the Black Sea represented a meaningful boundary. The concept of a separate Europe, an area that was culturally distinctive, was allowed to remain intact.

Using the political map of the mid-1980s, T.G. Jordan defined Europe on the basis of 12 traits measured for each country (Figure 6.6):

1. majority of population speak an Indo-European language;
2. majority of population have a Christian heritage;
3. majority of population are Caucasian;
4. more than 95 per cent of population are literate;
5. an infant mortality rate (IMR) of less than 15;
6. a rate of natural increase (RNI) of 0.5 per cent or less;
7. per capita income of US$7,500 or more;
8. 70 per cent or more of population are urban;
9. 15 per cent or less are employed in agriculture and forestry;
10. 100 km (62 miles) of railway plus highway for 100 km² (39 square miles);
11. 200 or more kg (441 lbs) of fertilizer are applied annually per ha of cropland;
12. free elections are permitted.

Figure 6.6 implies a regionalization; to proceed further would require a more explicit focus on variables such as language or religion. One particularly innovative delimitation of regions within the larger European region, by Jordan (1988: 402), combines three approaches: a north/south distinction, an east/west distinction, and a core/periphery distinction. Of course, recognizing such divisions is really to admit that Europe cannot be defined in terms of a single culture, an idea that is supported by the fact that Europeans have a long history of fighting each other on the basis of religion or language.

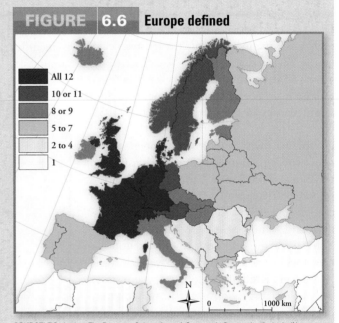

FIGURE 6.6   Europe defined

Legend:
- All 12
- 10 or 11
- 8 or 9
- 5 to 7
- 2 to 4
- 1

SOURCE: T.G. Jordan, *The European Culture Area: A Systematic Geography*, 2nd edn (New York: Harper & Row, 1988), 14.

A regionalization of the United States

Figure 6.7 is a regionalization of the United States explicitly based on the interesting concept of **first effective settlement**: 'Whenever an empty territory undergoes settlement, or an earlier population is dislodged by invaders, the specific characteristics of the first group able to effect a viable, self-perpetuating society are of crucial significance for the later social and cultural geography of the area, no matter how tiny the initial band of settlers may have been' (Zelinsky, 1973: 13).

Accordingly, Zelinsky demarcates five regions—West, Middle West, South, Midland, and New England. Only New England has a single major source of culture (England); the other four regions are further divided into various subregions, all of which have multiple sources of culture. Thus, the Midland is divided into two subregions; the South is divided into three subregions; the Middle West is divided into three subregions; and the West is divided into nine subregions. In addition, a number of sub-subregions are noted, along with three regions of uncertain status (Texas, peninsular Florida, and Oklahoma). As a cultural regionalization, this example is especially useful because of the single clear variable employed. Yet the simplicity of the approach does not disguise the complexity of the cultural landscapes revealed by the many subregions.

first effective settlement
A concept based on the likely importance of the initial occupancy of an area in determining later landscapes.

A regionalization of Canada

In our third North American example, Canada is divided into six regions (Figure 6.8). This regionalization employs several variables, but a basic cultural division is evident. The North is delimited according to political boundaries and can be subdivided into a Northwest that is relatively forested and has a mixed Native and European population and an Arctic North that is treeless, is populated mostly by Inuit, and has limited resource potential by southern Canadian standards. The British Columbia region is characterized by major differences in physical and human geography and has a resource-based economy; the same is true of the Atlantic and Gulf region, which also has the problems of low incomes and high unemployment rates. The Interior Plains is an agricultural region (cattle, wheat, and mixed farming), and its dispersed farmsteads and often regularly spaced towns reflect government surveying and planning before settlement. The core region of Canada is the Great Lakes–St Lawrence Lowlands, with most of the people, the largest cities, major industry, and intensive agriculture; this is the cultural and economic heartland of Canada. The largest region is the Canadian Shield, a sparsely inhabited area in which most of the settlements are resource towns.

The Making of Cultural Landscapes

So far, our discussion of cultural regions has been descriptive rather than explanatory. Now we will consider in more detail how regions of distinctive cultural landscape arise, beginning with a look at cultural variations over space and the ability of culture to affect landscape.

Although all cultures share certain basic similarities—all need to obtain food and shelter and to reproduce—they differ in the methods used to achieve those goals. As humans settled the earth, culture initially evolved in close association with the physical environment. Over time, however, cultural adaptations brought increasing freedom from environmental constraints. Gradually, as humans' ties to the physical environment loosened, their ties to their culture increased. One of the most important changes in human history was the shift (associated with the Industrial Revolution) to a capitalist mode of production, which largely

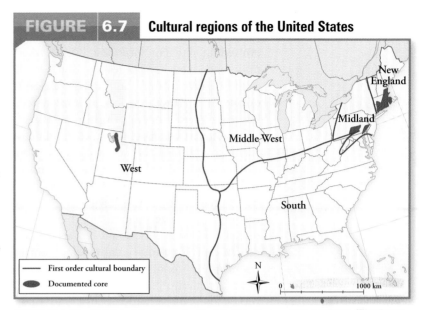

FIGURE 6.7 Cultural regions of the United States

— First order cultural boundary
⬮ Documented core

SOURCE: Adapted from W. Zelinsky, *Cultural Geography of the United States* (Englewood Cliffs, NJ: Prentice-Hall, 1973), 118.

destroyed our ties to the physical environment but created a new set of ties—to culture itself. Some human geographers see this change as part of an ongoing process that is effectively homogenizing world culture by minimizing regional variations; this view is derived most clearly from Marx (see Box 2.6).

FIGURE 6.8 Regions of Canada

SOURCE: Adapted from J.L. Robinson, *Concepts and Themes in the Regional Geography of Canada*, rev. edn (Vancouver: Talon Books, 1989), 17.

CULTURAL ADAPTATION

As humans settled the earth, different cultures evolved in different locations. Despite their many underlying similarities, each one developed its own variations on the basic elements of culture: language, religion, political system, kinship ties, and economic organization. As we seek to understand how different cultures create different landscapes, we must also ask how sound relationships develop between humans and their physical environment. This question is central to human geography: witness environmental determinism, possibilism, and related viewpoints. One answer centres on the concept of **cultural adaptation**. Humans are adapting continuously to the environment, genetically, physiologically, and culturally. Cultural adaptation may take place at an individual or at a group level—human geographers are interested in both—and consists of changes in technology, organization, or ideology of a group of people in response to problems, whether physical or human. Such changes can help us to deal with problems in any number of ways: by permitting the development of new solutions or by improving the effectiveness of old ones; by increasing general awareness of a particular problem or by enhancing overall adaptability. In other words, cultural adaptation is one of the essential processes by which sound relationships evolve between humans and land.

We noted earlier the idea that culture represents the human ability to develop ideas from experiences and subsequently to act on the basis of those ideas. Cultural adaptation thus includes changes in attitudes as well as behaviour. One current example of attitudinal change is our increasing awareness of the importance of environmentally sensitive land-use practices; an example of behavioural change would be the implementation of such practices. Needless to say, behavioural changes do not always accompany attitudinal changes—or even follow them.

Human geographers have proposed a number of theories to explain more precisely what cultural adaptation is and how it works. Among them are the following.

cultural adaptation
Changes in technology, organization, and ideology that permit sound relationships to develop between humans and their physical environment.

Culturally habituated predisposition

In their analysis of the evolution of agricultural regions, Spencer and Horvath (1963: 81) suggested that such regions could be seen as the 'landscape expression of . . . the totality of the beliefs of the farmers over a region regarding the most suitable use of land'. According to this view, cultures are particular beliefs, psychological mindsets, that result in a culturally habituated predisposition towards a specific activity and hence a specific cultural landscape; for example, the corn landscape of the early European-settled American Midwest reflected the commercial aims of incoming settlers.

Cultural preadaptation

According to another view, a cultural group moving into a new area may be preadapted for that new area: conditions in the source area are such that any necessary adjustments have already been made prior to the move or are relatively easy to make thereafter. Following this logic, Jordan and Kaups (1989) proposed that the American backwoods culture had significant Northern European roots, specifically in Sweden and Finland.

Core, domain, and sphere

Perhaps the most useful explanation is the core, domain, and sphere model (Meinig, 1965). According to this view, a cultural region or landscape can be divided into three areas—core (the hearth area of the culture), domain (the area where the culture is dominant), and sphere (the outer fringe)—and cultural identity decreases with increasing distance from the core. Figure 6.9 shows what most observers would agree is one of the easiest regions to delimit: the Mormon region of the United States (Box 6.4). But Meinig has successfully extended these ideas, in modified form, to other regions such as Texas and the American Southwest. Few attempts have been made to apply these ideas outside North America.

Each of the three views outlined above is, however indirectly, very much in the Sauer landscape school tradition. There is a consistent concern with the landscapes created by cultural groups and particularly with the material manifestations of culture: the visible landscape.

Two aspects of culture are particularly important in understanding our human world: language and religion. Not only are they important in themselves, but they are also good bases for delimiting cultures, and hence regions, and they affect both behaviour and landscape.

Language

Both language and religion have been of interest traditionally to cultural geographers, but neither has been recognized as a core concern. Park's observation (1994: 1) that 'the study of geography and religion remains peripheral to modern academic geography' applies also to the study of geography and language.

Yet—as we will see in our discussions of ethnicity (Chapter 7) and political states (Chapter 8)—language and religion often are the fundamental characteristics by which groups distinguish themselves from other groups. In the current chapter we will look at language and religion in terms of classification, origins, diffusion, regionalization, links to identity, and relations to landscapes. No attempt is made in either case to produce comprehensive human geographies. This is appropriate because, as noted, neither linguistic geography nor religious geography is a well-recognized subdiscipline of human geography; rather, both are cultural variables that relate to many different aspects of human geography. Both affect human behaviour and the building of landscapes, and both are inherently communal, helping members of language and religious groups share a sense of togetherness and belonging, as well as being inherently categorical, emphasizing differences and thus helping keep groups apart.

A CULTURAL VARIABLE

Language, probably the single most important human achievement, is of interest to human geographers for several reasons. *First*, language is a cultural variable, a learned behaviour that initially evolved so that humans could communicate in groups, probably for the purpose of organizing hunting activities. It is possible that early humans, concentrated in the area of origin, all spoke the same early language. According to McWhorter (2002), this brief phase of human unity occurred about 150,000 years ago in East Africa. However, as people moved across the earth, different languages developed in different areas offering new physical environmental experiences. From the beginnings of cultural evolution, therefore, language has been a potential source of group unity (and therefore of total-population disunity)—a topic of great interest to human geographers.

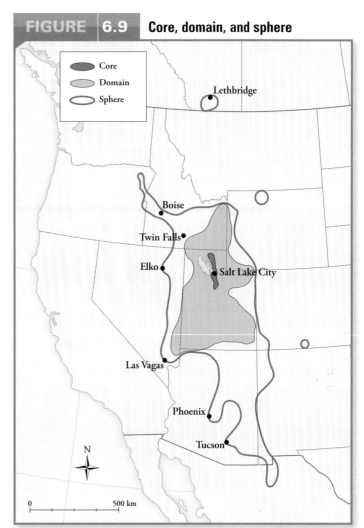

FIGURE 6.9   Core, domain, and sphere

SOURCE: Adapted from D.W. Meinig, 'The Mormon Culture Region: Strategies and Patterns in the Geography of the American West, 1847–1964', Annals, Association of American Geographers 55 (1965): 214, Fig 7 (MW/RAAG/P1887) Used with permission of Taylor & Francis Ltd., http://www.informaworld.com.

A *second* important feature of language is its usefulness in delimiting groups and hence regions. Language is the means by which a culture ensures continuity through time—and the death of a language is often seen as the death of a culture. It is for this reason that such determined efforts have been made to ensure the continuity, or even revival, of traditional languages in places such as Wales and Quebec. As a fundamental building block of nationhood, language is of particular interest to human geographers concerned with the plight of minority groups and the relationships between language and nationalism.

A *third* interest flows directly from the second. As suggested at the outset of this chapter, divisions between places and peoples are one outcome of culture, including language. For example, with reference to urban centres

Box 6.4   The Mormon Landscape

Members of the Church of Jesus Christ of Latter-day Saints, commonly known as Mormons, first arrived near Salt Lake in 1847 and subsequently settled in an extensive area that included part of southern Alberta. Most decisions concerning settlement expansion and landscape activities were made by the church leadership and reflected the fundamental character of the Mormon religion: decisions were made in response to the requests of church leaders and as one part of the larger experience of being a Mormon. Thus settlement followed an orderly sequence, in accord with larger church concerns and ambitions, and landscape change demonstrated a high degree of uniformity. More than most groups, Mormons emphasized co-operation, community, and economic success as means to establish and solidify their occupation of a region.

Settlement sites were usually selected by leaders, and the settlers either volunteered or were called by church leaders to move to each new location. The Mormon landscape that evolved has a number of distinctive features, all intimately related to the religious identity. Speaking in 1874, church president George Smith observed: 'The first thing, in locating a town, was to build a dam and make a water ditch; the next thing to build a schoolhouse, and these schoolhouses generally answered the purpose of meeting houses. You may pass through all the settlements from north to south, and you will find the history of them to be just about the same.'

Francaviglia (1978) describes the Mormon landscape in detail, based on extensive travels within the region. Mormon towns were laid out in approximate accord with the City of Zion plan as detailed in 1833 by the church's founder, Joseph Smith: there was a regular grid pattern, with square blocks, wide streets, half-acre lots that included backyard gardens, houses of brick or stone construction, and central areas for church and educational buildings. Towns laid out along these lines continue both to reflect and to enhance the Mormons' community focus, and represent a distinctive town landscape within the larger region of the mountain West.

The rural landscape of the Mormons is also distinctive, emphasizing arable agriculture rather than pastoral activities (which necessarily involve greater movement and lower population densities). Among the notable features of the farm landscape are networks of irrigation ditches, Lombardy poplars, unpainted fences and barns, and hay derricks. Together, these features serve both to distinguish the Mormon landscape from the surrounding areas and to impose a sense of unity on it.

The wide main street in Manassa, Colorado; the central temple in Cardston, Alberta; and a hay derrick at Cove Fort, Utah.

William Norton

in English-speaking Ontario and French-speaking Quebec, Mackay (1958) compared the expected interaction (calculated using accepted ideas that assume interaction relates to settlement size and the distance between settlements) and the actual interaction. Results showed that there was much less interaction between the two regions than there was within each region, and it is clear that language plays a role in lessening the expected interaction.

Our *fourth* interest in language concerns its interactions with environment, both physical and human. Spatial variations in language are caused in part by variations in physical and human environments, while language itself is an effective moulder of the human environment. The importance of language to the symbolic landscape links it to group identity, returning us to our first interest.

HOW MANY LANGUAGES?

Like many cultural variables—indeed, like culture itself—language began as a single entity (or at most a few different ones) and diversified into many. Over the long period of human life on earth, many languages have arisen and many others have died out. During the past several hundred years, however, a new language is a rarity while the disappearance of a language is commonplace. Of the roughly 7,000 distinct languages (not counting minor dialects) that existed 400 years ago, approximately 1,000 have disappeared, leaving about 6,000 languages today. Many more may vanish over the next few hundred years. Current estimates by UNESCO suggest that about 3,000 languages are endangered and that one language dies about every two weeks. None of these numbers are certain. For example, some sources put the current number of languages at closer to 7,000.

Languages die for two related reasons. First, a language with few speakers tends to be associated with low social status and economic disadvantage, so those who do speak it may not teach it to their children. In some cases this results in people choosing to speak a different language associated with economic success and social progress. Also, when a language has few speakers the specific reason for disappearance might be a natural disaster, such as drought or the spread of disease. Second, because globalization depends on communication between previously separate groups, it is becoming essential for more and more people to speak a major language such as English or Chinese. Already, a very few dominant languages effectively control global economics, politics, and culture. About 96 per cent of the world's population speak only 4 per cent of the world's languages. Table 6.4 identifies the eight languages spoken by the most people. Mandarin has more native speakers than any

other language, with Spanish ranking second. (Different sources often suggest different data on numbers of speakers because some sources focus on all speakers rather than native speakers; if this method of estimating is employed then the number of English speakers increases markedly from that noted in Table 6.4.)

DISAPPEARING LANGUAGES

There are varying reactions to the idea that perhaps half of our present languages will disappear during the current century. For many observers, language loss represents culture loss and is just as serious a threat as loss of biodiversity. First, each language might be understood as a window on the world, a distinct and different way of seeing and thinking. This point is highlighted by the fact that most languages include words that are effectively untranslatable. A survey of 1,000 linguists identified the most untranslatable word as *ilunga*, from the Tshiluba language spoken in the southeast of the Democratic Republic of Congo. In English this word means 'a person who is ready to forgive any abuse for the first time, to tolerate it for a second time, but never a third time'. Losing a language, a different way of seeing the world, might be described as a process of cultural forgetting. Second, loss of languages is an important practical issue because most languages include detailed knowledge—about local environments, for example—not available elsewhere. Three examples of the many disappearing languages are Eyak (Alaska), the last speaker having died in 2008, Haida (Haida Gwaii, British Columbia), with fewer than 65 speakers, and Tofa (Siberia), with fewer than 30 speakers.

Table 6.4	Numbers of Speakers, Major Languages, 2007
Language	Number of Speakers (millions)
Mandarin	937
English	340
Spanish	332
Hindi/Urdu	242
Portuguese	177
Bengali	171
Russian	145
Japanese	122

NOTE: The numbers in this table refer to 'first language' speakers.

SOURCE: Updated from World Almanac Education Group, *The World Almanac and Book of Facts 2005*. Copyright © World Almanac Education Group. All rights reserved.

Master weaver Delores Churchill poses for photos in her home in Ketchikan, in southeast Alaska. Churchill is one of few fluent Haida speakers in all of Alaska. The Haida Nation has less than 65 people who speak the Haida language, the average age is 70 to 80.

Farah Nosh/Getty Images

overlook questions about why just a few languages have been so spectacularly successful. A first explanation is migration; as noted below several of the European languages spread around the world as one component of colonial expansion. Closely related to migration is a second explanation, namely that some languages become successful when people find that speaking a particular language is to their economic advantage. A third explanation, as hinted above, is prestige, the number of speakers increasing if the language is associated with a culture that is in some way impressive, perhaps in military, artistic, economic, or religious terms. However, prestige is necessarily temporary, suggesting that the current popularity of English may not be a permanent state of affairs.

CLASSIFICATION AND REGIONS

Others see the loss of linguistic diversity as a sign of increasing human unity, countering the divisive tendencies identified at the beginning of this chapter. From a pragmatic perspective, it is notable that an endangered language is likely to be saved only if speakers are prepared to be bilingual or multilingual.

The current concern with endangered languages means that there is a tendency to

By studying language families—groups of closely related languages that suggest a common origin—scholars have found evidence of language evolution dating from perhaps 2.5 million years ago and evidence of language distribution patterns dating back some 50,000 years. Nevertheless, it appears that the distribution shown in Figure 6.10 just prior to the changes initiated by European overseas

---

FIGURE 6.10   **World distribution of language families before European expansion**

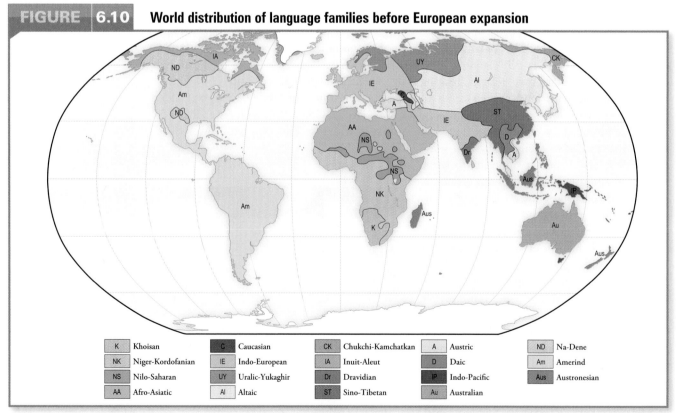

K	Khoisan	C	Caucasian	CK	Chukchi-Kamchatkan	A	Austric	ND	Na-Dene
NK	Niger-Kordofanian	IE	Indo-European	IA	Inuit-Aleut	D	Daic	Am	Amerind
NS	Nilo-Saharan	UY	Uralic-Yukaghir	Dr	Dravidian	IP	Indo-Pacific	Aus	Austronesian
AA	Afro-Asiatic	Al	Altaic	ST	Sino-Tibetan	Au	Australian		

expansion is a product of relatively recent migrations, during the last 5,000 years or so. Note that this map would be significantly different today because of the expansion of several of the Indo-European languages since the sixteenth century.

As human groups moved, their language changed, with groups that became separated from others experiencing the greatest language change. Something similar has happened over the past several hundred years as Europeans especially have moved overseas. Although obvious differences in dialect have developed (compare the varieties of English spoken in southern England, the southern United States, and South Africa, for example), no new languages have come into being as a result of these more recent movements, for two general reasons: (1) not enough time has elapsed for this to occur; (2) the groups that moved have remained in close contact with their original groups. This is not to say that languages cannot change significantly over a relatively short period of time, however: just compare the English of Shakespeare with that spoken today.

Table 6.5 provides a numerical summary of the number of speakers in each of the 19 language families. The largest family is the Indo-European, followed by the Sino-Tibetan. Interestingly, the Indo-European family is spatially dispersed today, while the Sino-Tibetan is essentially limited to one area. The difference reflects the fact that many Indo-European-speaking countries colonized other parts of the world, whereas Sino-Tibetan-speaking countries did not. Thus, the numerical importance of Mandarin is attributable simply to the large population of China (similarly, the numerical importance of Hindi reflects the large population of India), whereas the numerical importance of English, Spanish, and Portuguese is attributable to, among other things, colonial activity.

As shown in Figure 6.11, an interesting feature of the spatial distribution of languages is that temperate areas of the world have relatively few languages each typically with many speakers, whereas tropical areas have many languages each typically with relatively few speakers. One reason for this distribution is that agriculture spread throughout temperate areas with the resultant emergence of large-scale societies that gradually replaced smaller groups. Counting languages is fraught with problems; but Europe has about 200, the Americas about 1,000, Africa about 2,400, and the Asian-Pacific region about 3,200. Perhaps the most remarkable example of linguistic diversity is the country of Papua New Guinea, which has about 800 languages.

The Indo-European language family

Rather than review each of the language families in detail, we will investigate one family, Indo-European, and one component language, English. This choice is not arbitrary. English is the 'language of the planet, the first truly global language' (McCrum et al., 1986: 19).

The parent of all Indo-European languages, what we might call proto-Indo-European, probably evolved as a distinct means of communication in either the Kurgan culture of the Russian steppe region, north of the Caspian Sea, or in a farming culture in the Danube valley. The likely period of origin is between 6000 and 4500 BCE. Our knowledge of these origins has been gleaned largely from reconstruction of culture on the basis of vocabulary. The evidence suggests a culture that was partially nomadic, had domesticated various animals (including the horse), used ploughs, and grew cereals. The fact that contemporary

Table 6.5	Language Families	
Language Family	**Number of Languages**	**Number of Speakers (000)**
Indo-European	430	2,562,896
Sino-Tibetan	399	1,275,532
Niger-Kordofian	1,495	358,091
Afro-Asiatic	353	339,479
Austronesian	1,246	311,740
Dravidian	73	221,516
Altaic	64	145,069
Daic	57	50,000
Nilo-Saharan	197	34,953
Uralic-Yukashir	36	22,623
Amerind	583	18,000
Austric	4	7,000
Caucasian	38	5,000
Indo-Pacific	731	2,735
Khoisan	22	363
Na-Dene	41	180
Inuit-Aleut	9	85
Australian	224	35
Chukchi-Kamchatkan	5	14

SOURCE: Adapted and updated from M.C. Ruhlen, *A Guide to the World's Languages: Volume 1, Classification* (Stanford, Calif.: Stanford University Press, 1987). ISBN for 1991 edition: 0804718946, 9780804718943.

FIGURE 6.11   Linguistic densities

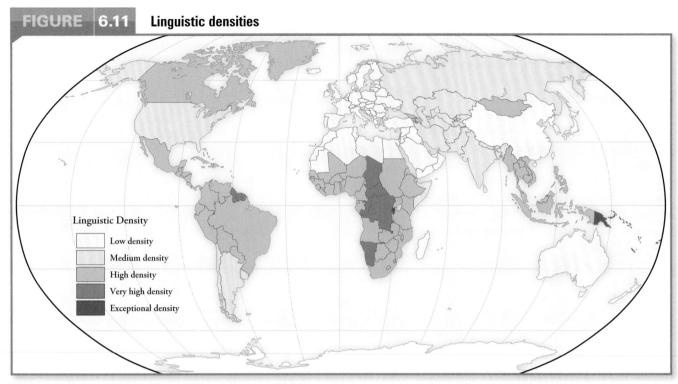

Linguistic Density

- Low density
- Medium density
- High density
- Very high density
- Exceptional density

SOURCE: Reprinted from *The UNESCO Courier*, April 2000, at: <www.unesco.org/courier>.

Indo-European languages have similar words for 'snow' but not for 'sea' places the original language in a cold region some distance from the sea.

As this cultural group dispersed, using the horse and the wheel, the language moved and evolved (Figure 6.12). Different environments and group separation gave rise to many changes. The earliest Indo-Europeans to settle in what is now England were Gaelic-speaking people who arrived after 2500 BCE. They were followed by a series of Indo-European invaders: the Romans in 55 BCE and the Angles, Saxons, and Jutes in the fifth century CE (Figure 6.13). The latter three groups introduced what was to become the English language, forcing Gaelic speakers to relocate in what is now the Celtic fringe. Surprisingly, little mixing took place between Gaelic and the incoming Germanic language. Old English, as it is called, was far from uniform, and was subsequently enriched with the introduction of Christianity (Latin) in 597, the invasions of Vikings between 750 and 1050, and the Norman (French) invasion of 1066. Each of these invasions led to a collision of languages, and English responded by diversifying. Today it has an estimated 500,000 words (excluding an equal number of technical terms); by contrast, German has 185,000 words and French 100,000.

Beginning in the fifteenth century, the English language, along with other Indo-European languages, spread around the globe. This spread encouraged the diversification of English, including the development of many American varieties. Not only is English the first language of many countries, but it is the second language of many others. In India, for example, perhaps 70 million people speak English as a second language: a vital unifying force in a country of many languages. In China, English is increasingly a requirement for admission to university. English is indeed becoming the global language, entirely replacing some minority languages and becoming increasingly necessary as a second language in many parts of the world, including Europe where it has become the de facto working language of the European Union. It is an accident of history that at the time when the need for a common second language became apparent in the twentieth century, an English-speaking country, the United States, was dominant.

LANGUAGE AND IDENTITY

For many groups, language is the primary basis of identity—hence the close links between language and nationalism, the desire to preserve minority languages, and even the various efforts that have been made to create a universal

FIGURE 6.12   Initial diffusion of Indo-European languages

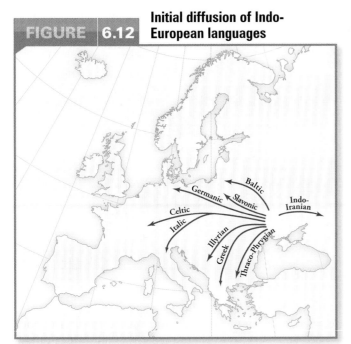

FIGURE 6.13   Diffusion of Indo-European languages into England

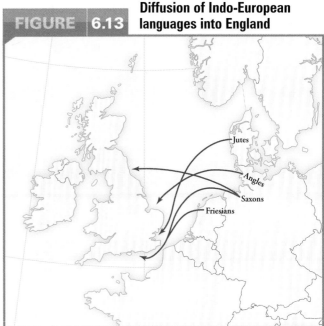

language. A common language facilitates communication; different languages create barriers and frictions between groups, further dividing our divided world.

Language and nationalism

The link between language and **nationalism** is clear. Rarely, however, are the boundaries of language regions clearly defined. This has resulted in many difficulties for those countries aspiring to declare their national territory on the basis of a common language. In medieval Wales there was a single word for 'language' and 'nation', and in present-day Ireland the survival of the Gaelic-speaking area is considered to be 'synonymous with retention of the distinctive Irish national character' (Kearns, 1974: 85). For two reasons language is often seen as the basis for delimiting a nation. First, a common language facilitates communication. Second, language is such a powerful symbol of 'groupness' that it serves to proclaim a national identity even where, as in Ireland, the language itself does not serve a significant communication function.

Before the nineteenth century, the boundaries between states and languages rarely coincided. France and the United Kingdom were single-language states, but most of Europe was politically divided on the basis of factors other than language. From the beginning, the rise of nationalism in the nineteenth century represented an effort to integrate language and state.

As a result of this effort, the state of Italy was established in 1870 and the state of Germany in 1871.

Multilingual states

Other areas in Europe, however, did not follow this trend. Switzerland is a prime example of a viable political unit with several official languages: 70 per cent of the people speak German, 19 per cent French, 10 per cent Italian, and 1 per cent Romansh (Figure 6.14). The political stability for which Switzerland is known reflects several factors, including the pre-1500 evolution of the Swiss state, its long-standing practice of delegating much governmental activity to local regions, and its neutrality in major European conflicts, the latter factors aided by the extreme geography.

More typical examples of **multilingual states** are Belgium and Canada (Box 6.5). Despite conscious efforts to follow the Swiss example, Belgium is politically unstable. Created as late as 1830 as an artificial state—a move that pleased other European powers but not necessarily the people who became Belgians—Belgium has not managed to remain detached from European conflicts and today is divided between a Flemish-speaking north and French-speaking south (Figure 6.15). Although it is a bilingual state, the two areas are regionally unilingual. The fact that the capital, Brussels, is primarily French-speaking, even though it is located in the unilingual

nationalism
The political expression of nationhood or aspiring nationhood; reflects a consciousness of belonging to a nation.

multilingual state
A state in which the population includes at least one linguistic minority.

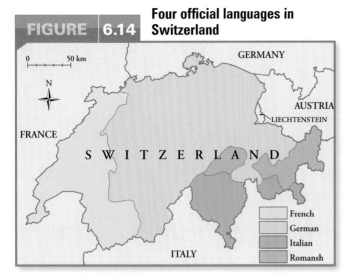

FIGURE 6.14 Four official languages in Switzerland

French
German
Italian
Romansh

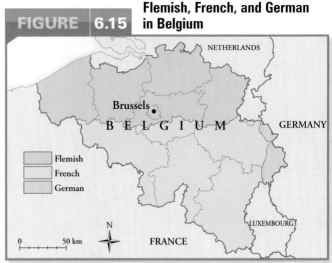

FIGURE 6.15 Flemish, French, and German in Belgium

Flemish
French
German

minority language
A language spoken by a minority group in a state in which the majority of the population speaks some other language; may or may not be an official language.

Flemish area, only aggravates a difficult situation: not only is Belgium not a bilingual country in practice, but it does not have a bilingual transition zone.

The situation is similar in Canada (Figure 6.16 locates French speakers in both Canada and the United States). There is no continuous language transition zone between English and French Canada—only a series of pockets, especially around the cities of Montreal and Ottawa and in Acadian areas of the Maritime provinces. This pattern suggests a parallel with Belgium (Box 6.5).

Minority languages

Any language is most likely to survive when it serves in an official capacity. **Minority languages** without official status typically experience a slow but inexorable demise. Thus, minority-language speakers often strive for much more than the mere survival of their language: they may press for the creation of their own separate state, using the language issue as justification.

Examples of minority languages include Welsh and Irish in Britain, Spanish in the United States, French—despite its official status—in

Box 6.5   Linguistic Territorialization in Belgium and Canada

Although the specific causes are quite different, the Belgian and Canadian language situations are basically similar. Belgium is clearly divided into two linguistic territories, while Canada appears to be approaching the same situation.

Since about the fifth century, the area that is now Belgium has been divided into a northern area of Flemish speakers (Flemish is the ancestor of modern Dutch) and a southern area where Walloon (French) is spoken. Thus, Belgium lies on both sides of the linguistic border separating the Germanic and Romance subfamilies of the Indo-European family. To expect a viable state to emerge in this context as late as 1830 was quite ambitious, particularly when the French speakers explicitly wished to join France. In fact, Belgium is an artificial state created especially by the British and Germans, neither of whom favoured territorial expansion for France. Thus, it comprises two distinct language areas with a minimal transition zone; it has attempted to follow the example of Switzerland, but without success.

Canada, settled by both French and English during its formative years, is also a bilingual state. With Confederation in 1867, the French formed a majority in Quebec and the English a majority in Ontario, Nova Scotia, and New Brunswick. Most of the remainder of Canada west of Ontario was settled either by English speakers or by immigrants from other countries who settled in English-speaking areas; immigration of French speakers has been very limited. Over centuries, interaction between French and English had given rise to distinct transition zones in western New Brunswick, southern Quebec, and eastern Ontario. However, analysis of local migration and interaction between the two groups (Cartwright, 1988) suggests that the zone is disintegrating into a series of pockets, and that the linguistic territorialization evident in Belgium since 1830 is becoming a reality in Canada as a consequence of spatially delimited language differences, despite the passage in 1969 of the Official Languages Act, which institutionalized French as one of two official languages in Canada.

FIGURE | 6.16 | French and English in North America

Legend:
- 10% or more of the population speaks French
- Majority of the population speaks French

Edmonton

Montreal

Boston

Atlantic Ocean

Pacific Ocean

New Orleans

N

0   1000 km

Canada, Basque in Spain, Hausa and other languages in Nigeria, and Cantonese in China. This is a highly condensed list; most countries in the contemporary world have at least one minority language. The consequences are varied. In Britain both Welsh and Irish continue to strive for survival and indeed independence (Box 6.6). In Spain the Basque language—one of very few languages in Europe that do not belong to the Indo-European family—is at the centre of a powerful and often violent Basque independence movement, although the principal terrorist group declared a ceasefire in 2006. In Nigeria the official language is English—a result of colonial activity and of the fact that, although Nigeria has several indigenous languages, no single one is predominant.

Communications between different language groups

Historically, some languages have played important roles even when they are not the first language of populations. In India, following independence, English joined Hindi as an official language. In parts of East Africa, Swahili is an official language; combining the local Bantu with imported Arabic, Swahili is an example of a **lingua franca**, a language developed to facilitate trade between different groups, in this case Africans and Arab traders. In some other colonial areas **pidgin** languages have developed as simplified ways of communicating between different language groups; they are especially common in areas that developed as trading centres, with imported slave populations, and in areas of plantation economies. Pidgins are common in Southeast Asia, where they are usually based on English, with some Malay and some Chinese. A pidgin that becomes the first language ('mother tongue') of a generation of native speakers is known as a **creole**. Creoles have larger vocabularies than pidgins and a more sophisticated grammatical structure. They are relatively common in the Caribbean region and vary according to whether the principal European component is English, Spanish, or

lingua franca
An existing language used as a common means of communication between different language groups.

creole
A pidgin language that assumes the status of a mother tongue for a group.

pidgin
A new language designed to serve the purposes of commerce between different language groups; typically has a limited vocabulary.

Box 6.6   The Celtic Languages

The Celts were one of the most important early groups to diffuse from the Indo-European core area. Beginning about 500 BCE, they spread across much of Europe. But after some 500 years of expansion, as they came into contact with other, more organized groups, the Celts gradually retreated into some of the more inhospitable and isolated areas of Western Europe. Today the remains of the once-large Celtic group live in four small pockets: western Wales, western Scotland, western Ireland, and northwest France. Each of these four areas still has some Celtic speakers: in Wales about 500,000 speak Welsh, in Scotland about 80,000 speak Gaelic, in Ireland about 70,000 speak Erse (Irish Gaelic), and in northwest France about 675,000 speak Breton.

In Britain the Celts were pushed to the western limits by the Anglo-Saxon in-movement. Over time, various Celtic languages, such as Cornish (southwest England) and Manx (Isle of Man in the Irish Sea), disappeared, leaving the four pockets already noted.

The remaining Celtic languages are in precarious positions, basically because the languages are not associated with political units. The Irish-language area, called the Gaeltacht, covered about 33 per cent of Ireland in 1850 and had perhaps 1.5 million speakers; today it covers about 6 per cent and has lost 95 per cent of its speakers. The first concerted efforts to save the Gaeltacht came in 1956. The basic assumption is that language is a key requisite for any group that aspires to retain a traditional culture. Consequently, the Irish government has, since 1956, actively encouraged retention of the language and cautious social and economic development for the rural Gaeltacht region. So far, the various government efforts to foster development have been quite successful and have not resulted, as some feared they might, in loss of the Irish language.

Welsh, a minority language intimately tied to traditional Welsh culture and a rural way of life in an increasingly Anglicized environment, is in a similar position. Welsh-speaking Wales is restricted to the extreme northwest and southwest, and the Welsh language is threatened, despite vigorous local efforts to preserve it. As a part of the United Kingdom, Wales is not in as strong a position as Ireland when it comes to implementing local development and language retention policies.

French; often the other components are the native Carib and imported Bantu.

Some countries are uncertain which of many languages spoken merit official status. In the South American country of Suriname, with about one-half million people, languages spoken include Dutch (it was a Dutch colony), Portuguese, English, and variants of Chinese, Hindi, Javanese, and six creole languages. Dutch is the official language and is taught in schools, but the main language of everyday communication is a creole, Sranan Tongo (meaning Suriname tongue). This creole is based largely on English. In general, people speak Dutch in formal settings and Sranan Tongo in informal settings.

A number of attempts have been made to promote the use of a single universal language; artificial languages have been invented for this purpose. In 1887, the most popular such language, Esperanto, was introduced, but failed to make a significant impact, perhaps because it had no ties to any specific tradition, culture, or environment although it was based on roots common to several European languages. For the same reason, most human geographers reject the idea of a universal language—even though they recognize that, in principle, a universal language could promote communication and understanding between groups, thus minimizing division and friction. In any case, it seems unlikely that any artificial language will succeed in playing such a role. Today, the best hope for a universal language rests with English.

LANGUAGE IN LANDSCAPE

Naming Places

Our discussion so far has emphasized the centrality of language to culture and group identity, and its effects on our partitioning of the earth. But language also plays a key role in landscape, in the form of place names or **toponyms**. We name places for at least two reasons. The first is in order to understand and give meaning to landscape. A landscape without names would be like a group of people without names; it would be difficult to distinguish one from another. Second, naming places probably serves an important psychological need—to name is to know and control, to remove uncertainty about the landscape. For these two reasons, humans impose names on all landscapes that they occupy and on many that they do not (the moon is a prime example).

Place names, then, are a significant feature of our human-made landscapes, often visible

toponym
Place name; evidence provided by place names can be crucial in a historical study of movement and settlement if other sources of information are unavailable.

in the form of road signs and an integral component of maps—our models of the landscape. Many place names combine two parts, generic and specific. 'Newfoundland', for example, has *Newfound* as the specific component and *land*—the type of location being identified—as the generic component.

Analysis of place names can provide information about both the spatial and the social origins of settlers. In addition, names such as Toivola (Hopeville) in Minnesota or Paradise in California tell us something about their aspirations. Place names typically date from the first effective settlement in an area. Because the Finns in Minnesota were among the later European settlers, their opportunity to name places was restricted to the local scale, whereas other groups, most notably the French and Spanish, were able to place their languages on the landscape at the regional scale; witness the profusion of French names in Quebec and Louisiana and of Spanish names throughout the American Southwest.

Place names can be an extremely valuable route to understanding the cultural history of an area. This is especially clear in areas where the first effective settlement was relatively recent, but is also the case in older settled areas. In many parts of Europe, for example, former cultural boundaries can be identified through place-name analysis. Jordan (1988: 98) provides an example from before 800, when the Germanic–Slavic linguistic border roughly followed the line of the Elbe and Saale rivers in what is now Germany. Although German-speakers began moving east after 800, evidence of the earlier boundary remains: place names are German west of the line and Slavic east of it. Similarly, throughout much of Britain it is possible to identify areas settled by different language groups by studying place names; in northeast England, for example, place-name endings such as *–by* are evidence of Viking settlement.

Renaming places

Place names also are about power. The former republic of Yugoslavia that is now called Macedonia came close to war with neighbouring Greece over the choice of name because Macedonia also is the name of a Greek province. In response to this political issue, the United Nations elects to call Macedonia by the much more convoluted name, the Former Yugoslav Republic of Macedonia (or FYROM).

One of the greatest confusions in the world today concerns international uncertainty over the names of countries and of many landscape features. Especially throughout much of Africa, Central Asia, and in India, place names are being changed as one part of a larger rejection of the colonial era. For example, Mumbai has replaced Bombay, and Iqaluit Bay has replaced Frobisher Bay. In some multilingual countries the situation is extraordinarily complex. In South Africa, which has 11 official languages, the same city is called Cape Town (English), Kaapstad (Afrikaans), and eKapa (Xhosa). Elsewhere in South Africa there has been opposition by white residents to some name changes, for example, changing the name of the town of Lydenburg, which was named by early Dutch settlers, to Mashishing, an African name. In Israel the city of Jerusalem (English) is called Yerushalayim in Hebrew and either Urshalim or Quds in Arabic. Sometimes the consequences of place-name uncertainty can be very damaging, as evidenced during the first Gulf War when bombs were dropped in an unintended location because of misunderstanding about place names. The UN even has a committee of geographers to discourage the use of **exonyms**, names given to a group or place by a group other than the people/place to which the name refers.

Renaming also occurs because previous names may now be seen as causing offence. A cape on Lake Ontario known as Niggerhead Point, because it was on the route of the underground railway aiding slaves to escape to Canada, was renamed Negrohead Point, and is now marked on New York state maps as Graves Point. For a somewhat different reason, Gayside in Newfoundland was renamed Baytona.

The 'Great American Desert'

Just as language is everywhere in landscape, it can also make landscape. A striking example is the nineteenth-century use of the term 'desert'—as in 'Great American Desert'—to describe much of what is now the Great Plains region. Most European North Americans knew nothing of the Plains until several expeditions in the early nineteenth century, including those of Zebulon Pike (1806) and Stephen Long (1823), returned with reports of a 'desert' that would surely restrict settlement. This misconception arose because individual explorers recorded what impressed them most, because the small areas of sand desert were indeed

exonym
A name given to people (or a place) by a group other than the people to which the name refers (or who are not native to the territory within which the place is situated).

impressive, and because the absence of trees did make the Plains a desert in comparison with the heavily forested East. Among the consequences was delayed settlement as migrants bypassed the Plains in favour of the west coast. Thus, the language used to describe a specific environment affected human geographic changes in the landscape.

LANDSCAPE IN LANGUAGE

Not only is language in landscape, but landscape is in language. First, because physical barriers tend to limit movement, there is a general relationship between language distributions and physical regions (this is evident on a world scale in Figure 6.10). Second, the vocabulary of any language necessarily reflects the physical environment in which its speakers live; hence there are many Spanish words for features of desert landscapes and few comparable English words. Similarly, many words are available to discriminate between different types of snow in the Inuit language and between the colourings of cattle in the Masai language. As languages move to new environments, new words are added to help describe these new environments; the word 'outback', for example, entered English only with the settlement of Australia. Finally, the human landscape is in language to the extent that language reflects class and gender; within a given language, word choices and pronunciations say a great deal about social origins. This area of research is now labelled 'sociolinguistics'.

Religion

A second fundamental cultural element is religion. Although the specific origins of religious beliefs are no less difficult to trace than those of language, the universality of religion suggests that it serves a basic human need or reflects a basic human awareness. Essentially, a religion consists of a set of beliefs and associated activities that are in some way designed to facilitate appreciation of our human place in the world. In many instances, religious beliefs generate sets of moral and ethical rules that can have a significant influence on many aspects of behaviour. Typically, a religion has a core set of beliefs and these find expression in many forms, including texts, rituals, everyday behaviour, symbols, and, of course, landscapes.

There are often major distinctions between women and men with regard to religious behaviour. Women usually make up the majority of the followers of a religion, and women play a critical role in teaching religion to their children. Further, most major religions—with the exception of Islam—have at least some female figures of worship (although the most important figures are generally male). Yet in many religions women are excluded from serving in a formal, structured role; in the Christian tradition, although women are now admitted to the ministry in many Protestant churches, they remain largely excluded from the Catholic and Orthodox traditions. Men have played the dominant official roles in Hinduism, Buddhism, Christianity, and Islam, to name only the largest organized religions.

Figure 6.17 identifies the origin areas of the four largest religions, as measured by numbers of adherents—Hinduism, Buddhism, Christianity, and Islam—while Figure 6.18 maps the contemporary distribution of these and other religions. These maps make two things especially clear. First, there are two major religious 'hearth' areas: Indo-Gangetic and Semitic. The Indo-Gangetic hearth has given rise to Hinduism, Buddhism, Jainism, and Sikhism, while the Semitic hearth has given rise to Judaism, Christianity, and Islam. Second, two of the four major religions, Christianity and Islam, have diffused over large areas, whereas Hinduism has not experienced significant spread and Buddhism has effectively relocated to China and other parts of Asia. These conclusions are confirmed by Table 6.6, which provides detailed data on the distribution of religious groups by selected major world regions. Christianity is the leading religion numerically, with 33 per cent of the world population declaring themselves, at least nominally, Christian, followed by Muslims (followers of Islam) at 20 per cent, Hindus at 13 per cent, and Buddhists at 6 per cent. A high 13 per cent are classed as non-religious.

Of course, using a single word to describe a religious identity can be misleading as all four of the major religions are divided into often very different branches—with differing beliefs and practices—that are in competition. These schisms arose as religions moved through landscapes, coming into contact with different cultural contexts so that new understandings emerged; also, oral traditions and ancient texts are easily subject to differing interpretations.

A useful classification of religions, favoured by geographers because it relates closely to

FIGURE 6.17 Hearth areas and diffusion of four major religions

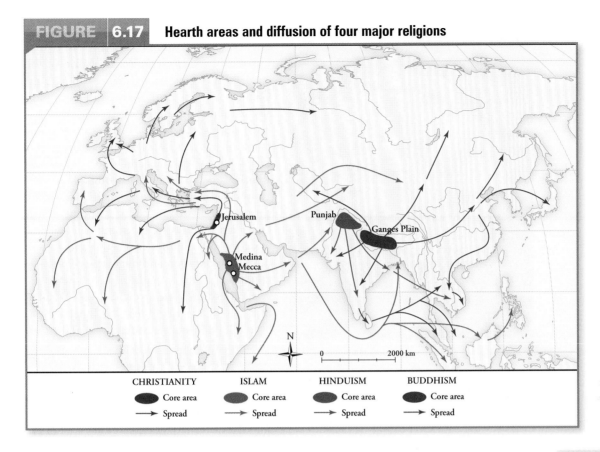

CHRISTIANITY	ISLAM	HINDUISM	BUDDHISM
Core area	Core area	Core area	Core area
→ Spread	→ Spread	→ Spread	→ Spread

spatial distributions, distinguishes between universalizing religions, which actively seek converts, and ethnic religions, which are closely identified with a specific cultural group. Although very useful, this twofold classification necessarily excludes hundreds of numerically small religions. For example, many cultures embraced a form of **animism**—in which a soul or spirit is attributed to various phenomena, including inanimate objects—especially before the diffusion of universalizing religions, notably Christianity and Islam.

animism
A general name for beliefs that attribute a spirit or soul to natural phenomena and inanimate objects.

FIGURE 6.18 World distribution of major religions

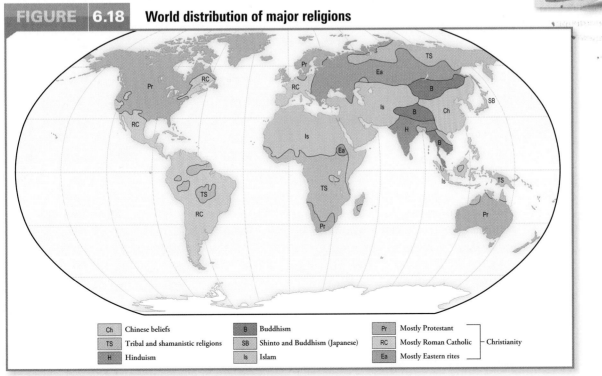

Ch	Chinese beliefs	B	Buddhism	Pr	Mostly Protestant
TS	Tribal and shamanistic religions	SB	Shinto and Buddhism (Japanese)	RC	Mostly Roman Catholic
H	Hinduism	Is	Islam	Ea	Mostly Eastern rites

Christianity

A Hindu devotee making the painful pilgrimage to Batu Caves, Kuala Lumpur, Malaysia.
Robert Churchill/iStock

ORIGINS AND DISTRIBUTION OF ETHNIC RELIGIONS

The principal ethnic religions emerged earlier than their universalizing counterparts. They are associated with a particular group of people and do not actively seek to convert others.

Hinduism

Of the several hundred religions in this category, the largest is Hinduism, which evolved in the Indo-Gangetic hearth—a lowland area of north India that drains into the Indus and Ganges rivers—about 2000 BCE. Hinduism, initially more correctly described as Vedism, was the first major religion to evolve in this area, from which it spread east down the Ganges and then south through India, eventually to dominate the entire region. As it diffused during the last few centuries BCE and much of the first millennium CE through an already diverse cultural landscape, core features absorbed and blended with other local religious beliefs. There were particular distinctions between Indo-Aryan cultures to the north, Dravidian cultures to the south, and numerous local hill cultures in central and eastern India. The diffusion and ongoing evolution of Hinduism means it might be better described as a variety of related religious beliefs rather than as a homogeneous religion. Indeed, variations of Hinduism are evident even at the scale of the local village community.

From India Hinduism was carried overseas, but it has not retained significant numbers of adherents outside of India. Today, it is an Indian religion associated with a country and a broadly defined cultural group. Hinduism has no dogma and only a loosely defined philosophy by religious standards. It is polytheistic (worshipping more than one god) and has close ties to the rigid social stratification of the **caste** system. Like many other religions, Hinduism has spawned numerous offshoots. Jainism, developed from the teachings of a sixth-century BCE holy teacher, rejects

caste
A social rank, based solely on birth, to which an individual belongs for life and that limits interaction with members of other castes.

Table 6.6	Major World Religions: Number of Adherents (thousands), 2008							
Religion	Africa	Asia	Europe	Latin America	North America	Oceania	World	%
Christianity	465,880	364,106	583,802	536,162	277,089	27,476	2,254,535	33.4
Islam	392,636	992,850	40,749	1,830	5,556	460	1,434,081	21.2
Hinduism	2,813	906,190	1,681	760	1,756	471	913,671	13.5
No religion	6,012	619,845	82,658	16,958	39,847	4,294	769,614	11.4
Chinese folk religions	38,5	385,861	312	186	747	150	387,294	5.7
Buddhism	165	377,515	1,792	767	3,504	575	384,318	5.7
Ethnic religions	116,125	147,571	1,153	3,654	1,567	343	270,413	4.0
Atheism	614	126,914	15,676	2,839	1,852	427	148,322	2.5
New religions	126	104,208	393	819	1,633	90	107,269	1.6
Sikhism	65	22,592	475	6	647	50	23,835	0.4
Judaism	130	5,750	1,850	1,046	6,212	108	15,096	0.2
Spiritism	4	0	143	13,348	168	7	13669	0.2
Baha'i faith	2,229	3,786	142	910	660	141	7,368	0.1
Confucianism	0	6,346	18	0	0	53	6,418	0.1

Source: *Time Almanac, 2009* (Chicago: Encyclopedia Britannica, 2009), 568–9.

Hindu rituals but shares many basic tenets of Hinduism, including the belief in reincarnation and *ahimsa* (the ethical doctrine that humans ought to avoid hurting any living creature). A more recent offshoot is Sikhism, a hybrid of Hinduism and Islam that arose about 500 years ago in the Punjab region of India.

Judaism

A second ethnic religion is Judaism. The first monotheistic religion (worshipping one single god), Judaism originated about 2000 BCE in the Near East, initially in the form of proto-Judaism. Following the Romans' destruction of Jerusalem in 70 CE the Jews were driven out of their homeland and eventually dispersed throughout Europe; the entire body of Jews living outside Israel, in Europe and elsewhere, is known as the Diaspora.

Judaism contains significant internal divisions reflecting theological and ideological differences. These differences developed as Jews moved, with Sephardic Judaism concentrated in Spain and Ashkenazic Judaism elsewhere in Europe. Sephardic Jews easily adapted to different areas, but Ashkenazic Jews tended not to integrate with larger non-Jewish, Christian, society. Undoubtedly this lack of integration was a twofold process, with Jews wishing to maintain their traditions, and with Christians being antagonistic towards Jews. Ashkenazic Judaism experienced a number of divisions, especially with the eighteenth-century rise of Hasidic Judaism in western Ukraine and the emergence of ultra Orthodox Judaism initially in Hungary.

Much Jewish movement, including a general eastward migration in Europe from about the fourteenth century onward and the late nineteenth- and early twentieth-century movements from Europe to North America, can be understood as a response to persecution. Anti-Jewish sentiment was most evident when it became a part of larger state policies in Germany in the 1930s. Only in 1948 was the long-term goal of a Jewish homeland achieved with the creation of the state of Israel.

Other ethnic religions include Shinto, the indigenous religion of Japan, and Taoism and Confucianism, both of which are primarily associated with China.

ORIGINS AND DISTRIBUTION OF UNIVERSALIZING RELIGIONS

Both Hinduism and Judaism have given rise to major universalizing world religions.

Buddhism

Buddhism was the first universalizing religion, an offshoot of Hinduism founded in the

The seventeenth Karmapa (left), at the Gyuto Ramoche Monastery near Dharmsala, India, in February 2000. One of the highest lamas in Tibetan Buddhism, Urgyen Trinley Dorje had escaped from Chinese-occupied Tibet in January and reached Dharmsala—the home of the Tibetan government in exile—after a 1,400-km trek over the Himalayas.

CP/AP photo/John McConnico

Indo-Gangetic hearth by Prince Gautama, who was born in 644 BCE. During his lifetime he preached in northern India, but after his death Buddhism was spread by missionary monks into other parts of India and then throughout much of Asia. There is debate concerning whether this diffusion of Buddhism occurred mostly before or after two versions of the religion evolved in the first few centuries CE—Mahayana Buddhism emerged in the Upper Indus valley, while Theravada Buddhism first displayed a clear identity in Sri Lanka. Some scholars contend that, by the third century BCE, Buddhist scriptures were codified, that the religion was effectively the state religion of a large India-wide empire, and that it was then carried to China after about 100 BCE and throughout most of Asia. But these dates are uncertain. What is clear is that the Mahayana version, which is more inclusive (syncretistic), diffused north from India into Central and East Asia, including China, Korea, Japan, Tibet, and Mongolia, while the conservative Theravada version, which does not seek to reconcile with other belief systems, moved into Southeast Asia, including Burma (Myanmar), Thailand, Cambodia, Laos, and Vietnam.

Perhaps surprisingly, this spread and growth did not result in Buddhism's acceptance as the religion of China, as both Taoism and Confucianism wielded important influences. Also, Buddhism in India merged with Hinduism, which also competed with it in some parts of Southeast Asia. Only in Burma and Thailand did Buddhism achieve the status of a state religion. The patterns of diffusion remain evident in the current distributions, with Mahayana Buddhism prevailing in Central and East Asia and Theravada Buddhism in Southeast Asia. Like Hinduism, Buddhism is characterized by numerous local variations as it blended easily with pre-existing local beliefs.

Christianity

A second universalizing religion, Christianity, developed about 600 years after Buddhism as an offshoot of Judaism in the Semitic hearth area. Christianity began when disciples of Jesus of Nazareth accepted that he was the expected Messiah. The religion spread slowly during his lifetime and more rapidly after his death as missionaries carried it initially to areas around Jerusalem and then through the Mediterranean to Cyprus, Turkey, Greece, and Rome. Although local variations soon appeared, these never assumed the same significance as was evident in Hinduism and Buddhism. This was because Christianity included the idea of a unified Church, because written details of the founding of the religion were soon available as the four gospels, and because there were deliberate efforts to formalize universal doctrines at a series of councils, the first of which was held in Nicea in 325 CE.

Christianity spread rapidly through the Roman Empire and was formally adopted as an official religion. However, the collapse of the empire and resultant political fragmentation meant there were competing Germanic and Roman influences, with many Germanic people practising a version of Christianity known as Arianism. A division was evident between a Latin-speaking western region and a Greek-speaking east. This west–east distinction became more significant with debate about the status of the bishop of Rome, later known as the Pope. Christians in the east rejected claims by bishops of Rome that they had authority over the entire Church. As a result of these differences, a major east–west division between the Roman Catholic (western) and Eastern Orthodox forms of Christianity occurred in 1054, and the line dividing these two remains the most basic religious boundary in contemporary Europe (see Figure 8.10). The two versions of Christianity competed for converts throughout the Slavic areas of Europe.

The Protestant Reformation of the 1500s produced a third major version of Christianity. The Reformation was an effort to reform Roman Catholic dogma, teachings, and practices, and also an attempt by some northern European leaders to challenge the wealth of the Catholic Church. Protestants established a number of independent churches, notably Lutheran, Zwinglian, Calvinist, and Anglican. Several of these churches, most notably Anglican, had strong national connections. Northern Europe became largely (though by no means exclusively) Protestant with all the religious diversity that this implied, but the Reformation did not affect Southern Europe, which remained predominantly Catholic.

Both Catholic and Protestant versions of Christianity spread to other parts of the world—the Americas, Africa, much of Asia, and Australasia—in the course of European movement overseas. The importance of Christianity in the contemporary world reflects not only the large number of its adherents and their wide spatial spread, but also the fact that Christian thinking has been a cornerstone of Western culture, affecting attitudes and behaviours at the level both of the group and of the individual. Christianity has proven to be the most influential religion in shaping the world today because it was carried overseas by European colonial powers and because it has been very adaptable and therefore effective at encouraging converts.

Islam

The third major universalizing religion, Islam, also arose in the Semitic hearth area; it is related to both Judaism and Christianity but has additional Arabic characteristics. Rather like Christianity, Islam is an all-encompassing world view, shaping both group and individual attitudes and behaviours. Founded by Muhammad, who was born in Mecca in 570, by the time of his death in 632 it had diffused throughout Arabia. Further diffusion was rapid as a result of Islamic political and military expansion. Arab Muslims created an empire that stretched west to include parts of the northern Mediterranean as far as Spain and much of North Africa, and east to include the areas of modern-day Iraq, Iran, Afghanistan, and Pakistan. Islam has also spread into much of Southeast Asia. In India, Muslims added to the already diverse religious identities and landscapes of Hinduism. Essentially, this historical expansion of Islam was limited to a vast west–east belt that barely penetrates into temperate or extreme tropical climates. The European expansion of Islam resulted in conflicts with Christians; in the sixteenth and seventeenth centuries, for example, there was a massive expulsion of Muslims from Spain, with many relocating in North Africa.

For several reasons, Islam has not experienced the same degree of divisiveness as the other two major universalizing religions, Buddhism and Christianity. The basic structure and content of Islam was widely accepted from the outset, as was the idea of the Islamic *umma*, as a group sharing basic beliefs, along with the idea of the obligatory *Haj* or pilgrimage. One notable division in Islam is between Sunni and Shia versions, which came into being with the death of Muhammad. Sunnis believed that the *umma* was responsible for electing Muhammad's successor, the person who would serve as leader, while Shiites believed that Muhammad had designated a

Muslim pilgrims throw stones at three pillars representing Satan on the second day of Eid al-Adha in Mina, near the holy city of Mecca, as part of the annual hajj pilgrimage to Mecca.

Mohammed Abed/AFP/Getty Images

specific person (his cousin and son-in-law). The break between these two groups, formalized in 680, thus revolved around the question of human leadership. Other schisms in Islam have involved the reassertion of fundamental beliefs and practices, and these are discussed in the following section.

Today, Sunnis, who represent about 90 per cent of the total Muslim population, dominate in Arabic-speaking areas, as well as in Pakistan and Bangladesh; Shiites are a majority in Iran and Iraq. For most of the history of Islam the Shiite–Sunni divide has been peaceful, at least partly because the two groups usually occupied different places; but in recent years both Iraq and Iran especially have experienced conflict, and the two countries fought a bitter war in the late 1980s.

Although Islam is often considered a religion of the Middle East, the largest Muslim populations today are in Indonesia, India, Pakistan, and Bangladesh. Islam was carried overland and overseas as part of a larger political expansion, but it has not been as adaptable as Christianity because Muslims see their religion as a way of life that encompasses social and political affairs, meaning that it is well-suited to some societies but not to others. Islam is a more prescriptive religion than Christianity as it comprises a universal code of behaviour that adherents accept and pursue. The body of Islamic law, the *shariah*, serves as a basis for the religion and for the political state. A significant tension in many parts of the Islamic world today concerns disagreements about state observance of the *shariah*, with revolutionary movements in some countries asserting that any departure from the Islamic law indicates that the government lacks legitimacy. This is a very sensitive issue in a globalizing world where ideas about universal human rights, as evident in UN documents, derive primarily from a Christian tradition.

RELIGION, IDENTITY, AND CONFLICT

The preceding account of the origins, spread, and growth of major religious groups has highlighted an important general fact about religious identity, religious territory, and conflict between religious groups: a person's sense of identity and community, and all that this implies, can often be closely tied to religion. Religions have competed, directly or indirectly, with each other, and with different versions of the same religion, as they have spread from source areas and become established in particular places. Indeed, sometimes such competition has resulted in conflict and the expulsion of the members of one group. In the case of competing universalizing religions and their many versions, this competition has included attempts at conversion. These circumstances, which can often be sources of tension, are now considered.

Most notably, our human tendency to identify with a specific religion has given rise to many military conflicts, from the medieval Crusades and the European religious wars of the sixteenth and seventeenth centuries to recent conflicts in Pakistan (Muslim–Hindu), Lebanon (Christian–Muslim), and Northern Ireland (Catholic–Protestant Christian). In some of these conflicts, religious differences might have been used as excuses for aggression that actually has other motives; many commentators see this as the case in Northern Ireland, arguing that the real cause of conflict was British involvement in Ireland. But there is no doubt that, in many instances, religion promotes mistrust of non-believers, an attitude that, combined with a general lack of understanding, may lead to hostility and conflict. Christian attitudes to, and sometimes forced expulsion of, Jews and Muslims were referred to in the preceding section.

The example of the area of Palestine is informative. Locations are sacred to Jews, Christians, and Muslims, and the region has long been contested space. Beginning in 1095 and continuing for about the next 200 years, the Crusades were attempts by Christian Europe to wrest control of Palestine from Muslim empires. More recently, especially since the creation of the state of Israel in 1948, conflict has intensified between Jews and Muslims concerning the legitimacy of the state itself as well as access to and ownership of many specific locations within the state.

As discussed in Box 6.7 and in Chapter 8, some contemporary scholars point to hostility between Islam and Christianity in particular as a major cause of conflict in the contemporary world, although other scholars consider this a gross oversimplification. Not surprisingly, this debate intensified following the terrorist attacks on New York and Washington in 2001. What is clear today is that just a few religious fanatics can take actions that have drastic consequences on very large populations. Only 19 people were physically involved in the terror attacks of 11 September 2001, but their actions contributed to American-initiated invasions of Afghanistan and Iraq. In turn, these invasions prompted many Muslims elsewhere to worry about the possibility of American invasion and to numerous expressions of anti-American sentiment. More generally, it appears that mutual misunderstanding contributes to disagreements and even outrage.

RELIGIONS AS CIVILIZATIONS?

One of the more contentious debates in the human geographic study of religion concerns the question of links between religions and civilizations. The term 'civilization' is not restricted to one group of people or way of life. Rather, it is used to refer to a variety of technological and related changes occurring in particular regions at particular times. In the twentieth century, scholars began working to identify and label different civilizations as they are now distributed around the world. Most recently, it has become common to define two civilizations in terms of their religious bases in Islam or Christianity and to interpret a person's belonging to one or the other as a fundamental basis for human identity. Other religions are not usually regarded as defining civilizations.

The debate over the degree to which religion and civilization are connected is especially important in a world currently experiencing serious frictions between Christian and Islamic identities. Equating Islam and Christianity

Box 6.7   Islamic and Christian Identities

As the brief account of religions as civilizations suggests, some of the differences between the two closely related religions of Islam and Christianity are assuming major importance in the early twenty-first century. But it is important to place these tensions in perspective by stressing that only a small fraction of the about 1.5 billion Muslims in the world are involved in violent activities against the Western world. The few Muslims who do subscribe to extremist views reflect an Islamic reformist ideology, a jihadist movement, propagated in Arabia in the 1740s by Muhammad ibn Abd al-Wahhab, known as Wahhabism, which is based on the claim that an adherent has a religious duty to kill someone who does not convert to his way of thinking. Wahhabism did not prosper in its extreme form, although it played a role in the creation of the Saudi state and in influencing jihadist movements that helped create a number of West African states, including in Mali, Niger, northern Nigeria, Cameroon, and Sudan. Indeed, Wahhabism has survived to this day, and is evident both in the Taliban in Afghanistan and in the movement led by Osama bin Laden (Allen, 2006).

To understand these religious outlooks, it is crucial to appreciate that most Muslim countries are profoundly Muslim, whereas most Christian countries are no longer profoundly Christian (although, in the United States especially, fundamentalist Christian ideas have had an increasing influence on social and political life). In the Muslim world, values that might compete with religion have not been as evident as in the Christian world. The influential and often controversial scholar of Islam, Bernard Lewis (2003), argues that the West sees the world as a system of nations, whereas Islam sees the world as a system of religions. A key idea in Islam is that religious unity is more important than tribal loyalties. Linked to this concept is the fact that Muslim countries have found it difficult to create successful democracies, although some Muslim countries with large populations are democratic, notably Indonesia, Malaysia, Bangladesh, and Turkey.

Part of the difference in viewpoint is that democracy implies that humans make laws, whereas more fundamentalist interpretations of Islam contend that the laws dictated by God to Muhammad cannot be changed. This belief creates close ties between religion and state in some Islamic countries, while these countries are seen by the Western world as overly rigid and inflexible. Saudi Arabia, for example, bans the practice of all religions excepting Islam and, as a Sunni Islam state, even looks unfavourably on other versions of Islam. At the same time, however, most Saudis are unsympathetic to fundamentalism and strongly oppose violence conducted in the name of Islam. For many Western observers these differing points of view within Islam appear to be contradictory. An unfortunate tendency is to treat the apparent illogic in dehumanizing terms.

On the other hand, Muslims sometimes view the West as practising double standards. The row that exploded in 2006 over the publication, initially in a Danish paper and subsequently in other European newspapers, of cartoons deemed by some Muslims as offensive to Muhammad was justified by many European commentators on the grounds of freedom of speech. But many Muslim observers were quick to point out that European countries typically have limits to freedom of speech on the grounds of national security and prevention of disorder. Most notably, 11 countries—Austria, Belgium, the Czech Republic, France, Germany, Israel, Lithuania, Poland, Romania, Slovakia, and Switzerland—have laws against denying the Holocaust, and, coincidentally, a British historian, David Irving, was jailed in Austria for denying the Holocaust at about the same time as the cartoon row erupted.

The cartoon row was just one specific expression of an underlying tension among Europeans and Muslims that also received expression in, for example, the French ban on head scarves and other displays of religious symbols in state schools.

Improved understanding of and respect for each other within the Islamic and Christian worlds is a prerequisite for a more stable and secure cultural and political world. A better understanding of Islam might develop in the Western world if there was fuller appreciation of some of the contributions made by Muslims. Just four examples of many: the first pin-hole camera was invented by Ibn al-Haitham after he realized that light entered the eye rather than leaving it; the crankshaft, which translates rotary into linear motion, was invented about 1200 by al-Jazari to raise water for irrigation; the windmill was invented in the seventh-century Islamic world; and the soap that we use today was perfected in the Islamic world.

A specific suggestion that might help major religions play a more positive role in the world today is the proposal, raised at a 2006 meeting of Jewish and Muslim leaders, for the creation of a world body with representatives from major religious groups—a form of United Nations of religions.

with civilizations can lead to some unfortunate consequences. One problem is that both are universalizing religions, which means that, strictly speaking, both hold it an intrinsic duty to spread their message of truth to all people: when these messages differ, mutual respect—even tolerance—may be difficult. Toynbee identified three 'civilizations' associated with two different forms of Christianity and with Islam. Obviously the civilization he called 'Western Christendom'—now perhaps more commonly known simply as 'the

West'—includes a great many non-Christian elements. In the case of Islam, however, we have only the one word, 'Islam', to designate both the religion and the civilization. As a result, people in the West frequently fail to distinguish between the two. In recent years this tendency has been especially common in the context of terrorist activity. Thus, terrorism is seen as representative of Islam itself—even if most Muslims would disclaim it as antithetical to their faith.

Religion and identity today

The importance of religion for individual and group identity today varies considerably. In much of the more developed world, especially in urban areas, secularism prevails such that religion is not central to human activity, and many people consider it irrelevant or marginal. During the twentieth century some countries actively rejected religion and replaced it with a political belief, such as communism, as a guide to beliefs and behaviour. Despite these examples of the declining relevance of religion, however, in reality most humans are unable to separate themselves entirely from religion, some because they choose to be actively religious and others because religion is one part of state identity. Indeed, there is a long-standing relationship with religion reinforcing state identity, as with the Church of Greece, Church of Sweden, Church of Norway, and Church of England. In Iran the state is defined in religious terms, and in Saudi Arabia religious leaders play a key role in formulating state policy. In other cases religion is one of the arguments for political separation, as with Sikhism in the Indian state of Punjab.

The role played by religious belief, or lack of belief, today continues to be a source of tension within and between societies. Some people consider their faith to be the basis of life and regard their own as the only true religion. This way of thinking is a contributing factor to the rise of fundamentalism in Christianity, Islam, and Hinduism especially. Others recognize their beliefs as but one of many possibilities and are less inclined to seek to impose their beliefs on others. Still others reject all religious belief. In many societies these different ways of thinking have implications for such sensitive issues as birth control, abortion, and capital punishment. Debates between religion and secularism have become a part of our everyday world.

Concern about the role of religion in public life is now quite pronounced in many countries, arising in often unexpected circumstances. For example, the 2004 appointment of Ruth Kelly—a staunch Roman Catholic and member of Opus Dei—as Education Secretary in the British government was clearly compromised by her strict beliefs on contraception, embryo research, cloning, and abortion. A subsequent 2006 appointment as head of a new local government department was similarly compromised by the fact that she felt unable to support many of the landmark social changes relating to homosexuality that were part of government policy. In another instance, in March 2009 controversy erupted when Gary Goodyear, Canada's Science Minister and one of those playing a key role in assigning funds for scientific research, invoked his religious beliefs and was less than clear when asked if he accepted evolutionary theory. More generally, gay and lesbian groups have sometimes met with huge religious opposition. For example, Catholics opposed the holding of a gay parade in Rome, home of the Vatican, in 2000, while a variety of religious groups successfully opposed 2005 and 2006 plans for a parade in Jerusalem—this parade eventually took place in 2007. This form of contestation of space is discussed further in the Chapter 8 account of sexuality.

Sometimes these debates are more explicitly between those with religious beliefs and those who question the existence of a god or gods. In London, England, a grassroots campaign began in 2008 in response to religious advertisements on London buses asking people to contribute about $10 to fund alternative atheist advertisements. Hugely successful, by early 2009 there were 'atheist' buses in cities in many European countries, Australia, the United States, and Canada. Toronto, Montreal, and Calgary were the first three Canadian cities to have buses with these advertisements. In some cities, including Calgary, religious groups have countered these slogans with their own bus advertisements. In many respects, this very public airing of opinions is a quite remarkable extension of what has more usually been an intellectual debate conducted by a relatively small group of people.

Not only do religions divide our world, as do languages, they also often encourage people to engage in what may appear, from a detached perspective, to be inappropriate behaviour;

recall James's comment quoted at the beginning of this chapter. In some cases, religion serves as an even more potent force for group unification than language. Religion is even more capable than language of resisting external influences. The North American experience of immigration and settlement by people from many different regions of the world suggests that religion is often the most lasting feature of a culture, retained long after language has been lost. The roles played by both religion and language in the creation, continuation, and, sometimes, collapse of political states will be considered in more detail in Chapter 8.

RELIGIOUS LANDSCAPES

Religion and landscape are often inextricably interwoven, for three principal reasons.

1. Beliefs about nature and about how humans relate to nature are integral parts of many religions.
2. Many religions explicitly choose to display their identity in landscape.
3. Members of religious groups identify some places and load them with meaning—these are called sacred spaces.

Religious perspectives on nature

In many cases an important function of religion is to serve an intermediary role between humans and nature—although the type of relationship favoured varies. Judaism and Christianity place God above humans and humans above nature. In other words, they have traditionally incorporated an attitude of human dominance over the physical environment that is reflected in numerous ways. Christians in particular have seen themselves as fulfilling an obligation to tame and control the land. Other religions, however, take a very different view. Eastern religions in general, as well as many Aboriginal belief systems, see humans as a part of, not apart from, nature, and both as having equivalent status under God. The result is a quite different relationship between humans and land.

Religious beliefs about human use of land, plants, and animals can have significant impacts on regional economies. For example, the pig is a common domesticated animal in Christian areas, but absent in Islamic and Jewish areas. The traditional Catholic avoidance of meat one day per week prompted European fishermen to sail across the North Atlantic to fish the teeming Grand Banks off the coast of Newfoundland long before Europeans settled in the New World. Viticulture diffused with Christianity because of the use of wine as a sacrament. Hindus regard cows as sacred (Box 6.8) and not to be consumed. These are among the most familiar examples of how beliefs influence human behaviour.

Different religions incorporate different beliefs and attitudes, and what is important to one group may not be important to another. When religion affects the way we use land, it can be a powerful cultural factor operating against economic logic.

Religious displays of identity in landscape

Many religions, especially when they are a minority group, may choose to reinforce their identity through settling in close proximity. The Mormon example discussed in Box 6.4 is a prime example, and there are many others in North America among smaller Protestant denominations such as the Amish and Mennonites. In Europe, the Jewish ghetto was commonplace in many cities and this preference was brought to North America.

Landscape is a natural repository for religious creations, a vehicle for displaying religion. Sacred structures, in particular, are a part of the visible landscape. Hindu temples are intended to house gods, not large numbers of people, and are designed accordingly; Buddhist temples serve a similar function. By contrast, Islamic temples (mosques) are built to accommodate large congregations of people, as are Christian churches and cathedrals; yet these different faiths, in large, modern cities such as Toronto, can affect landscape differently. Islamic temples, because they do not use the conventional seating of Christian churches and can accommodate many more people in the same amount of space, require larger parking lots outside the building (Hoernig and Walton-Roberts, 2006: 414–15). The size and decoration of religious buildings often reflect the prosperity, as well as the piety or devotion, of the local area at the time of construction.

When a religious building is proposed for an area where there are no similar buildings, friction often results as established residents can interpret it as a threat to traditional and generally accepted religious identity. There have been many recent cases in Europe of disputes over the building of mosques needed by the large immigrant Muslim population. A particularly sensitive case erupted in London when a large and secretive Islamic sect, Tablighi Jamaat, proposed construction of what would be the largest mosque in Europe, housing 70,000 worshippers, close to the financial district and the area designated as the main site of the 2012 Olympic Games. As of 2009, a final decision on construction remains to be made.

Religious identity is often displayed by adherents at the personal scale of the body. Sikh men and Ultra Orthodox Jewish men, to cite just two examples, are readily visible to others through choice of dress and beards. Muslim women often favour the wearing of veils or burkas, some Christians choose to wear crosses, while some Hindus are marked on the forehead.

Religion and sacred spaces

Most religions recognize a holy land: there is the promised land of Israel for Jews, all of India for some Hindu fundamentalists, western Arabia including Mecca and Medina for Muslims, and the larger Palestine area for Christians. As discussed earlier, several different religions sometimes value the same places, often with unfortunate results. Indeed, sacred places are not always treated with appropriate respect, especially when there are competing claims. The traditional site of Christ's crucifixion, the Church of the Holy Sepulchre in the old city of Jerusalem, was the site of a brawl between Greek Orthodox and Armenian monks in 2008, each group blaming the other for the incident. Six Christian sects share control of the church, and disagreements are not unusual.

Penitents on their knees on the approach to the shrine of Our Lady of Fatima, in Portugal. Roughly four million people visit the shrine each year.

Paul Preece/Alamy/GetStock

Many religions ascribe a special status to certain features of the physical environment. Rivers and mountains may be sacred places, including the Ganges for Hindus, the Jordan for Christians, and Mount Fujiyama in Shintoism. Human environments may also achieve sacred status, including Mecca for Islam, Varanasi for Hinduism, and the Vatican for Catholicism. Some sites attract diverse visitors: the Golden Temple at Amritsar, India, is sacred to Sikhs but also attracts many others, as do Lourdes in France and Westminster Abbey in London.

More generally, almost any religious addition to landscape—church, temple, mosque, cemetery, shrine—is sacred. With sacredness, come tourists. What might be called faith tourism is no longer a small niche market. In addition to the places noted, the following are just a few of the many sacred places attracting increased visitor numbers.

1. Tongi, north of Dhaka in Bangladesh, hosts an annual three-day gathering for millions of Muslims on the banks of the River Turag.
2. Sri Pada, a mountain in southern Sri Lanka, is sacred to Hindus, Buddhists, Muslims, and Christians, with each group having a particular reason; Buddhists, for example, believe that Lord Buddha ascended the mountain, leaving a footprint.
3. Uman, in central Ukraine, is where Rabbi Nachman of Breslov, an influential Hassidic Jew, died and was buried; the site attracts thousands of pilgrims during Rosh Hashanah.
4. Mount Kailash, in the remote western Tibetan Plateau of China, is sacred to Buddhists, Hindus, and other groups; it is believed that walking around this holy mountain erases a lifetime of sins.
5. Medjugorje, a small village in Bosnia, is where local youth claimed to see the Virgin Mary in 1981; subsequent sightings have been claimed, with the result that this is now a major location of faith tourism.
6. Djenne, in Mali, has long been a centre of Islamic learning and pilgrimage; the Grand Mosque is the tallest dried-earth building in the world.
7. Mount Athos, in Greece, is the oldest surviving monastic community in the world, and receives a strictly limited number of male visitors for brief stays.

In Calcutta, traditional Hindu temple architecture reflected the belief that temples are the homes of gods. Hence, because mountains are also traditional dwelling places of gods, temple towers were built to resemble mountain peaks, while small rooms inside the temples resembled caverns. In the late eighteenth century, these temples were replaced by flat-topped two- or three-storey buildings built next to the homes of the very rich. Finally, in this century, a series of new temple styles appeared as a result of the pressures of urbanization and related institutional processes. These styles are described in detail by Biswas (1984) and are a clear example of the effects of changing power relations.

Places to house gods or to gather for worship are typical features of most landscapes. In many Christian communities, the church is a religious and social centre serving many extra-religious functions. Although some religious groups actively reject such external expressions of religion as churches and create landscapes devoid of religious expression (Box 6.8), such cases are unusual.

Other religious practices that create a distinctive landscape include the construction of roadside shrines and the use of land for burying the dead. Hindus and Buddhists cremate their dead, but Muslims and Christians traditionally opt for burial, a practice that requires considerable space.

These brief comments do little more than highlight some of the many ways in which religious beliefs may be expressed in landscape. Additional examples of religion and landscape symbolism will be noted in the following chapter.

Cultural Globalization

The contents of this chapter reflect the importance to human geography both of regions—one of our three recurring themes—and of the culturally based divisions described by James at the beginning of the chapter. Today, however, those divisions may be reduced in the process of cultural globalization.

One way to conceive of cultural globalization is to imagine a process that began before the rise of the nation-state, when cultures and identities were essentially local. With the emergence of the nation-state, a second option became available: membership in a national culture. In this sense, nation-states can be seen as cultural integrators, bringing together various local identities in such a way that it became possible for individuals to understand themselves and their lives in both traditional local and newer national contexts—although (as we will see in Chapter 8) the transition from a singular to a dual cultural identity has not been easily accomplished in many parts of the world. Today, some suggest that we have reached a stage where a global cultural identity is developing.

Proponents of this argument see a homogeneous global culture replacing the multitude of local cultures that has been characteristic for most of human history (for an essentially Marxist interpretation of this phenomenon, recall Box 2.6). The culture that is diffusing is Western in character and largely derived from the United States. Among the mechanisms that allow this culture to spread spatially are various aspects of the mass media and consumer

Jerusalem: the Western or 'Wailing' Wall and Dome of the Rock mosque.
travelpixs/Alamy/GetStock

Five baptized Sikhs lead a procession at the Golden Temple in Amritsar, India, in 2003, marking the 337th anniversary of the birth of the tenth guru, Gobind Singh.

CP/AP photo/Aman Sharma

culture—newspapers, magazines, the Internet, music, television, films, videos, fast-food franchises, fashions. That these aspects of popular culture are being diffused around the world, and that they are influencing aspects of non-material culture such as religion and language,

is undeniable and will be discussed in further detail in Chapter 7. However, assessing their significance is not as easy as it might seem.

In fact, it appears unlikely that globalization can erase the power of local places and local identities—what Vidal called *pays* and *genres de vie*. Physical environments vary throughout the world, and there is no denying that the connections between physical and human geographies are often intimate. Most places still look different from other places. In some ways, at least, their inhabitants continue to behave differently from other people, and they still hold some attitudes, feelings, and beliefs that differ from those of other people. Despite clear evidence that the number of the world's languages is decreasing, the roughly 6,000 languages that still exist today are compelling examples of cultural variation from place to place. We are human, but significant differences remain between places and groups of people, resulting primarily from differences in cultural identity. Certainly, the world we live in is not homogeneous. Thus, it may be that a conflict is developing between parochial ethnicity and global commerce.

These comments reflect some of the uncertainties about precisely what globalization is and

Box 6.8   Religious Landscapes: Hutterites and Doukhobors in the Canadian West

Most landscapes include evidence of religious occupation. Places of worship in particular are visible even in an increasingly secular developed world. Unlike the Mormons (Box 6.4), the Hutterites and Doukhobors, in parts of the Canadian West, have not settled across a broad region, and neither have they aspired to express themselves in landscape. Hence, the impact of their religion on landscape is not quite so visible.

A number of Protestant Anabaptist groups emerged during the Reformation in Europe. All shared a belief in adult baptism and a literal interpretation of the Bible, but they varied in other areas of religious belief and practice. Three such groups, all of which immigrated to North America because of persecution in Europe, are the Hutterites, Mennonites, and Amish. Each expressed some aspects of their religious belief in landscape. Hutterites immigrated beginning after World War I and continue to live in the Canadian prairies, creating a landscape that reflects their belief in community and their desire for isolation from the larger world (Simpson-Housley, 1978). At the centre of each communal settlement are a kitchen complex and long houses, and around them are buildings used for economic functions; the two types of building are painted different colours. Hutterite communal settlements do not include

commercial stores or bars. The Hutterite farming landscape typically involves a greater diversity of activities than do neighbouring farms.

Doukhobors, like Hutterites, are a Christian sect. They broke from the Orthodox Church in the eighteenth century and were banished to the Caucasus region, where they built a flourishing community. As a result of persecution, they immigrated to Canada beginning in 1898. Doukhobors reject the 'externalities' of religion; their settlements are thus without houses of worship and there are no crosses or spires in the landscape. In short, the Doukhobor areas lack any religious symbolism. Doukhobors also believe in the equality of life; hence their settlements are communal and all work and financial matters have a group rather than an individual focus. Communal living arrangements required the construction of distinctive double houses that accommodate up to 100 people (Gale and Koroscil, 1977).

Although most Christian landscapes show evidence of religious symbolism and beliefs, such as houses of worship, cemeteries, and roadside shrines, some areas, settled by community-based religious groups, reflect group organization even when they do not include obvious symbolic features. Both Hutterites and Doukhobors belong in the second category.

what impacts it is having, as these were noted in Table I.2. Indeed, as noted in Table 6.7 and in Box 6.9, the three theses about globalization—hyperglobalist, skeptical, and transformational—posit quite different cultural consequences.

Thinking about cultural globalization obliges us to acknowledge that there are no uncontested and unidirectional processes at play in the contemporary world. Globalization, in the sense of an ever-increasing connectedness of places and peoples, is a fact, but it is not the only important fact. Some regional economies remain distinctive, and it is possible that some regional differences actually are being enhanced with the rise of regional trading blocs, as we shall see in Chapter 9. It is hard to believe that the cultural world is becoming uniform at a time when so many ethnic groups are reasserting their identities—at least partly in reaction against the declining importance of national political and cultural identities.

But there is a third, more emancipatory, way to think about the consequences of cultural globalization. If Western values in particular are spreading around the world, then those values include the basic ideals of liberal democracy, including freedom of speech and freedom of cultural expression. If we are free to be what we wish to be, then perhaps we will not choose either to be the same as others or to separate

Table 6.7	Globalization Theses and Cultural Geography		
	Hyperglobalist: The Global Era	Skeptical: Increased Regionalism	Transformationalist: Unprecedented Interconnectedness
	New global civilization.	Clashing cultures.	New global and local hybrid cultures.
	Homogeneous consumerist culture and global brands dominant.	Civilizational blocs entrenched and cultural identities differentiated and relativized.	Possibility of progressive cultural change, but westernization currently dominate.
	Universalization of cultural identities.		

SOURCE: Adapted from W.E. Murray, *Geographies of Globalization* (New York: Routledge, 2006), 353–4.

Box 6.9   Blurring Space and Time

Marshall McLuhan spoke of the coming 'global village' as early as the 1960s and geographers have long recognized that our increasing technological ability to overcome distance has resulted in what we might call a 'shrinking world', one in which places are closer together in terms of both space and time. Most of us live in one place, but through travel and broadcast and telecommunications media we are increasingly aware of many other places. Consequently, our experiences are a strange mix of near and far. Many of us also are finding that our social relations are being stretched, with face-to-face interactions less important as many of our contacts are with those who are physically distant.

With these contexts in mind, Urry (2000) introduces the idea of global or instantaneous time that involves:

- increasing disposability of products, places, and images;
- rapidly changing fashions, products, ideas, and images;
- the idea of an 'always-open' society and economy, with many employees expected to be available for flexible working hours;
- a decrease in the symbolic significance of family rituals, such as mealtimes.

And yet, perhaps local places and local times continue to matter. The distinguished human geographer David Harvey wrote:

The more global interrelations become, the more internationalized our dinner ingredients and our money flows, and the more spatial barriers disintegrate, so more rather than less of the world's population clings to place and neighborhood or to nation, region, ethnic grouping, or religious belief as specific marks of identity. . . . Who are we and to what space/place do we belong? Am I a citizen of the world, the nation, the locality? Not for the first time in capitalist history . . . the diminution of spatial barriers has provoked an increasing sense of nationalism and localism, and excessive geopolitical rivalries and tensions, precisely because of the reduction in the power of spatial barriers to separate and defend against others. (Harvey, 1990: 427)

Reflecting the skeptical thesis, Harvey raises the challenging possibility that there is a second way of thinking about cultural globalization. Instead of making the world more and more homogeneous, perhaps it is reinforcing the distinctiveness of local places and identities. Certainly globalization does not appear to mean the end of diversity or the imposition of a single global culture. On the contrary, the discussion of political geography in Chapter 8 will provide considerable support for the argument that localism is increasing rather than decreasing.

ourselves from them. Perhaps, rather, we will choose to value both our own personal rights and freedoms and those of others, including the inalienable right to improve their economic circumstances. Our future—wherever we are—may well be one of greater pluralism, more choices, and (most important) enhanced mutual respect. These comments—which are in close accord with the 1948 United Nations Universal Declaration of Human Rights—offer a more hopeful counterpart to the sombre scenario outlined at the beginning of this chapter.

CHAPTER 6 SUMMARY

OUR DIVIDED WORLD

The human world is divided physically and, more important, culturally.

CULTURE AND SOCIETY

The terms 'culture' and 'society' are difficult to define. Neither has a single universally accepted meaning, and sometimes the two are used interchangeably. For our purposes, 'culture' refers to humans' ability to knowingly change physical landscapes in directions suggested by experience, while 'society' refers to a cluster of institutionalized ways of doing things. Traditionally, North American geographers have used the term 'culture', whereas European geographers have used 'society'. This chapter and the one following reflect the importance of integrating the two concepts.

HUMAN SCALE OF ANALYSIS

Geographers can study humans at any scale, from all humans in the world to single individuals. Typically, however, cultural geography operates at the scale of a group of people with some recognizably common set of operating rules; hence geographers study language groups and religious groups.

CULTURAL EVOLUTION

Pre-agricultural groups, probably numbering between 10 and 30 individuals, used language, fire, and tools. Cultural variation prior to the development of agriculture was probably limited to features directly related to the physical environment. Agriculture and civilization were closely related; among the factors that may have contributed to the development of both are irrigation, social change, class conflict, population pressure, climatic changes, and basic physical geography. Once in place, agriculture generally is permanent, whereas civilization has not proved to be. Most civilizations have collapsed as a result of either natural disaster or conflict (internal or external).

CULTURAL REGIONS

Regions can be delimited on various scales. The world scale can provide a useful overview but lacks precision. North America can be usefully regionalized, from a European settler point of view, using the concept of first effective settlement. Europe is usefully defined by reference to specific traits.

CULTURAL LANDSCAPES

Cultural regions have distinct cultural landscapes because of the impact of culture on land and the variations in human–land relationships. 'Cultural adaptation' is cultural change in response to environmental and cultural challenges. Geographers studying cultural adaptation have proposed a variety of explanations, including (in addition to first effective settlement) culturally habituated predisposition, cultural preadaptation, and the core, domain, and sphere model.

LANGUAGES

Probably the single most important human achievement, language, is an essential key to understanding human groups, their attitudes, beliefs, and behaviours. Indeed, it is not far-fetched to assert that language is the single most appropriate indicator of culture.

As the basis for group communication, language is the earliest source of group unity and the means by which cultures continue through time. For many groups, language is culture. Today there are perhaps 6,000 languages in the world. Individual languages can be grouped into families—languages that share a common origin. The language family with the most speakers is Indo-European. Mandarin has more speakers than any other single language, but English is the most widespread and the nearest to a world language. Many languages are in danger of disappearing. Increasingly, English is assuming the status of a global language.

LANGUAGE AS IDENTITY

Languages create barriers between groups and facilitate the development of group identity. For many groups, language is the principal basis for a national identity. A characteristic of multilingual states is that they are less stable than unilingual states.

LANGUAGE AND LANDSCAPE

Place names—toponyms—are the clearest expression of language on landscape. Place-name studies help us understand early settlement. But places also are renamed, often for political reasons. Language also helps to make landscape in the sense that a place may become what it is named—as in the case of the Great American Desert. Both physical and human landscapes are reflected in language.

RELIGIONS

For many people, religion is the basis of life. Thus, religion is a useful variable for regionalizing and may be even more powerful than language in reinforcing group identity. Because different religions affect attitudes and behaviour in different ways, they help to distinguish one group from another and to promote group cohesiveness. Religions are usefully classified as ethnic or universalizing. A feature of the world today is the increasing tension between Islam and Christianity.

RELIGION AND LANDSCAPE

Many religions function as intermediaries between humans and nature. Often, specific physical environments are ascribed a special status as sacred spaces. Some religious beliefs can affect regional economies. Landscape also is a natural vehicle for religious expression.

CULTURAL GLOBALIZATION

Increasing evidence suggests that globalization is transforming our culturally divided world, mainly through the diffusion of Western-derived attitudes, beliefs, and behaviours.

QUESTIONS FOR CRITICAL THOUGHT

1. Why is culture our 'greatest human achievement' on the one hand and the source of so much conflict on the other hand?

2. Norton states that humans might choose to 'create new sets of [cultural] values, and thereby 're-engineer' ourselves. Can this be done? Should this be attempted? Justify your position.

3. Discuss the rationale underlying the 'Regions of North America' depicted in Figure 6.5. Do these regions exist in the minds of the people who live in them or are they figments of the geographer's imagination?

4. Why is language so important to culture? If a language disappears or becomes extinct, will that culture also disappear?

5. Why has religion been the basis of so much conflict throughout the course of human history?

6. Do you think cultural globalization (i.e., the homogenization of culture on a global scale) is a good thing? What would be the advantages and disadvantages of there being a single culture?

FURTHER EXPLORATIONS

Anthony, D.W. 2007. *The Horse, The Wheel, and Language: How Bronze-Age Riders from the Eurasian Steppes Shaped the Modern World*. Princeton, NJ: Princeton University Press.

A scholarly review of the emergence and impact of proto-Indo-European.

Austin, P.K., ed. 2008. *1,000 Languages: The Worldwide History of Living and Lost Tongues*. London: Thames and Hudson.

A product of the Endangered Languages Academic Program at the University of London, this comprehensive book is organized regionally and considers spoken and written language.

Francaviglia, R.V. 1978. *The Mormon Landscape*. New York: AMS Press.

A detailed description of one of the most distinctive cultural landscapes in North America.

Gade, D.W. 1999. *Nature and Culture in the Andes*. Madison: University of Wisconsin Press.

A fieldwork-based study of human–nature relationships in the Andean region; unified by commitment to the Sauer-inspired school of cultural geography.

Harrison, K.D. 2007. *When Languages Die: The Extinction of the World's Languages and the Erosion of Human Knowledge*. New York: Oxford University Press.

A careful consideration of the threats posed by language loss, described as a catastrophe of cultural forgetting.

Jordan-Bychkov, T.G., and B. Bychkova Jordan. 2002. *The European Culture Area: A Systematic Geography*, 4th edn. London: Rowman and Butterfield.

An excellent book. One of the few cultural geographies of a major world region; includes a highly original synthesis and analysis of language, religion, and other cultural traits.

Kent, R.B. 2006. *Latin America: Regions and People*. New York: Guilford.

Detailed overview of the regional geography of Latin America.

Leighly, J. 1978. 'Town Names of Colonial New England in the West', *Annals, Association of American Geographers* 68: 233–48.

One example of a study of toponyms used to analyze settlement history.

McColl, R. 2007. 'Costa Rica's Churches: Keys to Place Identity, Navigation, and History', *Focus* 50, 3: 30–6.

An informative article discussing some of the churches that serve as ideal markers of place, each with its own history, architecture, and identity.

Meinig, D.W. 1969. *Imperial Texas: An Interpretive Essay in Cultural Geography*. Austin: University of Texas Press.

One of several fine regional studies by this author; includes several original approaches to region and landscape analysis.

Monmonier, M. 2006. *From Squaw Tit to Whorehouse Meadow: How Maps Name, Claim, and Inflame*. Chicago: Chicago University Press.

The title of this wonderful book says it all. Maps legitimize place names, but changing public taste and political circumstances result in renaming and hence in different maps.

Ostergren, R.C., and J.G. Rice. 2004. *The Europeans: A Geography of People, Culture, and Environment*. New York: Guilford.

Detailed and informative account of European geographies, both past and present. Excellent regional geography textbook.

Robinson, A. 2009. *Lost Languages: The Enigma of the World's Undeciphered Scripts*. London: Thames and Hudson.

This book focuses on eight of the many languages with undeciphered scripts. Deciphering a script facilitates study of the people who spoke the language.

Rooney, J.R., Jr, W. Zelinsky, and D.R. Loudon, eds. 1982. *This Remarkable Continent: An Atlas of United States and Canadian Society and Cultures*. College Station: Texas A&M University Press.

An original attempt to map and discuss a wide variety of cultural traits.

Shortridge, J.R. 1977. 'A New Regionalization of American Religion', *Journal of the Scientific Study of Religion* 16: 143–53.

A comprehensive statistical regionalization of some religious differences.

Sopher, D.E. 1967. *Geography of Religions*. Englewood Cliffs, NJ: Prentice-Hall.

The first book-length study of this topic, full of useful ideas and facts.

Spate, O.H.K., and A.T.A. Learmouth. 1967. *India and Pakistan*, 3rd edn. London: Methuen.

Three major regions, subdivided into 37 regions, are identified by physical geographic and historical human identity criteria in this classic regional geography.

Spencer, J.E. 1978. 'The Growth of Cultural Geography', *American Behavioral Scientist* 22: 79–92.

One of many overviews, but especially useful to the new student of geography.

Stock, R. 2004. *Africa South of the Sahara: A Geographical Interpretation*. New York: Guilford.

Excellent regional geography text exploring all aspects of physical and human geographies and discussing successes and failures.

Wagner, P.L. 1974. 'Cultural Landscapes and Regions: Aspects of Communication', *Geoscience and Man* 5: 133–42.

An early statement regarding the importance of communication to our understanding of culture.

Warf, B., and P. Vincent. 2007. 'Religious Diversity across the Globe: A Geographic Exploration', *Social and Cultural Geography* 8: 597–613.

An examination of religious diversity at the global scale, showing that China, India, Russia, Japan, and Indonesia are among the world's most religiously diverse countries.

ON THE WEB

CIVILIZATIONS, LANDSCAPES, AND REGIONS

World Civilizations: Timeline and Overview

www.bbc.co.uk/religion/tools/civilisations/index.shtml

An informative account of world civilizations.

World Civilizations: Information Sources

www.wandertheglobe.com/ancient/

Provides a list of websites concerned with ancient civilizations; global in coverage.

Cultural Landscape Preservation

www.tclf.org/whatis.htm

Webpage of the Cultural Landscape Foundation. Focuses on the idea that cultural landscapes are a legacy for everyone. Proposes a useful classification of types with examples.

Landscapes as World Heritage

whc.unesco.org/en/culturallandscape

A UNESCO website detailing World Heritage sites designated as cultural landscapes. These are viewed as expressions of the long and intimate relationships between humans and the environments they occupy.

Library of Congress Maps

cweb2.loc.gov/ammem/gmdhtml/setlhome.html

Detailed information on maps of parts of the US showing human changes to physical landscapes as settlers established settlements, built transport routes, and named places.

LANGUAGE

World Languages

www.ethnologue.com

Comprehensive coverage of language families and all known languages of the world.

Place Names in Canada

geonames.nrcan.gc.ca/

This Natural Resources Canada site provides detailed information on place names in Canada.

Place Names in Europe

lazarus.elte.hu/~guszlev/euro/

Information on European place names using an interactive map. Also included is an exonym database.

RELIGION

Religious Adherents

www.adherents.com/Religions_By_Adherents.html

Data on major world religions with detailed discussions

Facts about Religions

www.religionfacts.com/big_religion_chart.htm

Interesting site that provides a comparison of religions with respect to basic beliefs and practices.

GLOBAL CULTURE

Global Cultural Diversity

portal.unesco.org/culture/en/ev.php-URL_ID=34321&URL_DO=DO_TOPIC&URL_SECTION=201.html

This part of the UNESCO website is a good starting point for seeking information on global cultural issues; focus on diversity.

Cultural Globalization

www.globalpolicy.org/globaliz/cultural/index.htm

Links to articles and documents concerned with the spread of a global culture based on Western capitalist ideas.

SOCIAL IDENTITIES AND LANDSCAPES

This chapter continues the discussion of landscapes as they reflect relationships between humans and land. However, the emphasis here is rather different. Specifically, although some of the material in Chapter 6 had conceptual, social, or symbolic overtones, the overall approach was empirical.

By contrast, this chapter is strongly theoretical. Following a reintroduction to 'culture' that emphasizes the plurality of cultures, we outline three types of society: feudalism, capitalism, and socialism. Then we offer some additional comments on Marxism and humanism and introduce the newer theoretical approaches of postmodernism and feminism.

Informed by these new theoretical perspectives, we return to landscape, this time as the spatial constitution of culture. A third cultural variable is introduced: like language and religion, ethnicity often is a fundamental factor in establishing and maintaining group identity. This account of geographies of difference, which examines the problematic term 'race', also considers gender and sexuality. Following a discussion of well-being, we consider folk and popular culture as they are reflected in landscape and as they relate to globalization processes. Next, the geography of tourism and recreation highlights how tourist activity often involves the commodification of peoples and places. The chapter concludes with a brief look at social engineering.

'Paris Las Vegas': this resort hotel/casino in the Nevada desert offers tourists a simulacrum of the quintessential symbol of France.
Das Fotoarchiv

Rethinking Culture

Chapter 6 restricted itself to a single view of culture—albeit an important one with strong roots in anthropology. In recent years, however, several other perspectives on culture have had a significant influence in human geography.

In the traditional interpretation, especially for Sauer and other members of the landscape school, culture was a given; accordingly, these geographers analyzed the impact of culture on landscape, especially as it was reflected in the material and visible landscape and the formation of regions. Sauer's view of culture as cause remained largely unquestioned in geography until about 1980. Increasing interest in both Marxism and humanism, as well as debate over the landscape view, encouraged a substantial rethinking of the concept of 'culture'. The term 'new cultural geography' distinguished the revised concepts of culture from the traditional landscape-school view introduced by Sauer.

A SYMBOLIC VIEW

With the rise of humanistic geography in the 1970s, a symbolic interpretation of culture became prominent in the discipline. This view broadens the concept of culture to more fully embrace non-material culture and to emphasize that individuals create groups through communication. This broader view of culture allows human geographers to consider topics beyond landscape: specifically, topics that fall under the general heading of what we have called the spatial constitution of culture. As P. Jackson and S.J. Smith explain it:

> Culture, in the sense of a system of shared meanings, is dynamic and negotiable, not fixed or immutable. Moreover, the emergent qualities of culture often have a spatial character, not merely because proximity can encourage communication and the sharing of individual life worlds, but also because, from an interactionist perspective, social groups may actively create a sense of place, investing the material environment with symbolic qualities such that the very fabric of landscape is permeated by, and caught

up in, the active social world. (Jackson and Smith, 1984: 205)

This interpretation of culture had been anticipated a decade earlier by P.L. Wagner (1975: 11): 'The fact is that culture has to be seen as carried in specific, located, purposeful, rule-following and rule-making groups of people communicating and interacting with one another.'

This shift in geographical thinking was in line with a major school of thought in sociology called **symbolic interactionism**. Essentially, this set of closely related theories, derived from the ideas of the American social philosopher G.H. Mead (1934), argues that humans learn the meanings of things through social interactions, and that our behaviour in any given situation is the product of our response to the perceived environment. In other words, interactions with other people provide us with meanings for things that we are then able to use to understand those things. Once we have acquired such an understanding, we use it to define the situations that we encounter and then act accordingly.

NEW CULTURAL GEOGRAPHY

Closely related to the symbolic view of culture is the idea that in fact there is no single, fixed entity called culture, but rather a plurality of cultures, understood as those values that members of human groups share in *particular* places at *particular* times. Considered from this perspective, cultures are not objects but mediums or processes—what Jackson (1989: 2) described as maps of meaning: the 'codes with which meaning is constructed, conveyed, and understood'.

This interpretation has led geographers to study many topics besides landscape, from previously ignored groups, new cultural forms, and 'otherness' to Eurocentrism and ideologies of domination and oppression. Most generally, human geographers now tend to base their questions about human identity on the logic of **constructionism** rather than the more traditional logic of **essentialism**. Whereas the essentialist view sees the characteristics that constitute identity as inherited and largely unchanging, the constructionist view stresses that those characteristics are socially made or acquired and that they

symbolic interactionism
A group of social theories that see the social world as a social product with meanings resulting from interaction.

constructionism
The school of thought according to which all our conceptual underpinnings (for example, ideas about identity) are socially constructed and therefore contingent and dynamic, not given or absolute.

essentialism
Belief in the existence of fixed unchanging properties; attribution of 'essential' characteristics to groups.

are contested in the sense that there are no unequivocal meanings: different characteristics are important in different places and at different times. The distinction is important philosophically, since the essentialist view is associated with empiricism, while the constructionist view is closely related to feminism and postmodernism (to be discussed later in this chapter). This shift in perspective is reflected here: in Chapter 6 we discussed the identity characteristics of language and religion from a primarily essentialist perspective, while this chapter discusses place, ethnicity, gender, and sexuality, and the following chapter discusses nationality, from a primarily constructionist perspective.

The rethinking of culture has opened a number of new directions for cultural geographers, some of whom, taking a postmodern perspective, have now rejected the landscape school's emphasis on the regional mapping of material and visible features of landscape on the grounds that such mapping assumes the existence of a single distinctive and unchanging cultural group constituting a culture. Instead, approaches are favoured that acknowledge the existence of a plurality of cultures, located in specific times and places. Among the topics for study suggested by recognition of multiple cultures are cultural identities and their links to place and issues of cultural dominance and subordination.

Along with these new views within human geography of culture has come a new understanding of landscape itself, including the natural landscape, as something that is socially constructed. Studies in this area focus on two aspects of landscape: symbolic (specifically, the meaning contained in landscape) and represented (landscapes as represented in literature and art, as well as the more usual visible and material landscapes). Such work often treats landscape as a text that is open to interpretation, and, recognizing the importance of images, includes among its research methods **iconography**: the description and interpretation of images to uncover their symbolic meanings (see Box 7.5). The idea that nature is socially constructed—in effect, that it is part of culture—has several important implications. We need to recognize that our understanding of nature is filtered through human representations of it, and that these representations vary with time and place. We also need to be aware that the representation of nature is never neutral: any such representation—and there may be several of them for any part of the natural world—is ideologically loaded, and it is part of the geographer's task to interpret them.

The Cultural Turn

One way to think about the rise of new cultural geography is to suggest that a 'cultural turn' has taken place. This phrase—which has been used in the context of recent changes in many of the social sciences and humanities—essentially refers to an increased appreciation of the importance of culture in understanding humans and their political and economic activities.

The impact of this cultural turn has been especially notable in studies of human identity and human difference and of the politics related to these matters. There is now a persistent questioning of traditional concepts, classifications, and categories such as those used in our discussions of language and religion in Chapter 6. Identity is increasingly being understood in terms of constructionism rather than essentialism, that is, as something socially created and therefore subject to ongoing change rather than something predetermined and fixed. This more sophisticated understanding of identity has prompted human geographers to examine the power relations between dominant groups and other groups, as well as the politics of difference. Some relevant key ideas about how geographic knowledge is constructed are noted in Box 7.1.

With these rather different views of culture and landscape in mind, we can now consider different types of society, specifically feudalism, capitalism, and socialism, and then evaluate various approaches to the subject matter of this chapter.

Types of Society

Our contemporary cultures reflect a long evolutionary process that has followed different routes in different parts of the world. This section summarizes the European transition from feudalism to capitalism and the experience of socialism. Using Marxist logic, each of

iconography
The description and interpretation of visual images, including landscape, in order to uncover their symbolic meanings; the identity of a region as expressed through symbols.

Box 7.1   Constructing Geographic Knowledge

Numerous theorists from various disciplines have contributed to the 'cultural turn' in geography and have introduced important concepts for understanding this turn towards a constructionist view. Most of the material considered here relates closely to the account of postmodernism later in this chapter, while several of the concepts are employed throughout the chapter, especially in the discussions of ethnicity, gender, sexuality, and contested landscapes.

A key idea is that of **hegemony**. According to Antonio Gramsci (1891–1937), any group that desires to attain power in a larger society needs to achieve a degree of cultural and intellectual hegemony, that is, the ability to determine the ruling discourse, including what questions are or can be asked. This allows the group to express its world view and to structure social and other institutions in accord with its goals. Gramsci also introduced the term *subaltern* to refer to those socially subordinate, marginalized, exploited, and oppressed groups that lack both the unity and the organization of more dominant groups that exercise **authority** and control. The term 'subaltern' has been used especially in the context of attempts to write about peoples and places from 'below', such as post-colonial literature produced by the colonized.

Closely linked to these ideas is the concept of discourse. Michel Foucault (1926–84) saw subjectivity as constructed within and through discourses, with discourse defined as a system, comparable to a language, that enables the world to be made intelligible. Discourses are important because they serve to legitimize a particular view of the world that then becomes part of the taken-for-granted world. Thus, **power** is practised through discourse, and dominant discourses affect how members of a society understand the world. It can be argued, for example, that the discipline of geography evolved within the late nineteenth-century discourses of imperialism, colonialism, and racism.

Building on these ideas, the cultural turn has highlighted the idea that research conducted by geographers may not be free from bias. For example, much previous work might be best described as *colonial research* essentially exploitive of those being studied. *Post-colonial research* explicitly premised on the rejection of colonial attitudes, ideas, and values is advocated instead; such research is inclusionary, seeking to empower those whose voices previously were unheard.

Another influential writer, Edward Said (b. 1935), introduced the related idea of **Orientalism**, a key post-colonial concept contending that the Orient is an invention of those who study it from outside, and that North America and Europe employ ideas from within their cultures, such as freedom, democracy, and individualism, to facilitate their conquest and domination of other regions. Thus, the Orient is a necessary European image of the **Other**, a construct of a dominant European discourse. The contemporary importance of these ideas is evident in the Chapter 6 account of Islam and Christianity, where it was noted that parts of the Christian world sometimes see the Islamic world in dehumanized terms, a tendency exacerbated by terrorist activities and by the ongoing conflict between Palestinians and Israelis. More than one point of view may be legitimate; it is important to be attentive to multiple voices.

The concept of **contextualism** acknowledges that a specific discourse is employed in human geography. We need to be aware of, and sensitive to, precisely how knowledge is being constructed—in other words, readers need to know who is conducting the research and what their agenda is. For this reason, some current work in human geography explicitly acknowledges the **positionality** and **situatedness** of the author, in short, who they are and what they believe. Expressed rather differently, the impossibility of producing universally valid and accurate representations of the world increasingly is accepted because any **representation** is actually an interpretation. Traditionally, it was assumed that a real world could be mirrored by geographers in their writings; this assumes a neutrality on the part of the author that is not feasible. Rather, geographers interpret the world through particular lenses, especially in the context of prevailing power relations and discourses. Furthermore, representations do not simply interpret the world; they also help to shape the world.

---

feudalism
A social and economic system prevalent in Europe, prior to the Industrial Revolution, in which land was owned by the monarch, controlled by lords, and worked by peasants who were bound to the land and subject to the lords' authority.

these societal types is an example of a mode of production.

FEUDALISM

Feudalism was a non-centralized system of governance and social and economic organization that developed in Northern Europe over the centuries following the collapse of the Roman Empire (*c.* 375). Under the feudal system, all land was owned by the king, who effectively delegated control of it to his warrior lords (vassals) in return for their military and political support. The 'direct producers' who worked the land were the peasants, who were permitted to live on the land in exchange for their labour and were subject to the legal and political control of their individual lords.

Feudalism has several important implications for both place and people. In a feudal society, people were defined by their social class and social mobility was extremely limited. Further, peasants were not free to choose whom they would work for; this was the situation that Marx would describe as exploitation.

CAPITALISM

In some Western European countries, as cities grew in size and number, as European overseas movement proceeded apace, and as feudalism began to decline, mercantilist economic policies were put in place. Beginning in the sixteenth century, these were essentially government policies favouring national economic unity, protectionism, increased global trade, and centralized political control. This mercantilist stage can be seen as sowing the seeds for the onset of capitalism, a new type of social and economic organization that began to emerge as early as the late sixteenth century and that was fully in place in many parts of Europe by the eighteenth century. The term used to describe this system, capitalism, first was popularized by Marxists in the late nineteenth century. Capitalism is characterized by the transformation of labour into a commodity that can be bought and sold and the separation of the producer (the worker) from the means of production, which are owned by the 'capitalist' class. Today, capitalism is the dominant form of economic and social organization, having diffused from Europe throughout most of the world (Box 7.2).

Among the distinctive characteristics of capitalism, according to Marx, are its capacity for self-expansion through ceaseless centralization and concentration of capital; continual technological changes to the production process; the cyclical nature of the associated process of development; and the divisions it creates between classes, leading to class conflict. In Marx's view, capitalism, like feudalism, exploits the peasant or working class, and because people under capitalism are not fully free, they are alienated (see Box 7.2).

Another important social theorist, Max Weber, saw capitalism in quite a different light, as an **ideal type**. He traced its origins to the religious ethic of Protestantism, the growth of cities, and the legal and political framework provided by the rise of a new type of nation-state.

Class

The concept of **class** is often employed by social scientists in discussions of capitalism. Specifically, sociologists use the term either as a structural category with respect to status—typically distinguishing lower, middle, and upper class—or as an expression of self-identity. The first interpretation has proven of little interest to human geographers, at least partly because the categories have different meanings in different parts of the world. The second interpretation ought to be of interest because different classes tend to locate in different areas—most apparent in the internal geography of cities—and because questions of self-identity are of increasing concern in the new cultural geography.

Despite these attractions, however, human geographers have used the concept infrequently and somewhat ineffectively. The principal exceptions are some Marxist geographers who consider the concept of class to be a more useful general concept than culture; for Blaut (1980), class also has the advantage of implying links between groups, with some classes exercising authority over other classes. It can be argued that cultures are discrete and separate entities, whereas classes are linked by power relations (although much of the new cultural geography interprets cultures, too, in terms of power relations). Viewed in this light, class involves patterns of dominance and subordination reflected in our attitudes and behaviours and in the landscapes we create. The links are clear between this interpretation of the concept of class and the Marxist philosophy outlined in Chapter 2.

SOCIALISM

In the twentieth century, a number of societies rejected capitalism in favour of **socialism**, a form of social and economic organization based on common ownership of the means of production and distribution of products. The common feature of all socialist endeavours is the opposition to capitalist individualism; as the term itself suggests, socialism focuses on community, equality, the well-being of society as a whole, and the vision of a classless society.

Multiple versions of socialism were proposed in the nineteenth century, communism representing the most extreme, social democracy the least. It is notable, however, that communism as it was eventually implemented in

---

hegemony
A social condition in which members of a society interpret their interests in terms of the world view of a dominant group.

authority
The power or right, usually mutually recognized, to require and receive the obedience of others.

power
The capacity to affect outcomes; more specifically, to dominate others by means of violence, force, manipulation, or authority.

Orientalism
Western views of the Orient, implying a view of the periphery from the centre; closely associated with post-colonial theory, especially the work of Edward Said.

Other
Subordinate group as seen by and contrasted to dominant groups; implies both difference and inferiority.

contextualism
Broadly, the idea that it is necessary to take into account the specific context within which any research is conducted.

Box 7.2   Capitalism Is Good, and Capitalism Is Bad

Of course, the key idea of capitalism—that making money is a good thing—had been around long before the rise of capitalism, sometimes viewed favourably and sometimes not. Some early Jewish and Buddhist stories noted the high status of merchants, while Hinduism, Confucianism, and Christianity often looked unfavourably on commerce.

Once in place as a social and economic system (a mode of production, using Marxist terminology), it proved highly successful at managing national economies through market forces rather than through state control.

Widely acknowledged as imperfect because of frequent evidence of corporate power abuse and the creation of social and economic inequalities (which became all too apparent with the onset of a global recession in 2008), capitalism seems to be a flawed system but also, perhaps, to be the best system available for managing economies. John Maynard Keynes famously observed: 'For my part, I think that Capitalism, wisely managed, can probably be made more efficient for attaining economic ends than any alternative system yet in sight, but that in itself it is in many ways extremely objectionable. Our problem is to work out a social organisation which shall be as efficient as possible without offending our notions of a satisfactory way of life' (Keynes, 1926: 52–3).

Capitalism has undergone several changes since Keynes wrote these words, but there is little evidence that the changes have ameliorated the negative social and economic components of the system. After World War II, **competitive capitalism** (as it is now usually called) changed considerably as a result of the rapid growth of major (often transnational) corporations and increased involvement by the state in the economy (often through public ownership). This is known as **organized capitalism**. Most recent evidence suggests that a further transformation is now underway: the term **disorganized capitalism** refers to a new form characterized by a process of disorganization and industrial restructuring. The

transition from organized to disorganized capitalism (also referred to as the transition from **Fordism** to **post-Fordism**) is in part a reflection of the malfunctioning of capitalism suggested by economic instability, social injustice, poverty, and unemployment. The significance of these transitions is evident in the accounts of economic restructuring in Chapters 9–14.

Another major failing of capitalism, according to Marx, is that it has a dehumanizing effect on individuals living within it, an effect that he termed **alienation**. Because they sell their labour and do not control the means of production, they lack control over their own lives. At the same time, the capitalist state uses democracy and guarantees of individual human rights to legitimize the maldistribution of political and economic power. It has even been suggested that the state itself evolved to legitimize capitalism and to prevent substantial popular opposition to the circumstances of alienation (Johnston, 1986: 176). As capitalism has spread across the globe, the people of states at all levels of the world system—core, semi-periphery, and periphery—have experienced alienation. Although capitalism is most detrimental in the periphery and least detrimental in the core (since the core's living standards are highest), the basic effects are the same.

The concept of alienation is central to many discussions of global problems. For example, the alienated individual can no longer interact directly with the natural world; all our relations with that world are in some way organized by forces that are not part of us or of nature. It also is important in everyday life with people experiencing frustration, powerlessness, and an overriding sense of helplessness, circumstances most fully developed within existentialist philosophy.

So, is capitalism good or is it bad? Many critics today hedge their bets, arguing that capitalism is sick, but then disagreeing as to whether the illness is curable or terminal. More usefully, perhaps, judgements as to whether capitalism is good or bad often conclude that it is the best of a set of unsatisfactory options, the least worst system.

---

positionality
The ideological preference and the identity of the researcher as this relates to the subjects of the research.

Russia and elsewhere differed significantly from Marx's own vision. Moreover, the late twentieth-century rejection of communism in the former USSR and throughout Eastern Europe has resulted in an expansion of the capitalist world. Today, the prospect of a transition from capitalism to socialism seems unlikely. For many people 'socialism' now signifies nothing more radical than the provision of certain basic welfare measures (such as state-funded medical care) within a basically capitalist economy.

Social Theory

DIVERSITY OF CURRENT APPROACHES

Social theory is not the property of any one discipline; all the social sciences ask questions about ways of life and human behaviour. During the middle part of the twentieth century, the social sciences were dominated by a positivistic approach based on ideas borrowed from the physical sciences and focused on the development of theory, the testing of hypotheses, and the creation of laws. Since the late 1960s, however, dramatic changes have occurred in our theoretical preferences. The weakening of the influence exerted by positivism has permitted the flowering of various alternative approaches linked both by their rejection of two positivistic claims—that social science can be value-neutral, and that the theory/hypothesis/law approach necessarily produces the best explanation—and by their acceptance of the idea that social science is an interpretive endeavour—in other words, questions of meaning and communication are relevant.

Several of these newer approaches trace their intellectual origins to earlier writers: for example, Marxism to Marx and humanism (phenomenology) to various nineteenth- and twentieth-century geographers. Others, notably postmodernism, are of more recent origin.

The current proliferation of theoretical approaches has provoked two quite different responses among practising social scientists. Some see it as a problem: if social theorists are unable to agree among themselves, what possible use can social theory serve for those actively engaged in research? Others see it as an invaluable aid in helping scholars to avoid the dogmatism that results from a single dominant approach. This is an important question—and one for which there is not a single correct answer.

HUMANISM AND MARXISM REVISITED

One topic of interest to contemporary human geographers is the social significance of space and place. Humanists are interested in the elements that combine to produce a sense of place, in the symbolism of landscape and in iconography. Those with a more Marxist perspective are interested in social inequality with respect to such variables as language, religion, ethnicity, class, gender, and sexuality, especially as they are reflected in space. Both groups recognize the importance of space in the constitution of social life. The Marxist perspective also introduces the social constructionist idea of nature as a product of culture, the idea that in addition to the natural world as conventionally understood, apart from humans, there is a second nature that emerged with industrial landscapes—nature transformed as the product of human labour. Much of the work conducted in the Marxist tradition is identified as **critical geography**, a term referring to the need for geographic studies to be emancipatory and not ideological.

Two more recent theoretical approaches now making important contributions to human geography are feminism and postmodernism. A third theoretical approach, structuration theory (see, e.g., Giddens, 1984), which developed from numerous earlier contributions to social theory, including Marxism, focuses on the capacities that permit people to institute, maintain, and alter social life, but this approach has proven to be of less value to human geography.

FEMINIST THEORY

There is no single body of feminist theory; rather, various schools of feminist thought are associated with larger bodies of theory such as liberalism, Marxism, socialism, or postmodernism. Nevertheless, all schools of **feminism** are united in their commitment to improving the social status of women and securing equal rights with men.

Women are systematically disadvantaged in most areas of contemporary life. The fundamental reason for this inequality is **patriarchy**: a social system in which men dominate women. Under the traditional division of labour, women were economically dependent on male breadwinners, and the few women who did work outside the home were paid less than men received for equivalent work. *Culture* is seen as a key factor in the construction of **gender** differences through various socialization processes. *Sexuality* and *violence* are seen as forms of social control over women. Finally, the *state* is seen as typically reinforcing traditional households and failing to intervene in cases of violence against women.

Of the several traditions of feminist thought and action, the oldest, dating back to the late eighteenth century, is *liberal feminism*, aimed at securing equal rights and opportunities for women. Of the more recent traditions, developed since the 1970s, two of the most important argue that the oppression of women cannot be corrected by superficial change because it is embedded in deep psychic and cultural processes that need to be fundamentally changed. *Radical feminism* contends that gender differentiation results from gender inequality, and that the subordination of women is separate

situatedness
An idea that rejects notions of researcher authority and impartiality—knowledge is not neutral and cannot be acquired in some detached and disembodied manner. Rather, all knowledge is partial and located somewhere.

representation
A depiction of the world, acknowledging impossibility to be exact as all such depictions are affected by the researcher's identity.

ideal type
A term introduced by the sociologist Max Weber to refer to a hypothetical norm used to deepen understanding of the real world through comparison.

class
A large group of individuals of similar social status, income, and culture.

socialism
A social and economic system that involves common ownership of the means of production and distribution.

competitive capitalism
The first of three phases of capitalism, beginning in the early eighteenth century; characterized by free-market competition and laissez-faire economic development.

organized capitalism
The second phase of capitalism, beginning after World War II; increased growth of major corporations and increased involvement by the state in the economy.

disorganized capitalism
The most recent form of capitalism, characterized by disorganization and industrial restructuring.

post-Fordism
A group of industrial and broader social practices evident in industrial countries since about 1970; involves more flexible production methods than those associated with Fordism.

Fordism
A group of industrial and broader social practices introduced by Henry Ford, including mass-production assembly line, higher wages, and shorter working hours.

from other forms of social inequality, such as those based on class. *Socialist feminism* similarly emphasizes gender inequality, but links that inequality to class; what this means is that both men and capital benefit from the subordination of women.

POSTMODERNISM

As the name implies, **postmodernism** emerged as a reaction to **modernism**, a general term used to refer to any number of movements beginning in the mid-nineteenth century that broke with earlier traditions. Modernism developed most fully in art and architecture, but social science methodology arose via the positivism that first appeared in the nineteenth century. It assumes that reality can be studied objectively and can be validly represented by theories (as in positivism), and that scientific knowledge is practical and desirable. Modernism is also closely linked to the Industrial Revolution and the rise of capitalism, emphasizing such classic liberal themes as the rationality of humans, the privileged position of science, human control over the physical environment, the inevitability of human progress, and a search for universal truths.

What was the postmodern reaction? Most contemporary human geographers would agree that postmodernism is an especially difficult body of ideas to understand. At least one reason is that postmodernism is, by its reactive nature, unstructured and ambiguous; anarchic concepts may be embraced. Another is that there are many versions of postmodernist theory, most of them developed in disciplines far removed from human geography—most notably architecture, literature, and other expressions of culture. Despite these difficulties, postmodernism is playing an increasingly important role in contemporary human geographic research.

The postmodernist alternative

Postmodernism rejects all the assumptions of modernism. Reality cannot be studied objectively because it is based on language. Rather, reality should be thought of as a **text** in which all aspects are related (it is 'intertextual'). Therefore, reality cannot be accurately represented. *Taken to the extreme*, this means that truth is relative and, for practical purposes, non-existent. Causality does not exist, and theory construction has no meaning.

So how does a postmodernist approach work? The emphasis is on the **deconstruction** of texts and the construction of narratives that do not make claims about truthfulness; such narratives tend to focus on differences, uniqueness, irrationality, and marginal populations. Deconstruction questions the established readings of a text and highlights alternative readings. Overall, postmodernism considers modernist claims to be arrogant, even authoritarian. 'Postmodernism and deconstruction question the implicit or explicit rationality of all academic discourse' (Dear, 1988: 271).

Among the principal attractions of postmodernism for contemporary human geography is its emphasis on cultural otherness, its openness to previously repressed experiences such as those of women, gays and lesbians, and, in general, those lacking power and authority. Later sections of this chapter will reflect this emphasis on the diversity of human experience.

Diverse postmodernisms

Not all postmodernism is quite as described above. In fact, the concept varies considerably between disciplines and even within them. Some who embrace postmodernism nevertheless continue in the progressive directions suggested by modernism, becoming involved in social movements or working to break down the barriers between researchers and subjects so that people are allowed to speak for themselves. These versions are relatively close to some earlier concepts of culture, such as symbolic interactionism, and some other social philosophies, such as humanism. That there is no single unequivocal version of postmodernism in human geography is not surprising, given that the central message of postmodernism is the importance of diversity. For critical responses to the postmodern approach from two different perspectives, see Boxes 7.3 and 7.4.

Other human geographers acknowledge the strengths of the postmodern perspective, such as its emphasis on cultural otherness, but express concern about the postmodern tendency to focus on topics that could be seen as trivial. For example, Hamnett (2003: 1) worries that human geography has become a 'theoretical playground where its practitioners stimulate or entertain themselves and a handful

It is hardly surprising that an approach such as postmodernism, which effectively undermines much previous academic work because of its insistence that there are multiple ways of reading any reality, will be subject to opposing views; nor is it surprising that human geographers who have worked primarily in a positivist tradition will be at the forefront of such criticism.

Brian Berry is one of North America's most distinguished human geographers and, according to the Social Science Citation Index, one of the world's most frequently cited geographers since the 1960s. In the course of reviewing a collection of essays edited by Abler, Marcus, and Olson (1992), *Geography's Inner Worlds: Pervasive Themes in Contemporary American Geography*, Berry (1992) noted the postmodern emphasis in several essays and took exception to the implications of such an emphasis for human geography. He expressed specific concern about the implications of two claims:

1. 'While scientific method remains accepted in physical geography, critical social theorists have raised challenges to the use of formal modeling in human geography.... Human geographers continue to be engaged in lively epistemological debate in which there is little consensus except for considerable negativism towards logical positivism' (quoted in Berry, 1992: 491–2).
2. 'Contemporary human geography reflects ... a postparadigm condition in which disciplinary practices and concepts appear, for good or bad, to have broken loose from any notion of disciplinary closure and unity' (quoted ibid., 492).

More important, Berry identified a fundamental contradiction between the editors' assertion that geography needs to be a 'coherent, synthetic, global discipline focused on human use of the earth' (quoted ibid., 493) and their warm approval of the postmodern diversity represented in the book:

The success of *disciplines* is that they think in ways *disciplined* by theory; for them, paradigm shifts occur because older theories are found wanting and are pushed aside by theories that offer better explanation. But postmodernism carries with it a New Age social vision that embodies not only ideas of holism (the interdependence of all systems, spiritual as well as material) and earth awareness (the interdependence of all things on earth, including humankind) but also a particular view of human rights (the rights of all individuals to chase their *sadhana* and live transformative lives) that, in academic practice, translates into a concept of inner meaning (geographer's inner worlds) that is fundamentally antitheoretic.... If this be geography's karma so be it, but the implications are not pleasant. Detached from the world of science, a generation of geographers will, like the regionalists, have little accumulative wisdom to share with others, simply disparate fragments—some insightful, some beautiful, but none coming together into the whole cloth of theories that provide order to inquiry and help frame practice. (Ibid.)

of readers, but have in the process become increasingly detached from contemporary social issues and concerns.'

Landscape as Place

Landscapes are not simply locations; they are also places in the sense that they convey meaning. This idea is reflected in the increasing numbers and sophistication of iconographic analyses (Box 7.5). The meaning that a place has is a human meaning, dependent on social matters. There is an important circularity here. Humans, as members of groups, create places; in turn, each place created develops a character that affects human behaviour. This circularity is characteristic of all landscapes, but is perhaps most relevant in the extreme cases—landscapes of the advantaged and the disadvantaged, the privileged and the underprivileged, insiders and outsiders, rich and poor, and men and women.

HUMANISTIC UNDERSTANDINGS OF PLACE

The humanistic concept of place is central to this chapter, and some extensions of that concept will aid our understanding of the material that follows. First, our experience of place and the meaning we attach to place are not simply individual matters: they are intersubjective (shared). This idea is in accord with our symbolic rethinking of culture. The most compelling example of the intersubjective character of place is the idea of home. Whatever specifically the word 'home' refers to—a dwelling, other people, the earth itself—for most people, a home is shared.

Second, as noted in Chapter 2, our experience of any place may be characterized

alienation
The circumstance in which a person is indifferent to or estranged from nature or the means of production.

critical geography
A collection of ideas and practices concerned with challenging inequalities as these are evident in landscape.

Box 7.4   A Marxist Response to Postmodernism

Another established geographic approach that takes exception to postmodernism is Marxism. David Harvey, the most influential Marxist human geographer since the early 1970s, explained his reservations in his book *The Condition of Postmodernity* (1989). The greatest concern for Marxists is the postmodernist claim that Marxism is unable to explain the growth of disorganized capitalism (as opposed to organized capitalism). Harvey firmly believes in the power of Marxist theory to explain these processes and events.

From a Marxist perspective, Harvey argues that postmodernism reflects 'a particular kind of crisis within . . . [modernism]':

one that emphasizes the fragmentary, the ephemeral, and the chaotic side (that side which Marx so

admirably dissects as integral to the capitalist mode of production) while expressing a deep scepticism as to any particular prescriptions as to how the eternal and the immutable should be conceived of, represented, or expressed. But postmodernism, . . . with its concentration on the text rather than the work, its penchant for deconstruction bordering on nihilism, its preference for aesthetics over ethics, takes matters too far. (Ibid.)

Most critically, Harvey (ibid., 117) sees the rhetoric of postmodernism as dangerous because it ignores 'the realities of political economy and the circumstances of global power'.

Although we obviously cannot resolve any of these debates about postmodernism, it is important to recognize some of the tensions that exist in contemporary human geography.

---

feminism
The movement for and advocacy of equal rights for women and men, and commitment to improve the position of women in society.

patriarchy
A social system in which men dominate, oppress, and exploit women.

gender
The social aspect of the relations between the sexes.

postmodernism
A movement in philosophy, social science, and the arts based on the idea that reality cannot be studied objectively and that multiple interpretations are possible.

either by topophilia, a positive attitude towards place, or by topophobia, a negative attitude towards place. These terms refer to the emotional attachment that exists between person and place, so that the specific feelings of topophilia or topophobia are related to specific characteristics of both the individual and the environment in question. Individual characteristics include personal well-being and familiarity with place. Environmental characteristics include the attractiveness of places as judged by individuals; for example, a pastoral setting may be attractive, while an urban slum may be unattractive.

A third extension of the place concept is the degree to which something in the environment exhibits the characteristics of place or placelessness. As we saw earlier in a Marxist context (Box 2.6), there is a tension present in all places between the local and the global. Aspects of an environment that are local in origin, such as a festival or an indigenous building style, enhance place identity, whereas those that are global in origin, such as chain department stores or an imported building style, contribute to placelessness. Even if all placeless areas do not look alike, they may, nevertheless, lack the local characteristics that make a place distinctive, comfortable for insiders, and perhaps incomprehensible to outsiders. Global characteristics are those that make a place similar to other places and therefore relatively placeless, perhaps less comfortable for insiders but more

comprehensible for outsiders (Entrikin, 1991). The concept of placelessness is an important one because it suggests that the more places become placeless, the more similar human experience becomes. The idea that the modern world is losing environments with distinctive local features and gaining environments with global features has implications for some of the subjects discussed in the rest of this chapter, especially vernacular regions, popular and folk culture, and tourism.

PLACES AS SOCIAL CREATIONS

One distinguished human geographer (Johnston, 1991: 67–8) argues that our focus should be on places as they are created socially. The thrust of his argument is as follows:

1. Place creation is a social act; places differ because people made them that way. Thus differences in physical environment are relevant, but are not a cause.
2. Places are self-reproducing entities because they are the contexts within which people learn. Thus, people are made by the places that they create.
3. Cultures in places do not exist separately from the individual members of the culture.
4. Places are not autonomous units; they interact with other places.
5. Places are often the deliberate creations of those in control.
6. Places are potential sources of conflict.

Box 7.5   The Iconography of Landscape

conography is the description and interpretation of visual images to uncover their symbolic meanings. To the extent that landscapes can be regarded as depositories of cultural meanings, it is possible to subject them to iconographic analysis. A successful iconographic analysis will reveal how a landscape is shaped by and at the same time shapes the regional culture of which it is a part. Two Canadian examples are briefly summarized here.

Osborne (1988) focused on the development of a distinctive national Canadian iconography both in artistic images of lands and peoples (notably the paintings of Tom Thomson and the Group of Seven) and in the various responses these images provoked. The declared aim of those artists was to assist in the development of a national identity, largely through their paintings of the Canadian North. During the first half of the twentieth century, they helped to create a distinctive image of Canada: 'rock, rolling topography, expansive skies, water in all its forms, trees and forests, and the symbolic white snow and ice of the "strong North"' (Osborne, 1988: 172). Although this image is clearly limited, and is now just one among a vast range of national images, it remains a compelling example of iconography.

A different application of iconographic logic can be seen in the examination, by Eyles and Peace (1990), of images of the city of Hamilton, Ontario, in their societal context, asking why these images appear as they do. Derived primarily from newspaper accounts, the images were varied, but they were dominated by a single—negative—theme: Hamilton as steeltown. This image most frequently arose in comparisons with neighbouring Toronto, in which Hamilton appeared as the blue-collar city of production and Toronto as the city of consumption. Because Hamilton is an industrial city, it was seen as a polluted landscape representing the past rather than the future, as an area of economic decline.

Hamilton at dusk.

Mike Grandmaison/All Canada Photos

Each of these six points will reappear in the discussions and examples presented later in this chapter.

REGIONS AND SENSE OF PLACE

The cultural regions discussed in Chapter 6 were formal regions; areas with one or more cultural traits in common, they were the geographic expressions of culture. Such regions are essentially the product of the geographer who analyzes people and place and proceeds to delimit regions. Vernacular regions are rather different in that they may or may not be formally defined: what matters is that it is perceived to exist by those living there and/or by people elsewhere. Such regions introduce the idea of the spatial constitution of culture, and these regions often are most clearly characterized by a sense of place. A formal region may also be a vernacular region; the Mormon landscape, for instance, can be defined formally (Francaviglia, 1978), but it generally is perceived also as a distinctive entity by Mormons and by others.

French Louisiana

In other areas the link is not so clear. The area of French Louisiana, which is the only surviving remnant of the vast French Mississippi Valley empire, is perceived as a distinctive vernacular region in which the population is associated with one group identity, namely Cajun. According to Trépanier (1991), however, this perception is not an accurate reflection of cultural reality. Four important French subcultures inhabit the area: white Creoles, black Creoles, French-speaking Indians, and the descendants of the Acadians who were expelled from Nova Scotia in 1755, now known as Cajuns. Furthermore, until recently, outsiders' perceptions of the Cajun group identity were largely negative, and only since the late 1960s has a process of image modification, which Trépanier (1991: 161) called 'beautification', been underway. The Creole identity, on the other hand, carried a positive image for both black and white French speakers. How are we to explain this anomaly?

modernism
A view that assumes the existence of a reality characterized by structure, order, pattern, and causality.

text
A term that originally referred to the written or printed page, but that has broadened to include such products of culture as maps and landscape; postmodernists recognize that there may be any number of realities, depending on how a text is read.

The decision to identify the area with the Cajun group was made by the state government. In 1968, the Louisiana legislature created the Council for the Development of French in Louisiana (CODOFIL) to gain political benefits by cultivating a French image, but this did not garner any real popular support. Once created, CODOFIL was determined to unify French Louisiana using the then-negative Cajun label rather than the more positive Creole label. Trépanier (ibid., 164) interpreted this effort as a way to guarantee a white identity for the French-speaking area. Once this decision was made, it became necessary to improve the popular image of the Cajun identity, which was achieved through publicity campaigns, including the organization of Cajun festivals. It helped enormously that a new governor who identified with the Cajun group was elected.

For many, French Louisiana now exists as a vernacular region with a Cajun identity that disguises the variety of French-speaking people and the fact that both the black Creoles and the French-speaking Indians reject the Cajun identity. As this example vividly illustrates, vernacular regions are much more than areas perceived to possess regional characteristics; they are also regions to which specific meanings and values are attached—they have social and symbolic identity.

Vernacular regions in North America

North American geographers have delimited many vernacular regions, typically by collecting information on individual perceptions. Probably the most elaborate survey was conducted by Hale (1984); it gathered 6,800 responses from such people as local newspaper editors. In order to identify true vernacular regions, any survey has to ensure that the responses received in some way represent those of average people.

Zelinsky (1980) avoided the survey problem by studying the frequency of a specific regional term and a more general national term in metropolitan business usage; for example, if the region in question is the American South, the incidence of the term 'Dixie' might be compared with that of the term 'American'. Figure 7.1 shows Zelinsky's findings: there are 15 named vernacular regions, an area that lacks regional affiliation, and two areas that are included in either one or both of two regions (these are the areas between the West and Southwest and between the Northwest and West). It is instructive to compare this map and Figure 6.7, the map of cultural regions based on the concept of first effective settlement, to see the extent to which the cultural region and the vernacular region are related.

A newspaper journalist, Joel Garreau, made an insightful contribution to the geographic

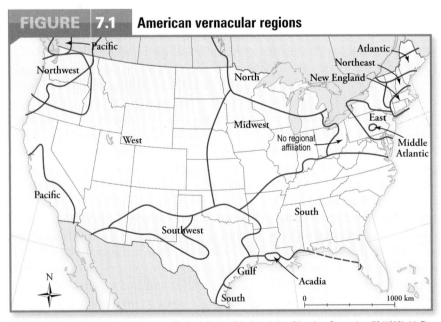

FIGURE 7.1   American vernacular regions

SOURCE: W. Zelinsky, 'North America's Vernacular Regions', *Annals of the Association of American Geographers* 70 (1980): 14, Fig 9 (MW/RAAG/P1889). Used with permission of Taylor & Francis Ltd., http://www.informaworld.com.

understanding of North America, first with a short article that proved very popular and then with his book, *The Nine Nations of North America*. Describing the inspiration for his work as 'a kind of private craziness', he collected information from fellow journalists who spent much time on the road concerning their impressions of places (Garreau, 1981: ix). Gradually, it became clear that most political divisions lacked meaning and that North America was better understood as a collection of nine nations, places that are full of meaning but that do not appear on any map. Certainly, Garreau produced a regionalization that is a useful way of thinking about North America. The nine nations are shown in Figure 7.2 and justified as follows (Garreau, 1981: 1):

> Forget the pious wisdom you've been handed about North America.
>
> Forget about the borders dividing the United States and Canada, and Mexico, those pale barriers so thoroughly porous to money, immigrants, and ideas.
>
> Forget the bilge you were taught in sixth-grade geography about East and West, North and South, faint echoes of glorious pasts that never really existed save in sanitized textbooks.

FIGURE 7.2 **The nine nations of North America: a journalist's perception**

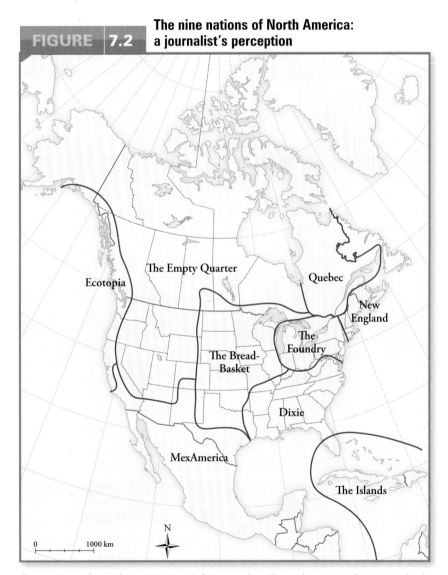

Geographers often refer to this thoughtful map when discussing vernacular or perceived regions. It is instructive to compare this map not only to Figure 7.2 but also to the divisions of North America suggested in Figure 6.5.

SOURCE: Adapted from J. Garreau, *The Nine Nations of North America* (Boston: Houghton Mifflin, 1981), following 204.

Having forgotten these and other ideas, Garreau invites the reader to consider how North America really works, identifies the nine nations, and notes that each nation is characterized by a particular way of looking at the world. Consider just two examples. First, it is not helpful to talk about Colorado because Colorado is three different places being part Bread-Basket, part Empty Quarter, and part MexAmerica. Second, San Francisco and Los Angeles are not best seen as two cities in California; rather, they are the capitals of two of the Nations, San Francisco of Ecotopia, and Los Angeles of MexAmerica.

Vernacular regions are often viewed more positively by those living within the region than by those outside. For many living in the American Bible belt, the name is a source of great pride; yet for some outsiders it is a term of derision. In other cases, if the name of the vernacular region has been imposed for purposes of tourism or commercial promotion, with no real roots in the region, it may mean little to either residents or outsiders.

Regional identity as place

If an area has an identity specific enough to be named, this suggests that the identity is meaningful to those using the name. In other words, vernacular regions are not simply locations: they are places. The name of the region conveys a meaning—or possibly more than one. Parts of the world possess such a powerful regional identity that the mere mention of the name conjures up vivid mental images. Most North Americans, regardless of religious affiliation, recognize that the term 'Holy Land' refers to the land bordering on the eastern Mediterranean. Other places have one meaning for one group and a quite different meaning for another. For country-music lovers, Nashville is likely to be most meaningful as the home of the Grand Ole Opry; for others it may be simply the capital of Tennessee. Similarly with literature, where vernacular regions associated with particular novelists or poets may be created and then marketed to interested tourists, as in the well-known British examples of Wordsworth country (the Lake District), Bronte country (the Yorkshire moors), and D.H. Lawrence country. In such cases, place blends with identity.

An especially vivid example of the links between regional identity and sense of place is the example of wine-producing regions.

More so than most crops, wine grapes and the wines produced from them vary from place to place. Even within a particular wine-producing region microclimate, soil, slope and related incoming radiation, methods of winemaking, and other details can vary such that the resulting wines taste quite different from one another. The key concept here is that of *terroir*, a French term that literally means ground or soil, but is better understood as referring to all of the local environmental characteristics, and even social characteristics, that have an effect on the wine. In a wonderful turn of phrase, Sommers (2008: 19) suggests that to taste wine is to 'taste geography'. A similar logic applies also to cheese-producing regions, especially those in Europe but also in parts of southern Ontario and Quebec, for example.

Pyschogeography

The term 'psychogeography' is used to refer to 'how people feel about, experience, paint themselves into the world and take that portrait back into themselves as literal parts of who they are and hence their well-being' (Stein and Thompson, 1992: 63). The aim in a psychogeographic study is to identify an internal self-image of a region, without reference to external perceptions (see Box 7.6).

Another way to think about psychogeography—vernacular regions that are clearly perceived by those living within them—is in terms of a **homeland**. According to Nostrand and Estaville (1993: 1), a homeland has four basic ingredients: people, place, sense of place, and control of place. 'Sense of place' refers to the emotional attachment that people have to a place, while 'control of place' refers to the requirement of a sufficient population to allow a group to claim an area as their homeland. Most homelands are associated with groups defined on the basis of common language, religion, or ethnicity.

Geographies of Difference

POWER, DIFFERENCE, AND INEQUALITY

Our discussions of social theory and related contemporary concerns in cultural geography make it clear that peoples and places are now being considered in a variety of conceptually sophisticated ways. Human geographers are

homeland
A cultural region especially closely associated with a particular cultural group; the term usually suggests a strong emotional attachment to place.

Box 7.6 | Psychogeography: The Sense of Oklahomaness

Understanding a regional cultural landscape requires consideration of regional identity, which includes popular attitudes that have developed through long experience in a landscape. In a study of regional self-awareness, Stein and Thompson (1992: 65) used the state of Oklahoma 'as an example of a community of meaning within a politically-defined territory'—what they call a cultural identity system. '"Oklahomaness" connotes what is distinctively Oklahoma; that is, the boundaries of the identity and the contents within it' so that 'Oklahoma is first and foremost a state of mind which springs from a common pool of self images, a community of meanings which lends the state much of its regional character' (ibid., 66).

What are the self-images, the popular attitudes, that shape the identity of Oklahoma today? First, there are images associated with Native Americans: the Trail of Tears resulting from the forced migration of the Five Civilized Tribes from the southeastern United States to Oklahoma; reservation lands taken from them by the United States government to give to white Americans. Second, there are images associated with a famous series of land rushes to the 'unassigned' land taken from Native Americans between 1889 and 1901: images of unbridled enthusiasm and opportunism summed up in the labels 'Boomers' and 'Sooners'.

As a result of these land rushes, Oklahoma came to occupy a special place in the American psyche; it is no accident that the musical show (later a film) *Oklahoma!* was so successful on Broadway from 1943 to 1948, a time of exuberant American nationalism.

A third component of Oklahoma's regional identity is the tragic and haunting image of the 'Dust Bowl' of the 1930s, associated with the folksongs of Woody Guthrie and so movingly described in John Steinbeck's novel, *The Grapes of Wrath*. Cowboys and open spaces combine to produce a fourth image, one that today implies an 'ideological statement of moral superiority over those with whom Oklahomans have compared themselves and felt inferior' (ibid., 73). Both of these images are linked by Stein and Thompson to a fifth: the symbolism of land and sky.

These images, along with a number of others (such as the college football mentality), are essential to understanding the state of Oklahoma as a meaningful cultural identity system; in effect, they are boundaries delimiting an area that has real meaning for those inside. These images are not limited to a single group of people and are in fact closely tied to relations between those inside and those outside the region. The identity of Oklahoma is continually changing as relationships with others change.

---

increasingly aware of the diversity of peoples and places as a result of the insights suggested by humanism and Marxism and, more recently, by feminism and postmodernism. Together, these varied approaches invite us to explore more than cultures as ways of life and the visible landscapes related to them (the geographic expressions of culture); they invite us to explore culture as a process in which people are actively involved and landscapes as places constructed by people (the spatial constitution of culture). Thus, a first reason for the interest in difference and inequality relates to changing approaches to human geography.

As we learned in Chapter 2, traditional regional geography was essentially empirical in emphasis, focusing on the description of peoples and places; it was not concerned with uncovering the power differences that we now recognize as underlying human geographies. Similarly, as we also saw in Chapter 2, the spatial analysis school was explicitly positivist in philosophy, taking it for granted that such work was by nature objective, unaffected by the identity of the human geographer conducting

the research. We now understand that such assumptions are often unwarranted. It was only about 1970, with the emergence of the new philosophical approaches based on humanism and, especially, Marxism that human geographers began to consider questions of meaningful human difference in areas such as access to resources and quality of life. In short, human geographers added a political dimension—a concern with power and inequality—to their studies.

A second reason why difference and inequality did not emerge as topics of interest before about 1970 was that until then the field of geography—like most other academic disciplines—was dominated by white, middle-class, heterosexual, able-bodied males. In the era when the dominance of patriarchal discourse was near-universal, questions of difference and inequality were rarely acknowledged.

About 1970, however, a new radical geography, sometimes called social geography, began to consider previously neglected aspects of human landscapes, such as the evidence of differences in power and related differences in well-being, incidence of crime, health, and availability of health care. Over time, such studies have broadened to include analysis of different qualities of landscapes—what we call elitist landscapes and landscapes of stigma—and of gender and sexuality as these relate to landscape and to economic development. Of central concern here is what we might call the quality of life. This is a very difficult concept to discuss objectively, for it can be interpreted only in some specific context.

As suggested above in the accounts of social theory, and as described in our discussion of types of society, significant social and spatial variations exist in the distribution of power and authority. Further, the logic of world systems theory (Chapter 5) relies on the argument regarding a dominant core and a subordinate periphery in the contemporary world. It is not surprising, then, that to understand peoples and places we must recognize the importance of the distribution of power and authority.

But, as suggested especially in Box 7.1 and in the account of postmodernism, there is also need to focus on difference. Put simply, 'space and place are intimately connected to . . . gender, class, sexuality and other axes of power; all geographic knowledges are situated, and location matters' (McKittrick and Peake, 2005: 43).

Building on ideas about hegemony and power introduced in Box 7.1 and also on the postmodern focus on previously repressed experiences, Tables 7.1 and 7.2 identify some of the key issues relating to geographies of difference. Table 7.1 suggests that difference is evident at various geographic scales, from the body through to the entire world, and Table 7.2 provides examples of inclusion and exclusion at different geographic scales. Together, these two tables are suggestive of the need for geographers to think critically about how spaces are lived in, experienced, contested, and resisted. The following accounts of geographies of difference focus on ethnicity, gender, and sexuality.

ETHNICITY

Ethnicity is often poorly defined; one researcher who examined 65 studies of ethnicity noted that 52 of them offered no explicit definition at all (Isajiw, 1974). The geographer K.B. Raitz argued that an **ethnic group** is any group that has a common cultural tradition, that identifies itself as a group, and that constitutes a minority in the society where it lives: 'ethnics are custodians of distinct cultural traditions. . . . the organization of social interaction is often based on ethnicity' (Raitz, 1979: 79). Thus, the group may be delimited according to one or more cultural criteria, but it is necessary that the group not be living in their national territory: Swedes in Sweden do not constitute an ethnic group, whereas Swedes in the United States do, because they identify themselves as one.

Although this definition is clear and reasonable, it is not universally employed. Indeed, it is important to recognize that 'ethnicity' is generally regarded as one of the most confusing terms in social science. The greatest confusion occurs when terms such as 'race' or 'minority' are used interchangeably with 'ethnic', and because of this confusion it is necessary to clarify what race is and, more important, what it is not.

The illusion of race

Today, scientific research is finally dispelling three long-standing myths about human

> **ethnic group**
> A group whose members perceive themselves as different from others because of a common ancestry and shared culture.

Table 7.1	Examples of Scales of Difference			
	Body	Home	Nation	Globe
Geographies of domination	Racialization, racism, heterosexism	Domestic violence, domestic labour	Colonization, genocide, apartheid	Imperialism, globalization, language, religion
Experiential geographies	Bodily geographies, such as transgendered bodies	Geographies of fear, geographies of fleeing or staying put	Geographies of diaspora and migration; critiques of nation	Geographies of fair trade, refugees; anti-globalization activism

SOURCE: Adapted from K. McKittrick and L. Peake, 'What Difference Does Difference Make to Geography?', in N. Castree, A. Rogers, and D. Sherman, eds, *Questioning Geography* (New York: Blackwell, 2005), 45.

Table 7.2	Examples of Scales of Inclusions and Exclusions			
	Home	Neighbourhood	Nation	Globe
Spaces of assimilation and/or exclusion	Master bedrooms, den	Ethnic neighbourhoods, gated communities	Public spaces such as parks, shopping malls	Trading blocs, language, religion
Spaces of containment/ internment/exile	Homeless shelters, homelessness	Ghettos	Concentration camps, refugee camps, prisons	Apartheid-like systems
Spaces of objectification	Women in home seen as housewives	Youth in malls seen as delinquent	Immigrants seen as drain on national welfare system	Women in less developed world seen as victims of development processes

SOURCE: Adapted from K. McKittrick and L. Peake, 'What Difference Does Difference Make to Geography?', in N. Castree, A. Rogers, and D. Sherman, eds, *Questioning Geography* (New York: Blackwell, 2005), 45.

evolution. The *first myth* concerns the idea of human evolution, or evolution generally, as some sort of ladder of progress. Evolution is not a history of steady progress to a finished product; it is not correctly seen as a ladder. Rather, to borrow a metaphor popularized by the distinguished evolutionary biologist Steven Jay Gould, evolution is best seen as a bush with endlessly branching twigs, each twig representing a different species. In the terms of this metaphor, the human species is 'a fragile little twig of recent origin' (Gould, 1987: 19).

A *second myth* concerns the supposed existence of several different human **races** (Gould, 1985; Montague, 1964). As modern humans, *Homo sapiens sapiens*, moved from Africa to different physical environments and splintered into distinct spatial groupings, adaptations developed not only in culture but also in certain body features or **phenotypes**. Relatively distinct phenotypes emerged: different skin colours, head shapes, blood groups, and so forth. Certain characteristics became dominant in certain areas, either because they were environmentally appropriate or as a result of chance genetic developments. The resulting variations in human physical characteristics are quite minor and do not in any way represent different types of humans (Box 7.7).

A *third myth* is the idea that races not only exist but can be classified according to their level of intelligence. We are fortunate today in

race
A subspecies; a physically distinguishable population within a species.

phenotype
Any physical or chemical trait that can be observed or measured.

Box 7.7 Species and Races

There is only one species of humans in the world, namely *Homo sapiens sapiens*. That all humans are members of the same species is biologically confirmed by the fact that any human male is able to breed with any human female.

All species, including humans, display variation among their members. Indeed, many plant and animal species exhibit sufficient variation that they can be conveniently divided into subspecies or races: geographically defined aggregates of local populations. Different races emerge within a species if groups are isolated from one another, in different environments, for the necessary length of time. As a consequence of selective breeding (the result of isolation), together with various adaptations to their respective environments, each group will eventually exhibit some genetic differences from every other group.

To what extent is it possible to delimit distinct races—geographically defined aggregates of local populations—in the human species? To answer this question, we need to identify three stages:

1. We know that *Homo sapiens sapiens* evolved by 195,000–150,000 years ago, probably only in Africa.
2. Over the course of many millennia, members of this species then moved across the surface of the earth, replacing archaic *Homo sapiens* and gradually separating into various groups; as a consequence of these movements and groupings, some selective breeding and adaptation to different environments occurred, giving rise to *some minor* genetic differences.
3. As a result of population increases and further migrations, groups that had been temporarily isolated began to mix with other groups.

During the second stage the groups of humans often known as Negroid, Mongoloid, Caucasoid, and Australoid appeared. But are these groups races? The answer is no, because none of these human groups was ever isolated long enough or completely enough to allow separate independent genetic changes to occur. Use of the word 'race' is therefore mistaken because, genetically speaking, humans cannot be divided into clear-cut stable subspecies. This fact was powerfully affirmed in 2000 when researchers from the Human Genome Project announced that all humans are 99.9 per cent identical at the genetic level and that, therefore, there is no genetic basis for 'race'.

Because race is a biological concept referring to a genetically distinct subgroup of a species, there are today no such things as races within our human species. There are some small genetic differences between members of the human species, such as skin colour and shape of eyes, as a consequence of the earlier period of isolation of groups in different environments. Biologically, however, these differences are not sufficient to justify the identification of separate races. Thus, the differences among people throughout the world are very few and very minor.

We are a biologically variable species, and our variability results from the same processes that produce variability in all living things: namely, natural selection in different physical environments. These biological facts are of basic importance. They immediately inform us, for example, that it is incorrect to equate a cultural group with a racial group. Thus there is a Jewish religion, but not a Jewish race. Similarly, there are Aryan languages, but not an Aryan race.

There are no subspecies—races—within our human species.

racism
A particular form of prejudice. Attributing characteristics of superiority or inferiority to a group of people who share some physically inherited characteristics.

that we have clear scientific evidence of the unity of the human race. Such knowledge was not available until relatively recently. The characteristic response to physical variations in humans, particularly different skin colours, has been to regard these as evidence of different types of humans with different levels of intelligence. Box 7.8 summarizes the history of the erroneous idea that the human population can be divided into distinct and unequal groups, usually called races. The reality is that all humans are members of the same species;

the only differences within the species are differences of secondary characteristics, such as skin colour (Kennedy, 1976).

Regardless of the fact of human biological unity, today the term 'race' commonly is used to set apart outsiders whose physical appearance does not accord with some generally accepted norm; the key divide is the most visible one: skin colour. The fact that races do not exist does not prevent the label from being applied, often with tragic consequences.

Box 7.8   A History of Racism

Box 7.7 explained the unity of the human species: there are no distinct subspecies or races within that species. Notwithstanding these biological facts, the concept of distinct racial groupings has long been popular in both lay and scientific circles. Ideas about the existence of races and the relative abilities of the supposed races have been central to many cultures.

Racism is the belief that human progress is inevitably linked to the existence of distinct races. Notions of racial variations in ability appear to have been common in most cultures. They flowered especially in Europe beginning about 1450 with the overseas movement of Europeans, which brought them into contact with different human groups, and new theories were later developed according to which these different groups came from different origins. This notion of multiple origins is incorrect.

By the eighteenth century, racial explanations for the increasingly evident variety of global cultures were standard. Philosophers such as Voltaire and David Hume perceived clear differences in ability between what were commonly seen as racial groups. Undoubtedly one of the key racist thinkers was Joseph Arthur de Gobineau, whose *Essay on the Inequality of Human Races* was published in 1853–5. Gobineau ranked races as follows: whites, Asians, Negroes. Within the whites, the Germanic peoples were seen as the most able. A second influential racist was Houston Stewart Chamberlain, but racist logic was apparent among many of the greatest scientists. Darwin wrote of a future when the gap between human and ape would increase because such intermediaries as the chimpanzee and Hottentot would be exterminated (see Gould, 1981: 36). As we have seen, the racial categories employed in all such discussions have no biological meaning.

One explanation for the popularity of racist thought is the fact that most cultures classify other cultures relative to themselves. The consequences of racist thinking are varied and considerable. Belief in the inferiority of specific groups has led to mass exterminations, slavery, restrictive immigration

policies, and, most generally, unjust treatment (the specific example of apartheid in South Africa is detailed in Box 7.9).

The Statement on Race of the American Anthropological Association (1998) summarizes the development and consequences of thinking based on 'race'.

'Race' . . . evolved as a worldview, a body of prejudgments that distorts our ideas about human differences and group behavior. Racial beliefs constitute myths about the diversity in the human species and about the abilities and behavior of people homogenized into 'racial' categories. . . . Given what we know about the capacity of normal humans to achieve and function within any culture, we conclude that present-day inequalities between so-called racial groups are not consequences of their biological inheritance but products of historical and contemporary social, economic, educational, and political circumstances.

Today, discussions about racism, both past and present, generate heated debate. For example, as noted in Chapter 5, discussions at the 2001 United Nations World Conference Against Racism, Racial Discrimination, Xenophobia, and Related Forms of Intolerance included arguments that rich countries should pay poor countries compensation for the abuses (such as slavery) inflicted on them in the past, and also arguments about whether or not Zionism is racism. On both sides of the colonial divide, of the 'privileged' West and the 'othered' Third World, politics often is central to discussions of racism. In 2009, a follow-up UN conference held in Geneva was boycotted by many Western countries, including Canada, over concerns about anti-Western and anti-Israel bias—of particular concern was that the President of Iran, who has both denied the Holocaust and made homophobic statements, was to address the conference. (Of course, these same Western countries did not want to be called to account at such a global conference for past and ongoing 'racialized' abuses against minorities in their midst.)

One example of the expression of a racist ideology in landscape was the apartheid landscape of South Africa, formally set in place beginning in 1948 and finally dismantled in 1994 (Box 7.9 and Figure 7.3).

Genocide

Tragically, groups seen as different from the majority within a society often find themselves labelled negatively and treated unequally. One extreme consequence of this circumstance is the targeting of a group for genocide. The UN Convention on Genocide was written following the Holocaust and sees victim groups in national, ethnic, racial, or religious terms. Following this definition, the sociologist Michael Mann (2004) uses terms such as 'classicide' or 'politicide' to cover other bases for mass killing. However, most scholars favour a broadening of the term 'genocide' to include groups defined politically, occupationally, or socially. Thus, genocide 'is a form of one-sided mass killing in which a state or other authority intends to destroy a group, as that group and membership are defined by the perpetrator' (Chalk and Jonassohn, 1990: 23).

A necessary precondition for genocide is a symbolic, and sometimes spatial, distancing or separation of one group—the perpetrator 'in-group'—and another, the victim 'out-group'. The victim group often is given a derogatory label, further emphasizing that they do not belong. Identities frequently are linked to place, and in most cases genocide is justified by the perpetrators in terms of their 'right' as a group to occupy a particular place without others present. Of course, for genocide to occur, the perpetrator group must be able to exercise power over the victim group. Totalitarian states have been especially prone to commit genocide because they function without real constraints on their ability to exercise power. The worst mass killings of the twentieth century were committed by dictatorial states in which most aspects of cultural, political, and economic life were controlled by the government. Genocide requires the participation of many people in the perpetrator group, and this level of participation is more likely to occur when the actions taken are formally authorized by the state, such that participating in mass killing is legitimized. Further, much evidence suggests that most of those who participate in genocide are 'relatively ordinary people engaged in extraordinary behaviors that they are somehow able to define as acceptable, necessary, and even praiseworthy' (Alvarez, 2001: 20). Most cases of genocide occur at times of severe national tension.

Twentieth-century examples of genocide include forced movements and murder of Armenians in Turkey during World War I; population purges and deportations in the USSR from the 1920s to the 1950s; the Holocaust perpetrated by Nazi Germany against the Jews, as well as against the Roma (gypsies), the disabled, and homosexuals; mass killings in Cambodia by the Communist Khmer Rouge between 1975 and 1979; killing of Kurds in Iraq in the 1980s; killing of Muslims by Serbs in the former Yugoslavia in the 1990s; and Hutu killing of Tutsis and moderate Hutu in Rwanda in 1994. In all of these cases, genocide is one part of larger efforts by one group to dramatically reshape the geography of peoples and places. Racist and nationalist ideas form the basis for categorizing populations into groups. Typically, victim groups are blamed for any and all social and economic problems and are seen as impediments to progress and prosperity. The perpetrators, using powerful metaphors of 'purifying' and 'cleansing', justify genocide on the grounds that victim groups are something less than human.

Ethnic identities

Some groups are generally regarded as minorities—either by themselves or by themselves and others—because they are in some way different from (and therefore sufficiently visible to be excluded by) the majority. Common bases for delimiting a minority are language, religion, ethnicity, perceived racial identity, and recent immigrant status, or any combination of these. The term 'ethnicity' is a useful one as it usually suggests some enduring collective identity through a shared history—a shared time as well as a shared space. Most groups who identify themselves as ethnic base their ethnicity on one or both of the two principal cultural variables discussed in the previous chapter: language and religion. Like language and religion, ethnicity is both inclusionary and exclusionary. Some people are defined as insiders because they share the common identity of the group, while others are seen as outsiders because they are different. Yet it is quite possible for an ethnic group to change its identity and behaviour over time.

Box 7.9   Apartheid

The first permanent European settlement in South Africa was established in 1652 to grow food to supply Dutch ships; population growth was slow until the nineteenth century. The colony became British in 1806, and from 1806 until 1948 South Africa was predominantly British in political character.

The history of South Africa since 1652 has been characterized by a series of conflicts between several relatively distinct societies. Segregation of different groups was characteristic of the South African landscape prior to 1948, but in that year legislation was passed to institutionalize the system known as 'apartheid', or 'separate development'. Thus, apartheid had roots in long traditions of social conflict and of relatively informal segregation. Understanding apartheid, however, clearly requires further consideration of the social group that formally institutionalized it: the Afrikaners.

The Afrikaners, as their name suggests, separated themselves from their European heritage, although their language is derived from Dutch and traditionally they have belonged to the Dutch Reformed Church. By the twentieth century they identified themselves fully with Africa, often being called 'the White tribe'. In 1948 the Afrikaners gained political control of a state made up of whites, blacks (of various linguistic and tribal groups), coloureds (people of mixed descent), and Asians. On gaining power, the Afrikaners instituted apartheid on the grounds that it was the best way to allow a socially fragmented country to evolve; their argument was that integration of different groups leads to moral decay and racial pollution, whereas segregation leads to political and economic independence for each group. Ten African homelands were created, four of which were theoretically independent (their independence was never recognized by the United Nations or by any government outside South Africa).

According to the Afrikaners, then, apartheid provided blacks with independent states and rights comparable to those of whites. In practice apartheid simply enabled the whites to maintain their own cultural identity and political power while exploiting black labour. Interestingly, the dismantling of the apartheid system, which began in the late 1980s and concluded with the 1994 election of the first government representative of the black majority, proceeded relatively peacefully. Today, South Africa is not immune from either internal ethnic tensions or xenophobic tendencies. There are, for example, many attacks on people from Zimbabwe and Mozambique who have moved into South Africa. Most commentators blame these attacks on economic and social problems, but Nelson Mandela noted: 'We cannot blame other people for our troubles' (quoted in Nolen, 2008).

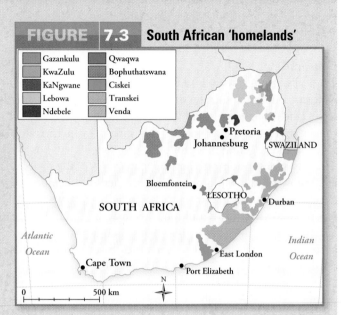

FIGURE 7.3   South African 'homelands'

Gazankulu	Qwaqwa
KwaZulu	Bophuthatswana
KaNgwane	Ciskei
Lebowa	Transkei
Ndebele	Venda

Apartheid on the national scale: the 'homelands' planned by the South African government in 1975.

During the apartheid era, desirable landscapes, such as this Indian Ocean beach north of Port Elizabeth, South Africa, were designated for 'whites only'. Similar signs were placed in many urban parks, while official buildings, such as post offices, had separate entrances for whites and other groups.

William Norton

'Ethnic' is a convenient term, partly because it defies explicit definition. Thus, human geographers often use it to identify and discuss cultural identities and cultural regions. Generally speaking, an ethnic region or neighbourhood is an area occupied by people of

Situated a little south of Madison, Wisconsin, in a region of rolling hills reminiscent of the Swiss alpine farmlands, the small town of New Glarus calls itself 'America's Little Switzerland'. The village was founded in 1845 by 108 Swiss immigrants, and succeeding generations have retained the community's Swiss-German language, vernacular architecture, and folk traditions. In this photo the Swiss architectural influence is evident in the church steeple as well as the chalet-style roof on the left. The decorative banners pair the symbolism of the Swiss flag with the American stars and stripes, and the 'Alpine' store on the right is clearly intended for the tourist market.

Photo courtesy of the New Glarus Chamber of Commerce

common cultural heritage who live in close spatial proximity.

Ethnic areas

Most immigrant ethnic groups, especially those moving into urban areas, experience an initial period of social and spatial isolation that may lead to low levels of well-being, relative deprivation, and the development of an ethnic colony, enclave, or **ghetto**. We can interpret these common initial experiences of deprivation and residential segregation as social and spatial expressions of outsider status. Often a local group identity is continually reinforced by **chain migration**, the process whereby migrants from a particular area follow the same paths as friends and relatives who migrated before them.

Despite often negative initial experiences, most new immigrant groups do eventually experience either **assimilation** or **acculturation**. Some groups that move into North American cities steadily lose their ethnic traits and assimilate, eventually becoming part of the larger culture. Other groups, however, are more likely to acculturate: to function in the larger culture but retain a distinctive identity. In both Canada and the United States, acculturation has been more common than assimilation.

One key factor determining whether or not assimilation occurs is the degree of residential propinquity: if group members live in close spatial proximity, then social interaction with the larger culture is limited and assimilation unlikely. Once again we see an intertwining of space and culture.

Immigrants moving into rural areas, especially as a group or in a chain migration, tend to retain aspects of their ethnic identity longer than do those moving into urban areas.

ghetto
A residential district in an urban area with a concentration of a particular ethnic group.

chain migration
A process of movement from one location to another through time sustained by social links of kinship or friendship; often results in distinct areas of ethnic settlement in rural or urban areas.

assimilation
The process by which an ethnic group is absorbed into a larger society and loses its own identity.

acculturation
The process by which an ethnic individual or group is absorbed into a larger society while retaining aspects of its distinct identity.

multiculturalism
A policy that endorses the right of ethnic groups to remain distinct rather than to be assimilated into a dominant society.

Many regional landscapes, especially in areas of European expansion, are characterized by distinct ethnic imprints, because of the relative recency of the occupation, because of the cohesion and homogeneity of the group, and because of the relative isolation of rural areas. Historical and cultural geographers often study the extent of cultural change that accompanies and follows settlement and the creation of landscapes (see McQuillan, 1993).

Multiculturalism

The extent to which groups assimilate also is related to state policies. Canada, for example, has had a policy of **multiculturalism** since 1971. The political logic for the introduction of multiculturalism was succinctly stated by then Prime Minister Pierre Elliott Trudeau: 'Although there are two official languages, there is no official culture, nor does any ethnic group take precedence over any other' (quoted in Kobayashi, 1993: 205). Versions of multiculturalism are in place in many other countries; but the idea and policy are contentious.

Some see multiculturalism as a vision of national identity based on pluralism; others see it as divisive, a form of 'have a nice day' racism. This is an especially challenging circumstance in many European countries today where it is often argued that multiculturalism facilitates extremism, especially Islamic extremism, because it tacitly encourages newcomers or those whose ethnicity is 'different' not to assimilate. Policy-makers in several European countries are seeking to identify and implement some alternative policy that promotes pluralism without creating divisiveness.

GENDER

Gender, like class, implies a distinction between power groups—in this case, dominant males and subordinate females. For feminist geographers, gender is a key category of analysis for two reasons: because it involves power relations and because it involves specific roles for males and females; that is, our identities are gendered.

That human geographers did not begin paying serious attention to gender differences until recently is regrettable, since the former insistence on discussing humans in general effectively meant ignoring all the ways in which, around the world, the lives and experiences of women and men are different. For example, it is still women who typically perform domestic work, which is unpaid, repetitious, and often boring, and the jobs available to women outside the home are often low-paid and low-skilled. In short, women and men tend to do different work, in different places, and to lead different lives, which include different visions of the world and of themselves.

Feminist geographers have usually accepted that gender is a social construction deriving largely from the natural category of biological sex. Gender is formed initially through the differential treatment of girls and boys and continues to be reinforced through the life cycle. This differential treatment is accompanied by different societal expectations of the values, attitudes, and behaviours of boys and girls.

But it is not sufficient to note that girls and boys are raised differently and expected to be different throughout their lives. They also are raised unequally, with boys socialized to be aggressive and to assume leadership roles, while girls are socialized to be passive and to be compliant followers. Such characteristics possibly have some initiating biological cause, but even if this is so, the socialization process clearly emphasizes and increases any natural differences and minimizes movement across the categories of male and female.

Overall, this process works to the advantage of men. In the context of work, for example, women are relegated to the private domain or to limited opportunities in the public domain. The differing employment experiences of women and men even in the more developed world were highlighted by 2006 data for the European Union countries showing that women achieve a better education overall but earn less and occupy fewer top jobs. Part of the explanation for this apparent contradiction is that women continue to perform more domestic work and more women hold part-time jobs. Data such as these are published regularly, often on the occasion of International Women's Day, 8 March. As noted, these differences are not explained in terms of biological differences (sex) but in terms of cultural differences (gender).

Gender in the landscape

Interestingly, gender differences are not always as apparent in the landscape as are some other differences, such as ethnicity and social class. Yet landscapes are shaped by gender, and they do provide the contexts for the reproduction of gender roles and relationships (Monk, 1992).

Numerous examples of landscapes, both visible and symbolic, reflect the power inequalities between women and men and demonstrate the embodiment of patriarchal cultural values and related **sexism**. In urban areas, statues and monuments reinforce the idea of male power by commemorating male military and political leaders: 'conveyed to us in the urban landscapes of Western societies is a heritage of masculine power, accomplishment, and heroism; women are largely invisible, present occasionally if they enter the male sphere of politics or militarism' (Monk, 1992: 126).

The design of domestic space is strongly influenced by ideas about gender roles. In many cases, including traditional Chinese and Islamic societies, areas for women and for men are separated, and those for women are often isolated from the larger world. In Western societies, the favoured domestic design has centred on the home as the domain of women and as a retreat from the larger world. Certainly, it has been usual in the Western world to locate the kitchen, designed as a separate area for the unpaid work of women, at the rear of the home, and some feminist scholars have interpreted this as devaluing women's work. Following this logic, a home can be understood as a site of unpaid domestic work, oppression, and sometimes violence.

Again, according to some feminist arguments, this same patriarchal logic influenced the expansion and morphology of city suburbs: men were seen as commuters and women as homemakers/consumers requiring ready access to shops and schools. It is now argued that this arrangement further disadvantages women, who may become isolated because of distance from town or who may struggle to cope with work at home and limited opportunities for paid employment in the suburbs. The morphology of the city therefore works to reinforce traditional gender roles and identities and serves as an obstacle to change. This is evident in the shopping mall. Because of increasing personal mobility, additional discretionary income, and conformity with gender expectations, women are the principal consumers at shopping malls. As a result, the typical shopping mall landscape reflects the gendered identity of consumers in the way the mall is laid out, in the way many individual stores display merchandise, and more generally in advertisements directed at women.

Gender and human development

Many expressions of gender in landscape typically disadvantage women, but perhaps the clearest evidence of gender inequality can be found in data on human development. Recall from Table 5.9 that the Human Development Index measures average development in a country, but note that it does not incorporate gender inequality as this is evident in development. Because of this failing, the Gender-related Development Index (GDI) was first calculated by the UN in 1995. This statistic uses the same criteria as the HDI except that it also captures inequalities between women and men for each criterion. The greater the gender disparity in basic human development, the lower is a country's GDI relative to its HDI. Table 7.3 provides details of this more sophisticated, gender-sensitive measure of human development for the 10 countries that rank highest. Of the 157 countries for which it was possible to calculate a GDI, those with the lowest value all were African, with Sierra Leone ranked at the bottom. The gender inequality evident in many African countries relates to relatively low education levels and poverty. The global pattern of gender-related human development is mapped in Figure 7.4.

Using a different data set that provides information for just 58 countries, Table 7.4 identifies the top five and the bottom five countries for gender equality. The list of the bottom five needs to be interpreted with care as so many countries are not included in the data set. China, the country with the largest population, ranked thirty-third on the list—China outlawed sexual harassment and established

sexism
Attitudes or beliefs that serve to justify sexual inequalities by incorrectly attributing or denying certain capacities either to women or to men.

Table 7.3	Gender-Related Human Development	
Country	**GDI Rank**	**GDI Value**
Iceland	1	0.960
Norway	2	0.954
Australia	3	0.953
Netherlands	4	0.947
Sweden	5	0.946
France	6	0.946
Canada	7	0.944
United Kingdom	8	0.942
Switzerland	9	0.941
Finland	10	0.939

SOURCE: United Nations Development Program, *Human Development Report*, 2008: International Cooperation at a Crossroads: Aid, Trade, and Security in an Unequal World Fighting Climate Change: Human Solidarity in a Changing World (New York: UNDP, Macmillan, 20075). By permission of Oxford University Press, Inc.

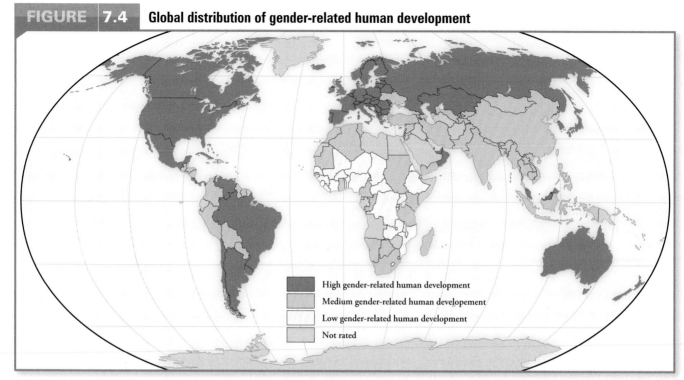

FIGURE 7.4   Global distribution of gender-related human development

High gender-related human development
Medium gender-related human developement
Low gender-related human development
Not rated

Notably, some countries that likely would rank low on this map are not rated.

Source: UNDP, *Human Development Report, 2007/2008. Fighting Climate Change: Human Solidarity in a Divided World* (New York: Macmillan, 2007), Table 28.

gender equality as a national goal as recently as 2005. India, the country with the second largest population, ranked fifty-third.

The annual *Human Development Report* published by the United Nations also provides much additional specific data on gender inequality. For example, Table 7.5 shows the five countries with the highest percentages of women holding seats in national legislatures. A 2008 election in Rwanda, at the top of the list, had women winning 44 of 80 seats (55 per cent). Women clearly are playing much greater roles in political life than in previous years. Most notable, perhaps, is the increasing number of women playing key leadership roles in political life. In the first two months of 2006, three countries elected female leaders—Liberia, Chile, and Jamaica. It is especially significant that Liberia's president is Africa's first female elected leader. But women also play important roles fighting for democracy—in Myanmar, Aung San Suu Kyi, who has been under house arrest for many years, is the leading campaigner for democratic rights in a country with one of the worst records in the world for arbitrary punishment, corruption, and injustice.

Human rights groups regularly highlight Saudi Arabia as a country that severely limits women's rights in order to maintain male control. In 2008 the New York-based Human Rights Watch reported that women have limited access to justice and may have to obtain permission from male relatives to work, travel, study, marry, or receive health care. More generally, speaking on the UN's International Day for the Elimination of Violence Against Women in 2008, the UN Secretary-General stated that one in every three women in the world will be abused, beaten, or forced into sex in her lifetime.

Eliminating gender inequality should be a fundamental goal of any civilized society. But many observers also see this as key to winning

Table 7.4	Gender Inequality		
Top Five Rank	Country	Bottom Five Rank	Country
1	Sweden	54	South Korea
2	Norway	55	Jordan
3	Iceland	56	Pakistan
4	Denmark	57	Turkey
5	Finland	58	Egypt

NOTE: This survey looked at 58 countries only. Different data collection and analysis result in different rankings between this table and Table 7.3.

SOURCE: World Economic Forum, Women's Empowerment: Measuring the Global Gender Gap (Geneva). At: http://www.weforum.org/pdf/Global_Competitiveness_Reports/Reports/gender_gap.pdf, pp. 8-9 .

the war on global poverty. This point was central to discussions in Chapter 4 concerning population growth, where the empowerment of women, especially improved education, was identified as critical for reducing fertility.

SEXUALITY

Sexuality—particularly sexualities other than heterosexuality—is studied as an expression of identity and as one way in which the dominant heterosexual landscape can be challenged. Work on diverse sexual identities and related landscapes has been conducted mostly within the larger framework of feminist geography, but researchers also employ **queer theory**, a rather controversial term that refers to a concern with all people who are seen as and/or who have been made to feel different, and who share a common identity on the fringes. Queer theory also emphasizes the fluidity and even hybridity of identities, and is concerned with empowering those who lack power. Terminology is important in discussions of sexuality, as certain terms tend to be loaded with emotive meanings. The terms 'gays' and 'lesbians' are commonly preferred by male and female homosexuals.

Consensus judgements about particular sexual behaviours usually reflect power relations, with most feminists arguing that the critical power relation in this context is patriarchy, which is in turn based on heterosexuality. It is now generally accepted that sexual behaviour is not simply the product of some instinctual drive designed to ensure the continuity of the species; rather, sexual activity is a multi-functional behaviour that can be fully understood only in a larger cultural context. This is why a particular sexual behaviour may be unacceptable in one cultural context and yet may be acceptable in another, and why cultural attitudes towards sexual orientation change through time. By the early twentieth century, the dominant view of homosexuality in the Western world was negative, a view reinforced by both the legal and medical establishments.

Today, attitudes are changing; for example, relationships involving same-sex partners are being legally recognized as legitimate and socially acceptable lifestyle choices, although many contrary views are evident, especially where religious fundamentalists are an influential presence. Countries in Northern Europe were the first to recognize same-sex unions. Nevertheless, in many parts of the world, lesbians

Table 7.5	Women in Political Life
Country	Seats in Parliament Held by Women (% of total)
Rwanda	55.0
Sweden	47.3
Finland	42.0
Norway	37.9
Denmark	36.9

SOURCE: United Nations Development Program, *Human Development Report, 2008: Fighting Climate Change: Human Solidarity in a Changing World* International Cooperation at a Crossroads: Aid, Trade, and Security in an Unequal World (New York: UNDP, Macmillan, 2005). By permission of Oxford University Press, Inc.

and gays continue to experience homophobia, homosexual acts remain criminalized, prejudice and discrimination are routine in many workplaces, and people in same-sex relationships have reduced rights to pensions and inheritance. Some of the least tolerant places in the world include Nigeria, Saudi Arabia, and Iran, where homosexuality can be punishable by death, and Jamaica, where male homosexuality carries a punishment of 10 years' hard labour. Even in relatively liberal countries, it is estimated that perhaps 50 per cent of gay men and lesbians do not identify as such in their workplace for fear of recrimination.

Sexuality in the Landscape

One of the ways in which sexuality is expressed in landscape, especially in large cities in the Western world, is through the gradual identification of a part of the city as being a residential and commercial area where lesbians and gays

sexuality
In some feminist and psychoanalytic theory, interpreted as a cultural construct rather than as a biological given. Aligned with power and control.

queer theory
Ideas developed in gay and lesbian studies and concerned with oppressed sexualities in terms of both social rights and cultural politics.

Women and men wearing traditional dress in Riyadh Saudi Arabia.
Jeremy Horner/Alamy/GetStock

Gays and lesbians parade in downtown Montreal at the close of week-long festivities. Public celebrations of sexual identity are now regular events in many large urban centres, where they are becoming major tourist attractions.

CP photo/La Presse/Alain Roberge

dominate. Early examples include Greenwich Village in New York, the Castro district in San Francisco, West Hollywood in Los Angeles, and Soho in London, England. Another expression of sexuality in landscape is a gay pride parade, march, or demonstration. These are now routine in many cities and are often major celebratory occasions attracting large tourist audiences. But they can be unsettling, and controversy often surrounds these displays of identity.

For example, as noted in Chapter 6, the 2005 parade in Jerusalem was opposed by some Jewish, Christian, and Muslim religious leaders. In Auckland, New Zealand, the first two gay pride parades took place outside of the area understood to be gay, and hence were seen as challenges to the larger heterosexual community—as expressions of resistance in landscape. The third parade, in 1996, took place within an area widely perceived to be gay. The fact that the 1996 parade was less controversial than the first two suggests that the identical behaviour may be interpreted by the majority differently, depending on whether it represents a landscape challenge to the heterosexual majority (see Johnston, 1997).

CONTESTING LANDSCAPES

Two important general ideas emerge from these discussions of difference that have focused on ethnicity, gender, and sexuality.

First, dominant groups have constructed landscapes that take certain characteristics—the characteristics of the dominant group—for granted. Thus, urban areas in the Western world have been constructed assuming heterosexual nuclear families, women dependent on men, and able-bodiedness. Because of these assumptions, those individuals and groups who do not conform to these societal expectations are perceived as different—and are often excluded or disadvantaged as members of society. When such people intrude into landscapes that were not constructed for them, the result is often controversial, both culturally and spatially. Indeed, as evident in the account of sexuality, a distinctive feature of the contemporary world is the unsettling effect that the expression of other identities and the crossing of spatial boundaries have on dominant groups.

The second general idea concerns the way some groups of people understand themselves in relation to others. In the case of apartheid, the dominant Afrikaners imposed their views of themselves and the indigenous black population on the latter. Like all imperialists, they found their justification in a logic that in the context of Asia and the Middle East has come to be known as Orientalism (see Box 7.1). According to this logic, which implies a relational concept of culture, the indigenous people of areas that Europeans wanted to colonize were 'others'—not only different from but less than the Europeans, who as 'superior beings' had a natural right to use them and their lands as they chose.

The same logic has been extended beyond the colonial context, as conventionally understood, to operate in more locally defined situations. In many cases, those colonized have responded by creating landscapes of resistance.

It is not surprising that places often become sites of conflict between different groups of people, as in the example of the apartheid landscape or gay pride marches. Throughout history, groups have endowed parts of the earth's surface with meaning: that is, they have created places. But the understanding of place held by one group may be different from that held by another group; hence the identity and ownership of places may be contested. Many examples of these landscapes of resistance are associated with new social movements—in support of the environment, social justice, ethnic separatism, and so on—and they are best interpreted as expressions of opposition to power. Other subcultures, especially those associated with youthful populations, express opposition to mainstream lifestyles and mainstream places. Such groups may choose to display their difference through the adoption of alternative lifestyles, including musical and clothing preferences, and through their attachment to places such as inner-city neighbourhoods.

Ethnicity, gender, and sexuality are not the only grounds for exclusion from the landscape created by the dominant group. The homeless, the unemployed, the disabled, and the elderly all find themselves in some way incapable of fitting into the dominant landscape. Indeed, a 2005 British survey concluded that ageism was a bigger problem than racism, sexism, or ableism. Some disadvantaged groups are more visible than others, and some are more controversial; but all are in some way oppressed because the landscapes created by dominant groups presume a set of identity characteristics that they do not possess. For example, dominant groups 'presume able-bodiedness, and by so doing, construct persons with disabilities as marginalized, oppressed, and largely invisible "others"' (Chouinard, 1997: 380).

Geographies of Well-being

The term **well-being** is used to refer to the overall condition of a group of people and to specific components of well-being—economic, social, psychological, and physical. Landscapes at various spatial and social scales differ all too widely in terms of the well-being of their occupants. On the global scale there are areas of feast and areas of famine; some children are born into poverty and others into affluence. On the regional scale, many landscapes

in areas of European overseas expansion have continued to evolve in response to the needs of the former imperial powers and the capitalist imperative to seek profit over equity. Needless to say, the results have been to the detriment of most indigenous inhabitants.

On a more local scale the landscapes of disadvantaged groups also reflect the fact that there are dominant and subordinate groups in any culture. Human geographers are paying increased attention to these landscapes, at a variety of scales, and to the social and power relations that lie behind them.

Much of the work done in these areas is labelled **welfare geography**; this is a general approach to a wide range of issues, but with an emphasis on questions of social justice and equality. Concern with the geography of well-being and welfare geography has encouraged research into such important issues as the geography of education (focusing on the often inequitable spatial distribution of services and facilities) and the geography of justice (focusing on the variations in the spatial availability of social benefits).

MEASURING WELL-BEING

In the 1960s, concern about social problems in the United States prompted the development of spatial social indicators as measures of well-being. In a pioneering study, Smith (1973) identified seven sets of indicators representing seven different components of well-being:

1. income, wealth, and employment;
2. the living environment, including housing;
3. physical and mental health;
4. education;
5. social order;
6. social belonging;
7. recreation and leisure.

Examining the extent to which quality of life differs for different groups in different places, Smith found extreme inequalities at all spatial levels in the US.

But do all people have the same needs and the same ideas of what is beneficial and what is harmful? One attempt to conceptualize along these lines argues that all of us share a basic goal, to avoid harm (Doyal and Gough, 1991). Achieving this basic goal requires satisfaction of two needs: physical health and the ability to make informed decisions concerning personal

welfare geography
An approach to human geography that maps and explains social and spatial variations.

well-being
The degree to which the needs and wants of a society are satisfied.

behaviour. These two in turn require that the following conditions be satisfied:

1. adequate supply of food and water;
2. availability of protective housing;
3. safe workplace;
4. safe physical environment;
5. necessary health care;
6. security while young;
7. relationships with others;
8. physical security;
9. economic security;
10. safe birth control and child-bearing;
11. required education.

It is, of course, one thing to list such needs and another to measure them effectively in such a way that the integrity of different groups of people is not lost. As we have seen, the United Nations has led the way in the measurement of well-being, variously called quality of life or human development. An important issue in this work is the existence of spatial inequalities, something that most of us take for granted. Chapter 5 highlighted some of the glaring differences between more and less developed world places and lives, but there are clear links also between where we live and our overall well-being at the national scale. An atlas of British identity (Thomas and Dorling, 2007) shows that where an individual lives can be a strong predictor of identity, including class, health, life expectancy, and family structure. In short, place provides advantages or disadvantages, life opportunities or barriers.

Indeed, our capitalist society and economy are predicated on inequalities. Capitalism allows a few to be very wealthy, many to be comfortable, and large numbers to live in poverty. This does not need to be the case. It is possible to strive for greater spatial equality so that at least basic human needs are met for all people in all places. Two important areas where spatial differences are clearly evident are crime and health and health care.

CRIME

Human geographers have addressed the study of criminal behaviour and activities from a range of philosophical perspectives, but with the central concern of relating crime to relevant spatial and social contexts (Evans and Herbert, 1989). Mapping the incidence of crime has helped identify places where criminal activity is most prevalent, such as inner cities and other areas of poverty, low-quality housing, high population mobility, and social heterogeneity. These essentially spatial, and often empiricist or positivistic, studies are accompanied by studies of the social environments of criminals, the victims of crime, and the geography of fear, all of which are more clearly Marxist or humanist in focus.

Two topics of particular concern at present are the relationship between areas of criminal activity and urban decline, and how our use of space is affected by our concerns about the likelihood of a criminal act (Box 7.10; also see Box 13.4 for discussion of crime in the city).

HEALTH AND HEALTH CARE

Susceptibility to disease, morbidity (illness), mortality, and health-care provision are closely related to questions of well-being and the quality of life, and all have been analyzed by human geographers (see Chapters 4 and 5). Regardless of spatial scale, there has been a tendency to focus research on the less privileged members of society to demonstrate how health issues relate to a wide range of social and economic conditions. Clearly, health problems are related to the physical environment, to social factors such as housing and living conditions, and to access to health services. Several studies have emphasized inequalities in availability of services (see Jones and Moon, 1987).

Canada's Native peoples offer a compelling example, as their health status and their access to care are much poorer than those of other Canadians. This situation is one aspect of a way of life characterized by poverty and low self-esteem, which result at least partly from a long period of oppression and, more specifically, from a residential school system that removed children from their parents and imposed an alien way of life. Today, many Native communities have high rates of physical and sexual abuse and of alcohol and substance abuse. As we saw in Box I.2, Native Canadians have higher rates of infant mortality and a lower life expectancy than do other Canadians. Major causes of death include suicide and tuberculosis (a disease related to poverty), while a common chronic condition is diabetes mellitus (indicative of poor diet).

THE GEOGRAPHY OF HAPPINESS

An interesting recent focus is on the idea of happiness. Notoriously difficult to measure, happiness is clearly related to, but different

Box 7.10   The Geography of Fear

Geographers have made considerable advances in the analysis of social and spatial variations in the distribution of fear. Compelling evidence suggests that the fear of crime is a major problem affecting our perceptions of certain areas and our behaviours in them (Pain, 1992; Smith, 1987; Evans, 2001). The problem is especially acute for women, the elderly, and other groups within society who are relatively powerless and vulnerable.

Fear of domestic violence, sexual harassment, and rape can limit women's access to and control of space within and outside the home and also imposes limitations on social and economic activities. A study of the geography of rape and fear in Christchurch, New Zealand, by Pawson and Banks (1993), demonstrates the extent of the problem. Christchurch has a population of 292,000, of whom 91 per cent are of European origin. Pawson and Banks had two goals: first, to determine the spatial and social incidence of rapes that were publicly reported in the press; second, to map the spatial and social distribution of the fear of crime. The results of their research showed a spatial distribution that emphasized the inner-city area and a general correlation with patterns of other criminal activity (such as burglary), areas of youthful population,

and rental homes. They demonstrated that younger women under 25 years are especially at risk from rapists; that about 50 per cent of the rapes occurred in the victim's home, not a public place; that most occurred at night during or immediately after the hours of social contact and at a time of minimal community surveillance; and that there was a marked pattern of seasonality, with fewer rapes in the winter months.

To map the spatial and social distribution of the fear of crime, a survey of about 400 persons was conducted. The results showed that although fear was widespread, there was a distinct spatial pattern, with the low-income, least stable, and high-rental areas reporting the greatest levels of fear. Socially, women and the elderly displayed the greatest concern, and fear was much greater at night than during the day. The evidence clearly showed that women and the elderly are often reluctant to use either their neighbourhoods or the city centre at night.

The conclusions of this research are generally in accord with those of several other studies. As this book has already emphasized in other contexts, we live in an unequal world. For women, the city at night may well be a landscape dominated by males.

---

from, the idea of well-being. Essentially, it is how we perceive ourselves, and thus cannot be measured objectively. It is not unusual in some countries for surveys to be conducted that ask us, to put it simply, whether we are very happy, just happy, or not so happy. There is even a regularly updated World Database of Happiness housed at Erasmus University, Rotterdam. Those who conduct these surveys and those who then interpret and publish their thoughts on happiness are at the forefront of what *The Economist* (2006: 13) humorously described as the 'upstart science of happiness', blending economics, psychology, and geography. Two general points that emerge from this work are, not surprisingly, that the rich report greater happiness than do the poor, and, perhaps surprisingly, that people in affluent countries have not become happier as they have become richer. The latter finding might be because: (1) for many people, happiness is having things that others do not have, so as others become wealthier then the already wealthy become less happy; (2) when people achieve a better standard of living they are unable to appreciate its pleasures.

Of particular interest to human geographers is research that focuses on the spatial distribution of happiness. The first world map of happiness at the country scale, published in 2006, showed Denmark first, followed by Switzerland, Austria, Iceland, Bahamas, Finland, Sweden, Bhutan, Brunei, and Canada. Least happy of the 178 countries included were DR of Congo, Zimbabwe, and Burundi (Figure 7.5). There is a close relationship between happiness and measures of health, prosperity, and education. These results are much as expected, and much of the interest in this work is on the detailed differences between countries, on how happiness might be more scientifically measured, and on whether or not individual responses reflect reality. With this latter point in mind, a geography of happiness project has been set up in Britain involving geography teachers travelling through Europe to determine how happy people are and whether these informal surveys accord with the published data.

ELITIST LANDSCAPES

Much of the discussion in this chapter, and elsewhere in this text, focuses on the different types of landscapes occupied by different

FIGURE 7.5   World distribution of happiness

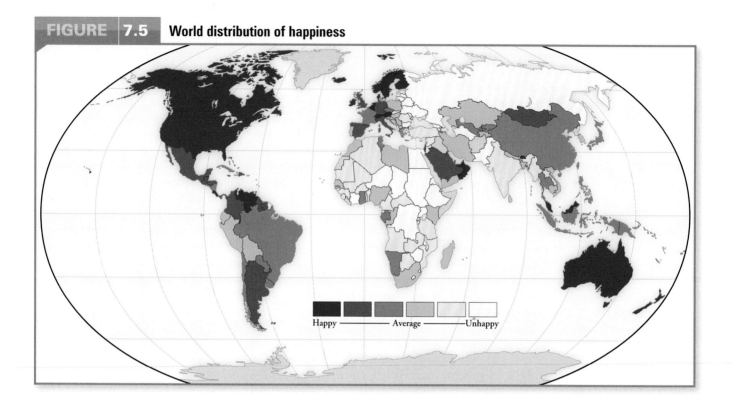

types of people. As we have seen, there are landscapes for the privileged and the less privileged, the advantaged and the disadvantaged; and they generally are labelled elitist landscapes and landscapes of stigma, respectively. There are obvious links between the theoretical questions of class and gender discussed above and these contrasting landscape types.

The existence of elitist landscapes is obvious, although they have seldom been analyzed. Geographers studying cities have traditionally recognized that all cities include a range of areas associated with different social classes; it is not unusual for the higher-class areas to be at a higher elevation than other areas, to be close to a lake, river, or ocean, or to be located so that they are not affected by the pollution generated by industry. In British cities, it is usual for higher-status areas to be spatially segregated and to include such features as golf courses and specialty retail areas. In Melbourne, Australia, an elite residential area has been a consistent feature of the urban landscape, one that moved steadily south as the city grew. In an important sense, identity and landscape are closely interwoven in these elite areas. A 2008 survey listed the richest streets in the world as measured

by property value. Ranked first was Avenue Princess Grace in Monaco, with apartments costing about US$50 million. Among the other rich streets are Severn Road in Hong Kong, Fifth Avenue in New York, Kensington Palace Gardens in London, and Avenue Montaigne in Paris.

Elite regions can also be identified. In Britain a long-standing distinction between south and north clearly reflects a distinction between privilege and lack of privilege. More obvious examples of privileged landscapes are tourist areas, especially in less developed countries (discussed below).

LANDSCAPES OF STIGMA

In recent years, human geographers have identified pariah landscapes, landscapes of despair, and landscapes of fear. Examples of pariah landscapes include many ghetto areas in cities and Aboriginal reserve lands in Canada and the United States. The experiences of discharged mental patients prompted the introduction of the term 'landscapes of despair', a vivid descriptor that applies equally well to pariah landscapes. One of the most publicized landscapes of despair in recent years is the inner-city landscape of

The planned community of Seaside, Florida, is an elitist landscape. An early example of the movement known as the New Urbanism, or neo-traditionalism, it was established in the early 1980s in a conscious effort to create a modern version of the idealized old-fashioned American town.

© Steven Brooke Studios

despair of homeless people. All such landscapes are expressions of exclusion, spatial reflections of social injustice (Box 7.11).

As already noted, violence is closely identified with specific locations. Many people must cope on a regular basis with landscapes of fear; elderly people and women are especially susceptible to violence, usually by males. Most people have mental maps identifying some areas as safe, others as unsafe—though certain 'safe' areas may actually be safe only for groups and/or during daylight hours. Large, open spaces, such as parks, are often perceived as unsafe, as are closed areas with limited exits, such as trains. The recent emergence of 'Neighbourhood Watch' schemes is a clear reflection of increased fear and a corresponding social interaction in many city areas. Landscapes of fear result in restricted use of public space.

Folk Culture and Popular Culture

The distinction between 'folk' and 'popular' culture adds another dimension to this account of the social and symbolic aspects of landscape that is also relevant to our earlier discussion of the concepts of place and placelessness.

The relative homogeneity of people in traditional folk cultures today survives mainly in rural areas among groups linked by a distinctive ethnic background and/or religion and language. Folk cultures tend to resist change and to remain attached to long-standing attitudes and behaviours. By contrast, popular culture is largely urban and tends to embrace change— even to the point of seeking change for its own sake, as in the case of fashion. Geographic studies of both folk and popular culture are plentiful

Box 7.11   Consigned to the Shadows

The title 'Consigned to the Shadows'—borrowed from an article in a geographical magazine (Evans, 1989)—could refer to many disadvantaged groups. In fact, it refers to people who are mentally ill. Those of us living in larger urban centres are probably accustomed to the occasional horrific news story about deplorable conditions in some home for the mentally ill. However, you may not have heard about the Greek island of Léros, variously described as a concentration camp, an island of outcasts, and a colony of psychopaths. Established in 1957, this institution off the Greek coast was home to some 1,300 mentally ill people, who were provided with little food, minimal medical care, and inefficient sanitation; when the odour became too unpleasant, they were hosed down. People with various disorders were grouped together without proper care, let alone hope of recovery. The European Union began a project to close the institution in 1990 but it took several years before the last inmates were deinstitutionalized.

Conditions on Léros may appall us today, but the island is not an extreme case historically. Mentally ill people have typically been separated from the majority with no real attempt to cure them. Indeed, until very recently, there has been no real understanding of mental illness. In the United States, until 1973, homosexuality was regarded as a mental disorder; in Japan, private mental hospitals now house about 350,000 people (one of the highest per capita rates in the world), at least partly because it is relatively easy to have a person so detained.

The problem of properly housing the mentally ill is increasing. The World Health Organization (WHO) predicts a dramatic rise in the numbers of people with serious mental disorders in the less developed world. This is not surprising, given that such illnesses can be caused by a wide range of factors, including medical and dietary factors. A recent WHO estimate places the number of mentally ill people worldwide at 100 million.

Homeless man collects bottles in Vancouver. Scenes such as this are increasingly common as many cities struggle to cope with growing numbers of people living on the street.

Ryan Koopmans/Alamy

in North America and Europe. At present folk studies still tend to be closely associated with the traditional landscape school of Sauer; however, the more theoretically informed methods outlined and employed in this chapter are likely to become increasingly prominent. Studies of popular culture are already being conducted in the traditional landscape school and in the newer arena of cultural analyses enriched by social theory. Although the boundary between 'folk' and 'popular' is far from clear-cut, it is possible to identify some basic characteristics of each.

FOLK CULTURE

In general, folk cultures are more traditional, less subject to change, and, in principle, more homogeneous than popular culture. (This third identifying characteristic is of limited value, however, since popular culture landscapes also can be homogeneous.) Folk cultures have pre-industrial origins and bear relatively little relation to class. Religion and ethnicity are likely to be key unifying variables, and traditional family and other social traditions are paramount. Folk cultures tend to be characterized by a rural setting and a strong sense of place.

This is not to say that elements of folk culture cannot be found in urban as well as rural areas. In large multicultural cities, attitudes and behaviours specific to many different ethnic groups not only survive but flourish, and in some cases have even become features of the broader popular culture—prime examples include the Caribbean festivals held every year in Toronto and London, England, and the dragon boat races mounted by the Chinese communities in various Canadian cities.

Domestic architecture, fence styles, and barn styles vary from one folk culture to another. However, the landscapes of individual folk culture groups, such as the Amish or Mennonites, are uniform and largely unchanging—a reflection both of central control and of individuals' desire to conform to group norms. Diet, too, reflects a preference for long-established habits, as well as the rootedness in an agrarian or maritime economy and lifestyle. The same is true of musical preferences: traditional styles associated with a particular region (e.g., western swing in north Texas and Oklahoma) are strongly preferred over the products of North American/global popular culture. Indeed, folk traditions in musical landscapes and cultural regions have been studied extensively.

POPULAR CULTURE

Trends in popular culture, including new attitudes and behaviours, tend to diffuse rapidly, especially in more developed areas where people have the time, income, and inclination to take part. Today, the diffusion of popular culture is an important part of the cultural globalization introduced in Chapter 6.

Jackson (1989) describes the rise of popular culture in nineteenth-century Britain, showing how the social activities of the working-class populations were constrained by more powerful classes. New places of entertainment (such as music halls) became major contexts of class struggle, as did activities such as prostitution. Many issues today similarly involve the control of space, and Jackson (1989: 101) argues that the 'domain of popular culture is a key area in which subordinate groups can contest their domination.' One of the better-known instances of social inequality that has expression in landscape relates to matters of sexuality. For example, in most large cities gay/lesbian neighbourhoods serve as powerful examples of the spatial constitution of culture.

Other geographic analyses of popular culture focus on specific landscapes (shopping malls, tourism, sports, gardens, urban commercial strips) and regions (musical regions, recreational regions). Shopping malls—artificial landscapes of consumption, located in most urban areas in the more developed world—provide a vivid illustration of the impact of popular culture on landscape (Box 7.12). Usually enclosed, windowless, and climate-controlled, catering to the commerce-driven consumer tastes of the moment, malls are characterized above all by their sameness. In fact, they exemplify the placelessness that, as we saw in Chapter 2, is one of the by-products of globalization (discussed in detail in Chapter 9). When completed in 1986, the West Edmonton Mall in Edmonton, Alberta, was the largest in the world, with 836 stores, 110 restaurants, 20 movie theatres, a hotel, and a host of recreational features (see Jackson and Johnson, 1991). As products of popular culture, shopping malls encourage the further acceptance of popular culture. Tourist landscapes, discussed more fully below, are another expression of popular culture. The emergence of themed areas is especially significant, with one current example being the planned Biblical theme park to be built by an evangelical Christian consortium near the Sea of Galilee.

Music

As with folk culture, there is much interest in studying musical landscapes and regions in popular culture contexts, and also a focus on links between music and identity. As identities evolve, cities sprawl, and landscapes change, so music changes. Much recent and contemporary music is closely tied to specific places, as with the Detroit Motown sound of the 1960s and the Seattle grunge music of the 1990s.

According to a 2008 study of United Kingdom musical tastes that analyzed sales, regional charts, and live performances, there are striking differences in musical preference from place to place. Folk is particularly popular in Scotland, jazz in the southwest, heavy metal in the Nottingham area, world music in Bristol, handbag house in Northern Ireland, reggae in Birmingham, and Eurodisco on the south coast. Some of these preferences are linked to immigration patterns and ethnic identity, while others are more difficult to explain. Of course, some musical preferences are far more widespread, even global, reflecting the globalization processes that contribute to cultural homogeneity.

Sport

Sport is another important part of popular culture. In fact, major organized sports, such as baseball, hockey, and soccer, originated in the late nineteenth century in association with the social changes initiated by the Industrial Revolution. Regions can be delimited in terms of sporting preferences among spectators and participants alike. Rooney (1974) discovered clear regional variations in the United States: for example, prior to the 1970s American professional football players came primarily from a cluster of southern states (Texas, Louisiana, Mississippi, Alabama, and Georgia). Geographers often see links between such preferences and vernacular region (four of the five states noted above can be identified as the American Deep South).

For some people, supporting a particular sports team is an important component of their everyday identity. College and high school sports teams in the United States often receive huge support from local people and from those who have graduated from these institutions. In Europe and Latin America affiliation with a soccer team can be a key part of many people's lives, and they have a passionate interest in how well their team performs. Identity at the national scale comes to the fore when there are major international competitions, most notably the World Cup of Soccer.

locale
The setting or context for social interaction; a term introduced in structuration theory that has become popular in human geography as an alternative to 'place'.

Recreation and Tourism

Recreation, including time spent with family and friends, or in reading, or participating in or watching sports, is now a significant part of many people's lives. Recreational activities include tourism, an industry generated by the more developed world but that operates both in the more and in the less developed worlds.

The Red River dragon boat races at the Forks in Winnipeg, Manitoba, Canada.

Terrance Klassen/Alamy/GetStock

The tourism industry has experienced dramatic growth since about 1960; according to the World Travel and Tourism Corporation it is now the largest industry in the world and is growing at a phenomenal rate—23 per cent faster than the overall world economy. This group claims that travel and tourism account for about $5.9 trillion of economic activity, which is about 10 per cent of global GDP. Globally, the number of tourists is estimated at about 700 million annually, and World Travel and Tourism, somewhat optimistically perhaps, expects that number to increase to some 1.5 billion by 2020. Of course, tourist expenditures and tourist numbers are very difficult to determine precisely and these numbers need to be viewed with caution.

Regardless, the importance of this industry cannot be measured simply in numerical terms. First, 'tourism is a significant means by which modern people assess their world, defining their own sense of identity in the process' (Jakle, 1985: 11). Second, because it is 'one of the most penetrating, pervasive and visible activities of consumptive capitalism, world tourism both reflects and accentuates economic disparities, and is marked by fundamental imbalances in power' (Robinson, 1999: 25). In other words, tourism is a means by which both tourist and host communities create their respective identities and emphasize their difference from one another. Because tourism offers significant opportunities for cultural contact, it creates opportunities for both cultural understanding and cultural conflict. In short, tourism is one of the means by which human geographies are created and recreated.

THE RISE OF TOURISM

In seventeenth-century Europe, the elite visited spas for medical purposes; later, the supposed health benefits of seaside resorts attracted a broader clientele. The idea that travelling might be for pleasure and also for broadening the traveller's horizons first emerged in Europe during the eighteenth century, especially as European nations developed their overseas empires and travellers and colonial officials published accounts of exotic **locales** (local social systems) and peoples.

Mass tourist travel, however, depended on the introduction of the annual vacation (as opposed to holidays for religious observance) negotiated between employer and workforce—and that was a product of the

Industrial Revolution. By the late nineteenth century the vacation proper was established and seaside resorts in Europe, especially Britain, were becoming playgrounds for the working classes. Beginning about 1960, changes in employment patterns that allow for more leisure time, additional discretionary income for many people, decreasing travel costs (especially air travel), and increasing numbers of retirees have helped to diversify tourism and make it a year-round industry. By the 1990s, the Internet allowed people to arrange all aspects of the vacation experience without using a professional travel agent, thus reducing overall costs.

The most recent trend in the tourist industry is the involvement of what are often called emerging economies; the rise of these economies is noted in the Chapter 9 account of globalization. These include what are sometimes called the BRIC economies (Brazil, Russia, India, and China) and also some countries in Southeast Asia and several of the Persian Gulf states.

TOURIST ATTRACTIONS

There appear to be six principal attractions for contemporary tourists:

1. good weather (usually meaning warm and dry);
2. attractive scenery (coastal locations are especially favoured);
3. amenities for such activities as swimming, boating, and general amusement;
4. historical and cultural features (old buildings, symbolic sites, or birthplaces of important people);
5. accessibility (increasingly, tourists travel by air; in general, costs increase with the distance travelled);
6. accommodation.

Locations with an appropriate mix of these features tend to be major tourist areas. In the more developed world, such areas usually are urban or coastal. In Europe, many of the most popular coasts for tourism are peripheral (for example, in Spain and Greece), although this is not the case in countries such as the United States and Australia.

Also, some landscape features have assumed prominence as tourist attractions because of their identification as World Heritage sites as defined by UNESCO. There are currently

English football fans in Trafalgar Square watch the quarterfinals of the World Cup.
Juliet Butler/Alamy

almost 900 such sites, and this list is regularly reviewed. One recent addition, in 2007, was the Rideau Canal.

MASS TOURISM

Understood as part of a larger Fordist economy, mass tourism is a form of mass consumption: it involves the purchase of commodities produced under conditions of mass production; the industry is dominated by a few producers; new attractions are regularly developed; and many supposedly different sites are essentially similar. Together, these characteristics make the mass tourist experience much the same everywhere, regardless of specific site. Further, as with other cases of mass production and consumption, the producers of tourist sites effectively determine the consumers' options, having the power to direct what is sometimes called the 'tourist gaze'.

While commercial producers create such attractions as coastal resorts, governments, too, can direct the tourist gaze by creating parks and other favoured locations, although the tourist element in such cases is sometimes quite small. For example, the 10 new national parks and park reserves that the Canadian government has created in the Canadian North since 1972 were intended primarily to preserve habitats and landscapes.

ALTERNATIVE TOURISM

The impact of tourism on local ways of life is contradictory. On the one hand, tourism

Table 7.6	Characteristic Tendencies: Conventional Mass Tourism vs Alternative Tourism	
Variable	Conventional Mass Tourism	Alternative Tourism
Accommodations		
Spatial pattern	high density/concentrated	low density/dispersed
Size	large-scale	small (local)-scale
Impact	obtrusive (modifies local landscape)	unobtrusive (blends into local landscape)
Ownership	non-local/big business	local/small business
Attractions		
Emphasis	commercialized cultural/natural	preserved cultural/natural
Character	generic, contrived	local, authentic
Orientation	tourists only	tourists and locals
Market		
Volume	high volume	low volume
Frequency	seasonal (cyclical)	year-round (balanced)
Segment	mass tourist	niche tourist
Origin	a few dominant markets	diverse origins
Economic impact		
Status	dominant sector	complementary sector
Linkages	non-local	local
Leakages	profit expatriation, high import level	low import level
Multiplier	low multiplier	high multiplier
Regulation		
Control	mainly non-local	mainly local
Amount	minimal	extensive
Basis	self-regulating free-market forces	public-sector emphasis
Priority	develop, then plan	plan, then develop
Motives (holistic/integrated)	economic (sectoral)	economic, social, environmental
Time frame	short-term	long-term
Development and tourist ceilings	no ceilings	imposed ceilings that recognize carrying capacities

SOURCE: D.B. Weaver, 'Ecotourism in the Small Island Caribbean', *Geojournal* 31 (1993): 458. Copyright © 1993 Springer. Reprinted with kind permission of Springer Science and Business Media.

ecotourism
Tourism that is environmentally friendly and allows participants to experience a distinctive ecosystem.

in places such as Phuket, Thailand, because protective mangrove forests and coral reefs had been cleared away to create playgrounds for visiting tourists and shrimp farms to feed European demand. Increasing awareness of the problems associated with conventional mass tourism has stimulated development of various 'alternative' types of tourism centred on unspoiled environments and the needs of local people. Table 7.6 lists some of the advantages of such forms.

Recent years have seen a striking upsurge of interest in 'environmentally aware' or 'sustainable' tourism, sometimes called **ecotourism**. Belize, in Central America, is one of the best-known ecotourist destinations and has hosted two major conferences on the topic. Tourism accounts for 26 per cent of Belize's gross national product and is becoming more important as prices for traditional cash crops, especially sugar cane, drop. Awareness of the fragility of coastal ecosystems led the government of Belize to encourage ecotourism; however, several observers have seen little difference between the developments designed for ecotourists in Belize and other, more traditional developments (Wheat, 1994: 18). In fact, specialized alternatives like ecotourism are not likely ever to attract the market that mass tourism does, although they may be able to make some contribution to local economies.

CREATING PLACES AND PEOPLES FOR TOURISTS

Culturally, the tourist industry can be seen as one more example of the dominance of the more developed world over the less developed. It can be argued that the activities of tourists both reflect and reinforce, perhaps even legitimize, existing patterns of inequality and the dominance of some groups over others. Increasingly, one reason for people to visit 'exotic' peoples and places is to experience cultural difference. Yet—not surprisingly—those peoples and places are rarely authentic; in many cases they have been constructed specifically to satisfy the tourist gaze as described above. Travel companies often play on the desire for novelty by using 'exotic' images and descriptions in their brochures. In fact, for many tourists, the places and peoples they visit are commodities, and—as with many other commodities—the advertising used to present them to prospective consumers relies more on fancy than on factual information.

encourages local crafts and ceremonies; on the other hand, it can destroy local cultures and dramatically change local environments. One commentator suggested that 'we are more prone to vilify or characterize mass tourism as a beast; a monstrosity which has few redeeming qualities for the destination region, their people and the natural resource base' (Fennell, 1999: 7). Indeed, the Asian tsunami of December 2004 had such a devastating impact

Human geographers are now analyzing images and representations such as those used in tourist brochures in order to unpack and understand the meanings they convey, and in many cases they are finding that such advertising stresses the power of the tourist in some dependent, often former colonial, area. Specifically, using the terminology earlier in this chapter, it is clear that 'tourist' places and peoples are being socially constructed—in terms of ethnicity, gender, food, and drink—to stress the difference between 'them' and 'us' and enhance the elements of mystery and exoticism that attract tourists. Figure 7.6 shows one way of understanding the relationship between the places produced by the tourist industry and the places consumed by the tourist, suggesting that the social construction of tourist places comprises two subsystems: place production by the tourist industry and place consumption by the tourists. Where these constructions are in agreement there is a zone of convergence.

As this discussion suggests, recreation and tourism are increasingly attracting the attention of human geographers interested in the new cultural geography. In particular, tourism is of interest as a process of consumption, a product of place construction, and a reflection of difference.

Falsifying place and time

Among the most distinctive tourist attractions today are **spectacles**, such as sporting events

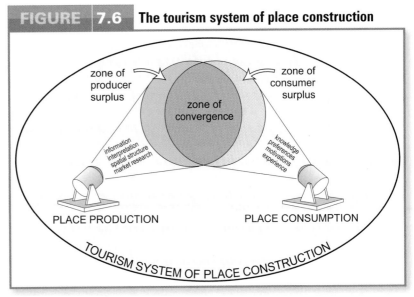

FIGURE 7.6 **The tourism system of place construction**

SOURCE: After M. Young, 'The Social Construction of Tourist Places', *Australian Geographer* 30 (1999): 386, Fig 3 (MW/CAGE/P1892) Used with permission of Taylor & Francis Ltd., http://www.informaworld.com.

or world fairs, theme parks, such as Disney World (Ley and Olds, 1988), and large shopping centres, such as the West Edmonton Mall (Shields, 1989; Hopkins, 1990; Jackson and Johnson, 1991). Many such manifestations of popular culture are specifically created by developers to appeal to tourists as consumers, and they lend themselves to a wide range of human geographic interpretations.

Attractions created where previously there was nothing to draw visitors are completely artificial. It is probably correct to say that what matters in this kind of venture is innovation,

spectacle
A term referring to places and events that are carefully constructed for the purposes of mass leisure and consumption.

The Governor's Palace, Colonial Williamsburg.

Paul Hakimata/Alamy/GetStock

creating something that people want—sometimes even creating a demand for something that no one had thought of wanting before. Two important attractions of artificial sites are that they offer a controlled and safe environment while purporting to be exotic, providing a glimpse into other cultures and places.

In many parts of North America, tourist attractions such as gambling facilities capitalize on a previously unsatisfied demand. In many respects, Las Vegas is the pioneer in this area, and it regularly reinvents itself to continue as a prime tourist destination. Most of the early hotels in Las Vegas offered little more than basic accommodation for gamblers and relatively tame nightclub performances, but in 1966 the first of the large theme hotels, Caesars Palace, opened, setting the pattern for subsequent development. Today the city attracts many families and business conferences in addition to gamblers.

One of the more interesting aspects of these essentially artificial tourist sites is that the experiences they offer are inauthentic—more about myths and fantasies than about reality. Most observers see this form of tourism as bound up with postmodernism, the current emphasis on consumption rather than production, and the commodification of people, place, and time. Perhaps nowhere is this seen more clearly than in Branson, Missouri, on the edge of the Ozarks, which has been transformed from a quaint town with one significant tourist attraction—Silver Dollar City—into a country music (and other performances) destination for millions of visitors that offers dozens of single-purpose theatres owned and/or operated by the formerly famous or near famous, from Boxcar Willie and Glen Campbell to Shoji Tabuchi, Yakov Smirnoff, and the ubiquitous Elvis Presley impersonators.

Even many supposedly 'real' tourist sites may have more to do with myth than reality. Heritage and other historical sites, for example, may be so thoroughly reconstructed and packaged that it is difficult for tourist consumers to know how true to the past their experience is. Colonial Williamsburg in Virginia sometimes is criticized on these grounds, as are many battlefields and heritage sites, at least partly because the meanings attached to the sites themselves are necessarily plural and contested. For some critics, our understanding of ourselves, our places, and our pasts increasingly is formed by the tourist industry.

TOURISM IN THE LESS DEVELOPED WORLD

Many of the favoured tourist destinations today are in the less developed world—Mexico, various Caribbean and Pacific islands, and some Asian and African countries. Some of these places do offer an appropriate mix of the six principal attractions noted above, but they may also be attractive for other reasons. For example, low labour costs usually mean that these areas can offer competitive rates, and some may represent—at least for Europeans or North Americans—an 'exotic' cultural experience. Certainly tourism is growing in the less developed world, especially in coastal areas, in response to increasing demand from the more developed world.

For most of the world's least developed countries, tourism generates the largest percentage of foreign exchange. Kenya has an especially well-developed tourist industry, and other African countries are seeking to follow a similar path. Hoping to capitalize on a long Atlantic coastline, lush forests, waterfalls, and wildlife, Angola has begun construction of several luxury hotels only a few years after the end of a civil war that left much of the country devastated.

This trend poses several problems for the less developed countries, however. Most important from the economic perspective is that dependence on tourism makes a country vulnerable to the changing strategies of the tour companies in more developed countries. Many destinations offer sun, sand, and surf, and those that have no distinctive attractions must keep their rates low if they are to draw clients away from their competitors. Even so, most less developed countries appreciate the benefits that tourism can provide: local employment, stimulation for local economies, foreign exchange, improvements in services such as communications. In short, in less developed countries the tourist areas are generally much better off than the larger economy. Today, tourism is a major growth industry in the tertiary sectors of several less developed countries, and current evidence suggests continued growth (Box 7.13).

Shangri-La

Not surprisingly, the tourist industry in China is becoming increasingly similar to that of countries in the more developed world. In particular, in addition to attracting tourists from other countries, domestic tourism is emerging.

Box 7.12   Tourism in Sri Lanka

Sri Lanka is representative of tourist destinations in the less developed world in that its tourist industry is problematic and uncertain. Long described as a paradise by Europeans (see Box 5.8), from a European perspective it is a logical area to be marketed as a tourist destination. There are scenic landscapes, including beaches, tropical lowlands, and mountains, as well as a rich and diverse cultural heritage. During the 1960s, the Sri Lankan government created the necessary infrastructure of hotels, roads, and airline facilities, and improvements and additions to this infrastructure have continued. The industry experienced spectacular growth, and by 1982 tourism was second only to tea as an earner of foreign exchange.

A period of decline between 1983 and 1989 reflected the ethnic conflict described in Box 5.8. Tourist numbers dropped by more than 50 per cent; northern and eastern Sri Lanka experienced the greatest losses in tourist numbers and, hence, the greatest economic damage. After 1989, the industry began to recover and attract visitors from elsewhere in Asia. By the mid-1990s, the industry seemed more secure because of strong government support and careful regulation (O'Hare and Barrett, 1993), but the hoped-for stability has not been achieved. Damage caused by the 2004 tsunami had an immediate negative impact with more than 100 hotels destroyed or badly damaged. More significant is ethnic conflict, meaning that the island is not seen as a desirable destination—indeed, by 2006, beach resorts catering to wealthy tourists from North America and Europe were functioning at only 20 per cent of capacity largely because of the ongoing conflict. Although the conflict came to a bloody end in 2009, ethnic tensions are not conducive to a successful tourism business. Another problem for many tourist businesses is a government regulation passed in 2005 that created a buffer zone within 100 metres of the coast—within this zone current businesses cannot repair damaged structures, nor can any new structures be built. However, this regulation was welcomed by environmentalists, and there is hope that an ecotourism industry focused on the shoreline environments might develop.

The Sinhalese ruler Parakramabahu (1153–86) built the city of Polonnaruwa, the site of this 14-m-long figure depicting the Buddha entering nirvana.

Euan White

Consider the example of northwestern Yunnan Province on the edge of the Tibetan Plateau in western China.

Designated a World Heritage site in 2003, here three of the great rivers of Asia—Yangtze, Mekong, and Salween—run parallel through dramatic gorges in a region that may be, according to UNESCO, the most biologically diverse temperate region in the world. Particularly interesting is that one small nearby town—Zhongdian—has been completely transformed to become a major tourist centre, attracting visitors especially from the cities of eastern China.

Capitalizing on the symbolism of the fictitious place in the novel by James Hilton, *Lost Horizon*, Zhongdian was renamed Shangri-La in 2001. The once-derelict buildings are now rediscovered as examples of traditional Tibetan architecture, and the town is a commercialized centre of Tibetan culture. Urban development proceeded rapidly, including hotel and airport construction and road improvements. Also nearby is Mount Kawagebo, one of the most sacred sites in Tibetan Buddhism. Many pilgrims walk around the mountain, a challenging trek that takes about two weeks. The combination of an outstanding nearby physical geography, a pilgrimage site, and the commodification of Tibetan cultural attractions encouraged 3 million tourists to visit Shangri-La in 2008.

Group Engineering in a Globalizing World?

Perhaps the real significance of culture is that it conditions our attitudes and behaviour. Do we then want to engineer ourselves? Is there value in attempts to change our cultural beliefs and practices? These difficult and sensitive questions provide a challenging conclusion to our discussions of culture.

In this chapter we have highlighted the value of perceiving landscapes or regions as places with social and symbolic content. We have pursued this idea using a wide variety of theoretical underpinnings and focusing on

how a number of important variables—such as gender and ethnicity—interact to form groups and related landscapes. A recurring feature has been the reality of inequality: unequal social groups living in unequal landscapes.

Groups in particular places have very different experiences; income levels vary, as do levels of health, education, and overall environmental quality. All human geographers are concerned about these issues, but the importance that any particular geographer attaches to them is clearly related to philosophical persuasion. The empiricist might emphasize the need to describe facts accurately; the positivist might focus on statistical precision and the development of theoretical explanations; the humanist might centre attention on the experiences of those living in such places; the Marxist might strive to explain issues, identify causes, and advocate change; the feminist might uncover the inequalities associated with the gendering of landscape; the postmodernist might focus on

World Waterpark in West Edmonton Mall. Built in three phases from 1981 to 1986, the mall is a prime tourist destination—a massive complex of department stores, shops, restaurants, recreation areas, amusements, and services that also includes a luxury hotel.

Courtesy West Edmonton Mall

Box 7.13 | Geography, Consumption, and Identity Formation

While I am watching television, wearing trainers, eating at a restaurant, shopping at the local supermarket, going out clubbing, or jetting off on holiday, I am not often consciously thinking about geography. Indeed, until recently, geographers largely ignored such activities, deeming them frivolous and peripheral, to be left to the disciplines of sociology or cultural studies (or left outside academia altogether). This neglect is now being redressed, however, and cultural geographers in particular are increasingly arguing that the taken-for-granted activities which fill our everyday lives should be seen as very important and therefore worthy of serious academic inquiry. (Jayne, 2006: 34)

Some human geographers occasionally express surprise, even concern, at the seemingly never-ending proliferation of subject matter that now includes advertisements, yard sales, clothing preferences, shopping malls, and much more. But, in an important sense, such concerns reflect a misunderstanding. Geographers, especially cultural geographers, have long studied the everyday lives of people and place, with many studies in the Sauerian tradition focusing on such seemingly 'esoteric' topics as covered bridges, agricultural fairs, log building construction, geophagy, and much more.

Indeed, it is not so much the choice of subject matter that is different, as evidenced by the fact that Sauerian and

new cultural geographers have shared interests in many aspects of consumption—food, drink, sports, and music to name a few—and in many aspects of identity formation as these relate to consumption. What is new is the philosophical underpinning of research. Whereas much earlier work on everyday lives and landscapes, including consumption habits and places of consumption, was essentially empiricist, current work is informed by a disparate body of social theory, much of which is postmodern in character. Consider, for example, the construction and reconstruction of shopping mall cultures. The best-known example of this phenomenon was the emergence of the much ridiculed Valley girls who identified themselves with the Glendale Galleria in California.

Most notably, current work typically acknowledges that the 'consumption' of places (for example, tourist heritage sites or resort settings), the 'consumption' of peoples (for example, indigenous or ethnic peoples), and the links between identity and consumption (as evidenced, for example, by musical, clothing, and food and drink preferences) are worthy of study because such consumption is constituted by economic, political, and cultural processes that vary from place to place. Most notably, for many researchers there is an interest in disentangling local processes, regional processes, and global processes. In short, human geographers today recognize that what we consume, when we consume, how we consume, and where we consume are inherently geographic and therefore merit study.

the writing of alternative geographies inspired by the experiences of previously excluded groups. Each of these approaches has merit, and we may hope that, together, they will help us to find solutions to the uneven distribution of well-being.

The previous chapter began with a series of powerful metaphors: 'mountains of hate, rivers of inflexible tradition, oceans of ignorance' (James, 1964: 2)—physical geographic terms used to describe aspects of our human condition. Language, religion, ethnicity, class, gender, and sexuality all have at some time in some place been used as justifications for war, cruelty, hypocrisy, or dogma. The goal of a single universal language remains elusive, although English approaches that need; the prospects for uniting diverse religious beliefs are negligible; ethnic groups continue to value traditions that in some cases may be inappropriate to modern societies; class distinctions may even be increasing; and the likelihood of eliminating the cultural implications of gender and sexuality

seems slight. All of these variables (including gender-divided cultures) are human constructions, and they are inevitable consequences of cultural evolution. Not only does each variable involve divisions within the human population, but divisive attitudes and behaviours are actively encouraged by many cultural groups; for religious groups especially, dogma is fact.

Human diversity, then, has a dangerously divisive aspect. We are at a point in human cultural evolution when we can give serious thought to social engineering: designing ourselves, and hence our future. It is therefore appropriate to ask whether such a project is desirable—to devise new, overarching values and moral standards that might bridge the divisions between us. These are heady issues. Today, the more developed world is actively questioning traditional value systems. Only 100 years ago, racial hierarchies were proposed by serious scholars and accepted by other serious scholars; even more recently, most people of European descent took the racial inferiority of others to be fact. Today, such theories are largely discredited and such attitudes are less common, though ethnic hatred still divides peoples within states.

A world without human diversity at both group and individual levels may be unimaginable.

Perhaps what we really need to achieve is a diversity that involves mutual respect between groups. Only then will our divided world be free of the 'mountains of hate, rivers of inflexible tradition, oceans of ignorance' that James spoke of.

CHAPTER 7 SUMMARY

SYMBOLIC INTERACTIONISM

Sees cultural groups as being created and maintained through communication; closely associated with humanistic concepts.

NEW CULTURAL GEOGRAPHY

A welcome addition to Sauerian cultural geography, based on a revised concept of culture, which introduces new approaches and much new subject matter, especially relating to issues of dominance and subordination. The new cultural geography often implies constructionist rather than essentialist interpretations of such topics as human identity.

TYPES OF SOCIETY

There are three types of society ('modes of production', in Marxist terminology): feudal, capitalist, and socialist. The feudal and capitalist types explicitly divide people into unequal groups or classes. Our contemporary socioeconomic world is dominated by capitalism.

CURRENT SOCIAL THEORY

There are several social theories to which human geographers can turn to facilitate their analyses of people and place. In addition to the humanistic and Marxist approaches introduced in Chapter 2, these include the approaches known as feminism and postmodernism. Human geographers are also actively using radical and socialist versions of feminist theory, and pursuing various aspects of postmodernism. Some of these additional conceptual inspirations reflect a 'cultural turn' in the social sciences and humanities.

PLACE

Today, landscapes are recognized as signifying systems that have meanings; we used the term 'place' in this sense. In addition, they are interpreted as the continually changing product of ongoing struggles between groups and between different attitudes and behaviours. The distinction between 'place' and 'placelessness' is particularly significant, as is that between the related ideas of local and global environments.

VERNACULAR REGIONS

Vernacular regions are perceived to exist by people living inside and/or outside them; they may be created institutionally and typically possess a strong sense of place. In North America, the map of vernacular regions closely resembles the map of regions delimited using the concept of first effective European settlement. A related area of interest is known as psychogeography. Some vernacular regions may qualify as homelands if they are especially closely identified with a distinctive cultural group.

GEOGRAPHIES OF DIFFERENCE

Geographers today understand difference through socially produced markers, such as ethnicity, gender, sexuality, able-bodiedness, and age (and also language and religion), and often focus on contested landscapes. Overall, these studies mark a dramatic break with the traditional focus initiated by Vidal and Sauer especially. In brief, we are seeing increasing interest in Marxism, humanism, feminism, and postmodernism; decreasing concern with the landscape per se; and growing acknowledgement of the role of human agency in both social relations and power relations. These newer concerns first emerged in the 1970s.

THE UNITY OF THE HUMAN RACE

Humans are members of one species—*Homo sapiens sapiens*. Early spatial separations of groups of humans facilitated the development of physical variations. These are of minimal relevance to our understanding of either people or places. The concept of race among humans has no basis in fact.

RACISM

Race may be an illusion, but racism is a fact.

ETHNICITY

Ethnic groups are often loosely defined; the key linking variable may be language, religion, or common ancestry. They are minorities and may be seen as set apart from the larger society, but over time many experience acculturation or assimilation. Canada is an officially multicultural society.

GENDER

Landscapes are gendered to reflect the dominance of patriarchal cultures: dominant men and subordinate women. Gender differences in well-being and the quality of life in general are especially evident at the aggregate scale using measures of human development.

SEXUALITY

Sexuality—particularly sexualities other than heterosexuality—is studied as an expression of identity and as one way in which the dominant heterosexual landscape can be challenged.

WELL-BEING

Human geographers are increasingly concerned with landscapes of resistance (constructed by those who are excluded from the landscapes constructed by and for dominant groups); topics such as crime and health as they reflect differences from place to place; and what might be called elitist landscapes and landscapes of stigma.

FOLK CULTURE

Folk cultures tend to retain long-standing attitudes and behaviours and thus are conservative of traditions passed on from one generation to the next both orally and by example. A wide range of landscape features and folk activities are typically studied in the manner of the Sauer landscape school.

POPULAR CULTURE

The term 'popular culture' is applied to groups whose attitudes and behaviours are constantly subject to change, in contrast to folk culture. Geographic studies of popular culture employ both traditional and more recent conceptual backgrounds and include analyses of landscapes and regions. Today, popular culture is usually discussed in the larger context of globalization processes.

THE TOURISM INDUSTRY

The largest industry in the world, tourism, is generated by the more developed world, but many favoured destinations are in the less developed world. Spectacles and theme parks are a distinctive form of tourist attraction. Many less developed countries take advantage of attractive climates and landscapes to cultivate a tourist industry, although their success can be affected by volatile political circumstances, as in the case of Sri Lanka. Tourist areas may benefit economically, but may also experience negative cultural and environmental consequences. Some countries, such as Belize, are encouraging ecotourism. Human geographers are increasingly interested in how the places and peoples consumed by tourists are represented.

FUTURE CULTURAL IDENTITY

Cultural variables are human constructions. They promote divisions between people and landscape that have often led to conflict between groups. Geographers now ask questions about our future cultural identity and the possibility of human engineering—deliberately creating new cultures.

QUESTIONS FOR CRITICAL THOUGHT

1. Outline the various theoretical perspectives on culture that underlie the 'new cultural geography'. How can/should we make sense of these competing theories? Is one theory better than another? Justify your position.

2. How can/should we balance theory and empiricism?

3. Do you think that there can be too much attention paid to theory and, by implication, not enough attention paid to real-world social issues and concerns.

4. What do geographers mean when they talk of 'landscape as place'? Why is this so important?

5. To what extent do vernacular regions exist in the minds of people who live in them? Are such regions merely constructs of geographers that bear little or no meaning to those who live in them?

6. Why are topics such as gender, race, and ethnicity discussed under the heading of 'geographies of difference'?

7. How can geographical perspectives on well-being be used to address issues of inequality at the local and global levels?

8. At the end of the chapter, Norton raises the issue of future cultural identities and the possibility of human engineering deliberately creating new cultures. What do you think about this? Is it a good idea? Justify your position.

FURTHER EXPLORATIONS

Amadeo, D., R.G. Golledge, and R.J. Stimson. 2009. *Person-Environment-Behavior Research*. New York: Guilford.

Overview of research, both conceptual and empirical, concerned with how space relates to people's everyday experiences and behaviour.

Atkins, P., and I. Bowler. 2001. *Food in Society: Economy, Culture, Geography*. New York: Oxford University Press.

Covers a wide range of topics, including food consumption, regional differences in food preferences, and links between food and gender.

Bale, J. 2003. *Sports Geography*, 2nd edn. New York: Routledge.

Material from a disparate literature; probably includes something to interest every geography student, regardless of philosophical persuasion or conceptual preference.

Blunt, A., et al., eds. 2003. *Cultural Geography in Practice*. New York: Oxford University Press.

Undergraduate text emphasizing how to do cultural geographic research; practical instruction in key methodologies and discussion of research that puts the methodology into practice.

Burton, R. 1995. *Travel Geography*, 2nd edn. London: Pitman.

A factually detailed overview of the world's geographical resource base for tourism and of the spatial patterns of world tourist activity.

Cartier, C., and A.A. Lew, eds. 2005. *Seductions of Place: Geographical Perspectives on Globalization and Touristed Landscapes*. New York: Routledge.

A collection of studies that employ recent innovations associated with the new cultural geography.

Gatrell, A.C. 2002. *Geographies of Health: An Introduction*. Malden, Mass.: Blackwell.

A comprehensive textbook covering basic concepts, philosophies and related methods of explanation, techniques, and a wide range of empirical discussions; includes discussion of health issues related to human impacts on environment.

Gleeson, B. 1999. *Geographies of Disability*. New York: Routledge.

An innovative book covering concepts and both historical and contemporary discussions.

Hamnett, C., ed. 1996. *Social Geography: A Reader*. New York: Arnold.

An edited collection of readings covering such topics as gender, class, race, and social justice.

Graves, J.L., Jr. 2004. *The Race Myth: Why We Pretend Race Exists in America*. New York: Penguin.

An evolutionary biologist demonstrates that racial distinctions are nothing more than social inventions.

Hall, R.E., ed. 2008. *Racism in the 21st Century: An Empirical Analysis of Skin Color*. New York: Springer.

Important text focusing on both the flawed concept of race and the reality of racism.

Hubbard, P. 2005. 'Places on the Margin: The Spatiality of Exclusion', in M. Philipps, ed., *Contested Worlds: An Introduction to Human Geography*. Burlington, Vt: Ashgate, 289–316.

Interesting overview of links between marginalized stereotyped groups and the places they occupy; focus is on urban examples.

Jenkins, M. 2009. 'Searching for Shangri-La', *National Geographic* 215, 5: 56–83.

Popular article, with excellent photos, about this region in western China.

Johnston, L., and R. Longhurst. 2008. 'Queer(ing) Geographies "Down Under": Some Notes on Sexuality and Space in Australasia', *Australian Geographer* 39: 247–57.

Insightful discussion of the importance of place in the production of queer geographies.

Johnston, R. 2006. 'The Politics of Changing Human Geography's Agenda: Textbooks and the Representation of Increasing Diversity', *Transactions of the Institute of British Geographers* 31: 286–303.

Thought-provoking article that highlights the role played by textbooks in introducing geography students to disciplinary diversity. Interesting in relation to any chapter in this text, but perhaps especially so here as there are so many new geographic ideas introduced.

Knight, D.B. 2006. *Landscapes in Music: Space, Place, and Time in the World's Great Music*. Lanham, Md: Rowman & Littlefield.

Pioneering work exploring links between orchestral music and landscape, a previously ignored field.

Krims, A. 2007. *Music and Urban Geography*. New York: Routledge.

A challenging text that applies Marxist theory to one aspect of urban cultural studies.

Minca, C. 2007. 'The Tourist Landscape Paradox', *Social and Cultural Geography* 8: 433–53.

A challenging article about how tourists conceptualize landscape; focuses on the Jamaa el Fna square in Marrakech, a UNESCO World Heritage site.

Mitchell, D. 2000. *Cultural Geography: A Critical Introduction*. Oxford: Blackwell.

A good overview of the new cultural geography with a thoughtful discussion of its origins.

Mitrašinovi, M. 2006. *Total Landscape: Theme Parks, Public Space*. Aldershot, UK: Ashgate.

Attractive academic book proposing that 'total landscapes' exist in the merging of material differences between natural and artificial landscapes; focuses on theme parks as places where consumer desires are created.

Moore, N., and Y. Whelan, eds. 2007. *Heritage, Memory, and the Politics of Identity: New Perspectives on the Cultural Landscape*. Aldershot, UK: Ashgate.

Addresses two key questions about links between landscape, cultural identity, and power: whose heritage is being remembered in landscape, and why their heritage?

Noble, A.G., ed. 1992. *To Build in a New Land: Ethnic Landscapes in North America*. Baltimore: Johns Hopkins University Press.

An excellent first book-length study of North American ethnic groups and their landscapes; largely traditional cultural geography in content, but with some humanistic overtones.

Norton, W. 2006. *Cultural Geography: Environments, Landscapes, Identities, Inequalities*. Toronto: Oxford University Press.

A textbook that incorporates and attempts to integrate the traditional landscape school of cultural geography and the new cultural geography.

Nostrand, R.L. 1992. *The Hispano Homeland*. Norman: University of Oklahoma Press.

A seminal contribution to homeland studies that identifies a distinctive group and their landscape through a reconstruction of their history.

Rosenau, P.M. 1992. *Post-Modernism and the Social Sciences: Insights, Inroads and Intrusions*. Princeton, NJ: Princeton University Press.

Clearly written account distinguishing between two principal versions of post-modernism that defines the key terms and issues.

Smith, D. 2008. *Penguin State of the World Atlas*, 8th edn. New York: Penguin.

An important source for many of the topics of interest to human geographers.

Stratford, E., ed. 1999. *Australian Cultural Geographies*. New York: Oxford University Press.

Writings on various relationships between people and place from the perspective of the new cultural geography; although the examples are Australian, they are relevant in a wider context.

Wallis, M., and A. Fleras. 2009. *The Politics of Race in Canada.* Toronto: Oxford University Press.

This important collection provides numerous insights into the fallacy of race yet the reality of racism with examples drawn from the Canadian context; a thought-provoking text.

Watson, A. 2008. 'Global Music City: Knowledge and Geographical Proximity in London's Recorded Music Industry', *Area* 40: 12–23.

Interesting article about the spatial clustering of the music industry in London, England, analyzing organizational connections at local and global scales,

Weiner, E. 2008. *The Geography of Bliss: One Grump's Search for the Happiest Places in the World*. New York: Twelve Books.

A colourful and entertaining book that reports on the author's travels and experiences as he searches for the elusive reasons why people in different places are happy or unhappy.

ON THE WEB

MODES OF PRODUCTION

Capitalism

www.capitalism.org/

A site that argues the case for capitalism as a system based on the rights of the individual.

Socialism

www.worldsocialism.org/

A site authored by the World Socialist Movement that argues the case for socialism as a system designed to make the world a better place for all.

THE MYTH OF RACE

Race: The Power of an Illusion

🔍 www.pbs.org/race/000_General/000_00-Home.htm

Excellent source of reasoned information about the idea of race; companion site to a PBS television series.

The Slave Trade

🔍 www.culture24.org.uk/history/people+%2526+ society/slavery+and+abolition/art50051

Website related to an exhibition at the Manchester (England) Museum. Discusses three key myths at the heart of racist thinking.

Apartheid Facts

🔍 www.africanaencyclopedia.com/apartheid/apartheid.html

Encyclopedia entry on the South African policy of apartheid.

Apartheid Photographs

🔍 www.unmultimedia.org/photo/subjects/apartheid.html

UN website with a substantial collection of photographs that help provide insight into the policy and practice of apartheid.

QUALITY OF LIFE

Gender

🔍 hdr.undp.org/en/statistics/indices/gdi_gem/

UNDP website with information on the gender-sensitive measure of human development.

Global Fund for Women

🔍 www.globalfundforwomen.org/cms/

The goal of this US-based site is to promote women's economic security, health, education, and leadership. Discusses issues, highlights success stories, and encourages involvement.

Happiness

🔍 worlddatabaseofhappiness.eur.nl/

Information on scientific research aimed at providing a world database on the subjective appreciation of life.

FOLK AND POPULAR CULTURE

American Folk

🔍 www.americanfolk.com/

A site concerned with all aspects of American folk and popular culture.

TOURISM AND RECREATION

Canada

🔍 parkscanada.gc.ca/

The Parks Canada site includes links to numerous discussions of issues related to tourism and recreation.

World Heritage

🔍 whc.unesco.org/

UNESCO provides full details of its World Heritage sites and regularly updated news items.

Global Tourism

🔍 www.world-tourism.org/

The World Tourism Organization, the leading international organization in the field of travel and tourism, serves as a global forum for tourism policy issues and a source of practical information and statistics.

POLITICAL IDENTITIES AND LANDSCAPES

One of the most basic divisions in the world today is the division into political states. Political states play an important role in individual and group behaviour. They also provide much of the data on which human geographers rely. A focus on political states follows logically from our discussions of language and religion (Chapter 6) and ethnicity (Chapter 7).

To explain how the world map came to look as it does today, we begin by exploring the subjects of nationalism, the nation-state, and colonialism. A discussion of geopolitics introduces some of the general models that geographers have used to facilitate understanding of world affairs. Considering why some states are relatively stable and others subject to stress, this section helps to explain why human geographers can make important contributions to understanding many of the problems that states encounter. Subsequent sections address the role of the state in everyday life, the rise of new political movements, and the geography of elections.

We conclude with a discussion of the geography of peace and war and the possible waning of the nation-state. Drawing on the understanding provided by Chapters 6, 7, and 8, this concluding section raises the big question—what will the future bring?—and outlines nine possible scenarios. This chapter reflects the subdiscipline of political geography—broadly understood as the study of the geographical manifestations of political phenomena.

Pakistani Rangers and Indian Border Security Force personnel perform the daily retreat ceremony at the India-Pakistan Wagah border.
Arif Ali/AFP/Getty Images

If human geography is about the role played by space in the conduct of human affairs, politics is about struggles for power: specifically, the power to exercise control over people and the spaces they occupy. Like many other animal species, humans have an inherent need both to delimit territory—an area inside which we feel secure against outsiders—and to apportion space among different groups. The creation of specific territories is the basis for political organization and political action. The political partitioning of space creates the most fundamental of geographic divisions: the sovereign state. Most states are recognized as such by other states; their territorial rights are typically respected by others; they are governed by some recognized body; and they have an administration that runs the state.

State Creation

For most people, the most familiar map of the world is one that shows it divided into states. Our familiarity with the concept of international boundaries reflects our recognition that security, territorial integrity, and political power are enmeshed in this pattern of states. Each state, large or small, is a sovereign unit. Today, there are more than 190 such states.

DEFINING THE NATION-STATE

'Nation' is not an easy word to define. As we have seen in Chapters 6 and 7, humans are divided into numerous cultures based on such variables as language, religion, and ethnicity; cultural affiliation is important not least because it provides people with an identity and a sense of community. The term **nation** is potentially applicable to all cultural groups. But not all cultural groups aspire to be nations, and a key question is why some cultures regard themselves as also possessing national identity while others do not.

Humans are divided also into formally demarcated political units known as states. The **state** is a set of institutions, the most important being the potential means of violence and coercion; a state also makes the rules that govern life within its territory, encouraging cultural homogenization.

A **nation-state** is a clearly defined large group of people who self-identify as a group (a nation) and who occupy a spatially defined territory with the necessary infrastructure and social and political institutions (a state). Each

nation-state is, in principle, a political territory including all members of one national group and excluding members of other groups. In practice, few nation-states fit that strict definition because the vast majority of them are not composed of just one national group.

Indeed, although it is often supposed that nations are in some way natural groupings of people, the reality is quite different, with most nations having been constructed rather than occurring naturally. Most countries in the world engage in an ongoing process of nation construction and reconstruction. In this sense it is useful to conceive of nations as 'imagined communities' (Anderson, 1983).

Nevertheless, the concept of the nation-state is important to us for two reasons. First, because the phrase is part of a dominant discourse (see our discussion of human geographic concepts in Chapter 2), commentators often use the term 'nation-state' to refer to the political unit of a state. Second, the concept of the nation-state has played, and continues to play, an important role in shaping the contemporary political world.

PREDECESSORS OF THE NATION-STATE

Our current political divisions are the products of a long evolutionary process that began with claims to territory and is now characterized by independent states, most of which aspire to be nation-states—that is, political groupings of people with a common identity. In Europe before about 1600, the sovereignty of most individual rulers depended on the allegiance of people rather than on control over a clearly defined territory; however, the classical Greek and the medieval Italian city-states were exceptions to this generalization.

Classical Greece was not one political unit but an amalgam of city-states. Both Plato and Aristotle gave thought to the size, location, and population of the ideal state. Plato favoured a self-sufficient and inward-looking state, while Aristotle favoured a self-sufficient but outward-looking state. The principal state in Greek times was the Persian Empire, a large and loosely structured unit built through conquest and dependent on efficient leadership. One of Aristotle's pupils, Alexander the Great, actively pursued the expansionist (outward-looking) policies favoured by his teacher, destroying the Persian Empire and creating an even larger, although short-lived, empire.

nation
A group of people sharing a common culture and an attachment to some territory; a term difficult to define objectively.

state
An area with defined and internationally acknowledged boundaries; a political unit.

nation-state
A political unit that contains one principal national group that gives it its identity and defines its territory.

From the retreat of the Romans until the sixteenth century, Europe experienced a succession of empires of which the major divisions were recognized on religious grounds (for example, between Christian and Islamic areas and between Christian and pagan areas). Most states in medieval Europe were feudal in structure (see Chapter 7). Their boundaries were uncertain, their spatial extent was constantly subject to change, a few individuals usually exercised political control, and there was a clear social hierarchy. Gradually, these feudal states were replaced by absolutist states that concentrated power in a monarchy; starting about 200 years ago, these states were replaced by nation-states.

Historically, then, the fact that a political unit exists does not mean that its boundaries delimit a national identity. Many states (such as city-states or tribal units) have been too small to qualify as nations, while others (such as empires) have been so large as to be multinational.

THE RISE OF THE NATION-STATE
Nationalism

To understand the rise of nation-states, we need to refer again to the concept of nationalism, which reflects the belief that a nation (cultural group, or large group of people who self-identify as a group) and a state (political unit) should be congruent. Nationalism assumes that the nation-state is the natural political unit and that any other basis for state delimitation is inappropriate. Thus, an aspiring nation-state will usually argue that:

1. All members of the national group have the right to live within the borders of the state.
2. It is not especially appropriate for members of other national groups to be resident in the state.
3. The government of the state must be in the hands of the dominant cultural group.

It is not difficult to understand how application of these arguments can lead to conflict, whether external (with other states) or internal (between cultural groups). Although this principle of nationalism is taken for granted in the modern world, as our brief historical account of states has indicated, it is a relatively recent idea. In retrospect it seems that nationalism was more often a romantic yearning than a coherent political program, and therefore must be a flawed—but vastly influential—concept.

Indeed, the English cleric W.R. Inge (1860–1954) summed up this contradiction rather acerbically: 'A nation is a society united by a delusion about its ancestry and by common hatred of its neighbours.'

Also important to our understanding of the rise of nation-states is the concept of **sovereignty** as it became invested in states. The idea that every state is a sovereign unit became established following the 1648 Treaty of Westphalia, which concluded a long period of European conflict. At that time it was accepted that the ruler of a realm had the authority to determine the religion of its people. Eventually, the concept of sovereignty would contribute to the emergence of independent states.

Explaining nationalism

The map of Europe as it was redrawn in 1815, following the Napoleonic Wars, depicted states demarcated on the basis of dynastic or religious criteria; no real attempt was made to correlate the national identities of populations with their states or their rulers. Indeed, such an exercise would have proved difficult because the cultural map of Europe was exceedingly complex and those cultural groups included many social divisions. Such an exercise would also have been inappropriate, because nationality was not of great importance at the time; people did not question boundaries based on dynastic, religious, or community differences. This raises an important question: Why, despite obvious problems of definition, did national identity emerge as the standard criterion for state delimitation during the nineteenth century? Here are five common theories:

1. Nation-states emerged in Europe in response to the rise of nationalist political philosophies during the eighteenth century.
2. Humans want to be close to people of similar cultural background.
3. The creation of nation-states was a necessary and logical component of the transition from feudalism to capitalism, as those who controlled production benefited from the existence of a stable state (a Marxist argument).
4. Nationalism is a logical accompaniment of economic growth based on expanding technologies.
5. The principle of one state/one culture arises from the collapse of local communities and the need for effective communication within a larger unit.

sovereignty
Supreme authority over the territory and population of a state, vested in its government; the most basic right of a state understood as a political community.

Police keep separatists (right) and federalists (left) apart during a protest in 1998 by English rights activists in Montreal demanding English signs at the Eaton's department store.

CP photo/Ryan Remiorz

In Europe, Germany and Italy were created in the nineteenth century in response to centrally organized efforts at political union of a national group. The transition came earlier for Britain and France. Britain evolved gradually throughout the eighteenth century, while in France monarchical absolutism was intact until the 1789 revolution; in these cases an early general correspondence existed between culture and state.

Nation-states in the contemporary world

Despite the transition to nation-states, our contemporary political map still includes many states that contain two or more nations. Among the multinational states are many African countries whose boundaries were drawn by Europeans without reference to African national identities. Although nationalistic ideology has been clearly articulated in the context of anti-colonial movements, no substantial changes to the European-imposed boundaries have been achieved to date.

Many multinational states are politically unstable, prone to changes of government and/or expressions of 'minority nation' discontent. Canada and Belgium are examples of politically uncertain binational states: both include more than one language group and both are experiencing internal stresses related to the differing political aspirations of these groups. States that approach the nation-state ideal, such as Italy and Denmark, are found primarily in Europe. States that incorporate members of more than one national group are not necessarily unstable; the United States has largely succeeded in creating a single nation out of disparate groups, while Switzerland is probably the prime example of a stable and genuinely multinational state.

EXPLORATION AND COLONIALISM

Clearly, not all states in recent centuries have aspired to the nation-state model. European countries in particular attempted to expand their territories overseas to create empires similar to those that existed in earlier periods in Europe and elsewhere. Such empires are multinational by definition, and were typically regarded as additions to rather than replacements for nation-states.

European countries began developing world empires about 1500, and most such empires achieved their maximum extent in the late nineteenth century. During the twentieth century, most of these empires have been dissolved. Figure 8.1 shows the British Empire as it was in the late nineteenth century, when large areas of the world were under British control. Today the British Empire is limited to a few islands, most of which are in the Caribbean and South Atlantic (Box 8.1).

Exploration

Most empires began as a result of exploratory activity (see Box 1.1 on the contentious term 'exploration'). In general, geographers have paid little attention to exploration—defined as the expansion of knowledge that a given state has about the world—other than to list and to describe events. One geographer, however, has derived a conceptual framework for the process of exploration, a framework that focuses on the links between events (Overton, 1981). As depicted in Figure 8.2, the process begins with a demand for exploration and may lead, in due course, to the 'development' of the new area. Such 'development' is likely to mean exploitation of the explored area as a source of raw materials needed by the exploring country. Britain, for example, viewed Canada as a source of fish, fur, lumber, and wheat, and Australia as a source of wool, gold, and wheat.

Colonialism

The way a new area was used by the exploring country usually was a function of perception and need. Economic, social, and political

FIGURE 8.1 The British Empire in the late nineteenth century

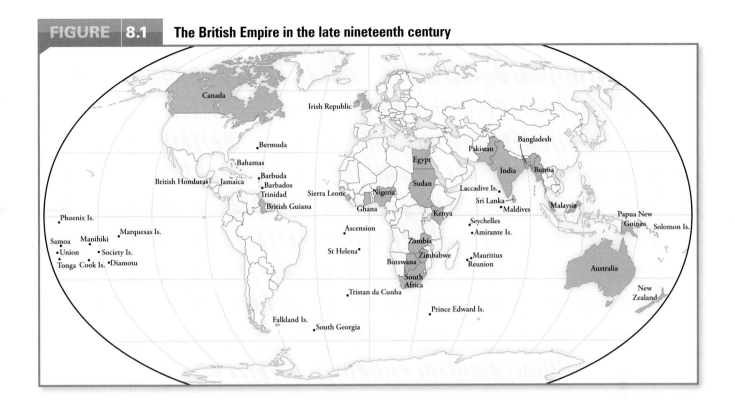

activity in explored areas that became colonies was determined by and for the exploring power. This was true of European empires and of the territories acquired by the United States from 1783 onward (Figure 8.3). In most cases, Aboriginal populations were eliminated or moved if their numbers represented a military or spatial threat. The economic activities that were encouraged benefited the central power. This is the process known as colonialism (see Chapter 5).

In recent European history, colonialism took the form of territorial conquest throughout the Americas, much of Asia (including Siberia), and most of Africa. Competition for colonies was a major feature of the world political scene from the fifteenth to the early twentieth centuries. The earliest colonial powers were Spain and Portugal, followed by other Western European states, the United States, and Russia. The remarkable extent of European colonialism is shown in Figure 8.4.

FIGURE 8.2 Principal elements in the process of exploration

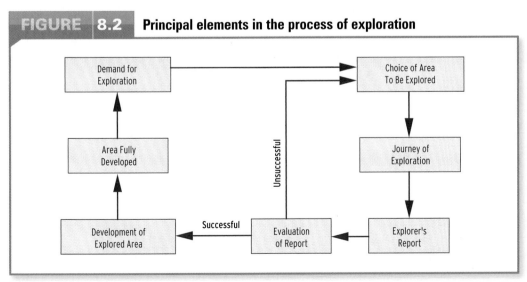

SOURCE: Adapted from J.D. Overton, 'A Theory of Exploration', *Journal of Historical Geography 7* (1981): 57.

FIGURE 8.3 **Territorial expansion of the United States**

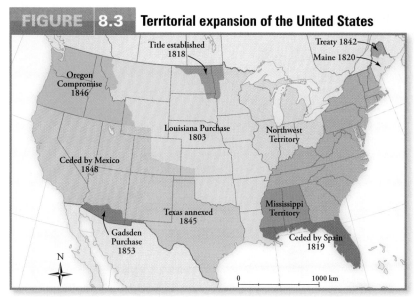

The phrase 'manifest destiny' reflected the belief that the new United States had an exclusive right to occupy North America. In 1803 President Jefferson purchased the Louisiana territory from France. Spain ceded the Florida peninsula in 1819, while war with Mexico resulted in the acquisition of Texas and the Southwest. The Gadsden purchase of 1853 completed the southern boundary of the US. The northern boundary along the 49th parallel was established in 1848. Subsequently, American interests expanded north to Alaska, purchased from Russia in 1867, and across the Pacific to include Hawaii.

Explaining colonialism

The reasons behind colonial expansion are complex; although often summarized as 'God, glory, and greed', they may also have included various changes in Europe involving the prevailing feudal system and economic competition at the time. Recall that, shortly before the beginning of European colonialism, China and the Islamic world were the two leading world areas; indeed, gunpowder, the mariner's compass, printing, paper, the horse harness, and the water mill all were invented in China. The fact that Europe was the region that initiated global movements appears to be related to the demands of the economic growth that began in the fifteenth century, as well as to internal social and political complexity, turmoil, and competition. Reasons for the colonial fever that swept Europe, especially in the nineteenth century, included the ambitions of individual officials, special business interests, the value of territory for strategic reasons, and national prestige. But perhaps the most compelling motive was economic: colonial areas provided the raw materials needed for domestic industries and additional markets for industrial products.

Colonialism effectively ended following World War II. The defeat of Japan and Italy immediately removed them as colonial powers, and most of the remaining parts of the British Empire achieved independence over the next few years. Three European powers—France, the Netherlands, and Portugal—tried to retain their colonies, but all were unsuccessful.

FIGURE 8.4 **European imperial coverage of the globe**

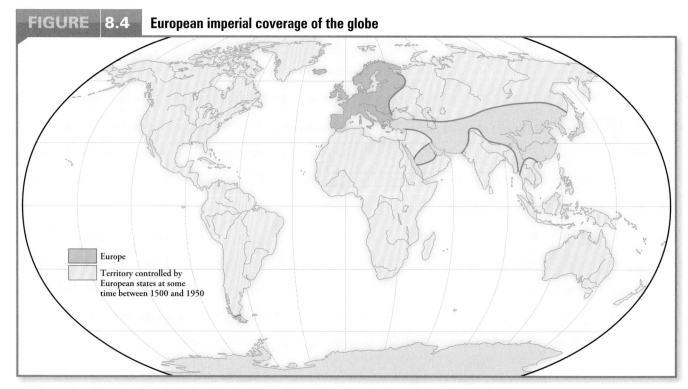

Europe

Territory controlled by European states at some time between 1500 and 1950

SOURCE: W.E. Murray, *Geographies of Globalization* (New York: Routledge, 2006), 82. Copyright 2006. Reproduced by permission of Taylor & Francis Books, UK

Box 8.1   Remnants of Empire

In recent years, two former colonies have changed their political status in accord with an agreed-upon timetable: Hong Kong was transferred from Britain to China in 1997, and Macao was transferred from Portugal to China in 1999. Today, the UN identifies 16 other political units, representing some 1.2 million people, still best described as colonies. Although some colonies are likely to become independent as economic circumstances change, others are likely to remain politically dependent for some time to come, for two general reasons.

First, some areas are perceived as particularly valuable to the colonial power. Both France and the United States use islands in the Pacific for military purposes, and the US is known to be reluctant to relinquish control over Guam in the Pacific and Diego Garcia (a British colony mostly populated by US military) in the Indian Ocean. Indeed, both the UK and the US have refused to co-operate fully with the UN as it aims to end all colonial situations.

Second, it appears that some colonial areas would simply not be viable as independent states. A prime example is St Helena, an island of 122 km² (47 square miles) with a population of 5,700 located in the South Atlantic Ocean (see Figure 8.1) that is perhaps best known as the location of Napoleon's exile and death.

First reached by the Portuguese in 1502, the island later became the object of competition between the Dutch and the British, and today it remains part of the much-reduced British Empire. The competition reflected St Helena's location, which made it valuable as a stopover on the Europe–India route. Over time, movements of slaves from East Africa and indentured labour from China, as well as British and some other European settlement, created an ethnically diverse population who supported themselves by providing food and other supplies for ships. By the late nineteenth century, however, technological advances such as steamships and refrigeration, and the opening of a new route using the Suez Canal in 1869, had made St Helena unnecessary as a stopover, and the economic problems soon became evident. With a very limited resource base, St Helena suffers from its small size and small population. For much of the twentieth century the agricultural industry was dominated by flax; at times, half the working population was employed in flax production. Today, however, agriculture is limited to livestock and vegetables produced only for subsistence purposes.

The future of St Helena is uncertain. It is isolated, dependent on financial support from Britain, and unlikely to become a tourist attraction. From the British perspective, it is no longer a prestigious overseas possession but a financial liability (Royle, 1991).

Decolonization

In recent years the number of states in the world has increased, from 70 in 1938 to more than 190 in 2009. Most of the new states have achieved independence from a colonial power. Many of them reflect national groupings, but some occupy areas around which Europeans drew boundaries for their own reasons.

As we have seen, the idea of a national identity originated in Europe, but it has been welcomed by colonial peoples who were discontent with colonial rule, aspired to independence, and suffered psychologically from foreign rule. The transition to independence has been violent in some cases and peaceful in others, depending in part on the way individual European powers responded as the independence movements in their colonies gathered momentum in the 1950s. Most of the former British colonies experienced a relatively smooth transition to independence, and almost all of them are now members of the Commonwealth, a voluntary organization of 54 nations. Some other former colonies, however, achieved independence only after long civil wars; examples include Algeria (France), Mozambique (Portugal), and the Congo region (Belgium).

Effects of colonialism

It appears that many colonies may have resulted in net losses to the national economies of the colonial powers, benefiting the stock exchange and a few individuals rather than state treasuries. But there were other benefits. For example, the vast expanses of Canada, Australia, and South Africa offered a way for Britain to reduce population pressure. The effects on the areas colonized are complex. As we saw in Chapters 4 and 5, many areas have experienced massive population growth and associated economic problems because of the colonial experience and related reductions in death rates. Many believe that **imperialism**—in which one powerful state or territory seeks to control another, weaker territory—has had extremely negative consequences for the less powerful area. This is one

imperialism
A relationship between states in which one is dominant over the other.

of the central arguments of the world systems theory introduced in Chapter 5.

A related set of ideas, dependency theory (introduced in Chapter 5), contends that African and Asian countries became poor as a result of their colonization, while the colonial powers advanced at their expense. Specifically, dependency theory claims that less developed countries depend for their survival on their links with dominant countries in the more developed world—links formed during the colonial period, when occupation by the imperialist powers led to the disintegration of indigenous cultures and the formation of the less developed world.

STATE CREATION: SOME CONCEPTUAL DISCUSSIONS

Philosophical discussions of state creation and expansion can be traced as far back as Plato and Aristotle; Strabo took a favourable view of the Roman Empire; ibn-Khaldun distinguished between the territories of nomadic and sedentary peoples; and Bodin and Montesquieu both favoured the emerging nation-state model. Three more recent discussions are outlined below.

Ratzel

The great early nineteenth-century German geographer Ritter viewed states as organisms. He saw cultures and states as progressing through a life cycle, an idea Ratzel would elaborate on. Ratzel's seven laws concerning the spatial growth of states (Box 8.2) are not the product of rigorous scientific logic; rather, they are generalizations based on observations

of a supposed ideal world. Central to Ratzel's thinking was the notion of the state as a living organism. Although this notion was criticized even in Ratzel's day because it reified (assumed the independent existence of) something that was actually a human creation, Ratzel's ideas have been highly influential, both conceptually and practically.

Jones

A rather different view of states was developed by S.B. Jones (1954) on the basis of some ideas put forward earlier by Gottman, Hartshorne, and the political scientist Deutsch. Jones proposed a chain of events beginning with some political idea and concluding with the creation of a political area (Figure 8.5). In this way, an idea of state creation may lead in due course to state creation. Between idea and area are a positive decision to implement the initial idea; movement of some variety following a political decision; and a field of activity, created by that movement, which ultimately leads to the delimitation of a political area. Jones's application of these ideas to the creation of the state of Israel is outlined in Box 8.3: the political idea is Zionism; the political decision is the Balfour Declaration of 1917; the principal movement is the immigration of Jews; the resulting field is settlement and government activity; and the political area is Israel.

Deutsch

A third concept was proposed by Deutsch (see Kasperson and Minghi, 1969: 211–20), who describes a process of state creation that can involve as many as eight stages:

| Box 8.2 | Laws of the Spatial Growth of States |

Ratzel's 'laws', published in 1896, are actually generalizations:

1. The size of a state increases as its culture develops.
2. The growth of a state is subsequent to other manifestations of the growth of a people.
3. States grow through a process of annexing smaller members. As this occurs, the human–land relationships become more intimate.
4. State boundaries are peripheral organs that take part in all transformations of the organism of the state.
5. As a state grows, it strives to occupy some politically valuable locations.
6. The initial stimulus for state growth is external.
7. States' tendency to grow continually increases in intensity.

The key idea is that a state grows as its level of civilization rises.

These laws present an interesting parallel with Ravenstein's laws regarding migration (Box 5.2). In neither case is the logic employed strictly deterministic.

1. transition from subsistence to exchange economy;
2. increased mobility leading to formation of core areas;
3. development of urban centres;
4. growth of a network of communications;
5. spatial concentration of capital;
6. increasing group identity;
7. rise of national identity;
8. creation of a state.

Both Jones and Deutsch emphasize evolution and focus primarily on human actions. Neither regards states as organisms. As we have seen, nations do not exist as conveniently packaged bundles of people waiting for state boundaries to be drawn around them (Taylor, 1989). Nations and states are human creations; explaining the existence of nations and national identities requires consideration of ethnic distributions and the political construction of nations. An example is contemporary Quebec, where aspirations for a separate political identity appear on occasion to be politically orchestrated.

These conceptual discussions bring us to one of the most exciting and provocative aspects of the study of our political world: geopolitics.

Introducing Geopolitics (and *Geopolitik*)

One influential interest within political geography, which studies the geographical manifestations of political phenomena, is **geopolitics**—the study of the relevance of space and distance to questions of international relations. Geopolitical discussions originated in the late nineteenth century, but manipulation of the concept by some German geographers and Nazi leaders in the 1930s and 1940s led others to reject it. Only since the 1970s has geopolitics revived and once again become a legitimate area of interest. Numerous discussions have centred on the 'new geopolitics', especially a critical geopolitics that places emphasis on the social construction of national and other identities. The emphasis on a culturally informed geopolitics is especially interesting in light of the new cultural geography introduced in Chapter 7.

The intellectual origins of geopolitics lie in the work of Ratzel, specifically in his seven

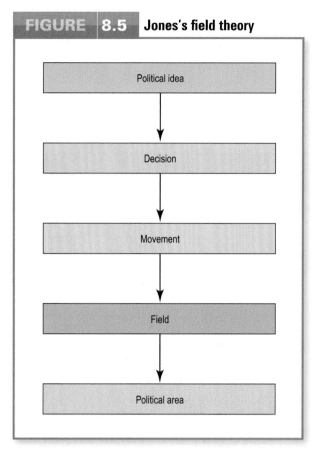

FIGURE 8.5   Jones's field theory

Political idea
↓
Decision
↓
Movement
↓
Field
↓
Political area

laws of state growth (see Box 8.2). The term 'geopolitics' was coined early in the twentieth century by a Swedish political scientist, Kjellen, who expanded on Ratzel to argue that territorial expansion was a legitimate state goal. Clearly, this argument is only one aspect of the larger field, the relevance of space and distance to questions of international relations. It was, however, this narrow interest that dominated geopolitical discussions until they were largely abandoned in the 1940s.

GEOPOLITICAL THEORIES

A leading British geographer, H.J. Mackinder, writing in 1904, was the first to formulate a geopolitical theory. Mackinder's **heartland theory**, which attempted to explain how geography and history had interacted over the past 1,000 years, has strong environmental determinist overtones, reflecting British concerns about perceived Russian threats to British colonies in Asia, especially India. Mackinder (1919: 150) contended that the Europe–Asia land mass was the 'world island' and that it comprised two regions: an interior 'heartland' (pivot area) and a surrounding 'inner or marginal crescent' (Figure 8.6):

geopolitics
The study of the importance of space in understanding international relations.

heartland theory
A geopolitical theory of world power based on the assumption that the land-based state controlling the Eurasian heartland held the key to world domination.

Who rules East Europe commands the Heartland;

Who rules the Heartland commands the World Island;

Who rules the World Island commands the World.

Mackinder was arguing that location and physical environment were key variables in any explanation of world power distribution. The theory is flawed because it overemphasizes one region (Eastern Europe) and because Mackinder could not anticipate the rise of

Box 8.3   The Jewish State

The Jews are a religious group with roots in the area that is now Israel. Jerusalem, in particular, is an important symbolic location for Jews, as it is for Christians and Muslims. During the time of Jesus, this area (known as Palestine) was under Roman rule. In 636 it was invaded by Muslims and from then until 1917 (with a brief Christian interlude during the Crusades), it was a Muslim country, specifically Turkish after 1517. Over the centuries, in a process known as the Diaspora, Jewish populations moved away to settle throughout much of Europe and overseas. At the end of World War I (1914–18), Syria and Lebanon were taken over by France, and Palestine and Transjordan (now Jordan) were taken over by Britain, under League of Nations (the predecessor of the United Nations) mandates. The Palestine mandate—specifically the Balfour Declaration of 1917—allowed for the creation of a Jewish national home but without damaging Palestinian interests. An impossible goal was being set because a different national group, Palestinians, already inhabited the space that was being assigned as a Jewish state.

In 1920 Palestine included about 60,000 Jews and 600,000 Arabs. Most of the Jews had arrived since the 1890s. Jewish immigration increased under the mandate, especially as persecution of Jews in Hitler's Germany intensified. After World War II, Jewish survivors struggled to reach Palestine, which by 1947 included 600,000 Jews, 1.1 million Muslim Arabs, and about 150,000 Christians (mostly Arab). Arab–Jewish conflicts were by then commonplace, and Britain, unable to cope, announced that it intended to withdraw.

The United Nations then produced a plan to partition Palestine into a Jewish state and an Arab state, with Jerusalem as an international city. The Arabs did not accept the plan, and the Jews declared the independent state of Israel (1948). Conflict immediately flared between Israel and nearby Arab states and between Jews and Palestinian Arabs, and has continued to the present. One long-standing issue of contention has been Israel's occupation of the West Bank (of the Jordan River) and the Gaza Strip, which it captured (from Jordan and Egypt respectively) during the Six Day War of 1967. Although Israel has established relations with Egypt (in 1979), the Palestine Liberation Organization (in 1993), and Jordan (in 1994), serious obstacles to peace remain. Among them are the absence of diplomatic relations between Israel and Syria, the uncertain status of Jerusalem, the continuing

presence and expansion of Jewish settlements in the West Bank, and, finally, ongoing terrorist actions against Israel by Palestinian militants and Israeli military actions against the Palestinians. The result is that Israel's borders are vulnerable and the state is inherently unstable. Also, about 5 million Palestinians live in historic Palestine under Israeli control—2.5 million in the West Bank, 1.47 million in Gaza, and 1.13 million in Israel. There are also about 2.8 million Palestinians in Jordan and about 2 million more elsewhere. In 2002 the United States acknowledged the need for an independent, democratic, and viable Palestinian state—an explicit acknowledgement of Palestinian identity—existing in peace and security with Israel. But progress in this direction has been uncertain, especially since the victory of Hamas, a militant Islamic group, in the 2006 Palestinian elections in the Gaza Strip and three weeks of fighting in early 2009. Today, there is no obvious resolution to the closely related problems of Israeli security and the creation of a Palestinian state. Indeed, some observers now advocate a single-state solution.

A Palestinian boy searches for his family's belongings in the rubble of a four-storey building at the Nusseirat refugee camp in the Gaza Strip, March 2003. The building was destroyed in a battle between Israeli troops firing from tanks and helicopters and dozens of Palestinian gunmen. Seven Palestinians, including a four-year-old child, were killed.

CP/AP photo/Karel Prinsloo

FIGURE 8.6 Mackinder's heartland theory

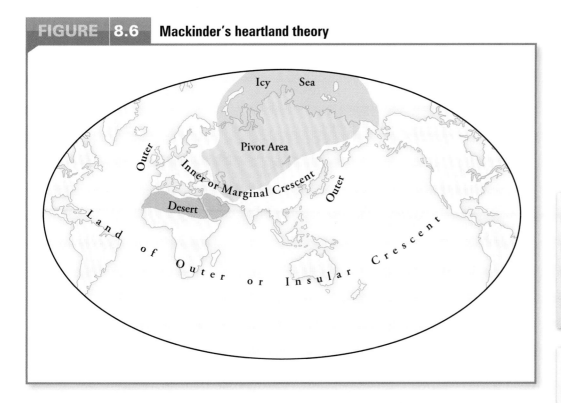

Geopolitik
The study of states as organisms that choose to expand in territory in order to fulfill their 'destinies' as nation-states.

rimland theory
A geopolitical theory of world power based on the assumption that the state controlling the area surrounding the Eurasian heartland held the key to world domination.

centrifugal forces
In political geography, forces that make it difficult to bind an area together as an effective state; in urban geography, forces that favour the decentralization of urban land uses.

centripetal forces
In political geography, forces that pull an area together as one unit to create a relatively stable state; in urban geography, forces that favour the concentration of urban land uses in a central area.

air power, but it did influence other writers in general and may well have exercised a continuing influence on the United States' policy in particular.

Certainly the earlier work of Ratzel, Kjellen, and Mackinder influenced the rise of *geopolitik* in the 1920s. ***Geopolitik*** is a specific interpretation of more general geopolitical ideas. It focuses on the state as an organism, on the subordinate role played by individual members of a state, and on the right of a state to expand to acquire sufficient *lebensraum* (living space). The individual most responsible for popularizing these ideas was Haushofer, a German geographer in Nazi Germany who was bitterly disappointed by the territorial losses Germany experienced as a result of World War I. Haushofer's academic justification for German expansion coincided with Nazi ambitions. The extent of Haushofer's influence on actual events is uncertain, but it is the case that the German interest in *geopolitik* resulted in a general disillusionment in the Western world with all geopolitical issues.

Despite this disillusionment, several scholars continued to present their views on the global distribution of power. Spykman, writing in the 1940s, built on Mackinder's 'inner or marginal crescent' (rimland) to develop a **rimland theory**, arguing that the power controlling the rimland could control all of Europe and Asia and therefore the world. As an American, Spykman saw considerable advantage in a fragmented rimland—a view that has certainly influenced post-World War II American foreign policy in Asia (Korea and Vietnam) and in Eastern Europe. Also during the 1940s, de Seversky stressed that the US needed to be dominant in the air to ensure state security.

CENTRIFUGAL AND CENTRIPETAL FORCES

In a consideration of the stability, or instability, of states, the geographer Hartshorne (1950) usefully distinguished between centrifugal and centripetal geopolitical forces. **Centrifugal forces** tear a state apart; **centripetal forces** bind a state together. When the former exceed the latter, a state is unstable; when the latter exceed the former, a state is stable.

The most common centrifugal forces are those involving internal divisions in language and religion that lead to a weak raison d'être or state identity. Other centrifugal forces include the lack of a long history in common (the case in many former colonies) and state boundaries that are subject to dispute. The most common centripetal force is the presence of a powerful raison d'être: a clear and widely accepted state identity. Other centripetal forces include a long state history and boundaries that are clearly delimited and accepted by others. The various centrifugal or centripetal forces are often closely related; for example, the presence of internal divisions generally indicates that groups do not share a long common history and that boundaries are not agreed upon. Closely linked, also, are stability and peace—and instability and conflict—within and between states.

BOUNDARIES

Boundaries mark the limits of a state's sovereignty. They are 'lines' drawn where states meet or where states' territorial waters end. A state's stability often reflects the nature of its boundaries.

The characteristics that give identity to a nation, such as language, are rarely as abruptly defined as are its boundaries. For this reason, many countries include at least one significant minority population. In principle, such situations are less common where the boundaries are *antecedent* (established before significant settlement began), because settlers moving into areas close to the boundary must acknowledge its presence. Boundaries of this kind frequently are geometric—as in the case of the US–Canada boundary west of the Great Lakes, which follows the forty-ninth parallel.

Other boundaries are *subsequent*: defined after an area has been settled and the basic form of the human landscape has been established. Such boundaries may attempt to reflect national identities (for example, the present boundary of France is an approximation of the *limites naturelles* of the French nation) or may totally ignore such distinctions (this is the case with most colonial boundaries). Many subsequent boundaries are continually subject to redefinition, as in Western Europe.

All boundaries are artificial in the sense that what is meaningful in one context may be meaningless in another. Rivers are popular boundaries, since they are easily demarcated and surveyed, but they are generally areas of contact rather than of separation, and so tend to make poor boundaries. Rivers may also be poor boundaries if they are wide and contain islands (e.g., the Mekong River in Southeast Asia) or if they repeatedly change course (e.g., the Rio Grande along the Mexico–US [Texas] border). Using a watershed as a boundary sounds logical and non-controversial, but what happens when glaciers shrink because of global warming and the watershed shifts? This is now happening along one stretch of the border between Italy and Switzerland where glacier melt has shifted the border by more than 10 metres. Alpine countries acknowledge that any line based on a glacier watershed is necessarily temporary, rendering the concept of a movable border a reality.

In some cases, groups have erected barriers not to demarcate boundaries but rather to serve as physical obstructions preventing others from entering. Well-known examples include the Great Wall of China, constructed in the third century BCE to prevent invasion from the north, and Hadrian's Wall, constructed across northern England in the second century CE to prevent invasion from Scotland. Along some parts of the US–Mexico border, fences are in place to restrict illegal immigration. In one exceptional instance—the Berlin Wall—the aim was to prevent people leaving the eastern sector for the west. In 2003 Israel began building an elaborate system of fences and military checkpoints along the border with the West Bank in the hope of reducing terrorist incursions.

Divided states

Some states have been divided into two or more separate parts; this situation increases the likelihood of boundary problems. A classic example was the partition of India as part of the decolonization of Britain's Indian empire. In some cases of partition the boundaries between the newly formed states have been especially inappropriate. Examples include Germany (East and West from 1945 until 1990) and Korea (North and South since 1945). In both cases, one nation was divided into two states for reasons that had more to do with the political wishes of other states than with the wishes of the people themselves. There are sound reasons to argue that artificial, externally imposed boundaries such as these are rarely long-lasting.

Building on these various geopolitical ideas, we next consider two seemingly contradictory trends in the contemporary world. First, there is much evidence to suggest that the number of states will continue to increase from the current about 190 because so many states have internal divisions, related especially to claims of ethnic nationalism. Second, there is also evidence that some states are willing to sacrifice aspects of their independent national identity as they join with other states to form some degree of commonality, as has been the case with the European Union. Both trends contribute to the creation of new maps of our political world.

Unstable States

As the account of state creation explained, concordance between nation (a group of people) and state (a political territory) is rare. Internal ethnic (usually linguistic and/or religious) divisions often occur within a state. Often these do not threaten the stability of the state, but in some cases, to varying degrees, they do.

In countries with significant internal divisions, any one of three general situations may threaten state stability. First, secessionist movements arise when nations within multinational states want to create their own separate states; examples include the Québécois in Canada, the Flemish and Walloons in Belgium, and the Welsh and Scottish nationalist movements in the United Kingdom. Second, in other cases 'nations within' may want to link with members of the same nation in other states to create a new state; the Basques in Spain and France and the Kurds in Iraq, Iran, Syria, Turkey, Armenia, and Azerbaijan are two examples (Box 8.4). Third, **irredentism** involves one state's seeking the return from another (usually neighbouring) state of people and/or territory formerly belonging to it. Irredentism

> **irredentism**
> The view held by one country that a minority living in an adjacent country rightfully belongs to the first country.

| Box 8.4 | The Plight of the Kurds |

According to some commentators, the biggest losers in the Middle East over the past 100 years have been the Kurds. Since the collapse of the Ottoman (Turkish) Empire at the end of World War I, other groups—Turks, Arabs, Jews, and Persians—have consolidated or even created their own states. Yet more than 20 million Kurds remain stateless, dispersed in Turkey (10.3 million), Iran (4.6 million), Iraq (3.6 million), Syria (1 million), and in smaller numbers in other neighbouring states. Their failure to achieve state identity is all the more surprising given that, in 1920, Britain, France, and the United States all acknowledged the need for a Kurdish state: 'No objection shall be raised by the main allied powers should the Kurds . . . seek to become citizens of the newly independent Kurdish state' (Article 64, Treaty of Sèvres, signed 20 August 1920, quoted in Evans, 1991: 34).

Why have the Kurds not succeeded in imposing their identity on a region to create a state? Part of the answer lies in the question of Kurdish identity itself. Most Kurds are Sunni Muslims, but not all are; their society, located in a mountain environment, is typically poor and divided tribally; their dialects are varied; and there is no one political party to speak for all of them. In short, it could be argued that the Kurds are united more by persecution than by any sense of shared identity. A second part of the answer is the fact that a Kurdish state would weaken the authority of all the states in which Kurds now live. To discourage the development of Kurdish nationalism, Turkey banned the use of the Kurdish language in schools until 2002 (when it abandoned that policy as part of its unsuccessful effort to gain entry to the European Union), and since the 1990s has resisted any suggestion of independence for Iraqi Kurds living next to its own Kurdish population.

The Kurds' preferred location for their own state would be in the mountainous border regions of Turkey, Iraq, and Iran.

Unfortunately, this is an area valued by existing states for its resources, which include oil deposits at Kirkuk. Iraq has used chemical warfare against the Kurds and, following the end of the 1991 Gulf War, mounted an extensive campaign to slaughter the Kurds in Iraq, which caused massive population movements into Turkey. Ironically, Turkey is far from a safe haven: in 1979, the country's prime minister stated: 'the government will defeat the disease (of Kurdish separatism) and heads will be crushed' (quoted ibid., 35). Political developments following the 2003 war in Iraq highlighted the clear divisions between Sunni, Shiite, and Kurdish groups and have led to a degree of Kurdish autonomy in that country, but the larger picture in the region is unclear and not promising.

Members of a Kurdish family prepare to leave their home in the Kurdish town of Chamchamal, near Kirkuk in northern Iraq, in March 2003. Local residents feared a chemical attack by Saddam Hussein's forces in the event of a US-led war.

CP/AP photo/Newsha Tavakolian

is commonplace today, in parts of the Balkan region, the Middle East, and sub-Saharan Africa, all regions where state boundaries were largely determined by other powers. In addition, as suggested in Table 5.8, many protracted refugee situations lead to irredentist claims.

To help clarify these general points, we now consider some of the details of internal divisions through reference to a wide variety of examples from Africa, Europe, the former USSR, and South Asia, as well as the case of Canada.

NATION AND STATE DISCORDANCE IN AFRICA

Discordance between nation and state is especially evident in Africa (Figure 8.7). In only a few instances does a distinct national group correlate with a state; the southern African microstates of Lesotho and Swaziland provide examples. Ironically, perhaps, it has been suggested that Lesotho might join South Africa (Lemon, 1996). The principal argument in favour of Lesotho's sacrificing its identity in

this way is that it is the only sovereign state in the world, apart from the microstates of San Marino and the Vatican, to be completely surrounded by a single neighbour.

Among the many African nations that once had related states but no longer do are the Hausa and Fulani nations in Nigeria, the Fon in Benin, and the Buganda in Uganda. The states of contemporary Africa are not products of a long African history but creations of colonialism that subsequently achieved independence. State boundaries, shapes, and sizes are colonial creations, reflecting past European rather than African interests. In consequence, Africa is characterized by political fragmentation. Some states have a high degree of contiguity (when there are many states, most states have many neighbours); some are very small in size and/or population; some have awkward shapes and long, often environmentally difficult, boundaries; and some lack access to the sea. Each of these difficulties is related to the colonial past, a past that ignored national identity in the process of creating states.

National identity was evident in many of the pre-colonial African states. Two early states in West Africa were Ghana and Mali (names later given to modern states covering different areas); both of these states succumbed to Islamic incursions in the eleventh century. States continued to grow and decline. Figure 8.8 shows maps for the sixteenth, eighteenth, and nineteenth centuries. The colonial impact is especially clear when we compare Figures 8.7 and 8.8; European neglect of African national identities is a prime cause of African nationalism and contemporary instability in many African states (Box 8.5).

The future of African political identities remains uncertain. The Organization of African Unity (OAU) was established in 1963 with the explicit goal of providing a common voice for the emerging independent African states. Sadly, it failed; most commentators found that in practice it did more to impede than to promote democracy, human rights, and social and economic growth. In 2002 the OAU was replaced by the African Union. There is real hope that this new organization will succeed in enhancing national identities, encouraging co-operation, and permitting Africa to speak with one voice in a global context. In a similar vein, some observers argue that the best way forward for Africa is through federalism, a linking of countries to facilitate co-operation.

FIGURE 8.7   African ethnic regions

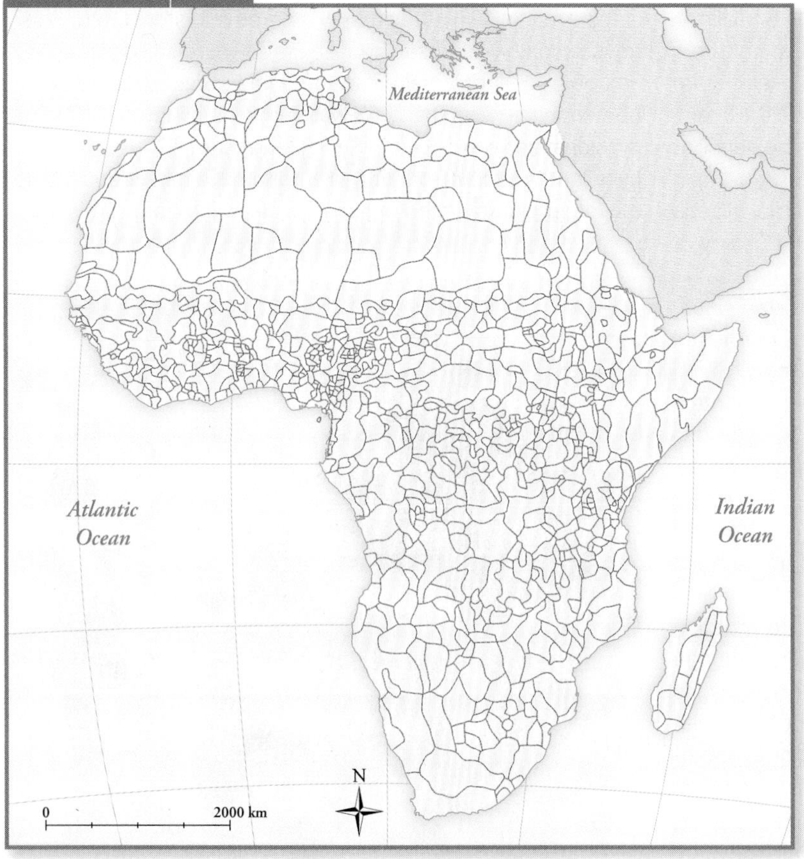

SOURCE: L.D. Stamp and W.T.W. Morgan, *Africa: A Study in Tropical Development*, 3rd edn (New York: John Wiley & Sons, 1972), 41. Copyright © 1972 John Wiley & Sons. Reprinted with permission.

Box 8.5 Tribes, Ethnicity, Political States, and Conflict in Africa

There is little evidence in Africa of pre-colonial state or tribal conflicts. By creating states reflecting their political and economic concerns, the colonial powers disrupted African identities and provided a basis for the many ethnic conflicts that are evident today.

At the root of recent and current African conflicts was the tendency of colonialism to privilege one group over others and thus to stimulate competition over resources. During the decolonization phase, it was not unusual for different groups to unite against a common enemy, the colonial power. But independence has in many cases exposed the fragility of such alliances; not surprisingly, African countries have not been able to construct nations out of different ethnic groups. Overall, African countries find it difficult to keep in place the many politically fragile states that Europeans created. Their task has not been facilitated by the many problems—including rapid population growth, inadequate food supplies, and declining prices for commodities—that they have encountered. Tragically, violence sometimes becomes widespread as ethnic groups compete for resources and political power. Unfortunately, it appears that once ethnic violence begins, it encourages additional and more widespread ethnic violence. The post-colonial era in Central Africa, therefore, has witnessed a cycle of ethnic killings.

In Central Africa, conflict erupted in Rwanda in 1994. Rwanda is situated in one of the most densely populated regions in Africa—the highland region surrounding Lake Victoria. Part of German East Africa in the late nineteenth century, both Rwanda and neighbouring Burundi were placed under Belgian administration at the end of World War I and became independent, as separate states, in 1962.

The two principal competing ethnic groups, Hutu and Tutsi, are not clearly defined; indeed, they speak the same language and share many cultural characteristics. About 90 per cent of the population in Rwanda are Hutu, Bantu-speaking herders and farmers (82 per cent in Burundi). Prior to the colonial period, they lived side by side with the minority Tutsi—who are pastoralists and probably of Hamitic descent—without significant conflict. But European colonialism created four countries—Rwanda, Burundi, the Democratic Republic of Congo, and Tanzania—each containing both Hutu and Tutsi

populations. Europeans then reinforced the ethnic division by favouring the Tutsis, whom they regarded as superior; hence today Hutus and Tutsis occupy different social positions. Differences between these two groups were not the cause of their recent conflicts. Rather, the causes included the lack of human development, high population densities, lack of agricultural land, and a collapse in the price of coffee, the region's most important commercial crop. The conflict assumed an ethnic character, but was not directly caused by ethnic hostility.

A related conflict began in the Democratic Republic of Congo (formerly Zaire) in 1998 and continues today with fighting between the Congolese army and ever-changing alliances of rebel militia groups. Estimates are that about 4 million have died since 1998, many because of untreated disease or starvation. The fact that no rail or road routes run east–west across the country makes it easier for warring groups to consolidate themselves in particular areas. In an attempt to facilitate a ceasefire, the UN deployed military personnel in 2000 and there are 17,000 in the country as of 2009—the UN's largest such mission. One reason for the conflict is to gain access to the country's vast mineral wealth.

There are conflicts elsewhere in Africa. In Kenya in 2007 the dominant Kikuyu ethnic group was perceived as being involved in election fraud to the disadvantage of other groups. Massive population displacement, property damage, and claims of ethnic killings resulted. Conflicts are ongoing in the Central African Republic and Sudan.

A combination of the 2002 formation of the new African Union and spreading democracy offers hope for the future. For example, in 2005 the African Union worked along with the Economic Community of West African States to develop a real show of unity among African nations as they succeeded in reversing a coup in Togo. Also, as noted in Chapter 5, there are continuing indications of the spread of democratic principles, with elections held in the Central African Republic in 2005 and in Mauritania in 2006, and evidence that Liberia, under the leadership of Africa's first female head of state, is undergoing a dramatic turnaround following a devastating civil war.

NATIONS AND STATES IN EUROPE

In Africa, then, present-day post-colonial states rarely reflect nationalist aspirations. The situation is different in Europe, where states correspond more closely to national identity. Nevertheless, most instances of internal division are based on language and/or religion. Rokkan (1980) identifies four functional prerequisites for the existence of a state—economy, political power, law, and culture—and argues that the tensions between European cores and their peripheries reflect the fact that peripheries have less political power, less developed legal structures, and less dominant cultures. He identifies two axes: a north–south cultural axis from the Baltic to Italy, and an east–west economic axis. The north–south cultural axis was Protestant (a religion that favoured national

FIGURE 8.8   African political areas in the sixteenth, eighteenth, and nineteenth centuries

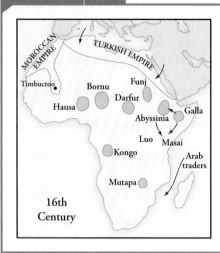
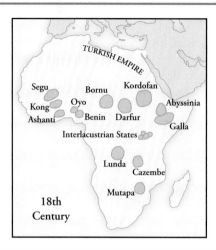
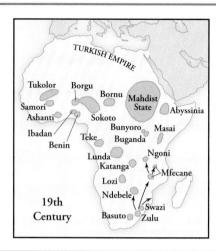

devolution
A process of transferring power from central to regional or local levels of government.

core/periphery
The concept that states are often unequally divided between powerful cores and dependent peripheries.

aspirations) in the north and Catholic (a religion that cut across national boundaries) in the south. Along the east–west economic axis, the west had key commercial centres (an important factor in state formation), but in the east the urban centres were economically weak (and hence unable to offer the resource base needed for the building of states). Arguing that the success of efforts at nation-building depended on cultural conditions, whereas the success of efforts at state-building depended on economic conditions, Rokkan proposed that the north–south cultural axis determined the former, while the east–west axis determined the latter. Hence, the most favourable conditions for building nation-states existed in the northwest.

Contemporary nationalist movements in Western Europe can thus be seen as relics of a Rokkan-type core–periphery cleavage in areas not yet homogenized into an industrial society with its related class divisions. In our present context, this view is certainly useful.

Separatist Movements in Europe
Regionalism and associated efforts to assert separate identities remain strong in Europe (Box 8.6). Figure 8.9 shows several areas with political parties whose policies are explicitly regional and nationalist. Some favour separation; others, self-government within a federal state structure, perhaps by means of **devolution** (the transfer of power from central to regional or local levels of government). In most of these European cases, language is a key factor. However, it is worth noting that most of the areas asserting a distinct identity are peripheral and economically depressed. Indeed, one of the principal causes of social unrest in general is the **core–periphery** economic spatial structure—a common feature of modern states. Rich cores and poor peripheries often exacerbate separatist tendencies in the peripheral areas. Thus, most of the areas of unrest shown in Figure 8.9 may feel disadvantaged not only in terms of national identity but also in terms of economic well-being. Many peripheral areas have a specialized or short-lived economic base that is particularly subject to economic problems such as unemployment.

Northern Ireland: Since partition in 1921, a large Catholic minority has sought to separate from the United Kingdom and integrate with

Liberian President Ellen Johnson Sirleaf, left, and World Bank President Robert Zoellick take part in a joint news conference at the World Bank headquarters in Washington announcing that Liberia had significantly reduced its foreign debt.

Pablo Martinez Monsivais/AP Photo

predominantly Catholic Ireland. The most significant movement towards resolution of the conflict came in 1998, when the 'Good Friday' agreement established a new framework to help decide future constitutional change. The 2005 disarmament commitment from the Irish Republican Army was a further positive step. Today, Northern Ireland continues to be a spatially segregated society, with Catholics and Protestants living in different areas of the city of Belfast, a fact that is unfortunately highlighted when Protestant groups choose to march through or close to Catholic neighbourhoods, especially on the Glorious Twelfth (12 July, commemorating the victory in 1690 of William III at the Battle of the Boyne over the Catholic army of the deposed James II).

Wales: Plaid Cymru, the Welsh national party, was formed in 1925 to revive the Welsh language and culture; it favours self-government for Wales. The devolution referendum of 1978 saw 12 per cent in favour and 47 per cent against a devolved assembly. However, in a 1997 referendum Welsh voters approved proposals—introduced by the British government—for a devolved National Assembly. Since 1999, the National Assembly has been responsible for the development and implementation of policy in a wide range of areas including agriculture, industry, the environment, and tourism.

Scotland: The Scottish National Party was formed in 1934 and achieved notable political success in 1974; by 1978, 33 per cent of voters favoured a devolved assembly, with 31 per cent opposed. As part of the same initiative that led to the creation of the Welsh National Assembly, proposals for a Scottish Parliament were introduced by the British government and a 1997 Scottish referendum approved its creation by a substantial majority. The Parliament's responsibilities are similar to those of the Welsh National Assembly. Whether the new Scottish and Welsh legislatures will lead to further separation remains to be seen. Today, the Scottish National Party is focusing attention on representation in the European Community.

Flanders and Wallonia: The two linguistically distinct regions of Belgium—Flanders is Flemish-speaking and Wallonia is French-speaking—were combined as a single state in 1830 and have never achieved effective national unity; indeed, the two areas legally

FIGURE 8.9 Principal areas with nationalist/separatist movements in Europe

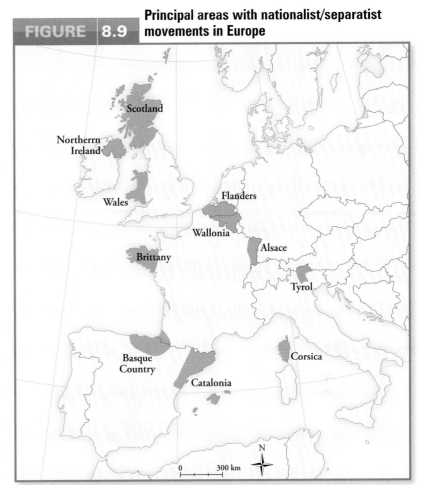

became unilingual regions in 1930 (see Figure 6.15). Today Belgium seems even less unified than in the past following a long-running constitutional crisis in 2007–8.

Brittany: Breton (a member of the Celtic family) is the original language of this region in northwestern France and is still spoken by 10 per cent of the population. Although various groups agree that culture and language in the region need to be defended, typically separation is not favoured.

Basque country: A distinctive language region divided by the French–Spanish border is the country of the Basques. One separatist organization practised terrorism for many years but declared a permanent ceasefire in 2006. Other Basque organizations have long striven for separate identity through democratic means. A Basque assembly was convened in 1977.

Catalonia: The Catalan language is being revived, and today 17 per cent of the people in this autonomous region in northeast Spain

Box 8.6   Conflicts in the Former Yugoslavia

From the mid-fifteenth to the mid-nineteenth centuries, the area that was to become the federal state of Yugoslavia was part of the Ottoman Empire. By the beginning of the twentieth century, some parts—notably Serbia—were independent, while others belonged to the Austro-Hungarian Empire. At the end of World War I, Yugoslavia was founded as a union of south (Yugo) Slavic peoples, although the name was not adopted until 1929. Following World War II, a socialist regime was established under the leadership of Tito, who dominated the political scene until his death in 1980, after which a form of collective presidency was established. The ethnically diverse area of Yugoslavia was organized as six socialist republics and two autonomous provinces (Table 8.1).

The ethnic diversity evident in Table 8.1 is further complicated by the major European cultural division between eastern and western Christianity that runs through former Yugoslavia as well as through Romania, Ukraine, and Belarus (Figure 8.10). It is therefore not surprising that this federal state was dissolved in 1991, following Serbia's efforts in 1988 to exert a greater influence in the federation by assuming direct control of the two autonomous provinces (Kosovo and Vojvodina).

The process of dissolution began in 1991 with the secession of Slovenia, which was quickly followed by that of Croatia. In Bosnia bitter conflict between Croatian Muslims

Table 8.1	Ethnic Groups in the Former Yugoslavia
Republic	**Ethnicity**
Serbia	85% Serbs
Croatia	75% Croats, 12% Serbs
Slovenia	91% Slovenes
Bosnia-Herzegovina	40% Slavic Muslims, 32% Serbs, 18% Croats
Montenegro	69% Montenegrins, 13% Slavic Muslims, 6% Albanians
Macedonia	67% Macedonians, 20% Albanians, 5% Turks
Province	
Kosovo	77% Albanians, 13% Serbs
Vojvodina	54% Serbs, 19% Hungarians

and Serbs introduced that terrible euphemism 'ethnic cleansing' (recall the discussion of genocide in Chapter 7). In 1999 violence flared in the southern province of Kosovo as Serbians and ethnic Albanians contested an area that has meaning to both groups; Western powers eventually intervened on behalf of the Albanian population.

Many commentators attributed that conflict to the particular effect of Kosovo on the Serbian national psyche, for it was in Kosovo that the Serb Prince Lazar was killed by Turks in 1389—a defeat that would lead to Serbia's annexation by the Ottoman Empire in 1459. Nevertheless, as a result of Prince Lazar's reported heroism in the face of adversity, many Serbs—or at least Serb leaders striving to reinforce Serbian identity—came to regard Kosovo as a sacred space. Moreover, according to the Serbs, the ethnic Albanian population of the region had sided with the Turks. Not surprisingly, the Albanian people of Kosovo see things differently: according to their traditions, Albanians were living in the region before the Serbs and fought with the Serbs against the Turks. When, in 1999, Serb forces entered Kosovo, purportedly to quell an armed independence struggle by the Kosovo Liberation Army, the conflict led to several instances of ethnic cleansing and massive refugee movements of ethnic Albanians.

Following this series of nationalist independence movements, several instances of ethnic slaughter, and a series of wars, what had been one country, Yugoslavia, became five countries: Slovenia, Croatia,

FIGURE 8.10   The former Yugoslavia

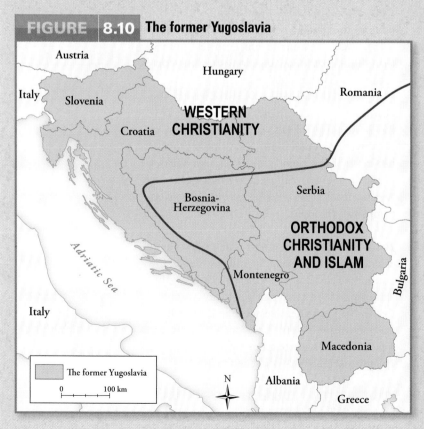

Continued

Bosnia-Herzegovina, Serbia-Montenegro, and Macedonia. Subsequently, in 2006 Montenegro voted to become independent from Serbia, a move that leaves Serbia 'landlocked and alone, a final nail in the malign nationalist dream of a "Greater Serbia" that fuelled such bloodshed in the 1990s' (*The Times*, 2006).

Finally, in 2008, Kosovo—with a 90 per cent ethnic Albanian population that has long sought independence—declared itself independent. The declaration was rejected by Serbia, essentially on the grounds that Kosovo was not a republic in the former Yugoslavia but rather a province of Serbia. But it was accepted by the United States and by some other countries (54 in total one year after the declaration), although described as a special case of separation not intended to set a precedent. As of 2009 debate continues about the legality of Kosovo as an independent state following a UN decision to refer the matter to the International Court of Justice.

Following the collapse of the Ottoman, Austro-Hungarian, and Soviet empires in the twentieth century, the demise of Yugoslavia might be seen as marking a symbolic end to efforts to bind multiple ethnic groups into artificial single entities. Identity—national, ethnic, linguistic, and religious—has been a central feature of recent conflicts in the former Yugoslavia. It was with this thought in mind that S.P. Huntington described the conflicts as one example of a 'fault line' war, where West and East, Christianity and Islam, and Catholicism and Eastern Orthodox struggle to coexist and even to survive.

One especially tragic component of the conflicts that erupted as Yugoslavia disintegrated was the practice of 'ethnic cleansing' or genocide. The most notorious example was carried out in 1995 at Srebenica—in a supposedly safe area under the control of UN peacekeeping forces—where 23,000 Bosnian Muslim women and children were expelled by Serbs and over 7,000 Muslim men and boys murdered, in some cases after being blindfolded and bound (Wood, 2001).

Ethnic tensions have also surfaced within the larger region of the Balkans. Since the collapse of their Communist regimes, Bulgaria has felt threatened by its Turkish population and Romania by its Hungarian population.

---

speak it. A Catalan assembly was convened in 1977. A historic agreement was reached in 2006 with the Spanish government accepting that Catalans have a right to be described as a nation and also giving the region some powers over taxation.

Corsica: One militant group seeks full independence from France, and today the French government perceives Corsica as a real threat to national unity. Schools are now permitted to teach Corsican. However, a 2003 referendum narrowly rejected an autonomy offer from France.

Other areas in Western Europe that recently have seen movements for autonomy or independence include Galicia (northwest Spain), Andalusia (southeast Spain), southeast Belgium (German speakers), Alsace (a German-speaking region in France), the Jura region (France–Switzerland), Tyrol (northern Italy), and Sardinia.

Thus, although language is the principal centripetal force in many Western European states, there are numerous specific centrifugal forces. Much the same can be said of pre-1989 Eastern Europe and the former USSR, where nationalist sentiments were rarely publicized. Events since 1989 have highlighted the national problems of states such as Romania, where a Hungarian minority has been repressed; Bulgaria, where the majority attempted to assimilate Turks in their efforts to create a united Bulgaria; and Yugoslavia, where Serbian nationalism has proved to be highly destructive (Box 8.6).

The 2006 referendum that approved the independence of Montenegro raises the possibility of an increasingly fragmented Europe as

Exiled Uighur leader Rabiya Kadeer punches the air during a 2009 protest by Australian Uighurs outside the Chinese Consulate in Melbourne. Kadeer is calling on the Australian government to push for an international inquiry into the alleged atrocities in Urumqi, the capital of Xinjiang province.

Paul Crock/AFP/Getty Images

other groups seek independent status. A concern here is that such aspirations may increase political instability in already unstable regions, especially in the former Soviet Union.

THE FORMER USSR

As discussed in Box 8.7, the USSR was not a typical state; rather, it was the last great world empire. In fact, in 1918 the three Baltic republics (Latvia, Lithuania, and Estonia), along with Ukraine, Byelorussia, Georgia, Azerbaijan, and Armenia, all fought for independence from the Russian empire; in 1990 conflict flared once again between Christian Armenians and Muslim Azerbaijanis in what was then the Soviet Republic of Azerbaijan;

and between 1988 and its collapse in 1991, the Soviet Union experienced considerable separatist pressure from the three Baltic republics and from southern republics with significant Islamic populations.

Today, Georgia, one of the former republics within the USSR, is being challenged by two regions along the border with Russia that, with Russian support, are seeking to separate from Georgia. Abkhazia and South Ossetia were the sites of major conflict in the early 1990s and again in 2008. Abkhazia is seeking independence, while the principal goal of the population in South Ossetia is to unite with North Ossetians living in a neighbouring autonomous republic of the Russian Federation.

Box 8.7   The Collapse of the USSR

The Union of Soviet Socialist Republics was never really a union; rather, it was an empire tightly controlled by one of the republics, Russia (Figure 8.11). In this sense, it differed little from its predecessor, the pre-1917 Russian empire.

Russian expansion from the small core area around Moscow usually is explained in terms of a continuing search for good agricultural land, seaports, and borders that could be easily defended against regular invasions by Vikings, Poles, Germans, and Mongols. To the north, the Russians built the White Sea port of Archangel by 1584; to the west, they established St Petersburg in 1703. Meanwhile, Russia was expanding to the east, across Siberia; Russians crossed the Urals in 1582, reached the Sea of Okhotsk at the edge of

FIGURE 8.11   The former USSR

Continued

the Pacific by 1639, claimed Alaska in 1741, and during the nineteenth century established fur-trade settlements as far south as California. To the south, Russia moved down the rivers Don, Dnieper, and Volga, acquiring Ukraine and gaining access to the Black Sea by 1800. During the nineteenth century, Russia moved into the area between the Black Sea and the Caspian Sea.

Russia retreated from North America in 1867, when it sold Alaska to the United States, but the empire that the new Communist government inherited in 1917 nonetheless was massive. It expanded even farther with the suppression of local independence movements and the re-annexation of several areas that had broken away from the Russian Empire after World War I, including Ukraine, Moldova, Estonia, Latvia, and Lithuania.

From 1917 until the collapse of the USSR in 1991, considerable numbers of ethnic Russians moved into other republics of the Union, which adopted policies favouring the Russian language and suppressing the practice of religion. The collapse of the USSR, as part of the sweeping changes that occurred in Europe in the late 1980s and early 1990s, left a difficult and uncertain legacy of ethnic tensions and frequent ethnic conflict. Table 8.2 provides data on ethnic populations in the former republics of the USSR, which are now 15 independent states, and highlights the high percentages of Russians in most of these states. The creation of the Commonwealth of Independent States in 1991 was an attempt by Russia to maintain links with the newly independent states, but it has proven ineffective.

Table 8.2	Ethnic Groups in the Former USSR		
Country (former republic)	Population (millions)	Titular Nationality (%)	Principal Minorities
Russia	141.9	82	Tatars 4%
Ukraine	46.2	73	Russians 22%
Uzbekistan	24.4	71	Russians 8%
Kazakhstan	15.4	40	Russians 38%, Ukrainians 5%
Belarus	9.7	78	Russians 13%
Azerbaijan	7.7	83	Russians 6%, Armenians 6%
Tajikistan	6.2	62	Uzbeks 23%, Russians 7%
Georgia	5.4	70	Armenians 8%, Russians 6%, Azeris 6%
Moldova	4.1	65	Ukrainians 14%, Russians 13%
Turkmenistan	4.8	72	Russians 9%, Uzbeks 9%
Kyrgyzstan	4.7	52	Russians 22%, Uzbeks 12%
Armenia	3.8	93	Azeris 2%
Lithuania	3.4	80	Russians 9%, Poles 7%
Latvia	2.3	52	Russians 35%
Estonia	1.3	62	Russians 30%

There are problems also in the Central Asian area where the collapse of the USSR resulted in the five republics of Kazakhstan, Uzbekistan, Tajikistan, Turkmenistan, and Kyrgyzstan emerging as independent countries. Each contains a significant minority population of Russians—in Kazakhstan it was as high as 38 per cent, but is declining as a result of out-migration.

One group closely related to the majority populations in these new countries and with similar aspirations for independent state status seems unlikely to see those aspirations fulfilled in the immediate future. The Uighurs, who live in the northwestern Chinese province of Xinjiang, are part of the same larger group to which the people of the five new Central Asian countries belong. This group conquered much of Asia and Europe before converting to Islam in the fourteenth century; at that time the larger region was one of great wealth and included such important centres as Tashkent and Samarkand. But by the mid-nineteenth century, the entire region had been conquered, either by Russia or, in the case of the Uighur group, by China. Today only the Uighurs have not achieved independence. They launched a new round of independence claims in 1997, following the death of long-time Chinese leader Deng Xiaoping. However, so many Han Chinese immigrants have moved into the region that the Uighurs now make up only about half of the total Xinjiang population of some 16 million. The future of the Uighurs remains uncertain.

SOUTH ASIAN CONFLICTS

Ethnic and other tensions in South Asia are reflected in a highly unstable political landscape. Among the factors behind those tensions are a complex distribution of ethnic groups and associated religious rivalries (especially, but not exclusively, Hindu–Muslim), the caste system, and the 1947 departure of the British as colonial rulers (Zurick, 1999).

South Asia had been the 'jewel in the crown' of the British Empire. With the withdrawal of the British in 1947, the two new states of India and Pakistan were born, and in 1948 Ceylon became independent (it would be renamed Sri Lanka in 1972). However, the goal of creating national identities has proved elusive. The intention in planning for the formation of the new states was that India would be predominantly Hindu and Pakistan predominantly Muslim, with the boundaries to be imposed (on the basis of 1941 census data) where none had existed before, dividing several well-established regions—notably Kashmir and the Punjab in the northwest and Bengal in the northeast. One immediate result was the movement of about 7.4 million Hindus from Pakistan to India and 7.2 million Muslims from India to Pakistan. It is estimated that several million people died in the chaotic redistribution of population that ensued. The new state of Pakistan was divided into two parts, West and East, with roughly equal numbers of people separated by 1,600 km (1,000 miles) of Indian territory. The capital was located in West Pakistan, first at Karachi, then at Islamabad, and—perhaps inevitably—relations between the two sections were poor. By 1971 a large part of the Pakistani army was in East Pakistan fighting Bengali guerrillas. In that year the eastern section seceded, creating the independent state of Bangladesh.

Today conflict is endemic in much of South Asia (see Figure 8.12). Five areas of tension are located in the Himalayan Mountains alone. Most notably, the northwestern region of Kashmir is the subject of ongoing struggles between India, which officially controls the entire state, and Pakistan, which in practice administers the upper portion of it and also lays claim to the predominantly Muslim Kashmir Valley to the south; it is widely feared that these tensions could erupt into war. At the same time, multiple Kashmiri separatist movements are demanding independence

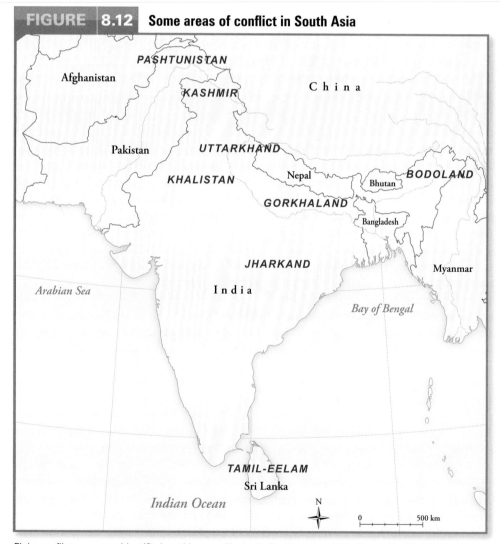

FIGURE 8.12   Some areas of conflict in South Asia

Eight conflict areas are identified on this map: Kashmir, Pashtunistan, Uttarkhand, Gorkhaland, Bodoland, Khalistan, Jharkand, and Tamil-Eelam.

The northeastern part of Sri Lanka, the disputed territory and the site of much of the fighting during the civil war.

Design Pics Inc./Alamy/ GetStock

from both India and Pakistan. Northwest of Kashmir in Pashtunistan, Pathan tribes are divided between Afghanistan and Pakistan but make strong claims for statehood. South of Kashmir, Pahari-speaking people favour creation of a separate state they call Uttarkhand, while in the eastern Himalayan region of Gorkhaland some 10 million ethnic Nepalese aspire to statehood and have been involved in several uprisings in recent years. Farther still to the east, in Bodoland—where India borders Myanmar and China—a number of groups live independent both of one another and of central state control.

Meanwhile, south of the Himalayas in the Punjab, Sikhs aspire to independence from India and the creation of their own state, Khalistan. The struggle for control led to serious violence in the 1980s, but recent years have seen Sikhs turning to more peaceful political activity. Farther east, numerous groups collectively known as the *adivasi* (indigenous peoples) assert their right to a state known as Jharkand. Finally, Sri Lanka was for many years in a state of civil war between the Buddhist Sinhalese majority, who run the government, and the Hindu Tamils, who want their own independent homeland, Tamil-Eelam. The first Tamils arrived on the island around the eleventh century, but a second group, known as Indian Tamils, was taken there by the British in the late 1800s to work on tea plantations. Conflict first flared in the early 1900s when

Tamils moved onto Sinhalese land, prompting Sinhalese resentment and political and other discrimination. Sinhalese nationalism became more assertive over time, and Tamils mounted many organized terror campaigns between the 1970s and 2009, when they finally surrendered to government forces. It is estimated that about 80,000 people died during the almost 30-year civil war. Although that war has ended, feelings of resentment and claims for a separate political identity are unlikely to go away (see Box 5.9).

The above account covers only a smattering of the conflicts that plague South Asia. In some cases ethnic groups war against the state; in others, ethnic groups war against one another; and in yet other cases the conflicts have international dimensions. There are, for example, continuing tensions between India and China (although a border agreement was signed in 2005) and between India and Pakistan (although recent developments suggest that these two nuclear-armed powers are increasingly able to co-operate—notwithstanding tensions relating to the Mumbai terror attacks in 2008). Further complicating the scenario are ongoing Hindu–Muslim differences, the discovery of a Hindu terror cell in 2008, and the complexities of the caste system.

INTERNAL DIVISIONS IN CANADA

Are there internal divisions in Canada? It is a huge country of 9.8 million km² (3.8 million square miles), 33.7 million people, and large unsettled areas. Much of the land is inhospitable to humans; sixteenth-century French explorer Jacques Cartier described the southern coast of Labrador as the land that God gave to Cain. Further, Canada comprises 10 provinces and three territories, with marked regional variations. Ways of life range from traditional to very modern. It is a plural society made up of Aboriginal peoples, the two founding European cultures (English and French), and a host of other national groups. Together, these physical and human factors have produced strong regional feelings—a major cause of instability. In recent years, the unity of Canada has been threatened not only by Quebec (French-language) separatism (which suffered a setback with the 2003 election of a Liberal provincial government in Quebec City), but by various western Canadian separatist movements as well (Box 8.8).

Groupings of States

As noted, the contemporary world is characterized by two divergent trends. In addition to the many cases outlined above of regions and peoples actively seeking to create their own independent states, some groups of states are actively choosing to unite, sometimes even at the cost of sacrificing aspects of their sovereignty.

EUROPEAN INTEGRATION

The principal example of such a voluntary grouping was established following the end of World War II in 1945—the date that marked the end of the old Europe. Various moves towards European union resulted in the 1957 formation of the European Economic Community (EEC) by France, Belgium, the Netherlands, Luxembourg, Italy, and West Germany. These six states had already created a European Coal and Steel Community and a European Atomic Energy Community, both of which merged with the Economic Community in 1967 to form a single Commission of the European Communities. A common agricultural policy was adopted. Other European countries that did not favour such close integration formed the European Free Trade Association in 1960.

Gradually, the EEC, now the European Union, assumed dominance, with Britain, Denmark, and Ireland joining in 1973, Greece in 1981, Portugal and Spain in 1986, and Austria, Sweden, and Finland in 1995 (Norwegians voted against joining in 1994; indeed, perhaps

Box 8.8   Regional Identities and Political Aspirations

Many regional populations lack a sense of belonging as part of a political unit with which they are unable to identify, and the desire to be an independent political unit is at the head of the agenda for many groups. This desire is understandable, given the importance of states in the contemporary world: 'they behave very much like Greek gods, so much so that, given the great powers which they wield, they need to be taken very seriously' (East and Prescott, 1975: 1).

Is it the existence of local and regional nationalisms a good thing or a bad thing? Of course, there is no correct answer; but the question is still worth thinking about. The current world political map, as we know all too well, is not sacrosanct. Changes occur, and often for very good reasons. It may seem that regionalism is in fashion today. But is the trend towards the creation of new states, reflecting ever more localized identities, something that should be encouraged? The world as a whole is an uncertain and somewhat unpredictable environment. On the one hand are the processes of globalization and movement towards the integration of some states; on the other are numerous groups arguing that they are different from others and therefore should be separate from them. Most states in which certain groups are seeking some degree of independence react negatively to the possible loss of territory, population, and prestige.

Is it better for a group to look outward and encourage cosmopolitanism or to look inward and encourage ethnicity? Or is it possible to accomplish both these goals simultaneously? Does support for a sub-state level of nationalism result in an ethnocentric world view and encourage lack of respect for others, or does it allow for increased self-respect and a better understanding of others? Although there are no agreed-upon answers to such difficult questions, the literature in social psychology suggests strongly that simply classifying people into groups prompts a bias in favour of one's own group (Messick and Mackie, 1989). Such biases may arise even between groups that are interdependent and co-operative, as is the case with English- and French-speaking Canadians.

Canada is not the only country in which different groups hold different opinions of movements to achieve some degree of autonomy within a larger state.

A US border patrol agent drives along the US–Mexico border in Jacumba, California, as prospective immigrants wait to attempt an illegal crossing.

CP/AP photo/Susan Sterner

the most surprising fact about the EU is that neither Norway nor Switzerland, two countries with very high per capita incomes, has opted to join.). The most dramatic expansion of the EU took place in 2004 with the addition of 10 countries, bringing the total to 25. Eight of the new members—Estonia, Latvia, Lithuania, Poland, the Czech Republic, Slovakia, Hungary, Slovenia—are in Eastern Europe. The remaining two are the Mediterranean islands of Malta and Cyprus. The latter poses a political problem because it was partitioned after an attempted Greek coup in 1974, and since then a buffer zone policed by the United Nations has divided the Greek-controlled Republic of Cyprus from the northern area, which is controlled by Turkey and not internationally recognized.

When these 10 countries were accepted for membership, three others—Romania, Bulgaria, and Turkey—were considered not ready to join. Romania and Bulgaria joined in 2007, although with some initial membership restrictions, meaning that the number of members in 2009 was 27. However, the case of

Turkey is seen as problematic as the country has a record of human rights violations. Likely, the remaining Baltic countries and several in Eastern Europe will join within the next few years. It is also possible that the EU will expand into North Africa, which would raise complex questions about 'European' identity. Also, in response to the prospect of national financial collapse in 2009 related to the larger global economic crisis, Iceland was fast-tracked to join the EU.

Why have so many sovereign states been willing to sacrifice some components of their independence? Probably the most important reason is the inherent appeal of a united Europe in a world dominated before 1991 by the US and USSR and now by the US alone. Today the 27-member EU has a population of more than 500 million—larger than the total population of the three countries (Canada, the US, and Mexico) linked by the North American Free Trade Agreement (NAFTA). During the formative years of the EU, the United States was seen as a real economic threat and the Soviet Union as a real military threat. Thus, it seemed that there was a place for Europe in the world—but not for some patchwork of European states. As the proponents of union saw the two superpowers, both were large; both occupied compact blocks of territory with a low ratio of frontier to total area; both had substantial east–west extent that increased the area of comparable environment. Both were mid-latitude countries with large populations and densely settled core areas, and both had a wide variety of natural resources. Individual European countries did not share any of those crucial characteristics—but a

united Europe would. Unlike the US, however, a united Europe would be multinational, and a voluntary multinational state might be more stable than an involuntary one such as the empire of the former USSR.

A radically different, indeed controversial, view of Europe's future was proposed by L. Kohr (1957). Arguing that aggression is a result of great size and power, Kohr suggested that Europe would be a more peaceful place if, instead of uniting, it divided into many small ethnic states as shown in Figure 8.13. Such an arrangement would minimize the problems associated with borders and national minorities

FIGURE 8.13   European ethnic regions

This map needs to be read carefully and critically. As the text discussion notes, Kohr suggested that sometimes things do not work because they are too big, and, little over a decade after World War II, he advocated a return to a pre-modern European world. He is sometimes described as a radical decentralist, unsympathetic to larger globalization tendencies. The map is not a definitive statement—indeed, it might be considered highly problematic in that whether these regions ever really existed in some meaningful way is debatable. The map is included here to encourage critical thought about the merits, or drawbacks, of a political world comprised of many small, ethnically based, political territories.

SOURCE: L. Kohr, *The Breakdown of Nations* (Swansea: Christopher Davies, 1957).

and, in effect, return Europe to its medieval form. Although at present Europe appears to be moving in a different direction, evidence is increasing of regional authority within existing states.

Indeed, even as Europe is becoming increasingly integrated, the profusion of states and independence movements within it suggests that fragmentation might be more characteristic of its component parts. Since the end of World War II, certainly, political tensions in Europe have centred on intranational rather than international disputes. We are confronted with a complex scenario in which individual states are integrating with one another at a time when many of those same states are experiencing serious internal stresses.

OTHER GROUPINGS OF STATES

Apart from the EU, which has its own parliament, elected representatives from the member countries, and a rotating executive, the principal groupings of states, with the exception of the British Commonwealth, the African Union, and the Organization of American States, are essentially limited to establishing and maintaining trade blocs, such as NAFTA and ASEAN, noted earlier. Other such groupings include the Economic Community of West African States, the Latin American Integration Association, MERCOSUR (Mercado Común del Sur, a customs union of Argentina, Brazil, Paraguay, Uruguay, and Venezuela), the Organization of Petroleum Exporting Countries (see Chapter 14 for a detailed discussion of this grouping of countries), Asia–Pacific Economic Co-operation (APEC), and the Maghreb Union. These are considered in Chapter 9 in the context of regional economic integration. In Eastern Europe an economic group named Comecon (the Council for Mutual Economic Assistance) began meeting secretly in 1949, but was terminated in 1991 following the collapse of the Communist bloc. Finally, there is the prospect of a single world government. Since 1945, the United Nations has striven to assist member states in limiting conflict. Today, most countries are members, and states in the less developed world now make up the majority.

The Role of the State

Most of the world's people are citizens of a specific state and, as such, are subject to its laws; they have limited power—if any—to change the state significantly, and they are spatially tethered. Our everyday life is irrevocably involved with government, for all states 'are active elements within society, providing services which are consumed by the public' (Johnston, 1982: 5). Political units correspond spatially to cultural, social, and economic units, and human geographers often find it useful to define regions on a political basis. The state is a factor affecting human life; it has evolved with society and is a key element in the mode of production. Because of the state's crucial role in the lives of people and places, it is important to recognize that there are many different ways in which a state can be governed.

FORMS OF GOVERNMENT

The two principal political philosophies today are capitalism and socialism, but a number of related or alternative ideas are also important to us. Fundamental to the capitalist form of government is **democracy**: rule by the people. Democracy implies five features: regular free and fair elections to the principal political offices; universal suffrage; a government that is open and accountable to the public; freedom for state citizens to organize and communicate with each other; and a just society offering equal opportunity to all citizens. Democracy was important for a period in classical Greece, but only in the nineteenth century did it reappear as a major idea, and only in the twentieth century was it generally approved and commonly practised.

Monarchy is rule by a single person. Constitutional monarchies do survive in some countries, but generally without any real power. Britain's monarchy legitimizes a hierarchical social order; those of the Netherlands, Denmark, and Norway are more democratic. **Oligarchy** is rule by a few, usually those in possession of wealth. The term was introduced by Plato and Aristotle; the favoured term today is 'elite'.

Government by **dictatorship** implies an oppressive and arbitrary form of rule established and maintained by force and intimidation. Military dictatorships are common in the contemporary world; an example is North Korea. Fascism is an extreme form of nationalism that, especially in Europe between 1918 and 1945, provided the intellectual basis for the rise of political movements opposing governments whether they were capitalist or socialist. In Italy and Germany, fascist parties achieved

Democracy
A form of government involving free and fair elections, openness and accountability, civil and political rights, and the rule of law.

monarchy
The institution of rule over a state by the hereditary head of a family; monarchists are those who favour this system.

oligarchy
Rule by an elite group of people, typically the wealthy.

dictatorship
An oppressive, anti-democratic form of government in which the leader is often backed by the military.

anarchism
A political philosophy that rejects the state and argues that social order is possible without a state.

Maoism
The revolutionary thought and practice of Mao Zedong (1893–1976), based on protracted revolution to achieve power and socialist policies after power is achieved.

federalism
A form of government in which power and authority are divided between central and regional governments.

power through a combination of legal means and violence.

The political philosophy of **anarchism** may emphasize either individualism or socialism and was advocated by two notable nineteenth-century geographers, Kropotkin and Réclus. It rejects the concept of the state and the associated division of society into rulers and ruled.

SOCIALIST LESS DEVELOPED STATES

'Socialism' is an imprecise term, but we can identify two general characteristics of socialist regimes:

1. They aim to remove any and all features of capitalism, especially private ownership of resources, resource allocation by the marketplace, and the class structure associated with them.
2. They have the power, in principle, to make substantial changes to society.

Recent experiences in the socialist developed world of Eastern Europe have made it clear that neither of these characteristics has met with popular approval.

Nevertheless, socialism has had a major impact in the less developed world. Several African countries have attempted to combine tradition with socialist ideals; perhaps the most successful has been Tanzania. In Latin America, Cuba's socialist system has survived since 1959 in defiance of the dominant capitalist model, although changes began in 2008 when ill-health prompted Fidel Castro to step down in favour of his brother, Raul. Also, socialist movements have played important roles in mobilizing the poor in various South American countries, notably in Uruguay, Bolivia, and Venezuela, with the election of leftist governments.

But in Asia socialism has been able to exert its most significant and continuing influence (Hirsch, 1993). Today there are socialist or Communist governments in four major Asian countries: China, Laos, Vietnam, and North Korea. Of these, only North Korea still maintains a traditional hard-line, secretive, closed system. China, in particular, seems clearly to be moving towards a form of capitalism and towards more democratic institutions. Interestingly, North Korea today is run by a Communist 'royal family'; before his death in 1994, Kim Il Sung groomed his son, Kim Jong Il, as heir. North Korea remains a closed

country, although it made meaningful contact with democratic South Korea in 2000 and with China in 2006, and in recent years closely monitored visits between relatives in the two Koreas have been arranged. However, North Korea's underground testing of a nuclear weapon in October 2006 served, at least initially, to isolate it even further from the wider world.

One particular version of socialism is based on the revolutionary thought and practice of Mao Zedong, the leader of the peasant revolution in China that led to the 1949 creation of the People's Republic of China. **Maoism** has two components: a strategy of protracted revolution to achieve power, and the practice of socialist policies after the revolution succeeds. Both aspects continue to be influential in other parts of the world.

In many less developed countries, socialism has a strong anti-colonial, nationalist component. Unlike the former socialist states in the developed world, which are largely urban and industrial, most of these states are firmly rural in character. Although the details vary, it is fair to say that in these states, even more than in capitalist states, individual behaviour is often determined by larger state considerations. Fertility policies, for example, are likely to be rigorous, as in China. Perhaps most important, central planning—that is, planning by the central government—means considerable state involvement in people's everyday lives.

SUBSTATE GOVERNMENTS

In many states, political authority is not entirely centralized at the national level. The political geographer Paddison (1983) has distinguished three levels of centralization (Figure 8.14):

1. Unitary: The most centralized form of government; local governments are used by the central state to organize the political hinterland.
2. Federal: The least centralized; examples include Australia, Canada, and the United States. One purpose of **federalism** is to prevent one level of government from dictating to another.
3. Compound unitary: Midway between the federal and unitary types, these systems devolve substantial powers to subnational governments, but less power than in the federal case. There are two types: Type I has established regional governments, and Type

FIGURE 8.14 World distribution of state types

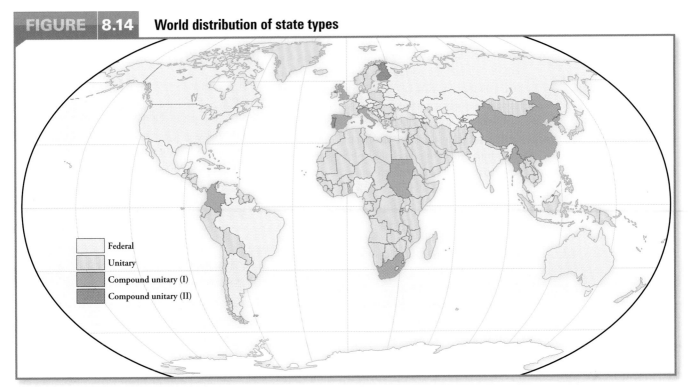

Federal
Unitary
Compound unitary (I)
Compound unitary (II)

This map reflects the distribution of states in the early 1980s, before the collapse of the Soviet Union.

SOURCE: R. Paddison, ed., *The Fragmented State: The Political Geography of Power* (New York: St Martin's Press, 1983), 32.

II involves small (usually peripheral) areas that maintain some distinct identity while retaining ties to the state.

According to this classification, most of the large-population states (except China) are federal, although the most common type generally is unitary. Any understanding of how a particular state works must therefore consider all appropriate levels of government.

More generally, an increasing trend towards decentralization means that various administrative, political, and financial responsibilities are being transferred from national to subnational governments. In South Africa, the government is hoping that decentralization will help to unify a country so long divided by apartheid legislation. In both Ethiopia and Bosnia-Herzegovina, decentralization has been introduced to accommodate various ethnic tensions. Increasing decentralization in the federal state of India is in evidence, with several states devolving powers to local governments. Finally, even the unitary state of China is also moving in this direction.

EXERCISING STATE POWER

In capitalist countries, state power is exercised through various institutions and organizations (Clark and Dear, 1984); this **state apparatus** includes the political and legal systems, the military or police forces to enforce the state's power, and mechanisms such as a central bank to regulate economic affairs. Significant spatial variations occur in the management of this state apparatus, as well as in public-sector income and spending and the provision of **public goods**, including services such as health care and education. Inequalities in the distribution of public goods often reflect government efforts to influence an electorate prior to an election (a popular American term for this behaviour is 'pork barrelling'). Clearly, the form of government and the particular political philosophy favoured directly affect the manner in which state power is exercised.

One of the most critical issues concerning the power exercised by individual states is the need for international co-operation in solving global environmental problems. As we saw in Box 3.2 on the 'tragedy of the commons', state power is needed to ensure that private-sector industries do not harm the environment. In the same way, some international authority is needed to ensure that individual states behave responsibly. But as yet there is no such international power—only the occasional meeting of states convened to address particular problems.

state apparatus
The institutions and organizations through which the state exercises its power.

public goods
Goods that are freely available to all or that are provided (equally or unequally) to citizens by the state.

There are at least two dilemmas here (Johnston, 1993). First, governments in more developed countries may be unwilling to protect the global environment when such actions will result in losses of jobs and wealth (and therefore votes) in their own states. Second, governments in less developed countries contend that they cannot afford to implement environmentally appropriate policies and practices, that their practices are not the principal causes of environmental problems, and that as they develop they should not be put at a disadvantage vis-à-vis the lack of stringent policies in place when the more developed countries first set out on the path of industrialization.

The politics of protest: Social movements and pressure groups

The relationship between a state and its population is always subject to change. In extreme cases, a substantial proportion of the people may reject the state's authority altogether—a situation that can lead to significant internal conflict. An example is Northern Ireland, where many citizens believe that the British state has no legitimate authority over them. In other cases, the people may accept the legitimacy of the state itself but oppose the specific governing body; in such cases, efforts to replace the people in charge with others deemed more appropriate may lead to civil war. Even in states that are relatively stable and peaceful, however, usually several groups are striving to influence government policy in various areas.

In many parts of the more developed world today, it is not uncommon for state governments to face protest from groups with specific agendas; such 'beyond the state' social movements have been organized around a vast range of concerns, from racism and labour issues, to women's issues, environmental conservation, gay and lesbian rights, and nuclear weapons, to housing conditions and urban redevelopment.

Social movements are collective endeavours to instigate change, characterized by active participation on the part of their members and—unlike the state—a relatively informal structure (Carroll, 1992; Wilkinson, 1971). Specific details vary, of course, but it is not uncommon for social protests to begin following the failure, from the perspective of the interested group, of earlier attempts at persuasion and collaboration. Nevertheless, social protest does not always lead to social change.

gerrymandering
The realignment of electoral boundaries to benefit a particular political party.

Elections: Geography Matters

Much attention has been paid to the question of voting since E. Krebheil's (1916) analyses of geographic influences in British elections. The focus tends to be on spatial and temporal variations in voting patterns and on such causal variables as environment, economy, and society. In any analysis of elections, geography is an important factor: in the boundaries of voting districts, voting behaviour, government activity, and larger world issues.

LEGITIMACY OF ELECTIONS AND VOTER TURNOUT

Some elections are correctly described as free—meaning there is a multi-party system, universal adult suffrage, public campaigning, secret ballots, and no voter fraud—but others lack one or more of these characteristics and are better described as compromised. An analysis of recent elections in 154 countries identified 39 as compromised, mostly in African countries but also in Asia.

Voter turnout varies enormously. Recent compromised elections in Turkmenistan, Rwanda, and Russia reported voter turnouts in excess of 95 per cent. Free elections in Australia regularly have 95 per cent turnouts, at least partly because Australia is one of about 30 countries where voting is compulsory—but this requirement is not rigorously enforced in all cases and in other cases the penalty for not voting is not significant (in Australia it is a $20 fine). Countries that often have a very low voter turnout include several African and southeastern European countries. The lowest-ever voter turnout for a federal election in Canada was just under 60 per cent in 2008.

CREATING ELECTORAL BIAS

In 1812 the governor of Massachusetts, Elbridge Gerry, rearranged voting districts (Figure 8.15) to favour his party. Today, **gerrymandering**—a word coined by Gerry's political opponents—refers to any spatial reorganization designed to favour a particular party. It aims to produce electoral bias in one of two ways: either by concentrating supporters of the opposition party in one electoral district, or by scattering those supporters so that they cannot form a majority anywhere. Racial gerrymandering has been especially

FIGURE 8.15 The original 'gerrymander'

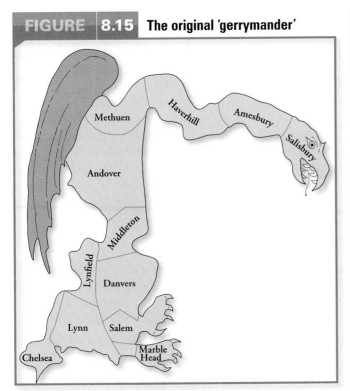

This apportionment was Intended to concentrate the vote for one party in a few districts. The term (introduced in the *Boston Gazette*, 26 March 1812) reflected the name of the governor who signed the law—Gerry—and the supposedly salamander-like shape of the new configuration.

SOURCE: R. Silva, 'Reapportionment and Redistricting', *Scientific American* 213, 5 (1965): 21. Picture Collection, The Branch Libraries, The New York Public Library, Astor, Lenox and Tilden Foundations.

FIGURE 8.16 Gerrymandering in Mississippi

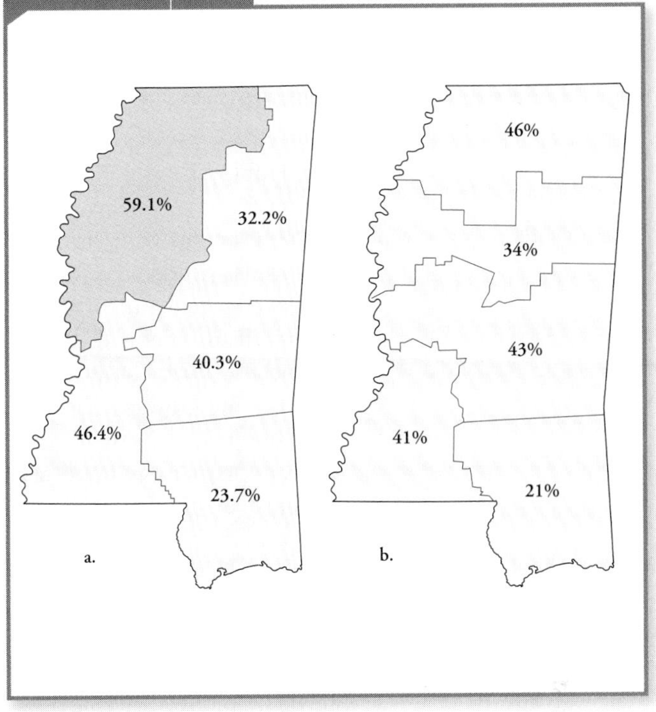

Congressional districts in 1960s Mississippi with the percentage of black voters noted: (a) the pre-1966 districts with one of the districts having a majority of black voters; (b) the post-1966 districts that guaranteed a minority of blacks in all districts—an example of deliberate gerrymandering.

SOURCE: Adapted from J. O'Loughlin, 'The Identification and Evolution of Racial Gerrymandering', *Annals, Association of American Geographers* 72 (1982): 180, Fig 1 (MW/RAAG/P1885). Used with permission of Taylor & Francis Ltd., http://www.informaworld.com.

common. Figure 8.16 shows how congressional district boundaries in Mississippi were manipulated in the 1960s: prior to 1966, blacks formed a majority in one of the five districts, but the redrawn boundaries ensured that blacks were a minority in all districts. Gerrymandering was especially prevalent in the early 1990s, when several US states fiddled with their electoral borders to create some districts dominated by black voters and others dominated by whites. In both cases the intention was to benefit the party in power (the Republicans) by reducing the impact of the black vote, which has traditionally gone mostly to the Democrats: in the first instance, by ensuring that many black votes would be wasted (since the winning candidate would be elected by a huge majority) and in the second, by ensuring that black voters would be too few to elect the candidate of their choice.

Gerrymandering is a deliberate effort to produce an electoral bias; yet it has only recently been deemed a violation of the Constitution in the United States. A second method, particularly useful to parties that appeal to rural voters, is **malapportionment**. Rural areas are typically much less densely populated than urban areas; thus, boundaries are drawn so that rural electoral districts contain a relatively small number of voters, who are thus able to exert a much greater influence than the same number of opposition voters in a large urban district.

Clearly, then, electoral bias is not difficult to produce. But it can also be produced quite unintentionally. Under what circumstances can electoral boundaries be considered fair? Some observers argue that a district ought to be a meaningful spatial unit in a human geographic context; others argue that districts ought to be as diverse as possible. Probably all that can be agreed on is that the number of voters in each district should be as close to equal as possible and that districts should be continuous.

malapportionment
A form of gerrymandering, involving the creation of electoral districts of varying population sizes so that one party will benefit.

VOTING AND PLACE

It is not uncommon to assert that class is a dominant influence on voting behaviour. Thus, in Britain there is a distinction between the Labour Party, traditionally representing the views of workers against those of the capitalist employers, and the Conservative Party, representing the employers. A similar distinction can be made in the US between the Democrats (worker-based) and Republicans (employer-based). In Canada the situation is less clearly defined: although the New Democratic Party (NDP) and the Liberals are more oriented towards the working class than the Conservatives, ethnic and regional divisions can play a more important role than class in Canadian politics. Recent election results in Northern Ireland, Scotland, and Wales also suggest a tendency for voting to be linked to ethnic or nationalist interests.

But is class really a key explanation of voting behaviour? Do people from the same class but different regions vote similarly? We have already noted that nationalism can disturb this relationship. Place also matters. Geographic research has clearly demonstrated that where a voter lives is crucial. This is especially well documented by Johnston (1985), who used data from the 1983 British election to show that national trends cannot simply be transferred to the local scale. Four types of local influences on voting can be identified:

1. *Sectional effects:* 'a long-standing geographical element of voting whereby differences in local and regional political culture produce spatial variations in the support given to the various political parties' (ibid., 279). Sectional effects have been clearly demonstrated in presidential elections in the US. Long-standing geographic cleavages require any successful candidate to build an appropriate geographic coalition; these cleavages are so factored into the electoral college system that, to be elected president, votes alone are not enough—they must be in the right place.

2. *Environmental effects:* for example, in the 1983 British election it was found that the higher the level of unemployment in an area, the more successful the Labour Party candidate was. Candidate incumbency is another environmental factor: candidates who already hold their seats usually attract more votes than challengers do.

3. *Campaign effects:* 'Vote-switching was more common in the safer seats and the longer-established parties (Conservative and Labour) did better in the marginal constituencies' (ibid., 287).

4. *Contextual effects:* individuals may be influenced in their voting decisions (as they are in other behaviours) by their social contacts—friends, neighbours, relatives, and fellow employees with whom they are in contact. The emphasis on social contacts here is in general accord with the 'symbolic interaction' interpretation of culture noted in Chapter 7.

Voting, then, is influenced by both class and place. Thus, any successful political party needs to develop a strong social and spatial base, and any meaningful analysis of elections needs to consider both factors.

The Geography of Peace and War

'In geographical terms, this planet is not too small for peace but it is too small for war' (Bunge, 1988).

A 2005 report produced by the Human Security Centre at the University of British Columbia shows that wars are less frequent today than in the past and have had fewer casualties. There has been a decline in all forms of political violence, excepting terrorism, since the early 1990s. Principal reasons for the reduction in conflicts include intervention by the UN, the end of colonialism, and the end of the **Cold War**.

CONFLICTS

The concepts of war and peace were not clearly distinguished in European thought until the eighteenth century, when capitalism and nation-states emerged. In this period war became a state activity with the creation of professional armies and the elimination of private armies, and war became a temporary state of affairs, with long periods of peace between conflicts. Between the French Revolution and World War I, wars were waged between nation-states but were generally territorial; after 1918 conflicts became more ideological—variously between communism, fascism, and liberal democracy. Only since

Cold War
The period of confrontation without direct military conflict between Western (led by the US) and Communist (led by the USSR) powers that began shortly after the end of World War II and lasted until the early 1990s.

the end of the Cold War have non-Western cultures begun to play a major role in global politics. Since 1945, most conflicts have been centred in the less developed world. Before 1945, most casualties were caused by global wars; after 1945, most casualties were caused by relatively local wars.

The contemporary world contains many traditional enmities. Most of these relate to ethnic rivalries and/or competition for territory. Conflicts between peoples, a feature of human life for eons, have become increasingly formalized and structured over time as the number of independent states has increased. Indeed, some authors believe that aggression is a natural human behaviour (see Box 8.9). Since 1945, the United Nations has offered member states the opportunity to work together to avoid conflict, and in Korea several countries fought together under the UN flag to resist aggression.

Conflicts may be grouped in five categories: (1) traditional conflicts between states; (2) independence movements against foreign domination or occupation; (3) secession conflicts; (4) civil wars that aim to change regimes; and (5) action taken against states that support terrorism.

Category 1 conflicts since 1945 include three India–Pakistan wars, four wars involving Arab states and Israel, and the Vietnam War. Category 2 conflicts have arisen primarily as a consequence of decolonization; examples include the conflicts in the Belgian Congo (1958–60), Mozambique (1964–74), and Indonesia (1946–9). Category 3 conflicts, over secession, include the struggles of Tibet (1955–9), Biafra (1967–70), and Philippine Muslims (1977–present). Category 4 conflicts include the civil wars in China (1945–9), Cuba (1956–9), Bolivia (1967), and Iran (1978–9). Category 5 conflicts include the US-led invasion of Afghanistan in 2001 and the 2003 war on Iraq. However, in both of the latter cases terrorism was probably only one of several motivations for the US invasions. In brief, the world may not be at war globally, but it is certainly not at peace with itself.

Conflicts between states (Category 1) have the greatest potential to disrupt the human world. Geographers have focused on various theories of international relations to explain such conflicts. Among the principal variables in such theories are power, environment, and culture. Much of the related empirical work is strongly quantitative.

Civil wars

Although the disruption caused by civil wars may be somewhat less than that caused by violent conflict between states, there is good reason today to be especially concerned about them. Most wars now are civil wars—there were approximately 25 ongoing as of 2009—and most continue for years. Governments in the more developed world often assume that

Box 8.9   Naturally Aggressive?

A fundamental debate in social science concerns a particular human characteristic: aggression. Are we naturally (genetically) aggressive or do we learn (through acculturation) to be aggressive?

Certainly, aggression is a normal part of the world we live in. It operates at many spatial and social scales and takes a variety of forms. But there are cultural variations in aggression: some cultures display considerably more aggression than others.

Desmond Morris believes we are naturally aggressive. In his popular book *The Naked Ape* (1967), he argues that humans fight for three reasons—to establish dominance, to defend territory, and to defend family—and, hence, that aggression is biologically programmed. In *On Aggression* (1967) another popular writer, Konrad Lorenz, presents a similar view, developed on the basis of a detailed study of animal behaviour.

More recent studies suggest that when different group identities prevail and disagreements arise, war is a default position. Because nature is unkind (the argument goes) humans have evolved to be suspicious of difference. The implications of this claim are enormous, suggesting as they do that any heterogeneous society needs to encourage a common culture and not to celebrate diversity because this, biologically, will only result in conflict.

Ashley Montague, by contrast, does not see aggression as genetically controlled. In *The Nature of Human Aggression* (1976), he argues that early human societies, far from being aggressive, were characterized by co-operation; it was with the advance of technology that humans became increasingly violent. For Montague, humans are neither 'naked apes' nor 'fallen angels'.

Officers of the Royal Ulster Constabulary inspect the debris left by a bomb that killed 28 people and injured 220 in the market town of Omagh, Northern Ireland, in August 1998. A telephone warning had encouraged police to move people closer to the blast area.

CP/AP photo/Alastair Grant

such wars are rooted in ancestral ethnic and religious hatreds and that little can be done to prevent them. Yet, a 2003 World Bank study suggests that this assumption is not necessarily valid. Rather, the study found that the principal cause of civil war is lack of development. As countries develop economically, they become progressively less likely to suffer violent conflict, and this in turn makes further development easier to achieve. By contrast, when efforts at development fail, a country usually is at high risk of civil war, which will further damage the economy, increasing the risk of further war. The fact that many civil wars are linked to natural resources, as discussed below, suggests that they can be a curse rather than a blessing, with much national wealth that might be used to foster human development instead being directed to fund wars as groups within a country fight over access to oil, metals, minerals, and timber. Angola, where groups fought over oil and diamonds from 1975 until 2002, and Sudan, where fighting from 1983 to 2005 was related to oil, are prime examples.

Civil wars cause population displacement (as discussed in the Chapter 5 account of refugees), mortality, and poverty among local populations, and have spillover effects on neighbouring countries. But they also have global impacts. For example, in a country engaged in civil war, some areas are impossible to control

for illegal activities such as drug production or trafficking—about 95 per cent of the world's hard drugs are produced in countries experiencing civil war. Tragically, civil wars may be prolonged because a few people benefit financially from such activities, and this is one of the reasons why a state might be appropriately described as failed.

FAILED STATES

It has become commonplace, at least since the end of the Cold War, to suggest that some countries have failed because they are either critically weak or no longer functioning effectively—they are either ungoverned or misgoverned. Indeed, it can be argued that there are three types of state in the contemporary world: premodern states are those that have failed; modern states are governed effectively; postmodern states are those where national sovereignty is being voluntarily dissolved, as in the member countries of the European Union.

Failed states pose both national and global problems. In Chapter 5 several examples of failed states were discussed in the context of the discussion of refugees. The global security implications of state failure are apparent. Such states may serve as safe havens for terrorists and illicit drug production (Afghanistan), and/or allow pirates to operate freely in busy shipping lanes (Somalia).

Drug production, movement, and conflict

Several Latin American countries are struggling to cope with drug production and movement. Large quantities of cocaine are produced in and regularly shipped from the Andean region, forcing peasant farmers from the land, prompting gang wars, and compromising state institutions. In Mexico an estimated 6,000 people were killed in drug-related conflict in 2008, prompting a suggestion by the US Joint Forces Command that Mexico was close to becoming a failed state.

A recent development in drug trafficking is movement from Colombia in South America to West Africa and then to Europe. This route is favoured because trafficking through the Caribbean has become more difficult as a result of more intensive policing. In West Africa, the small country of Guinea-Bissau might be labelled the world's first narco state after the arrival of Colombian drug cartels in 2005. Already a failed state following a series of conflicts, Guinea-Bissau, with no prisons and few police, was a logical location for the storage of cocaine prior to shipment to Europe. Of course, the demand in Europe and North America fuels most of the production, movement, and related conflict.

TERRORISM

Defining terrorism

Terrorism is not quite so easy to define as might be expected. Attempts to reach international agreement on a definition date back to 1937 when the League of Nations proposed the following: 'All criminal acts directed against a State, and intended or calculated to create a state of terror in the minds of particular persons or a group of persons or the general public'. This proposal was not formally accepted. More recently, in 1999, the United Nations developed two statements, proclaiming that it:

1. *Strongly condemns* all acts, methods and practices of terrorism as criminal and unjustifiable, wherever and by whomsoever committed;
2. *Reiterates* that criminal acts intended or calculated to provoke a state of terror in the general public, a group of persons or particular persons for political purposes are in any circumstance unjustifiable, whatever the considerations of a political, philosophical, ideological, racial, ethnic, religious or other nature that may be invoked to justify them.

Less formally, terrorism is the use or threat of violence by a group against a state or other group, with the general goal of intimidation designed to achieve some specific political outcome.

Terrorists or freedom fighters?

Difficulty in defining 'terrorism' reflects a fundamental contradiction, namely, that many people labelled as terrorists by those whom they attack self-identify as freedom fighters and are considered as such by those whose interests they represent. In recent decades, most terrorist activities were a response to national policies. For example, the Irish Republican Army conducted a campaign against British rule in Northern Ireland, various Palestinian groups fought against Israeli occupation of specific territories, and ETA (Euskadi ta Askatasuna, meaning 'Basque Fatherland and Liberty') fought for an independent Basque homeland

Two farmers harvest opium from poppy plants near Kandahar, Afghanistan, in May 2002. According to the UN drug control agency, growers took advantage of a power vacuum created by the US-led war and the collapse of the Taliban to re-establish Afghanistan as the world's largest producer of opium.

CP/AP photo/Victor R. Calvano

in northern Spain and southwest France. Terrorism, then, is a well-established and popular strategy employed by many groups over the years, with numerous countries—including England, Ireland, France, Spain, Israel, Iraq, Pakistan, India, Indonesia, and Thailand—being especially vulnerable. Cohen (2003: 90–1) identifies roughly 100 states that have been exposed to terrorist activities since the end of World War II. The contrast with peaceful civil disobedience, most famously employed by M.K. Gandhi in his opposition to British rule in India, is clear.

Global terrorism

Terrorism is essentially a regional, not a global, phenomenon. But many commentators consider that the nature and goals of terrorism changed with the 2001 attacks on the United States by the militant Islamic group, Al-Qaeda. These attacks and the organization behind them are clearly of a different order to previous terrorist activities. Al-Qaeda was founded by Osama bin Laden in Afghanistan in 1988, moving to Sudan in 1991, and returning to Afghanistan in 1996. The organization has links with many other groups worldwide and is responsible for an explosion at the World Trade Center in New York (1993), attacks on US military in Somalia (1993) and Saudi Arabia (1996), explosions in the US embassies in Kenya and Tanzania (1998), an attack on a US warship in Yemen (2000), the 11 September attacks on the US (2001), an explosion at a synagogue in Tunisia (2002), attacks on nightclubs in Bali (2002), bombings in Saudi Arabia (2003), suicide attacks in Morocco (2003), separate attacks on synagogues and British interests in Turkey (2003), train bombings in Madrid (2004), killings in Saudi Arabia (2004), and bombings in London (2005). More generally, Osama bin Laden issued a 'fatwa' in 1998 calling for attacks on Americans. Of these, the most devastating, both in terms of the numbers killed and in terms of the response, were the 11 September attacks in the US.

Much of the immediate reaction to these attacks was couched in the context of the long history of conflict between Islam and Christianity, and employed the 'clash of civilizations' terminology introduced by Huntington (discussed below). Others, however, found that view superficial and argued that 'Islamic' terrorism is the work of individuals and small extremist groups—not representative of Islam itself (see Chapter 6 for an account of the character of Christian–Islamic conflict). Certainly, organizations like Al-Qaeda do not reflect larger religious identities, and it is an error to associate terrorism with Islam in general. Indeed, 34 of the states identified by Cohen as being exposed to terrorist activities are Islamic states subject to attack from within by extremist Islamic groups—a fact that raises further questions about the logic of the 'clash of civilizations' scenario.

Coping with terrorist organizations is especially difficult because—unlike states—they are not spatially tethered components of the political landscape. This does not mean, however, that they function in a geographic vacuum. Most terrorist groups are supported, at least informally, by states—hence the American-initiated invasions of Afghanistan and Iraq following the 9/11 attacks, although in the latter instance no linkages were ever proven, any more than weapons of mass destruction (another American government rationale for the invasion of Iraq) were found. Certainly, the recent upsurge of extremist Islamic terrorism has dramatically changed global geopolitics, and it is not surprising that some commentators see this as a principal feature of the world today and one likely to continue for the foreseeable future, especially as many analysts in the US and throughout the West point out that the American 'War on Terror' in fact has created terrorists and the environment conducive to terrorism.

CONFLICT OVER RESOURCES

As we saw in Chapter 3, human activities can have serious implications for resource availability. Today, some commentators note that conflict over resources is central to several ongoing or recent civil wars and predict that resource shortages will play a key role in future conflicts (e.g., Klare, 2001). Perhaps most obviously, there may be increased competition for access to oil and gas, water, and commodities such as timber, copper, gold, and precious stones. In such a scenario, the areas containing those resources—many of which are located in contested and unstable areas of the less developed world—would be major sites of conflict.

In the case of oil and gas, areas vulnerable to conflict include the Caspian Sea basin, the South China Sea, Algeria, Angola, Chad, Colombia, Indonesia, Nigeria, Sudan, and Venezuela, not to mention the Persian Gulf region (it has been widely suggested that one US motive for the 2003 war in Iraq was to gain control of the region's oil reserves). In the case of water, the most vulnerable areas might be those where a major water body is shared between two or more states, as in the case of the Nile (Egypt, Ethiopia, Sudan, and others), Jordan (Israel, Jordan, Lebanon, Syria), Tigris and Euphrates (Iran, Iraq, Syria, Turkey), Indus (Afghanistan, India, Pakistan), and Amu Darya (Tajikistan, Turkmenistan, Uzbekistan). In time, the Great Lakes shared by Canada and the US could become a source of greater contention that is now the case. Also, there may be competition for diamonds in Angola, the Democratic Republic of Congo, and Sierra Leone, for emeralds in Colombia, for gold and copper in the DR of Congo, Indonesia, and Papua New Guinea, and for timber in many tropical countries.

Claiming the Arctic

One cause of tension in many parts of the world is disagreement about precisely what territory belongs to a country and therefore who has the right to access natural resources in these contested areas. The example of the Arctic region has come to the fore in recent years.

Canada claimed sovereignty over the entire Arctic Archipelago of North America in 1895, but others see the matter differently. In particular, the United States and other shipping nations regard the Northwest Passage as an international waterway, while Russia, Norway, Denmark, and the US also have sectoral claims to parts of the High Arctic. These differing views were of relatively little importance until about 2005 when diamond mining, oil exploration, pipeline development, and navigation became more imaginable and feasible because of global warming. Some estimates suggest that the Northwest Passage will be a reasonable route for ocean-going ships within about a decade. Significantly, the journey between Asia and Europe is about 7,000 km shorter through the Passage than through the Panama Canal.

Legally, Canada's claim to about one-third of the Arctic region appears well-founded, but this claim could be weakened if Canada does not maintain a significant presence in the region. Hence, several military exercises, flag-raising events, and scientific expeditions have been conducted in recent years. Most notably, in 2008 Canada began a survey of parts of the Arctic seabed intended to demonstrate that Canada's continental shelf extends through much of the Arctic region, and in 2009 a diplomatic campaign was initiated to forcefully and repetitively identify Canada as a major Arctic power.

THE GEOGRAPHY OF NUCLEAR WEAPONS

As of 2009, there are nine nuclear powers. The United States, Russia, China, France, and Britain are also the five permanent members of the UN Security Council and are all committed to the Nuclear Non-Proliferation Treaty. The other nuclear powers are India, Israel, Pakistan, and, as recently as 2006, North Korea. There is much concern today about the nuclear intentions of North Korea, and also of Iran. Three other countries—Algeria, Saudi Arabia, and Syria—are suspected by the West of having nuclear intentions, but no weapons programs have been identified. Two countries, Iraq and Libya, recently ended their nuclear programs, while Belarus, Kazakhstan, and Ukraine all scrapped the weapons inherited from their inclusion in the former USSR.

As discussed in Chapter 3, nuclear power is an increasingly important source of energy; unfortunately, however, it is also the first step along the road to the production of nuclear weapons. Thus, as more countries develop the technology for nuclear power, it seems likely

Canadian Prime Minister Stephen Harper and Defence Minister Peter MacKay watch military exercises in Frobisher Bay in August 2009.

Adrian Wyld/Canadian Press

that the world will confront more crises about the nuclear weapon intentions of those countries. Of course, it is important to ask: do countries that currently have nuclear weapons have any moral authority to say that other countries cannot have those weapons?

Geographers also have helped to clarify possible consequences of nuclear war. In Britain, Openshaw, Steadman, and Greene (1983) published an extensive series of estimates of probable casualties following any nuclear attack. Bunge (1988) published a provocative yet penetrating *Nuclear War Atlas*, and other geographers have explored the possible climatological effects of a nuclear exchange (Elsom, 1985). Nuclear war would mean national suicide for any country involved, and a large-scale nuclear war would affect all environments. Large areas of the northern hemisphere would likely experience sub-zero temperatures for several months, regardless of the season. Low temperatures and reductions in sunlight would adversely affect agricultural productivity. A nuclear winter would result in many deaths from hypothermia and starvation. It is not necessary to belabour these points. Let us consider instead the role that geographers can play in influencing public awareness and alerting the makers of public policy to the folly of nuclear war.

Geographers have two important responsibilities: to teach people to love the land and the peoples of their state, and to teach people not to hate and fear other states. Undoubtedly, human geographers are in an enviable position to achieve such goals. They have studied the environmental and human consequences of nuclear war. They have used cartography (long a highly effective propaganda tool) to teach, and so are able to identify any deliberate territorial biases. And their study of landscapes, places, and their interrelationships can help to develop the global understanding so desperately lacking in our contemporary world.

Our Geopolitical Future?

One of the most interesting, and most important, questions we might ask concerns the character of the geopolitical world as it is unfolding today. As we have seen throughout this chapter, that world is a complex mix of relative stability and great uncertainty. Some of the political changes described here have followed quite predictable directions, while others have seemed quite unexpected. It would not have been difficult to predict the eventual decolonization of Africa, for example, as it took place over the decades after 1945. On the other hand, most observers were not prepared for the mass expressions of discontent throughout Eastern Europe that led to the end of the Cold War, or for the peaceful termination of apartheid in South Africa. So, with some trepidation, we ask: what might our geopolitical future have in store? Many answers to this question have been suggested; this section will discuss five intertwined possibilities.

DEMOCRACY AND THE POSSIBILITY OF PERPETUAL PEACE

Is it possible that we are approaching the beginning of what Kant called 'perpetual peace'? One school of thought believes that we have seen the end of major wars between states, although local wars will continue. This 'end of history' thesis is based on the idea that, with the termination of the Cold War in the early 1990s, the principles of liberal democracy are increasingly accepted around the world, and that liberal democracies do not wage war against one another (Fukuyama, 1992). This may be so, but the many local and regional conflicts that have erupted since the end of the Cold War make it difficult to see the phrase 'perpetual peace' as applicable to the contemporary world. Nationalist, populist, and fundamentalist movements have proliferated in recent years, with resulting conflicts in many areas, including the Persian Gulf region, the former Yugoslavia, Afghanistan, Somalia, and Central Africa. On the other hand, in recent decades, democracy continues to gain ground (Table 8.3). Figure 8.17 maps democracy by country globally for 2008.

Around the world, citizens are increasingly unwilling to accept authoritarian rule: since 1980 alone, more than 30 military regimes have been replaced by civilian governments. Challenges to dictatorship may be internal, external, or both. Some of the most remarkable political events of recent years reflect the role played by 'people power', with images of large crowds or single individuals peacefully protesting undemocratic governments an icon of the contemporary world. The Philippines in 1986, much of Eastern Europe in 1989, Serbia in 2000, Ukraine in 2004, and Nepal in 2006 all are examples of the power of mass protest. Also, since 1990, the number of countries that have ratified the six main human rights conventions and covenants has risen from about 90 to about 150.

But, according to some sources, this spread of democracy is showing signs of being arrested in recent years. The influential American lobby group, Freedom House, has identified some disturbing reversals; most notably, according to this organization, in 2007–8 democracy weakened in Russia, Pakistan, Bangladesh, Sri Lanka, the Philippines, Venezuela, Kenya, and Nigeria.

Table 8.3	Global Trends in the Spread of Democracy, 1977–2007		
	Free	Partly Free	Not Free
1977	43	48	64
1987	58	58	51
1997	81	57	53
2007	90	60	43

NOTE: This table employs the standard terminology used by the Freedom House lobby group, identifying each country as free, partly free, or not free.

SOURCE: A. Puddington, 'Findings of *Freedom in the World 2008*—Freedom in Retreat: Is the Tide Turning?', 2008, at: <www.freedomhouse.org/template.cfm?page=130&year=2008>.

Of course, the spread of democracy does not preclude conflict. Democratic states frequently experience challenges from within, especially from ethnic or regional groups that feel marginalized in some way. Indeed, acceptance of democratic principles may encourage challenges to national identity. Identities—whether at the level of the nation-state itself or of a national group within the state—are not fixed but dynamic, continuously playing themselves out in a complex negotiation. For many states there is a fairly obvious basis in language and/or religion and also some sense of a shared history, but many individuals have several competing identities, such as belonging at once to a local community, a larger region, and a state. Democratic states also may experience the power of civil disobedience.

From a Western perspective, the most significant reasons to doubt the likelihood of perpetual peace were the September 2001 terrorist attacks on the US and subsequent attacks in London and Madrid. The United

Members of the Mexican Federal Police and Mexican Army patrol the streets of Ciudad Juarez as part of the Chihuahua Joint Operation in the northern state of Chihuahua, Mexico. In June 2009, about 2,400 people had been killed since the beginning of the year despite the deployment to Ciudad Juarez of more than 36,000 government security forces, including 8,500 soldiers.

Omar Torres/AFP/Getty Images

FIGURE 8.17 Global distribution of freedom, 2008

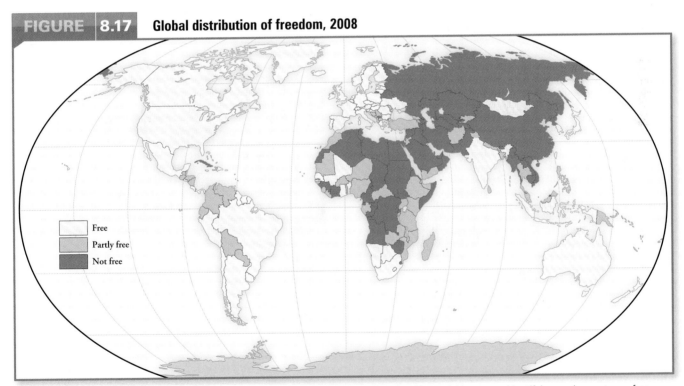

Free

Partly free

Not free

Like Table 8.3, this map employs the standard terminology used by US-based Freedom House lobby group, identifying each country as free, partly free, or not free. The details shown on this map need to be interpreted with caution. The concept of democracy is notoriously difficult to define, being understood differently in different parts of the world.

SOURCE: Freedom House, at: <www.freedomhouse.org/template.cfm?page=363&year=2008>.

States responded first with a military campaign in Afghanistan designed to replace the Taliban government, which was sympathetic to Al-Qaeda, and then with an invasion of Iraq that toppled the regime of Saddam Hussein.

CLASH OF CIVILIZATIONS

Another possibility is what Huntington (1993: 22) describes as a clash of civilizations or cultures:

> It is my hypothesis that the fundamental source of conflict in this new world will not be primarily ideological or primarily economic. The great divisions among humankind and the dominating source of conflict will be cultural. Nation-states will remain the most powerful actors in world affairs, but the principal conflicts of global politics will occur between nations and groups of different civilizations. The clash of civilizations will dominate global politics. The fault lines between civilizations will be the battle lines of the future.

Huntington (1996) identifies nine major cultures: Western, Sinic, Buddhist, Japanese, Islamic, Hindu, Orthodox, Latin American, and African (Figure 8.18). It is instructive to compare these with the regions proposed by Toynbee, discussed in Chapter 6, and the cultural regions depicted in Figure 6.4, and to reflect again on the quotation from James that introduced Chapter 6. It can be argued that conflict will be based on cultural divisions for six basic reasons:

1. Cultural differences—of language, religion, and tradition—are more fundamental than differences between political ideologies. They are basic differences that imply different views of the world, different relationships with a god or gods, different social relations, and different understandings of individual rights and responsibilities.
2. As the world becomes smaller, contacts will increase and awareness of cultural differences may intensify.
3. The ongoing processes of modernization and social change separate people from

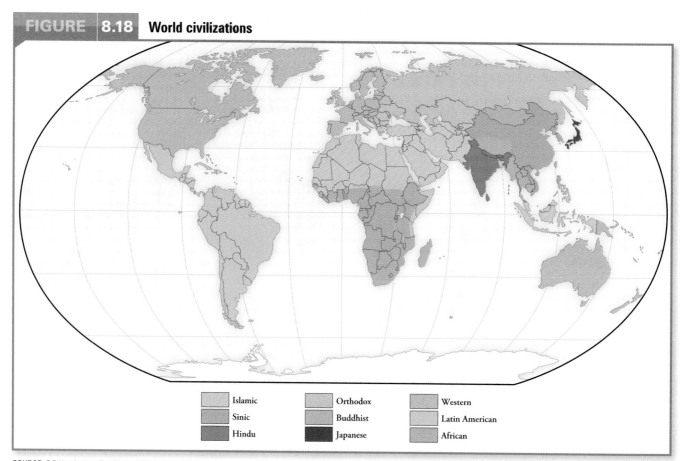

FIGURE 8.18   World civilizations

Islamic   Orthodox   Western

Sinic   Buddhist   Latin American

Hindu   Japanese   African

long-standing local identities and weaken the state as a source of identity. Increasingly, various fundamentalist religions are filling the gap, providing a basis for identity.

4. The less developed world is developing its own elites, with their own—non-Western— ideas of how the world should be.
5. Cultural characteristics, especially religion, are difficult to change.
6. Economic regionalism is increasing and is most likely to be successful when rooted in a common culture.

The argument that cultural difference will be the principal basis for future global conflicts proved both controversial and influential. Many commentators interpreted the 2001 terror attacks on the United States using this logic, while in this chapter and the two preceding chapters we have already discussed numerous regional and local conflicts rooted in cultural differences, especially of language and religion. On the other hand, Huntington ignored the fact that many conflicts occur within rather than between so-called civilizations; consider the many occasions when Europeans have fought each other and the instances of Muslims fighting other Muslims.

WORLD ORDER—OR DISORDER?

During the Cold War period the world was ideologically divided between the democratic capitalist states belonging to the US-dominated North Atlantic Treaty Organization (NATO) and the Communist states belonging to the USSR-dominated Warsaw Pact. This bipolar division no longer exists: NATO has expanded from its original 12 member countries to a total of 28 in 2009, and all of the new members are former Communist states in Eastern Europe. Other new members may be admitted in the near future.

The first stage in this geopolitical transition was Poland's installation of a non-Communist government—approved by the USSR—in 1989. Events followed with remarkable speed: the collapse of Communist governments elsewhere in Eastern Europe, the symbolic breaching of the Berlin Wall on 9 November 1989, the reunification of Germany in 1990, and the collapse of state communism in the USSR in 1991. These events ushered in a new and uncertain geopolitical world order with just one major world power (the United States) in the early twenty-first century.

Attempting to make sense of these changes, some political geographers focused on conventional geostrategic issues; de Blij (1992) used Mackinder's heartland model, while Cohen (1991) saw Eastern Europe as a 'Gateway Region' between maritime and continental areas. Other political geographers, however, proposed a geo-economic view of the world (O'Loughlin, 1992). They saw the world as consisting of three core regions— North America, Western Europe, and East Asia—each of which is undergoing a process of integration: the North American Free Trade Agreement (NAFTA), the European Union (EU), and the Association of Southeast Asian Nations (ASEAN) (the economic implications of these alliances are discussed in Chapter 9).

However, a world with just one major power, the United States, seems unlikely to last much longer. Indeed, it is widely accepted that the economic, military, and political dominance of the United States is in decline while other powers are strengthening their position. The National Intelligence Council in the United States sees China, India, and Russia as growing in influence. Also, of course, the power of the Islamic world is increasingly evident. It seems unlikely that the United Nations will be able to assert more authority in this changing world.

POLITICAL GLOBALIZATION

In many respects, a scenario of political globalization is the most complex, overlapping as it does with several other scenarios. For many observers it also seems the most plausible. There is much evidence to support the claim that a process of political globalization has been underway for quite some time, although as suggested in Table 8.4, the significance of this is uncertain.

Table 8.4	Globalization Theses and Political Geography	
Hyperglobalist: The Global Era	**Skeptical: Increased Regionalism**	**Transformationalist: Unprecedented Interconnectedness**
Market-led global governance. Nation-state replaced by 'natural' region states. Boundaries dissolve and sovereignty surrendered to global market.	Core-led regionalism. Sovereignty surrendered to regional groupings designed by powerful nation-states. Boundaries retrenched.	Multi-layered governance at three scales: global, national, local. Nation-state remains central. Centralization and devolution at same time.

SOURCE: Adapted from W.E. Murray, *Geographies of Globalization* (New York: Routledge, 2006), 353–4.

Political globalization is typically seen as one component of the overall globalization first noted in the Introduction, which will be more fully examined in Chapter 9. Thus, improvements in technology, specifically communications technology, and the spread of increasingly dominant transnational corporations are producing an integrated world economy in which, as trade increases, consumer behaviour patterns become more and more similar. The idea of political globalization, however, also incorporates two explicitly political circumstances.

First, following the end of the Cold War, the new world political order meant that, in principle, the United Nations finally was in a position to play the role for which it was created. Previously, it could be argued, both the US and the Soviet Union actively sought to prevent the UN from fulfilling its mandate. However, it might also be argued that the UN faces many challenges today at least partly because it continues to be dominated by the five powers that have permanent seats and vetoes on the Security Council.

Second, globalization not only relies on the spread of democracy but also tends to promote it. Because democratic states tend to form closer links with one another than with other political forms, states that want to trade with democratic states will be persuaded to become more democratic. Perhaps more important, globalization may also promote peace if the 'perpetual peace' scenario is correct in suggesting that democracies do not wage war with one another.

It can be argued that the process of political globalization has been underway for a very long time: in 1500 BCE the world contained perhaps 600,000 autonomous polities, and the long-term trend has been towards political integration. On the other hand, developments in the twentieth century suggest that, in the short term at least, the trend may be towards more rather than fewer political units. The end of World War I brought the collapse of several empires and the rise of several new states. The end of World War II ushered in a period of previously unheard-of co-operation between European states and a 'Cold War' that divided much of the world into two camps, American and Soviet. The decades since 1945 have seen some apparently anti-globalizing trends as colonial areas have achieved independence; approximately 95 new states have come into

being this way, and the collapse of the Soviet Union and the disintegration of Yugoslavia were responsible for the creation of about 19 more. Additional new states may be formed as a result of the trend towards decentralization discussed earlier in this chapter.

Decline of the state

Because political globalization challenges the integrity of the individual state, it entails a reduction in the importance of the role played by states. By the early 1990s, Ohmae (1993: 78) was arguing that the nation-state 'has become an unnatural, even dysfunctional unit for organizing human activity and managing economic endeavor in a borderless world', and suggesting that on 'the global economic map the lines that now matter are those defining what may be called "region states"'. A region state is a natural economic area. It may be part of a nation-state, as in the case of northern Italy, or it may cross boundaries, as in the case of Hong Kong and south China before 1997 (when China regained control of Hong Kong). Its principal economic links are global rather than national. In the Canadian context, it might be suggested that British Columbia forms a region state with the American West; other parts of Canada also have north–south relationships with areas of the US. In Canada, as elsewhere, the rise of region states has significant implications for national unity.

Certainly, the role of the state is changing. But perhaps states themselves will be able to adjust to the changing circumstances resulting from political globalization. In the past, many states sought power as measured by territory and population; today, by contrast, states seek markets. The challenge for the state in the twenty-first century may well be to integrate with other states without sacrificing national identity. On the other hand, the nation-state may prevail after all, since people tend to resist change as long as they are reasonably comfortable. Thus most Canadians have wanted Quebec to remain part of Canada, if only because of the insecurity that its secession would entail.

TRIBALISM: NOT ONE WORLD, BUT MANY?

Our final scenario focuses on globalism's opposite number: tribalism, or the tendency for groups to assert their right to a state separate from the one they occupy. Those who see tribalism as gaining ground point to secession movements, civil war, terrorism, and cross-border ethnic conflict fed by the resurgence of local identities. Support for this scenario can be found in two other trends.

First, as we have seen, groups of people with some common identity are increasingly emphasizing their distinctiveness, often at the expense of the states to which they 'belong'. During the 1990s a number of new states were created as, for example, the former Czechoslovakia divided peacefully into the Czech Republic and Slovakia, and the former Yugoslavia disintegrated in bitter conflict (see Box 8.6). At least some of the new states created in this way may turn out to be more meaningful, culturally, than their predecessors. In any event, the potential for the creation of additional new states is considerable.

Second, in some states, such as Canada and Australia, the principle of multiculturalism is widely accepted and cultural pluralism is valued. It is possible that acceptance of multiculturalism will lead to a weakening of central authority as 'tribal' identities become stronger.

UNCERTAIN TIMES

This overview of five possible scenarios suggests that future generations are likely to look back on the early twenty-first century as a period of uncertainty. We may hope that they will also see it as a transitional period in which—however gradually—conflict declined, democracy spread, and the chances improved for people around the world to live without being oppressed, discriminated against, or persecuted. In truly human geographic terms, we all have a right to have a home and to feel at home in the world.

CHAPTER 8 SUMMARY

DIVIDING TERRITORY

Human beings partition space. The most fundamental of the divisions we create is the sovereign state. Today there are more than 190 such states in the world. Almost all people today are subjects of a state. Loosely structured empires were typical of Europe before 1600. The link between sovereignty and territory emerged most clearly in Europe after 1600.

NATION-STATES

In principle, a nation-state is a political territory occupied by one national group. Despite the importance of nationalism and the territory–state–nation trilogy, there are many binational or multinational states today. Numerous scholars have attempted to explain how states are created: Ratzel viewed states as organisms; Jones developed a field theory involving a progression from idea to area; and Deutsch outlined an eight-stage sequence. Nationalism is the belief that each nation has a right to a state; only in the twentieth century did nationalism become a prevalent view.

EMPIRES: RISE AND FALL

World history includes many examples of empire creation and collapse. European empires evolved after 1500 and typically disintegrated after 1900, greatly increasing the number of states in the world. Imperialism is the effort by one group to exert power, via their state, over another group located elsewhere. The colonies created in this way quickly became dependent on the imperial states. The most compelling motivation for colonialism was economic.

GEOPOLITICS

Geopolitics is the study of the roles played by space and distance in international relations. Early geopolitical arguments by Ratzel and Kjellen included the idea that territorial expansion was a legitimate state goal. This idea, in the form of *geopolitik*, dominated geopolitics until the 1940s when it was discredited for being too closely associated with Nazi expansionist ideology. Other important geopolitical theories include Mackinder's heartland theory and Spykman's rimland theory. In the early 1990s the Cold War, which had started in the late 1940s, ended and a geopolitical transition began.

BOUNDARIES OF STATES

Stability may be closely related to boundaries. State boundaries are lines that can be depicted on maps, whereas national boundaries are usually zones of transition. Antecedent boundaries precede settlement; subsequent boundaries succeed settlement.

STATES: INTERNAL DIVISIONS

The stability of any state is closely related to the relative strength of centrifugal and centripetal forces within it. A key centripetal force is the presence of a powerful sense of shared identity; a key centrifugal force is the lack of a strong shared identity—perhaps because of the presence of minority groups that, in fact, may wish to secede. In Africa, discordance between nation and state is common where state boundaries were imposed by former colonial powers without reference to national identities. European states typically have greater internal homogeneity, although many countries do experience conflict over minority issues, some of which might be explained by reference to core/periphery concepts.

The collapse of the former USSR, an ethnically diverse empire, is one example of instability; the disintegration of Yugoslavia, a country with several republics and a deep divide between Christians and Muslims, is another. Canada's internal divisions are related to a variety of physical and human factors. Many of the peoples asserting a distinct identity occupy areas peripherally located in their states and economically depressed.

GROUPINGS OF STATES

There are two opposing trends in our political world: minority groups increasingly aspire to create their own states, while states increasingly desire to group together, usually for economic purposes. The EU is the principal instance of the second trend. As of 2009 it includes 27 countries, but is scheduled to expand further.

SUBSTATE GOVERNMENTS

Political authority is decentralized in federal states. Many of the states with large populations are federations. Like state boundaries, substate boundaries often do not reflect physical or human differences.

THE POWER OF THE STATE

There are numerous types of states and related political philosophies. States perform many important functions, both internally, with regard to the distribution of public goods, and externally, in their relationships with other states, especially regarding conflict and environmental concerns.

ELECTIONS

Geography is a central factor in elections. Some electoral boundaries are deliberately drawn to create an electoral bias using gerrymandering or malapportionment. The supporters of different political parties are often spatially segregated. Place has a significant influence on voting behaviour.

PEACE AND WAR IN THE TWENTY-FIRST CENTURY

Peace and war, with their obvious implications for both people and place, are inherently geographic. Geographers can help us understand the causes and consequences of conflict and can contribute to increased understanding between states. Even countries not at war are heavily committed to military and related expenditures. Although the geopolitical future remains uncertain, most scholars agree that conflict will continue. The world may not be at war globally, but it is certainly not at peace with itself.

QUESTIONS FOR CRITICAL THOUGHT

1. What role did colonialism play in creating the world political map of today?

2. Geopolitics is the study of the relevance of space and distance to questions of international relations. How has the 'relevance of space and distance' changed over the past half-century? In other words, are space and distance more or less important today than they were 50 years ago?

3. It has been said that Canada has too much geography and not enough history, while the former Yugoslavia has too much history and not enough geography. What does this mean?

4. What similarities and differences are there in comparing the various separatist movements in Europe? Why do you think there are so many such movements in Europe?

5. How important will 'groupings of states' (e.g., the European Union) be for the geopolitical stability of the world in the future?

6. Are you optimistic or pessimistic about the future political stability of the world? Why?

7. To what extent is conflict due to the 'fact' (theory) that humans are naturally aggressive?

FURTHER EXPLORATIONS

Agnew, J. 2002. *Making Political Geography*. London: Arnold.

An informative basic political geography text, covering all the topics addressed in this chapter; especially useful on the reasons why combining geography and politics aids our understanding of the contemporary world.

Archer, J.C., et al. 1988. 'The Geography of US Presidential Elections', *Scientific American* 259, 1: 44–51.

An important article explaining that the US electorate is divided by enduring geographic cleavages such that any winning candidate must build a geographic coalition.

Blacksell, M. 2006. *Political Geography*. New York: Routledge.

A critically informed introduction to political geography.

Boyd, A. 2001. *An Atlas of World Affairs*, 10th edn. London: Routledge.

An indispensable account of our contemporary political world; companion atlases deal with Africa, North America, and the European Union.

Bunge, W. 1973. 'The Geography of Human Survival', *Annals, Association of American Geographers* 63: 275–95.

A thought-provoking and deeply passionate piece by a concerned and often polemical writer.

Cohen, S.B. 2003. *Geopolitics of the World System*. Lanham, Md: Rowman & Littlefield.

A substantial volume that covers issues from a pragmatic perspective, identifying a hierarchy in the world system (geostrategic realms, geopolitical regions, national states, quasi-states, and territorial subdivisions), along with other features; includes predictions for the future.

De Blij, H. 2005. *Why Geography Matters: Three Challenges Facing America— Climate Change, the Rise of China, and Global Terrorism*. New York: Oxford University Press.

This informative and readable book includes a detailed discussion of terrorism.

Funk, M. 2009. 'Arctic Landgrab', *National Geographic* 215, 5: 104–21.

Informative account of the several competing claims to the Arctic region, with a detailed map.

Gottman, J. 1973. *The Significance of Territory*. Charlottesville: University Press of Virginia.

An original essay by a well-known political geographer.

Gray, C.S., and G. Sloan, eds. 1999. 'Special Issue on Geopolitics, Geography and Strategy', *Journal of Strategic Studies* 22, numbers 2 and 3.

An interesting collection of often challenging revisionist interpretations of traditional geo-political thinkers, such as Mackinder, Mahan, and Haushofer, along with a number of pio-neering new analyses on various topics.

Hobsbawm, E. 2007. *Globalisation, Democracy, and Terrorism*. London: Little, Brown.

A collection of essays from a leading British political historian that expresses both per-plexity and pessimism about our global future.

Kasperson, R.E., and J.V. Minghi, eds. 1969. *The Structure of Political Geography*. Chicago: Aldine.

A pioneering collection of 40 articles; includes some works by Ratzel, Mackinder, Whittlesey, Hartshorne, and Wallerstein.

Khanna, P. 2008. *The Second World: Empires and Influence in the New Global Order*. New York: Random House.

Discusses the decline of the US and the related rise of Europe and China as major powers able to influence what is happening in the 'second world'.

Knight, D.B. 1982. 'Identity and Territory: Geographical Perspectives on Nationalism and Regionalism', *Annals, Association of American Geographers* 72: 514–31.

A clear discussion of some of the difficulties of defining nation and state, as well as the related problems of group delimitation.

Kuus, M. 2009. 'Political Geography and Geopolitics', *Canadian Geographer* 53: 86–90.

Valuable discussion of contributions made by Canadian political geographers to the subfield of geopolitics in recent years; emphasizes human agency.

Monmonier, M. 2001. *Bushmanders and Bullwinkles: How Politicians Manipulate Electronic Maps and Census Data to Win Elections*. Chicago: University of Chicago Press.

An overview of how contemporary American politicians are using cartography to help them adjust electoral district boundaries to their advantage.

O Tuathail, G., and S. Dalby. 1998. *Rethinking Geopolitics*. London: Routledge.

A study of critical geopolitics, focusing on approaches that are not state-centred and that are suitable in contemporary situations; includes some insightful discussions of the relevance of culture and identity.

Pepper, D., and A. Jenkins, eds. 1985. *The Geography of Peace and War*. Oxford: Blackwell.

An excellent set of readings focusing on such issues as the Cold War, the arms race, nuclear war, and peace education.

Smith, N. 2005. *The Endgame of Globalization*. New York: Routledge.

Employing a Marxist-inspired political econ-omy perspective, Smith dissects attempts by the US to ensure that global superpower status is achieved and maintained.

Storey, D. 2004. 'Geography and National Identity', *Geography Review* 17, 5: 28–33.

Very readable article that covers basic ideas about nations and nationalism with refer-ence to current examples.

Tyner, J.A. 2009. *War, Violence, and Population*. New York: Guilford.

Insightful discussion of how states use violence to control and administer space; includes case studies to elucidate general concepts.

Zakaria, F. 2008. *The Post-American World*. New York: W.W. Norton.

Zakaria, a former managing editor of *Foreign Affairs*, discusses what seemed unthinkable just a few years ago, a global future with an enfeebled US.

Zielonka, J. 2006. *Europe as Empire: The Nature of the Enlarged European Union*. New York: Oxford University Press.

An interesting account of the growth of the EU, arguing that it reflects successful foreign policy but that it will not become as integrated as the US.

ON THE WEB

GLOBAL POLITICAL GEOGRAPHY

World Politics

🔍 www.chass.utoronto.ca/~dwelch/
netlinks.html

Links to information on world politics.

Global Geopolitics

🔍 globalgeopolitics.net/

In-depth articles and reports on global issues
and specific countries or regions; hosted by a
non-commercial educational research centre.

Global Issues

🔍 www.globalissues.org/issue/65/
geopolitics

This part of the Global Issues website includes
a wide variety of articles on various geopolitical
issues.

CONFLICTS

Many websites concerned with specific conflict situations have a particular perspective. To
access a balanced viewpoint the best sites to
research are those of major news organizations
such as the BBC and CNN. The following site is
also useful.

World Conflicts Today

🔍 www.worldconflictstoday.com/?a_
username=GCPSS&a_password=
GCPSS

Although aimed at pre-university students,
this website provides some sound basic
information.

PEACE AND SECURITY

United Nations

🔍 www.un.org/english/

Home page of the United Nations

NATO

🔍 www.nato.int/cps/en/natolive/index.htm

The home page for the North Atlantic Treaty
Organization; excellent source of news and
statistics.

Democracy

🔍 www.freedomhouse.org/template.cfm?
page=1

Freedom House reports, from an American
bias, on democracy and freedom globally.

Elections

🔍 www.electoralgeography.com/new/en/

This site reports on elections worldwide.

A GLOBALIZING WORLD

This chapter builds on the distance concept introduced in Chapter 2. We begin with the challenging idea that distance is now less important, as a human geographic concept, than it was in the past. An outline of five kinds of distance (physical, time, economic, cognitive, and social) is followed by an overview of diffusion—a concept that has already played an important role in our discussions of migration, cultural regions, and colonial activity, and that will resurface with reference to the spread and growth of both agriculture and industry. Here we identify three approaches to diffusion research: cultural geography, spatial analysis, and political economy. The next section considers transportation as one means of overcoming distance, specifically the modes available for the movement of goods: by water, railway, road, and air. We then turn to trade, a second and related means of overcoming distance; the factors affecting trade; various theories of trade; and efforts at regional integration—a discussion that leads to our final two sections.

Many commentators believe that globalization is transforming the human world from a collection of connected but often very different places to a single entity in which differences are greatly reduced, if not eliminated. We will discuss this theory in a long-term context. Finally, we will take a detailed look at the processes involved in economic globalization, especially the rise of transnational corporations and the changing technologies of information transmission. This final discussion is an essential preface to our discussions of agriculture in Chapter 10, settlement in Chapters 11, 12, and 13, and industry in Chapter 14.

Indians gather during a meeting to discuss rights of indigenous peoples at the WSF.
Andre Penner/AP/The Canadian Press

Distance Friction and Frictionless Distance

Why are things located where they are? This is one of the key questions that human geography seeks to answer. The standard assumption has always been that location—whether of people, farms, towns, industries, or anything else—is not random, but follows a pattern. Sometimes the phenomena of human geography are close to one another and sometimes they are separated by great distances, but 'spatial regularity' means that we are able to make sense of those locations. For example, as we saw in Chapter 5, at the most basic level global population distribution is closely related to physical geography: some parts of the world have been better suited than others to support human life.

With this idea of spatial regularity in mind, one way to think about human geography is as 'a discipline in distance'—a phrase used by the distinguished British geographer Wreford Watson (1955) in explicit acknowledgement of the way humans use space and distribute their activities as a way of adjusting to distance.

Why is distance important in this way? Because overcoming it requires effort, and, as the American sociologist Zipf (1949: 6) explained, 'an individual's entire behavior is subject to the minimizing of effort.' Applied to human movement, this **principle of least effort** suggests that location decisions are made to minimize the effort required to overcome what we call the friction of distance. Human geographers agree that many of our current distribution patterns reflect the friction of distance (Box 9.1). We saw in Chapter 6 how the first law of geography—that near things are more closely related than distant things—applied to cultural regions. Similarly, Chapters 10–14 will present many more examples of the relationships between one set of geographic facts and another: the choice of crops grown by farmers and access to market; the choice of residential location and workplace location; industrial sites and the locations of resources and markets. In every case, the friction of distance has played a central role.

Today, however, it seems that in many instances this friction of distance is decreasing, perhaps even to the point where some distances, at least, will become frictionless. If that is true, is human geography still a discipline in distance? The answer is yes—and no.

On the one hand, even a cursory glance at the world today shows that distance still matters, since in most cases overcoming it still takes some effort. On the other hand, that same glance is likely to show that in many instances distance now matters somewhat less than it did in the past. The reason for the declining importance of distance as a factor determining where we locate things is that friction decreases as communication technologies improve: what is increasingly important today is the way locations are integrated with high-speed communication networks. For most of us, the most potent examples of an essentially frictionless distance are e-mail and the Internet—technologies that make almost instantaneous contact with faraway places a routine matter. Estimates for 2009 suggest that more than 50 per cent of the world's population use a mobile phone and almost 25 per cent use the Internet. There is, of course, a digital divide evident between rich and poor countries, with Sweden, South Korea, and Denmark leading the way in the use of information and communications technology and sub-Saharan African countries being the least connected.

Highlighting the changing importance of distance, this chapter moves from the topic of diffusion to transportation, trade, and, finally, economic globalization. The friction associated with distance will become noticeably less as we progress through these topics. Nevertheless, for most of us, distance continues to play a central role in everyday life. You only have to think about your own daily life—travelling between home and school, visiting friends, choosing which restaurant or movie theatre to go to—to realize that the friction of distance and the principle of least effort are still important considerations in many respects.

Concepts of Distance and Space

Different interpretations of distance and space are appropriate for different activities. For example, in some cases distance may be best measured in terms of time or money, and by extension the same is true of sets of distances: spaces. Our discussions of distance and space are traditionally couched in terms of Euclidean geometry. This tradition has been described as a 'container view of space' (Harvey, 1969: 208), since it sees space as a framework in which

---

principle of least effort
Considered to be a guiding principle in human activities; for human geographers, refers to minimizing distances and related movements.

geographic facts are located and geographic events occur. However, it is important to note that there are many different types of geometry. Riemann geometry, for example, is more appropriate than Euclidean geometry when we are concerned with a spherical surface, such as the surface of the earth, because in Riemann space the shortest distance between two locations is a curved line, not a straight one. The characteristics of a particular geometry can be described in terms of the path of minimum distance between two points: geodesics. For example, when we say that economic and time distances are different, we are saying that economic space has a different geodesic from that of time space. Five different but related concepts of distance and space—physical, time, economic, cognitive, and social—are of interest here.

PHYSICAL DISTANCE

The spatial interval between points in space is the physical distance. It is often measured with reference to some standard system and is therefore a precise measurement. Sometimes, however, physical distance is measured in less precise terms. In ancient India, the standard unit of distance measurement was the *yojana*: the distance that the royal mode of transportation—an elephant—could travel from dawn to dusk (Lowe and Moryadas, 1975: 22). Even in advanced technologies, imprecise measurements may be used; in North American cities

it is common to measure distance by reference to numbers of city blocks.

The shortest travel distance between points is often not a straight line. In a grid-pattern city, it is a series of differently oriented straight lines (Figure 9.1). In other instances, the physical distance between two points may be related to the direction of travel, as in the case of vehicle movement in a one-way street system (Figure 9.2).

TIME DISTANCE

For some movements, especially those of people as opposed to materials or products, time is important; hence the preferred route may be the quickest rather than the shortest. Time distance is related to the mode of movement, traffic densities, and various regulations regarding movement. Figure 9.3 depicts **isochrones**—lines joining points of equal time distance from a single location—for the city of Edmonton. Clearly, travel time is not directly proportional to physical distance; if it were, the isochrones would be equally spaced concentric circles. Figure 9.4 presents an imaginative extension of these ideas depicting Toronto in (a) physical space and (b) time space. The time–space map shows space stretching in the congested central area and shrinking in the outlying areas—a direct consequence of the greater time needed to travel a given distance on congested as opposed to freely flowing routes. This pattern is not fixed:

> **isochrones**
> Lines on a map of equal travel time from a given starting point. One example of an isoline. Isolines generally allow map readers to infer change with distance and to estimate specific values at any location on the map.

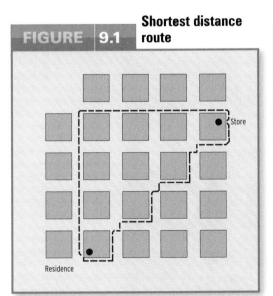

FIGURE 9.1 Shortest distance route

A non-straight line, shortest distance route.

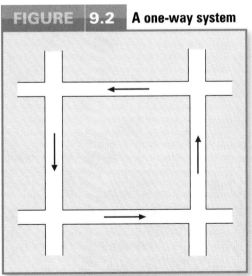

FIGURE 9.2 A one-way system

To travel a block, the distance covered may be one block or three, depending on direction.

FIGURE 9.3   Time distance in Edmonton

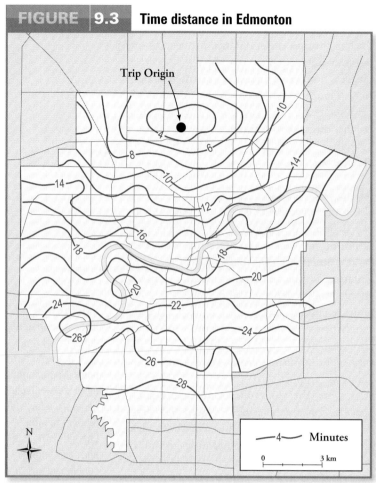

Trip Origin

N

—4— Minutes

0   3 km

SOURCE: Adapted from J.C. Muller, 'The Mapping of Travel Time in Alberta, Canada', *Canadian Geographer* 22 (1978): 197–8.

time–space convergence
A decrease in the friction of distance between locations as a result of improvements in transportation and communication technologies.

the extent of stretching and shrinking varies according to time of day and day of the week (rush hour on a business day generates the most stretching).

Converging locations

The idea that travel times typically decrease with improvements in the technology of transport has been labelled **time–space convergence** (Janelle, 1969). We can conceive of locations converging on each other; these are locational changes in relative space (recall that in Chapter 2 we described relative space as subject to continuous change). It is possible to calculate a convergence rate as the average rate at which the time required to move from one location to another decreases over time. For example, Janelle (1968) calculated the convergence rate between London, England, and Edinburgh, Scotland, from 1776 to 1966 as 29.3 minutes per year. More generally, it took Magellan three years (1519–22) to circumnavigate the globe. Today we can fly around the world in less than two days. With the laying of the first telegraph cable across the North Atlantic seabed in 1858, the 'distance' between Europe and North America was reduced from weeks to minutes, and now information can be transmitted around the world almost instantaneously.

ECONOMIC DISTANCE

Movement from one location in space to another usually entails an economic cost of one kind or another. Thus 'economic distance' can be defined as the cost incurred to overcome physical distance. As with time distance, there is not necessarily a direct relationship between physical distance and other measures. If we consider the movement of commodities, for example, costs frequently increase in a

Landing the telegraph cable at Heart's Content Bay, Newfoundland, from the *London Illustrated News*, 8 Sept. 1866.
Library and Archives Canada/Robert Dudley/C-004484

step-like fashion and the cost curve is convex (see Figure 14.5). Similarly, in some cities taxi fares are determined not by physical distance but by the number of zones crossed.

For many industries and other businesses, cost distance is of paramount importance. There is considerable logic to the notion that economic activities should be mapped in economic space, not physical or container space.

In Britain in 1950 the average person travelled about 8 km (5 miles) a day, mostly on foot. Today this average has increased to about 45 km (28 miles), most of it not on foot. Similar changes have taken place throughout the more developed world. Much of the increase in distance travelled can be attributed to improvements in transportation networks and the growth of suburbs that often require commuters to cover greater distances. Distance travelled is closely linked to economic status—people who regularly overcome large distances for business and pleasure are sometimes described as 'hypermobile'.

COGNITIVE DISTANCE

Spatial cognition is a matter of individuals' capacity to be precise in determining locations and distances. Each individual has a different interpretation of reality—that is, of actual locations and actual physical distances. Understanding individual interpretations may be essential if we are to understand human spatial behaviour. Indeed, cognitive distance may be a better explanation for human movement than physical distance—but how can it be measured? Physical, time, and economic distance each can be measured precisely, but cognitive distance cannot. Instead, cognitive distances are determined by asking respondents about the relative proximity of various locations to one given location. The importance of cognitive distance to behaviour is then assessed by analyzing the discrepancies between physical and cognitive distance. Also of interest are the reasons why cognitive distances differ from physical ones.

Most research has demonstrated that cognition affects both our measures of distance and our larger spatial comprehension. Cognition itself is affected by the individual and social characteristics of humans. Most of us are able to be relatively precise about distances and locations in the area that we know well, but as physical distance increases, cognitive distance becomes less precise. In addition,

FIGURE 9.4 | Toronto in physical space and time space

(a) Physical space

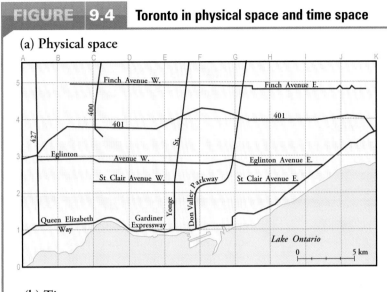

(b) Time space

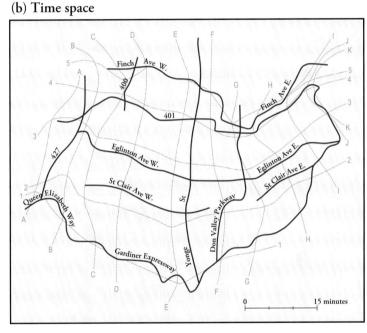

SOURCE: G.O. Ewing and R. Wolfe, 'Surface Feature Interpolation on Two-Dimensional Time–Space Maps', *Environment and Planning A* 9 (1977): 430, 435. Pion Limited, London.

most of us judge favoured locations to be closer than locations we dislike. In North America, for example, many people feel that the west coast is an attractive place to live and hence may judge it to be closer than it actually is, whether they measure the distance in physical, temporal, or economic terms. (For more on mental maps, see Chapter 5, especially Figure 5.2.)

Although it is clear that cognition affects behaviour in space, including movement, geographers are uncertain whether any specific causal relationships exist. It remains to be determined whether cognitive distances and spaces can be used to predict spatial behaviour as well as spatial preferences.

SOCIAL DISTANCE

The concept of social distance is complex, and the sociological literature offers a variety of interpretations. **Proxemics** is the study of social distances as defined by individuals in particular circumstances (for example, in social interactions such as conversation), and is typically a subject for social psychologists, not geographers (Hall, 1966). At the group scale, the correspondence between social and physical distance has been studied with special reference to residential distributions. Following the lead of urban sociology, geographers have analyzed the links between social status (often interpreted as class) and residence. Decisions about residence result in the creation of relatively distinct social spaces. One way to measure social distance is by the amount of interaction between groups—for example, the degree of intermarriage. Social

distance reaches its greatest extent in cultures with rigid hierarchical caste systems.

TRANSFORMATIONS OF DISTANCE

In geographic studies, the application of physical distance in reasoning may often be inappropriate; the relevant measure may be temporal, economic, cognitive, or social. This is one reason why it may be useful to transform physical distance data (others have to do with certain statistical procedures). As an example, we turn to the classic work of the Swedish geographer Hagerstrand (1967).

Hagerstrand recognized that movement intensity decreased sharply with increasing distance from the focal point because potential movers were less familiar with distant places than with nearer ones, and thus he mapped the logarithms of the physical distances (Figure 9.5). The focal point for his analysis was the small town of Asby, Sweden, located at the centre of a map representing the information space of Asby residents. The cartographic result of this logarithmic transformation is that nearby distances appear exaggerated, while distances farther away seem to shrink (to appreciate the displacement of the other locations identified, compare this map of Sweden with a more conventional one). Many types of human interaction, as well as human movement, exhibit the same negative relationship to distance. Figure 9.5 may well be the correct cognitive map. From a practical cartographic viewpoint, a logarithmic-transformed map creates room for the most information where it is most needed, at the shorter distances.

A rather different type of transformation is the **topological map**. The correct spatial order of locations is maintained, but space itself is elastic because distance and direction can be altered. An excellent example of a topological transformation is a map of the London underground system (Figure 9.6). If we assume that most of us view distance not in precise quantitative terms but rather in relative terms (such as 'less than' or 'greater than'), some topological transformations may be highly relevant to the geography of movement—perhaps more relevant than the original untransformed physical-distance data.

GRAVITY AND POTENTIAL MODELS

Many geographic studies of movement—whether of people, things, or ideas—between

---

proxemics
The study of personal space, the invisible boundaries that protect and compartmentalize each individual.

topological map
A diagram that represents a network as a simplified series of straight lines.

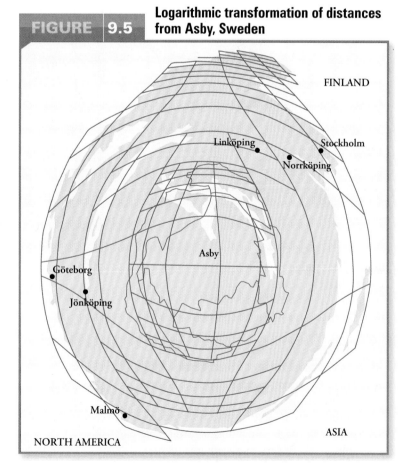

FIGURE 9.5 **Logarithmic transformation of distances from Asby, Sweden**

FINLAND

Linköping   Stockholm
        Norrköping

Göteborg

Jönköping

Asby

Malmö

NORTH AMERICA   ASIA

FIGURE 9.6 The London underground system

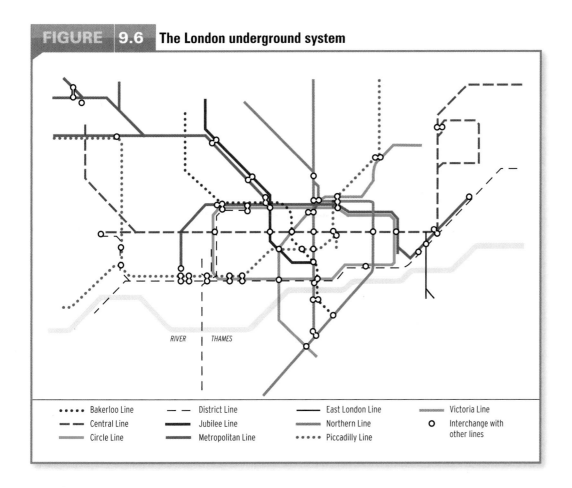

RIVER | THAMES

••••• Bakerloo Line	— — District Line	—— East London Line	Victoria Line
– – Central Line	—— Jubilee Line	Northern Line	O Interchange with other lines
Circle Line	—— Metropolitan Line	••••• Piccadilly Line	

urban locations have found a close link between the quantity of movement on the one hand and both the physical distance between locations and the sizes of locations on the other. There is a clear parallel here with Newton's law of gravitation, and geographic work that applies this logic is often labelled **social physics**. Of course, all such work recognizes that individual humans make choices and are not simply responding to the variables of size and distance, but this does not necessarily detract from the merits of employing this approach to study movement at an aggregate rather than an individual scale. The larger context for this logic was addressed in the Chapter 2 account of positivism, while critiques arise mostly from a humanistic perspective. Human geographers often study movement using **interaction theory**. Two versions are now described.

Gravity model

The gravity model describes the relationship between quantity of movement and distance as follows:

$$I_{ij} = \frac{M_i \times M_j}{D_{ij}^b}$$

where

I_{ij} is the interaction (movement) between locations i and j

M_i and M_j are the sizes (usually population sizes) of locations i and j

D_{ij} is the distance between locations i and j

b is a distance decay function

As noted, this formulation is a human equivalent to the gravity concept in physics. It proposes that the quantity of movement between two locations increases as their size increases: in other words, the larger $M_i \times M_j$ is, the greater the movement. It also proposes that the quantity of movement between two locations decreases as D_{ij}—the distance between the locations—increases. Empirical evidence suggests that the impact of distance is not uniform and that D_{ij} needs to be raised to some power: thus D_{ij}^b, where b varies, depending on circumstances, from as low as 0.5 to as high as 3.0. Note that a

social physics
An approach to aggregate human movement and interaction based on the assumption that such phenomena are analogous to processes in the physical sciences.

interaction theory
A body of theories explaining movements of goods and people between locations.

high value of *b* implies considerable distance friction, while a low value of *b* implies the reverse. As noted in the opening comments to this chapter, transportation and technological advances cause the value of *b* to decrease; that is, they decrease the frictional effects of distance (Box 9.1).

This valuable model was introduced by the nineteenth-century social scientist H.C. Carey. Carrothers quoted Carey as follows:

> Man, the molecule of society is the subject of Social Science. . . .The great law of *Molecular Gravitation* [is] the indispensable condition of the existence of the being known as man. . . . The greater the number collected in a given space the greater is the attractive space that is there exerted. . . . Gravitation is here, as everywhere, in the *direct* ratio of the mass, and the *inverse* one of distance. (Carrothers, 1956: 94)

We have already encountered one of the first uses of this concept in our discussion of migration. In 1885, Ravenstein used a variation of the model when analyzing migration flows between English cities. Other applications appeared from the 1920s onward and included a 'Law of Retail Gravitation' (Reilly, 1931).

In summary, the gravity model assumes that interaction—in the form of the movement of people or goods, telephone communication, or whatever else—takes place between locations. It also assumes that the greater the size of the location (such as an urban centre), the greater the interaction will be; in so doing, it treats all individuals as the same. Of course, this assumption is overly simplistic, since individuals are not equal in ability to generate interaction. Nevertheless, the gravity model is attractive because it is simple, it has parallels in physical science, and it has provided useful descriptions.

Potential model

The gravity model is concerned with relationships between pairs of points. The potential model is conceptually similar to the gravity model, but is concerned with the influence that many points have on one point. It can be expressed as follows:

Box 9.1   The Tyranny of Distance

The Tyranny of Distance is the title of a history of Australia by the eminent Australian historian Geoffrey Blainey. Blainey (1968: 2) begins: 'in the eighteenth century the world was becoming one world but Australia was still a world of its own. It was untouched by Europe's customs and commerce. It was more isolated than the Himalayas or the heart of Siberia.' The twin ideas of distance and isolation frame the processes of change in Australia in the European era. The four principal 'distances' are:

1. distance between Australia and Europe;
2. distance between Australia and nearer lands;
3. distance between Australian ports and the interior; and
4. distance along the Australian coast.

The distance concept proposes new ways of explaining why Britain sent settlers to Australia in 1788 and why the colony was at first weak but gradually succeeded. Similarly, it aids understanding of Chinese immigration in the 1850s and later immigrations from Southern Europe. Within Australia, the distances from coast to interior and along the coast are closely linked to economic and other change. The first

Australian staple, wool, was effectively able to overcome the distance from pasture land to port and the distance from port to Europe. Wool opened up much of southeastern Australia and was followed briefly by gold and then wheat. These products, along with technological innovations, combined to overcome the problems of distance.

Blainey is acutely aware that distance was not the only relevant variable—climate, resources, European ideas, and events were also important. Similarly, he continually acknowledges that distance is not a simple concept: it is fluid, not static, and can be measured in many ways. In later writings, Blainey has noted that distance is also a key to understanding the history of Australian Aboriginals.

Sensibly used, the distance concept can help us to understand the unfolding of life in many parts of the world. Blainey's work is eminently geographical.

Although the distance concept has not been explicitly applied by Canadian historians, it is central to a leading thesis in Canadian history. This 'Laurentian' thesis, proposed by Harold Innis, places the Canadian historical experience in the context of staple exports and communications that—as in Australia—effectively overcame the tyranny of distance.

$$V_i = \sum_{j=1}^{n} \frac{M_i}{D_{ij}^b}$$

where

V_i = potential at location i

M_i = size of location i

D_{ij} = distance between locations i and j

b = a distance decay function

Thus if a town, i, is located in a region with other towns, then town i has some potential for interaction with other towns. The potential at i is an aggregate measure of the influence of all other $j = 1$ to n places. As in the gravity model, distances may be raised to some power to accommodate the friction of distance. This model, too, has been extensively used by human geographers.

As noted, studies such as these, in which physics analogies are applied to social phenomena, are known as social physics. Although not all human geographers are comfortable with an approach that treats humans as molecules and places as masses, some human geographers contend that 'while it may not be possible to describe the actions and reactions of the individual human in mathematical terms, it is quite conceivable that interactions of groups of people may be described this way' (Carrothers, 1956: 201).

Extensions of the basic models

The gravity and potential models are simplifications of complex phenomena, and, quite appropriately, numerous variations and extensions of the basic models have been proposed. Two closely related sets of models are worth noting here. First are the entropy-maximizing models introduced by Wilson (see Wilson and Bennett, 1986). The term **entropy**, which originated in thermodynamics, refers to the disorganization in a system: entropy-maximizing models identify the most likely state of a system—that is, how many geographic facts are located in certain areas, given certain constraints. They have been extensively used for modelling spatial interaction. Second are location-allocation models, which can be used to determine the optimal location—that is, the location that minimizes movement and other costs associated with facilities, such as hospitals.

LINKS TO REGIONAL SCIENCE

In the late 1940s, the hybrid discipline of regional science developed in response to dissatisfaction with regional economic analysis. Largely the creation of the American economist Walter Isard, who wrote two major works (1956, 1960), it favoured the use of neo-classical economic theory and statistical techniques. This influential new discipline proved attractive to the many human geographers who favoured spatial analysis as the dominant approach to their discipline. Regional science is important in the present context because it proved to be one of the principal sets of ideas through which social physics procedures were accepted and practised by geographers such as Warntz (e.g., 1957). Since about 1970, in response to the humanist and Marxist critiques of positivistic spatial analysis, interest in regional science and social physics has waned considerably. As will become evident in Chapters 10, 12, 13, and 14, contemporary geographers are more attracted to approaches such as those based on political economy.

Diffusion

Spatial patterns are rarely static. To understand present patterns, it is often helpful to analyze past patterns; similarly, understanding current circumstances may help us to predict the future. We have frequently had occasion to acknowledge (at least implicitly) this point—and the importance of time generally—in this book.

Diffusion is best interpreted as the process of spread in geographic space and growth through time. Migration—the movement of people—can be regarded as a form of diffusion, although the term is more often used in the context of some particular **innovation**, such as a new agricultural technique. Diffusion research has a rich heritage in geography, and has been associated with three approaches in particular: cultural geography, spatial analysis, and political economy.

DIFFUSION RESEARCH IN CULTURAL GEOGRAPHY

Until the 1960s, most diffusion research took place under the general rubric of historical and cultural geography in efforts to understand cultural origins, cultural regions, and cultural landscapes. It is specifically associated with the landscape school initiated by Sauer. Typical studies focused on the diffusion of particular material landscape features, such as housing types, agricultural fairs, covered bridges,

innovation
The introduction of a new invention or idea, especially one that leads to change in human behaviour or production processes.

entropy
A measure of the disorder or disorganization in a system.

place names, or grid-pattern towns. The usual approach was to identify an origin and then describe and map diffusion outwards from it; the work of Kniffen (1951) is a good example.

Various issues emerged from this research. In many cases, debate centred on the question of single or multiple invention, the numbers and locations of hearth areas, or the problem of determining source areas for agricultural origin and subsequent diffusion. The importance of ethnic or social characteristics in relation to particular groups' receptivity to innovations was frequently acknowledged. For example, a detailed study of cigar tobacco production in the United States that focused on the significance of ethnicity—'In each of the tobacco producing districts, tobacco culture came to be identified with hard work, clever farming techniques, economic independence, and with an ethnic group' (Raitz, 1973: 305)—found, in Wisconsin, a very close relationship between people of Norwegian descent and tobacco production.

Diffusion continues to be a central concern in cultural geography.

DIFFUSION RESEARCH AND SPATIAL ANALYSIS

The character of diffusion research changed substantially following the work of the innovative cultural geographer, Hagerstrand (1951, 1967), who pioneered the use of models and statistical procedures. This shift was associated in North America with the rise of spatial analysis as a major approach in human geography after about 1955. Although the specific interests of these researchers were quite different from those of more traditional cultural geographers, the central concern was unchanged: to study diffusion as a process effecting change in human landscapes. The importance of chance factors was explicitly acknowledged by Hagerstrand in his use of a procedure known as Monte Carlo **simulation**, which allowed for the likelihood of any given acceptance of an innovation to be interpreted as a probability. Spatial analysis-oriented diffusion research also introduced a number of themes that we might call empirical regularities, because they are consistently observable in geographic analyses. Four of these empirical regularities are outlined below.

simulation
Representation of a real-world process in an abstract form for purposes of experimentation.

Neighbourhood effect

An innovation such as a new farming technology will likely be adopted first close to its source and later at greater and greater distances. This is the neighbourhood effect—a term that describes the situation where diffusion is distance-biased. In general, the probability of new adoptions is higher for those who live near the existing adopters than it is for those who live farther away. The simplest description of this situation, of course, is a diagram showing a series of concentric circles of decreasing intensity with increasing distance—similar to the ripple effect produced by throwing a pebble into a lake.

The neighbourhood effect occurs in those circumstances where an individual's behaviour is strongly conditioned by the local social environment. It is most common in small rural communities with limited mass media communication, where the most important influences on behaviour are personal relationships—in traditional or *Gemeinschaft* societies (see Chapter 11). The logarithmic transformation illustrated in Figure 9.5 describes such circumstances. The neighbourhood effect is likely to be least evident in urban settings and in circumstances of improving transport and communication technologies that reduce distance friction.

Hierarchical effect

The hierarchical effect is evident when larger centres adopt first and subsequent diffusion spreads spatially and vertically down the urban hierarchy. Thus the innovation jumps from town to town in a selective fashion rather than spreading in the wave-like manner associated with the neighbourhood effect.

This effect is likely to occur in circumstances where the receptive population is urban rather than rural, and is increasingly evident as transport technology improves (Box 9.2).

Resistance

Of course, reception of an innovation does not guarantee acceptance. Resistance to innovation is a sociological phenomenon. The greater the resistance, the longer the time before adoption occurs. Resistance, too, shows a spatial pattern: urban dwellers typically are less conservative than their rural counterparts.

S-shaped curve

Probably the best-supported empirical regularity is the S-shaped curve. When the cumulative percentage of adopters of an innovation is plotted against time, the typical result is an

S-shaped curve (see Figure 2.8). This curve describes a process that begins gradually and then picks up pace, only to slow down again in the final stages.

A simple conceptual framework

A simple framework for conceptualizing innovation diffusion is derived from a variety of sources, especially from Rogers (1962).

The process of innovation diffusion has four components: the innovation, a population of potential adopters, the communication process, and the adoption process. The *innovation* is assumed to be a desirable advance upon earlier procedures. For such an innovation to be diffused and either accepted or rejected, a *population* is necessary. At any given time during the process of diffusion, there are two classes of individuals in the population: adopters and non-adopters. There are three explanations for the presence of non-adopters: they have not heard of the innovation; they have heard and are in the process of making a decision; or they have heard and have made the decision to reject it. In addition, a third category of individuals, called change agents, is often relevant. These are people who promote the innovation, but who may be regarded as external to the system; for convenience, they are considered to be neither adopters nor non-adopters.

The *communication process* is the process by which an innovation, or knowledge relating to

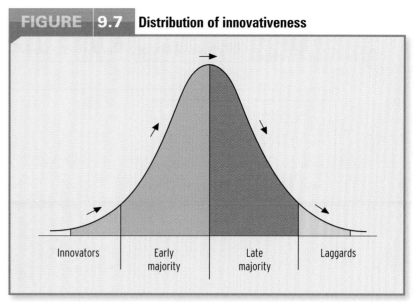

FIGURE 9.7 Distribution of innovativeness

Innovators | Early majority | Late majority | Laggards

As the innovation is adopted over time, an S-shaped curve emerges.

it, is diffused. It operates when either a change agent or an adopter communicates with a non-adopter. Communication may take two basic forms: pair-wise (individual to individual) or mass (individual or organization to group). Adoption of an innovation is rarely immediate upon receipt of information; it takes time. The *adoption process* may be described as follows: awareness—interest—evaluation—trial—adoption. Individuals vary in their innovativeness and may be classified as innovators, early majority, late majority, and laggards (Figure 9.7).

Box 9.2 Cholera Diffusion

In the nineteenth century, North America experienced three major cholera pandemics: in 1832, 1848, and 1866. Cholera is generally spread by water contaminated with human feces, but it can be carried also by flies and food. Climatically, its diffusion is promoted by warm, dry weather.

Each of the three pandemics diffused differently as the urban and communication systems changed. The 1832 outbreak spread in a neighbourhood fashion, the 1866 outbreak in a hierarchical fashion. The 1848 outbreak demonstrated aspects of both types of diffusion (Pyle, 1969). Figure 9.8 maps the spread and growth of the disease for the 1832 and 1866 pandemics.

The 1832 pandemic occurred at a time when both urbanization and the communication system were limited. Urban centres were few and small, and movement was largely by water—coastal, river, and canal. Data on cholera spread show a neighbourhood process. After initial outbreaks near

Montreal and New York City, the disease moved along distance-biased, not city-size-biased, paths. The 1866 pandemic occurred in the context of a very different urban and communication system: eastern North America was by then well served by rail, and an integrated urban system was developing. These changing circumstances allowed cholera to diffuse hierarchically. From New York City it travelled along major rail routes to distant second-order centres, as well as along a number of other minor paths. In many cases, cholera took much longer to reach areas close to the urban areas where it started than it did to reach more distant large cities.

This example is a useful reminder of the importance of economic and other infrastructures to any diffusion process. In general, a neighbourhood effect is more likely in circumstances of limited technology and on a local scale, while a hierarchical effect is more likely in a developed technological context and in a larger area.

Continued

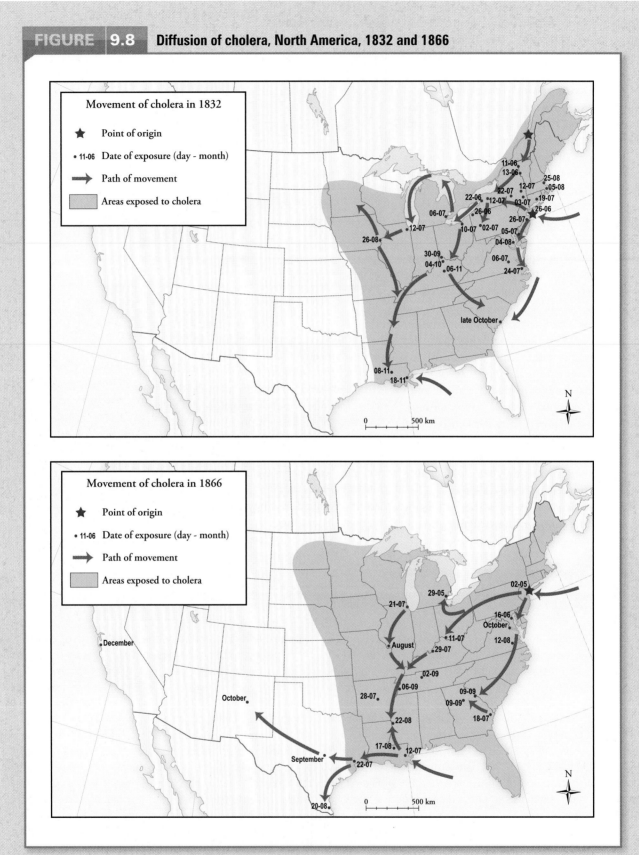

FIGURE 9.8   Diffusion of cholera, North America, 1832 and 1866

SOURCE: G. Pyle, 'The Diffusion of Cholera in the United States in the Nineteenth Century', *Geographical Analysis* 1 (1969): 63, 72. Published by Blackwell Publishing Ltd.

The global diffusion of products such as Coca-Cola is facilitated by the more general diffusion of American cultural and economic characteristics. These two images are from Guadaloupe and Malaysia.

V. Last, Geographical Visual Aids

The pattern of innovation diffusion is assumed to be equivalent to the pattern of adopters, and the adoption process is assumed to be a learning process. The entire process, then, is envisaged as a stimulus–response situation, with receipt of information as the stimulus and the decision to adopt or reject as the response. Adoption or learning may occur as a result of reinforcement, insight, or both.

This framework recognizes processes of communication and adoption that allow an individual to be in one of three states:

1. *No knowledge:* the state prior to the successful completion of the communication process.
2. *Knowledge:* the state after receipt of the information, but prior to the decision- making.
3. *Adoption:* the state after the decision to adopt is made.

Like any conceptual framework, this one deliberately simplifies a complex reality and is intended merely to aid in diffusion research.

DIFFUSION RESEARCH AND POLITICAL ECONOMY

For many researchers, the framework outlined in the preceding section represents yet another example of the dehumanizing approach characteristic of the spatial analysis school. Accordingly, after about 1970 a third approach to diffusion research was developed that focused not on the mechanics of the diffusion process but rather on the explicitly human consequences of diffusion. Although there are various aspects to this third approach, it is perhaps most frequently associated with Marxist perspectives (discussed in more detail in Chapter 10).

Any diffusion process adds to existing technological capacity, but it also affects the use of resources in some way. For example, some innovations result in significant time savings and hence change the daily time budgets of individuals; others actually demand more time, as in the case of the introduction of formal schooling. Analyzing such innovations means analyzing cultural change.

Contemporary research into innovation diffusion also considers the extent to which there may be a spatial pattern of innovativeness related to a wide range of social variables such as relative wealth, level of education, gender, age, employment status, and physical ability. These variables go far beyond the broad spatial variables we have noted before now, such as rural–urban or ethnic differences. Not surprisingly, such research also tends to consider overarching social, economic, and political conditions over which most individuals exercise little or no control.

In a study of the diffusion of agricultural innovations in Kenya since World War II, Freeman (1985) discovered that the diffusion process could be greatly affected by the pre-emption of valuable innovations by early adopters. The difference between the two curves shown in Figure 9.9 reflects the influence of an entrenched elite who, as early adopters, saw advantages in limiting adoption by others. From their positions of social authority, members of the elite were able to take political action, lobbying for legislation to prevent further spread of an innovation or limiting access to essential agricultural processing facilities. This pattern was observed in Kenya for three agricultural innovations (coffee, pyrethrum, and processed dairy products).

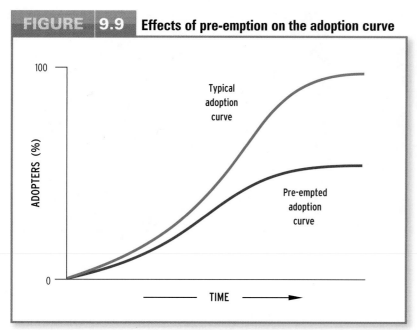

FIGURE 9.9 **Effects of pre-emption on the adoption curve**

This example makes it clear that the process by which an innovation spreads can be just as important as the innovation itself.

THE DIFFUSION OF DISEASE

Geographers concerned with mapping and modelling disease diffusion have used a variety of procedures reflecting diverse philosophical perspectives. The discussion of cholera diffusion in Box 9.3 exemplifies the spatial analysis tradition in describing the diffusion process. In addition, research on the diffusion of infectious diseases has sometimes used mathematical modelling techniques (e.g., Cliff et al., 1986; Cliff and Smallman-Raynor, 1992).

Mapping has been particularly useful with respect to the diffusion of AIDS; Gould (1993) demonstrates the value of maps in understanding the spread of the disease, and Smallman-Raynor, Cliff, and Haggett published an AIDS atlas (1992). Work such as this emphasizes the complexity of the diffusion process, identifying cultural, social, behavioural, economic, political, and transportation factors that have played a part in past and present distribution patterns of AIDS. Thus physical and human geographers are contributing to our understanding of AIDS through their interest and expertise in space, movement, and diffusion. Indeed, epidemiology—the study of the incidence, diffusion, prevention, and control of disease—is in large measure a geographical discipline.

Overcoming Distance: Transportation

Humans have continually striven to facilitate movement across the surface of the earth by reducing distance friction and the costs of interaction. One result has been the evolution of transport systems. Complex components of the human landscape, transport systems use specific modes (such as road, rail, water, or air) to move people and materials, to allow for spatial interaction, and to link centres of supply and demand. The fact that in much of the less developed world transport systems are largely deficient has a direct economic effect, limiting activities such as commercial agriculture and mineral production.

In some countries, a transport system has been constructed to encourage national unity and open up new settlement regions. It is not a coincidence that this has occurred in Canada and Russia, the two largest countries in the world, both with large areas of inhospitable environment. Canada's transcontinental railway, completed in 1886, served as a crucial centripetal factor and encouraged settlement of the prairie region. The trans-Siberian railway between Moscow and Vladivostok has been operating since the beginning of the twentieth century.

TRANSPORT GEOGRAPHY

Geographic study of transport is not as well developed as the studies of agriculture, settlement, and industry discussed in the following chapters. Nineteenth-century geographers such as Ratzel and Hettner viewed transport routes as landscape features and as factors related to more general landscape changes, and the early twentieth-century French school of human geography regarded transport as a critical component of the geography of circulation. Nevertheless, transport geography remained largely unexplored until the 1950s, when several geographers published studies of particular modes of transport. The most significant developments came with the rise of spatial analysis; they involved quantitative and modelling studies, many of which had a planning orientation. Even today, transport geography has a notably positivistic flavour, and has been relatively unaffected by humanist, Marxist, and other types of social theory (Hoyle and Knowles, 1992).

EVOLUTION OF TRANSPORT SYSTEMS

Three processes characterize the evolution of transport systems: intensification, or the filling of space; diffusion, or spread across space; and articulation, or the development of more efficient spatial structures. However, changes in transport systems do not occur at a steady pace. Just as agriculture, settlement, and industry have experienced times of revolutionary (as opposed to evolutionary) change, so has transportation. In fact, the changes that have occurred in transportation reflect many of the same basic forces that have prompted revolutionary change in other aspects of the human geographic landscape. In short, transport systems change in response to (1) advances in technology and (2) various social and political factors. A review of the evolution of transport networks in Britain will clarify these generalizations.

Britain

The first organized transport system in Britain was created by the Romans between 100 and 400 CE for political and military reasons. Unlike the earlier system of local routes intended simply to permit social interaction between settlement centres, the Roman system focused specifically on facilitating rapid movement between London and the key Roman centres. Once the Romans withdrew from Britain, their system fell into disuse because it did not serve the needs of the local population. Only in the seventeenth century did a new national system emerge, reflecting the emerging national interest; this system took the form of carriers and coach services. By the early eighteenth century, two innovations diffused rapidly across Britain: turnpike roads (an organizational innovation) and navigable waterways, especially canals (a technological innovation). In the nineteenth century, the technological innovation of railways appeared, and in the twentieth century, the road system responded to the new technology of automobiles. Finally, air transport added a new dimension to the overall transport system.

Turnpike roads, charging tolls, transferred the cost of road maintenance from local residents to actual road users. No new technology was involved, and thus, in principle, a turnpike system could have been set up at any time. Beginning about 1700, turnpikes diffused rapidly. During the early industrial period, *canals* played an important role in serving mines and ironworks. After 1790 they became the dominant transport mode, and until about 1830 they played an important role in industrial location decisions and related population distributions. Canals were tied to industry, since they were best suited for moving heavy raw materials, whereas turnpikes were better suited to the movement of people and information. Thus, the two new systems were largely complementary rather than competitive.

Neither of these eighteenth-century developments compared to railways in long-term impact. *Railways* dominated British transport from about 1850 to about 1920. The first railway was completed in 1825. The early railway network was relatively local and related to industry, but after 1850 railways linked all urban centres. Railway construction in Britain was organized and financed by small groups of entrepreneurs with minimal government intervention. Elsewhere in Europe and overseas, governments typically played a larger role in railway construction.

Each of these three innovations—turnpike roads, canals, and railways—displays the characteristic S-shaped growth curve (Figure 9.10): a slow start, rapid expansion, and a slow completion period. The pattern by which transport systems evolved in Britain is fairly typical. Today, most parts of the developed world have transport systems that include navigable waterways, railways, roads, and air traffic, and are continually changing in response to technology and demand.

Theories of transport system evolution

Several human geographers have studied the evolution of transport systems from a theoretical perspective. For the less developed world, Taafe, Morrill, and Gould (1963) proposed four stages of development (Figure 9.11):

1. A series of ports are scattered along an entry zone (most likely a coastline); there is no real network (Fig. part a).
2. A few of the early ports prosper and develop networks inland (parts b, c).
3. A number of feeder routes and lateral connections arise (parts d, e).
4. High-priority linkages emerge (part f).

SOURCE: Physical and Social Environment E. Pawson, Transport and Economy. The Turnpike Roads of Eighteenth Century Britain. London and New York: Academic Press, 1977, 13.

FIGURE | 9.10   Diffusion of transport innovations in Britain, 1650–1930

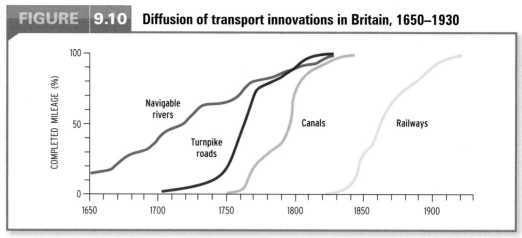

This generalized model usefully complements detailed empirical studies and has inspired a substantial body of research (Box 9.3). A second and rather different theoretical focus concerns the complex relationship between transport change and larger issues of economic change. Transport is both cause and effect and must be considered in analyses of, for example, agricultural, settlement, and industrial change.

NETWORKS

A network consists of specific locations, or *nodes*, linked together by routes to form some interconnected system. Nodes are both sources and destinations of all types of movement; they may be individuals or aggregates of individuals in the form of specific businesses or settlements. In the latter two cases, nodes are spatially fixed and their distributions can be described as dense, sparse, clustered, dispersed, and so forth. Whereas the central place theory discussed in Chapter 11 is closely concerned with settlements as nodes and their relative locations, in this chapter we are more concerned with the routes that facilitate movement between nodes. *Routes*—the channels along which interaction occurs—are sometimes spatially fixed, as in the case of roads and railways, and sometimes more flexible, as in the case of air and sea routes.

Geographers have long been interested in the characteristics of routes, especially their locations. Lines of communication are an essential factor in all economic geography: farmers and their products must travel to market and industrialists must move materials and products. In one sense, then, routes are the skeleton on which the flesh of human geography is built. 'Communications are themselves a part of the landscape' (Appleton, 1962: xviii). As geographers, we ask our usual questions about where and why.

FIGURE | 9.11   Evolution of a transport network

Box 9.3   Port System Evolution

A study of the ports of Ghana concluded that the system developed from 'a highly unstable scattering of numerous primitive surf-ports with restricted hinterland links' to 'a gradual concentration of traffic and the emergence of a stable system with dependence on two efficient deep-water ports' (Hilling, 1977: 104).

In this Ghanaian example, transport development and port development are linked:

1. First there was a long period (1482–1900) of primitive surf-port operations corresponding to the first scattered-ports stage of the Taafe, Morrill, and Gould (1963) model. Surf-ports—necessary because there were no good natural harbours—usually involved a pier extending beyond the surf zone.

2. Between about 1900 and 1928 the transport system penetrated inland and two ports, Sekondi and Accra, achieved dominance. Both of them acquired additional port facilities.

3. After 1928 the interior transport system developed a series of interconnections, while deep-water ports were constructed at the two most suitable sites, Tema (near Accra) and Takoradi (near Sekondi).

4. Finally, a high-priority route system developed, focusing on the two deep-water ports, and most of the smaller ports lost trade.

As a consequence of these changes, Ghana now has a contemporary, cost-efficient port and a transport system.

Graph theory

Networks are most easily analyzed by using graph theory. Although they involve simplification, graphs make objective analysis possible. Geographers were introduced to graph theory by one of the pioneers of theoretical geography, William Garrison (1960). Lowe and Moryadas (1975: 79) provide an example of the procedure. Figure 9.12a shows Martinique, a mountainous island in the West Indies with a circuitous road network; Figure 9.12b converts the network to a graph. The graph simplifies the network by excluding incidental characteristics but retaining the essential topology. The elements of the graph are nodes and routes—elements that can be used to generate a series of measures such as the number of nodes and routes. Kansky (1963) introduced a large number of other graph theoretic measures such as the beta index, the number of linkages per node:

$$beta = r/n$$

where

n = number of nodes
r = number of routes

A beta value greater than one suggests that alternative routings exist between some pairs of nodes. For the Martinique road network, beta = 1.49. Beta is generally higher in more developed countries than in less developed ones. A second measure introduced by Kansky (1963) is the gamma index: the ratio between the actual and the possible number of routes (minimum 0, maximum 1): gamma = $r/3(n - 2)$. For the Martinique road network, gamma equals 0.52. This measure is often expressed as a percentage; 0.52 is converted to 52 per cent, meaning that the network has a 52 per cent level of connectivity.

The beta and gamma measures indicate the utility of graph theory and are used extensively. Nevertheless, graph theory shares the drawbacks of all such simplified procedures; information is lost (deliberately), and highly variable nodes and routes are treated as though they were identical. Many networks have a hierarchical structure, in which the nodes that are close together are especially well connected.

MODES OF TRANSPORT

In some cases, different modes of transport that may have evolved at different times, and under very different human geographic circumstances, combine to create an integrated system. In other cases, route duplication and a general lack of co-ordination may combine to create a system that as a whole is somewhat incoherent, although each specific mode may be a perfectly logical system in itself. Each of the various transport modes has advantages and disadvantages.

Water transportation

Movement on water is a particularly inexpensive means of moving people and goods over long distances because water offers minimal resistance to movement and because waterways

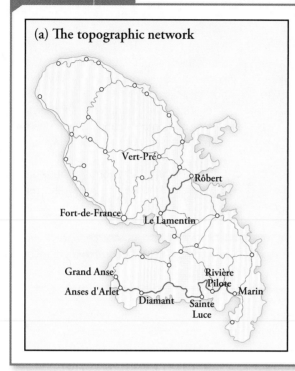

(a) The topographic network

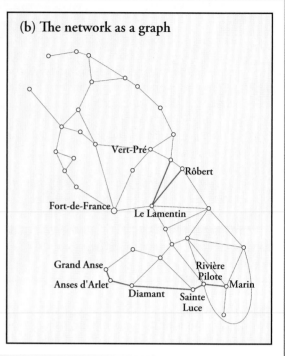

(b) The network as a graph

SOURCE: J.C. Lowe and S. Moryadas, *The Geography of Movement* (Boston: Houghton Mifflin, 1975), 80. Copyright © 1975 by Houghton Mifflin Company. Reprinted with permission.

are often available for use at no charge. But it is slow, may be circuitous, and not all locations are accessible by water. In the case of inland waterways, topographic features such as variations in relief can be a major obstacle. In some areas weather conditions—especially freezing and storms—may cause particular problems for water transport: in northern Manitoba, for

A section of the Panama Canal.
Central America/Alamy/GetStock

example, the port of Churchill on Hudson Bay is open only from mid-July to mid-November.

The opening of the Suez (1869) and Panama (1914) canals resulted in dramatic changes to travel distances. The Suez Canal reduced the distance between the Indian and Atlantic oceans (Asia and Europe), while the Panama Canal reduced the distance between Pacific and Atlantic oceans (Figure 9.13). Canals continue to be a critical component of many continental transportation systems. In Europe, the Rhine–Main–Danube Canal was completed in 1992, allowing for a continuous water link between the North Sea and the Black Sea.

Railway transportation

Land transportation includes railways and roads. Railways are the second least expensive form of transportation, after water, and are suitable for moving bulk materials when waterways are not available. However, in some areas local topography makes railway construction (and maintenance) expensive, if not impossible. The most significant railway construction in recent years has taken place in Australia, China, and Europe. In Australia, a railway line between Adelaide and Darwin that was first promised in 1911 was finally completed in 2004; this railway between north and south is a powerful symbolic achievement. In China, a new railway line, including more than 1,000 km of fresh track, was completed in 2005 between Golmud in the far west and Lhasa in Tibet. In Europe, the 'Channel tunnel' between England and France, opened in 1994, is in fact three tunnels—two for trains and one for access by maintenance workers. In the Scandinavian region a series of rail and road links (including tunnels and bridges) link Sweden and Norway through Denmark to Germany (Scanlink, completed in 2000). In recent years several train tunnel projects have been undertaken in the Alpine region, on the routes between Italy and Germany. The most notable of these is the Loetschberg rail tunnel, the longest in the Alps at 34 km (21 miles). Also under construction is the 57 km New Gotthard rail tunnel scheduled to open in 2015.

Road transportation

Roads accommodate a range of vehicles and typically are less expensive to build than railways; hence they are often favoured in regional planning and related schemes. Most of the internal movement within Europe is by road, with inland waterways second and railways third. Construction of bridges often reduces road distances. A recent example is the 2006 completion of a second major bridge, both road and rail, over the Orinoco River, which makes it much quicker for Brazilian goods to reach Venezuela's Caribbean seaports.

Air transportation

Transportation by air is the most expensive, but it is rapid and is usually favoured for small-bulk, high-value products. Air transport is particularly subject to rapid technological change and to political influence. Even small countries tend to favour maintaining a national airline for reasons of status and tourist development. In many cases airport construction and maintenance are funded by governments.

Containers

A major development in ocean and land transportation is the technology of the shipping container. Developed in the 1950s by Malcolm P. McLean, an American businessman who ran a trucking company, containers being used today are estimated to number as many as 300 million. The widespread use of containers has resulted in dramatically reduced transportation costs, despite the requirement of an entirely new way of handling freight, including new storage facilities, different cranes, changes at ports, and changes to trains and trucks. Theft is reduced, the storage and sorting of breakbulk cargo is eliminated, and time taken to load and

---

FIGURE | 9.13   **The impact of the Suez and Panama canals on ocean travel distances**

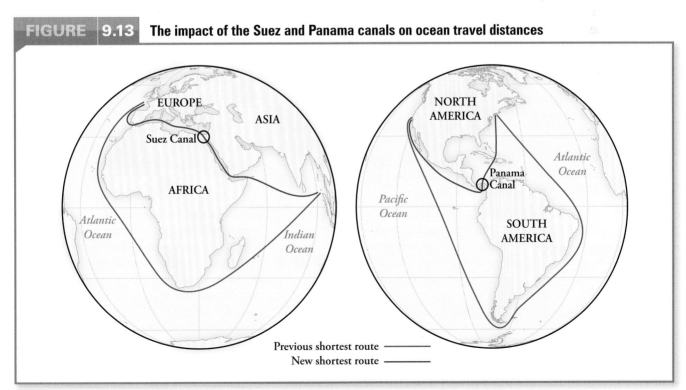

Previous shortest route ———
New shortest route ———

SOURCE: Adapted from, Rodrigue, J-P., C. Comtois, and B. Slack. 2006. The Geography of Transport Systems. New York: Routledge, Figure 1.12, p.21.

Passengers arriving from London at the Gare du Nord in Paris. The train that makes the Channel run is called the Eurostar.

V. Last, Geographical Visual Aids

• Opening of the world's longest sea-crossing bridge in 2008, a 36-km, 6-lane highway, across Hangzhou Bay that halves the travel time between two busy ports, Ningbo and Shanghai.
• Use of bullet trains in many places, including between Beijing and its nearest port, Tianjin, and a planned bullet train link between Beijing and Shanghai.
• Ongoing road and rail improvements throughout the country.

The fact that China is a Communist country, with secretive planning and no significant public input, means that such major infrastructural change can happen at great speed and without meaningful consideration of the impact on people's lives and on the environment. Consider, for example, that it took about the same length of time to conduct a public inquiry into the proposed construction of Heathrow's Terminal 5 as it did for China to build the new Beijing airport.

Overcoming Distance: Trade

Trade can take place if the difference between the cost of production in one area and the market price in another will at least cover the cost of movement.

In principle, domestic trade is identical to international trade. The major difference is that it is less likely to be hindered by human-created barriers. International trade is playing an increasingly important role in the world, as it has grown much more rapidly in recent years than international production. This confirms the apparent trend towards international integration in the global economy.

FACTORS AFFECTING TRADE

The movement of goods from one location to another reflects spatial variations in resources, technology, and culture. The single most relevant variable related to trade is distance. Like all forms of movement, trade is highly vulnerable to the friction of distance—a friction that decreases as technology advances. Other relevant variables include:

unload is halved. Most container ships carry about 4,000 containers although a few carry over 10,000. The largest ships are unable to use the Panama Canal (a plan to widen the Canal to accommodate these new ships was announced in 2009), and can access only a few major ports (such as Los Angeles, Rotterdam, Hong Kong, and Singapore). The important role played by containers in facilitating economic globalization is difficult to exaggerate, and the subtitle of a recent book on container shipping, 'How the Shipping Container Made the World Smaller and the World Economy Bigger', is not inappropriate (Levinson, 2006).

Transportation in China

The rapid growth of the Chinese economy and its ongoing integration into the global economy, discussed in Chapter 14, has necessitated major improvements in all aspects of China's transport system. Between 2001 and 2005 more was spent on transport than in the previous 50 years and, notwithstanding the slowdown in the global economy that began in 2008, it is likely that massive expenditures will continue. Examples of recent transport improvements include the following.

• The completion of a new Beijing airport terminal prior to the 2008 Olympic Games. Built in only four years by 50,000 workers, the terminal is 3 km long and the floor space is 17 per cent larger than all five of London Heathrow's terminals.

1. the specific resource base of a given area (needed materials are imported and surplus materials exported);

2. the size and quality of the labour force (a country with a small labour force but plentiful resources is likely to produce and export raw materials); and

3. the amount of capital in a country (higher capital prompts export of high-quality, high-value goods).

All these variables are closely linked to level of development, and in fact trade between more and less developed countries is often an unequal exchange: the former export goods to the latter at prices above their value, while the latter export goods to the former at prices below their value. Moreover, the less developed countries are inclined to excessive specialization in a small number of primary products, or even a single staple. More developed countries may also specialize, of course, but they do so in manufactured goods. This difference reflects the technologically induced division of labour between more and less developed worlds. The fact that the less developed countries are typically dependent on the more developed countries as trading partners is central to the world systems theory discussed in Chapter 5. Today the majority of world trade moves between developed countries, although the rise of the newly industrializing countries (see Chapter 14) is affecting trade flows. For a simplified explanation of commodity flows, see Box 9.4.

TRADE THEORIES

It may be true, as Isard (1956: 107) wrote, that 'trade and location are as the two sides of the same coin.' Yet from a geographic viewpoint, trade theory excludes the central concept of location theory, namely distance. Quite literally, trade theory typically considers countries as though there were no space between them. Integrating location theory and trade theory seems a difficult task, not least because trade theory often lacks empirical support.

Classical trade theory originated in the work of economists such as David Ricardo (1773–1823) and focuses on the natural or historically created differences between the countries specializing in production and related trade; its key variable is labour. Trade is explained by the law of comparative advantage as follows: in the absence of barriers to trade, a country will specialize in the production and export of those commodities it can produce at a comparatively low cost and import those goods that can be produced at a comparatively lower cost in other countries.

Neo-classical trade theory extends classical theory by using geometric techniques and adding further concepts such as the notion of external economies, which helps to explain why trade can be conducted between areas that are similar in most respects. External economies are the benefits that an industry enjoys as a result of the expansion of the industry to include related activities.

Box 9.4 Explaining Commodity Flows

The study of commodity flows has long been a topic of concern to geographers. In North America, the most influential writer on this topic has been Edward Ullman. Ullman was especially interested in transportation, and even defined geography as the study of spatial interaction (apparently after hearing a sociologist define sociology as social interaction). On the basis of detailed studies, Ullman (1956) concluded that commodity flows could be reasonably explained by reference to three variables:

1. *Complementarity*. Different places have different resources. When a surplus of a certain commodity (resource) in one area is matched by a deficit of the same commodity in a second area, this complementarity permits the commodity to flow from the first area to the second. This logic applies at all spatial scales.

2. *Intervening opportunity*. Complementarity can result in product movement only if no other area between the complementary pair is an area of either surplus or deficit. If such an intervening opportunity exists, it will either serve the deficit area or receive the surplus because it is closer to both areas than they are to each other.

3. *Transferability*. Commodity flows are affected by the ease with which a commodity can be transferred. A product that is difficult or expensive to move will be less mobile than one that is easier or less expensive.

Together, these three factors provide a simplified explanation of commodity flows at various times and at a multitude of spatial scales.

Modern trade theory extends the range of variables. Whereas classical theory focused on labour, modern theory also considers such variables as production, capital, land, and entrepreneurship, comparing countries in terms of their overall endowment with these factors. Applied to a historical topic such as the growth of trade associated with the transition from feudalism to capitalism, modern theory might focus on new producers seeking ever-larger markets as an example of entrepreneurship.

REGIONAL INTEGRATION

The fact that trade occurs across international boundaries means that it can be regulated. In fact, most governments actively intervene in the importing and exporting of goods. For example, they may set up trade barriers to protect domestic production against relatively inexpensive imports. The most popular form of regulation involves the imposition of a **tariff**.

As introduced in Chapter 8, regional integration—the linking of separate states in some form of economic union—has become important since World War II. The basic motive behind integration is the opportunity for increased trade and the expansion of potential markets. Until recently, most integration was quite limited, but today there is compelling evidence of a widespread desire among separate states to unite economically, and consequently the world is increasingly divided into trade blocs. In many cases economic integration also involves some degree of political integration. There are five stages in the process of integration (Conkling and Yeates, 1976: 237):

1. The loosest form of integration is the free trade area, consisting of a group of states that have agreed to remove artificial barriers, such as import and export duties, to allow movement and trade among themselves. Each state retains a separate policy on trade with other countries. Two major examples of free trade areas are the North American Free Trade Agreement (NAFTA), consisting of Canada, Mexico, and the United States and established in 1994, and the Association of Southeast Asian Nations (ASEAN), originally comprising Brunei, Indonesia, Malaysia, the Philippines, Singapore, and Thailand and established in 1967.

2. The second stage in international economic integration is the creation of a customs union, in which member states not only remove trade barriers between themselves but impose a common tariff barrier. In this case there is free trade within the union and a common external tariff on goods from other states.

3. A common market has all the characteristics of a customs union, and in addition allows the factors of production, such as capital and labour, to flow freely among member states. Members also adopt a common trade policy towards non-member states.

4. There are two levels of integration beyond the common market that groups of states may achieve. An economic union is a form of international economic integration that includes a common market and harmonization of certain economic policies, such as currency controls and tax policies.

5. Finally, at the level of economic integration member states have common social policies and some supranantial body is established with authority over all. The extent to which individual states may be prepared to sacrifice national identity and independence to achieve full economic integration is not yet clear.

Figure 9.14 maps some of the major trade blocs. At present, the most developed example of regional economic integration is the European Union (EU), which established a single European market in 1992. The importance of transportation as a means of encouraging integration is evident in the several ambitious projects recently completed or underway (the Channel tunnel, Alpine tunnel routes, and Scanlink). As the process of regional integration continues, states left outside the major groupings may suffer economically (Cleary and Bedford, 1993). This is one reason why many countries—including countries in Western Asia and North Africa that would generally be considered non-European—aspire to join the EU.

ASEAN today consists of 10 countries—Brunei, Cambodia, Indonesia, Laos, Malaysia, Myanmar, the Philippines, Thailand, Singapore, and Vietnam—and encourages free trade between members. It is also moving towards free trade with China and strengthening trading connections with Japan and India. Economic progress in this area was significantly impeded by the Asian economic crash of 1997.

MERCOSUR, the South American economic grouping of Argentina, Brazil, Paraguay,

tariff
A tax or customs duty on imports from other countries.

FIGURE 9.14 Selected economic groupings of countries

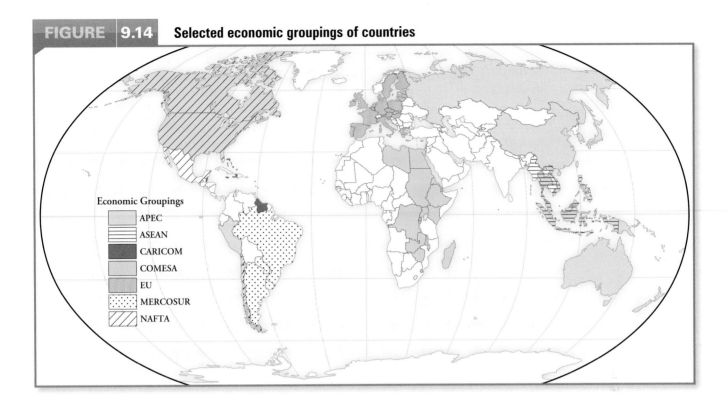

Economic Groupings
- APEC
- ASEAN
- CARICOM
- COMESA
- EU
- MERCOSUR
- NAFTA

and Uruguay formed in 1991, is a customs union that in the past has experienced real difficulties because of financial crises in Argentina. With oil-rich Venezuela becoming a full member in 2009, Bolivia likely to gain full-member status in the near future, Chile, Peru, Colombia, and Ecuador as associate members at present, and Mexico expected to move from observer to associate status, the region might well follow the route taken by the EU. One potential stumbling block is the risk that further integration might imply some loss of national sovereignty, although a majority of MERCOSUR states have rejected the formation of a Free Trade Area of the Americas (FTAA) proposed by the United States and perhaps view their participation in this Latin American customs union as not only economically important but also ideologically significant.

It seems likely that African countries also will move towards some form of integration in the near future, especially with the 2002 creation of the African Union (Chapter 8). Among the problems that those countries face are a generally low level of economic development, by global standards, and the fact that one of the few areas in which African countries are most competitive—agriculture—is the one major area still subject to protective tariffs (see Box 9.5) in the West. Integration of all areas into the emerging global economy and the

role of trade and of trade blocs in the contemporary globalizing world are considered later in this chapter.

Towards One World

The rest of this chapter is devoted to globalization. We will begin by establishing the broad context necessary to place the phenomenon in perspective, before addressing some specific aspects of economic globalization. This is not, of course, our first discussion of globalization, nor will it be our last. But it is a pivotal one.

Much of the content of this chapter so far has anticipated a discussion of economic globalization, especially the accounts of the possible declining role of distance, the development of container shipping, and the rise of trade blocs. For most observers, globalization is primarily economic. In this chapter we will examine economic globalization in more detail, and the next five chapters will build on this material in discussions of agriculture, settlement, and industry. In each case we will see how the friction traditionally associated with distance has been reduced with the advent of globalization. First, however, to help us place globalization in the big picture, reinforce our current understanding, and shed some additional light on why our world is the way it is—and why it is constantly changing—let us take a brief journey through time.

SHAPING THE CONTEMPORARY WORLD

Geography is ambitious, as befits a discipline with the fundamental goal of 'writing about the earth'. Accordingly, it is sometimes important to step back and acknowledge how many factors, past and present, have shaped the world we see today. What fundamental things do we need to know in order to understand contemporary human geographies?

Physical geographies (Appendix 1)

Regardless of our technological achievements, human and physical geographies continue to be closely related. This is true whether we are thinking at the scale of the world as a whole (e.g., determining global population distribution and density), or at the local, personal level (e.g., deciding what clothes to wear depending on the season). In physical geographic terms, all parts of the world are not the same, and different environments are often associated with different human geographies. Human geographers may no longer subscribe to any rigid notion of environmental determinism, but it is still essential to take physical geographies into account in our analyses.

Life on earth (Appendix 2)

We are not alone on the earth. Humans are only one of many life forms, and we cannot consider ourselves in isolation. Just as we need to take into account basic physical geographies, so we need to consider our relationships with plants and other animals. The fundamental lesson of ecology is how completely we depend on other organisms—from the trees that produce oxygen to the food that we eat to the bacteria that enable us to digest it.

Human origins (Appendix 2)

Current evidence suggests that *homo sapiens sapiens* first appeared in Africa only about 195,000 years ago and eventually spread around the world in a process of diffusion closely related to physical geography. The fact that, biologically speaking, humans are a single species exposes the fallacy of the concept of race. Since the emergence of *homo sapiens sapiens*, human evolution has been primarily cultural, not biological.

Culture and diversity (Chapters 6 and 7)

We may all belong to one biological species, but human culture is astonishingly diverse. Culturally, humans have evolved in a seemingly infinite number of directions—as evidenced by our many languages, religions, and identities—and this is what makes our world such a divided place. Whether or not you agree with James's interpretation of cultural diversity as having created 'one world divided', you will probably find at least some truth in the idea that cultures set up barriers that are often difficult to cross.

Origins of agriculture (Chapters 3 and 10)

The transition to agricultural life that began about 12,000 years ago, often known as the agricultural revolution, was a critical technological advance that permitted population growth, new forms of social organization, the beginnings of urbanization, and the development of specialized occupations. Because urbanization largely depended on agriculture, many of our cities are located in agricultural areas. During the long period since domestication began, agricultural technologies have diversified and advanced, allowing humans to settle new areas and achieve increasing yields.

European overseas movement (Chapters 1, 5, and 8)

For various reasons, Europeans moved themselves, their cultures, and their economies across large parts of the world beginning in the fifteenth century. This process was not really completed until the late nineteenth century, when European states carved up much of Africa among themselves. During this period of approximately 450 years many of the characteristics of our contemporary world were put in place. Most notably, according to world systems theory (Chapter 5), European expansion was responsible for the emergence of the two worlds we know today: the less and the more developed. It also promoted the spread and growth of Christianity, of several European languages, and of Eurocentric racism.

Social and economic organization: Modes of production (Chapters 2 and 7)

Modes of production are important because they both enable and constrain human activities;

in any particular mode, there are some things that can be done and some things that cannot. In the contemporary world the dominant mode of production is capitalism, which involves what is often described as the ceaseless accumulation of capital.

States (Chapter 8)

States are the basic building blocks that most of us have in mind when we think about world events. The nation-state may be declining in importance, as we saw in Chapter 8, but it remains the case that most people in the world are citizens of a particular state and are tethered to it, with little opportunity to move permanently to another state.

The Industrial Revolution (Chapter 14)

The term 'Industrial Revolution' refers to a series of technological changes, beginning in England between about 1750 and 1850, that involved the large-scale use of new energy sources, especially coal, and the introduction of new machines that necessitated the construction of factories. These factories generated urban growth, and industrial cities replaced earlier forms of settlement. New transport links—canals, railways, and roads—were essential for moving raw materials to factories and finished products to markets. Agriculture became increasingly mechanized, and the proportions of the total labour force engaged in agriculture declined rapidly. In much of Europe this was also a period of large-scale overseas migration.

The twentieth century

Many dramatic changes took place during the twentieth century. The fundamental difference between the world of the early twentieth century and today is the increase in the number of people—from 1.6 billion to 6.8 billion in little over 100 years. Furthermore, each of us is expected to live longer—in the more developed world life expectancy increased from 45 to 75 years, and in the less developed world from 25 to 64 years. The way that we make our living has changed as well. In 1900, about 90 per cent of the world's people practised agriculture: this figure has fallen below 50 per cent, and approximately 50 per cent of us now live in cities (in Canada, 80 per cent of the population are urban dwellers). Urban living encourages smaller families and new conceptions of appropriate family structures. In addition,

forms of discrimination, especially racism and sexism, have diminished in both intensity and impact throughout much of the world over the past century, and technological advances have brought marked changes in our ability to move from place to place and to transmit information. Finally, the political geographic world experienced dramatic shifts during the twentieth century. The world of 1900 was dominated by colonialism, but changes over the century ultimately resulted in a world without empires. The vast overseas possessions of such countries as Britain and France have disappeared, as have the land-based empires of Austria-Hungary, Russia, and the Ottoman world. One consequence of decolonization has been a remarkable increase in the number of states in the world, from about 50 in 1900 to more than 190 today.

Perhaps of greatest import, the economic organization of the world has changed markedly. One broad conclusion to be drawn is that we are now able to talk intelligibly on a global scale as well as in local and regional terms—an idea that lies at the heart of many accounts of our contemporary world. Indeed, it is often asserted—by human geographers and others—that globalization, understood broadly as an ever-increasing connectedness of both people and places, is the greatest challenge facing humans at the present time.

Globalization

Because the processes of globalization are still unfolding, it is not necessarily easy to recognize all of them, let alone to understand all their implications. The rest of this chapter will address the most potent aspect of the phenomenon: economic globalization. This form of globalization involves both technological change, especially the instantaneous transmission of information, and organizational change, especially the rise of transnational corporations able to 'slice their way at will through national boundaries and render all state policy-makers redundant' (Dicken, 1992: 48).

Economic Globalization

To begin, it is important to point out that, for most commentators, economic globalization is not the same thing as economic internationalization. The internationalization of economic activity began with the development first of empires and then of economic links—especially through the movement of capital and goods—between colonial powers and their colonies. But this process of internationalization did not extend to production, which continued to take place in a single national territory. Trade between countries is not globalization. Only when production and distribution are no longer contained by national boundaries, and this change has been brought about primarily by the rise of transnational corporations, can we properly speak of globalization. As Figure 9.15 shows, the increasing economic importance of transnationals means that the production process is now organized *across* rather than *within* national boundaries. Table 9.1 provides a useful introduction to this account, highlighting as it does the three globalization theses noted in the Introduction and links to economic geography.

THE GLOBAL ECONOMIC SYSTEM

To understand how transnational corporations emerged, it may be helpful to review recent changes in the global economic system. Certainly, economic co-operation between states has ebbed and flowed over the past century. In much of the Western world, including Canada, the 1920s and 1930s were a period of trade barriers and limits on foreign investment.

The beginnings of what we now think of as economic globalization came at the end of World War II, when the United States became the world's principal economic power and was able to initiate the Bretton Woods agreement of 1944. This agreement led to the creation of two important economic institutions: the International Monetary Fund (IMF) and the World Bank (initially called the International Bank for Reconstruction and Development). The IMF was to provide short-term assistance to countries whose currencies were tied either

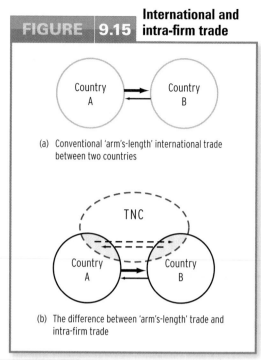

FIGURE 9.15   International and intra-firm trade

(a) Conventional 'arm's-length' international trade between two countries

(b) The difference between 'arm's-length' trade and intra-firm trade

Conventional international trade and intra-firm trade within a transnational corporation.

SOURCE: P. Dicken, *Global Shift: The Internationalization of Economic Activity*, 2nd edn (London: Paul Chapman, 1992), 49. Copyright © Sage Publications Ltd., 1992. Reprinted by permission.

to gold or to the US dollar, and the World Bank was to provide development assistance to Europe after the war. An important component of the economic order introduced at Bretton Woods concerned international capital movement, which was to be regulated primarily through national systems of exchange controls, with the US dollar playing the key role.

Another feature of the post-war economic system was free trade. This was the concern of a third international institution, created in 1947: the General Agreement on Tariffs and Trade (GATT), intended to reduce barriers to trade between countries. Between 1947 and 1994 there were eight rounds of GATT trade talks; the final ('Uruguay') round, which ended in 1994, accomplished the most sweeping liberalization of trade in history (see Table 9.2).

However, the economic order introduced in the 1940s under the leadership of the United States has changed significantly in two ways. First, the movement of capital is now virtually immediate (because it is achieved electronically), and almost unregulated. This represents a dramatic change from the Bretton Woods system. The principal geographic expression of this new form of capital movement is the rise of world cities, most notably New York, Tokyo,

Table 9.1	Globalization Theses and Economic Geography		
	Hyperglobalist: The Global Era	**Skeptical: Increased Regionalism**	**Transformationalist: Unprecedented Interconnectedness**
	Market is omnipresent and transnationals powerful. Networks replace core–periphery division. Activity disembedded from nation-states. Deregulation and convergence.	Trading blocs formed. Core–periphery division increased. Activity embedded in nation-states. Transnationals reflect national strategies. Regulation and divergence.	Networks and structural patterns coexist. Nation-states and transnationals govern markets. Re-regulation and convergence and divergence at the same time.

SOURCE: Adapted from W.E. Murray, *Geographies of Globalization* (New York: Routledge, 2006), 353–4.

and London, that dominate global patterns of trade, communication, finance, and technology transfer (these are discussed in Chapter 11). In additional changes from the Bretton Woods system, the roles of the IMF and World Bank have been expanded, and in 1995 the GATT was replaced by the even more powerful World Trade Organization (WTO) (Box 9.5).

Second, as we saw above, there are now several close-knit regional organizations that are essentially discriminatory and protectionist in their approach to trade. Some commentators even suggest that we are moving towards a world composed of three regional trading blocs that will function not only as economic unions (Figure 9.16) but as political ones as well, eventually replacing states. This remains to be seen. In any event, in addition to the EU, NAFTA, and ASEAN, the list of regional organizations today includes the Caribbean Community (CARICOM), the Central American Common Market (CACM), the Commonwealth of Independent States (CIS), the Economic Community of West Africa (ECOWAS), the Central African Customs and Economic Union (UDEAC), the Maghreb Union, the Southern African Development Co-ordination Conference (SADCC), the Common Market for Eastern and Southern Africa (COMESA), and Mercado Común del Sur (MERCOSUR). On the other hand, the Asia-Pacific Economic Co-operation forum (APEC), comprising 18 countries (including the three largest economies in the world—the United States, Japan, and China), is attempting to move even further along the road of trade liberalization. (Figure 9.14 maps some of the principal regional trading blocs.)

Political changes in the late 1980s and early 1990s also had significant economic repercussions. With the collapse of communism in the former Soviet Union and Eastern Europe and the end of apartheid in South Africa, several new participants entered the world trade arena. And the country with the largest population in the world, China, is adopting increasingly capitalist economic policies.

In any event, these political changes are not directly related to the economic globalization we are discussing here. In that context, the principal organizational change is the rise of transnational corporations (corporations that operate in two or more countries), while the principal technological changes are taking place in the realm of communications and information flow.

Table 9.2	From Protectionism to Free Trade: A Chronology
Date	**Events**
1947	GATT founded in Geneva with 23 members; some tariff cuts.
1948	First GATT meeting, held in Havana, Cuba.
1949	Second round of talks, held in Annecy, France; 500 tariff cuts; 10 new countries admitted.
1950–1	Third round, held in Torquay, England; 8,700 trade concessions; four new countries admitted.
1956	Fourth round, held in Geneva, Switzerland; extensive tariff cuts.
1957	The Haberler Report outlines new guidelines for GATT; problems facing less developed countries considered; impact of new European Economic Community considered.
1960–2	'Dillon round' (named after US Under-Secretary of State) held in Geneva; 4,400 tariff concessions.
1964–7	'Kennedy round' negotiations, also in Geneva, depart from the product-by-product approach used in previous talks and adopt an across-the-board method of cutting tariffs for industrial goods; 50 per cent tariff cut in many areas.
1973–9	'Tokyo round', involving 99 countries; reduction of import duties and other trade barriers by industrial countries on tropical products exported by developing countries; average tariff on manufactured goods produced by nine largest markets cut from 7 per cent to 4.7 per cent.
1982	Ministerial meeting in Geneva reaffirms validity of GATT rules for the conduct of international trade and commits to combatting protectionist pressures.
1986–94	'Uruguay round'; cuts in industrial tariffs, export subsidies, licensing and customs valuation; first agreements on trade in services and intellectual property.
1995	WTO established as a replacement for GATT.
1996	First WTO ministerial conference, in Singapore.
1998	WTO ministerial conference in Geneva.
1999	WTO ministerial conference in Seattle; major anti-globalization protests by a wide variety of interest groups.
2001	WTO ministerial conference at Doha, Qatar, provides mandate for negotiations on a range of subjects, and other work including issues concerning the implementation of the present agreement. Mandate places emphasis on assisting less developed countries through further liberalization of global trade.
2003	WTO ministerial conference (Cancun, Mexico); there are now 144 members, including China. In September an alliance of poorer countries, led by China, India, and Brazil, challenges the right of the US and EU to determine the details of trade deals; when talks collapse, each side blames the other.
2004	WTO meeting (Geneva, Switzerland); some progress made in the areas where the Cancun meeting failed; barriers to international trade reduced with wealthy countries agreeing to cut the subsidies they give farmers for exports.
2005	WTO ministerial conference (Hong Kong); there are now 149 members (Russia has observer status but is not yet a full member); agreed procedures designed to reduce tariffs further and to remove many agricultural subsidies
2006–7	Attempts to implement the 2001 Doha mandate to expand free trade with emphasis on helping less developed countries continue to stall at meetings in Geneva (2006) and Potsdam (2007).
2008	Talks to implement the Doha mandate have now stalled three times since 2001. There are now 153 members of the WTO. Onset of a global recession in 2008 likely to affect future negotiations.

FIGURE 9.16   The contemporary geo-economy

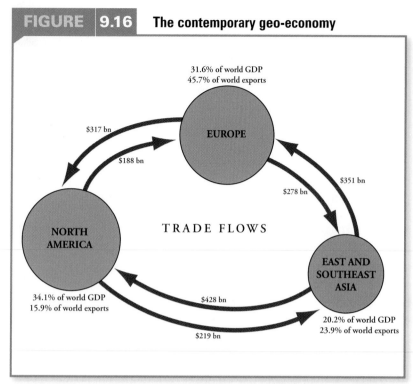

Dicken emphasizes that the power structure of the contemporary world economy is based on three regions: North America, the European Union, and East and Southeast Asia. Each contains one of the three most important and powerful global cities—London, Tokyo, and New York—identified in Table 11.12. Clearly a key consideration in the creation of these trading blocs is spatial proximity. Dicken writes: 'The "triad" appears to sit astride the global economy like a modern three-legged Colossus, constituting the world's "mega-markets" and "sucking in" more and more of the world's production, trade and direct investment.' This situation represents a significant change from the mid-twentieth century, when European colonial empires were the crucial regions and the international division of labour was a simple core–periphery structure.

SOURCE: Adapted from P. Dicken, *Global Shift: Mapping the Changing Contours of the World Economy*, 5th edn (New York: Guilford: 2007), 39, Figure 2.5.

TRANSNATIONAL CORPORATIONS

Transnational corporations are distinguished from earlier business organizations by several important characteristics. Most fundamentally, they are able to command and control production and sales at a global scale and usually can relocate production and other facilities with relative ease. They are also functionally integrated and able to benefit from geographic variations in capital, knowledge, labour, resources, national regulations, and taxes.

Table 9.3 provides sales data for the 10 leading transnational corporations, with GDP data from selected countries for comparative purposes. Only 21 countries have a GDP larger than the sales of the top transnational, Wal-Mart. Remarkably, if we consider sales data and GDP as comparative measures of an economy, then the top 100 economies in the world in 2005 include 53 countries and 47 transnationals.

Because transnationals are able to adjust their activities not just within countries but also between them, they are able to take advantage of variations in factors such as land costs and labour costs at both scales. They undermine national economies because they are able to organize movements of information, technology, and capital between countries and, to a considerable extent, site production and profit in countries with relatively low wages and taxation.

Different transnationals carry out different proportions of their production at home and abroad. For several of the large automobile

foreign direct investment
Direct investment by a government or multinational corporation in another country, often in the form of a manufacturing plant.

Anti-WTO protestors in Hong Kong, 2005.
Ron Yue/Alamy/GetStock

Box 9.5   The World Trade Organization: Friend or Foe?

Formed in 1995 to replace the GATT as the guardian of international trade, the WTO has become closely associated with globalization. In 2006 it had 149 member countries, and at least 30 more were hoping to join (see Table 9.2). The WTO's mandate includes facilitating smooth trade flows, resolving disputes between countries, and organizing negotiations; it is the only international body with such authority (Hoad, 2002). WTO decisions are absolute and binding on member countries. The GATT and the WTO have been spectacularly successful: world trade is increasing rapidly, and tariffs are on average about one-tenth what they were in 1947. Today, protectionism remains significant in just one important area: agriculture.

In late November–early December 1999, a WTO meeting of member countries' trade ministers in Seattle was overwhelmed by well-organized protest demonstrations in which labour organizations, environmental activists, consumer groups, and human rights activists shared their concerns about the circumstances and presumed consequences of economic globalization.

The concerns are many and varied. Some argue that the WTO is too secretive and that its discussions should be held in a more public context, involving a much broader range of participants. Some accuse the organization of being a front for corporate interests, especially influential transnationals. Some see it as benefiting the wealthy, more developed countries and neglecting the poor of the less developed world (certainly the most powerful players are the European Union, Japan, and the US). Some point out that the WTO is inconsistent in promoting unrestricted movement for some goods while permitting restrictions on the movement of others. In particular, these critics object to the continuation of agricultural subsidies in more developed countries that allow those countries to export artificially low-priced produce to the less developed world while preventing the importation of produce from less developed countries. Perhaps most fundamentally, some consider it ironic that the WTO's wealthiest member states continue to resist the unrestricted movement of labour in the form of international migration. With this contradiction in mind, Massey (2002) points out that many politicians and business leaders in the more developed world take different stances depending on the issue under discussion: if that issue is free trade they support the idea of a world without borders, but if the issue is international migration they support the nationalist idea of separate states with defensible borders.

The WTO rejects most of these criticisms, arguing that the best way to address global inequality is through trade liberalization, so that free trade becomes fair trade. In many respects the WTO is the most democratic of international organizations, having a one member–one vote system and serving as a forum for discussion between governments, the vast majority of which are democratically elected; Rigg (2001) notes that many of the WTO's opponents are more secretive, less accountable, and less democratic than the organization itself. There is probably some truth on both sides of the debate, with the WTO acknowledging the need for some internal reforms and many critics accepting that the WTO is necessary in some form.

---

companies, about 50 per cent of employment is foreign, whereas for the Swiss-based Nestlé and the UK-based BAT Industries, the foreign-employment figure is closer to 90 per cent.

Most of the transnationals are based in the US, Europe (especially the UK and Germany), or Japan. But others have their home bases in the newly industrializing countries in Asia, or even in countries (such as Canada) that are more usually regarded as host countries for **foreign direct investment** (FDI)—a term that refers to the investment of capital into manufacturing plants located in countries other than the one where the company is based. In general, countries that are major sources of FDI are also major hosts. Indeed, one way to identify economic globalization is to focus on the amount and location of these flows of capital, which have grown even more rapidly than international trade since World War II. In most cases both the investors and the targets of their investment are located in the more developed countries (Table 9.4). Global FDI rose to an all-time high in 2007, but declines will occur, presumably temporarily, as the global economy began to shrink in 2008.

Typically, a transnational will have its headquarters in the more developed world but locate its manufacturing activities in the less developed world, where wages are lower. There is, then, a division of production referred to as the **international division of labour** (see Chapter 14, especially Figures 14.12 and 14.13, which show the percentages of national labour forces in industrial and service employment, respectively).

Since 1945, trade has increased much more rapidly than production—a clear indication of how much more interconnected the world has become. Most of the products dealt in by transnationals fall into two groups: either they are technologically sophisticated, or they

international division of labour
A term referring to the current tendency for high-wage and high-skill employment opportunities, often in the service sector, to be located in the more developed world, whereas low-wage and low-skill employment opportunities, often in the industrial sector, are located in the less developed world.

Table 9.3	Revenue Data for the Top 10 Transnationals (2008) and Gross National Income (GNI) for Selected Countries (2007)	
Transnational or Country	**Revenues or GNI (US$ billions)**	
South Africa	455	
Wal-Mart	379	
Exxon Mobil	373	
Belgium	369	
Greece	362	
Royal Dutch/Shell Group	356	
Austria	317	
BP	291	
Toyota Motor	230	
Bangladesh	213	
Chevron	211	
ING Group	202	
Denmark	198	
Total (French oil company)	187	
General Motors	182	
ConocoPhillips	179	
New Zealand	111	

SOURCES: World Bank, *World Development Indicators*, 2008; Fortune Global 500, 2008.

TECHNOLOGICAL CHANGES

Of all the space-shrinking and time-compressing technologies that have been developed in recent years, the most important to economic globalization are those that facilitate the almost instantaneous transmission of information regardless of distance. These include advances in the mass communications media and in the use of cables, faxes, and satellites.

Other technological changes relate especially to the production process, the principal change being from Fordism to post-Fordist 'flexible production'. As we will see in Chapter 14, 'flexible production' means that manufacturers are able to produce specific goods less expensively in small quantities, rather than having to mass-produce as during the era of Fordism.

Globalization: Good or Bad?

You may not be surprised to learn that this textbook will not answer the question in the heading above. Rather, it will provide some food for thought, in the hope that you will use it to clarify your own thinking. As with environmental issues (Chapter 3) and the state of the world generally (Chapter 5), many people express strong opinions on globalization; and, as with those other topics, opinions often reflect ideological positions—personal views

involve mass production and mass marketing. In addition, much of the recent growth in trade reflects the increasing role of services—commercial, financial, and business activities—conducted by transnationals.

Activists in Hong Kong called for a boycott of McDonald's in August 2000 after it was reported that in Shenzen, China, children as young as 14 were working 16-hour days for just 1.50 yuan (less than 30 cents) an hour making promotional toys for the fast-food chain. The characters under the 'golden arches' read 'exploiting workers'.

of the world and how it ought to be. Let us begin with the arguments put forward by those who oppose globalization.

OPPOSING GLOBALIZATION

Why is opposition to globalization so widespread? There are several interrelated reasons.

In the broadest terms, globalization is seen as benefiting the more developed world at the expense of the less developed. More specifically, it is seen as benefiting those few countries and private investors within the more developed world where transnational corporations are based, at the expense of many (specifically working) people, especially those in the less developed world. There are also more specific criticisms, however, many of which are directed at the World Trade Organization in its capacity as guardian of world trade. For example, when the members of the WTO meet, they tend to insist on closed discussions. By excluding other interested parties, such as labour unions and environmental and anti-poverty groups, the WTO gives the impression that it is undemocratic and working to benefit only its members—or, indeed, since the WTO now includes less as well as more developed countries, only a select few of those.

Certainly the movement towards free trade—or what critics describe as selective free trade—can be interpreted as serving the interests of the major economic powers, such as the US, Japan, and the European Union, at the expense of the less developed countries that are only now becoming industrialized. Figure 9.17 shows that the rich countries of the world continue to dominate trade, with about 15 per cent of the world population accounting for about 66 per cent of world exports. It is worth noting that when the major powers themselves were industrializing, in the nineteenth and early twentieth centuries, they all enjoyed the benefits of protectionist trade policies, government intervention, subsidies designed to encourage domestic industrial growth, and few if any restrictions on environmental degradation.

It is often charged that globalization is contributing to the ever-widening gap between the rich and the poor. Certainly it is reasonable to suspect that if those who are already rich are in a position to dictate the rules of the game, they may become richer at the expense of the poor. On the other hand, the gap between rich and poor is not a new phenomenon. Since the beginning of European overseas expansion and

early capitalism, those countries and groups of people with the power to exercise control over others have benefited at the latter's expense; this is the basic logic of the world systems and dependency theories (Chapter 5).

Table 9.4	Foreign Direct Investment Flows by Selected Regions and Countries, 2007 (US$ millions)	
Region/Country	FDI Inflows	FDI Outflows
World	1,833,324	1,996,514
European Union	804,290	1,142,229
United States	232,839	313,787
Africa	52,982	6,055
Latin America and the Caribbean	126,266	52,336
Asia	319,333	194,663
China	83,521	22,469

SOURCE: United Nations Conference on Trade and Development, *World Investment Report: Transnational Corporations and the Infrastructure Challenge* (New York: United Nations, 2008), Annex Table B.1, 255–6.

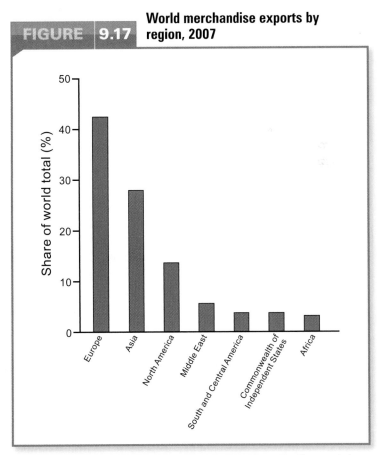

FIGURE 9.17 World merchandise exports by region, 2007

This graph shows that merchandise exports are dominated by Europe, Asia, and North America. The total value of world exports in 2007 was US$13,619 billion, almost double the 2003 value of US$7,375.

SOURCE: World Trade Organization, *Statistics: International Trade Statistics, World Trade Developments in 2007* (2008), Table 1.6, at: <www.wto.org/english/res_e/statis_e/its2008_e/its08_world_trade_dev_e.htm>.

A vegetarian holds a banner that reads in Portuguese 'One kilogram of meat equals 10 square kilometres of forest plus 15 million litres of water' during a protest at the 2009 World Social Forum in Belem, Brazil. The World Social Forum is an annual gathering to protest the simultaneous World Economic Forum in Davos, Switzerland.

Silvia Izquierdo/AP/The Canadian Press

For some critics, the most serious problem associated with globalization is that it gives priority to export-centred economies, with the result that the merits of locally sustainable economies are downplayed. Exploitation of resources in the less developed world to satisfy demand in the more developed world is likely to have a negative impact on local economies. There are related worries about the environmental consequences of globalization, particularly with respect to resource extraction. Increasingly, local landscapes are being treated as global commodities, and local communities are losing their traditional resource bases.

In addition to these primarily economic issues, specific concerns have been expressed about the rise of supranational organizations that could replace nation-states and democratic processes as the basic units of the political world; increasing cultural homogenization and the loss of local cultural identity; and social consequences in areas such as gender equality.

Finally, the criticism that may be the most profound is also the most uncertain. It has been suggested that globalization is unnatural in that many of the processes associated with it do not reflect any natural pattern of economic change—rather, they are the products of carefully designed economic strategies. Of course, similar criticism could be directed at the earlier protectionist phase. The extent to which the current situation represents a departure from the past can be debated, but public opinion in general is suspicious of globalization.

The extent of the opposition seems undiminished each year. One indicator is the annual World Social Forum, first held in 2001, and various related regional forums, at which anti-globalization activists meet to talk and plan strategy. Four of the first five annual meetings were held at Porto Alegre (Brazil), with the 2004 meeting in Mumbai. More recent meetings have been held in other places, including at three sites in 2006 (Caracas in Venezuela, Bamako in Mali, and Karachi in Pakistan), in Nairobi in 2007, by numerous local organizations in 2008, and in Belem (Brazil) in 2009. In general, participants reject the claim that globalization is in some way inevitable and argue that 'Another World is Possible'.

SUPPORTING GLOBALIZATION

But are the reasons for opposing globalization the only ones worth thinking about? Most of those who support globalization do so for business-oriented reasons. However, Johan Norberg defends globalization for reasons that are essentially moral. In his book *In Defence of Global Capitalism* (2001), Norberg argues in favour of capitalistic economic globalization on the grounds that it represents the best hope for eliminating poverty and fulfilling human potential. This moral perspective makes Norberg's argument compelling for many people. Using data from many countries, including the two largest, China and India, he argues that lowering trade barriers and investing capital in less developed countries helps to reduce global poverty, and hence is not only defensible but morally imperative.

As an example of the way transnational investment in less developed countries increases wages, Norberg points to a Nike shoe manufacturing plant in Vietnam: when it first opened, in the early 1990s, workers walked to the factory; three years later they used bicycles; three years after that they used scooters; and finally they are beginning to use cars. Although Norberg differs from many other proponents of globalization in favouring the free movement of people as well as goods, in general he maintains that improvements in food supply, education, equality, human rights, and gender issues show that globalization is making the world a better place for most people.

To the extent that global capitalism increases opportunities and wealth for the world's poor, obviously it is a positive thing. The problem, of course, is that globalization is far too complex a set of processes for us to be sure that all its consequences will be so benign. As is so often the case, reasonable people may well take very different positions, and many will come down somewhere in the middle, recognizing the benefits that globalization can bring but also aware of the constant effort that will be required to prevent or mitigate adverse consequences.

One thing seems clear, even to supporters of globalization—the benefits are unevenly distributed. Most of the benefits have accrued to countries in East and South Asia, especially China, while sub-Saharan Africa and, at least for a few years, the countries of the former Soviet bloc have lost rather than gained. Part of the explanation lies in specific circumstances. China, for example, has a long tradition of trading and is embracing a new capitalist approach. The Soviet bloc, in contrast, was sheltered for decades from free-market forces and the region has not found it easy to embrace the emerging economic order. In the case of sub-Saharan Africa, numerous problems of conflict, inadequate infrastructure, and some cases of corrupt government all work together to limit economic opportunities.

Some recent changes are likely to exacerbate these circumstances. Consider, for example, that in 2005 all quotas and restrictions on clothing imports were abolished, meaning that world trade in textiles was fully liberalized. This liberalization, the outcome of a 40-year process, will mean that some of the smaller producers, such as the Philippines, Bangladesh, and Mexico, will suffer, while China and India will significantly increase their exports, especially to the huge US market.

The 2008 and 2009 G20 summits held in Washington and London, respectively, were perhaps another striking sign that globalization is changing the established economic order in a positive way. Global economic summits used to involve the G7 countries, then the G8. The fact that a meeting of a group of 20 countries—including China, India, Russia, Brazil, and Saudi Arabia—was held is an indication that a changing cast of characters is responsible for discussing global economic issues. Consider, also, that there is increasing

evidence of what are often called emerging national economies playing larger roles in the global economy. In 2008 the *Fortune 500* list of companies included 62 from Brazil, Russia, India, and China (sometimes called the BRIC economies).

In some instances, the consequences of change are less clear. In late 2005, the EU agreed to reduce the subsidies paid to Europe's sugar farmers in accord with demands from the WTO after formal complaints from Australia, Brazil, and Thailand—the EU had been paying producers three times the world price. In principle, reduction of subsidies will benefit poor countries that produce sugar, but the situation is more complicated because 18 former colonies of Europe—countries such as Fiji, Barbados, and Mauritius—also had access to the subsidies.

Many supporters of globalization acknowledge these problems but remain convinced that the more general trend is one of trade liberalization leading to rising incomes and longer-term development for the world's poor. Two pieces of evidence in favour of this claim are that, in 2005, the emerging economies of the world (not the rich countries) produced a little more than 50 per cent of world output measured at purchasing-power parity and also accounted for more than 50 per cent of the increase in global GDP. It is possible to interpret these statistics as indicative of the biggest shift in economic strength since the rise of the United States about 100 years ago. In short, more and more countries are industrializing.

A factory in Indonesia producing Nike footwear.
Irene Slegt/Panos Pictures

A rather different understanding of globalization from those evident in these accounts of opposition and support is noted in Box 9.6.

A Global Village?

Today, space is shrinking, time is being compressed, and national borders are becoming less important than they used to be. These transformations are particularly evident in four recent developments:

1. New markets have arisen: global markets in services; global consumer markets with global brands; deregulated, globally linked financial markets.
2. New participants have emerged: most important among these are transnationals, the WTO, and the several regional trading blocs; there have also been numerous corporate mergers and acquisitions.
3. Faster and less costly means of communication have developed: the Internet, cellular phones, and fax machines facilitate the transmission of information, and there

are also ongoing technological advances in the movement of goods by air, rail, and road.
4. New policies and practices are being put into place concerning matters such as human rights and environmental issues.

But do all these ongoing changes mean that the world we live in, the lives we live, and the landscapes that we occupy are changing? Specifically, do we now all live in one global village? Certainly many people have suggested that this is the case, but it is important not to exaggerate the point. The reality is that, for most of the people in the world, life at the local, regional, and national levels still matters most. For example, use of the Internet is high in the more developed world but it is much lower in the less developed world. Similarly, crossing national borders is much more difficult for unskilled than for skilled workers.

Globalization is a reality, but by no means an uncontested one; we may think of the world as a global village, but the truth is that most of its people are not yet villagers.

Box 9.6 Globalization in Retreat?

Rather than debate the pros and cons of globalization, some prefer to focus on related but rather different issues. Thomas L. Friedman (2005), the noted *New York Times* columnist, asserts that the world is now what he describes as 'flat', a result of improved communications and the free movement of capital, meaning that the global economy is, for the first time, a level playing field as societies around the world conform to free-market principles and practices. Further, Friedman argues that globalization is intensifying.

But John Ralston Saul (2005), a Canadian philosopher and novelist, sees the world quite differently, going so far as to suggest that globalization, seen essentially not as a historical process of innovation and technological change but rather as a neo-conservative ideology, is now in retreat. Saul (2005: 270) writes: 'At the most basic level of societal knowledge, we do know that Globalization—as announced, promised and asserted to be inevitable in the 1970s, '80s and much of the '90s—has now petered out. Bits and pieces continue. Other bits have collapsed or are collapsing. Some are blocked. And a flood of other forces have come into play, dragging us in a multitude of directions.'

Rather than Friedman's flattening of the world, Saul sees a series of regional dislocations and inequalities resulting from the various successes and failures of globalization. This image of the world prompts Saul to write a book that often

reads like a manifesto designed to encourage the reader to become involved, to care about depopulation in the heartland of North America, about the inexorable spread and growth of Wal-Mart, and above all about the plight of the poor in the less developed world, all of which are understood to be evidence of the failings of globalization. Not only is it in retreat, it has failed to make the world a better place.

Saul dates the onset of globalization from 1971 (when the United States went off the gold standard and settled on a floating dollar to which most other currencies were pegged), claiming that it emerged fully grown. The heyday was in the mid-1990s, by which time trade liberalization had proceeded apace, global markets were dominant, and the WTO was in place to preside over the new global economy. But all was not well, as evidenced by the Asian financial crisis of 1997–8 and by the explosion of anti-globalization activity that began in earnest in Seattle in 1999. Building on these facts, Saul contends that globalization is now in retreat, with a key defeat being the US decision to invade Iraq without the support of other major powers if necessary. Thus, it is suggested that globalization might reverse because of an increasingly nationalistic US foreign policy.

But this thought-provoking and highly readable book does leave largely unanswered the critical question: what does the future hold if globalization is in retreat? Certainly, a return to a world dominated by nation-states seems unlikely.

CHAPTER 9 SUMMARY

A DISCIPLINE IN DISTANCE

The spatial location of geographic facts is not random: humans typically choose to minimize movement in order to minimize the frictional effects of distance.

TYPES OF DISTANCE

Physical distance is the spatial interval between two points that are usually (though not necessarily) measured with reference to some standard system. Time distance is often more important than physical distance, especially where movement of people is concerned. The valuable concepts of plastic space and time–space convergence reflect the fact that time distance between any two locations is a function of changing circumstances. 'Economic distance' is the cost incurred in overcoming physical distance; it is often the paramount consideration for industries and other businesses. 'Cognitive distance' reflects individual perceptions of distance and is especially difficult to measure. 'Social distance' is particularly relevant to residential distributions—and the creation of distinctive social spaces.

GRAVITY AND POTENTIAL MODELS

Movement between locations is affected by both the distance to be covered and the population size of the locations—a human equivalent to the gravity concept in physics. The 'potential model' is similarly derived from physics and expresses the potential that a particular location has for interaction with other locations. Both models are attractive simplifications of complex real-life situations.

THE DIFFUSION PROCESS

Following the pioneering work of Hagerstrand, diffusion research in geography changed from a study of landscape features to a study of process. Four empirical regularities are the neighbourhood effect, the hierarchical effect, resistance, and the S-shaped curve. Most recently, diffusion studies have focused less on the process and more on the human consequences of diffusion, especially its potential to cause spatial inequalities.

TRANSPORT SYSTEM EVOLUTION

Transport systems are both cause and effect of other economic aspects of landscape. Major changes are associated with political expansion and technological advances. In the case of Britain, the most dramatic changes came with the Industrial Revolution: turnpike roads, navigable waterways, railways, and finally an improved road network. Theoretical approaches to the evolution of transport systems focus on the identification of stages that are related to larger geographic issues.

NETWORKS AND MODES

Transport networks are made up of nodes and routes and are often analyzed in a simplified form using graph theory. Each of the principal modes of transport—water, land, and air—has some advantages and some disadvantages.

TRADE

Although trade is clearly related to the distance between locations, the three principal branches of trade theory all exclude any reference to distance. Trade is also related to resource base, labour force, and capital. Today, much world trade is regulated by various agreements between countries concerning economic integration.

SHAPING OUR WORLD

To understand our contemporary world we need at least a basic understanding of the forces that have shaped it, many of which are touched on elsewhere in this book. Among them are physical geography, human dependence on other life forms, human origins and early movements, cultural evolution, agricultural technologies, European overseas movement, diverse modes of production, the rise of states, industrial technologies, twentieth-century changes, and current globalization trends.

ECONOMIC GLOBALIZATION

Production is no longer contained by national boundaries, a change brought about primarily by the rise of transnational corporations. Further, since 1945 trade has increased much more rapidly than production—a clear indication of how much more interconnected the world has become. Although the reality of economic globalization seems clear, the details of its impact are contested. For example, on the one hand, a new global economy is emerging in response to the internationalization of capital, production, and services; on the other hand, there are several major examples of regional integration and associated regional protectionism. According to some observers, trade liberalization is leading to rising incomes and longer-term development for the world's poor. However, one thing that is certain is that globalization is meeting considerable opposition from a variety of groups.

QUESTIONS FOR CRITICAL THOUGHT

1. As globalization continues to transform the human world, can we expect fundamental geographic concepts such as space and distance to become less significant as determinants of human spatial behavior? Why?

2. How might the gravity model and the potential model be rewritten so as to reflect the (apparent) decrease in the importance of distance attributed to globalization? Are there other factors (apart from physical distance) that could be incorporated into these models to make them more realistic?

3. Why is diffusion such an important concept in geography? How has globalization influenced diffusion processes?

4. What factors do you think will shape globalization and international relations in the next century?

5. What are the advantages and disadvantages of economic globalization? Are you, personally, better or worse off as a result of economic globalization? Do we all live in one global village, as Norton postulates at the end of the chapter?

FURTHER EXPLORATIONS

Brook, T. 2007. *Vermeer's Hat: The Seventeenth Century and the Dawn of the Global World*. London: Bloomsbury Press.

Suggests that transculturation, the movement of ideas and things from one culture to another, followed European overseas movement, leading to our globalizing world.

Chanda, N. 2007. *Bound Together: How Traders, Preachers, Adventurers, and Warriors Shaped Globalization*. New Haven: Yale University Press.

Argues that globalization has proceeded for centuries, but now it is simply more visible.

Chorley, R.J., and P. Haggett. 1970. *Network Analysis in Geography*. New York: St Martin's Press.

An account of many of the topics raised in this chapter, with a consistent focus on spatial analysis.

Clark, G.L., M.P. Feldman, and M.S. Gertler, eds. 2000. *The Oxford Handbook of Economic Geography*. Oxford: Oxford University Press.

Covers a wide range of topics in economic geography; includes overviews of global economic integration and changes.

Cohen, D. 2006. *Globalization and Its Enemies*. Cambridge, Mass.: MIT Press.

Thoughtful discussion of climatic and other geographic conditions that might impact economic growth.

Gatrell, A.C. 1983. *Distance and Space: A Geographical Perspective*. Oxford: Clarendon.

A clear account of distance and spatial concepts with many examples of appropriate analyses.

Hagerstrand, T. 1967. *Innovation Diffusion as a Spatial Process*, trans. A. Pred. Chicago: University of Chicago Press.

A seminal work integrating the cultural and spatial analytic approaches to diffusion; the postscript, by Pred, is an excellent summary of diffusion research before 1967.

Haggett, P. 2000. *The Geographical Structure of Epidemics*. Oxford: Clarendon.

Written by one of the most original of twentieth-century geographers; examines epidemics using diffusion concepts and provides valuable insights into infection patterns.

Hanson, S., and G. Giuliano, eds. 2004. *The Geography of Urban Transportation*, 3rd edn. New York: Guilford.

A comprehensive collection that covers a wide range of urban transportation topics, but with emphasis on the US.

Murray, W.E. 2006. *Geographies of Globalization*. New York: Routledge.

A balanced discussion of the topic with comprehensive coverage of economic, political, and cultural trends.

Ritzer, G. 2004. *The Globalization of Nothing*. Thousand Oaks, Calif.: Sage.

An entertaining and informative study premised on the idea that the grand narrative of globalization is a transition from something to nothing.

Rodrigue, J.-P., C. Comtois, and B. Slack. 2006. *The Geography of Transport Systems*. New York: Routledge.

Sound overview covering all aspects of transport geography.

Stiglitz, J.E. 2006. *Making Globalization Work*. New York: W.W. Norton.

Contends that globalization holds much promise for a better world, but that this is not being realized because the current economic order favours rich countries.

Vance, J.E. 1986. *Capturing the Horizon: The Historical Geography of Transportation*. New York: Harper and Row.

A major conceptual and factual work dealing with the growth of transport networks.

ON THE WEB

EKISTICS

World Society for Ekistics

www.ekistics.org/

Informative site explaining ekistics as an interdisciplinary approach to the needs of human settlements.

DISEASE DIFFUSION

John Snow

www.ph.ucla.edu/epi/snow.html

A site hosted by the UCLA Department of Epidemiology and devoted to the work of John Snow, one of the first medical geographers; includes a link to Snow's famous London cholera map.

Geography of Health

www.makingthemodernworld.org.uk/learning_modules/geography/05.TU.01/

Detailed information on the geography of health and disease with emphasis on spread.

TRADE AGREEMENTS

NAFTA

www.international.gc.ca/trade-agreements-accords-commerciaux/agr-acc/nafta-alena/index.aspx

Canadian government website with details of NAFTA.

European Union

europa.eu/index_en.htm

The home page of the EU; presents news and statistics.

ASEAN

www.aseansec.org/

A good source of news and detailed statistics.

ECONOMIC GLOBALIZATION

Global Policy Forum

www.globalpolicy.org/globaliz/econ/index.htm

Information on economic globalization, from a critical perspective, examining how globalization might be resisted or regulated to promote sustainable development.

International Forum on Globalization

www.ifg.org/

The IFG brings together leading activists concerned with the consequences of globalization processes.

World Trade Organization

www.wto.org/index.htm

Contains wealth of information on WTO as well as trade and related issues.

AGRICULTURAL LIVES AND LANDSCAPES

Agricultural geography—the identification, measurement, and explanation of spatial variations in agricultural activities—is generally considered to be a branch of economic geography. The other principal branches (settlement and industrial) are discussed in Chapters 11, 12, 13, and 14. It may be helpful to think of these five chapters as forming a unit. Economic geography obviously is affected by economic globalization. However, cultural and political factors also play very important roles. Cultural factors such as religious beliefs and ethnicity may influence the varieties of crops or animals produced in certain areas, and state intervention in the form of quotas and subsidies also has a direct effect.

This chapter begins with an examination of the way the subdiscipline of economic geography has changed in recent years. Drawing on the spatial analysis tradition—the third of our three recurring themes—we then examine why agricultural activities are located where they are, and discuss in detail a classic theoretical argument that many geographers have found useful.

Next, the origins and evolution of major agricultural activities are explained and nine principal agricultural regions around the world are identified. This is followed by a discussion of the global restructuring of agriculture. The chapter concludes with a look at changes in food consumption patterns and preferences. The final two sections reflect the current interest in agriculture as one part of a wider production and consumption system that includes inputs, farming, product processing, wholesaling, retailing, and food preferences.

Spring planting, potato field in Kinkora, Prince Edward Island, Canada.
Barrett and MacKay/All Canada Photos

Inventing, and Reinventing, Economic Geography

Geography is an eclectic and fashion-prone discipline. The attention span . . . for major theoretical or methodological perspectives is rather short-lived. For many a geographer, it is very hard to keep up with the endless re-formulations of spatial or geographical perspectives and theoretical influences. (Swyngedouw, 2000: 41)

The subjects of this chapter and the four that follow—focusing on agriculture, settlement, and industry—represent the core of what is usually labelled 'economic geography'. Although that label remains appropriate, our understanding of these traditionally 'economic' subjects is increasingly being informed by cultural, social, and political perspectives. As the quotation above suggests, a feature of human geography as a discipline is its frequent shifts in theory and methodology, one of its great strengths but also an occasional weakness.

Human geography is indeed an eclectic discipline that welcomes new approaches. In Chapter 2 alone, we saw how it incorporated four very different philosophical approaches:

- empiricism (1900–mid-1950s);
- positivism (mid-1950s–*c.* 1970)
- humanism (*c.* 1970–);
- Marxism (*c.* 1970–).

In Chapter 7 we considered at length two more approaches:

- feminism (1970s–);
- postmodernism (late 1980s–).

Reflecting the 'cultural turn' of recent years (see Chapter 7), most of our newer approaches have built on versions of humanist, Marxist, feminist, or postmodern ideas.

With this chronological summary of approaches in mind, let us now review what we might call the invention of economic geography—and its ongoing reinvention.

DESCRIPTIVE REGIONAL ECONOMIC GEOGRAPHY

As the name suggests, 'descriptive regional economic geography' is the traditional empirical approach outlined in Chapter 1. Until about the mid-1950s, most economic geography focused on economic activities within a regional framework. Geographers mapped and described land use, identifying regions on the basis of their major economic activities. Some of this work was general in character, focusing on a dairy belt or a manufacturing region, for example, while in other cases the ultimate goal was to delimit appropriate regional boundaries. Such studies remain a central part of geography, as evidenced by the accounts of world agricultural regions in this chapter and world industrial regions in Chapter 14.

SPATIAL ANALYSIS

From the vantage point of the early twenty-first century, spatial analysis dominated economic geography for only a short time, but it represented a conceptual, theoretical, and quantitative revolution. Abandoning traditional regional geography, with its focus on the empirical and idiographic (unique and particular), for the spatial analysis school, with its nomothetic approach (stating universal laws), geographers in the mid-1950s were inspired by developments in other social sciences, especially economics.

As a discipline, economics is concerned with many issues relevant to human geography: from the allocation of limited resources (such as the land and soil needed for agriculture or the minerals and power supplies needed for industry), to exchange transactions and the accumulation and distribution of wealth (issues central to human geographic analysis of settlement and transportation). Especially relevant have been the classical and neo-classical traditions in economics, since these were the ones that interested geographers of the spatial analysis school as they shifted their attention from region to location and interaction.

Of course, this revolutionary shift in focus owed much to economics, but it also made a valuable contribution to economics in the form of a new hybrid discipline called regional science (introduced in Chapter 9). For geographers, one obvious shortcoming of economics was that many of its theories seemed to have been constructed 'in a wonderland of no spatial dimensions' (Isard, 1956: 25). As they set out to rewrite economics in spatial terms, human geographers, not surprisingly, turned first to theories that already included a spatial dimension.

Positivist theory

Much of the rest of this book is concerned with location—specifically, the location of human activities and the interactions between locations. Inspired by classical and neo-classical economics, **location theory** explains geographic patterns at any given place and time, while interaction theory (introduced in Chapter 9) explains movements between locations.

As we saw in Chapter 2, theories are sets of interrelated statements that explain reality—for example, patterns of economic activity. Theories are usually generated inductively, from apparent facts; then hypotheses are deduced from them and tested in the real world (see Figure 2.1). Perhaps the greatest virtue of theories developed following the scientific method is their rigour: once a theory is generated, the process of hypothesis deduction and verification will not (in principle) vary with the researcher's abilities. Verified hypotheses eventually assume the status of laws, which can later become initial statements in other theories.

By definition, a theory is merely a simplification of complex reality, and may even be untrue. Yet even a theory that proves not to be true can still offer insight and facilitate further work. Indeed, advocates of positivism argue that any science—including human geography—depends equally on fact (subject matter) and theory (the means of explaining the subject matter).

We will examine three important examples of positivist theory in this chapter and in Chapters 11 and 14. All three are closely linked to the classical and/or neo-classical economic traditions. The first example, addressing the question of why agricultural activities are located where they are, was developed by an economist, Johann Heinrich von Thünen, but uses distance as the key explanatory variable. The second, discussed in Chapter 11, seeks to explain why urban settlements are located where they are; this theory was developed by a geographer, Walter Christaller, inspired by the von Thünen theory. Finally, in Chapter 14 we will encounter a theory explaining why industrial activities are located where they are; like von Thünen's theory, this was developed by an economist (Alfred Weber), but uses distance as the key explanatory variable.

When positivist-type theory was introduced to human geography, some critics argued that theories had no place in geography because all locations, indeed all geographic facts, are unique. While this is literally correct, it is misleading: however unique individual locations may be, it is always possible to recognize general types of location.

Theories explaining economic location have typically been **normative** in nature—concerned with what ought to be. By definition, then, they do not purport to explain reality itself, but rather what reality would be like in some hypothetical ideal situation. For example, our principal agricultural location theories are based on the concept of the 'economic operator' (Box 10.1).

Evaluating spatial analysis

Research in the area of spatial analysis was innovative and challenging. For the first time, human geographers became excited about conceptual advances, constructing theory, and the possibility of testing theoretical predictions quantitatively. By the early 1970s, as we have seen, such work was widely criticized as a version of human geography that lacked the human element. As a result, spatial analysis and the regional science associated with it were very influential for only a brief period. Nevertheless, these approaches left an important legacy, and they still play an important role in some contemporary work.

REINVENTING ECONOMIC GEOGRAPHY

Since the early 1970s, economic geography, like the larger discipline of human geography, has not been dominated by any single approach. However, most work in the field has been linked in some way to one or more of the post-positivist philosophical developments, especially variants of Marxism, feminism, and postmodernism.

Marxism

The emergence of Marxist perspectives in human geography was linked to social and economic issues, including the civil rights and anti-Vietnam War movements, increasing unemployment and inflation, and the declining health of many industrial areas. Together, these developments encouraged an interest in various forms of radical geography. The pioneering statement was a book by Harvey (1973; see Box 12.2) that served to introduce geographers to the Marxist political economy described in Box 10.2.

location theory
A body of theories explaining the distribution of economic activities.

normative
Focusing on what ought to be rather than what actually is; in normative theory, the aim is to seek what is rational or optimal according to some given criteria.

Box 10.1 | The Economic Operator Concept

The **economic operator** concept (also known as **rational choice theory**) is a normative concept according to which each economic operator minimizes costs and maximizes profits as a result of perfect knowledge and a perfect ability to use such knowledge in a rational fashion. This theory has its roots in classical economics, but is one of the basic propositions of the neo-classical school. There is, of course, no such person as an economic operator in the real world—none of us is blessed with the required 'omniscient powers of perception' and 'perfect predictive abilities' (Wolpert, 1964: 537).

What makes this unrealistic concept useful in economics and geography is that it allows us to generate hypotheses that are not encumbered by the complexities of human behaviour. In fact, the economic operator represents one extreme on a continuum of possibilities that range from optimization to minimal adaptation. We know that in reality most human behaviour falls between the two extremes and can be described as **satisficing behaviour**—aimed at satisfying the individual rather than optimizing the situation. The concept of satisficing behaviour is a basic alternative to rational choice theory.

economic operator
A model of human behaviour in which each individual is assumed to be completely rational; economic operators maximize returns and minimize costs.

rational choice theory
The theory that social life can be explained by models of rational individual action; an extension of the economic operator concept to other areas of human life.

satisficing behaviour
A model of human behaviour that rejects the rationality assumptions of the economic operator model, assuming instead that the objective is to reach a level of acceptable satisfaction.

Those working from a Marxist perspective strenuously reject the supposed value neutrality of spatial analysis, arguing that such an approach only perpetuates the status quo. Rather, the forces of capital accumulation and related social structures are responsible for the creation and ongoing recreation of the economic geographic landscape. The issues on which Marxist-inspired geographers focus—poverty, unemployment, industrial decline, uneven regional development—had been largely ignored by spatial analysts concerned primarily with explaining locations.

By the 1980s, however, much Marxist research was being criticized for focusing on the geographic outcomes of the large-scale processes of capital accumulation and not taking enough interest in political contexts and human intentionality. Some criticisms built on structuration theory (see Chapter 7) to highlight the importance of localities as the appropriate scale for analyzing the space economy of capitalism. Others reflected humanist and especially feminist perspectives.

Feminism

The central argument behind feminist critiques of Marxist work is simply that gender is just as important as class, if not more so. Notably, many feminists contend that gender analysis does much more than add an extra set of questions to the discussion of a subject like economic geography: it quite literally transforms it. McDowell (2000) suggests three examples:

1. Feminist economists and economic geographers have shown that the work done by women is often ignored. In particular, domestic labour—including caring for children or elderly family members—is unpaid and widely unacknowledged as work. Further, national economic statistics do not take women into account: for example, data on income and health typically reflect the overall circumstances of the household and do not capture differences between household members in regard to such issues as the allocation of resources.

2. In the labour market, assumptions about women tend to limit them to a relatively narrow range of occupations coded as feminine and typically seen as inferior.

3. Finally, it is possible to interpret features of the economic landscape—for example, workplace practices—not in terms of Marxist political economy but rather in terms of gender identities.

The body of research into these and related issues is substantial and growing. In addition, some feminist geographers draw on postmodern theory, notably when they focus on the social construction of gendered identities.

Postmodernism

On the one hand, postmodernism rejects the traditional empiricist faith that human geographers can provide factually reliable accounts of the world. On the other, it rejects the assumption—central to positivism and

Box 10.2 | Marxist Political Economy

The Marxist perspective gained adherents when it became clear that neither the empiricist nor the positivist approach offered a meaningful account of the agricultural changes that began in the 1970s. Neither classical description nor location theory offered any conceptual insight into issues such as the industrialization of agriculture, the rise of agribusiness, state intervention, or the social impacts of technological change.

According to Marx, capitalism is contradictory, characterized by social tensions and conflicts. The essential argument is as follows. A capitalist society is based on the circulation of capital—through production, exchange, and consumption—and this circulation leads to economic growth, or what is often called the ceaseless accumulation of capital. Because capital circulation involves the movement of money, goods, and labour, it creates, and continually recreates, geographies of production, consumption, and interaction. The requirement that circulation leads to economic growth implies that lack of growth, or decline, is untenable. Indeed, for the capitalist system to work, labour must produce more value in the production process than it receives in the form of wages, so that business owners can appropriate that 'surplus value' as their profit. For Marx, the accumulation of capital by owners at the expense of workers inevitably leads to class conflict.

The capitalist system, as Marx describes it, has three key geographic implications: social and spatial divisions of labour (e.g., along class and gender lines); competition among owners for space, resources, and economic infrastructure; and, most generally, inherent instability (e.g., periods of depression and inflation).

Marxist discourse has two important advantages for the study of contemporary agricultural geography: it blurs the lines between agricultural and other economic activities, especially industry, and it integrates the agricultural experiences of the more and of the less developed worlds. A specific key insight revealed by the Marxist perspective is that capitalist development in the broad agricultural and food system has increased dramatically as non-farm interests such as banks and the food-processing industry have gained increasing control of the agricultural production process.

In recent years, the Marxist-inspired political economy approach has provided a valuable perspective on agricultural change, especially in the context of an increasingly globalized world and food system. The approach has been widely employed in studies of agriculture in the less developed world, especially as it has been merged with a more traditional ecological (humans and land) approach as discussed in Chapter 3, and can be summarized as follows:

1. At the regional level, human geographers attempt to integrate the multiple social and ecological relations involved in the organization of agriculture and land use.
2. This approach permits a focus on the politics of place, especially regarding differences in power between groups and places (as discussed in a different context in Chapter 7).
3. It has clear implications for the design and implementation of rural development policies.
4. There is recognition of the need to involve local populations as well as governments and organizations in the planning process.
5. The interplay of social practices (agency) and political—economic conditions (structure) is often considered, as outlined in the account of structuration theory in Chapter 7.
6. In many cases the environmental impacts of agricultural activity can be addressed in political terms.

In this way a political ecology perspective can add significant new dimensions to more traditional human ecological approaches. Box 10.13 discusses an application of this type of analysis that provides a good indication of its procedures and merits. Other examples include studies by Zimmerer (1991), integrating regional political ecology concepts, structuration theory, a politics of place, and production ecology in an analysis of agricultural change in highland Peru, and by Grossman (1993: 346), who used a political ecology perspective to 'highlight not only the impact of political-economic relationships on resource-use patterns but also the significance of environmental variables and how their interaction with political-economic forces influences human-environment relations'.

Of course, the political ecology viewpoint has much to say also about the environmental impacts of agricultural activities (see Chapter 3), especially with regard to the unequal distribution of those impacts. For example, the hurricane that devastated parts of Central America in October 1998 did not affect all agricultural operations equally. In fact, those farms using traditional methods suffered much less soil erosion and crop damage than did those using modern chemical-intensive methods. More generally, there is good evidence that although modern agricultural methods are usually more productive than traditional methods, environmentally they are less friendly. The four most common types of environmental damage caused by agriculture are particularly dangerous for the less developed world: soil degradation, soil and water pollution, water scarcity, and reduction of genetic diversity.

Marxism—that theory can be relied on to explain reality. In both cases, postmodernism has had an enormous influence on the work of geographers and other academics and practitioners, for it has demanded new ways of viewing and of representing the world. We have already noted some of the implications of these ideas in Chapter 7, especially the concern with listening to previously repressed voices. The sheer diversity of the postmodern endeavour appeals to many geographers. Postmodern theory has suggested new directions for research—for example, focusing on consumption rather than production, or on the commodification of people and place, and both of these directions are evident in the accounts of agriculture in this chapter, of settlement in Chapter 13, and of industry in Chapter 14. The interest in research directions such as these is one component of 'the cultural turn' as it has affected economic geography. Today it is widely acknowledged that cultures, at all scales, affect economic activity and that economic activity in turn is one aspect of culture and of cultural change.

In conclusion, it is particularly important to note that postmodernists see geographic theory and practice as social activities, inevitably influenced by the geographer doing the work. Thus they demand that geographers situate themselves in relation to their work. Barnes (2001: 557) aptly describes this demand as reflecting the reality that all geographic writing is necessarily a 'view from *somewhere*'—an observation that may recall the humanistic critique of positivism noted in Chapter 2.

SUMMARY

This account of changing approaches to the subdiscipline of economic geography mirrors the larger history of human geography since 1900 outlined in Chapter 1. That history is further developed here because economic geography most clearly illustrates these changing approaches, which in turn are mirrored in the contents of this chapter and the four that follow it:

1. Any introductory account of economic geography needs to provide basic factual information, therefore, the empiricist regional approach continues to be an essential starting point.
2. Spatial analysis remains important, at least partly because such research addresses a fundamental geographic question: Why are things—farms, settlements, industries, or anything else—located where they are?
3. Marxist ideas have inspired a variety of important questions and related approaches, with particular emphasis on the role of class within a capitalist mode of production.
4. Feminist ideas have introduced important challenges to economic geography, questioning the validity of basic assumptions and suggesting new directions for research that acknowledge gender differences.
5. Postmodernism (like other ideas associated with 'the cultural turn') adds to the diversity and complexity of economic geography, enriching the theoretical background and opening the way to new and original directions in research. In particular, postmodernism has suggested a new focus on consumption and commodification.

One way to summarize the evolution of economic geography over the past century is to note how its focus has broadened, from the strictly economic (in the empiricist and spatial analysis traditions), to the economic and political (in the Marxist tradition), to the economic, political, and cultural (in the feminist and postmodern traditions).

This detailed analysis of the connections between apparently separate types of human geography may seem an unnecessary complication in an introductory textbook, but it reflects the world we live in, a world where economic, political, and cultural matters are irrevocably intertwined. One of our challenges in the rest of this book will be to analyze the three principal topics of traditional economic geography—agriculture, settlement, and industry—from multiple perspectives, taking advantage of the diverse approaches now at our disposal.

The Agricultural Location Problem

We begin our account of agriculture with a basic geographic question. Why are specific agricultural activities located where they are? Our answer will centre on economic issues, but first it will take into consideration several other factors—physical, cultural, and political—all of which are interrelated.

SOME PHYSICAL FACTORS

The agricultural landscape is typically made up of individual farms, and the details of the landscape reflect the decisions of farm owners, managers, and workers. Such decisions may lie anywhere along the continuum identified in Box 10.1—from optimization through satisficing behaviour to minimum adaptation—and reflect a variety of factors.

Physical factors are important influences in agricultural decisions and hence in the creation of agricultural landscapes. Animals and plants are living things and require appropriate physical environments to function efficiently. Farmers have two options: either to ensure that there is a match between animal and plant requirements and the physical environment, or to create artificial physical environments by, for example, practising irrigation or building greenhouses. Some farmers are prepared to create artificial environments for specific agricultural purposes, but may choose not to engage in activities that are environmentally detrimental. Because many environments are suitable for more than one crop, the agricultural decision can be based on other factors; similarly, those other factors can play a role in decision-making when the demand for a product is lower than the quantity that can be produced.

Many geographers believe that the best way to approach agricultural issues is to think in terms of ecosystems. Low-technology agricultural systems are structured along the same lines as natural ecosystems, while higher-technology systems are so complex, involving so many different types of circumstances, that to attempt to isolate one variable in explaining agricultural practice would be seriously misleading. The multiplicity of factors is examined more closely in the following sections.

Climate

Climatic factors are the main physical variables affecting agriculture. Plants have specific temperature and moisture requirements, although ongoing plant hybridization has succeeded in extending these requirements. Optimum temperatures vary according to the plant cycle, but growing-season temperatures between 18°C (64°F) and 25°C (77°F) are often required. Low temperatures result in slow plant growth, while short growing seasons may prevent crops from reaching maturity. In many areas, frost may result in plant damage. Moisture is crucial to plant growth; too little or too much can damage plants. In some areas, especially semi-arid zones, rainfall variability is a problem. Many of the world's major wheat-growing areas (such as the Canadian prairies) are highly vulnerable in this respect; Figure 10.1 indicates the relationship between wheat yield and mean annual rainfall at a time before chemical fertilizer was widely used. Animals also have water requirements, especially dairy cattle; sheep are much more adaptable to water limitations.

Soils and relief

Two other important physical variables are soils and topographical relief. Soil depth, texture, acidity, and nutrient composition all need to be considered. Shallow soils typically inhibit root development; the ideal texture is one that is not dominated by either large particles (sand) or small particles (clay). Most crops require neutral or slightly acidic soils. The most crucial nutrients are nitrogen, phosphorus, and potassium. Soil fertility needs to be maintained if regular cropping is practised; methods include fallowing, manuring, crop rotation, and use of chemical fertilizers.

Relief (the shape of the land) also affects agriculture, specifically through slope and altitude. The angle, direction, and related insolation (exposure to sun) of slopes determine both the probability of soil erosion and the use of machinery. Generally, the flatter the land

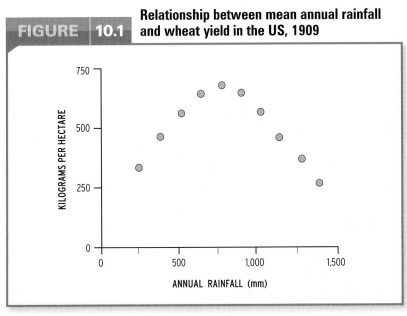

FIGURE | 10.1 **Relationship between mean annual rainfall and wheat yield in the US, 1909**

SOURCE: Adapted from O.E. Baker, 'The Potential Supply of Wheat', *Economic Geography* 1 (1925): 39.

the more suitable it is for agriculture. Altitude affects temperatures: in temperate areas, the mean annual temperature falls 6°C (11°F) for each 1,000 m (3,280 feet) above sea level.

SOME TECHNOLOGICAL, CULTURAL, AND POLITICAL FACTORS

In the past, agricultural location was generally considered to be determined by physical conditions:

> The Corn Belt is a gift of the gods—the rain god, the sun god, the ice god, and the gods of geology. In the middle of the North American continent the gods of geology made a wide expanse of land where the rock layers are nearly horizontal. The ice gods leveled the surface with glaciers, making it ready for the plow and also making it rich. The rain god gives summer showers. The sun god gives summer heat. All this is nature's conspiracy to make man grow corn. (Smith, 1925: 290)

Today we have advanced well beyond environmental determinism, and we recognize the important roles that technological, cultural, and political factors can play.

Technology

During about the past 300 years especially, advances in plant and animal breeding and, most recently, in biotechnology have improved agricultural productivity. A particularly effective

View of vineyards in the Okanagan, British Columbia, Canada.
laughing mango/iStock

approach has been to reduce the competition between crops and various pests for both light and nutrients through use of pesticides and fungicides. Also important is increasing energy inputs through use of fertilizers. The dramatic consequences of the development of nitrogen fertilizers and also current advances in biotechnology are considered in more detail later in this chapter.

Religion and ethnicity

One important cultural factor noted in Box 10.1 is that farmers are not profit maximizers. Farmers in the more developed world typically favour security and a relatively constant income over a life dedicated to the unlikely goal of profit maximization, and poor farmers, in the more developed and less developed worlds alike, typically maximize product output for subsistence rather than profit. Even these preferences can be interpreted as consequences of cultural attitudes towards an essentially economic issue. The influence of culture is more obvious, however, in the case of religion and ethnicity.

Group religious beliefs may favour specific agricultural activities because of the value placed on either the activity or the product of the activity. Christianity, for example, values wine made from grapes, which is used in the sacrament of Holy Communion. The Church's demand for wine encouraged the spread and growth of viticulture, including its introduction to California by early Christian missionaries. Other agricultural activities are negatively affected by religious beliefs. Pigs are taboo in Islamic areas, while Hindus and Buddhists believe it is wrong to kill animals, especially cattle.

Immigrants in a new land often continue to adhere to the agricultural practices of their homelands. Hence, an area settled by various ethnic groups often presents a patchwork appearance. North American examples abound. One geographer has explained the location of cigar tobacco production in the United States by reference to ethnic variables (Raitz, 1973). Another described two communities in south-central Illinois, one of German Catholic ancestry and one of non-Catholic British ancestry (Salamon, 1985). Although these communities are only 32 km (20 miles) apart and have similar soils, they differ substantially in farm size and organization. The German farms are small and diversified, including dairy, hogs, beef, and grain; the British farms are monocultural in grain. Arguing that these differences in farming practices stem from different attributes, Salamon contrasts the 'yeoman' Germans and the 'entrepreneurial' British (Table 10.1).

As Table 10.1 suggests, a variable such as landownership can affect farmers' behaviour, especially with respect to innovation. Owner-operated farms may be handled quite differently from rented farms or farms in socialist economies where decision-making is centralized.

The state

Political decisions affect much human behaviour, including the behaviour of farmers. Governments may have many reasons for deciding to influence farmers' behaviour, but in the contemporary developed world the most common motive is the desire to support an activity that, as measured by income, is in decline. In such cases government intervenes by fixing product prices and providing financial support for specific farm improvements such as land clearance (Box 10.3 and Figure 10.2). In the less developed world, governments provide assistance that enables farmers to adopt new methods and products.

The contemporary agricultural system is affected not only by state policies but by national and international trade legislation. Canada, the United States, Japan, and Europe all have marketing boards, quota requirements, government credit policies, and extension services to provide information to the farm community, all of which affect the agricultural landscape. Perhaps the best-known examples of interventionist policies come from the United States and the European Union, where in some cases farmers have been paid not to grow crops.

Some basic economic factors

The spatial patterns of agricultural activities are the end product of complex physical, cultural, and political variables. But human geographers agree that the most important variables generally—not necessarily in a specific location—are those labelled economic. Agricultural products are produced in response to market demand for them; thus farming is subject to the basic laws of supply and demand. Figure 10.3 depicts characteristic supply and demand curves. Since supply increases and demand decreases when the price received increases, an equilibrium

Satellite image of the Great Plains: bread basket of North America.
© Google Earth

Table 10.1	Contrasting Farming Types in Illinois
Yeoman	**Entrepreneur**
Goals	
Reproduce a viable farm and at least one farmer in each generation returns	Manage a well-run business that optimizes short-run financial returns
Strategy	
Ownership of land farmed preferred Expansion limited to family capabilities Diversify to use land and family most creatively Manage the most efficient operation possible	Ownership plus rental land to best utilize equipment Ambitious expansion limited by available capital
Farming Organization	
Smaller than average operations Animals plus grain, crop variety Land fragmentation Landowners often operators Expansion of community territory	Larger than average operations Monoculture cash grain Land consolidation Landowners frequently absentee Community territory stable
Family Characteristics	
Intergenerational co-operation Parents responsible for setting up son/heir Many children, non-farmers, live nearby Parents responsible for intergenerational transfer Early retirement geared to succession by children	Intergenerational competition Incumbent upon son/heir to set up self Often all children leave farming Heirs responsible for intergenerational transfer Retirement geared to personal desires
Community Structure	
Village central focus of community Community loyalty Population relatively stable Strong church attachment Farmers involved in village	Village declining Weak community attachment Population diminishing Church consolidations Farmers uninvolved in village

SOURCE: S. Salamon, 'Ethnic Communities and the Structure of Agriculture', *Rural Sociology* 50 (1985): 326.

Box 10.3   Government and the Agricultural Landscape

The Canada–United States border region between the Great Lakes and the Rocky Mountains provides an excellent example of how agricultural landscapes evolve in response to government policies. Around 1960, despite similar climatic, soil, relief, and drainage conditions, the landscapes differed considerably because the US government's agricultural policy was interventionist and its Canadian counterpart was much less so. By the 1970s, however, American intervention had decreased, while Canadian intervention had grown.

Figure 10.2 shows dramatic differences in crop and livestock combinations around 1960. The wheat allotment program in the United States meant that much wheat land was converted to raise barley, while the National Wool Act of 1954 encouraged sheep-raising through a guaranteed price.

By the 1970s, however, the United States had moved away from support programs towards relatively market-oriented policies. By contrast, Canada had come to favour greater intervention to ensure adequate incomes for farmers. As a result of these changes, the differences between the two countries were reduced and the landscapes changed accordingly. The transboundary differences are significantly fewer today than in the 1950s and 1960s.

SOURCE: Adapted from H.J. Reitsma, 'Crop and Livestock Production in the Vicinity of the United States—Canada Border', *Professional Geographer* 23 (1971): 217, 219, 221, Fig 1 (MW/RTPG/P1891). Used with permission of Taylor & Francis Ltd., http://www.informaworld.com.

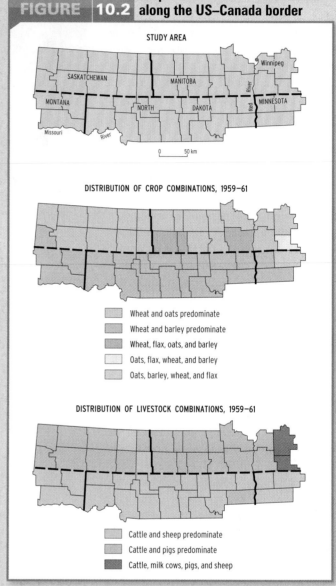

FIGURE 10.2   Crop and livestock combinations along the US–Canada border

STUDY AREA

DISTRIBUTION OF CROP COMBINATIONS, 1959–61

- Wheat and oats predominate
- Wheat and barley predominate
- Wheat, flax, oats, and barley
- Oats, flax, wheat, and barley
- Oats, barley, wheat, and flax

DISTRIBUTION OF LIVESTOCK COMBINATIONS, 1959–61

- Cattle and sheep predominate
- Cattle and pigs predominate
- Cattle, milk cows, pigs, and sheep

commercial agriculture
An agricultural system in which the production is primarily for sale.

subsistence agriculture
An agricultural system in which the production is not primarily for sale but is consumed by the farmer's household.

price (p) can be identified at the intersection of the two curves. The ideal economic world of **commercial agriculture** is occupied by profit maximizers (economic operators) who respond immediately to any price changes.

But this type of response is neither feasible nor desirable for farmers in either the more or the less developed world. In the more developed world, farmers also value stability and independence, and may be willing to sacrifice profits accordingly; they may also choose to respond to decreasing prices by increasing rather than decreasing supplies. Farmers in the less developed world also tend not to be profit maximizers, and their behaviour may well be rational. The aim of **subsistence agriculture** is to produce the amount of product required to meet family needs; profit maximization has no meaning for such farmers. The number of subsistence farmers worldwide has decreased significantly since about 1800 and continues to decrease. Today, human geographers studying agricultural location focus on commercial farmers.

FIGURE 10.3 Supply and demand curves

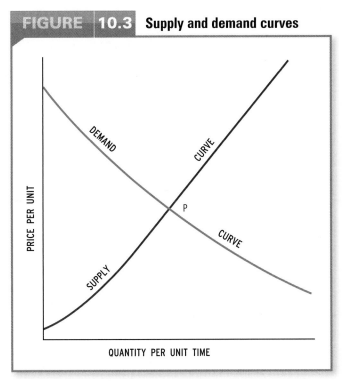

FIGURE 10.4 Rent-paying abilities of selected land uses

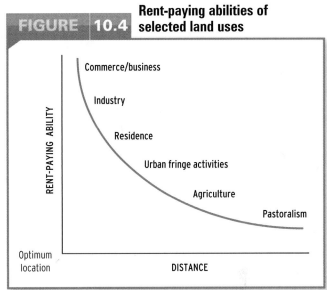

COMPETITION FOR LAND

Although several theories have been developed to explain observed variations in the spatial patterns of farming activities, they all focus on the fact that different activities compete for use of any location. Such competition arises because it is not possible for all activities to be carried out at their economically optimal location, and different activities may have identical optimal locations. The obvious question is how to determine the most appropriate use of a particular piece of land. Conventionally, land is assigned to the use that generates the greatest profits; thus, in a general sense, we can determine a hierarchy of land uses based on relative profits. But there is another way of viewing this situation. The greater the profits a particular use generates, the more that use can afford to pay for the land. The maximum amount that a given use can pay is called the **ceiling rent**. Figure 10.4 indicates the relative rent-paying abilities for a variety of land uses.

We can now identify the basic premise of much location theory: *the competition among land uses, a competition fought according to rent-paying abilities, results in a spatial patterning of those land uses.*

THE CONCEPT OF ECONOMIC RENT

Agricultural location theorists use the above premise in combination with the concept of **economic rent**, which explains why land is or is not used for production. Land is used for production if a given land use has an economic rent above zero (Box 10.4). This concept can be related to one or more variables, but especially to measures of land use by the economist who formulated the economic rent concept, David Ricardo (1772–1823); however, the economist Johann Heinrich von Thünen (1783–1850) regarded distance as more important.

VON THÜNEN'S AGRICULTURAL LOCATION THEORY

Von Thünen was a German economist and landowner who published two major works. His treatise *The Frontier Wage* identified appropriate wages for agricultural workers. *The Isolated State*, published in 1826 (Hall, 1966), reflected von Thünen's interest in economics (he was inspired by the work of Adam Smith) and his 40 years as manager of an estate on the north German plain.

The problem

To tackle the complex question of which agricultural activities should be practised where, von Thünen used a highly original method of analysis (Johnson, 1962: 214). He deliberately excluded several factors known to be relevant and proceeded on the basis of a framework that he called 'the isolated state':

economic rent
The surplus income that accrues to a unit of land above the minimum income needed to bring a unit of new land into production at the margins of production.

ceiling rent
The maximum rent that a potential land user can be charged for use of a given piece of land.

Box 10.4 | Calculating Economic Rent

It is important to clarify some details of the economic rent concept, notably how it is calculated and its spatial implications. Economic rent can be calculated as follows:

$$R = E(p - a) - Efk$$

where

R = rent per unit of land (dependent variable)
k = distance from market (independent variable)
E = output per unit of land
p = market price per unit of commodity   ⎫
a = production cost per unit of commodity ⎬ parameters
f = transport rate per unit of distance per ⎭
       unit of commodity

This equation is a straightforward linear relationship between rent and distance. Rent is a function of distance with the details of the relationship determined by $E(p - a)$, which gives the intercept (on the rent axis), and Ef, which determines the slope of the line.

What are the spatial implications of economic rent as calculated above? Because economic rent declines with increasing distance from market, it eventually becomes zero, and when more than one agricultural activity is practised, a series of economic rent lines will appear, as in Figure 10.5. Where lines cross, one activity replaces another as the more profitable. The result is that agricultural activities are zoned around the central market and a series of concentric rings emerges.

The simple concept of economic rent as a function of distance results in a situation of spatial zonation.

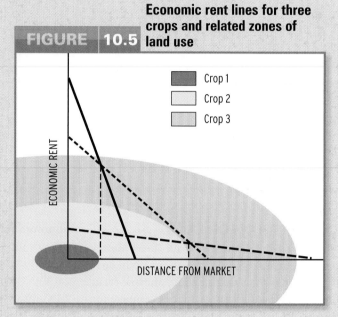

FIGURE 10.5 Economic rent lines for three crops and related zones of land use

Crop 1
Crop 2
Crop 3

ECONOMIC RENT

DISTANCE FROM MARKET

Assume a very large city in the middle of a fertile plain which is not crossed by a navigable river or canal. The soil of the plain is uniformly fertile and everywhere cultivable. At a great distance from the city the plain shall end in an uncultivated wilderness by which the state is separate from the rest of the world.

The assumptions
The isolated state model allowed for greatly simplified descriptions. Its assumptions may be elaborated as follows:

1. There is only one city, that is, one central market.
2. All farmers sell their products in this central market.
3. All farmers are profit maximizers—economic operators.
4. The agricultural land around the market is of uniform productive capacity.

5. There is only one mode of transportation by which farmers can transport products to market.

In effect, von Thünen was not so much excluding as holding constant such key variables as physical environment, humans, and transport. Only one key variable—distance from market—was allowed to vary. This was a distinctive and highly original contribution, and was, of course, in accord with his definition of economic rent, which centres on distance from market.

The answer
Given the above assumptions, von Thünen asked, 'How will agriculture develop under such conditions?' (Johnson, 1962: 214). The conditions are those of a controlled experiment isolating a single causal variable, distance from market. The answer is straightforward. Agricultural activities are located in a series

of concentric rings around the central market, one zone for each product for which there is a market demand. To determine the location and size of each product zone, von Thünen used data from his own estate on production costs, market prices, transport costs, and so on. *The principle of concentric zones emerges from the concept of the isolated state, while the number, size, and content of zones is a function of particular places and times.*

Combining the isolated state concept, his own data, and the economic rent concept, von Thünen came to two conclusions: (1) zones of land devoted to specific uses develop around the market, and (2) the intensity of each specific land use decreases with increasing distance from the market.

The crop theory

The first conclusion, the crop theory, is summarized in Figure 10.6, which illustrates how product location is affected by perishability and weight as they affect transport cost. Remember that while the specific activities depicted in this figure reflect north Germany in the early nineteenth century, the principle of zones applies generally. In zone 1 are market gardening and milk production. Both fresh vegetables and dairy products are perishable, give high returns, and have high transport costs. Accordingly, they have a steep economic rent line reflecting a high *Ef* value (see Box 10.4) and a high intercept on the rent axis reflecting a high $E(p - a)$ value. In zone 2 are forestry products (used for fuel and building). Forestry could command a location close to market because of the high transport costs involved in moving such bulky products. Rye, the principal commercial crop, is located in zones 3, 4, and 5. The differences between the zones reflect differences in the intensity of rye cultivation: zone 3 uses a six-year rotation, zone 4 uses a seven-year rotation, and zone 5 uses a three-field system. Livestock ranching—producing butter, cheese, and live animals—is located in zone 6. Beyond zone 6 is wilderness that can be brought into production if spatial expansion is needed in the future to meet increased product demand.

The intensity theory

Von Thünen's second conclusion is known as the intensity theory. Here von Thünen's own production data indicated that, for any given product, the intensity of production decreases with increasing distance from market. The easiest example to identify is the cultivation of rye in zones 3, 4, and 5, but the principle also applies inside specific zones.

VON THÜNEN'S CONTRIBUTION

In light of our discussion of theory and scientific method in Chapter 2, we need to ask to what extent von Thünen's work represents a true theoretical contribution. Although it does begin with a series of assumptions, they are not valid assumptions, or axioms, from which we can deduce hypotheses with confidence. Rather, they are simplifying assumptions that reduce the complexity of the problem. Hence, we know from the outset that von Thünen's hypotheses—the crop and intensity theories—cannot be literally correct. On the other hand, we also know that they serve the purpose intended: that is, they describe an ideal state (interestingly, the original title of the work). The two hypotheses describe a state of affairs against which we can compare real-world agricultural patterns. They describe what ought to be—a normative description. As in many economic models, a complex world is simplified

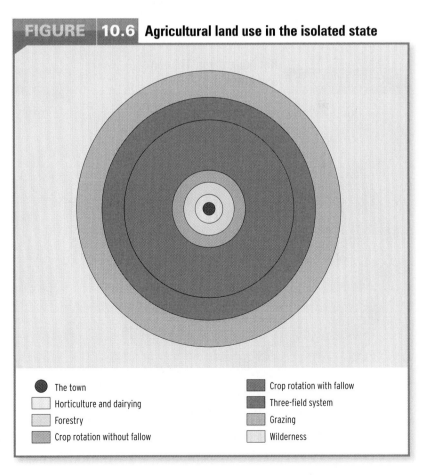

FIGURE 10.6 Agricultural land use in the isolated state

- ● The town
- Horticulture and dairying
- Forestry
- Crop rotation without fallow
- Crop rotation with fallow
- Three-field system
- Grazing
- Wilderness

to aid understanding; the spatial hypotheses produced are examples of ideal types.

Labelling von Thünen's work as a theory is also complicated by the inclusion of real-world data in the procedures. It is, nevertheless, an excellent example of a type of theory to which geographers have been greatly attracted.

SOME MODIFICATIONS AND EXTENSIONS

Von Thünen was well aware that his work was a simplification. Accordingly, he modified several assumptions when elaborating on the basic hypotheses. The assumption of a single mode of transportation to market was relaxed with the introduction of a navigable waterway serving to elongate zones (Figure 10.7). Relaxing the assumption of a single market and allowing a transportation network including roads resulted in a more complex pattern of zones (Figure 10.8). von Thünen also acknowledged that differential land quality would mean intensified use of the better areas.

Despite its age and some obvious weaknesses, von Thünen's work remains the fundamental answer to the question of agricultural location for at least two reasons. First, his decision to simplify the issue by excluding certain variables was a highly original contribution; this basic method has been used by many other location theorists, as we will see in Chapters 11 and 14 especially. Second, in both of von Thünen's concepts, economic rent and agricultural location, the key geographic variable is distance as measured in terms of transport cost.

Human landscapes in general are clearly related to distance, regardless of how we choose to measure it—straight-line distance, transport-cost distance, time, or any other format. Relationships between distance and spatial patterns, such that some spatial regularity is evident, lie at the heart of much theory and analysis.

Distance, Land Value, and Land Use

Numerous geographers tested the von Thünen hypotheses or the general relations with distance in the context of agricultural patterns. The seminal work in this area is Chisholm's (1962) remarkably thorough pioneering study of the scale ramifications of the concepts and an excellent survey of world evidence pertaining to the general issues.

FIGURE | 10.7 **Relaxing a von Thünen assumption**

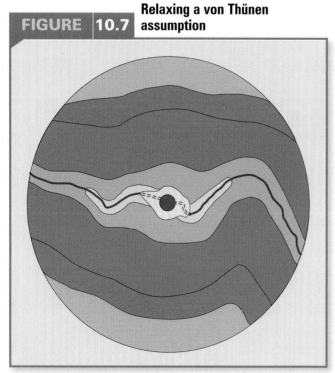

A navigable waterway elongates the agricultural zones.

FIGURE | 10.8 **Relaxing two von Thünen assumptions**

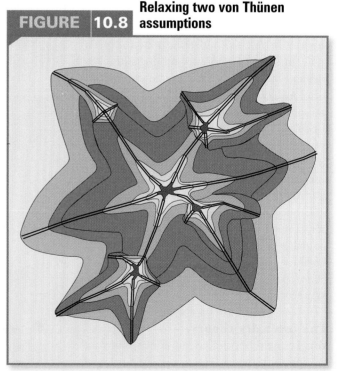

Multiple markets and a transport network create a more complex model.

HISTORICAL REGIONAL STUDIES

Von Thünen's own work is a historical empirical study, and it is hardly surprising that it is applicable in a general historical context. In fact, he was not the first to identify the existence of patterns in land-use zones. Two descriptions, one in 1576 and the other in 1811, identified four land-use zones close to London, England, namely zones of clay pits, cattle pastures, market gardening, and hay.

Many analyses of agriculture in nineteenth-century North America highlight the von Thünen principles in a spatial and temporal framework. One study of an area of Wisconsin centred on Madison found that the area close to the expanding centre was dominated by wheat production from 1835 to 1870, by market gardening from 1870 to 1880, and by dairy farming after 1880, demonstrating that zonal change for any one period becomes temporal change as the urban centre increases in size (Conzen, 1971). A study of western New York state focused on the related idea that a prolonged period of falling transport costs is associated with increasing regional specialization; Leaman and Conkling (1975) showed that the expanding transport network of roads and canals allowed a frontier area, initially akin to von Thünen's wilderness, to rapidly become first an area of wheat cultivation and then an area of regional specialization related to access to market.

The southern Ontario frontier

Both of the American studies noted above found that areas experiencing increased settlement are transformed from non-agricultural areas into areas of subsistence agriculture and finally into a series of commercial agricultural types. Similar conclusions were reached in a study of nineteenth-century southern Ontario that focused explicitly on testing the von Thünen hypotheses (Norton and Conkling, 1974).

The area analyzed lay north of the then-emerging centre of Toronto and can be regarded as one segment of the market area of that centre. Using data for 1861, the relationship between distance and land value is measured at $r = -0.6502$, producing an r^2 of 0.4228 and suggesting that 42 per cent of the spatial variation in agricultural land values is explained by distance to Toronto (Box 10.5 and Figure 10.9). The authors then follow the theoretical

logic one step further by relating land values to a series of variables: distance to Toronto, distance to smaller regional market centres, distance to major lines of communication, and a measure of land capability for agriculture. The statistical result is $R = 0.7367$, producing an R^2 of 0.5427, which means that 54 per cent of the spatial variation in land values is explained by the set of independent variables.

Study of the 1861 pattern of land use showed an area divided into two basic zones: (1) close to Toronto a zone of fall wheat, peas, and oats, with production oriented to the Toronto market, and (2) an outer zone with much unoccupied land, used for spring wheat, potatoes, and turnips—an area anticipating commercial development. The incipient zonation in place by 1861 was to become fully realized in the later nineteenth century. This Ontario study also analyzed the intensity hypothesis and showed that the intensity of the key commercial crop (fall wheat) decreased with increasing distance from Toronto.

A twentieth-century regional study

Griffin (1973) showed that Uruguay possesses many of the essential characteristics of the isolated state, and Montevideo has the key attributes of the theoretical central city. Actual land use there (Figure 10.10) closely resembles the von Thünen model of ideal land use. The first zone is used for horticulture and truck farming, followed by a zone of dairying, and finally a cereal zone, despite significant variations between model and reality in variables such as soil fertility, transport efficiency, and ethnicity.

STUDIES IN LESS DEVELOPED WORLD REGIONS

More surprising than these historical examples is considerable evidence suggesting von Thünen-type spatial patterns in less developed countries, where much agricultural activity is subsistence, not commercial, in orientation. A regional pattern of land use is often in fact a mix resulting from the different agricultural decisions made by relatively well-off commercial farmers and relatively less well-off subsistence farmers—groups that necessarily have different economic outlooks. The commercial farmers are concerned with profits and hence with rationalizing land use according to economic constraints, whereas

Box 10.5 Correlation and Regression Analysis

'A continuous theme in geographic research is that of analyzing the degree and direction of correspondence among two or more spatial patterns or locational arrangements' (King, 1969: 117). A traditional approach to the problem is to use map overlays—a procedure facilitated today by geographic information systems. Another approach is to apply a set of quantitative procedures known as correlation and regression analysis.

By simple correlation and regression, we may determine the degree of association between two variables. The correlation coefficient, r, is a statistical measure of the relationship, where r may vary between –1 (a perfect inverse relationship) and +1 (a perfect direct relationship). Squaring the r value gives a coefficient of determination, r^2, which may be expressed as a percentage and interpreted as the percentage variation in one variable explained by the other variable. A related set of procedures, known as multiple correlation and regression analysis, allows one variable to be related to two or more variables. In this case, the correlation coefficient is identified as R.

The procedure by which these results are obtained is as follows. The linear equation, $y = a + bx$, is used where

y = dependent variable
x = independent variable
a = y intercept
b = slope of the line

The economic rent equation is of this form: $R = E(p - a) - Efk$. This linear equation is the simplest equation for predictive purposes. A method known as least squares is used to calculate the best-fit line on the graph of x plotted against y. Once the best-fit line is calculated, we can measure how well it fits. If all points fall on the line, it is a perfect fit and $r = -1$ or $+1$; more typically points are not located on the line and r lies between -1 and $+1$. In cases where y does not vary with changes in x, r is close to 0 (Figure 10.9).

Procedures are also available to test the significance of r. All of the results reported in the text are statistically significant.

FIGURE 10.9 Scatter graphs, best-fit lines, and r values

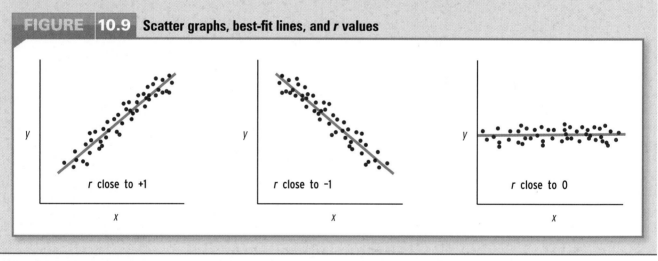

r close to +1   r close to –1   r close to 0

the subsistence farmers are limited in their activities by lack of capital and inadequate size of holdings.

Despite these complications, von Thünen-type patterns may emerge. Horvath (1969) analyzed land use *c.* 1964 in the vicinity of Addis Ababa, Ethiopia. Although this highland area is more varied, physically and culturally, than the von Thünen model assumes, basic von Thünen patterns were evident: an area of eucalyptus forest around the city that was used for fuel and building material (an orientation outward along roads was evident); vegetable cultivation within that forest, rather than in a separate zone; and a zone of mixed

farming that included commercial and subsistence activity beyond the first forest and vegetable zone. Horvath concluded that this was an example of incipient zonation, similar to the southern Ontario example above. Unfortunately, political circumstances in contemporary Ethiopia have made continuation of this analysis difficult.

STUDIES AT CONTINENTAL AND WORLD SCALES

Von Thünen's logic is relevant even at the continental and world scales. Several geographers have proposed concentric patterns of land use for North America, Europe, and the world.

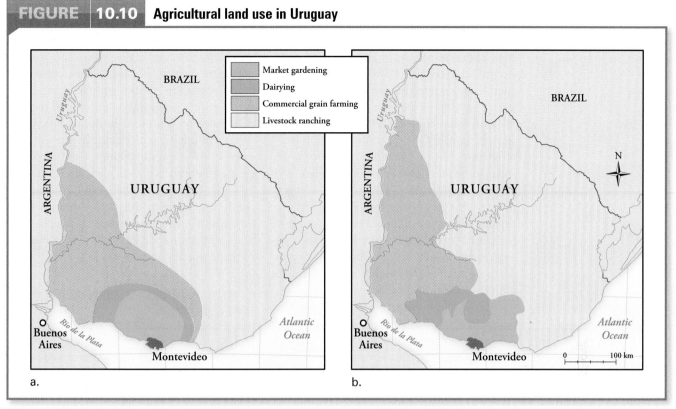

FIGURE | 10.10 Agricultural land use in Uruguay

Market gardening
Dairying
Commercial grain farming
Livestock ranching

a.

b.

(a) as predicted by von Thünen theory; (b) actual.

When Muller (1973) analyzed the nineteenth-century United States he found an expanding system of concentric rings. Jonasson (1925) described European agriculture in terms of an inner ring of horticulture and dairying, followed by zones of less and less market-oriented activity. On the world scale, Schlebecker (1960) traced the evolution of a world city from Athens to Western Europe to Western Europe–eastern US.

The growth of such a world market and the related expansion of the supply area can be analyzed by reference to import data. Over time, the mean distances over which agricultural imports have been moved have increased (Table 10.2). This spatial expansion of the export system has not been haphazard: it is clearly zonal in character. The changing details of the import system reflect a modified von Thünen model. Peet (1969) has detailed this situation specifically for wheat, showing that, in many instances, the arrival of wheat was preceded by extensive animal rearing (the Canadian prairies are an exception). In all cases, wheat cultivation was associated with major population growth, transport expansion, and the rise of an urban network. Spatially, wheat imports to Britain came first from Britain itself, then the Baltic region, followed by the United States, western Canada, and Australia.

STUDIES AT A LOCAL SCALE

Von Thünen's theory has also been tested on the scale of individual villages and farms. In such analyses, the concept of distance is equated less with transport cost than with minimization of movement in relation to time expended. Various investigations of villages

Table 10.2	Average Distances from London, England, to Regions of Import Derivation (miles)				
	1831–5	1856–60	1871–5	1891–5	1909–13
Fruit and vegetables	0	521	861	1,850	3,025
Live animals	0	1,014	1,400	5,680	7,241
Butter, cheese, eggs	422	853	2,156	2,590	5,020
Feed grains	1,384	3,266	3,910	5,213	7,771
Flax and seed	2,446	5,229	4,457	6,565	6,275
Meat and tallow	3,218	4,666	6,018	8,093	10,056
Wheat and flour	3,910	3,492	6,758	8,286	9,574
Wool and hides	4,071	14,207	16,090	17,811	17,538

NOTE: The distances shown here are for imports only; British agricultural production continued to be a major source in the London markets.

SOURCE: After S. Leonard, 'Von Thünen in British Agriculture', *South Hampshire Geographer* 8 (1976): 28.

have tested modified versions of the crop and intensity theories and confirmed their relevance. Von Thünen-type patterns are also evident in the immediate vicinity of large cities. According to Sinclair (1967), even though the considerations affecting land use close to cities are not those identified by von Thünen, the spatial consequences in terms of zonation are very similar. One such consideration is the prospect of urban expansion, specifically the anticipation of such expansion and hence of profits to landowners. The closer the land is to the city, the greater the anticipation of urban expansion and hence the lower the incentive for capital investment in the land. This argument helps us understand why so many areas on the outskirts of cities are either vacant or put to some clearly temporary use.

Overall, the message is clear. The methods pioneered by von Thünen—specifically, the deliberate simplifications producing an ideal world against which reality can be assessed—are of enormous value to agricultural geographers and, as we shall see in later chapters, to urban and industrial geographers.

Domesticating Plants and Animals

A discussion of agriculture on the world scale could focus on any number of topics: the influence of physical environment; the effects of regional population change; distances from markets; and so on. Perhaps the most useful approach, however, begins with the acknowledgement that the contemporary pattern of world agriculture is the still-changing product of a long history: 'The imprint of the past is still clearly to be seen in the world pattern of agriculture. To understand the present, it is essential to know something of the evolution of the modern types of agriculture' (Grigg, 1974: 1). To put it another way—as we did in the Introduction to this book—human decision-making, past and present, in diverse cultural contexts, has largely shaped and continues to shape our agricultural world.

EARLY DOMESTICATION AND DIFFUSION

The most fateful change in the human career can be said to have occurred . . . when the transition from foraging to farming began. During the preceding 150,000 years anatomically modern humans had successfully colonized almost all habitable and accessible areas and in so doing had learned to subsist, as 'hunter gatherers', on a great diversity of plant and animal foods. The most fundamental and far-reaching consequence of the agricultural revolution in the early Holocene was that it enabled more food to be obtained, and more people to be supported, per unit area of exploited land. It thus facilitated long-term sedentary settlement and the maintenance of larger and more complex social groups, which in turn enabled urban society to develop. (Harris, 1996: ix)

Agriculture originated in the domestication of plants and then animals. A domesticated plant is deliberately planted, raised, and harvested by humans; a domesticated animal depends on humans for food and, in many cases, shelter. As a consequence, domesticated plants and animals differ from their non-domesticated counterparts. From the human perspective, the domesticates are superior in that, for example, they bear more fruit or provide more milk: they have been deliberately engineered for these characteristics through selective breeding over long periods.

What was the first species to be domesticated? Where and when? How and why? In all

Disappearing farmland in Richmond Hill, Ontario, north of Toronto. Scenes like this are increasingly common around the world as residential developments encroach on valuable agricultural land.

Dick Hemingway Photographs

likelihood agriculture began some 12,000 years ago—a 2006 discovery in the Jordan Valley of figs that were grown through human intervention is dated between 11,200 and 11,400 years ago. From an initial centre (or, more probably, centres), agriculture diffused to other areas, gradually replacing the main pre-agricultural economic activities, hunting and gathering. Although it was traditionally thought that agriculture first emerged in a few Asian centres, most notably Southwest Asia (present-day Iraq), more recently it has been accepted that it evolved independently in several centres in Asia (east, southeast, and southwest), Africa (northeast and south of the Sahara), the North American Midwest, Central America, western South America, and Southern Europe. Figure 10.11 maps the likely centres of domestication and suggests directions of early agricultural diffusion (compare this map with the ideas summarized in Figure 6.1 that shows the shape of continents). It seems likely that, in each place, people began domesticating plants and animals in response to specific circumstances.

We are learning more and more about plant and animal domestication with advances in genetic research. Just as the study of genes is uncovering answers to specific questions about human origins, so it is illuminating many details of the domestication process. Most notably, such research is providing support for the idea that many plant and animal species were domesticated more than once, in different places at different times. Cows, pigs, sheep, goats, yaks, and buffalo each were domesticated at least twice, dogs at least four times, and horses on even more occasions. What this means is that many different groups of people came up with the idea of domesticating other animals. It also suggests that the reason some animal species are not domesticated is not that no one tried, but rather that they proved unsuitable candidates (typically, the animals most susceptible to domestication already lived in groups and had some form of hierarchical organization). Another recent (2009) discovery is that horses were first domesticated earlier than previously thought: in what is now northern Kazakhstan, horses were ridden, and used for meat and for milk, by about 3500 BCE. Horses represent a highly significant domestication as they greatly increase human mobility.

It is important to appreciate that there is nothing inherently difficult or complicated about the process of domesticating plants and animals; no knowledge of genetics is needed. The process simply makes use of artificial selection as opposed to natural selection. Natural selection results in the survival and reproduction of those plants and animals that

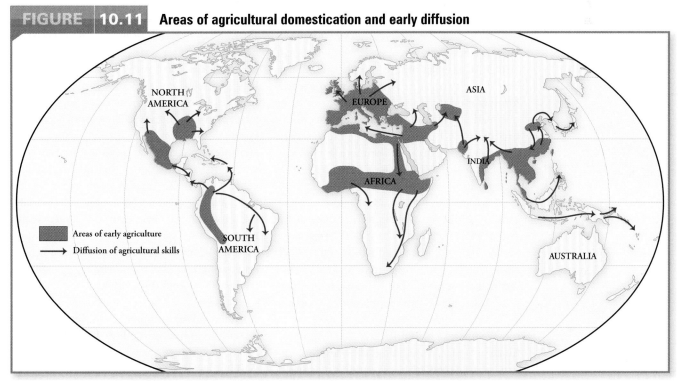

FIGURE | 10.11 Areas of agricultural domestication and early diffusion

SOURCE: P.N. Stearns, M. Adas, S.B. Schwartz, and M.J. Gilbert, *World Civilizations: The Global Experience* (New York: Pearson, 2007), 3.

are best able to cope in a particular environment. Artificial selection involves humans allowing certain plants and animals to survive and breed because they possess features judged desirable by humans—for example, plants with bigger seeds or animals that are less aggressive. The individual members of plant and animal species that humans favoured would reproduce and pass on the favoured characteristics, while the individual members with less desirable traits would be eliminated gradually.

POSSIBLE CAUSES OF DOMESTICATION

The idea that pre-agricultural societies lived a desperate and harsh life searching for food is discredited. Indeed, it is now suggested by some researchers that domestication was something that humans resorted to only when they encountered difficulties with food supply. But precisely how and why humans became food producers is debated.

Various explanations have been proposed. According to Sauer (1952), casual experimentation with plant and animal breeding probably began in a well-endowed environment that permitted a more sedentary way of life and a certain amount of leisure time. In his view, Southeast Asia, especially wooded hilly areas away from possible floods, was the most likely hearth area.

Perhaps the most satisfactory explanation combines a number of relevant variables: demographic, environmental, and cultural. The key idea, not yet proven, is that either climate change or population pressure, or a combination of the two, might have prompted a search for new food supplies. This hypothesis is based on the idea that although the process of domestication is not especially difficult, it can require a great deal of work and, therefore, is not especially attractive. Most societies did not need to increase food supplies because they had strategies to keep the population below carrying capacity. However, a change in climate at the end of the Pleistocene Era might have caused certain areas to become unusually rich in food resources, which would have made a more sedentary way of life possible and, hence, permitted an increase in population. Eventually, a larger population would have necessitated movement into marginal areas and competition for space, prompting the development of new strategies for food supply: specifically, domestication. Boserup (1965) and Binford

(1968) are among the authors who have proposed such a scenario.

Whatever the details, by *c.* 500 BCE agriculture had become a major economic activity. Grigg (1974: 21–3) has identified five core areas and agricultural types evident by this date (Box 10.6).

World Agriculture: A Brief History of Technological Changes

Contemporary agricultural landscapes are the outcome of a long evolutionary process. Since domestication began, plants and animals have diffused widely, far beyond the few areas where the wild varieties of the original species evolved. Wheat, for example, had diffused as far west as Ireland and as far east as China by about 2000 BCE, and this spread also meant the diffusion of related technologies and ways of life generally. One of the most important technologies was the plough, possibly first used about 4000 BCE. A major innovation took place in China in about 300 BCE with the invention of the horse collar that allowed the animal to pull a greater weight.

Many agricultural landscapes today emerged in response to the European overseas movement and/or to the demands of new population concentrations. In Europe, for example, some agricultural crops and methods prior to 1500 were the result of diffusion from the Southwest Asian hearth areas; after 1500, two American crops, potatoes and corn, were widely adopted, and other agricultural changes reflected increased market size; these included intensification, expansion of cultivated area, and increasing commercialization. Commercialization was facilitated by technical advances in transportation, particularly after the mid-nineteenth century. Agricultural commercialization at that time was merely one component of substantial growth and change in the larger world economy.

Five principal technological advances have transformed, or are transforming, agricultural landscapes around the world:

1. A second agricultural revolution associated with the onset of the Industrial Revolution in the eighteenth century.
2. The development of nitrogen fertilizers in the early twentieth century.

Box 10.6 Agricultural Core Areas, c. 500 BCE

1. Southwest Asia

Basic crops: Wheat, barley, flax, lentils, peas, beans, vetch.
Basic animals: Sheep, cattle, pigs, goats.

Subtypes: Irrigated farming in the Nile Valley, Tigris and Euphrates, Turkestan, and Indus Valley; dry farming (without irrigation) was the basic Southwest Asian complex mix of agricultural activities; Mediterranean agriculture was a mix of cereals and tree crops such as figs, olives, and grapes; in Northern Europe, oats and rye were added to basic crops.

2. Southeast Asia

Two types originated on the Southeast Asian mainland. Tropical vegeculture involved growing taro, greater yam, bananas, and coconuts, and raising pigs and poultry; a form of shifting agriculture. ('Shifting agriculture', practised in tropical forest areas, involves regular movement from one cut-and-burned area to another because of the rapidly declining soil fertility resulting from crop cultivation; also known as swidden agriculture, or slash-and-burn agriculture.) Wet rice cultivation gradually displaced the first type.

3. Northern China

Based on local domesticates (e.g., pigs, foxtail millet, soy beans, and mulberry) and imported domesticates (e.g., wheat, barley, sheep, goats, and cattle).

4. Africa

Two agricultural types, both shifting: tropical vegeculture (based on yams), and cereal cultivation (millets and sorghum).

5. America

Two agricultural types, both shifting: root crops and corn/squash/beans complex. No livestock.

This listing of five core areas and identification of numerous crops and animals must not disguise the fact that just three of the domesticated plants—wheat, rice, and corn—are of overwhelming importance in that, together, they have provided most of the calories that enabled populations to increase. Wheat was the oldest and most widespread of these.

Since c. 500 BCE, links have been established between areas and types of agriculture. Of particular relevance are those agricultural changes prompted by European expansion from the fifteenth century onward.

---

3. The 'green revolution' that began in the mid-twentieth century.
4. The biotechnology revolution that began in the late twentieth century and that, despite much opposition, is proceeding apace today.
5. The ongoing transition in some areas from ploughing the soil prior to planting to use of no-till strategies.

A SECOND AGRICULTURAL REVOLUTION: ENGLAND AFTER 1750

The most radical changes in agricultural activities since the beginning of domestication have occurred in the larger European world, helping to create what is today 'the great gulf in productivity between the present agricultures of Asia, Africa and much of Latin America, and western Europe, North America and Australia' (Grigg, 1974: 53). In most cases, those changes did not begin until the seventeenth century and it was only in the mid-nineteenth century that commercial agriculture became established on a large scale and the present pattern of agricultural activity began to evolve. The link between industrial progress and agricultural progress seems clear.

Thus, a period of significant agricultural change—significant especially because of its links with other technological, social, and economic changes—began in England about 1700. Sometimes described as a second agricultural revolution, it involved a series of changes:

1. development of new farming techniques, including introduction of fodder crops and new crop rotations;
2. increases in crop output because of improvements in productivity;
3. introduction of labour-saving machinery, most notably the 1701 invention in England of a seed drill that enabled much higher harvesting rates for a given number of seed sown and, later in the nineteenth century, the tractor; and
4. the ability to feed a growing population.

The first three of these represented changes in what Marx called the forces of production. But the new 'revolution' also involved changes in the relations of production associated with the decline of the feudal system and the emergence of capitalism. Among the institutional changes was the imposition of 'enclosure': the subdivision of large, open arable fields and large areas of pasture or wasteland into small fields divided by hedgerows, fences, or walls. Enclosure was part of a major social transformation from communal to private property rights. A second and related institutional change was the creation of farms that typically were rented from large landholders by capitalist farmers who employed labourers to work the land.

This second agricultural revolution—perhaps better described as an agrarian revolution—was one of the principal means by which England was able to begin supporting a growing population. 'The transformation of output and land productivity enabled the country to break out of a "Malthusian trap", allowing the population to exceed the barrier of 5.5 million people for the first time. Rising labour productivity ensured that extra output could be produced with proportionately fewer workers, so making the industrial revolution possible' (Overton, 1996: 206).

NITROGEN FERTILIZERS

Perhaps the most important development after the second agricultural revolution was the introduction of nitrogen fertilizers. Nitrogen is an essential nutrient for the cereal crops that have been staples for most people in most parts of the world since the beginnings of agriculture, and is added naturally to soil through rainfall. But rainfall alone does not add enough nitrogen, and it was soon discovered that growing a cereal crop on the same land year after year impoverished the soil and ultimately reduced yields. The favoured solution was to plant legumes along with the cereal crops, to replenish the soil's nitrogen content: peas and lentils in the Middle East, beans and maize in the Americas, soy and mung in Asia, and peanuts in parts of sub-Saharan Africa. Other strategies to prevent soil impoverishment included crop rotation and leaving land fallow for a season.

By the nineteenth century, the demand for cereals was increasing so rapidly that concerted efforts were made to find new sources of nitrogen. One readily available source in much of Europe was horse manure. In Asia human waste was used. Another source was discovered about 1843—the island of Ichaboe off the coast of Southwestern Africa was covered in about 8 m (26 feet) of guano (seabird excreta), all of which was removed by 1850. Another source was discovered in 1850 off the coast of Peru, on a few arid Pacific islands with whole cliffs made of guano. Guano mining quickly proved a very profitable business there as well, and it is estimated that some 20 million tons of guano were transported by ship from these islands to Europe between about 1850 and 1870. Once these sources were depleted, traders began exporting a fossil nitrate called 'caliche', found in desert areas in Chile, for use as a source of nitrogen. Even with the help of these organic fertilizers, however, crop yields were still too limited to permit human populations to grow beyond a maximum density of about 5 per hectare. Indeed, one of the principal constraints on population growth during the nineteenth century was the lack of a reliable source of nitrogen sufficient to enable farmers to increase their cereal yields significantly. This need was filled by two chemists who recognized that the vast ocean of air that surrounds us contains large quantities of nitrogen (Smil, 2001).

The solution was ammonia synthesis—a process that converts atmospheric nitrogen to ammonia, which is then used to produce synthetic nitrogen fertilizers. This 'Haber–Bosch process' was discovered in the early twentieth century, and nitrogen fertilizer production began on a commercial scale in 1912. Nitrogen fertilizers were used widely by the 1950s, and today this process produces about 2 million tons of ammonia each week, providing almost

Chiapas, Mexico: A Tzeltal Indian adds organic fertilizer to his shade-grown coffee in the Lacandon Jungle.
Danita Delimont/Alamy

100 per cent of the inorganic nitrogen used in agriculture. China is the single largest producer. In an important sense, humanity today is dependent on nitrogen fertilizers, and this technological breakthrough, more than any other, permitted world population to rise from 1.6 to 6.1 billion in the twentieth century. 'In a very literal sense, without Haber's invention, half the people alive today could not have been born' (Harman, 2004: 36).

At the same time, nitrogen fertilizers have been associated with environmental damage—including soil and water contamination, increasing soil acidity, and the release of nitrous oxide (a potent greenhouse gas) into the atmosphere—as well as increased risk for some cancers. Concerns such as these have prompted many to argue for a return to organic farming (see Box 10.7).

THE 'GREEN REVOLUTION'

Another major agricultural advance in the twentieth century was what came to be known as the 'green revolution'—the rapid development of improved plant and animal strains and their introduction to the economies of the less developed world. As we have seen, such improvements had been ongoing in the Western world since the eighteenth century, as one aspect of larger economic and technological changes. Until the 1960s, however, the less developed world did not benefit from these changes, and there were many instances, especially in Asia, of widespread hunger, malnutrition, and dependence on food aid. The initial breakthrough came as a result of scientific work in the more developed world.

In the 1960s, the Rockefeller and Ford foundations co-operated in the creation of an agricultural research system designed to transfer technologies to the less developed world. This system produced genetically improved strains of two key cereal crops, rice and wheat—often labelled high-yielding varieties, or HYVs—and eventually other crops such as beans, cassava, maize, millet, and sorghum also were improved. Although the new strains tend to have lower protein content than their predecessors, they produce higher yields, respond well to fertilizer, are more disease- and pest-resistant, and require a shorter growing season. Besides genetic improvements, the 'green revolution' involved expanded use of fertilizers, other chemical inputs, and irrigation. In many cases, the adoption of HYVs and other technologies quickly doubled crop yields—hence the name 'green revolution'.

The new strains and technologies allowed some farmers to grow enough not only to subsist but to market surplus production, raising farm incomes and thus stimulating the rural non-farm economy as well. Increasing incomes and lower prices led to better nutrition, with higher calorie consumption and more diversified diets. These positive impacts have been most evident in Southeast Asia and the Indian subcontinent (Box 10.8).

Yet despite undoubted successes, the results have not always met expectations. As with efforts to introduce population control policies, full acceptance has sometimes been prevented, or delayed, by cultural and economic factors. Peasant farmers are usually conservative and reluctant to abandon traditional practices. To persuade them to accept new plant strains, governments have sometimes found it necessary to offer subsidies and guaranteed prices. Thus, perhaps inevitably, the agricultural changes associated with the 'green revolution' have had some unfortunate consequences, and criticisms are plentiful.

One set of criticisms focuses on economic circumstances. New strains have been most readily adopted by farmers who are better off, who have some capital and relatively large holdings, with the result that in some cases improvements have led to the displacement of poor tenant farmers. Further, many of the poorest tenant farmers have been prevented from adopting improved strains because they cannot afford the required fertilizer and pesticides. This has led to increased inequality and exacerbated poverty in some regions. A related criticism is that new strains and techniques have been accompanied by unnecessary mechanization, which has reduced rural wages and increased unemployment.

A second set of criticisms focuses on undesirable environmental consequences. Excessive and sometimes inappropriate use of fertilizers and pesticides has resulted in water pollution, unwanted damage to insect and other wildlife populations, water shortages caused by increased irrigation, and some serious health problems.

Finally, a third type of criticism focuses on the broader social and political implications of the 'green revolution'. Specifically, there is a serious risk that imported knowledge, presumed to be expert, will be accepted at the

Box 10.7 | Organic Farming

Critics of contemporary conventional farming strategies argue that they should be replaced by more sustainable and ecologically more appropriate methods. Collectively, the alternative methods they propose differ from conventional farming in three ways (Atkins and Bowler, 2001: 68):

1. Alternative methods are less centralized: there are more farms, and marketing is local and regional rather than national and international.
2. Alternative methods emphasize the community rather than the individual: the focus is on co-operative activity, using labour rather than technology when appropriate, and giving due consideration to all costs, material and non-material, as well as moral values.
3. Alternative methods are less specialized: farming is seen as a system of related activities rather than as a series of individual components.

In short, alternative farming involves a completely different value system, in which farming is regarded as a way of life and not simply a matter of producing a product for the market.

Organic farming is the best known of these methods. Although it remains a minor component of the agricultural sector in most countries, it is viewed favourably by many and is becoming increasingly popular in many Western countries. Using 2006 data, organic agricultural land is 13 per cent of all agricultural land in Austria, 12 per cent in Switzerland, and 9 per cent in Italy. Several other European countries have between 7 and 4 per cent, Australia has 3 per cent, Canada has 1 per cent, and the United States and Japan have less than 1 per cent. In total in 2006, Canada had 3,571 organic farms, most of which grow field crops (typically buckwheat, barley, wheat); others grow fruits and vegetables.

It appears that buyers of organic food, and also of organic non-food products, are motivated primarily by personal health considerations but also by concern for the environment. Organic farmers may spend somewhat less than regular farmers on purchased inputs such as pesticides and fertilizers, but their production levels are significantly lower, and therefore their prices are higher. The farmers who pioneered organic production, and the consumers who supported them by purchasing their more expensive products, were motivated by real environmental and ethical concerns about conventional farming. As the demand for 'healthier' food increases, however, it seems clear that a significant part of the organic sector today is motivated primarily by profit. With increasing numbers of organic producers turning to more conventional marketing strategies—including supplying national or even international markets—organic

agriculture is not always what consumers believe it to be. In effect, producers who adopt the 'organic' label without the philosophy may be hijacking it. Sales data for 2008–9 in the United Kingdom suggested that sales of organic food fell dramatically, by up to one-third, probably because of the recession.

Meanwhile, some observers are questioning whether foods produced using organic methods really are better for consumers than more conventional foods. Some say that what the organic sector offers is simply a lifestyle package designed to make consumers feel good about paying higher prices for their food. It is also noteworthy that yields under organic cultivation are almost 50 per cent less than under conventional strategies and, in a world of continuing population increases and therefore increased demand for food, organic methods might be a luxury that cannot be afforded (Goodall, 2008). To produce all the food we produce now but to do so organically would require several times more agricultural land than is currently used.

A recent development with some similarities to the original organic farming movement is the movement for 'fair trade'. Still in its infancy, this is a grassroots social movement that appeals directly to people's sense of justice. The idea behind fair trade is that consumers pay a guaranteed price plus a small premium to groups of small producers supplying commodity goods such as coffee, tea, chocolate, and fruit. Prior to the dramatic increases in food prices in 2007, supporters of this movement hoped that fair trade would counter one of the evils often attributed to globalization, namely that the world price of many commodities had barely risen in 20 years, with the result that many small producers in the less developed world operated at a loss. However, a more significant concern today might be that much of the markup on 'fair trade' products goes to the retailer rather than the farmer.

Organic bell peppers grown in an Israeli greenhouse in the Jordan Valley.
PhotoStock-Israel/Alamy/GetStock

Box 10.8 | The 'Green Revolution' in India

For about 20 years after gaining independence in 1947, India depended on food aid, but 'green revolution' technologies have permitted steady increases in food production—although rapid population growth has offset some of the impact of these increases. You will recall from Chapter 4 that the Indian population is now over 1 billion and that India will shortly overtake China as the country with the largest population in the world. Part of the success of the 'green revolution' in India can be attributed to the presence of the Indian Agricultural Research Foundation (IARF), a government-funded organization that was established in 1905. With the adoption of the first new strain of rice in the late 1960s, funding to the IARF increased, and new technologies and a supportive infrastructure combined to put India at the fore-front of 'green revolution' advances.

The IARF is headquartered in Delhi and has nine regional research centres, a large library, and a modern mechanized farm. Since the late 1960s its principal goals have been to facilitate the spread and growth of 'green revolution' technologies developed elsewhere while conducting its own research into the development of higher-yielding and more disease-resistant varieties, genetic enhancement, breeding for multiple cropping and intercropping, analyzing the links between photosynthesis and productivity, and improving the nutritional value of crops. This work has resulted in the development of 64 new crop varieties since the 1960s. In addition the IARF has worked on related projects, including irrigation strategies, research on fertilizers and pesticides, and the use of manure.

Because India is a large country, diverse both physically and culturally, different strains and technologies are needed in different areas. For example, different physical settings need different environmentally friendly fertilizers. Efforts to develop monsoon-dependent crops continue, but so far without success. Overcoming cultural resistance is another problem, especially in areas of subsistence farming. Feeding the refugees, who continue to move into some of the densely populated areas—including Bangladeshis escaping floods and Afghans fleeing political uncertainty in their homeland—is an additional challenge for Indian agriculture (Slatford and Fishpool, 2002).

---

expense of the local knowledge that has been acquired and successfully applied over centuries. This downgrading of local knowledge is one component of **neo-colonialism**, which in turn may be seen as one aspect of globalization. On balance, it is clear that the 'green revolution' has helped to prevent possibly catastrophic food shortages. However, it is also clear that—ironically—adoption of new crop strains sometimes increases the dependency of the less developed country, as use of new technology increases the need to import fuel, fertilizer, and pesticides, and exposes the adopters to new risks, from environmental damage to rising oil prices. These problems notwithstanding, many observers see a need for additional 'green revolution' strains and technologies in sub-Saharan Africa.

BIOTECHNOLOGY: ANOTHER AGRICULTURAL REVOLUTION?

As a result of the green revolution, world grain production has increased spectacularly—about 45 per cent between 1970 and 1990. In the future, however, it seems likely that the most significant gains will be made through developments in the area of biotechnology.

Tissue culturing and DNA sequencing are two of the biotechnological methods that have helped to develop improved plant and animal varieties. Perhaps the most important—and controversial—aspect of biotechnology, however, is the alteration of the genetic composition of organisms, including food crops. As early as the 1970s, researchers discovered how to change the genetic structure of a plant by inserting genes (DNA) from another source to give the plant some desired new characteristic. An early example was the creation of a frost-free tomato by adding genes from a cold-water fish, the flounder. Other crops have been modified to make them less vulnerable to pests, or weeds, or drought, or salty soils, or diseases. In other cases, modification has focused on making a plant more nutritious.

The adoption of GM crops has been spectacularly successful since they were first commercially available in 1995, with large areas of grain cultivation having already switched to genetically modified varieties: by 2007 about 114 million hectares (282 million acres) of land were growing GM crops, a huge increase from the 2000 figure of 28 million hectares (69 million acres). The principal GM crops are

neo-colonialism Economic relationships of dominance and subordination between countries without equivalent political relationships; often develops after political colonialism ends and the former colony achieves independence, but may also occur without prior political colonialism.

soybeans, corn, cotton, canola, and potatoes. Most of the area under GM crops is in the United States, with more than 33 per cent of US corn and 55 per cent of US cotton coming from genetically modified varieties. Argentina, Brazil, India, and China all are increasing GM crop production. About 90 per cent of soybeans in Argentina now come from a genetically modified variety that is tolerant to a specific herbicide, which allows for better weed control and higher yields. General Foods has made a caffeine-free coffee bean, while scientists in the UK have developed a rice strain that produces about 20 times more beta carotene that the human body converts into vitamin A. Other products are expected to become available soon.

But despite all the promise, and indeed all the successes, biotechnology brings with it much controversy (Box 10.9). Genetically modified crops met with a mixed reception from the outset, with public opposition so significant in Europe that several countries banned already approved GM crops, actions that were opposed by the United States, Canada, and Argentina and that are being debated by the World Trade Organization. As of 2009, GM crops are rare in Europe, although Portugal, Germany, and France are showing some interest in an insect-resistant corn. The European Union currently restricts the importation of many genetically modified products on the grounds that the consequences for consumers and for the environment are uncertain. In January 2000, following a debate in Montreal, a global agreement was reached concerning safety rules for genetically modified products. This Biosafety Protocol to the UN Convention

Box 10.9   For and Against Genetically Modified Crops

From the perspective of food production and the world food problem, the greatest benefits of genetic modification are likely to come with the creation of crops resistant to virus infections. Some observers also claim that foods will be produced that offer enhanced nutritional and possibly even medicinal qualities. Proponents argue that GM crops increase yields without decreasing biodiversity through further forest clearance and decrease pesticide use and costs. So far, though, the main emphasis has been on engineering varieties that can resist specific herbicides (such as Monsanto's Roundup)—a development that is of less interest to consumers than to herbicide producers and farmers. It is not difficult to see why Monsanto might want to create plants resistant to its own herbicide. Indeed, much opposition focuses on the fact that a small number of chemical corporations (especially Monsanto, Dow, and Du Pont) dominate the GM industry.

But most of the concern expressed relates to possible risks to human health and to the environment, and increasing numbers of interest groups, especially environmental groups and some consumers, are expressing concern about GM crops. Indeed, GM crops appear to represent everything that environmental groups dislike, at least partly because they are promoted and marketed by companies such as Monsanto.

As we have seen, some favour a return to organic farming, which rejects the use of synthetic fertilizers and pesticides and promotes closer links between the producer and the consumer. On balance, the evidence for health risks does not seem to be significant, but several studies show gene flow from a modified plant to another plant that could possibly lead to some form of 'superweed'. It seems fair to say that scientists have insufficient understanding of this possible problem at the present time. Of course, it is also fair to say that many of the possible problems with GM crops are no different in principle to problems associated with conventional crops.

More generally, some commentators argue that the real problems of food production are not to be solved by technological advances but rather by solutions addressing issues of spatial and human inequality (Chapter 5). Indeed, in Africa, where the need for increased food production is very real, only South Africa is showing substantial interest in GM crops. Nevertheless, it may not be an exaggeration to suggest that developments in biotechnology could prove to be the most significant changes in agricultural production since the first animals and plants were domesticated. Some observers argue that the world faces a choice: either accept GM crops and thus help feed the world, especially through the introduction of drought- and disease-resistant varieties of such staple crops as rice and potatoes, or face the prospect of continuing famines in poor areas of the world. Recall, however, that discussions in Chapter 5 highlighted other reasons for famines, especially inefficient governments.

An important general point to bear in mind when debating the pros and cons of GM crops is this: the humans who first domesticated plants and animals through selective breeding were causing genetic changes. The essential difference today is that modern genetic engineering is able to go much further much more quickly than conventional selective breeding. The technology is very different and the time factor hugely so, but the outcomes are similar.

on Biodiversity includes rules designed to protect the environment from damage.

NO-TILL: A QUIET AGRICULTURAL REVOLUTION?

Ploughing, or tilling, soil is one of the oldest of farming techniques. Turning the soil over either after a harvest or prior to planting is typically seen as beneficial because it helps remove weeds and residue from the previous crop as well as aerates the soil. But increasing evidence suggests that in many places it may not be necessary and, indeed, may be detrimental. When ploughing, farmers usually turn over between about 6 to 10 inches of soil, whereas no-till farming involves making a groove no more than 3 inches deep in which to plant seeds. Ploughing thus results in much more soil disturbance and can be a major cause of soil erosion.

No-till farming was the norm until the invention of the plough and the use of domesticated animals to pull ploughs. Ongoing improvements to plough design facilitated the expansion of agriculture globally, including in many grassland areas such as the North American prairies, where turning the sod required such technology. But does this mean that ploughing continues to be necessary? According to many critics, the answer is no, with no-till practices being feasible in many different physical environments. No-till farming also is suited to most crops, with the notable exceptions of wetland rice and root crops such as potatoes.

Huggins and Reganold (2008: 75) summarize six principal benefits of no-till agriculture and seven trade-offs, and weighing these up they argue for no-till farming. The benefits are that it:

- reduces soil erosion;
- conserves water;
- improves health of soil;
- lowers fuel and labour costs because of less tillage;
- reduces sediment and fertilizer pollution of nearby water bodies;
- sequesters carbon.

The trade-offs are:

- transition from conventional to no-till farming may be difficult because many other changes are needed;

- needed equipment is costly;
- there is a heavier reliance on herbicides;
- there may be some unexpected changes in weeds and disease;
- at first more nitrogen fertilizer may be needed;
- germination may be slower and there may be a reduction in yields.

Table 10.3 describes what is involved in no-till farming, in conventional tillage farming, and in the intermediate strategy of conservation tillage (often called reduced tillage) farming in the specific case of corn–soybean crop rotation in the Corn Belt region. It is notable that no-till requires the fewest passes over a field.

Today, no-till farming is practised on only about 7 per cent of global cropland, with most such farming in the United States, Brazil, Argentina, Canada, and Australia. Adoption rates are much lower in Europe, Asia, and Africa. One reason why Europe lags is the lack of government policies promoting no-till, while in less developed countries the changes required are often too expensive to initiate.

Most of the no-till agriculture in Canada is in the Prairie region, where it has increased dramatically since the 1980s. Figure 10.12 shows the percentage of the land area of the

Table 10.3	Comparing Tillage Strategies		
# of Passes Over a Field	No-Till	Conservation Tillage	Conventional Tillage
1	Apply herbicide	Till with chisel plow, burying up to 50 per cent of crop residue	Till with moldboard plow, burying up to 90 per cent of crop residue
2	Plant	Till with field cultivator	Till with disc to smooth the ground surface
3	Apply herbicide	Plant	Till with field cultivator to prepare the seedbed for planting
4	Harvest	Apply herbicide	Till with harrows to smooth seedbed
5		Till with row cultivator	Plant
6		Harvest	Apply herbicide
7			Till with row cultivator
8			Harvest

SOURCE: D.R. Huggins and J.P. Reganold, 'No Till: The Quiet Revolution', *Scientific American* 299, 5 (2008): 74.

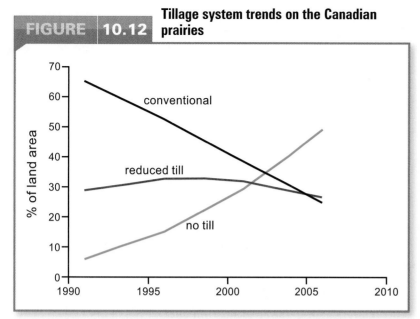

FIGURE 10.12 Tillage system trends on the Canadian prairies

SOURCE: Agriculture and Agri-Food Canada. 2009. Flexibility of No Till and Reduced Till Systems Ensures Success in the Long Term 2009, http://www4.agr.gc.ca/AAFC-AAC/display-afficher.do?id=1219778199286&lang=eng. Reproduced with the permission of the Minister of Public Works and Government Services Canada, 2009.

prairies under the three different tillage systems for each of the last four agricultural census years.

Huggins and Reganold (2008: 77) conclude that no-till, although 'not a cure-all', is a sustainable strategy that is able to 'deliver a host of benefits that are increasingly desirable in a world facing population growth, environmental degradation, rising energy costs and climate change.'

World Agriculture Today: Types and Regions

The noted American geographer Whittlesey identified nine major regional types of agriculture, and Figure 10.13 shows the global distribution of these nine types as elaborated upon by Grigg (1974); it includes a twofold division of one of the types and identifies those areas with little or no agriculture.

Even a cursory review of Figure 10.13 highlights the relationships between agriculture and climate (Figure A1.7), between agriculture and generalized global environments (Figure A1.8), and between agriculture and population distribution and density (Figure 5.1). The figure also suggests the complex links between areas of production and areas of consumption.

PRIMITIVE SUBSISTENCE AGRICULTURE

One of the earliest agricultural systems, 'primitive subsistence' or 'shifting' agriculture, is today practised almost exclusively in tropical areas. It involves selecting a location, removing vegetation, and sowing crops on the cleared land. Land preparation is minimal, and little care is given to the crops. Typically, agricultural implements are limited and livestock are not normally part of the system, although fowl and pigs may be present. After a few years, the land is abandoned and a new location is sought. In this farming system, land is not owned by an individual or family but rather by some larger social unit such as the village or tribe. Shifting agriculture has traditionally been for subsistence purposes, that is, for local consumption. Today, however, in addition to their subsistence crops many such cultivators also produce a different crop for sale so that they have at least some market orientation.

Superficially, shifting agriculture appears to be wasteful and indicative of a low technology. Yet, evidence suggests that it is a highly appropriate farming method. Not only is it an effective way of maintaining soil fertility in humid tropical areas, and well suited to circumstances of low population density and ample land, but it provides adequate returns for minimal capital and labour inputs. Shifting agriculture, then, can be explained by reference to a number of variables beyond economic considerations, including environment and population numbers.

WET RICE FARMING

Most of the rural population of East Asia is supported by wet rice farming, an intensive type of agriculture that requires only a small portion of the total land area—farms are small, perhaps only 1–2 ha (2.5–5 acres) and subdivided into fields—but large amounts of human labour. There are many versions of wet rice farming, but in all instances the crop is submerged under slowly moving water for much of the growing period. Wet rice farming is restricted to specific local environments, most notably flat land adjacent to rivers. Low walls delineate the fields and hold the water. Because this technology minimizes soil depletion, continuous cropping is often possible, producing multiple harvests each year. Cultivation techniques changed little until the green revolution of the 1960s, when some

FIGURE 10.13 World agricultural regions

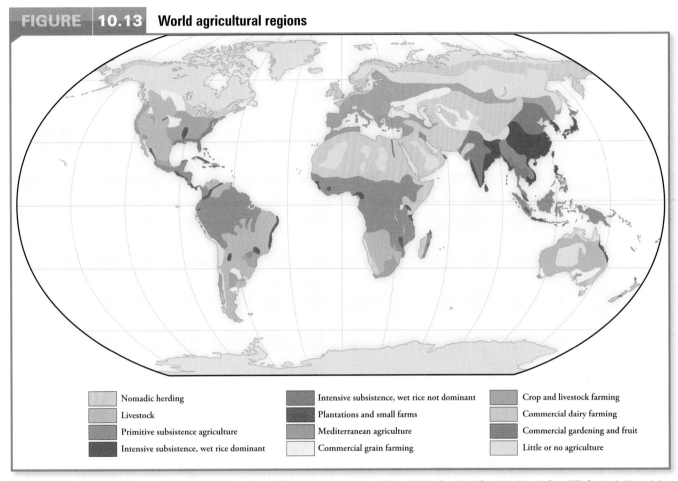

Nomadic herding

Livestock

Primitive subsistence agriculture

Intensive subsistence, wet rice dominant

Intensive subsistence, wet rice not dominant

Plantations and small farms

Mediterranean agriculture

Commercial grain farming

Crop and livestock farming

Commercial dairy farming

Commercial gardening and fruit

Little or no agriculture

SOURCE: D.B. Grigg, *The Agricultural Systems of the World: An Evolutionary Approach* (New York: Cambridge University Press, 1974), 4. Copyright © Cambridge University Press 1974. Reprinted with permission of Cambridge University Press.

areas began using new rice varieties, fertilizers, and pesticides. There are close links between wet rice farming, high population densities, and flatland environments (wet rice fields on terraced hillsides are characteristic of South China and some areas of Southeast Asia). In some areas UN-sponsored projects train farmers to use an integrated crop management system designed to conserve biodiversity and to reduce dependence on chemicals (Rogers and Chew, 2005).

PASTORAL NOMADISM

Traditionally practised in the hot/dry and cold/dry areas of Africa, Arabia, and Asia, pastoral nomadism is declining today as a result of new technologies and changing social and economic circumstances; however, it remains an important type of agriculture in Somalia, Iran, and Afghanistan. Pastoral nomads are subsistence-oriented and rely on their herds for milk and wool. Meat is rarely eaten, and the livestock—cattle, sheep, camels, or goats—is rarely sold. Usually one animal dominates,

A protestor dressed as a tomato crossed with a fish leads a demonstration in January 2000, calling on Ottawa to require labelling of genetically modified foods.

CP photo/Fred Chartrand

although mixed herds are not unknown. Pastoral nomadism likely evolved as an off-shoot of sedentary agriculture in areas of climatic extremes where regular cropping was difficult. Continually moving in search of suitable pastures, these groups achieved considerable military importance until the rise of strong central governments brought them under control (Box 10.10).

MEDITERRANEAN AGRICULTURE

Mediterranean agriculture can be designated as a type because it is associated with a particular climate (mild, wet winters and hot, dry summers) and because it has played a major role in the spread and growth of agriculture in the Western world. Traditionally, it has three components: wheat and barley, vine and tree crops (grapes, olives, figs), and grazing land for sheep and goats. Whereas both wheat cultivation and grazing are extensive land uses and low in productivity, vine and tree crops are intensive. Typically, all three activities are practised on all farms. Spatially, this type of agriculture evolved in the eastern Mediterranean. It had spread throughout the larger region by classical times, and then was exported to environmentally suitable areas overseas, such as California, Chile, southern South Africa, and southern South Australia. Population growth and technological changes have led to several variations, with a general increase in irrigation and a decline in extensive wheat cultivation.

MIXED FARMING

The four agricultural types discussed so far are generally subsistence in orientation (Mediterranean agriculture might be described as mixed subsistence/commercial); the remaining five are commercial. Mixed farming prevails throughout Europe, in much of eastern North America, and in

Box 10.10   Pastoral Nomadism in the Sahara and Mongolia

Pastoral nomads in the Atlantic Sahara region of Northwest Africa are declining in numbers for a variety of interrelated reasons, including government persuasion, economic change, drought, and war (Arkell, 1991). For more than 700 years, the Sahrawis, the indigenous population of the region, survived in a difficult desert environment by migrating vast distances to locate water and pasture for their herds of camels and goats. Even the imposition of artificial colonial boundaries in Northwest Africa, dividing the traditional territory of the Sahrawis, did not prevent them from continuing their nomadic way of life. Only in the 1960s did economic changes, in the form of phosphate exploitation and related urbanization, prompt many of the herders to seek wage employment in urban areas.

More damaging changes occurred in the 1970s as a consequence of drought and war. The war began when Spain ceded its colonial interests in Western Sahara in 1975, leaving Morocco and Mauritania to compete for the territory, and continued after a guerrilla group (backed by Algeria) claimed to represent the interests of the Sahrawis in Western Sahara. Many Sahrawis fled into Algeria, and cross-border movements between the latter and those remaining in Western Sahara are now virtually nil, as Morocco has constructed a series of defensive walls (sand-and-rubble parapets) around much of Western Sahara. The 1989 creation of the five-country Maghreb Union has not yet helped to resolve the conflict, and the future of the nomadic pastoralist Sahrawis remains uncertain.

In the popular imagination, perhaps the most famous pastoral nomads were those from Mongolia who struck fear into the fifth-century Roman Empire and who, from the thirteenth century on, founded dynasties in China, Persia (Iran), and India. The physical environment that supported those people is the Mongolian steppe, a fertile grazing region. It is likely that several animal species, including sheep, goats, camels, and horses, were domesticated on the steppe. The present political state of Mongolia owes its existence to the conflict between pastoralists and the expansionist Chinese agrarian society. Part of the Chinese Empire for two centuries, Mongolia achieved independence in 1911 and then became a Communist state under Soviet tutelage in 1924. Communism was abandoned in 1990, and today Mongolia is a liberal democracy. The pastoral economy remains, but it is subject to many stresses.

Perhaps the greatest stress is the contradiction between a free-market economy and public ownership of land in the pastoral areas. The current trend is towards private ownership, and a 2002 law giving Mongolians the right to own plots of land in urban areas seems likely to be extended to the rural pastoral landscape. On the surface this seems a positive step, for, as our discussion of the 'tragedy of the commons' in Box 3.3 demonstrated, public ownership can lead to environmental degradation. On the other hand, private ownership could be a disaster, undermining the 'best practices' developed over centuries. Mongolia, like any other part of the world, is a changing environment and the consequences of change are necessarily uncertain.

other temperate areas of European overseas expansion. It is clearly a variant of the earliest farming in Southwest Asia, involving both crops and livestock. The transition to modern mixed farming involved a series of changes, such as the adoption of heavier ploughs and a three-field system (in the early Middle Ages), the reduction of fallow and the use of root crops and grasses for animal feed (c. the seventeenth century), and general intensification (mid-nineteenth century). Most of these changes have been related to population and market pressures.

Contemporary mixed farming is intensive and commercial (being closely associated with large urban areas) and integrates crops and livestock. The principal cereal crop varies according to climate and soil—it may be corn, wheat, rye, or barley—while root crops are grown for animal and human consumption. Crop rotation is standard, because it has both environmental and economic advantages, aiding soil fertility and minimizing the impact of price changes. As with the other farming types identified, there are many versions of mixed farming. Differences are particularly significant between Western Europe and the American Midwest, which is characterized by larger farms, especially high productivity, and advanced technology.

DAIRYING

Specialization in dairying is closely related to urban market advances in transportation and to other technological changes that began in the nineteenth century. It is particularly associated with Europe and areas of European overseas expansion. Farms are relatively small and are capital-intensive. The sources of the dairy products marketed in a given region often reflect distance from market, with those farms close to market specializing in fluid milk and those farther away producing butter, cheese, and processed milk. On the world scale, those dairy areas close to population concentrations—in Western Europe and North America—specialize in fluid milk, while relatively isolated dairy areas—New Zealand, for example—focus on less perishable products. Fluid milk producers are making increasing use of the feedlot system, in which cattle feed is purchased rather than grown on the farm. In such cases dairying operations begin to look more like factory systems than the traditional image of an

Pineapple plantation near Pital, Costa Rica.
imagebroker/Alamy/GetStock

agricultural way of life. This development is most evident in North America.

PLANTATION AGRICULTURE

A number of crops required for food and industrial uses can be efficiently produced only in tropical and subtropical areas. For this reason, as Europeans moved into the non-temperate world, they developed what became known as plantation agriculture. Plantations produce crops such as coffee, tea, oil-palm, cacao, coconuts, bananas, jute, sisal, hemp, rubber, tobacco, groundnuts, sugar cane, and cotton for export to Europe and North America, primarily. This type of agriculture is extremely intensive, operating on a large scale using local labour (usually under European supervision) and producing only one crop on each plantation.

Because plantation agriculture evolved within the context of colonialism, it has profound social and economic implications and has been seen as exploiting both land and labour. However, the fault here lies with colonialism, not plantation agriculture in itself. Because the plantation economy was and is so labour-intensive, the first plantations, established in the fifteenth century by the Portuguese in Brazil to produce sugar cane, often used slave labour imported from West Africa; then, following the abolition of slavery, indentured labourers were brought from India and China (see Box 5.3). As a consequence of the worldwide search for labour, alien peoples were introduced to many areas and cultural conflict continues.

Because plantations are designed to serve the needs of distant areas, plantation companies are often multinational, that is, they operate in more than one country. Many such companies are also characterized by what is known as vertical integration: the same company that produces the crop also refines, processes, packages, and sells the by-products. Thus, although political colonialism is largely gone, it has in many instances been replaced by economic colonialism.

RANCHING

Commercial grazing—ranching—is generally limited to areas of European overseas expansion and is again closely related to the needs of urban populations. Cattle and sheep are the major ranch animals, since beef and wool are the products most in demand. Much of the European expansion into temperate overseas areas in North and South America, Australia, New Zealand, and South Africa was motivated by the desire to expand a grazing economy. Ranching is usually a large-scale operation because of the low productivity involved, and hence is associated with areas of low population density. In some areas the ranching economy has evolved into a ranching and live-stock-fattening economy; the most notable example is the American Great Plains. Previously, cattle were shipped to the Corn Belt states for fattening, but the late-twentieth-century proliferation of cattle feedlots within the Great Plains allows beef production to be vertically integrated in the region. This trend appears likely to continue; ranchers, especially in semi-arid areas of the temperate world, are at present experiencing difficult economic times.

LARGE-SCALE GRAIN PRODUCTION

In large-scale grain production the dominant crop is wheat, often grown for export. Major grain producers are the US, Canada, and Ukraine. Farms are large and highly mechanized. This type of production evolved in the nineteenth century to supply the growing urban markets of Western Europe and eastern North America, where in many cases it displaced a ranching economy. In some areas, such as Ontario in the mid-nineteenth century (McCallum, 1980: 4), wheat became the staple crop:

> A more classic case of a staple product would be difficult to imagine. More specialized in wheat production than the farmers of present-day Saskatchewan, Ontario farmers of the mid-nineteenth century exported at least four-fifths of their marketable surplus. Close to three-quarters of the cash income of Ontario farmers was derived from wheat and wheat and flour made up well over half of all exports from Ontario.

Wheat and other grain staples were the engines of economic growth in other areas. Today, most of those areas are more diversified, and some rely more on mixed farming than on grain production. On the Canadian prairies the transition from a predominantly wheat economy to a more diversified farming economy was evident even in the 1980s, reflecting both global trends and national economic strategies (Carlyle, 1994; Seaborne, 2001).

Harvesting canola near Ponoka, Alberta (south of Edmonton), August 2002. A severe drought that summer reduced crop yields by two-thirds.

CP photo/Adrian Wyld

Globally, as we saw in Chapter 9, world commerce is increasingly liberalized and competitive, and as a result governments are being forced to reconsider their interventions into the agricultural sector. Particularly important for farmers on the Canadian prairies was the 'Crow rate', named for the Crow's Nest Pass Agreement of 1897, under which the Canadian Pacific Railway had guaranteed western grain farmers a special low rate 'in perpetuity'. In 1984 the terms of this agreement were changed and the 'rate' became a direct federal subsidy, which in 1995 was eliminated altogether. In recent decades non-grain crops have become increasingly important, among them oil seeds, especially canola and flax, and such specialty crops as peas, lentils, mustard seed, and canary seed (Figure 10.14). The hog industry is also growing rapidly: in Saskatchewan, hog numbers increased by almost 50 per cent between 1996 and 2001. Overall, traditional areas of large-scale grain production are undergoing significant changes as the need to adjust to changing global, national, and regional circumstances becomes evident. In turn, changes in production lead to a multitude of other changes (Box 10.11).

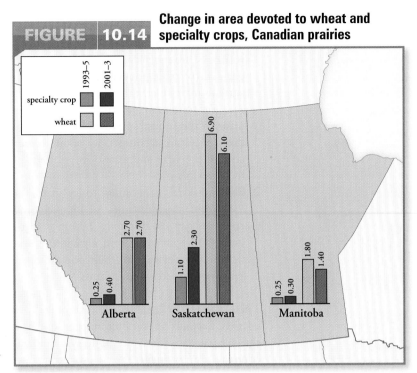

FIGURE 10.14 Change in area devoted to wheat and specialty crops, Canadian prairies

NOTE: Numbers indicate millions of hectares.

SOURCE: N. Fadellin and M.J. Broadway, 'Canada's Prairies: From Breadbasket to Feed Bunker and Hog Trough', *Geography* 91 (2006): 85. Reprinted by permission of the Geographical Association, www.geography.org.uk

Global Agricultural Restructuring

For almost all of human history, food was consumed in the same location where it was produced. Today, however, especially in the more developed world, producers and consumers are no longer in the same location. Agriculture is just one component of larger systems of production and consumption that are often affected by regulations that have a global impact (e.g., trade agreements), by the activities of transnational corporations, and by various production, processing, and marketing networks. As noted in Box 10.2, these important changes are usefully addressed through a political economy approach, not least because they contribute to uneven development as capital moves into those places where profits will be highest.

Notwithstanding globalization processes, there are notable differences between the agricultural economies of the more and less developed worlds, especially concerning the use of capital. In the more developed world much capital is invested in machinery and farm inputs, which means less need for human labour. But in the less developed world, most notably in tropical regions as evident in the account of world agricultural regions, much less capital is invested and agriculture continues to be relatively labour-intensive.

A process central to many of these changes in the agricultural industry (and other industries, as we will see in Chapter 14) is **restructuring**. In a capitalist economy, this term can refer to changes in either the type of capital invested in (for example, from human labour to machines) or the movement of capital resulting from changes in technology or labour relations. In agriculture, restructuring may involve spatial changes in agricultural activities, movements of capital from one level of the production process to another, or changes in the organization of production. Most simply expressed, restructuring can be seen as a process of ongoing adjustment to changing circumstances.

AGRICULTURE IN THE WORLD ECONOMY

The relative importance of agriculture in an economy decreases with economic growth (Grigg, 1992). This can be demonstrated as follows:

1. For pre-industrial economies, data for the World Bank category of low-income

> **restructuring**
> In a capitalist economy, changes in or between the various components of an economic system resulting from economic change.

countries suggest that today about 32 per cent of their gross domestic product (GDP) is derived from agriculture, while about 68 per cent of their population is engaged in agriculture; similarly, the limited data available for contemporary industrial countries prior to the Industrial Revolution indicate that agriculture accounted for about 40 per cent of the GDP and about 70 per cent of the workforce.

2. For industrial economies today, data for the World Bank category of high-income countries suggest that only 2.8 per cent of their GDP is derived from agriculture and that only 5 per cent of their population is actively engaged in agriculture.

These figures exaggerate the extent of the relative decline of agriculture because the agricultural data for contemporary industrial countries do not include food-processing and related activities, but the general picture is nevertheless clear: industrial and service activities have replaced agriculture as the most important sectors of national economies (Figure 10.15). Furthermore, agricultural incomes are lower on average than non-agricultural incomes, and this gap is widening all the time.

There are considerable spatial variations in the timing of the relative decline of the agricultural sector. The decline began in Britain, the home of the Industrial Revolution, where the agricultural percentage of the total labour force fell below 50 as early as the 1730s; in other European countries the decline to below 50 per cent had occurred by the 1840s; and in much of Africa and Asia it has not yet occurred.

CHANGES TO THE FOOD SUPPLY SYSTEM

The food supply system is becoming increasingly complicated. Whereas the von Thünen theory focuses on horizontal relations on one level only (that of farmers), today there are not only horizontal relations on several levels but also vertical relations between different levels. As Figure 10.16 shows, the food supply system now integrates activities at five levels—inputs, farmers, processors, distributors, and consumers—and is increasingly globalized. In recent years, the greatest change in this system has been the expansion of the purchasing and selling power of the processing and distributing sectors. Changes to the food supply system mean that geographers are directing much attention to the idea of a commodity network that links production, distribution, and consumption, and that operates in particular environmental, economic, cultural, and political contexts.

At least three significant ongoing changes in the food supply system are related to globalization processes (Marsden, 1997).

- Globalization, such as corporate manufacturing of food, is accompanied by national-scale processes, for example, state policies governing product prices.

Box 10.11 Canadian Farmers: Fewer and Older

The trend towards fewer and older farmers in most parts of the more developed world is well established and is one aspect of larger economic changes related to the changing nature of employment (a topic discussed more generally in Chapter 14). In Canada, the total number of farm operators continues to fall, declining by 7.1 per cent in the five years between 2001 and 2006. Although the decline in numbers of farmers has not reduced food production—with larger farms, those that remain are producing more than ever before—it is having a significant impact on rural life and the health and perhaps even the survival of many small towns. Indeed, many farmers today do not live on their farms but commute to their work from city residences.

The decline from 2001 to 2006 was neither new nor unexpected. On the prairies, rural populations grew rapidly until the 1930s, when the numbers began the decline that has continued into the twenty-first century. Meanwhile, as the farm landscape is gradually depopulated, the infrastructure of railway lines, grain elevators, small towns, schools, and so forth is being dismantled, and the large amounts of capital required to survive in an increasingly high-risk environment mean that more and more agricultural land is now in the hands of corporations and huge farm conglomerates.

In short, it appears that the small family farm is a thing of the past. Most small farmers now require some off-farm income to support themselves, and the median age of farmers is steadily increasing across Canada; in Manitoba it rose from 46 to 52 in 10 years (1996–2006). To survive, some small farmers are moving into specialized areas—in Manitoba, some focus exclusively on raspberries or other soft fruits—or activities such as organic farming (see Box 10.7).

FIGURE 10.15 Percentage of labour force in agriculture by country, 2007

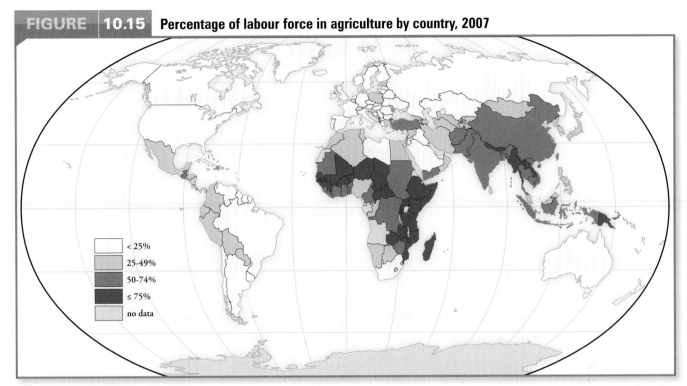

< 25%
25–49%
50–74%
≤ 75%
no data

This map provides a good indication of the importance of agricultural activity globally. The range is enormous. In many of the more developed countries less than 25 per cent of the labour force works in agriculture; in both Canada and the United States only about 3 per cent do so. In some less developed countries the proportion is more than 75 per cent; in Niger and Mali, more than 90 per cent of the people are agricultural workers. The correlation between development level and agriculture is far from perfect, however; several countries in Latin America have values below 25 per cent. The world average is 49 per cent. There are often significant differences between the genders: in Canada, 2 per cent of employed women and 4 per cent of employed men work in agriculture; in Bangladesh the corresponding percentages are 77 for women and 53 for men.

SOURCE: Updated from United Nations Development Program, *Human Development Report* (New York: Oxford University Press, 1996), tables 16 and 32. By permission of Oxford University Press, Inc.

- Globalization, such as the spread of standardized food items, is accompanied by localization, for example, an increase in local organic production (with the outcome sometimes called 'glocalization').
- Increasingly, the non-farm sector of the food supply system is controlling what people are able to consume and eat; this sector includes consumer agencies and food safety organizations.

Agricultural change in the more developed world has gone through four distinct phases since World War II: mechanization, chemical farming, food manufacturing, and biotechnology. The 1950s brought increasing mechanization, which raised yields and reduced labour requirements, and the 1960s saw the advent of large-scale chemical farming with nitrogenous fertilizers, herbicides, fungicides, and pesticides. These first two stages modernized agriculture. The 1970s witnessed significant growth in the sector dealing with the manufacture of purchased agricultural inputs and in the processing and distributing sectors (Figure 10.16), while the 1980s brought biotechnological advances in plant varieties and livestock breeds. The last two stages industrialized agriculture.

Smith (1984: 362) demonstrates that these changes have had a 'marked imprint on the landscape' in Quebec. With increasing vertical integration, retailers are assuming control of some processing, and processors are influencing farm operations; thus farmers themselves are losing some of their capacity for independent decision-making. At the processing level, plants are now fewer and larger, and there is greater locational concentration. Similarly at the retailing level, a few large organizations tend to dominate the market.

Agricultural change in the less developed world involves global and local forces playing themselves out in very different circumstances, as evidenced by agricultural change in Bhutan (Box 10.12) and by a land-use conflict in Côte d'Ivoire (Box 10.13).

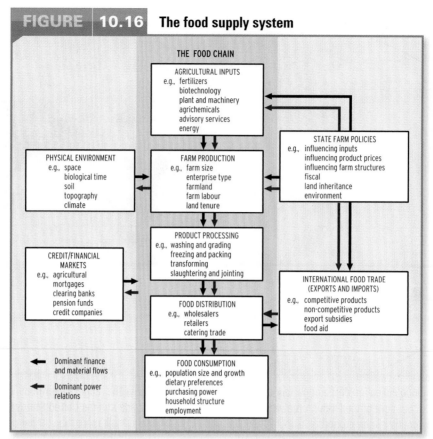

FIGURE 10.16　The food supply system

SOURCE: M. Pacione, ed., *Progress in Agricultural Geography* (Beckenham, Kent: Croom Helm, 1986).

INDUSTRIALIZING AGRICULTURE

For many human geographers, the most fascinating—most 'geographical'—aspect of agriculture is that it is carried out at the interface between humans and the physical environment. Agriculture, unlike most other economic activities, is constrained by physical geography, and its success, as measured by productivity or return on investment, is beyond the control of the farmer. But these physical constraints do not prevent the industrialization of agriculture. Rather, they guide that industrialization in particular directions.

Today, agriculture in the more developed world can be justifiably regarded as an industry, one in which the farmer is no longer the sole or even the primary decision-maker. According to Friedmann (1991: 65), 'the distinction between agriculture and industry is no longer viable, and should be replaced by the conception of an agri-food sector central to capital accumulation in the world economy.' The industrialization of agriculture has become especially evident since 1945. Both in the more developed world and increasingly in some parts of the less developed world,

agriculture has typically (but not always) become more intensive, more specialized, and more spatially concentrated.

Intensification, specialization, and spatial concentration are evident throughout much of the farming economy of the United States, for example, in the production of rice, sugar cane, tobacco, peanuts, and cotton in the South, the production of corn, soybeans, pork, and beef in the Midwest, and beef production in the West. The resulting increased production means that exports are of greater importance than previously.

Today, agriculture is one of the biggest industries in the world, employing about 1.3 billion people, although the numbers employed in the more developed world are decreasing. Agriculture produces about US$1.3 trillion worth of goods per year, and world output of food continues to increase. Nevertheless, farmers are always obliged to strive to be more competitive, more concerned with the environment, and more aware of changing consumer tastes. Some farmers are even entering the information age—for example, using GPSs (see Chapter 2)

Box 10.12   Agricultural Change in Bhutan

Bhutan is a mountainous land-locked kingdom between China and India that is inhabited by people of Tibetan or Hindu Nepalese origin and is currently striving to emerge from its geographic isolation. Only 46,500 km² (17,800 square miles) in size, in 2008 it had a population of 0.7 million, only 31 per cent of which was urban; a CBR of 30; a CDR of 7; and an RNI of 2.3 per cent.

Agriculture, focusing on rice and livestock, is limited to about 3 per cent of the land, but is nevertheless the most important sector of the economy. Not surprisingly, farming operations are small in scale and oriented towards subsistence, sustainability, and household self-sufficiency. It appears that this system works relatively well, for although Bhutan is a poor country, poverty at the individual or household level is rare.

Recently, however, the government, with support from elsewhere, has been attempting to introduce changes in Bhutan's agricultural economy—specifically, to increase exports by using imported 'green revolution' improvements, such as higher-yielding varieties of rice and wheat.

Unfortunately, it appears that the crop varieties imported by Bhutan need pesticides and fertilizers and may not be well suited to the local environment—yet they are displacing cereal varieties that, while lower-yielding, are much more appropriate. Similarly, the government has adopted new policies to promote the production of two cash crops, cardamom (a spice) and potatoes, which appear to be environmentally and culturally inappropriate. There is an important moral to this discussion. Blind, or at least short-sighted, acceptance of new technologies, without careful evaluation of their consequences, may be disastrous. Any change is best preceded by a trial period and a careful appraisal of the broader cultural implications (Young, 1991).

Farmers from Wangdi, Bhutan, threshing rice in the traditional way.

Anwar Hossain

to monitor soil and moisture conditions on a two-hectare square of land.

Changing Camembert cheese

One impact of the industrialization and globalization of agriculture is highlighted by the story of Camembert cheese, a story that can be repeated for many other food products.

Until the mid-nineteenth century, Camembert, like most other cheeses, was sold in only a few local markets. But the extension of a rail link allowed Camembert producers in Normandy to sell their product in the much larger Paris market with travel time reduced from three days to just six hours, while the use of a new light round wooden box meant the cheese could travel without being damaged. To

increase sales further, producers also worked towards a product that matched the expectations of consumers, with perhaps the most significant change being pasteurization, which had become widespread by the 1950s. They began to produce a more standardized cheese in terms of appearance, with the popular white Camembert product replacing by the 1970s earlier cheeses more varied in colour. In short, by the early twenty-first century Camembert cheese had been gradually stabilized and standardized in terms of taste and appearance, characteristics that in the past reflected local peoples and places.

But small local producers continued to campaign against this standardization and, in 2008, won a landmark victory when a French

Box 10.13   Peasant–Herder Conflict in Côte d'Ivoire

The intersection of global, national, and local processes in a less developed world context is evidenced by a study of conflict between two different ethnic and agricultural groups in the West African state of Côte d'Ivoire. This also provides a good example of how a political economy approach, as discussed in Box 10.2, can be applied to a traditional land-use problem (Bassett, 1988).

In the northern Côte d'Ivoire region, beginning about 1970, uncompensated crop damage by cattle in peasant fields gave rise to ethnic and economic tensions between Senufo farmers and Fulani herders, who accounted for 33 per cent of the country's beef production. By the mid-1980s, relations between the two groups had deteriorated to the point of open hostilities: 80 herders were killed in 1986. Some political candidates inflamed the situation by making election promises to banish the Fulani from Côte d'Ivoire. However, the official government policy is to encourage herders so as to reduce dependence on imported beef.

What is the basic cause of this land-use conflict? Traditionally, human geographers would have pointed to a declining resource base relative to population. But this answer does not fully explain why some people lose while others gain land; nor does it explain why conflicts occur in such land-abundant areas as northern Côte d'Ivoire. A more helpful answer would focus on the politics of land use and the human ecology of agricultural systems.

In this case, the key is to determine which of the two groups is able to exert greater political power. The answer to this question will likely influence future land-use patterns. The Fulani are new to the area, while the Senufo are indigenous, but current trends give rights to minorities as well as to established majorities (this was not the case in colonial times). The basic determinants of the conflict are outlined in Table 10.4. Note that the table includes both political ecological variables and traditional human ecological variables.

This example of the value of thinking about the intersection of global, national, and local processes highlights the always important geographic question of scale. A basic premise of political ecology research is the necessity to consider multiple scales in order to understand how places are disrupted and regions transformed.

Table 10.4	Determinants of Peasant–Herder Conflicts in Northern Côte d'Ivoire		
Ultimate Causes	**Proximate Causes**	**Stressors**	**Counter-risks**

Ivorian development model:
• surplus appropriation by
  – foreign agribusiness
  – the state
• livestock development
  policies
 	
Savanna ecology
 	
Fulani immigration
	Low incomes and beef consumption
	
	Insecure land rights
	
	Intersection of Senufo agriculture and Fulani semi-transhumant pastoralism
		Uncompensated crop damage
		
		Political campaigns
		
		Theft of village cattle
			Compensation
			
			Fulani expansion
			
			Crop and cattle surveillance
			
			Corralling animals at night

SOURCE: T.J. Bassett, 'The Political Ecology of Peasant–Herder Conflicts in the Northern Ivory Coast', Annals, Association of American Geographers 78 (1988): 456, Table 1 (MW/RAAG/P1889). Used with permission of Taylor & Francis Ltd., http://www.informaworld.com.

government committee decided that traditional methods of making Camembert, using raw milk, had to be maintained.

STATE INTERVENTION

Much contemporary agricultural activity can be understood only in the larger context of state agricultural policies and negotiated trading agreements. At first, states tried to respond to the inevitable fluctuations in production by regulating prices and marketing, with the ultimate goal of achieving some stability in agricultural production. In recent years, states have become increasingly involved in agriculture because improvements in productivity have led to the oversupply of domestic markets, reducing both product prices and farm income. Inadequate income for farmers is the principal reason behind government intervention in the form of income support.

In the more developed world, state policies aim to ensure the security of food supplies, price and income stability, protection of consumer interests, and regional development.

Such policy objectives may be achieved through guaranteed prices, import controls, export subsidies, and various other amendments to the capitalist market system, such as marketing boards. Canada, for example, has price-support policies, income-support policies, and supply-management policies—all of which are subject to change as trading agreements become increasingly important. In the less developed world, most state policies are aimed at increasing productivity and achieving higher farm incomes and dietary standards.

As we saw in Chapter 9, international trading agreements, including the GATT (replaced by the WTO in 1995) and the Common Agricultural Policy (CAP) of the EU, have emphasized the need to reduce state support for agriculture, with a notable recent example being the 2005 reduction of subsidies paid to Europe's sugar farmers following demands from the WTO. The issue of state support for domestic agricultural production is very sensitive, especially in the EU, the world's leading importer and second largest exporter of food. Globally, the principal issue relates to the fact that state support means other countries, particularly those in the less developed world, are unable to compete in state-supported markets. Nationally, a concern with state support is that it needs to reconcile the interests of consumers and the incomes received by farmers.

AGRIBUSINESS

The term 'agribusiness' refers to the activities of transnational corporations. Plantation agriculture was the first example of agribusiness, and even today most such businesses concentrate on production in less developed areas for markets in more developed areas. Table 10.5 suggests the extent to which agriculture in the less developed world is controlled by transnationals operating from and marketing their products in the more developed world.

Agribusiness has been widely criticized, most often on the grounds that, as Table 10.5 clearly shows, it is a form of economic colonialism, with little concern for the social, economic, and environmental consequences of activities in the producing area. More specifically, agribusiness often contributes to the creation of an agricultural economy vastly different from the local producing economy; benefits foreign investors and local elites rather than peasant groups; occupies prime agricultural land; imports labour that may lead to local

ethnic conflict (see Box 5.3); uses casual, part-time, and seasonal labour; and is reluctant to take responsibility for negative impacts on the physical environment.

GLOBAL TRADE IN FOOD

Two causes of current agricultural change are the ever-increasing demand for food and the concentration of people in large urban centres. Given the differences in population growth rates between the more and less developed worlds, movements of food products from areas of surplus to areas of deficit, through both sales and aid, are bound to increase.

The global trade in food is a relatively recent phenomenon, associated with the Industrial Revolution and, in particular, technological advances in ocean transportation and refrigeration. Most of this trade involves imports to the more developed world of products from the less developed world, notably coffee, tea, and cocoa.

At least half of the world's people rely on one of three cereal crops: wheat, corn, and rice. In many cases these crops are produced for

Table 10.5	Multinationals and Crops in the Less Developed World	
Crop	Companies	Observations
Cocoa	Cadbury-Schweppes Gill and Duffus Rowntree Nestlé	These companies control 60–80% of world cocoa sales
Tea	Brooke Bond Unilever Cadbury-Schweppes Allied Lyons Nestlé Standard Brands Kellogg Coca-Cola	Combined, hold about 90% of tea marketed in Western Europe and North America
Coffee	Nestlé General Foods	Control 20% of world market
Sugar	Tate and Lyle	Buys about 95% of cane sugar imported into the EU
Molasses	Tate and Lyle	Controls 40% of world trade
Tobacco	BAT R.J. Reynolds Philip Morris Imperial Group American Brands Rothmans	Combined, control over 90% of world leaf tobacco trade

SOURCE: Adapted from C. Dixon, *Rural Development in the Third World* (London: Routledge, 1990), 18.

local consumption, but there is also substantial global trade in them. Indeed, most of the foods exported from more developed countries are cereals; food exports are dominated by five countries—first and foremost the United States, but also Canada, Argentina, France, and Australia—while distribution of the surpluses is concentrated in five private corporations. All five of these corporations are international giants with diverse economic interests. The principal importer of grain in recent years has been the former USSR, but several Asian countries are increasing their imports, notably China, South Korea, Indonesia, the Philippines, and Japan.

GENDER RELATIONS

The significance of gender relations in the agricultural restructuring process is increasingly apparent, especially to those employing a political economy approach to the study of agriculture. In the more developed world, there is interest in the growing discrepancy between traditional gender relations and identities in farming communities and those in society as a whole. The trend for women to leave farming is generally considered a reflection of the conservative gender relations associated with farming communities.

In the less developed world, there is increasing recognition of the roles played by women, past and present. In the West African country of Gambia, for instance, women have played important roles in traditional subsistence agriculture and in recent initiatives in horticultural production. The smallest state in mainland Africa (11,300 km²/4,360 square miles), Gambia is a narrow strip of land bordering the Gambia River and encircled by Senegal except for a short Atlantic coast. In 2008 it had a population of 1.6 million, a CBR of 38, a CDR of 11, and an RNI of 2.7 per cent. The climate is subtropical, and wooded savanna covers most of the area. There are five principal ethnic groups, and some 85 per cent of the population is engaged in agriculture. Groundnuts are its one cash crop and account for about 90 per cent of exports.

While women play an important role in rice cultivation, this does not generate income. They have also been enthusiastically involved in recent government initiatives in dry-season horticultural production of commercial crops such as chilies, eggplants, and okra and subsistence crops such as potatoes and tomatoes (Barrett and Browne, 1991). If such initiatives increase women's incomes, they are clearly of great value; in fact, several of them yield well and generate needed income. Typically, however, such efforts face two problems.

First, the new schemes are based not on the local knowledge of horticulture but on imported (or at least non-local) knowledge. Second, they may not be environmentally or economically sustainable as this production often involves tree removal that in turn leads to reductions in the fertility and moisture-holding capacity of the soil. Once again, there is evidence of government efforts to improve agricultural circumstances, but such efforts are probably misguided. In Bhutan and Gambia, long-established agricultural practices are ignored and new imported practices are favoured. It is not difficult to understand why governments are tempted by the new, but it is also important that the new learn from the old. The agricultural practices that have survived, perhaps little changed, for generations are usually grounded in economic and cultural realities; they may need to be improved, but this does not mean that they need to be replaced.

THE CASE OF CHINA

Economic liberalization began in 1978 in China when the country was first opened to international trade and foreign investment. Agriculture was the first sector of the economy to be reformed and modernized, in the 1980s (Morrish, 1994). Since the collectivization of agriculture following the revolution of 1949, all farm operations had been run by groups rather than by individuals, to allow the Communist government to exert maximum control over production. Some 60,000 rural communes, organized by officials, were the key social and economic units before 1979. Under the less rigorous system in place since 1979, families work together on a voluntary basis; farmers contract to use government-owned land, agreeing to deliver a specified amount of produce to the government, and are allowed to sell any surplus. The result of this reform was a steady increase both in yields (about 6 per cent per year) and in incomes. Many peasant farmers are now opting not to produce staple crops but are diversifying into new specialty crops.

Much remains to be achieved. Although approximately 65 per cent of the labour force work in agriculture, this sector contributes only about 32 per cent of China's GDP.

Food, Consumption, and Identity

Much of this chapter has emphasized the political component of the traditionally economic study of agriculture. In this final section, by contrast, we address the cultural component that has become such an important aspect of human geography in recent years. In the context of agriculture, the 'cultural turn' introduced in Chapter 7 is reflected by an interest not only in production but in consumption.

YOU ARE WHAT YOU EAT—AND WHERE

Historically—as geographers such as Vidal recognized in the early twentieth century—the kinds of food consumed in various parts of the world usually reflected the available resources. The food that people consumed represented one of the closest links between them and their environment, as well as one of the principal means by which groups of people were distinguished one from another. Today, despite the undeniable impact of processes of interaction and globalization, this is still the case in many parts of the world, as Figure 10.17 shows.

Considering the history of food in terms of consumption and the consequences of production practices (rather than those practices in themselves), Fernandez-Armesto (2001) identified eight key stages:

- Sometime before 150,000 years ago, humans began to use fire to cook food. This new technique for changing food—existing techniques included burying, drying, masticating, and rotting—had significant social consequences, providing a new focus for human interaction.
- In the *second* stage, groups began to see food as representing more than just something to eat. Table rituals and dietary taboos began to appear in recognition of the other dimensions associated with food, especially the moral and religious dimensions.
- In the *third* stage food became associated with social status: people in positions of authority were able to eat not only more food than others but also different kinds.
- The *fourth* and *fifth* stages together constitute what we earlier described as the first agricultural revolution: the domestication of plants and animals. These two stages—sometimes separate and sometimes related—represented a fundamental change, as humans began producing food, not merely collecting it. Plant domestication opened the way to new forms of social organization, greater population density, and the rise of urban centres.

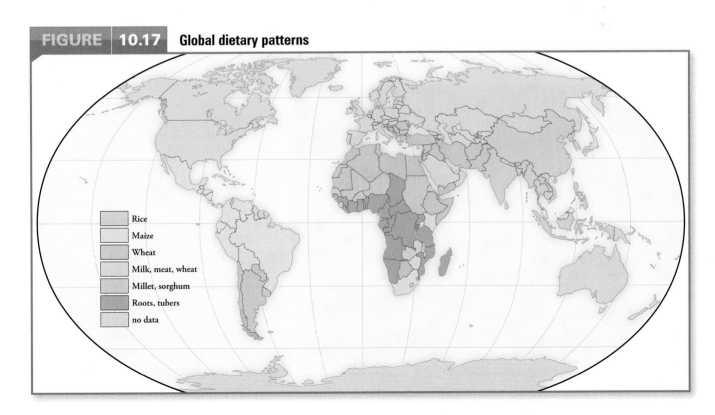

FIGURE 10.17 Global dietary patterns

Rice
Maize
Wheat
Milk, meat, wheat
Millet, sorghum
Roots, tubers
no data

- The *sixth* stage was a global movement of cultures in which food played several important parts. First, food has been one of the most important motivations for both overland and overseas movement over the past 1,500 years. Indeed, demand for salt and spices played a major role in the development of transport routes and cultural contact. The wholesale movements of people over the following centuries involved widespread diffusion of animals and crops in all directions. Potatoes especially fuelled economic development, feeding the poor and underpinning the Industrial Revolution in Europe.
- The *seventh* stage is perhaps best known under the label 'McDonaldization', with globalization of food accompanied by homogenization. Globalized foods have been so successful because they have travelled the routes already laid out by American power, capital, and culture. But much more has been involved than the spread of American fast food to other parts of the world. In North America, for example, 'ethnic' foods—from countries such as Italy, Mexico, China, India, and Thailand—have become popular.
- The *eighth* and most recent stage is the current trend towards fast food that is the first real shift away from the social interaction that was initiated in the first stage. To the extent that fast food makes it unnecessary for members of a family to sit down to a meal together, it may well contribute to

the fragmentation of the family. It is also possible that fast food is contributing to a general decline in health, not only because its nutritional value may be poor but because, without the discipline and camaraderie associated with the shared table, the temptation to overeat can be hard to resist. Certainly, the current increase in obesity in North America seems to be closely related to the increasing popularity of fast food.

CONSUMPTION AND PLACE

Today, food consumption habits remain key indicators of human identity, reflecting regional characteristics as well as individual and group tastes. Building on the ideas promoted by Sauer, much traditional cultural geography was concerned with regional and local foodways, particularly what were called 'ethnic preferences'—that is, the preferences of the specific group under study. By contrast, much contemporary cultural geography addresses the complex intertwining of consumption and identity, taking a particular interest in the political implications of consuming 'ethnic foods'—that is, foods associated with groups other than the one under study. The popularity of 'ethnic food' in the latter sense may well reflect a modern version of the historic association between food and social status. In Britain in 2001, the Foreign Secretary reported that the new national dish was neither fish and chips nor roast beef, but chicken tikka masala, a dish of Indian origin. Some commentators have suggested that this cross-cultural phenomenon perpetuates the former colonial relationship—although one could equally well interpret it as representing a reversal of that relationship.

An interesting example of using a consumer preference to highlight regional political identity arose in 2008. Darjeeling tea—often described as the champagne of teas—is grown in Gorkhaland, in India's northern West Bengal state, an area where ethnic Gorkhas (or Gurkhas as they have often been called) are campaigning for a separate political unit (this is one of the several Indian cases of claims for linking ethnic identity and territory noted in Chapter 8). Estate owners in the region are being pressured into adding the phrase 'The flavour of Gorkhaland' to packets of Darjeeling tea. The idea is that such labelling would highlight Gorkha ethnic and regional identity globally.

Happy Valley Tea Plantation in Darjeeling, India.
dbimages/Alamy/GetStock

Another consumption-related topic is the increasing popularity of specialty foods and drinks explicitly identified with a particular place (e.g., maple syrup as a distinctive product of eastern Canada). Through advertising, often directed at potential tourists, a conscious attempt may be made to construct images of product quality and the related place. This development is perhaps best seen as an extension of the long-standing association between products such as wine and cheese and particular places.

Eating away from home is increasingly popular, especially in North America. The remarkable variety of restaurants available, especially in urban areas, makes it possible not only to eat dozens of different sorts of food but to experience something of dozens of different cultures. Some places to eat and drink have become what are known as 'third places', that is, places other than home or work. Bars and coffee houses especially play important roles in many peoples' daily lives.

CHAPTER 10 SUMMARY

ECONOMIC GEOGRAPHY

Much of our behaviour that creates human landscapes has strong economic motivations. Economics inspired much of the new human geography (often called spatial analysis) that came to the fore in the 1960s. Contemporary studies of agriculture, industry, settlement, and planning take various approaches and may incorporate economic, cultural, and political content. Cultural and political content is increasingly being emphasized.

THE ROLE OF THEORY

Location theory explains geographic patterns, while interaction theory centres on movements between locations. Theories are valuable because of their rigour and simplicity. Many of the theories used by geographers are normative, describing what the theorist thinks ought to be rather than what is. The concept of the economic operator assumes that the goal of any economic activity is the maximization of profit in the context of perfect knowledge and a perfect ability to use such knowledge in a rational fashion.

THE LOCATION OF AGRICULTURAL ACTIVITIES

Many factors can influence the location of agricultural activities. For some geographers, an ecological approach, considering all factors, is the most appropriate. Other geographers focus on particular factors, depending on the specific issue in question. Physical factors include climate, soils, and relief. As living things, animals and plants require appropriate physical environments, either natural or human-constructed. Cultural, social, and political factors also play a role. A farmer's religious and ethnic background affects his or her decisions, as do the policies of various levels of government.

Economic factors include the basic principles of supply and demand.

ECONOMIC RENT

The concept of economic rent is central to agricultural location theory. Land uses compete for locations because it is not possible for all activities to occupy their economically optimal locations. Instead, land is devoted to particular uses according to a spatial pattern that reflects their differing economic rents (itself a specific measure of rent-paying ability). Economic rent can be based on one or more variables; Ricardo based such values on land fertility, while von Thünen used distance from market.

VON THÜNEN THEORY

In 1826 von Thünen, a German economist, published a landmark theory of agricultural location. A normative theory that holds all variables constant except distance from a central market, it generates two hypotheses: (1) zones of land use develop around the market; and (2) the intensity of each specific use decreases with increasing distance from the market. Von Thünen proposed particular contents of the zones for his time and place (early nineteenth-century north Germany). This theory and its variations have stimulated much geographic research.

EMPIRICAL ANALYSES

Studies of the relationships (1) between land value and distance and (2) between land use and distance can be conveniently subdivided into those with a historical focus, those in less developed areas of the world, and those on various scales ranging from the world to individual farms. Overall, these analyses confirm the value of the von Thünen theory as a basis for investigating agricultural patterns.

AGRICULTURAL ORIGINS AND CHANGE

Agriculture originated in the domestication of plants and animals, a long process that began about 12,000 years ago, probably in several different areas, and gradually diffused. Today pre-agricultural activities, such as hunting and gathering, are marginal. Major technological changes include a series of advances in eighteenth-century England, the twentieth-century development of nitrogen fertilizers, and the green revolution in the less developed world. Most recently, changing the genetic composition of food crops may be the greatest change in agricultural production since domestication began.

REGIONS AND TYPES

Today there are nine principal types of agriculture, each occupying a particular region or regions. Three types are classified as 'subsistence' (shifting agriculture, wet rice farming, pastoral nomadism); a fourth type, Mediterranean agriculture, is 'mixed subsistence/commercial'; and the remaining five are 'commercial' (mixed farming, dairying, plantations, ranching, large-scale grain production). All but one type can be traced back thousands of years; the exception is plantation agriculture, which developed as a result of European overseas expansion. The modern versions of dairying, ranching, and large-scale grain production are very different from their antecedents.

AGRICULTURE IN THE WORLD ECONOMY

Relative to other economic activities such as industry, agriculture has experienced a continuous decline in importance as measured by labour force and contributions to GDP.

TRADE IN FOOD

Food is an important component of world trade. Much food is sent as aid to poor countries. The former USSR and several countries in Asia are major importers of grain. The United States is the principal exporter.

AGRICULTURE IN THE LESS DEVELOPED WORLD

China is currently experiencing major changes in the organization of agricultural activities as part of a transition from strictly communistic practice. In some other areas, such as Bhutan, green revolution technologies have caused environmental problems. Recently, human geographers have employed a political ecology approach, combining traditional human ecology with political economy to address specific problems such as land-use change, land-use conflict, and the involvement of women in agriculture.

AGRICULTURE AS AN INDUSTRY

In many respects agriculture today functions as an industrial activity with several linked levels: input producers, farmers, processors, wholesalers and retailers, and consumers. National and international policies exert considerable influence on agricultural activities at all levels. Partly because of this organizational complexity, traditional von Thünen theory is inadequate to explain the spatial distribution of agricultural activities today.

CONSUMING FOOD

Historically, the kinds of food consumed in various parts of the world primarily reflected available resources. Today much interest is focused on trends in consumption: the act of consuming is seen as a way of distinguishing oneself from others, and perhaps even of establishing superiority.

QUESTIONS FOR CRITICAL THOUGHT

1. How has the subdiscipline of economic geography philosophically changed since the 1950s?

2. Summarize the general factors that affect where specific agricultural activities are located. How important are physical factors?

3. How important or relevant is von Thünen's theory in understanding patterns of agricultural production today?

4. Why does Harris describe the transition from foraging to farming as 'the most fateful change in the human career'? Do you agree? Why?

5. What role did the Industrial Revolution play in the development of agriculture?

6. What agricultural advances have been made in the 'green revolution'? What criticisms have been directed at the 'green revolution'?

7. How has globalization affected world patterns of agricultural production in the late twentieth and early twenty-first centuries?

8. If the shelves of Canadian grocery stores can be stocked with food produced in other countries, why should we support Canadian farmers?

FURTHER EXPLORATIONS

Atkins, P., and I. Bowler. 2001. *Food in Society: Economy, Culture, Geography*. New York: Oxford University Press.

Covers a wide range of topics, including food consumption preferences.

Baker, O.E. 1925. 'Geography and Wheat Production', *Economic Geography* 4: 389–434.

One of a series of articles by this early agricultural geographer, many of which exemplify a regional approach; see *Economic Geography* between 1925 and 1933 for other examples.

Barker, G. 2006. *The Agricultural Revolution in Prehistory: Why Did Foragers Become Farmers?* New York: Oxford University Press.

An excellent global overview of one of the most important questions of human cultural change.

Bellwood, P. 2005. *First Farmers: The Origins of Agricultural Societies*. Malden, Mass.: Blackwell.

An excellent overview of the origins of agriculture, in this case focusing on the roles of population and language dispersal whereas Barker (above) favours analysis of local changes.

Boisard, P. 2001. *Camembert: A National Myth*, trans. R. Miller. Berkeley: University of California Press.

Insightful study of one cheese that shows how an authentic local product has been industrialized, homogenized, delocalized, and pasteurized—in short, ruined—by modernity.

Clark, T. 2007. *Starbucked: A Double Tall Tale of Caffeine, Commerce and Culture*. New York: Little, Brown and Company.

Entertaining book that discusses Starbucks' role in cultural homogenization, American cultural imperialism, and the exploitation and impoverishment of poor farmers.

Coleman, W., W. Grant, and T. Josling. 2004. *Agriculture in the New Global Economy*. Cheltenham, UK: Edward Elgar.

A thorough analysis of agricultural changes at global and regional scales.

Conzen, M.P. 1971. *Frontier Farming in an Urban Shadow*. Madison: State Historical Society of Wisconsin.

A detailed analysis of agriculture in an area experiencing the effects of rapid settlement and commercialization.

Found, W.C. 1971. *A Theoretical Approach to Agricultural Land Use Patterns*. London: Arnold.

A sound theoretical overview.

Gilson, J.C. 1989. *World Agricultural Changes: Implications for Canada*. Toronto: C.D. Howe Institute, Policy Study 7.

A valuable survey focusing on government policies and needed reforms.

Goddard, T., M.A. Zoebisch, Y.T. Gan, W. Ellis, A. Watson, and S. Sombatpanit, eds. 2008. *No Till Farming Systems*. Bangkok: World Association of Soil and Water Conservation, Special Publication No. 3.

A global collection by academics, technical specialists, and farmers identifying the practice of and need for no-till farming systems.

Grigg, D.B. 1984. *An Introduction to Agricultural Geography*. London: Hutchinson.

An excellent textbook dealing with factors affecting agricultural activities.

Harris, D.R., ed. 1996. *The Origins and Spread of Agriculture and Pastoralism in Eurasia*. Washington: Smithsonian Institution Press.

A series of case studies; comprehensive and balanced coverage reflecting current understanding.

Ilbery, B.W. 1985. *Agriculture: A Social and Economic Analysis*. Toronto: Oxford University Press.

A well-written basic textbook.

Jakle, J.A., and A. Sculle. 1999. *Fast Food: Roadside Restaurants in the Automobile Age*. Baltimore: Johns Hopkins University Press.

An entertaining scholarly overview of the proliferation of roadside eateries in the twentieth-century United States, authored by a geographer and a historian.

McWilliams, J.E. 2005. *A Revolution in Eating: How the Quest for Food Shaped America*. New York: Columbia University Press.

A book concerned with five geographic regions, focusing on the establishment of regional preferences before the American Revolution.

Pacione, M., ed. 1986. *Progress in Agricultural Geography*. London: Croom Helm.

Covers a wide range of topics such as theory, classification, diffusion, government policies, and marketing.

Rangan, H., and C.A. Kull. 2009. 'What Makes Ecology "Political": Rethinking Scale in Political Ecology', *Progress in Human Geography* 33: 28–45.

Detailed consideration of political ecology stressing the need to understand how different scales intersect.

Reader, J. 2008. *Propitious Esculent: The Potato in World History*. New York: Heinemann.

Domesticated in the Andes and introduced to Europe in the sixteenth century, potatoes were being hailed as a wonder food by the eighteenth century. This book tells their story.

Robbins, P. 2004. *Political Ecology: A Critical Introduction*. Malden, Mass.: Blackwell.

Readable text with some very useful summaries of agricultural studies.

Robinson, G.M. 2004. *Geographies of Agriculture: Globalization, Restructuring and Sustainability*. Toronto: Pearson Education.

Comprehensive text with substantial coverage of conceptual and empirical topics.

Sauer, C.O. 1952. *Agricultural Origins and Dispersals*. New York: American Geographical Society.

A classic study by the most prominent twentieth-century cultural geographer.

Shephard, E., and T.J. Barnes, eds. 2000. *A Companion to Economic Geography*. Malden, Mass: Blackwell.

A comprehensive overview of economic geography; includes accounts of history and diverse approaches, along with a focus on production, resources, social issues, and circulation.

Smith, A.F. 2004. *The Oxford Encyclopedia of Food and Drink in America*, 2 vols. New York: Oxford University Press.

Comprehensive coverage of the topic with much detail on vernacular food items with a national or regional identity.

Winter, M. 2005. 'Geographies of Food: Agro-food Geographies—Food, Nature, Farmers and Agency', *Progress in Human Geography* 29: 609–17.

Informative progress report on recent work in the areas of food and nature and also on food and politics.

ON THE WEB

WORLD AGRICULTURE

Food and Agriculture Organization

🔍 www.fao.org/

This United Nations agency offers relevant facts and figures and topical discussions on all matters related to agricultural activity and food supply; also very useful for Chapters 3 and 5.

WTO

🔍 www.wto.org/english/tratop_e/agric_e/agric_e.htm

This WTO website provides details on agricultural agreements and related matters.

World Agricultural Forum

www.worldagforum.org/

The World Agricultural Forum initiates dialogue among those who can impact agriculture. It focuses on sustaining the lives and livelihood of the world's population by meeting the growing needs for food, fuel, and fibre.

AGRICULTURE IN THE LESS DEVELOPED WORLD

2008 World Development Report

web.worldbank.org/WBSITE/EXTERNAL/TOPICS/EXTARD/0,,menuPK:336688~pagePK:149018~piPK:149093~theSitePK:336682,00.html

This World Bank web page has links to various agriculture-related topics, including the World Development Report. Titled *Agriculture for Development*, this report provides a detailed discussion of agricultural investment and needs in the less developed world.

Overcoming Rural Poverty

www.ifad.org/

The International Fund for Agricultural Development, a specialized agency of the United Nations, was established as an international financial institution in 1977 following the food crises of the early 1970s that primarily affected the Sahelian countries of Africa. One important insight is that the causes of food insecurity and famine were not so much failures in food production, but structural problems relating to poverty and to the fact that the majority of the developing world's poor populations were concentrated in rural areas. Also useful for Chapter 5.

SUSTAINABLE AGRICULTURE

World Sustainable Agriculture Association

www.bcca.org/services/lists/noble-creation/wsaa.html

The goal of this association is to promote sustainable farming, food distribution, consumption, and biocycling systems throughout the world.

Solutions to Poverty

www.ifpri.org/

The mission of the International Food Policy Research Institute is to identify and analyze policies for meeting the food needs of the developing world in a sustainable manner. Also useful for Chapter 5.

Agricultural and Environmental Education

www.world-agriculture.com/

The intent of this website is to promote agricultural lifestyles that respect the natural world and that facilitate human relations.

Organic Farming

ec.europa.eu/agriculture/organic/home_en

European Commission website with much useful discussion of organic farming.

GENETICALLY MODIFIED CROPS

Greenpeace

www.greenpeace.org.uk/gm

This Greenpeace website offers a highly critical view of genetically modified crops, arguing that they pose a serious threat to biodiversity and to human health.

Monsanto

www.monsanto.com/biotech-gmo/asp/default.asp

This Monsanto website presents information about the benefits of genetically modified crops as described by farmers and researchers.

SETTLEMENT PATTERNS

This chapter focuses on patterns of rural and urban settlement, with particular emphasis on (1) the relationship between urban centres and the rural areas that surround them, and (2) the creation of networks, or systems, of urban centres. Among the many specific questions it addresses are the following:

- How do the urbanization experiences of more and less developed regions of the world differ?
- What are mega-cities, and where are they located?
- What differences can be found between dispersed and nucleated patterns of rural settlement?
- How do patterns of rural settlement evolve, and what theories of rural settlement location are available to help answer this question?
- What sorts of social and economic change do rural areas face?
- What is the concept of the 'rural idyll'?
- When and why did urban settlements first appear, and how did they change during the pre-industrial and industrial periods?
- What are some of the links between the capitalist mode of production and urbanization?
- Why are urban centres located where they are?
- How does Christaller's central place theory parallel von Thünen's theory?
- Are there networks of world cities?
- In what sense do world cities function as command centres in an increasingly globalizing world?

Dawson City, Yukon.
Gary Cook/Alamy

A number of the concepts and empirical analyses presented in this chapter are relatively new. In particular, the discussion of world cities represents a marked departure from the urban geography studied just a few decades ago. Whereas the latter emphasized national urban systems and hierarchies, today there is an additional focus on world systems and hierarchies as these are emerging in response to, and as one component of, the ongoing processes of economic, cultural, and political globalization. Of particular interest to geographers is the way new communication technologies are liberating people and businesses from the need to be located near one another in order to work together.

One way to highlight how much things have changed is to identify some of the new concepts being used to refer to relations between cities. Taylor and Lang (2004) list 50 such concepts, many of which include the words 'global', 'world', or 'international'. Most of these new concepts are concerned with the global reach of economic and other activities that are controlled from cities, or with relations between cities around the world. Of these concepts, two of the most useful are 'world city hierarchy' and 'world city network'.

Permanent settlements began to form with the introduction of agriculture. Today only a minority of human societies depend on hunting and gathering or agricultural activities that involve regular movement from place to place. Permanent settlements developed in association with specific economic activities, initially producing food for subsistence, later distributing surpluses, and eventually developing into trade and industrial centres. Principally agricultural settlements today are usually called rural; those that are principally non-agricultural are called urban.

This division between rural and urban reflects a long-standing geographic tradition. Rural settlement geography is often thought to have originated with the early twentieth-century studies by Vidal and other European geographers that focused on settlement patterns and their links to agriculture. Urban settlement geography is more recent in origin; the first English-language texts were published in the 1940s.

An Urbanizing World

Globally speaking, humans crossed the line from predominantly rural to predominantly urban in 2007. This passage is unlikely to be reversed, and in the future it may well be viewed as a significant turning point both in human settlement and in the relations between humans and land more generally. Although many of the major cities in the world have very long histories, for all of the time that humans have been on earth until 2007, the fundamental need either to find or to grow food meant that the majority of people lived in rural rather than urban settings. Villages arose with the beginning of farming, but even then most people still needed to be close to their crops and animals. When the first large permanent settlements began to develop about 3,000 years ago, it was because of improvements in farming technologies and the expansion of trade networks, which freed some people from the need to produce food.

For several thousands of years after the emergence of the first urban centres, the growth of urban populations was slow. It was not until the onset of the Industrial Revolution that the pace of **urbanization** really picked up as peasants left their rural homes to find work in the new factories. The rapid growth of cities is a recent phenomenon that has followed very different directions in the more and the less developed parts of the world. These different directions need to be understood in the context of recent significant increases in total world population and projected future increases.

The recent transition from rural to urban dominance has been very rapid. In 1800 urban dwellers were only 3 per cent of the world population. In 1900 only 14 per cent lived in a town or city, and there were just 16 cities in the world with more than one million people. But today one in every two people is an urban dweller, and the number of cities with populations of more than a million, already above 400, is increasing rapidly, with 500 expected by 2015.

DEFINING URBAN CENTRES

Before we consider recent and projected urban population trends, it is important to recognize two critical definitional issues. Why? Because some of the seemingly precise percentages cited in this discussion might be quite different if large-population countries such as China and India adjusted their definitions of 'rural' and 'urban'. For example, we note below that 53.6 per cent of the world population will be

urbanization
The spread and growth of cities.

urban by the year 2015. However, this level of precision is reliant on specific definitions and therefore in a sense is not fully justified. In some respects it might be fairer to say that the percentage is between 50 and 55.

There are two issues here. First, as Table 11.1 shows, different countries have different definitions of what constitutes an urban centre. If a country changes its definition of 'urban' in some way—for example, by raising or lowering the minimum number of residents required for a settlement to qualify for urban status—then the percentages of urban and rural dwellers in that country will change accordingly. And if the country is one with a very large population, the effect on the urban–rural proportions for the world as a whole may be significant. Consider the case of China. As Table 11.1 points out, the size of the urban population in China depends on how non-agricultural population is defined, and the decisions about that are made by local committees in towns and townships. Those decisions vary from one committee to another and are subject to change. It is also notable that urban data for China are sometimes unreliable, at least partly because growth is so rapid in some centres that the most recently available data may be out of date as soon as they are available (some aspects of Chinese urbanization are discussed more fully in Chapter 13).

Second, there is more than one way to define a specific urban centre, with the result that population counts for that centre can often vary widely. Consider the case of Toronto, which has three different ways of defining itself. In 2006 Metropolitan Toronto covered about 640 km² and had a population of 2.6 million; the Census Metropolitan Area of Toronto covered about 5,500 km² and had a population of 5.1 million; and the Greater Toronto Area covered more than 7,000 km² and had a population of 5.6 million. In the case of Tokyo, the urban population can range from about 8 million if only the central city is included to more than 40 million if the entire National Capital Region is included.

These two issues, the first involving the definition of 'urban' (for example, in terms of the cutoff number for either population density or total community population used to distinguish between rural and urban) and the second involving the way individual urban centres are demarcated, significantly affect the numbers and percentages contained in the following

Table 11.1	Some Definitions of Urban Centres
France	Communes containing an agglomeration of more than 2,000 inhabitants living in contiguous houses or with not more than 200 m between houses, and communes of which the major part of the population is part of a multi-communal agglomeration of this nature.
Portugal	Agglomerations of 10,000 or more inhabitants.
Norway	Localities of 200 or more inhabitants.
Israel	All settlements of more than 2,000 inhabitants, except those where at least one-third of the heads of households, participating in the civilian labour force, earn their living from agriculture.
Canada	(a) 1976: Incorporated cities, towns, and villages of 1,000 or more inhabitants and their urbanized fringes; unincorporated places of 1,000 or more inhabitants having a population density of at least 390 per km² and their urbanized fringes. (b) 1981: Places of 1,000 or more inhabitants having a population density of 400 or more per km².
United States	Places of 2,500 or more inhabitants and urbanized areas.
Botswana	Agglomerations of 5,000 or more inhabitants where 75% of the economic activity is of the non-agricultural type.
Ethiopia	Localities of 2,000 or more inhabitants.
Mexico	Localities of 2,500 or more inhabitants.
Argentina	Populated centres with 2,000 or more inhabitants.
Japan	Cities (SHI) having 50,000 or more inhabitants with 60% or more of the houses located in the main built-up areas and 60% or more of the population (including their dependants) engaged in manufacturing, trade, or other urban types of business. Alternatively, an SHI having urban facilities and conditions as defined by the prefectural order is considered urban.
China	Population living in areas under the jurisdiction of cities and towns. Cities are established with the approval of the central government. Towns are: (1) the location of county-level government agency; or (2) a township with a total population of less than 20,000, where the non-agricultural population of the location of a township government exceeds 2,000; or (3) a township with a total population of more than 20,000, where the proportion of the non-agricultural population to the total population of the location of a township government is greater than 10%; or (4) a remote area, mountainous area, small-sized mining area, small harbour, tourism area, or border area with a non-agricultural population of less than 2,000.
India	(1) All places with a municipality, corporation, cantonment board, or notified town area committee, etc. (2) A place satisfying the following three criteria simultaneously: (a) a minimum population of 5,000; (b) at least 75 per cent of male working population engaged in non-agricultural pursuits; and (c) a density of population of at least 400 per km².

NOTE: This table confirms that there is no simple and agreed-upon definition of 'urban centre' on the basis of population size, and emphasizes the danger of comparing data for different countries.

SOURCES: H. Carter, *The Study of Urban Geography*, 4th edn (London: Arnold, 1995), 10–12; <www.iiasa.ac.at/Research/LUC/ChinaFood/data/urban/urban_8.htm>; <http://72.14.207.104/search?q=cache:mDsAr85KWvEJ:www.censusindia.net/2001housing/metadata.pdf+>.

account, and it is therefore important to recognize that different definitions would produce quite different statistics. In fact, no two lists of the world's largest cities are likely to be identical. The largest city in the world may be Tokyo, and it may also be Mexico City—it all depends on how 'urban area' is defined. In this account most of the data used are derived from a single source, the United Nations Department of Economic and Social Affairs/Population Division.

URBANIZATION IN MORE AND LESS DEVELOPED REGIONS OF THE WORLD

In the more developed world, the Industrial Revolution beginning about 1750 provided the impetus for rapid growth in the number and size of urban centres, partly because of the increase in people moving from rural to urban areas and partly because of the rapid increase in total population during the second and third stages of the demographic transition. Growth involved expansion of the urban area and the creation of **suburbs** and, increasingly, a process of urban sprawl. By 2008, 74 per cent of the population of the more developed world was urban and only 26 per cent rural. The urban proportion is expected to increase to 82 per cent by 2030.

In the less developed world the urbanization process has been and continues to be very different. Rapid urban growth started later, in many cases as recently as the mid-twentieth century, and the process was even more dramatic, with urban centres acting as magnets for the rural poor during a period of exponential growth in the total populations in these regions. In fact, there is a close relationship between the rapid urbanization of the less developed world and rural poverty. In some cases, cities in the less developed world grew through rural-to-urban migration; in other cases, rural settlements were (and continue to be) transformed into urban centres. Both processes often involve the addition of slums at the margins of a settled area. By 2008, 44 per cent of the population of the less developed world was urban and 56 per cent rural. The urban proportion is expected to increase to 57 per cent by 2030. The social, environmental, and economic implications of this dramatic urban growth in many parts of the less developed world are discussed in Chapter 13.

Figure 11.1 shows the relative numbers of rural and urban dwellers in the world as a whole from 1950 projected through to 2030. The rapid convergence of the two lines between 1950 and 2007 is clear, as is the subsequent rapid divergence, with the urban percentage reaching 61 by 2030. Figure 11.2 looks at this trend in more detail, distinguishing between the more and less developed worlds through the same time period, showing that the greatest increases are occurring, and will continue to occur, in the years ahead, in the urban centres of the less developed world. As the figure shows, the number of urban dwellers in the less developed world is expected to exceed the number of rural dwellers by about 2017.

Table 11.2 provides data for the more developed and less developed world on the average annual rate of change for urban, rural, and total populations over two periods, 1950–2000

> **suburb**
> An outer commuting zone of an urban area; associated with social homogeneity and a lifestyle suited to family needs.

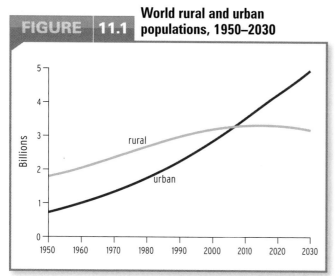

FIGURE 11.1 World rural and urban populations, 1950–2030

SOURCE: United Nations Department of Economic and Social Affairs, Population Division, *World Population Prospects: The 2003 Revision, Data Tables and Highlights* (New York: UN, 2004), 9. The United Nations is the author of the original material.

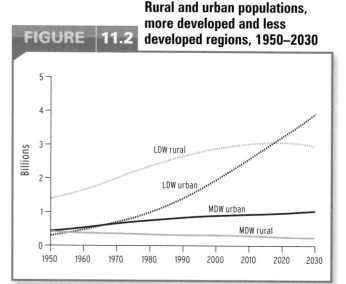

FIGURE 11.2 Rural and urban populations, more developed and less developed regions, 1950–2030

SOURCE: United Nations Department of Economic and Social Affairs, Population Division, *World Population Prospects: 2003 Revision, Data Tables and Highlights* (New York: UN, 2004), 11. The United Nations is the author of the original material.

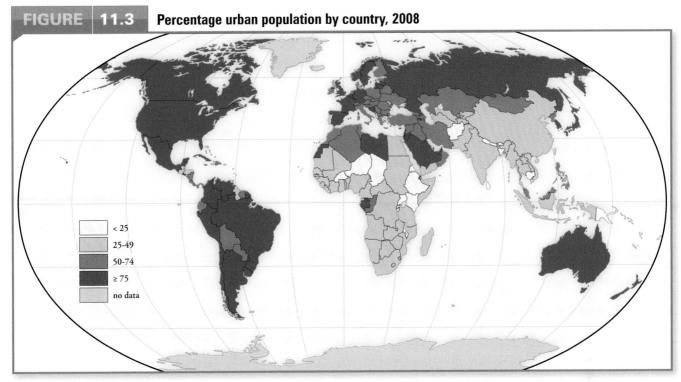

FIGURE 11.3 Percentage urban population by country, 2008

Legend:
< 25
25–49
50–74
≥ 75
no data

SOURCE: Population Reference Bureau, *2008 World Population Data Sheet* (Washington: Population Reference Bureau, 2008).

and 2000–30. As these data show, the rates of change are declining in all cases. Especially notable are:

- the relatively high projected growth rate for the urban population of the less developed world;
- the negative projected growth rate for the rural population of the more developed world; and
- the fact that the projected urban population growth rate is about twice the projected total population growth rate.

With reference to the last point above, the projected world urban growth rate of 1.83 per cent means that the urban population will double in about 38 years. In other words, most of the world population growth that will occur by 2030 will be concentrated in urban areas.

OTHER REGIONAL VARIATIONS

The differing figures for the more and less developed regions of the world show how misleading world averages can be. But there are also some notable regional differences within these two categories (Figure 11.3). In the less developed world, Latin America and the Caribbean were already highly urbanized in 2008, with about 77 per cent urban

population compared to 38 per cent in Africa and 42 per cent in Asia. All these percentages will increase by 2030, with Latin America and the Caribbean reaching 85 per cent, Africa 54 per cent, and Asia 55 per cent. In the more developed world, the European urban population is projected to increase between 2008 and 2030 from 71 to 80 per cent, while in North

Table 11.2	Population Growth Rates (Total, Urban, Rural), 1950–2000 and 2000–2030	
	Annual Rate of Change (%)	
	1950–2000	2000–2030
Total Population		
World	1.76	0.97
More Developed World	0.77	0.13
Less Developed World	2.10	1.15
Urban Population		
World	2.72	1.83
More Developed World	1.45	0.47
Less Developed World	3.73	2.29
Rural Population		
World	1.17	−0.03
More Developed World	−0.43	−1.05
Less Developed World	1.46	0.06

SOURCE: Adapted from United Nations Department of Economic and Social Affairs/Population Division, *World Population Prospects: The 2003 Revision, Data Tables and Highlights* (New York: UN, 2004), 4.

America the projected increase is from 79 to 87 per cent. Urban settlement in Canada is discussed in Box 11.1.

LOCATIONS OF CITIES

Most large cities are located either on coastlines or on navigable rivers with access to a sea. Only the Pacific coast of South America tends to lack large cities, because mountains close to the coast cut off ready access from and to the large agricultural hinterlands that might generate further growth for coastal cities. The only major region with many large inland cities is Europe, which includes several interior agricultural regions, is divided into many countries, and has several long rivers, notably the Danube and Rhine. Two large countries with many non-coastal cities are Canada and Russia; in both cases coastal regions suitable for settlement are limited but there are large interior areas with major river systems and, in the case of Canada, the Great Lakes.

One important reason why large cities tend to develop on a coast or a navigable river is that they represent natural breaks in transportation. As hubs where different modes of transport come together, they have obvious communications advantages, and through time these locations attract a wide range of economic activities.

MEGA-CITIES OR MANY CITIES?

The urban population is clearly increasing rapidly, both in absolute numbers and in relation to the rural population. Where is all this urban growth going? Is it concentrated in a handful of mega-cities (cities with populations of more than 10 million) or spread across many smaller cities? The answer is both. The populations of mega-cities are increasing, and new mega-cities are appearing, but most of the additional urban population growth is taking place in smaller cities, some with fewer than 500,000 people.

Table 11.5 provides further detail on projected urban growth, showing comparative data for 2003–15 for five categories of city size, and distinguishes between more and less developed regions; note that it is difficult to make detailed projections beyond 2015. With just the one

> **donut effect**
> A term that refers to a pronounced difference in the growth rates between a core city (slow growth or no growth) and its surrounding areas (faster growth).

Box 11.1   Changing Settlement in Canada

As in most of the world, urbanization is continuing in Canada. Changes in total, urban, and rural population between 1950 and 2000 and projected through to 2030 are shown in Table 11.3 and graphed in Figure 11.4. From 60.8 per cent in 1950, Canada's urban population for 2030 is projected to reach 87.2 per cent.

The urban centres increasing most in population are those located relatively close to the US border, those attracting immigrants both from elsewhere in Canada and from overseas, and those with economies based on manufacturing or services. Newfoundland and Labrador is the only region in Canada that is experiencing a declining urban population. A notable feature of Canadian urbanization is the continuing concentration of urban population in four areas:

- The Golden Horseshoe region extending from Oshawa through Toronto to St Catharines in southern Ontario (an area with 22 per cent of the Canadian population).
- Montreal and the adjacent region.
- The Lower Mainland of British Columbia.
- The Calgary–Edmonton corridor.

These four highly urbanized regions accounted for about 51 per cent of the Canadian population in 2001, up from 41 per cent in 1971, and this percentage continues to increase.

Another useful way to describe urbanization in Canada is with reference to Census Metropolitan Areas (CMAs). A CMA is a very large urban area (the urban core) together with the adjacent urban and rural areas (the urban and rural fringes) that have a high degree of social and economic integration with the core. To be defined as a CMA the urban core population must be at least 100,000 based on the previous census (Statistics Canada, 1999: 183).

Overall, in 2006 about 68 per cent of Canada's population lived in the 33 census metropolitan areas (Table 11.4 and Figure 11.5). In recent years, the CMA with the strongest growth rate is Calgary, while there is also substantial growth in Toronto, Vancouver, Edmonton, Ottawa–Gatineau, and Oshawa. In some CMAs, such as Regina, Greater Sudbury, Saguenay, and Saint John, population is relatively stable or declining slightly. A phenomenon known as the **donut effect**, where population in the central area is growing more slowly than in the surrounding area, is evident in some CMAs, most notably Regina and Saskatoon. This circumstance is discussed more generally in Chapter 13. In addition, the projected decrease in rural population is not simply a reflection of rural-to-urban migration but also indicates anticipated urban sprawl and rural growth, so that locations and households presently classified as rural will have been enveloped by urban growth and expansion and thus be reclassified.

Continued

FIGURE 11.4 — Rural and urban populations, Canada, 1950–2030

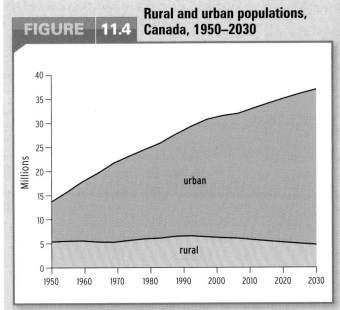

SOURCE: Produced using data from United Nations Department of Economic and Social Affairs, Population Division, *World Population Prospects: The 2002 Revision and World Urbanization Prospects: The 2003 Revision*. At: <esa.un.org/unup>.

Table 11.3 — Canada: Total, Urban, and Rural Population (thousands), 1950–2030

Year	Total Population	Urban Population	Percentage Urban	Rural Population	Percentage Rural
1950	13,737	8.356	60.8	5,381	39.2
2000	30,769	24,429	79.4	6,340	20.6
2015	34,133	28.667	84.0	5,467	16.0
2030	36,980	32,251	87.2	4,729	12.8

Source: Adapted from United Nations Department of Economic and Social Affairs, Population Division, *World Population Prospects: The 2002 Revision and World Urbanization Prospects: The 2003 Revision*. At: <esa.un.org/unup>.

FIGURE 11.5 — Canada: Census Metropolitan Areas, 2006

Table 11.4 — Canada: Census Metropolitan Area Populations (thousands), 2006

Toronto (Ont.)	5,113.1
Montreal (Que.)	3,635.6
Vancouver (BC)	2,116.6
Ottawa–Gatineau (Ont.–Que.)	1,130.8
Calgary (Alta)	1,079.3
Edmonton (Alta)	1,034.9
Quebec (Que.)	715.6
Winnipeg (Man.)	694.7
Hamilton (Ont.)	692.9
London (Ont.)	457.7
Kitchener (Ont.)	451.2
St Catharines–Niagara (Ont.)	390.3
Halifax (NS)	372.9
Oshawa (Ont.)	330.6
Victoria (BC)	330.1
Windsor (Ont.)	323.3
Saskatoon (Sask.)	233.9
Regina (Sask.)	195.0
Sherbrooke (Que.)	187.0
St John's (NL)	181.1
Barrie (Ont.)	177.1
Kelowna (BC)	162.3
Abbotsford (BC)	159.0
Greater Sudbury (Ont.)	158.3
Kingston (Ont.)	152.3
Saguenay (Que.)	151.6
Trois-Rivières (Que.)	141.5
Guelph (Ont.)	127.0
Moncton (NB)	126.4
Brantford (Ont.)	124.6
Thunder Bay (Ont.)	122.9
Saint John (NB)	122.4
Peterborough (Ont.)	116.6

SOURCE: Population and dwelling counts, for census metropolitan areas, 2006 and 2001 censuses - 100% data , http://www12.statcan.ca/english/census06/data/popdwell/Table.cfm?T=205&SR=1&S=3&O=D&RPP=33 , author department, year of work published. Reproduced with the permission of the Minister of Public Works and Government Services Canada, 2009.

Table 11.5	Urban Population by Size Class of Urban Settlement, 2003 and 2015	
	Percentage Distribution	
	2003	**2015**
World		
Urban area	48.3	53.6
10 million or more	4.5	5.0
5–10 million	2.8	3.7
1–5 million	11.0	12.7
500,000–1 million	5.0	5.0
Less than 500,000	25.0	27.2
Rural area	51.7	46.4
More Developed World		
Urban area	74.5	77.3
10 million or more	7.2	8.2
5–10 million	4.4	5.0
1–5 million	17.6	17.7
500,000–1 million	6.4	6.8
Less than 500,000	38.8	39.6
Rural area	25.5	22.7
Less Developed World		
Urban area	42.1	48.7
10 million or more	3.8	4.3
5–10 million	2.4	3.5
1–5 million	9.5	11.7
500,000–1 million	4.7	4.6
Less than 500,000	21.7	24.6
Rural area	57.9	51.3

SOURCE: Adapted from United Nations Department of Economic and Social Affairs, Population Division. *World Population Prospects: The 2003 Revision, Data Tables and Highlights* (New York: UN, 2004), 6.

exception (the 500,000–1 million category) percentage increases are projected across the range of city sizes. Most notable is the fact that roughly half of all urban dwellers live in cities of fewer than 500,000 people; this proportion is similar for more and less developed regions both in 2003 and in the 2015 projection.

The increasing number of mega-cities, and their increasing size, is highlighted by the following facts:

- In 1950 there were just two mega-cities: New York with 12.3 million people and Tokyo with 11.3 million.
- By 1975 there were four mega-cities: Tokyo with 26.6 million people, New York with 15.9, Shanghai with 11.4, and Mexico City with 10.7.
- By 2003 there were 20 mega-cities, and that number is expected to rise to 22 by 2015.

Table 11.6 provides data on the number and size of mega-cities in 2003 and projections for 2015, and Figure 11.6 maps the two patterns. Significantly, the vast majority of mega-cities are located in less developed countries. In 2003, only five of the 20 are in the more developed world (Tokyo, New York, Los Angeles, Osaka–Kobe, and Moscow). Of the other 15, nine are in Asia, four in Latin America, and two in Africa. The basic pattern is projected to change only in detail by 2015, with the two additional cities being Istanbul and Paris. Perhaps the most notable change between 2003 and 2015 is the substantial growth anticipated for the West African city of Lagos and the Asian cities of Mumbai (Bombay), Delhi, Jakarta, Dhaka, and Karachi.

Although less than 10 per cent of the world's urban population currently lives in cities of 10 million or more, and this percentage will not increase significantly by 2015, these mega-cities are of particular interest to geographers for two reasons. First, three of these places—Tokyo, New York, and London (sometimes a fourth place, Paris, is included)—play significant roles in controlling the global economy (this topic is addressed later in this chapter). Second, many of the other mega-cities are seen as representative of the global urban future because they are located in less developed countries and beset with problems of inadequate infrastructure, congestion, pollution, crime, and poverty (this topic is addressed in Chapter 13).

Rural Settlement

Before about 1950, when urban issues began to move to the fore, rural settlement issues were at the centre of human geography. In this section we will focus on three aspects of rural settlement: patterns of settlement; changes in the rural way of life; and the transition in land use from rural to urban.

RURAL SETTLEMENT PATTERNS

Patterns of rural settlement range widely between the extremes of dispersion (random or uniform) and nucleation (clustering; see Figure 2.4), depending on physical environment, culture, social organization, political influences, and economic activities. An important component of rural settlement landscapes is the pattern of fields, a landscape feature that often contributes greatly to the character of a place.

Mega-cities, 2003 and 2015

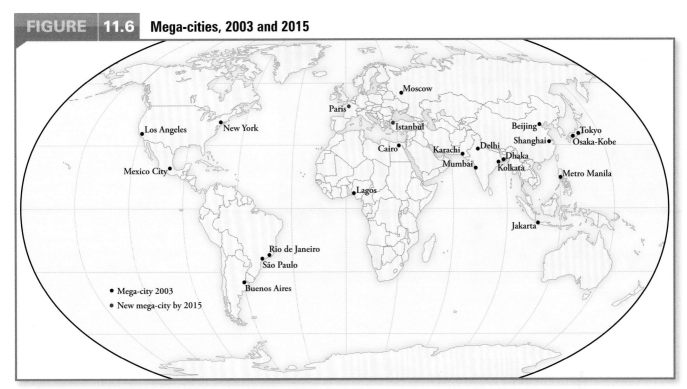

Mega-cities are urban agglomerations with more than 10 million people. Population size data for each of these cities are included in Table 11.6.

SOURCE: United Nations Department of Economic and Social Affairs, Population Division, *World Population Prospects: The 2003 Revision, Data Tables and Highlights* (New York: UN, 2004), 7. The United Nations is the author of the original material.

Table 11.6 | **Cities with More Than 10 Million People, 2003 and 2015**

	2003			2015	
Rank	City	Population (millions)	Rank	City	Population (millions)
1	Tokyo, Japan	35.0	1	Tokyo, Japan	36.2
2	Mexico City, Mexico	18.7	2	Mumbai (Bombay), India	22.6
3	New York, USA	18.3	3	Delhi, India	20.9
4	São Paulo, Brazil	17.9	4	Mexico City, Mexico	20.6
5	Mumbai (Bombay), India	17.4	5	São Paulo, Brazil	20.0
6	Delhi, India	14.1	6	New York, USA	19.7
7	Kolkata (Calcutta), India	13.8	7	Dhaka, Bangladesh	17.9
8	Buenos Aires, Argentina	13.0	8	Jakarta, Indonesia	17.5
9	Shanghai, China	12.8	9	Lagos, Nigeria	17.0
10	Jakarta, Indonesia	12.3	10	Kolkata (Calcutta), India	16.8
11	Los Angeles, USA	12.0	11	Karachi, Pakistan	16.2
12	Dhaka, Bangladesh	11.6	12	Buenos Aires, Argentina	14.6
13	Osaka–Kobe, Japan	11.2	13	Cairo, Egypt	13.1
14	Rio de Janeiro, Brazil	11.2	14	Los Angeles, USA	12.9
15	Karachi, Pakistan	11.1	15	Shanghai, China	12.7
16	Beijing, China	10.8	16	Manila, Philippines	12.6
17	Cairo, Egypt	10.8	17	Rio de Janeiro, Brazil	12.4
18	Moscow, Russia	10.5	18	Osaka–Kobe, Japan	11.4
19	Manila, Philippines	10.4	19	Istanbul, Turkey	11.3
20	Lagos, Nigeria	10.1	20	Beijing, China	11.1
			21	Moscow, Russia	10.9
			22	Paris, France	10.0

The long-lot settlement pattern of the St Lawrence lowlands was established in the French colonial period.

V. Last, Geographical Visual Aids

Dispersion

Humans are social animals, and throughout the world people have usually lived in close proximity to one another. Why, then, are many parts of Western Europe today characterized by dispersed rather than nucleated rural settlements? The answer lies mainly in the economic and social changes associated with the eighteenth-century transition from feudalism to capitalism. The development of this pattern reflected the new emphasis on individual ownership of land. In England, for example, between about 1750 and 1850 the rural landscape was dramatically transformed by the process of enclosure (see Chapter 10), which included land consolidation, new field boundaries, and the construction of dispersed dwellings. Government played a leading role in this transformation. Like many other aspects of the human landscape, field patterns can be seen as spatial expressions of power relations: the open pre-enclosure pattern reflected feudal social relations, while the enclosed pattern reflected individual ownership under a capitalist system.

Not surprisingly, the enclosure process gave rise to social discontent. In addition to losing their homes and livelihoods some people experienced a sense of loss of place and personal and community identity. These losses were sensitively expressed by the working-class poet John Clare (1813–64), who experienced the enclosure process at first hand while living in the village of Helpston in Northamptonshire, north of London. In 'Lament of Swordy Well', Clare wrote: 'The bees flye round in feeble rings/And find no blossom bye/Then thrum their almost weary wings/Upon the moss and die/Rabbits that find my hills turned oer/ Forsake my poor abode/They dread a workhouse like the poor/And nibble on the road.'

In North America, dispersion was established as the norm from the outset of European settlement because in most cases the government surveyed the land and divided it into discrete lots before building even began. As a result, geometric patterns soon became standard. Perhaps the most familiar example of a government survey comes from New France, where the authorities laid out long, narrow lots (rangs) at right angles to the St Lawrence that gave all settlers access to water. However, the nucleated settlement pattern that this arrangement produced was an exception.

More commonly, in English-speaking North America the government survey systems encouraged dispersed settlement; typically, each settler was granted a one-quarter section and required to construct a dwelling on it. This system was set in place in 1785, when the United States land survey laid out a baseline west to the Pacific from the point where the Pennsylvania border meets the Ohio River. For the government, the advantage of this type of survey was that it provided an orderly way to allocate land. A similar system was followed on the Canadian prairies, although in this area the landscape included some nucleated settlements as well (Box 11.2).

Nucleation

Historically, the rural landscape was dominated by nucleated settlements. At least part of the reason was the basic human need to communicate and co-operate with others; being with other people is important for security, social life, religious activities, and the regular exchange of goods and services. Other reasons for the development of nucleated settlements include a scarcity of good building land (for example, in areas that experience regular flooding), a need to defend the group against others, the need for group labour to construct and maintain a particular agricultural feature (such as terraces on steep slopes or irrigation systems), and political or religious imperatives (as in the case of the Israeli kibbutz or the Mormon village).

Box 11.2 | Rural Settlement in the Canadian Prairies

The rural landscape of the Canadian prairie region includes both dispersed and nucleated farm settlements. This pattern reflects both the policies of the central government and the wishes of some major ethnic groups.

The settlement that began in the 1870s was preceded by a series of numbered treaties that removed Native title to land and by a government survey that allowed for sectional settlement and dispersed farmsteads (Figure 11.7). Similar arrangements prevailed in many other areas of European overseas expansion, such as southern Ontario, much of the United States, and much of Australia. The Canadian government allocated some areas for settlement and others for grants to companies (especially railway companies) so they could generate income by selling to settlers. Dispersed settlement was the norm, whether the land was obtained

FIGURE 11.8 | Examples of nucleated rural settlement patterns

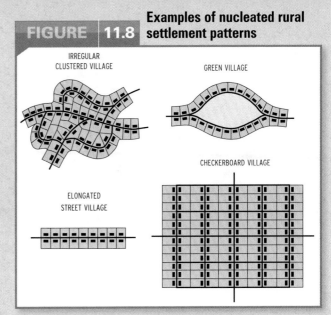

directly from the government or indirectly from a company. Most immigrants to the prairies accepted this arrangement—and the isolation that came with it—even though many had been accustomed to village life.

The principal exceptions to the dispersed settlement pattern were the farm villages established by specific ethnic groups. Doukhobors and Mennonites from Europe, Mormons from the western US, German Catholics, and Jews chose nucleation for religious and/or social reasons, to maintain a unified group. In these cases, the government rules requiring sectional dispersed settlement were waived on request. These nucleated settlements are an infrequent but distinctive feature of prairie rural settlement. Most of the farm villages were of the elongated-street variety (see Figure 11.8), although the Mormons favoured a square grid plan.

FIGURE 11.7 | Township survey in the Canadian prairies

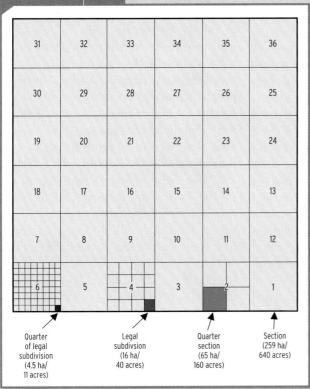

A typical township of 36 square miles (93 km^2) is divided into 36 sections of 1 square mile (640 acres/259 ha) each, and subdivided into four quarter sections of 160 acres (64 ha) each. Settlers were usually assigned a quarter section. Close to urban centres, quarter sections might be divided into four legal subdivisions of 40 acres (16 ha) each.

Even today, nucleation has been favoured over dispersion throughout much of the rural world. In China, the rural population is about 63 per cent of that country's total population. Most of those people lived in nucleated villages. Although many of these villages have populations of several thousand, they are best described as rural settlements since the inhabitants are farmers. The fields that surround these settlements are often very small.

Nucleated settlements are not uncommon in some parts of North America. The New

Nucleated long-lot settlement, west of Quebec City.

V. Last, Geographical Visual Aids

**counter-
urbanization**
A process of population
decentralization that
may be prompted
by several factors,
including the high
cost of living in
cities, improvements
in personal spatial
mobility, industrial
deconcentration, and
advances in information
technologies.

England landscape reflects the early settlers' desire to live close together, to reinforce group cohesion, and to provide some protection against attack. Similarly, the landscape of the Mormon cultural region in Utah and parts of adjacent states is characterized by nucleation as a consequence of the central planning practised by the Mormon Church (see Box 6.4).

The physical layout of nucleated rural settlements varies widely; see Figure 11.8 for examples. An irregularly clustered settlement suggests unplanned growth over time. Examples of regular settlements that may or may not result from some form of planning include elongated-street, green, and checkerboard settlements. Nucleated rural settlements range in size from a few dwellings (a hamlet) through to villages with populations of perhaps 25,000 in some parts of the world (as we have already noted, the distinction between rural and urban varies from country to country).

CHANGING PATTERNS OF RURAL SETTLEMENT

Rural depopulation

A significant trend in rural areas today is depopulation as a result of the spatial concentration of economic activities in urban areas and the move over the past half-century from family farming to factory farms. Strong evidence indicates a shift in the national economies of more developed countries from the agricultural sector to the manufacturing and service sectors (this topic is discussed in greater detail

in Chapter 14; see Table 14.5 for some relevant global data). Depopulation is especially prevalent in the vast central agricultural region of North America, and a compelling example is Saskatchewan (see Box 10.11). People are abandoning small towns and villages because the numbers working in agriculture are shrinking, and without the community support and purchasing power that farm families represent, there is no longer any reason for many small settlements to exist. This decline has been evident for some time: the number of small towns and villages in the province dropped from 317 in 1961 to 86 in 2002.

At the beginning of the twentieth century, settlers were moving to the prairies and nucleated centres were developing around the grain elevators established at 16-km (10-mile) intervals along railway lines. By the mid-1930s, however, the settlement process had effectively ended. The attractions of rural life gradually diminished as agricultural technology changed after World War II, becoming capital-intensive and requiring less labour, and as the recession of the 1980s reduced farm incomes. Today, small settlements in Saskatchewan are becoming less viable, both economically and socially. With declining farm populations, towns lose people, schools close, and businesses leave. A short distance northeast of Saskatoon, the former community of Smuts is a ghost town. The final blow for the 40 residents of Mendham (northwest of Swift Current), who had already lost their fight to keep schools open, came in 2001 when the Roman Catholic church that had served the community for 87 years was closed. Allan, a little south and east of Saskatoon, was founded by German families in 1903 and grew impressively at first, but it too has faltered; a school and a grain elevator still function, but continuing decline seems inevitable. These changes may well be normal consequences of larger global economic processes and national policies, but we may still mourn the losses that the disappearance of these towns represents.

Not all of Saskatchewan's small towns are losing population. Those that are thriving, however, usually have some economic base other than agriculture: Eastend, in the southeast of the province, has benefited from the discovery of dinosaur remains; Leader, near Mendham, is developing ecotourist facilities; Rouleau, south of Regina, is developing a small tourist industry as the setting for the popular television show, *Corner Gas*; and several towns remain viable because they are close to potash mines.

More generally, for Canada as a whole, a detailed analysis by Keddie and Joseph (1991) suggests that the period 1981–6 was one of sustained urban growth comparable to that of the 1960s. During this five-year period, the rate of urban growth in Canada was 5 per cent, compared to 0.8 per cent for rural growth. Table 11.7 summarizes the percentage change in regional rural population for the period 1981–6. Note that the figures vary substantially between regions: whereas the prairies experienced a rural population loss of 3.1 per cent, the Atlantic region's rural population grew by 3.5 per cent—a good example of the dangers of using data for large areas.

Counter-urbanization?

Urban growth is proceeding apace throughout the world. Or is it? At least until about 1970, the data were clear: people moved from less to more densely populated areas. Yet in the 1970s some parts of the more developed world saw a decline in this movement—and possibly even a countermovement. In some areas there was evidence of rural repopulation, a trend that gave rise to the term **counter-urbanization** (Hall et al., 1973). According to some researchers, this trend is the result of industrial changes in a less spatially concentrated economy. Other possible explanations include an increasing appreciation for rural lifestyles, especially among the growing numbers of seniors, and the rapid proliferation of new communication systems that reduce the need to be spatially close to services.

Significantly, the Canadian study that produced the data in Table 11.7 was prompted by evidence of some rural population growth in the 1970s, but no such trend was evident for the first half of the 1980s. Any rural repopulation in Canada during the 1970s was a 'fleeting and regionally specific phenomenon' (Keddie and Joseph, 1991: 379). Certainly, the period from 1981 through to 2001 showed significant rural population decline. Table 11.8, which provides data by province for the period from 2001 to 2006, suggests that this decline in rural population generally, and rural farm population specifically, is lessening somewhat. Much of the rural population change evident in Table 11.8 is related to small-town growth close to urban centres and many of the new rural residents

Rouleau, Saskatchewan, found a new lease on life playing 'Dog River', the fictional town that served as the setting for the comedy series *Corner Gas*. Prime Minister Stephen Harper (far right) visited the set in August 2006.
CP photo/Troy Fleece

Table 11.7	Percentage Change in Rural Population by Region, Canada, 1981–1986
Region	**Aggregate % Change**
Atlantic	3.5
Quebec	–0.1
Ontario	3.4
Prairies	–3.1
British Columbia	–1.1
Canada	0.8

SOURCE: Adapted from P.D. Keddie and A.E. Joseph, 'The Turnaround of the Turnaround? Rural Population Change in Canada, 1976 to 1986', *Canadian Geographer* 35 (1991): 371.

Table 11.8	Rural Population Change in Canada, 2001–2006	
Province	**Rural Population**	**Rural Farm Population**
Newfoundland and Labrador	–1.6	–8.6
Prince Edward Island	0.1	–14.0
Nova Scotia	1.4	–2.4
New Brunswick	–1.3	–9.2
Quebec	5.3	–7.8
Ontario	3.5	–5.4
Manitoba	4.2	–9.6
Saskatchewan	–3.0	–12.1
Alberta	3.7	–8.7
British Columbia	0.7	–2.7
Canada	2.7	–7.9

SOURCE: Statistics Canada, Farm population and total population by rural and urban population, by province, (2001 and 2006 Census of Agriculture and Census of Population), http://www40.statcan.gc.ca/l01/cst01/agrc42a-eng.htm. Reproduced with the permission of the Minister of Public Works and Government Services Canada, 2009.

Urban sprawl encroaches on a farm property south of Calgary.

CP photo/Larry MacDougal

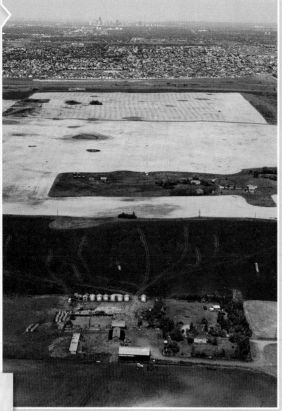

exurbanization
The movement of households from urban areas to locations outside the urban area but within the commuting field.

population decline was most evident in areas at greater distances from urban centres.

In Europe, which never had extensive areas of agricultural land, creeping urbanization is leading inexorably to the elimination of what might be thought of as the traditional countryside. With urban spread, the related development of transportation networks and nodes, and a declining agricultural sector, it increasingly seems that the urban never really ends and the rural never really begins. A report by the Campaign to Protect Rural England (2005) suggests that the English countryside will have all but disappeared by 2035.

The rural–urban fringe

Transition zones between rural and urban areas often have multiple land uses and varied social and demographic characteristics. Indeed, the absence of clear boundaries between rural and urban areas is a distinctive feature of contemporary land use and lifestyle. One way to look at this situation is to focus on the centrifugal forces that encourage urban land uses to expand into the fringe belt—a process that has come to be known as deconcentration, in contrast to the concentration more traditionally associated with urban development. The term **urban sprawl** is often used to describe the deconcentration that involves low-density expansion of urban land uses into surrounding

work in these urban centres. Examples of such growth in the 2001–6 period include Sylvan Lake near Red Deer (36.1 per cent growth) and Strathmore near Calgary (34.2 per cent growth). Not surprisingly, rural and small-town

Box 11.3   Exurbanization in Southwestern Ontario

In many areas, deconcentration is taking the form of residential movement to the rural–urban fringe. Such movement is not simply a matter of the extension of existing suburbs as it does not involve a steady outward expansion of urban areas but rather scattered and often relatively isolated pockets of settlement. Does it perhaps represent a form of counter-urbanization, or rural repopulation? The fragmented landscapes of exurbanization are most evident in North America, especially in the western United States where many of these new places are experiencing dramatic growth. This case study looks at southwestern Ontario.

For Woodstock in southwestern Ontario, Davies and Yeates (1991) show that both perspectives are relevant. Focusing on **exurbanization**—the movement of households from urban areas to locations outside the urban area but within the commuting field—they distinguish between exurbanite households located in dispersed rural settlements

(54 per cent of the total in the area around Woodstock) and exurbanite households located in villages (46 per cent of the total in the area around Woodstock). As shown in Table 11.9, 'the exurban-rural group is part of the new urban-to-rural migration pattern subsumed within the idea of counter-urbanization, while the exurban village group is part of the long-standing trend of suburbanization related to a search for more housing space and privacy at lower prices' (ibid., 186). This conclusion is based on the relatively high valuation that the exurban-village group places on lower taxes and lower house prices and on the relatively high valuation that the exurban-rural group places on the amenity attractions of rural areas. Both groups place a high value on privacy, larger houses, more land, the attractive rural landscape, a better quality of life, and less crime.

Another way to think about these trends is in terms of the push and pull factors introduced in the discussion of migration in Chapter 5.

rural areas (Box 11.3). (Urban sprawl is discussed more fully in Chapter 12 in the larger context of current changes to urban settlement patterns.)

In much of the world, the transition from rural to urban land uses can be a cause for concern. One particularly sensitive example is the Niagara Peninsula region in southern Ontario (Figure 11.9). This region is one of only two important soft-fruit-producing areas in Canada, but it is close to large urban areas and contains a number of urban centres. Although concerns about urban encroachment were first raised in the 1950s, few efforts were made to limit urban growth until the 1970s because it was generally considered an unmixed good. As a result of community pressure and subsequent government action, rulings in the 1980s identified substantial areas that could be used only for agriculture or related purposes. These rulings represented a significant reversal of attitude.

CHANGING RURALITY

A rural way of life

Numerous studies in diverse parts of the world have established that there are important differences between rural and urban ways of life. In 1887, for example, the sociologist Ferdinand Tönnies (see Cater and Jones, 1989: 170–1) distinguished between *Gemeinschaft* (community), human association based on close personal contact, as in village settings, and *Gesellschaft* (mass society), human association that is depersonalized, as in urban settings. It is true that, historically, rural settlements have been characterized by spatial isolation and limited spatial mobility—circumstances that may well tend to bring their inhabitants into closer social contact with one another. However, the notion that members of rural communities therefore enjoy close informal social relations may be misleading in some areas. Throughout much of Europe the typical social framework in a rural setting was (and in many areas still is) a rigid social hierarchy.

The seminal work on differences between rural and urban areas was done by C. Wirth (1938), who found that increases in the size and density of populations result in increased anonymity, further division of labour, social heterogeneity, and enhanced mobility. Wirth thus argued that urban areas were distinguished by a distinctive way of life. The lifestyle differences between rural and urban are generally

Table 11.9	Factors Influential in the Decision to Move to Exurban Locations around Woodstock, Ontario	
Factor	Exurban–Rural % Important	Exurban–Village % Important
Better schools	34.4	42.4
Lower taxes	33.3	53.8
Less commuting	26.2	24.7
Friends and relatives	25.1	31.0
More privacy	95.1	82.3
Lower house prices	43.7	65.2
Larger house	63.9	58.9
Better services	15.3	19.0
More land	80.9	75.9
Proximity to urban area	48.6	51.3
Attractive landscape	79.2	72.8
Better quality of life	85.8	86.0
Less crime	56.8	72.2

SOURCE: S. Davies and M.H. Yeates, 'Exurbanization as a Component of Migration: A Case Study in Oxford County, Ontario', *Canadian Geographer* 35 (1991): 177–86. Published by Blackwell Publishing Ltd.

well documented, although R.E. Pahl (1966) has suggested that, in some cases, the social differences between rural and urban areas are unclear; this lack of clarity is increasingly evident because of the twin processes of rural depopulation and creeping urbanization.

These two continuing processes have substantially modified many rural areas. Rural–urban differences, especially in the more developed world, are decreasing. Improved rural services are the norm, and the proximity of urban employment opportunities for many rural dwellers reduces the distinctiveness of the rural way of life. Even in those rural areas far from urban centres, access to similar mass media ensures a certain uniformity in preferred lifestyle. To put it bluntly, many rural communities, particularly those on the urban fringe that lack any pretense of homogeneity, are no longer the supportive, self-sufficient, and stable places we imagine they were in earlier times. But economic considerations have also played an important role: some areas that have experienced depopulation, such as the Canadian prairies, have lost basic services as their market bases have declined, and any repopulation that may have occurred has not reversed this trend. The prevailing ethic in the more developed world is that of the market economy, and this is not conducive to the establishment of services in rural areas with low population densities.

urban sprawl
The largely unplanned expansion of an urban area; typically discontinuous, leaving rural enclaves.

Gemeinschaft
A term introduced by Tönnies; a form of human association based on loyalty, informality, and personal contact; assumed to be characteristic of traditional village communities.

Gesellschaft
A term introduced by Tönnies; a form of human association based on rationality and depersonalization; assumed to be characteristic of urban dwellers.

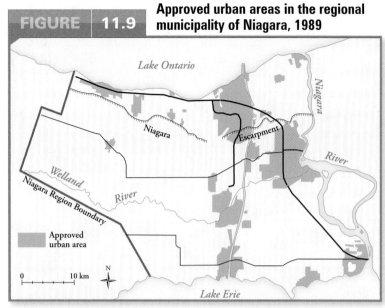

FIGURE 11.9 Approved urban areas in the regional municipality of Niagara, 1989

The Niagara region (population 370,000) is at the western end of an urban region, commonly called the Golden Horseshoe, containing 5 million people. North of the Niagara Escarpment, a fertile plain—the Niagara fruit belt—edges Lake Ontario.

SOURCE: Adapted from H.J. Gayler, 'Changing Aspects of Urban Containment in Canada: The Niagara Case in the 1980s and Beyond', *Urban Geography* 11 (1990): 377.

Images of rural life

The word 'rural' often conjures up images of idyllic life, in contrast to what is widely seen as the 'urban nightmare'. As Table 11.10 suggests,

Table 11.10	Rural and Urban Myths
Rural Idyll	**Urban Nightmare**
Nostalgic/part of national identity	Lacking identity
Traditional	Modern
Problem-free	Crime, poverty, homelessness
Close-knit/friendly	Anonymous/lonely
Better environment	Urban decay
Place of play	Place of fear
Simpler/more natural	Polluted/congested/dirty
Healthy	Unhealthy
Rural Anti-idyll	**Urban Dream**
Backward	Progressive
Unsophisticated	International/cosmopolitan
Unfriendly	Diverse/freedom to express yourself
Environmentally damaged	Architectural achievement
Dull/boring	Exciting/recreational
Poorly provided with services	Shopping/administrative centres
Sleepy	24-hour city

SOURCE: Adapted from R. Yarwood, 'Beyond the Rural Idyll: Images, Countryside Change and Geography', *Geography* 90 (2005): 24.

however, it is possible to see rural life as the antithesis of idyllic and urban life as the ideal. It is pointless to ask which image is fact and which is fiction. The 'important question is not "is it true?" but "whose truth is it?"' (Short, 1991: xvi; see also Yarwood, 2005).

Of course, not all the differences noted in the table are based solely on subjective impressions. For example, a 2005 Scottish study found some marked differences in health between rural and urban dwellers (Iversen et al., 2005). Most notably, rural dwellers are less likely to have respiratory problems such as wheezing and asthma or skin conditions related to allergies.

Some research into rural societies in the more developed world is being conducted under the general heading of the 'cultural turn' outlined in Chapter 7. This research suggests that in most of the more developed countries 'rural society has been subject to a process of cultural colonization in that the dominant images of rural life have been formulated by (middle class) urbanites and projected on to the countryside' (Cater and Jones, 1989: 194). Applying this logic to rural England, it might be argued that the 'idyllic' image detailed in Table 11.10 was constructed by white, middle-class males as an integral part of the larger English identity. Interestingly, in recent debates about fox hunting—a pastime closely associated with 'English-ness'—most rural dwellers and people who associate rurality with an 'English' way of life remained sympathetic to the hunt despite widespread public criticism.

More generally, the traditional image of rurality excludes at least some 'others', such as people of colour, who do not conform to the imagined norm. The basic logic of this argument is that, like all social constructions, rurality necessarily has exclusionary qualities. Some human geographers are now studying the experiences of 'others' in rural areas. One conclusion they have reached is that—contrary to nostalgic myths—the usual pattern is one of urban dominance and rural dependence.

Rural gentrification

An additional complication when discussing rural areas is that the people who have moved to rural areas in recent years have generally not been seeking one single 'rural lifestyle'. Some are commuters, some are retirees, some are those who could not afford to live in cities, and some are people seeking alternatives to the perceived unpleasantness of urban life. These newcomers are from a wide range of social backgrounds and economic circumstances, and they have a wide range of attitudes. Unlike earlier settlers, they do not form a single cohesive group with shared aspirations and goals for their rural way of life.

In both Europe and North America, there is evidence of rural **gentrification**: changes in rural landscapes resulting from settlement by relatively well-off people who are choosing rural over urban areas for reasons related to lifestyle preferences, and who in the process are changing the meaning of 'rural'. All too often, newcomers find themselves in conflict with local people, especially when questions arise about what rural communities are and need to become. The new rural settlement involves growth and change, developments that locals often oppose. In Europe, would-be gentrifiers are typically seeking the features outlined in Table 11.10 under the 'rural idyll' heading, whereas in North America the rural ideal involves a beautiful physical landscape, recreational opportunities, proximity to wilderness, and reasonable living costs. Some locations may also be associated with more specific attractions, such as a 'western' lifestyle; realtors in Missoula, Montana, for instance, are selling not only homes, but also dreams and getaways, with 'country style comfort . . . room for horses, . . . unique log home[s], with private natural setting[s] . . . rural yet minutes from the city' (Missoula County Association of Realtors, quoted in Ghose, 2004: 537). It is clear that, along with land developers and financial organizations, they are playing an active role in promoting rural gentrification.

As this discussion of gentrification, and indeed the larger account of rural areas, suggests, considerable change is taking place in some rural areas. Human geographers are beginning to investigate these changes in the context of globalization processes and even are talking about the 'global countryside'. What this term highlights is that it is not appropriate to conceive of rural areas as in some way separate from the larger social and economic processes that are transforming other geographic landscapes, a circumstance that was evident in the Chapter 10 account of agriculture.

Situated between Lakes Erie and Ontario, the Niagara region has a gentle climate that is well suited to grapes.
CP photo/Steve White

The Origins and Growth of Cities

Since about 1750, the world has experienced the rise of capitalism, a rapidly increasing population, a proliferation of states, many new technologies, and growth in both the numbers and the size of urban centres. By 1850, the major world cities were concentrated in the newly industrializing countries, a pattern that continued well into the twentieth century. Most of these major cities were located either in Europe or on the eastern seaboard of the United States.

In recent decades, however, the majority of the world's cities (particularly the largest ones) have been located in former colonies in the less developed world. Before 1950, cities in the less developed world were typically transportation centres or colonial government centres that formally excluded most indigenous people. With independence, cities in the less developed countries became population magnets, offering employment and wealth to some but only minimal benefits to the majority.

URBAN ORIGINS

The city–civilization link is an ancient one; the two words share the Latin root *civitas*. Civilizations create cities, but cities mould

gentrification
A process of inner-city urban neighbourhood social change resulting from the in-movement of higher-income groups.

civilizations. James Vance (1990: 4) quotes Winston Churchill: 'We shape our houses but then they shape us.' The earliest cities probably date from about 3500 BCE and developed out of large agricultural villages. It is no accident that the emergence of cities coincided with major cultural advances, such as the invention of writing. From the beginning, cities have been 'the chief repositories of social tradition, the points of contact between cultures, and the fountainheads of inspiration' (Smailes, 1957: 8).

The origins of urbanization and **urbanism** —the urban way of life—are debatable. It seems likely that most cities originated in one of four ways. First, we know that cities were initially established in agricultural regions, and that city life did not become possible until the progress of agriculture freed some group members from the need to produce food. Thus in some areas the first cities probably reflected the production of an *agricultural surplus*, possibly as a result of irrigation schemes. The earliest cities of this type were located in the Tigris–Euphrates region (present-day Iraq), the Nile Valley (Egypt), the Indus Valley (Pakistan), the Huang Valley (China), Mexico, and Peru, and were associated with the rise of agriculture in these regions (see Figure 6.3). These cities were the residential areas for those not directly involved in agriculture. They were small in both population and area. Populations generally ranged from 2,000 to 20,000, although Ur on the Euphrates and Thebes on the Nile may have numbered as many as 200,000. In 2001 a significant discovery was made on the coast of Peru. The city of Caral, located about 200 km north of Lima, has been dated as early as 2627 BCE—about 1,000 years before any other city in the Americas. The inhabitants of Caral appear to have supported themselves by practising irrigation, growing squash, beans, and cotton, but (interestingly) not corn.

A second group of cities were probably established as *marketplaces* for the exchange of local products. Some early cities were located on navigable waterways, others on long-distance trading routes: Venice was a prime example of a port city, while Baghdad was a port in the desert.

Third, some cities may have started as *military, defence,* or *administrative centres.* The Greeks built cities in their colonial areas as centres from which to exercise control; in many cases, they planned the city in advance as a grid pattern with a central space and a series of routes intersecting at 90° (this early example of planning is discussed in a larger context in Chapter 12). The Romans, who knew that urbanization was the key to controlling conquered areas, followed the pattern established by the Greeks as their own empire expanded. Many North African, Middle Eastern, and European cities trace their origins to either Greek or Roman times.

Fourth, some cities arose as *ceremonial centres* for religious activity. Chinese urban locations were selected using geomancy, an art of divination through signs derived from the earth; once an auspicious site was located, the city was laid out in geometric fashion. Chinese cities were square in shape, reflecting two fundamental beliefs: that the earth was square and that humans should be part of nature rather than dominate it (Wheatley, 1971).

PRE-INDUSTRIAL CITIES

The term 'pre-industrial city' may be misleading to the extent that it suggests that all cities were the same before the Industrial Revolution; in fact, there were notable differences between cities in different parts of the world, reflecting a wealth of cultural differences. It is a useful term, however, because it emphasizes the significance of the changes that accompanied industrialization. Before the advent of capitalism and the Industrial Revolution, most cities were concerned with marketing, commercial activities, and craft industries, and also served as religious and administrative centres. As new features in the human landscape, cities developed a new way of life—a new social system and a more diverse division of labour (Wirth, 1938). Political, cultural, and social changes followed the economic change from pre-agricultural to agricultural societies. Rural and urban settlements were complementary, the former producing a food surplus and the latter performing a series of new functions. From the beginning, then, cities have been functionally different from surrounding rural areas.

Inside the pre-industrial city

In modern industrial cities, different areas serve different functions. By contrast, homes, workshops, markets, and other functions in the pre-industrial city were located in a relatively haphazard fashion. The principal evidence of any planning inside the city was a basic division between elite and other areas. According to Sjoberg (1960), in all pre-industrial cities,

---

urbanism

The urban way of life; associated with a declining sense of community and increasingly complex social and economic organization as a result of increasing size, density, and heterogeneity.

regardless of time and place, the elite occupied the central core, which was also the economic, cultural, and political focal point. Vance (1971) argued that in Europe the **guild** system was more crucial to urban differentiation than the feudal system; thus there was land-use zoning in the sense that workers in a particular craft were spatially concentrated.

Pre-industrial urban growth

In Europe the process of urbanization faltered after the collapse of the Roman Empire in the fifth century as population movement and trade declined, and it did not resume for some 600 years. It began to revive in the eleventh century as commerce picked up, political units gained increasing power, population increased, and agricultural technology advanced. Old cities were revived and new ones were established throughout Europe, especially in areas well located for trade, on the Baltic, North Sea, and Mediterranean coasts. Urbanization was encouraged in the seventeenth century by the development of **mercantilism**: an economic philosophy arguing that governments should play an active role in economic activities in order to help states attain their maximum economic potential. Thus mercantilism encouraged colonial expansion and city growth.

Meanwhile, throughout the centuries of Europe's decline, stagnation, and rebirth, cities in China and the Islamic world flourished. It was only with the beginning of European colonial activity that traditional cities outside Europe began to experience substantial change.

Pre-industrial cities, then, followed the development of agriculture and soon took on distinctive functions and characters. The typical city functioned both as support for an agricultural population and as a trading centre—a function greatly enhanced by the development of mercantilism. All this would change dramatically with the advent of industrialization.

INDUSTRIAL CITIES

Beginning in England about 1750, the Industrial Revolution brought a series of changes that altered human landscapes. The principal economic change was related to the emergence of a new merchant and entrepreneurial middle class who opposed the state's mercantilist-inspired interference and the limits it put on their own profits. With the rise of the merchant class, mercantilism faded and the

Tourists visit the ruins of the ancient city of Caral in Peru.
AP photo/Karel Navarro

final remnants of the feudal society and economy disappeared. The new economic model of capitalism emphasized growth, profits, and a market economy. Capitalism, combined with a series of technological advances, particularly in the area of new energy sources, led to the Industrial Revolution and the emergence of the industrial city.

Capitalism and urban growth

Much recent city growth is related to the processes of economic globalization introduced in Chapter 9. Each of the three phases of capitalism outlined in Chapter 7 created new urban industrial geographies because each involved changes in what was produced, how, and where:

1. The first phase, competitive capitalism, may have begun as early as the late sixteenth century and extended to the middle of the twentieth. It was characterized by free-market competition between essentially local business activities, with little government regulation or interference. As this phase evolved, the scale of business activity increased, markets became more regional, and labour markets became more organized. From the eighteenth century onward, competitive capitalism contributed to the growth of towns as the centres of industrial and related economic activity, and by the early twentieth century, it became closely associated with Fordism. Together, mass

guild
An association of people who share the same trade or job skill. Historically, guilds were formed to protect shared interests and to ensure some uniformity of practice.

mercantilism
A school of economic thought dominant in Europe in the seventeenth and early eighteenth centuries that argued for the involvement of the state in economic life so as to increase national wealth and power.

production and mass consumption further contributed to the process of ongoing urban growth.

2. The second phase, organized capitalism, is usually considered to have begun after World War II and was associated with the development of larger companies, many of which eventually came to function as transnationals. Notable features of the latter part of this phase are the economic globalization processes outlined in Chapter 9, the rise of service activities outlined in Chapter 14, and an increasing emphasis on consumption rather than production.

3. The third phase is variously known as advanced, late, or disorganized capitalism. With the success of organized capitalism and the emergence of new consumer preferences, producers found it necessary to seek out smaller niche markets for more unusual products, which then required special retail outlets and the flexible production systems associated with post-Fordism. More generally, the continuing emphasis on consumption means continuing change for urban areas as new shopping areas and specialty retail outlets are established.

Industrial urban growth

As noted earlier, urbanization has been one of the key phenomena of the industrial age, involving the spatial movement of large numbers of people and major changes in social life. Industrial cities grew with remarkable speed in key resource locations, especially on coalfields. One of the first industrial areas in Britain was Coalbrookdale, northwest of Birmingham, where Abraham Darby first smelted iron ore with coke instead of charcoal in 1709. After 1760, the steam engine became available and old industrial activities, especially textile and metal production, were transformed. The importance of coal for power and transportation resulted in major new concentrations of activity, machinery, capital, and labour in the new industrial cities.

Urban growth proceeded apace in the new industrial areas—in Britain from about 1750 onward and then in Western Europe, the United States, parts of Southern Europe, Russia, and Japan by the end of the nineteenth century. The rapidity of this growth is illustrated by the United Kingdom, where the urban population increased from 24 per cent in 1800 to 89 per cent by 2005. Since World War II, many parts of the less developed world have also experienced industrialization.

ECONOMIC BASE THEORY

The changes described earlier have both economic and social causes. Generally, the economic causes are considered to be the financial benefits that accrue from city growth—reductions in the costs of assembly, production, and distribution. According to this view, cities are the most economically rational settlements. **Economic base theory** reduces urban economies to two interdependent sectors, basic and non-basic. The basic sector comprises all those activities that produce goods and services for sale outside the city; the non-basic sector, all those activities that produce goods and services for sale inside the city. Thus basic activities generate income for residents of the city, while non-basic activities circulate income within the city. This distinction leads to two important conclusions: (1) the larger the city, the less dependent it is on the basic sector, and (2) the larger the city, the more it is able to grow. As we will note in Chapter 13, this simple distinction is less useful today because of the rise of many new advanced service activities, such as computing, legal, and financial services, that serve both city and global economies and are therefore both basic and non-basic.

Explanations for city growth based on social causes revolve around the fact that groups offer security and that cities develop once the necessary social structures are in place. It seems reasonable to conclude that both economic and social causes play a role.

Locating Urban Centres

Why are urban centres located where they are? Various answers to this question have already been noted. In most cases location reflects function: provision of services to rural populations, military strategy, or trade. Occasionally, the location of a set of centres can be explained by reference to government planning: for example, it has been suggested that the settlement pattern on the Nile, c. 1317–1070 BCE, was determined by the state in order to maximize control of populations (Church and Bell, 1988). But such situations are not common. The most fully articulated arguments are (1) a group of related historical generalizations and

---

economic base theory
A theory that tries to explain the growth or decline of particular regions or cities in terms of 'basic' and 'non-basic' economic activities: 'basic' goods and services are those produced for sale outside the city or region.

(2) the influential central place theory developed by Christaller. The historical models refer essentially to more developed regions and especially to the United States. Central place theory is, in principle, relevant worldwide.

HISTORICAL EXPLANATIONS
The mercantile model

According to Vance's mercantile model (1970), the initial growth of an urban centre is the product of external factors such as long-distance trade. The case of North America illustrates that model:

Stage I   Initial search of the North American coasts to gather information relevant to European interests.

Stage II   Initial testing of the quality of resources discovered and, in some cases, use of specific resources as staple products.

Stage III   Settlement at key points, both coastal and inland, to facilitate export of products.

Stage IV   Expansion of the economic region by establishing urban locations, usually at key transport sites on waterways.

Stage V   Emergence of the key trading locations as cities.

Close links between the mercantile model and what is known as the staple model of economic growth indicate that the former is effectively an offshoot of the latter. The staple model is often used by economic historians, especially for areas such as Canada and Australia, to describe situations where the exploitation of one or more key resources lays the groundwork for further economic growth. Canada's staples, in approximately chronological sequence, have been fish, fur, timber, wheat, and minerals. For Australia the sequence was wool, gold, and wheat.

Reactions to Vance (1970) have been mixed. Sargent's (1975) study of Arizona towns confirmed the model, Earle (1977) found it inadequate to explain the pattern of early colonial towns in North America, and Meyer (1980) also considered it inadequate.

Stages of pre-modern development

Focusing more on the European than the North American experience, Rozman's model (1978) identified seven stages of pre-modern development (Table 11.11):

1. A pre-urban stage involved only unspecialized settlements.
2. Cities locate separately from one another and have only weak links to rural areas.
3. An administrative hierarchy develops.
4. High levels of centralization based on administration prevail.
5. Commercial centralization emerges, as do some periodic markets.
6. Increased commercial activity results in five or six levels of cities.
7. A national marketing system emerges.

This model identifies the number of settlement levels—groupings by population size—associated with each stage. For England, the second stage evolved in the second century BCE, the third stage was bypassed, stage four evolved in the first century CE, stage five in the tenth century, stage six in the twelfth century, and stage seven in the sixteenth century. Although this model is extremely broad, it has the clear advantage of being less time- and place-specific than the earlier formulations.

Comparison of these various urban-location models with von Thünen's theory of agricultural location (Chapter 10) reveals substantial differences. Although von Thünen's work was stimulated by a specific set of facts, it largely succeeded in explaining agricultural location in general. By contrast, historical explanations of urban location are clearly limited to individual cases and do not offer any general explanations. Fortunately, however, one urban-location theory, like von Thünen's, has general applicability: central place theory.

Table 11.11	Seven Stages of Pre-modern Urban Development	
Stage	Number of Levels	Characteristic
1	0	Pre-urban
2	1	Tribute city
3	2	State city
4	2, 3, or 4	Imperial city
5	4 or 5	Standard marketing
6	5 or 6	Intermediate marketing
7	7	National marketing

SOURCE: Adapted from G. Rozman, 'Urban Networks and Historical Stages', *Journal of Interdisciplinary History* 9 (1978): 79.

Winterset, Iowa, USA. The town is an urban centre that performs a series of functions for the surrounding rural population.
Patti McConville/Alamy/GetStock

CENTRAL PLACE THEORY

Von Thünen's most important contribution to geography is the concept of the isolated state: the use of a grossly simplified ideal situation to observe the effects of a particular variable. In von Thünen's study of agricultural location, the variable highlighted was distance expressed as transport cost. The German geographer Walter Christaller (1893–1969) used a similar 'isolated state' model to observe the role played by distance in the location of urban centres.

Although the influence of von Thünen is clear, Christaller's seminal work was perhaps even more directly influenced by Alfred Weber, a German economist who introduced theory to industrial geography (see Chapter 14) at a time when it was mostly descriptive. Like von Thünen and Weber, Christaller was a pioneer. The English title of his seminal work, originally published in 1933 but not translated into English until 1966, is *Central Places in Southern Germany* (Christaller, 1966). At the time of publication there was little interest in the book, but since 1945 it has proven highly influential in theoretical and empirical research, especially in Scandinavia and the United States. During the 1960s, Christaller's work came to the fore as one component of the positivist-inspired spatial analysis movement that swept English-language geography at that time.

In 1940 a second German academic, the economist August Lösch (1906–45), published *The Economics of Location*, which included

and expanded on many of Christaller's ideas. Although the present discussion of central place theory draws on the work of both theorists, it is largely couched in the language introduced by Christaller. Box 11.4 provides an interesting insight into the approach employed by this major theorist.

Assumptions

Theories begin with assumptions that simplify the key issue under investigation. The assumptions of Christaller's central place theory are similar to those used by von Thünen:

1. The land surface is a flat, endless plain. Christaller introduces the assumption of flatness to avoid the complications implied by a physically variable landscape; similarly, the plain is endless in order to avoid any complications associated with boundaries.
2. This uniform plain has a uniform distribution of rural population; furthermore, all members of this population have identical behaviour and purchasing power. Christaller introduces this assumption to avoid the complications of human variations.
3. A homogeneous transport surface allows equal ease of movement for all members of the population in all directions. Christaller introduces this assumption to avoid the complications of a network of roads and other lines of communication.
4. Finally, it is assumed that the above three conditions together describe some preliminary stage in landscape evolution, and that all subsequent evolution is related to the growth of cities as service centres.

Four key concepts

Christaller developed his theory using four key concepts:

1. The *central place*: an urban centre that performs a series of functions for the surrounding rural population. In fact, a central place originates in response to rural demands for functions.
2. The *range* of a good or service: the maximum distance that people are prepared to travel to obtain a particular good or service.
3. The *threshold*: a measure of the minimum number of people required to support the existence of a particular function.
4. *Spatial competition*: the idea that central places compete with each other for customers.

Box 11.4 | Theory Construction by Christaller

In a brief article originally published in German, Walter Christaller explained how he developed central place theory. The explicit references to von Thünen and Weber are illuminating. Christaller wrote:

Besides geography and statistics, I was also interested in sociology, which at that time (in 1913), when I was beginning my studies, had begun to exist as a new scholarly discipline. . . . At that time I witnessed the efforts of my teacher Alfred Weber in Heidelberg to create an industrial location theory.

. . . I continued my games with maps: I connected cities of equal size by straight lines, first of all, in order to determine if certain rules were recognizable in the railroad and road network, whether regular traffic networks existed, and, second of all, in order to measure the distances between cities of equal size.

Thereby, the maps became filled with triangles, often equilateral triangles (the distances of cities of equal size from each other were thus approximately equal), which then crystallized as six-sided figures (hexagons). Furthermore, I determined that, in South Germany, the small rural towns very frequently and very precisely were 21 kilometres apart from each other. This fact has been recognized earlier, but had been explained as being due to the fact that these cities were stopover places for long distance trade traffic, and that in the Middle Ages the distance daily by a cart was about 20 kilometres.

My goal was staked out for me: to find laws, according to which number, size and distribution of cities were determined.

. . . It was clear to me from the very beginning that I had to develop a theoretical schema for my regional investigation—a schema which, as is customary in national economics, is set up by isolating the essential and operative factors. It was, thus, as in the case of von Thünen's Isolated State, abstracted from all natural and geographic factors, and, also, partly to be abstracted from all human geographic factors. It had to be accepted as a symmetrical plain, without obstructions such as rivers or mountain ranges, with a uniformly distributed population, in order to then determine where, under such conditions, the site of a central city or market could form. I thus followed exactly the opposite procedure that von Thünen did: he accepted the central city as already having been furnished, and asked how the agricultural land was utilized in the surrounding area, whereas I accepted the inhabited area as already having been furnished, and subsequently asked where the city must be situated, or, more correctly, where should the cities be situated. Thus, I first of all, as is said today, developed an abstract economic model. This model is 'correct' in itself, even if it is never to be found in the reality of settlement landscape in pure form: mountain ranges, variable ground, but also variable density of population, variable income ratios and sociological structure of the population, historical developments and political realities bring about deviations from the pure model. In the theoretical portion of my investigation, I thus did not satisfy myself with setting up a model for an invariable and constant economic landscape (thus, for a static condition)—but instead I also tried to show how the number, size and distribution of the central places change, when the economic factors change: the number, the distribution and the structure of the population, the demands for central goods and services, the production costs, the technical progress, the traffic services, etc.

. . . I was able to find surprising concurrences between geographical reality and the abstract schema of the central places (the theoretical model) especially in the predominantly agrarian areas of North and South Bavaria.

SOURCE: W. Christaller, 'How I Discovered the Theory of Central Places: A Report about the Origin of Central Places', in P.W. English and R.C. Mayfield, eds, *Man, Space and Environment*. (New York: Oxford University Press, 1972), 601–10.

These four concepts apply at various scales and can be most easily illustrated by reference to a single business, such as a bakery. What conditions need to be satisfied for a bakery to operate profitably? First, there must be enough people wishing to purchase bakery products. For every central function, there is some minimum level of demand, which varies depending on the specific function (relatively low for a bakery, relatively high for a dentist); this is the threshold concept. The second condition to be satisfied involves the distance that people are prepared to travel to shop at the bakery, given certain prices. Third, people willing to purchase at the prices offered must be located within the range of the bakery. Finally, the people within the range who represent the demand must not have access to other competitive bakeries. All four of these conditions apply equally to individual business and the collection of businesses clustered in an urban centre.

Hexagons

In principle, to ensure the most efficient use of space, central places should be located as shown in Figure 11.10. Central places that compete with each other for the purchasing power of rural populations will tend to locate so as to produce a regular arrangement (Box 11.5). This triangular lattice represents the most economical use of space. Within a central place system, the ideal shape of each market area should be a circle; however, as Figure 11.11 shows, a set of circles will mean either that some areas are not served or that some areas are served twice. To completely cover an area without any overlap, the ideal shape is the hexagon; the nearest geometric figure to a circle, the hexagon has the greatest number of sides and provides total coverage without duplication.

Hierarchies

Central places, of course, have more than one central function. Bakery products are not the only ones purchased by rural populations. Furthermore, different central functions have different threshold populations and different ranges. We find, then, that some central places have many functions and are correspondingly large, while other central places have relatively few functions and are correspondingly small. Christaller identified seven such levels of settlement and showed that the highest-order settlements (the largest) will be few in number, while the lowest-order settlements (the smallest) will be more numerous. But the spatial regularity is not disturbed by these differences because for any given order, settlements are equidistant from one another. Figure 11.12 shows a two-order situation. Thus central place theory suggests a hierarchy of settlements and a nested hierarchy of trading areas.

Three principles

We have just outlined what Christaller called the marketing principle. He also proposed the transportation and administrative principles. In each of the three models, central places are located equidistant from each other. Although one of these three principles may be more important than the others in any given region, the effects of all three tend to be roughly equal.

The *marketing principle* invokes the principle of least effort: people will always go to the nearest centre that provides the required service. Christaller called this a $k = 3$ arrangement, with k representing the number of settlements at a given level in the hierarchy served by a central place at the next highest level. Figure 11.12 shows that each high-order centre serves three low-order centres (that is, one-third of each of the low-order centres that surround it, plus itself as a low-order centre). Note that higher-order centres include all the functions of lower-order centres.

The *transportation principle* allows for as many large centres as possible to be located on communication lines between centres. (Notice that we are now amending one of the initial assumptions.) The result is shown in Figure 11.13; there is a $k = 4$ (the centre itself plus one-half of each of six lower-order centres) system in place.

The *administration principle* assumes that a central government would not risk subdividing

Box 11.5   Nearest Neighbour Analysis

Geographers may describe the spatial pattern of urban centres (or of any other phenomena represented on a dot map) using a statistical procedure called nearest neighbour analysis. With this method a point pattern may be described as clustered, random, or uniform (see Figure 2.4).

Nearest neighbour analysis indicates the degree to which an observed distribution of points deviates from what might be expected if the points were randomly distributed in the same area. A single statistic, R_n, is calculated.

R_n equals distance observed (average distance between each point and its nearest neighbour) divided by distance expected:

$$\frac{1}{2\sqrt{\text{density of pattern}}}$$

If $R_n = 1.0$, the two distances are identical and the pattern is random. If $R_n < 1.0$, the observed distance is less than the expected distance, indicating that the points are relatively clustered. If $R_n > 1.0$ (to a maximum of 2.15), the points are relatively equally spaced—as proposed by central place theory.

Although a useful statistical procedure, nearest neighbour analysis needs to be employed at an approximate spatial scale (recall Figures 2.6 and 2.7). When only the first nearest neighbour of each point is used, a repeated pattern of two or more closely spaced points, occurring at large spatial intervals, gives a low value of R_n, although the pattern may be dispersed. Thus it is not uncommon to incorporate second-order and third-order neighbours or to confirm the results with an alternative procedure.

FIGURE 11.10 **A triangular lattice**

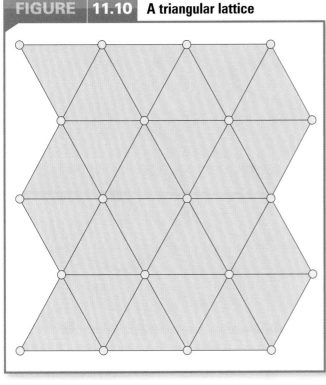

A central place system: the marketing principle

FIGURE 11.12

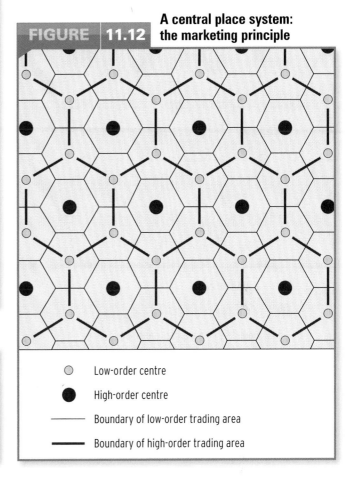

○ Low-order centre

● High-order centre

— Boundary of low-order trading area

━ Boundary of high-order trading area

FIGURE 11.11 **Theoretical trading areas**

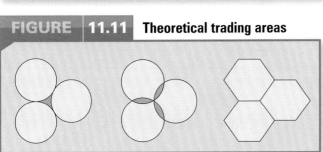

trading areas or regions. Thus the hierarchy is built up through the addition of whole regions. Figure 11.14 depicts the arrangement of centres and shows that a $k = 7$ situation applies: that is, a centre serves itself plus six lower-order centres. Box 11.6 offers some evidence to support the general hypotheses proposed in central place theory.

Other authors, especially Lösch (1954), have contended that Christaller's three principles are simply special cases of a more general situation. Extending this argument produces spatial arrangements that are much more complex than those proposed by Christaller.

The rank-size rule

Closely linked to central place theory, though separate in origin, is the rank-size rule, first proposed in 1913. This rule simply establishes a numerical size relationship between centres in a region. The rule is as follows:

$$P_r = P_1/R$$

where

P_r = population of centre r

P_1 = population of largest centre

R = rank size of centre r

Thus the population of a centre is inversely proportional to the rank of that centre. The largest centre in a region is named the **primate city**.

Geographers have noted that some regions conform to a rank-size distribution pattern, in which the second largest city is one-half the size of the largest and so forth, while other regions conform to a primate distribution pattern, in which one centre is more than twice the size of all other centres. This difference can be explained in various ways. For example, rank-size city distributions tend to occur in large countries, countries with a long history of urbanization, and countries

primate city
The largest city in a country, usually the capital city, which dominates its political, economic, and social life.

FIGURE 11.13 A central place system: the transportation principle

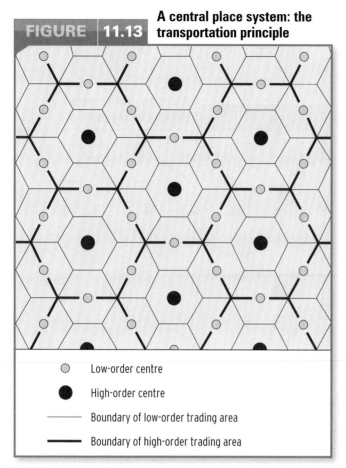

- ○ Low-order centre
- ● High-order centre
- —— Boundary of low-order trading area
- —— Boundary of high-order trading area

FIGURE 11.14 A central place system: the administration principle

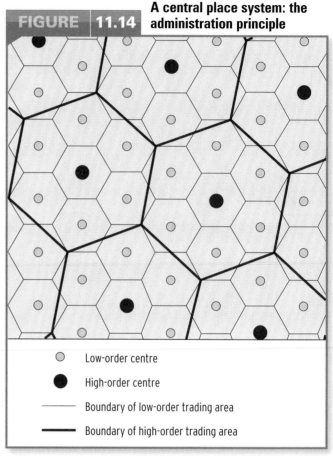

- ○ Low-order centre
- ● High-order centre
- —— Boundary of low-order trading area
- —— Boundary of high-order trading area

In this instance, the low-order centres are located in the middle of each of the six sides and not at the corners of the hexagons of the high-order centres; hence each low-order centre is linked to two, not three, high-order centres.

that are economically and politically complex. Primate city distributions tend to be found in small countries, those with a short history of urbanization, those with simple economic and political structures, and those at the centre of colonial empires.

World Cities

We now return to four facts discussed earlier in this chapter. First, as of 2008, 74 per cent of people in the more developed world and 44 per cent of people in the less developed world live in urban centres. Second, both percentages are increasing, especially in the less developed world. Third, the population of the world's mega-cities (more than 10 million people) is increasing and new mega-cities are appearing. Fourth, most of the projected additional urban population in the coming years will go to other cities, including some with fewer than 500,000 people today. These are significant changes, but

they are not the only ones occurring on the global scale.

As we have discussed throughout this chapter, cities are both the cause and the consequence of larger economic and social transformations, from the agricultural revolution through to the Industrial Revolution and related capitalist mode of production. But there is more.

Since the 1980s a global hierarchy and network of cities that serve as the basic nodes, or command centres, of the globalizing world economy has emerged. The first suggestion along these lines was made in the 1960s by Hall (1966); other important empirical and conceptual analyses include those by Friedmann (1986, 2002), Sassen (1991, 2002), and Taylor (2004).

Most of the world cities are located in more developed regions, since these are the core areas for the world economy. These centres are variously known as 'world cities' or 'global

Box 11.6 | Testing Central Place Theory

The essential hypotheses of central place theory—relating to the spacing and size of centres and to the shape and size of market areas—have been evaluated many times and for many different regions around the world. Together, these studies tend to confirm the basic hypotheses. Many of the early classic studies are described in detail by Berry (1967).

In evaluating central place theory, it is important to remember that it considers urban centres purely as service centres. It does not take into account urban centres established in response to, for example, some specific transportation or resource need. Thus Cape Town in South Africa began as a Dutch settlement on the route between Europe and Asia; the location decision had nothing to do with a surrounding rural population's demand for services. Similarly in northern Canada, many towns such as Sudbury, Ontario, are located to exploit a specific resource—again without any stimulus from a surrounding rural population. Nevertheless, analyses show that central place networks allow for such factors as they evolve; a modified version of a central place network can evolve from non-central place beginnings.

One of the most interesting criticisms of central place theory concerns the extent to which a recognition of hierarchies is legitimate or whether it results from the application of convenient but non-specific terms, such as 'hamlet', 'village', and 'town'. There is no doubt that the relationship between number of functions and urban size varies, but there may be a continuous functional relationship rather than a series of discrete stages.

A good example of a formal test of the hypothesis relating to urban centre spacing and size is Brush (1953). Brush identified three levels of settlement in southwest Wisconsin: hamlets, villages, and towns. There were 142 hamlets, 73 villages, and 19 towns; hamlets averaged two functions, villages 18, and towns 42; for hamlets, the average distance to the next centre was 8.8 km (5.5 miles); for villages, 15.8 km (10 miles); and for towns 33.9 km (21 miles). This research strongly supports Christaller's work.

It is difficult to exaggerate the importance of central place theory in the geography of urban centres and the spatial analysis school. Not only is it a major conceptual contribution, but it has helped to elevate urban geography in importance and stimulated further developments in location theory.

---

cities'—the two terms tend to be used interchangeably. It is often claimed that there are three cities at the top of the hierarchy, one for each of the three more developed regions of the world: London in Western Europe, New York in North America, and Tokyo in Pacific Asia. As shown in Table 11.12, this was the case in the late 1980s. (In some accounts, Paris is identified as a second Western European world city.) Today, however, the idea of three (or four) world cities oversimplifies what is now a complex and very dynamic situation. In this section we will consider cities at several levels of the hierarchy and the way they combine to form a global urban network.

WHAT IS A WORLD CITY?

There is no single way of identifying a world city. Still, geographers generally agree that great size is not the only qualification for world-city status. Table 11.6 and Figure 11.6 referred to mega-cities, not world cities. Rather, the factors that contribute to world-city status are mostly economic and include both management and production activities. As the following discussion suggests, these new world cities, 55 of which can be

categorized into three levels determined by the extent of their central place in world economic affairs (see below), can be best thought of as post-industrial locations.

Table 11.12	Three World Cities		
	Assets	Capital	Net Income
100 Largest Banks			
Tokyo	36.5	45.6	29.0
New York	8.6	8.8	20.8
London	4.2	5.7	13.2
All three	49.3	60.1	62.3
25 Largest Security Firms			
Tokyo	29.6	42.9	72.6
New York	58.6	50.0	22.0
London	11.1	4.9	2.8
All three	99.3	97.8	97.5

NOTE: This table shows the share in 1988 of the world's 100 largest banks and 25 largest security firms that are located in Tokyo, New York, and London, and confirms their status as the three pre-eminent world cities. These three cities also rank 1, 2, and 3 when we consider the locations of headquarters of major multinational firms (New York has 59, London has 37, Tokyo has 34); the next most important is Paris with 26.

SOURCE: S. Sassen, *The Global City: London, New York, Tokyo* (Princeton, NJ: Princeton University Press, 1991), 178–9. Copyright © 1991 Princeton University Press. Reprinted by permission.

producer services
Activities that offer a wide range of services to multinational and other companies that need to respond quickly to changing circumstances, including banking, insurance, marketing, accountancy, advertising, legal matters, consultancy, and innovation services; in recent years, the fastest-growing sector of national economies in most of the more developed countries.

Economic characteristics

First, world cities are the headquarters of transnational companies. As a result they influence global patterns of trade, communications, finance, and technology transfer. One specific reason why transnational companies prefer to locate in major cities is that this facilitates intimate interaction with capital markets, governments, and media. The urban locational choices made by transnational companies reflect a general shift in many national economies from manufacturing to service economies; hence older industrial cities may suffer as new service centres emerge.

Second, world cities are key centres for financial institutions and **producer services** such as banking, insurance, marketing, accounting, advertising, and legal matters. Through these institutions and services world cities are able to manage and co-ordinate global economic power. World cities serve as control centres for capital in the new international division of labour, in which company headquarters are located in one place and the related manufacturing activities are located elsewhere (often in areas of the less developed world where labour costs are low). In their capacity as home to transnational company headquarters and producer services, world cities function as global control centres, mediating between the global economy and specific nation-states. World cities are also marketplaces where both corporations and governments are able to purchase financial and other producer services.

Third, world cities are **gateway cities**. Necessarily, a world city will be a transportation and communications centre, with efficient intra-urban movement and various interregional and international links. This aspect of world cities is evident in the emerging geography of the Internet, which is widely dispersed in use but at the same time favours some places over others, with a commercial concentration in world cities (Zook, 2005). Certainly, e-commerce and e-business are encouraging further growth of and concentration of services in world cities.

As this account of economic characteristics implies, world cities are closely related to globalization trends. The components of globalization that are especially relevant to cities include: restructuring of the international financial system; globalization of property markets; investments by transnationals that prompt industrial dispersion; the rise of

gateway city
A city located at a key point of entry to a major geographic region or country, often a port or major rail centre, through which goods and people pass and in which several different cultural traditions are absorbed and assimilated.

service activities; 'stretching' of social relations as the links between widely separated locations intensify; and increased travelling and networking (Olds, 2001).

Cultural characteristics

In addition to these essentially economic features, world cities are also cultural centres, implicated in and responding to cultural globalization trends. They are usually culturally heterogeneous, serving as home to diverse ethnic identities, and are major players in such cultural activities as film, television, publishing, and sport. World cities are likely to be sites of various and varied spectacles and events, such as world fairs, theme parks, and major sporting events.

Political characteristics

Politically, the relationship between world cities and levels of government is complex. Typically, world cities are home to national and/or subnational governments. These cities also tend to receive support from governments as they strive to compete with other world cities, as evidenced by the national government support that London received for its successful bid to stage the 2012 Olympic Games. In some cases a city plays a distinct political role. Consider the unusual example of The Hague, in the Netherlands. Although not the official capital city of the country, this city is a political centre and has achieved significant stature as home to several international organizations concerned with law, peace, and justice, including the International Court of Justice and the International Criminal Court. Similarly, New York, though not even the capital of New York state, is the site of the United Nations.

WHY WORLD CITIES?

The emergence of world cities is generally understood as one response to and component of the process of economic globalization and the declining friction of distance, as discussed in Chapter 9, and also a response to and component of the diminishing role of the state, as discussed in Chapters 8 and 9. In particular, the globalization of economic activity has increased the quantity and complexity of economic transactions, thereby encouraging, indeed requiring, the growth of already existing institutions and the emergence of many new players. Economic globalization also involves increased demand for a wide range of services, including information technologies. Especially significant in this context is the massive growth of the Internet (Figure 11.15). A more detailed discussion of information technologies and cities is included in Chapter 12 in the context of movement inside the city.

THE HIERARCHY OF WORLD CITIES

As noted, there is general agreement that there are three or four principal world cities. They are London, New York, Tokyo, and, in some listings, Paris. But there is great value in moving beyond this basic identification to provide a more empirically justified and detailed statement.

Table 11.13 outlines the results of one comprehensive analysis that lists and ranks cities on the basis of the level of advanced producer services, specifically accountancy, advertising, banking/finance, and law. The result is the identification of 55 world cities grouped into three levels—10 at the first (Alpha) level, 10 at the second (Beta) level, and 35 at the third (Gamma) level. Not surprisingly, these 55 are located primarily in more developed world countries that function as the leading arenas of economic globalization. Also not surprisingly, the four leading world cities identified in this analysis are London, New York, Paris, and Tokyo.

More particularly, as shown in Figure 11.16, the Alpha world cities are evenly distributed across the three globalization arenas of Western Europe (four cities), North America (three cities), and Pacific Asia (three cities). Of the 10 Beta world cities, three are in Western Europe, two in North America, and two in Pacific Asia (counting Sydney, Australia, as Pacific Asia). The other three Beta cities are outside of the main globalization arenas, with two in Latin America and one in Eastern Europe. The three main globalization arenas again dominate for the 35 Gamma world cities (11 in Western Europe, eight in North America, and eight in Pacific Asia), while Latin America (four cities) and Eastern Europe (three cities) are confirmed as minor globalization arenas. Of particular note, only one city in sub-Saharan Africa—Johannesburg—is included in this listing of 55 world cities, and only one other city in sub-Saharan Africa (Cape Town, also in South Africa) shows any evidence, by the criteria used, of world-city status.

From a Canadian perspective it is notable that four Canadian cities are listed. Toronto is one of the 10 Beta world cities, Montreal is one

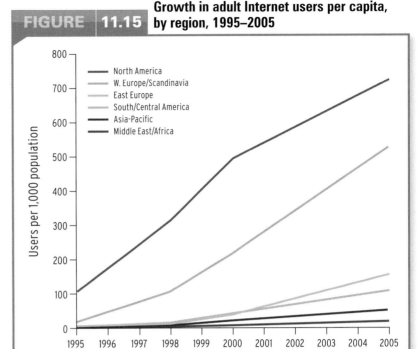

FIGURE 11.15 Growth in adult Internet users per capita, by region, 1995–2005

SOURCE: T.F. Golab and A.C. Regan, 'Impacts of Information Technology on Personal Travel and Commercial Vehicle Operations: Research Challenges and Opportunities', Transportation Research, Part C: Emerging Technologies Vol 9, Issue 2 (2001): 87-121. Copyright © 2001 Elsevier Science Ltd.

of the 35 Gamma world cities, while Vancouver shows some evidence of world city formation, and Calgary shows minimal evidence of world city formation.

Are all cities world cities?

It might be helpful to think of all cities as world cities in the sense that all cities are connected with other cities. Certainly, some cities

The Millennium Footbridge across the River Thames in London, England, with St Paul's Cathedral in the background.

imagebroker/Alamy/GetStock

Table 11.13	Inventory of World Cities

A. Alpha World Cities (full-service world cities)

  12: London, New York, Paris, Tokyo

  10: Chicago, Frankfurt, Hong Kong, Los Angeles, Milan, Singapore

B. Beta World Cities (major world cities)

  9: San Francisco, Sydney, Toronto, Zurich

  8: Brussels, Madrid, Mexico City, São Paulo

  7: Moscow, Seoul

C. Gamma World Cities (minor world cities)

  6: Amsterdam, Boston, Caracas, Dallas, Düsseldorf, Geneva, Houston, Jakarta, Johannesburg, Melbourne, Osaka, Prague, Santiago, Taipei, Washington

  5: Bangkok, Beijing, Montreal, Rome, Stockholm, Warsaw

  4: Atlanta, Barcelona, Berlin, Budapest, Buenos Aires, Copenhagen, Hamburg, Istanbul, Kuala Lumpur, Manila, Miami, Minneapolis, Munich, Shanghai

D. Evidence of World City Formation

  Di: *Relatively strong evidence*

  3: Athens, Auckland, Dublin, Helsinki, Luxembourg, Lyon, Mumbai, New Delhi, Philadelphia, Rio de Janeiro, Tel Aviv, Vienna

  Dii: *Some evidence*

  2: Abu Dhabi, Almaty, Birmingham, Bogotá, Bratislava, Brisbane, Bucharest, Cairo, Cleveland, Cologne, Detroit, Dubai, Ho Chi Minh City, Kiev, Lima, Lisbon, Manchester, Montevideo, Oslo, Riyadh, Rotterdam, Seattle, Stuttgart, The Hague, Vancouver

  Diii: *Minimal evidence*

  1: Adelaide, Antwerp, Arhus, Baltimore, Bangalore, Bologna, Brasilia, Calgary, Cape Town, Colombo, Columbus, Dresden, Edinburgh, Genoa, Glasgow, Gothenburg, Guangzhou, Hanoi, Kansas City, Leeds, Lille, Marseille, Richmond, St Petersburg, Tashkent, Tehran, Tijuana, Turin, Utrecht, Wellington

NOTE: Cities are ordered in terms of values ranging from 1 to 12. Within each class, cities are listed alphabetically. The values are produced by considering global service centres in terms of their provision of corporate services. The four services considered are accountancy, advertising, banking/finance, and law. For each of these four, cities score 3 for prime centre status, 2 for major centre status, and 1 for minor centre status, meaning the highest possible score is 12. The specific rationale for the divisions into classes is detailed in the article. This table was developed as part of the substantial research program associated with the Globalization and World Cities (GaWC) Study Group & Network, at: <www.lboro.ac.uk/gawc/index.html>. The authors describe it as the 'GaWC inventory'.

SOURCE: J.V. Beaverstock, R.G. Smith, and P.J. Taylor, 'Research Bulletin # 5: A Roster of World Cities', *Cities* 16 (1999): 456. Copyright © 1999 Elsevier. Reprinted with permission.

are more powerful than others—economically, culturally, politically, and technologically—but this fact does not dispute the importance of linkages for all cities. Alpha world cities are linked to other cities and influence those other cities, but they are, in turn, necessarily influenced by other cities. Interactions work both ways. What is happening in Mumbai needs to be understood in terms of what is happening in London, and vice versa. As noted in Chapter 9, there is no geography without connections.

FROM HIERARCHIES TO NETWORKS

World cities can be understood as occupying levels in a hierarchy but, of course, these cities are not isolated one from another. Rather, they form a network, with cities as nodes linked through transnational flows of capital, knowledge, information, commodities, economic activities, and people (especially people as high-value labour). World cities are also linked through the activities of transnationals and through arrangements between governments and cultural institutions. The principal agents of connectivity are firms, especially transnational companies and producer service firms, and hence the most important of these links are economic, with some links organized in cyberspace. Accordingly, one of the most useful ways to think about a world-city network is in terms of service firms that offer their services to users in many cities.

Recognizing networks highlights the fact that world cities cannot be seen in isolation; what happens in one world city is often inextricably tied up with experiences elsewhere. This is an obvious geographic fact that cannot be neglected and is one aspect of the account in Chapter 9 of the reduction of distance friction and associated time-space convergence. Networks function especially through telecommunications technologies such as telephone, fax, e-mail, computer networks, and the Internet. The major world cities are the key nodes in the global network; indeed, it can be said that the major world cities are moving closer together. For some commentators, one consequence of these emerging global networks is the formation of information-rich and information-poor regions within national economies.

A possible problem with the focus on networks of world cities is that most conclusions derive essentially from a consideration of financial and producer services. It can be argued that researchers are identifying connections between firms rather than connections between cities (Robinson, 2005: 758). Further, evidence suggests that connectivity between world cities retains a significant regional dimension, with European, North American, and Pacific Asian networks being evident. Figure 11.17 shows one simplified way of thinking about these regional networks, suggesting general spheres of influence for each

FIGURE 11.16 World cities

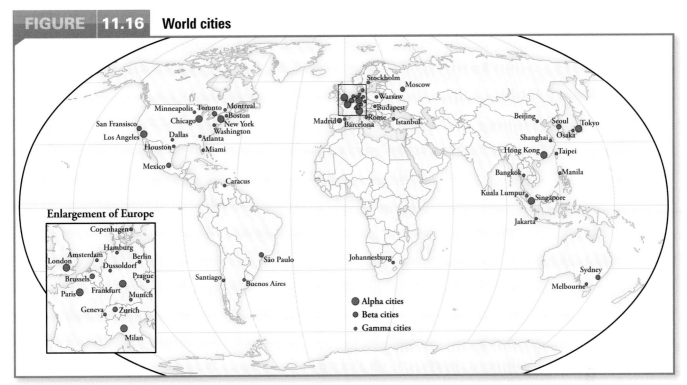

SOURCE: J.V. Beaverstock, R.G. Smith, and P.J. Taylor, 'Research Bulletin #5: A Roster of World Cities', Cities 16 (1999): 456. Copyright © 1999 Elsevier Science Ltd.

FIGURE 11.17 World cities and spheres of influence

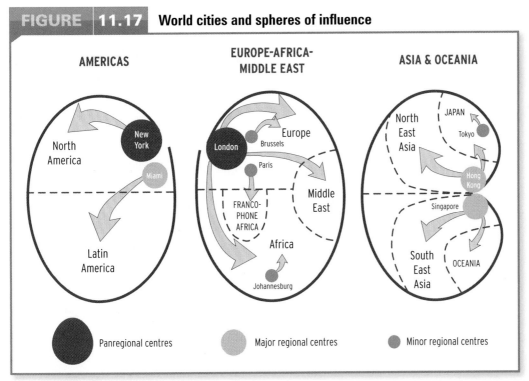

This diagram is derived from a detailed analysis of information about the location of business headquarters, subsidiary offices, and capital flows that identifies 55 world cities of varying degrees of importance, and questions many of the general assumptions about world cities. The patterns shown in this figure are quite different from some of the complex hierarchies described in other studies.

SOURCE: P.J. Taylor, 'World Cities and Territorial States under Conditions of Contemporary Globalization', Political Geography 19 (2000): 25. Copyright © 2000 Elsevier Science Ltd.

of these three, as well as some regional centres. Notably, two of these cities are not located in the regions for which they serve as centres (Miami serves Latin America, while London serves Africa and the Middle East), and one of those two is not an Alpha-level world city. However, as discussed in Box 11.7, perhaps the single most important global connection is that between two world cities separated by an ocean.

MORE AND MORE WORLD CITIES?

As noted, the first listings of world cities in the 1980s proposed three such places—New York, London, and Tokyo. As the concept of world cities has become more explicit, more of these locations have been identified and also hierarchies and networks have been proposed, a reflection both of improved understanding but also of our ever-changing world.

Many scholars envisage reinforcement of the new geography of centrality evident in the emergence of a limited number of key world cities that, among them, effectively manage the global economy. But there are other possibilities.

Most notably, although the principal world cities will remain important, many corporations may choose to locate their headquarters in suburban peripheral areas or perhaps even in smaller cities. Indeed, this trend can be thought of as a logical outcome of a world in which the friction of distance is significantly lessened, especially because of improved communication and transportation technologies.

Following this logic, perhaps it is the concentration of activities in a limited number of cities that is rather puzzling, not the dispersal. Certainly, dispersal is already evident in the global securities industry, previously concentrated in New York and London but increasingly scattered. Also, some major corporate headquarters favour locations outside world cities: the world's largest retailer, Wal-Mart, is based in Bentonville, Arkansas, and Berkshire Hathaway, a major American investment firm headed by Warren Buffet, is headquartered in Omaha, Nebraska. More generally, the fastest-growing areas in North America tend to be highly suburbanized cities in the American South and West, not only Atlanta, Houston, and Dallas, but also Charlotte and San Antonio.

Box 11.7   A Capital of the World?

Is there, or might there be in the future, a capital city of the world? Thinking along these lines suggests two possibilities.

First, it is possible to extend the world city network idea to suggest that in some instances two or more cities become so intertwined that they can in some respects be seen as one place. The most frequently suggested example is Nylon, with New York and London understood as the two poles of a single transatlantic urban centre. Conceiving of these two world cities as one effectively answers our question, with Nylon being the capital city of the world. A principal justification for this one-city claim is that, in addition to financial and business links, the two poles have shared interests especially in such popular culture activities as film, music, theatre, television, and publishing. It can also be argued that Paris and London are increasingly functioning as one city, especially since the 2007 opening of a new rail link, which means that the two are only 135 minutes apart.

There are many other suggestions for such linked cities, including Hong Sing (Hong Kong and Singapore in Asia), Amsfurt (Amsterdam and Frankfurt in Western Europe), Boslanta (Boston and Atlanta in the United States), and Manbrum (Manchester and Birmingham in England). As with the example of Nylon, these suggestions are based essentially on measuring the size of a city as a node in a network (the site of the city) and measuring the connectivity of a city in a network (the situation of the city).

But there is a second and very different answer to our question. Nylon may dominate today, but Chinese cities are growing and globalizing so rapidly since China opened its markets in the 1980s that some observers see either Beijing or Shanghai emerging as the dominant world city. Consider Shanghai: it has more skyscrapers than New York, is the regional headquarters for about 500 transnational companies, accounts for 30 per cent of China's exports, attracts 25 per cent of the foreign investment to China, and produces 20 per cent of China's manufacturing output. The world's largest container port is being constructed on an island that is 30 km offshore with a planned linking bridge. The recent explosive spatial growth of Shanghai has involved displacement of over one million rural dwellers—a circumstance possible only in a non-democratic country—while about four million workers have relocated to the Shanghai region from elsewhere in China. Such dramatic growth is not continuous (it was lessened by the onset of a global recession in 2008) and cannot continue in the long term, but the argument that China will become home to the capital of the world has merit.

Other cities, such as Fort Lauderdale, Boise, and Reno, are gaining business and professional service employment opportunities.

This trend is evident at least to some degree throughout the world. For example, in Europe, there are increased financial service activities in the northern English town of Leeds. It is clear also that the largest cities in the less developed world, such as Mumbai and Mexico City, have many problems that may discourage additional economic activity. In India, smaller centres, such as Bangalore and Jaipur, and suburban locations of large cities such as Mumbai are often preferred locations, while in Mexico smaller centres such as Chilango, Guadalajara, and Monterrey are increasingly favoured locations. The rising number of telecommuters in many parts of the world is also part of this larger trend.

In general, principal causes for this shift away from the centres of large cities are the lower cost of locating in less densely populated locations and the social benefits of more family-friendly settings, especially since such locations are not necessarily a disadvantage in an age of telecommunications. Whether or not this trend of deconcentration will prevail over the long term remains to be seen, but as we saw in Table 11.5, cities of less than 500,000 are home to a substantial part of the urban population. In addition, recall that Table 11.13 includes a total of 55 world cities and a further 68 that show evidence of world-city formation. Thus, in some respects and for some global firms, there may be an optimal size of city, with adequate

Shanghai, China.
SCPhotos/Alamy

or optimal supportive services, that does not easily equate with 'bigger is better'.

Acknowledging that there are many world cities introduces the idea that, rather than focusing attention on the dominant few, a better way to understand this outcome and component of economic globalization is in terms of what happens to all cities because of globalization. In short, the important question to ask may not be which cities are dominant, but, rather, how are all cities changing in our globalizing world. Perhaps, as Taylor (2004: 42) suggests, there is no such thing as a non-global city? The value of thinking in this way is evident in the following chapter.

CHAPTER 11 SUMMARY

COUNTING URBAN POPULATIONS

Two points are relevant here. First, the precise way in which urban and rural dwellers are defined varies by country. Any changes in these definitions necessarily change the numbers of each group in a country and, if the country has a large population, also impacts notably on global statistics. Second, the spatial extent of many cities can be measured in several ways, meaning that there are often very different statements as to how many people live in a city.

MORE AND BIGGER CITIES

The first permanent settlements appeared with the domestication of plants and animals. Dramatic urban growth was initiated in more developed countries with the beginnings of industrialization about 1750. The experience of less developed countries has been quite different, with substantial growth typically beginning after about 1950 as cities acted as magnets for the rural poor. Globally, humans crossed the line from predominantly rural to predominantly urban in 2007. The projected urban growth rate is 1.83, meaning that the world urban population will double in 38 years. About one-half of the world urban population live in cities of less than 500,000. The 20 mega-cities, each with over 10 million people, are home to about 10 per cent of the world urban population.

PATTERNS OF RURAL SETTLEMENT

Rural settlement patterns range from dispersed to nucleated. Nucleation, the oldest and most common pattern, may be regular or irregular. Dispersion is less common; in Europe it dates from the early eighteenth century, and it is typical of many temperate areas of European overseas expansion.

RURAL DEPOPULATION

Today, evidence indicates both rural depopulation and repopulation. Depopulation is normal, especially in extensive agricultural regions such as the Canadian prairies. Small towns are losing population and services as one component of larger global changes. There is some evidence of a process of counter-urbanization, with urban dwellers relocating to rural areas. In much of Europe there is every likelihood that rural areas will disappear during the next few decades. In some areas, urban expansion into rural areas is resulting in the loss of distinctive agricultural land uses, as in the Niagara Peninsula of Ontario.

THE RURAL WAY OF LIFE

Especially in the more developed world, differences between rural and urban ways of life (*Gemeinschaft* and *Gesellschaft*) are decreasing. Traditionally, rural areas have often been viewed in a positive light compared to urban areas. But contrasting images exist of both the rural and the urban, and no one image of either place is entirely accurate. A process of rural gentrification is occurring in many of the more developed countries as urban residents relocate to rural areas because of perceived benefits of rural life.

WHY SETTLE IN URBAN CENTRES?

Permanent settlements were initially associated with economic changes, specifically the production of an agricultural surplus; they also reflect social needs. The earliest cities developed about 3500 BCE as successors to large agricultural villages. The earliest cities (established before about 1750) and many cities in less developed areas can be conveniently labelled pre-industrial: they functioned primarily as marketing and commercial centres. Urbanization was stimulated by the mercantilist philosophy that flourished in Europe from the beginning of European overseas expansion until the emergence of capitalism. With the rise of industrialization and capitalist philosophy, cities increased in number and size at a rapid pace. The locations of the industrial cities that emerged at this time were often determined by the location of the resources they depended on. For their inhabitants, ways of life in these cities represented a marked departure from earlier experiences.

ECONOMIC BASE THEORY

The basic sector of an urban economy consists of the multitude of activities that produce goods and services for sale outside the urban area; the non-basic sector consists of all those activities that produce goods and services for sale inside the city. This distinction is not so clear in the case of producer services that serve both local and larger populations.

EXPLAINING URBAN LOCATIONS

Many attempts have been made to explain why cities are located where they are. Traditionally, attention was focused on specific cities, but increasingly it is shifting to networks, or systems, of cities. Most of the historical explanations that have been offered are most applicable to areas of European overseas expansion and identify a series of stages; examples include studies by Vance, Borchert, and Muller. Rozman identified seven stages of pre-modern urban development. The most influential and persuasive explanation, however, is the central place theory expounded by Christaller and Lösch. This theory, conceptually similar to the work of von Thünen, was published in 1933 and translated into English in 1966. It proved to be a major stimulus to research during the 1960s heyday of the spatial analytic approach. A series of simplifying assumptions, combined with four key concepts (central place, range, threshold, and spatial competition), generates three related principles—regarding marketing, transportation, and administration—concerning the size and spacing of cities. Many tests of these hypotheses have been conducted, confirming the basic logic of the approach.

THE RANK-SIZE RULE

This rule states that, in a given country, the population size of a city is inversely proportional to the rank of that city, where the largest city is rank 1.

WORLD CITIES?

Since the 1980s a global hierarchy and a global network of cities has become increasingly evident. Initially, Tokyo, London, New York, and possibly Paris were recognized as world cities, but it is now usual to recognize many more and to divide them into categories. A key feature of the emerging network of world cities is that connections are critical. In some instances, it is argued that two cities, such as New York and London, effectively function as one place in some respects. Most of the world cities are in more developed regions of the world. In general terms, they are understood in the larger context of globalization processes. Economically, world cities are typically home to transnational companies, are the sites of financial institutions and producer services, and function as gateway cities. Culturally, they are important centres, and are frequently the sites of world fairs and major sporting events. Politically, these cities are usually home to one or more levels of government. As one aspect of ongoing changes in the human geography of the world, world cities not only show evidence of the concentration of activities inside those cities, but also of dispersal of activities to city margins and to other smaller cities.

QUESTIONS FOR CRITICAL THOUGHT

1. Compare urbanization (level and rate) in less developed and more developed nations. In what ways are urban problems in less developed nations different from those in more developed nations?

2. What are the implications of the proliferation of cities with more than 10 million people (see Table 11.6)?

3. What are the causes and consequences of urban sprawl? What can be done to address sprawl-related problems?

4. How important is the 'rural way of life' to our nation? Should it be preserved? Why? What is lost as the rural way of life disappears?

5. Discuss the relevance of Christaller's central place theory to our understanding of: (1) contemporary urban systems, and (2) the retail structure of cities.

6. Why are New York City, Tokyo, and London classed as 'world cities'? What would it take for a city like Toronto to become a true 'world city'?

7. We have recently reached the point where one-half of the world's population is classed as urban. What are the implications of the continued rapid growth of the urban population on a global scale?

FURTHER EXPLORATIONS

Amen, M.M., K. Archer, and M.M. Bosman, eds. 2006. *Relocating Global Cities: From the Center to the Margins*. Lanham, Md: Rowman & Littlefield.

A distinctive contribution to the literature on global cities that focuses on cities that are not necessarily usually seen as having global characteristics.

Berry, B.J.L. 1981. *Comparative Urbanization: Divergent Paths in the Twentieth Century*. London: Macmillan.

An insightful overview of numerous urban issues by a leading geographer.

Brenner, N., and R. Keil, eds. 2006. *The Global Cities Reader*. New York: Routledge.

Substantial volume with 50 chapters on such topics as the concept of a global city, global city formation, the research focus on global cities in more developed countries, political and cultural aspects of global cities, and theoretical advances.

Bunting, T., and P. Filion, eds. 2006. *Canadian Cities in Transition: Local Through Global Perspectives*, 3rd edn. Toronto: Oxford University Press.

A major contribution to Canadian urban geography, this book looks at cities as systems and at the city as a system. A well-conceived and structured volume that is also full of detailed information on the full range of urban geographic topics.

Carter, H. 1983. *An Introduction to Urban Historical Geography*. London: Arnold.

A thorough review of urbanism and urbanization through time; varied international content.

Cloke, P., and J. Little, eds. 1997. *Contested Countryside Cultures: Otherness, Marginalization, and Rurality*. London: Routledge.

An interesting collection of readings focusing on the social construction of rurality and the related exclusion of particular identities—'others'; examples include new-age travellers in Britain and lesbian groups in the US.

Florida, Richard. 2008. *Who's Your City: How the Creative Economy Is Making Where to Live the Most Important Decision of Your Life*. Toronto: Random House Canada.

This easy-to-read exploration of the global geography of cities highlights what cities need to do to attract what the author describes as the creative class.

Gayler, H.J. 1990. 'Changing Aspects of Urban Containment in Canada: The Niagara Case in the 1980s and Beyond', *Urban Geography* 11: 373–97.

A detailed article focusing on a specific case study and the wider implications of urban expansion.

Kaplan, D.H., J.O. Wheeler, and S.R. Holloway. 2004. *Urban Geography*. New York: Wiley.

Detailed and informative text covering material relevant for both this and the following two chapters. The examples are predominantly American.

Knox, P.L., and P.J. Taylor, eds. 1995. *World Cities in a World System*. Cambridge: Cambridge University Press.

An excellent book that provides a thorough and often innovative account of the contemporary urban world, setting the discussion of world cities in the Wallerstein world systems framework.

Larsen, S.C., C. Sorenson, D. McDermott, J. Long, and C. Post. 2007. 'Place Perception and Social Interaction on an Exurban Landscape in Central Colorado', *Professional Geographer* 59: 421–33.

Detailed study of an exurban landscape that shows the growth of social networks and the presence of distinct place perceptions.

McCarthy, J. 2008. 'Rural Geography: Globalizing the Countryside', *Progress in Human Geography* 32: 129–37.

Review article covering topics such as the purchasing of second residences in rural areas, exurban and related inroads into rural areas, and environmental changes.

Marshall, J.U. 1989. *The Structure of Urban Systems*. Toronto: University of Toronto Press.

The most significant work to appear on the subject of central place theory since the 1960s; includes some informative Canadian examples.

Pacione, M. 2005. *Urban Geography: A Global Perspective*, 2nd edn. New York: Routledge.

A substantial and comprehensive text-book addressing urban geography generally. Useful for this and the following two chapters.

Soule, D.C., ed. 2005. *Urban Sprawl: A Comprehensive Reference Guide*. Westport, Conn.: Greenwood Press.

Detailed discussion of land use and other components of urban sprawl, including role played by immigration.

Woods, M. 2007. 'Engaging the Global Countryside: Globalization, Hybridity and the Reconstitution of Rural Place', *Progress in Human Geography* 31: 485–507.

Introduces the concept of a global country-side through a discussion of changing rural-ity, stressing the roles of negotiation and configuration rather than domination and subordination.

ON THE WEB

WORLD URBANIZATION

Population Reference Bureau

> www.prb.org/

The website of the Population Reference Bureau includes much useful information on world urbanization trends and references to relevant studies.

Population Division of the United Nations

> esa.un.org/unup/

The Population Division of the United Nations similarly provides much up-to-date information on world urbanization.

UN Habitat

> www.unhabitat.org/categories.asp?catid=9

Home page of the UN Human Settlements program, which is mandated to promote socially and environmentally sustainable urban centres.

CANADIAN SETTLEMENTS

Statistics Canada

> www12.statcan.ca/census-recensement/2006/dp-pd/prof/92-591/index.cfm?Lang=E

This Statistics Canada web page has links to individual community profiles using 2006 census data.

Ottawa Region Rural Settlement Strategy

> www.ottawa.ca/residents/public_consult/rural_settlement/index_en.html

Detailed documentation of research into rural development, groundwater resources, and agricultural development in the Ottawa region.

WORLD CITIES

Globalization and World Cities Study Group and Network

> www.lboro.ac.uk/gawc/index.html

This valuable source for researching world cities provides much data and discussion.

Photographs of Cities

> www.worldcityphotos.org/

More than 12,000 photographs of large cities around the world.

Cities

> www.mapsofworld.com/cities/

This site is a directory of major cities.

URBAN FORM AND GOVERNANCE

Cities around the world all are different, and yet they display some fundamental underlying similarities, at least partly because they are shaped by similar processes. The distribution of urban land uses, for example, typically relates to distance from the city centre and is also directionally biased. All cities today are governed and, at least to some degree, planned in accordance with particular and often changing political, social, and economic objectives. Also, globalization processes are today prompting the rise of new urban areas and influencing the location of many urban activities, typically encouraging new urban development outside of traditional urban areas in suburban or even more distant places. With these varied processes in mind, this chapter addresses the following questions:

- How are cities spatially organized?
- What conceptual formulations are used to understand the internal structure of cities in the more developed world?
- Do cities in different regions of the world have different characteristic internal structures?
- Is there a transition from modern to postmodern cities and, if so, what does this imply?
- How are cities governed, and how are city governance and planning related to political ideology?
- To what extent is the geography of cities a reflection of planning processes?
- How are the parts of a city linked together by transportation networks, and what roles are played by public and private means of transportation in the city?
- Are cities growing outward, or are new centres emerging?

Ottawa at night.
Ivy Images

Recall the comment at the outset of Chapter 11 concerning the relative newness of some urban concepts and related empirical analyses. In the previous chapter most of the new concepts were concerned with world cities and related city networks as these are emerging in response to and as one component of globalization processes. In this chapter, our focus is on two related trends, one involving population spread beyond city areas and the other involving dispersal of urban functions away from established city centres and towards suburbs, fringe locations, and even rural areas. Certainly, the urban environments that people live in today are often very different from those of just a few years ago.

Again, as for the content of Chapter 11, Taylor and Lang (2004) list a sample of 50 relevant new concepts—among the most useful for us here are those of 'edge city' and 'metropolitan suburb'.

Throughout the discussions in this chapter, it is evident that cities are highly complex organizational and institutional forms, serving as centres for social interaction, cultural activities, economic development, political activity,

and technological change. Further, it is evident also that changing economic, cultural, social, political, and technological circumstances combine to create changing urban areas and urban experiences. Not surprisingly, then, our discussions of urban areas and the urban way of life are conceptually diverse, reflecting a wide range of economic and social theories inspired by positivism, Marxism, humanism, feminism, and postmodernism. For example, in the 1960s researchers in the spatial analysis school used simple models to focus on the internal structure of cities. Since then, the focus has shifted to other topics, including links between the capitalist mode of production and the social and economic geography of the city, and also the way the city is perceived by those who live there. In addition, attention has been directed to the rise of post-suburbia and the post-industrial city.

This transition in conceptual approaches—what might be called a process of inventing and reinventing economic geography—is discussed in general terms at the beginning of Chapter 10, while Box 12.1 notes the transition as it applies to urban areas in the more developed world.

Box 12.1 Introducing Intra-Urban Theory

Urban theory has undergone a series of changes. The first important theoretical focus developed in the 1920s. The Chicago School of urban ecology regarded the city as a social organism and interpreted change by drawing ecological analogies. For example, the most important process was the competition for urban space, which resulted in spatial segregation of groups. Other explanations of the internal structure of the city also centred on a group's dominance of an area and the invasion of an area by a group that then succeeded some other group in that area. A related school of thought, which played a leading role in the 1960s, centred on the concept of neo-classical urban land rent; these von Thünen-type theorists interpreted urban areas in terms of the rational 'economic operator'.

Both of these approaches were criticized from various perspectives for their failure to explain many of the changes taking place in the modern city, especially those associated with the larger processes of transition from Fordist to post-Fordist modes of production (Chapter 7) and the related processes of economic and social restructuring (Chapter 9). Political economists pay special attention to the conflicting interests of, on the one hand, financiers concerned with capital accumulation and, on the other hand, inner-city residential and small-business interests. In human geography,

this Marxist approach is closely associated with the work of David Harvey (see Box 12.2).

Another focus, sometimes labelled 'urban managerialism', is derived from the work of the sociologist Max Weber and places particular emphasis on the state and other institutions. Humanistic concepts have also been applied to the study of urban areas, especially in the context of interpretive social research.

Most recently, there has been a movement towards some form of postmodern account of the city. A basic contention of postmodernism, as discussed in Chapter 7, is that all theories are to some degree repressive, because they privilege one explanation above all others. Postmodernists therefore question the legitimacy of all modern urban theories. This movement might be seen as a radical break from earlier theoretical traditions, requiring as it does entirely new understandings of what cities are and how they function.

All of these approaches are discussed in this chapter. In addition, two excellent accounts of urban theory that include appropriate examples of all the above types are those of Bassett and Short (1989) and Cooke (1990). The urban geography texts by Pacione (2001) and Kaplan, Wheeler, and Holloway (2004) also include discussions of these approaches and related examples.

Explaining Location inside the City

Human geographers analyze capitalist development in urban areas in terms of the major agents involved. The process by which urban land use changes or by which rural land becomes urban may involve many participants: governments, rural landowners, speculators, architects, developers, builders, subcontractors, real estate agents, urban homeowners, and various financial and legal consultants. To understand the urban development process, it is essential to recognize that although most of these participants stand to benefit in one way or another from a change in land use, their motives vary widely. Governments at various levels establish planning policies that constrain or facilitate the process of land-use change; rural landowners are primarily concerned with the agricultural productivity of land; speculators aim to buy land at a low price and sell at a high one; architects, developers, builders, and subcontractors together provide basic infrastructure and a finished dwelling and are motivated by the need to make a profit; real estate agents mediate between sellers and buyers and have a vested interest in promoting an active property market; financial and legal consultants, including mortgage providers, earn fees for their services. The interrelated activities of these various participants ensure that the city is a dynamic place, never at rest, and therefore difficult to describe in simple terms.

An important first concern when we consider the geography of the city is to describe and explain where people live and in what numbers. Interestingly, the distribution of population densities tends to be remarkably consistent for all urban areas, bearing little relation to the specifics of internal structure. Densities typically decline as a negative exponential function of distance from the centre of the city (see Figure 2.5).

A basic urban land-use model can be conceived similarly, operating on the same principles as von Thünen theory with different potential land uses having different economic rent lines. Thus land uses with the greatest values of economic rent at the city centre will have steep rent lines and occupy land adjacent to the centre. Following this logic, in principle a city is made up of a series of concentric zones. A typical arrangement might have businesses at

Land for sale in Brisbane, Australia.

V. Last, Geographical Visual Aids

the centre with industrial and then residential uses at increasing distances from the centre (see Figure 10.4). The value of land also characteristically decreases with increasing distance from the centre. A popular diagram of this urban land value surface that also allows for a transportation network is shown in Figure 12.1.

FIGURE 12.1 Urban land values

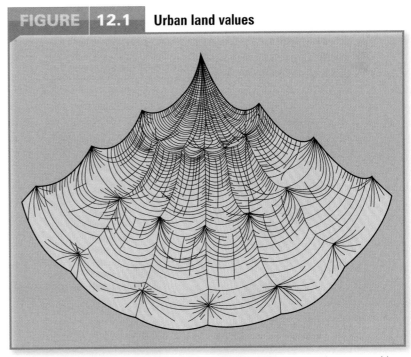

This classic 'circus tent' schematic diagram is often used by geographers to provide a generalized representation of urban land values. The peaks in the diagram are areas of high value, and the intervening areas are lower value. High-value areas are commercial or industrial locations on transport routes; low-value areas are residential or vacant.

ZONES, SECTORS, AND NUCLEI

Responding to the challenge of generalizing about where land uses are located inside the city, geographers and others initially focused attention on three classic models. These models of the internal structure of urban areas are shown in Figure 12.2.

Concentric zone model

The *concentric zone* model (a) was proposed by the pioneering Chicago sociologists (Box 12.1), most notably Park, Burgess, and McKenzie, who applied ecological concepts from biology and botany to urban areas, (Park and Burgess, 1921).

This von Thünen-type model is an extension of economic rent logic. The central business district is at the city centre; the second zone is a transitional area comprising original industries and older houses; and the third, fourth, and fifth zones are all residential, becoming increasingly affluent with increasing distance. This model, like von Thünen's, is a combination of theoretical logic and first-hand experience of the real world. As a descriptive device, it applies best to North American cities. It proved especially useful in the context of changing residential character with the claim that neighbourhoods change because of the operation of ecological processes. Specifically, a process of competition between ethnic groups prompts invasion of desirable locations; a sequence of invasions leads to a succession of land uses, with a dominant land use occupying the desired location at any given time.

This model has several limitations. First, there is the explicit use of physical science concepts—such as those of competition, invasion, succession, and dominance—that derive from plant and animal ecology. Second, ethnic groups are identified solely in terms of area of origin. Third, the premise that ethnic groups wish to remain separated from each other is clearly a generalization. Despite these limitations, this model has proven highly influential and remains of value as a general descriptive device. Of course, it is also important to appreciate that the model is normative, aiming to describe an ideal situation given certain assumptions, not to describe reality. The explicit identification of land-use types and of zones remains an important contribution to our understanding of the geography of the city.

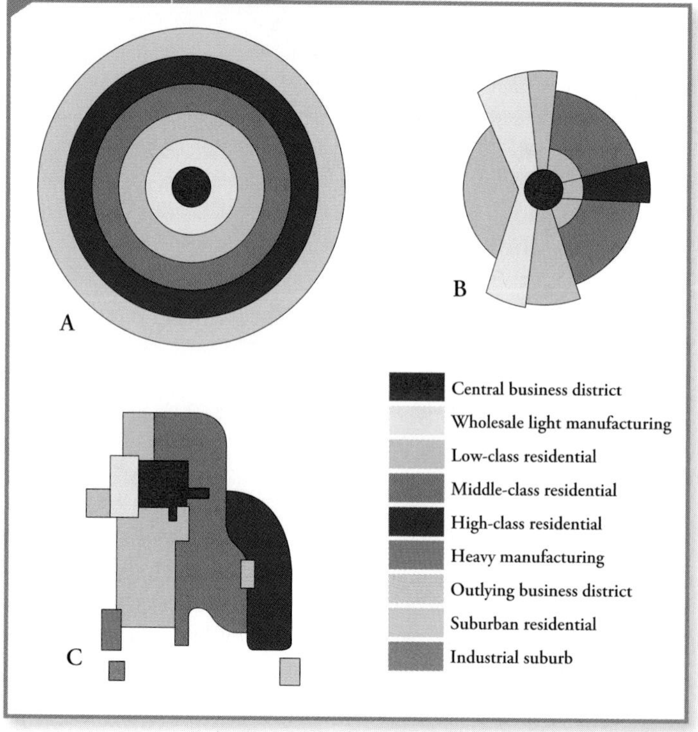

Sector model

The sector model of urban land use (b) was developed by Hoyt (1939) in response to criticisms of the concentric zone model following extensive empirical analyses of urban land markets. It is based on the claim that internal structure is conditioned not solely by distance from the city centre, but also by the location of routes that radiate outwards from the centre. This inclusion of both distance and direction is an improvement on the earlier model.

The model assumes that cities initially have a mix of land uses close to the centre but that, as growth occurs, each of these uses gravitates towards a particular sector. Part of the reason sectors develop is that specific uses are attracted to a particular line of communication and this prompts other uses to locate elsewhere. Once established in a sector, specific uses dominate that sector to the exclusion of other uses. Industrial uses, for example, tend to repel residential uses. Like the concentric model, this is a descriptive formulation that aims to provide a generalized statement of city structure.

Multiple nuclei model

The multiple nuclei model (c) incorporates a third variable in addition to distance and direction—the presence of several discrete centres in the urban area. Developed in 1945 by Harris and Ullman (1945), this model can take many

FIGURE 12.2

Three models of the internal structure of urban areas

Central business district
Wholesale light manufacturing
Low-class residential
Middle-class residential
High-class residential
Heavy manufacturing
Outlying business district
Suburban residential
Industrial suburb

forms in reality—the figure shows only one example. The number of nuclei varies according to city size and details of development. For North American cities, five general land uses were proposed: central business district, wholesaling and light manufacturing area, heavy industrial area, residential areas, and suburbs.

This model proved very attractive to geographers and influenced much subsequent research. In retrospect, it might be seen as an inspired formulation as it clearly anticipates many of the late twentieth-century changes that cities experienced, especially the decline of central business districts and the growth of suburban business and retailing centres as these are discussed later in this chapter.

REVISING THE CLASSIC MODELS

Urban researchers tested these models and also proposed alternative formulations. Working from a British rather than a North American perspective, Mann (1965) revised the concentric zone and sector models. In Mann's model, the higher-class residential areas are always located in a specific sector west and/or southwest of the city. This feature of many British cities is in response to the prevailing westerly/southwesterly winds that cause smoke to drift towards the east and northeast sectors of the city, which, in turn, are home to industrial uses and working-class housing.

A useful revision of the classic models is that by White (1987), who aims to update their arguments to suggest the general form of an early twenty-first-century city in the more developed world. Although this model is in response to changes affecting the internal structure of the city—changes such as deindustrialization and the rise of a service economy, decentralization of retailing activities, increased government intervention, increased automobile use, and related suburban expansion—it maintains the earlier emphasis on distance and also on direction and nuclei. There are seven components to his model (Figure 12.3).

1. The core area, which continues to function as the heart of the city, is the site of government, financial, and business offices. There is less retailing than previously. This zone grows upward rather than outward.
2. Surrounding the core is a zone that was previously light industrial and warehousing (as in the Park, Burgess, McKenzie model). In some cities this second zone is now stagnant,

The location of this Toronto Chinatown—in and around the city's long-time garment manufacturing district—is typical of many such ethnic enclaves.
Dick Hemingway Photographs

but in others it has benefited from business and residential investment.

3. Most cities include several areas of low-quality housing, characterized by poverty and often occupied by minority ethnic groups. Most of these areas are adjacent to the second zone, although some are located elsewhere.

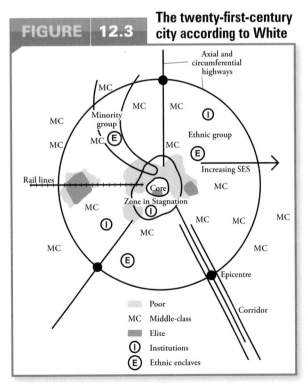

FIGURE 12.3 | The twenty-first-century city according to White

SOURCE: White, Micheal J., Figure 7.1, "The Shape of the Late Twentieth Century Metropolis." In *American Neighborhoods and Residential Differentiation*. © 1987 Russell Sage Foundation, 112 East 64th Street, New York, NY 10021. Reprinted with permission.

4. Much of the rest of the city consists of middle-class residences, often divided into relatively distinct neighbourhoods. Although this is the largest area of the city, it is not continuous; rather, it is interrupted here and there by the three remaining components.

5. Scattered throughout the middle-class area are elite residential enclaves. As with the poor areas, some of these are close to the city centre and some are suburban.

6. Also scattered throughout the middle-class area are various institutional and business centres such as hospitals, malls, and industrial parks.

7. Finally, most cities are growing outward along major roads, and peripheral centres may develop that will be largely independent of the original city form.

This model is helpful because it provides a set of generalizations but is more flexible than the three classic models of internal structure.

This discussion highlights the unsurprising fact that most models of urban internal structure rely primarily on distance, incorporating other variables such as direction and multiple nuclei to more closely approximate reality. All of the models were developed in the context of more developed world countries, especially the United States. But research and the building of theoretical models has also been pursued in regard to cities in the less developed world. These models also rely primarily on distance, adding direction and multiple nuclei and some other factors as judged appropriate. Several of these models are now described.

MODELLING THE LATIN AMERICAN CITY

Prior to colonial European settlement, there were important urban centres in the region of Central America reaching from central Mexico to Honduras, and also in the South American Andean and Pacific coast regions extending from Ecuador to Bolivia. Europeans transformed many of these centres and also built many others. Functionally, most of the cities belonged in one of five categories: agricultural, mining, industrial, commercial, or administrative. Regardless of function, the internal structure of these cities is quite similar.

As discussed later in this chapter, these cities were planned to include a central plaza with streets radiating outwards in a rectangular grid pattern. Wealthy Europeans lived close to the city centre while the city outskirts attracted Aboriginals seeking employment—this basic structure is reminiscent of Sjoberg's model of the pre-industrial city as described in the previous chapter and remains largely unchanged. Of course, Latin American cities have expanded and diversified dramatically since colonial times, and attempts to model internal structure reflect both colonial foundations and subsequent developments.

Figure 12.4 shows one model of the Latin American city. The key features are:

- The city remains centred on the plaza, with land uses varying according to distance from the centre and to direction.
- The centre has both modern and traditional components.
- A commercial spine focused on a major avenue, with high-quality residences adjacent, runs from the centre to a suburban mall.
- An industrial sector culminates in a suburban industrial park.
- A perimeter road connects the outer regions of the city.

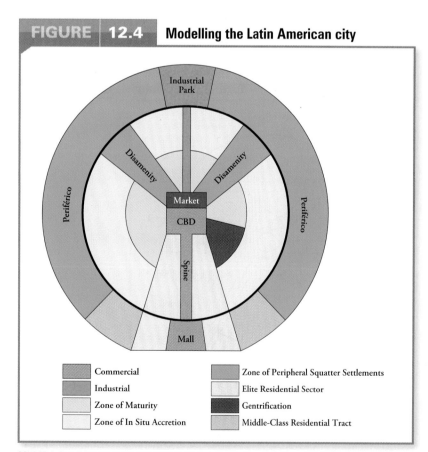

FIGURE | **12.4**   **Modelling the Latin American city**

Legend:
- Commercial
- Industrial
- Zone of Maturity
- Zone of In Situ Accretion
- Zone of Peripheral Squatter Settlements
- Elite Residential Sector
- Gentrification
- Middle-Class Residential Tract

SOURCE: L. Ford, 'A New and Improved Model of Latin American City Structure', *Geographical Review* 83 (1996): 438. Copyright © 1996 The American Geographical Society.

Squatter shacks outside the township of Soweto, south of Johannesburg, South Africa.

frans lemmens/Alamy

- In general, residential quality declines with increasing distance from the centre.
- The zone of maturity is relatively high-quality residential, with areas undergoing improvement in the form of gentrification (discussed in the following chapter).
- The zone of in situ accretion is a mixed and changing residential zone.
- There are sectors of very low-quality housing (labelled as disamenity) and also an outer zone of squatter settlement.

MODELLING THE AFRICAN CITY

The relatively simple models that have provided such useful generalizations for cities in the more developed world and also, to a lesser degree, for Latin American cities are not easily accomplished elsewhere in the world. It is particularly difficult to generalize about African and Asian cities as there are numerous regional traditions and varied colonial imprints. Seven types of African cities were noted by O'Connor (1983).

- Indigenous: there were important cities especially in West Africa and in what are now Ethiopia and Zimbabwe; Addis Ababa is an example.
- Islamic: located mostly in the Saharan region, and including Timbuktu.
- Colonial: located throughout much of the extensive area colonized by Europeans, often as capitals, mining towns, or trading centres; Kinshasa is an example.

- European: some colonial towns, such as Nairobi in the Kenyan highlands, were intended primarily for Europeans.
- Dual: comprising at least two of the previous four types, perhaps on opposite sides of a river as in the case of Khartoum-Omdurman.
- Hybrid: again, comprising at least two of the first four types but with the different parts integrated; Lagos is an example.
- Apartheid: politically, this type is no longer formally in place although many cities in South Africa continue to display the basic pattern of ethnic segregation that was the hallmark of the apartheid system (discussed in Chapter 7).

These different types are not readily generalized into a single model. Indeed, as urbanization proceeds apace, this simple classification is becoming less helpful.

Nevertheless, it remains the case that the internal structure of large African cities today displays variations according to distance and direction, as do most cities throughout the world. There is usually a commercial centre and, increasingly, development of high-quality housing close to the centre. Outer zones of squatter settlement are also usual.

MODELLING THE ASIAN CITY

There are numerous indigenous urban traditions in Asia and a varied colonial imprint. Regardless of location and date and reason for foundation, most Asian cities have an internal structure that reflects both distance and direction. Four of the many examples of city structure are described here.

In Southwest Asia and North Africa most cities display a dominant Islamic influence, being characterized by a major central mosque. A bazaar close to the mosque and an irregular street pattern are also usual features of cities in this region.

In South Asia there are two principal city types—indigenous and colonial. Many indigenous cities developed along important trade routes and are centred on a bazaar that today often offers a wide variety of retail outlets and accommodations. Close to the bazaar are high-quality residences, and housing quality generally declines with increasing distance. There are also enclaves scattered throughout the city for minority ethnic groups and people of low caste.

Most of the major colonial cities in South Asia are coastal and are usually in strategic political or economic locations. As shown in Figure 12.5, the centre was a military fort with adjacent open space. Beyond the open space were several sectors, including an administrative area, a business district, a housing area for the indigenous population, and an area of European settlement. Commercial nuclei centred on bazaars. The outer fringe of the city comprised a series of residential areas. Today, most of these cities are surrounded by squatter settlements.

As a final Asian example, Figure 12.6 diagrams the generalized internal structure of the Southeast Asian city, which has experienced a variety of cultural influences including Indian, Chinese, Arab, and European. Ethnic diversity dominates this model. There is a port zone, as well as a series of other zones and sectors. The overall zonal pattern is one of declining housing quality with increasing distance, while the sectors include commercial zones distinguished according to ethnicity. The outer zones include both suburbs and areas of squatter settlement.

RETHINKING THE CITY

The model formulations described so far, especially the three classic models, do have some significant limitations. Most notably, they are essentially positivist with roots in the neo-classical economics tradition. Not surprisingly, then, they are subject to the same sort of criticisms as are other similarly motivated models. Three important criticisms are: (1) they tend to ignore human decision-making in favour of the economic operator concept; (2) they claim to be value-free, but in effect they legitimize the capitalist status quo; and (3) they lack substantial consideration of social justice issues.

As noted, the concentric zone model focuses on distance, the sector model on distance and direction, and the multiple nuclei model on distance, direction, and multiple centres. The models proposed by Mann and White build on this earlier work. Similarly, the models developed for other regions of the world are closely related to the classic models, with the basic focus on distance, direction, and nuclei. To the extent that all of these models simplify a complex reality, they are generalizations; none of them purports to be universally correct. Thus, although they are valuable as descriptions against which real-world structures can be assessed, these models do not adequately explain the internal structure of urban areas. There is a general failure to answer (or even to ask) questions about processes, about the causes of particular spatial configurations. Further, the three classic models are essentially economic, largely ignoring social, cultural, and political variables. Even models for

| FIGURE | 12.5 | **Modelling the Asian colonial city** |

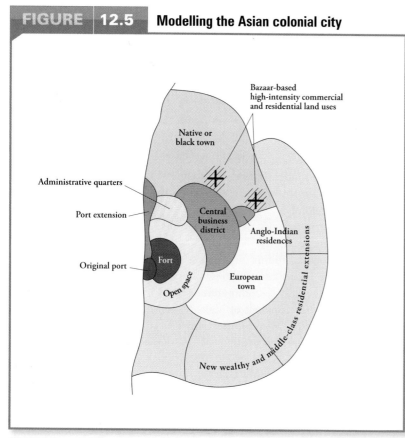

SOURCE: S. Brun and J. Williams, *Cities of the World*, 2nd edn (New York: Harper and Row, 1993), 360.

FIGURE 12.6 Modelling the Southeast Asian city

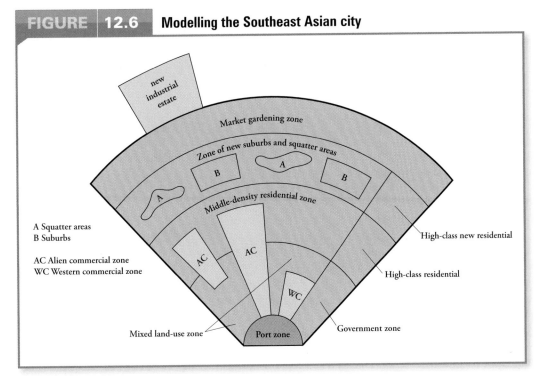

SOURCE: T. McGee, *The Southeast Asian City* (New York: Praeger, 1967), 128.

other world regions that explicitly incorporate cultural and social variables can be criticized for their scholarly dependence on the classic models (Yeoh, 1996: 5).

In the United States some empirical work added to the classic formulations by arguing that land uses relate to social variables. An early example of this work was that of Firey (1947), who analyzed two neighbourhoods in Boston. But the most important response to these models came in the form of a political economy perspective.

A political economy perspective

Political economy is derived from Marxist concepts as described in Chapter 2 and outlined in Box 10.2. Introduced into geography in the 1970s, political economy aims to understand the structural relations that create landscapes. The neo-classical assumption of the economic operator is rejected. Instead, the imperative of accumulating capital that is central to the capitalist mode of production is seen as the basic motivation for human behaviour. Critically, this means that the value of the goods produced by workers must exceed the labour costs involved in their production. This circumstance—the need for capitalists to extract surplus value from labourers—leads to conflict and financial crises. For example, one possible crisis is that of under-consumption:

this arises if workers are not paid enough to purchase the products they produce. Box 12.2 describes the work of one leading geographer who advocates, develops, and applies a political economy perspective.

The paradigmatic city

'The paradigmatic city may be defined as the city that displays more clearly than other cities

The town of Ipoh was founded by Malaysia's British rulers in the 1890s. The colonial influence is apparent in the imposing architecture of its railway station.

V. Last, Geographical Visual Aids

Box 12.2   A Marxist Interpretation of the Urban Experience in a Capitalist World

David Harvey is a human geographer who, after producing a major methodological work on positivism in 1969, has focused primarily on arguing the merits of a Marxist philosophy for understanding the urban way of life (Harvey, 1969, 1973, 1982, 1989, 1996, 2000, 2003, 2005). According to Harvey, Marxism offers a broad theoretical perspective (a meta-theory or higher-level theory) that allows geographers to approach such diverse issues as the built environment, urban economy, and local urban culture.

Using Marxist theory and adding spatial factors, Harvey (1982) argued that uneven spatial development is a necessary accompaniment of capitalism—what might be called 'accumulation by dispossession'. This argument applies on various scales, including that of the urban area. Thus urbanization is seen as one more aspect of the uneven spatial development that occurs in a capitalist context. The spatial generality of these ideas justifies use of the term 'meta-theory'.

Understanding urban places requires appreciation of both the flows of capital and surplus value through systems of cities and the class relations characteristic of a capitalist society. Explaining residential differentiation, for example, requires recognizing that the more powerful classes have resources that allow them to acquire land of their choosing. In Marxist terms, the explanation lies in the superstructure. It is not appropriate, from this perspective, to see residential differentiation as resulting from the preferences of all people; it should be seen as resulting from the preferences of a powerful few.

More specifically, this Marxist perspective identifies the city as the site of three circuits of capital. The primary circuit concerns the structure of relations in the production process, such as the manufacture of goods for sale; the secondary circuit concerns investments in fixed capital, such as property development; and the tertiary circuit concerns investment in science and technology that helps to increase production. According to this theory, the transfer of capital from the primary to the secondary circuit is one cause of suburban expansion: thus, in effect, restructuring capital can lead to spatial restructuring. For some human geographers, Marxist theory provides critical insights into the forces behind the production of the built environment of the city.

---

dual city
A postmodern circumstance; a city with areas occupied by the rich and powerful and areas occupied by the poor and powerless. Divisions may reflect ethnic, religious, and gender differences.

hybrid city
A postmodern circumstance. A city lacking conventional communities, containing instead new cultural categories including cultural hybrids.

cybercity
A postmodern circumstance. A city with parts that are so closely connected that location is less important than in the past.

the fundamental features and trends of the wider urban system' (Nijman, 2000: 135). For much of the twentieth century the paradigmatic city was the industrial city of Chicago and much of the urban geographic research accomplished fell under the general rubric of what was known as the Chicago School. Chicago was home for the sociologists who developed the concentric zone model, and Chicago and other similar Midwest American cities were the most intensively studied. In much of this work there was an implicit assumption that Chicago was the norm. More recently, however, it has been claimed that Los Angeles is now the paradigmatic city. This switch from industrial Chicago to post-industrial Los Angeles, referred to in Box 12.1 as a radical break, reflects the changing face of capitalism from competitive through organized to disorganized forms (using the Glossary terms introduced in Chapter 7).

This concept of the paradigmatic city is useful. Urban theorists today focus less on the old industrial city and more on the emerging post-industrial and postmodern city.

Postmodern urban theory

Contemporary cities in the more developed world typically reflect twentieth-century growth and planning practices, industrialization, and modernity in general. Today, however, a number of new and different processes are operating—what might be described as postmodern processes. The impact of postmodernism is particularly evident in architectural efforts to reproduce local vernacular building styles or to incorporate portions of old buildings in new developments. Similarly, more suburban office clusters than modernist office towers are being built.

Some human geographers have suggested that what urbanism is experiencing today may be not just an evolution but a transformation within the context of a postmodern break from earlier cultural, economic, and political trends. Dear and Flusty (1998: 50) pose the following questions: 'Have we arrived at a radical break in the way cities are developing? Is there something called a postmodern urbanism, which presumes that we can identify some form of template that defines its critical dimensions?' Their answers are in the affirmative. Arguing that most previous research into the urban landscape has been based on the Chicago School of urban ecology, with its emphasis on concentric zones of land use, they propose a new research model based on a southern California-style urbanism that emphasizes global–local link-ages (a reversal of the traditional relationship between hinterland and centre) and a ubiqui-tous social polarization. More specifically, the new Los Angeles School is distinguished from the Chicago School in three ways (Dear, 2005: 340).

- The idea of a central core is replaced by urban peripheries that organize the much changed and weakened core.
- Change in the city no longer relates to indi-vidual-centred agency but rather to global and corporate-dominated connectivity.
- A linear and evolutionist paradigm is replaced by a non-linear and chaotic process.

Accordingly, the proposed Los Angeles model depicts globalization and restructur-ing processes as the driving forces behind the new urbanism. Because the postmodern city includes populations that are not only cultur-ally heterogeneous but also politically and eco-nomically polarized, it is characterized by great disparities in wealth and styles of living. There are theme park areas, service-oriented edge cities, private housing areas based in common-interest developments, fortified zones, and mean streets dominated by gangs. A distinctive feature of this landscape is its fragmented char-acter (Figure 12.7).

It is notable that this new body of ideas, like much other postmodern work, requires a new vocabulary to express its ideas. Thus we have the concepts of 'world city' and of exurban-ization as noted in the previous chapter, and also the concepts of 'dual city', 'hybrid city', and 'cybercity'. A **dual city** is characterized by marked social polarization, by landscapes of privilege and landscapes of deprivation; a **hybrid city** is characterized by fragmentation and cultural hybridity; a **cybercity** is charac-terized by connections that reduce, perhaps even eliminate, the friction of distance.

The internal structure of many cities today, including the edge cities discussed later in this chapter, clearly departs from the classic mod-ernist models. The postmodern model may not be a sufficient description, but it is certainly suggestive of the complex chaos that is the contemporary urban landscape, especially in and around some American cities.

Of course, the idea of a postmodern urban-ism is not without critics. Some have suggested that such ideas stress style over substance,

that the basic premise of a transition from a modern to a postmodern epoch is question-able, and that the case of Los Angeles is not paradigmatic but rather represents an excep-tion to the norm. Certainly urban landscapes today include such new landscapes as **gated communities** designed to maintain privilege and exclude disadvantaged others. Yet at the same time many other components of the larger urban landscape suggest that there is at least as much continuity as there is change today. In any event, for the beginning human geographer postmodern ideas are an indication of the need for constructive re-evaluation of earlier ideas—and of the challenges to be faced in any attempt to make sense of contemporary urbanism.

> **gated community**
> A high-quality residential subdivision or community with access limited to residents and other authorized people such as domestic workers, tradespeople, and visitors. Often surrounded by a perimeter wall, fence, or a buffer zone such as a golf course.

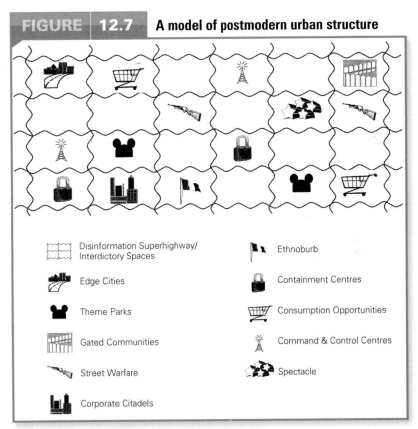

FIGURE 12.7 **A model of postmodern urban structure**

Legend:
- Disinformation Superhighway/ Interdictory Spaces
- Edge Cities
- Theme Parks
- Gated Communities
- Street Warfare
- Corporate Citadels
- Ethnoburb
- Containment Centres
- Consumption Opportunities
- Command & Control Centres
- Spectacle

The style and content of this diagram reflect the consequences of what is called keno capitalism (in a game of keno there is a game card with a numbered grid; during the course of the game, some squares are marked and others are not). Particular urban developments occur as capital is placed in particular locations in the city in an almost random manner. There is no apparent relationship between what is happening in one place and in any other place. The places, or interdictory spaces, are affected by the disinformation superhighway. 'Conventional city form, Chicago-style, is sacrificed in favor of a non-contiguous collage of parcelized, consumption-oriented landscapes de-void of conventional centers yet wired into electronic propinquity and nominally unified by the mythologies of the disinformation superhighway' (Dear and Flusty, 1998: 66).

SOURCE: After M. Dear and S. Flusty, 'Postmodern Urbanism', *Annals, Association of American Geographers* 88 (1998): 66, Fig 4 (MW/RAAG/P1886). Used with permission of Taylor & Francis Ltd., http://www.informaworld.com.

Overall, theorizing about the internal structure of the city is important because the generalized descriptions both reflect and direct empirical studies. With these theories in mind, we now focus attention on the closely related topics of the government of cities and cities as they reflect intentional planning decisions.

Governing the City

Prior to the mid-nineteenth century, governments were rarely active in managing cities, preferring to let economic and social processes unfold without much political interference or direction. By about 1850, social problems, especially relating to housing and disease, were evident in many industrializing European and North American cities, and these problems encouraged the beginnings of formal urban governance intended to manage the city more efficiently and fairly. Accordingly, notwithstanding the continuing role played by private enterprise, an awareness of urban politics is essential to understanding how cities grow and how they function today. There are two processes at play here.

National, regional, and local governments play a key role in determining the provision of services, thus contributing to the social geography of the city.

Urban development typically requires substantial public investment and can in turn create wealth for developers and other business interests. It is hardly surprising, then, that there are close links between politicians and business people.

The role played by local urban governments cannot easily be separated from both the number of levels of government and the larger national political ideology. In some countries, such as the United Kingdom and New Zealand, there are just two principal levels of government, national and local. In other countries, such as Canada, the United States, Germany, and Australia, there are three levels, national, regional (usually called state or provincial), and local. This is the same distinction as was made in Chapter 8 between unitary and federal systems of government.

The larger national political ideology varies between countries. This section is concerned principally with cities in the democratic, capitalist, more developed world. In a politically centralized country, such as China, local government circumstances can be quite different. Larger national political ideology also changes through time in any given country. For example, during the 1980s political thinking in many countries began to question the value of state involvement, preferring instead to encourage private enterprise to play a greater role; this change can be described as moving from liberal and socially progressive urban policies towards a more conservative laissez-faire approach.

URBAN GOVERNMENT

Although local urban governments are a relatively low rung on the governance ladder, they do many things. Much of the infrastructure of a city—including streets, sewers, lighting, and the water supply system—is publicly owned, meaning that local politicians and civil

One of numerous gated houses in the wealthy community of Jupiter Island, Florida.

AP photo/Steve Mitchell

servants are responsible for ensuring adequate provision of these services. They may also be responsible for public safety, health, education, and recreation. In terms of cost, education is often the largest expense. Money to fund activities is raised largely through property taxes and through income provided by other levels of government. Local governments typically intervene in property and labour markets, regulating the activities of the private sector through land-use zoning and building controls, and resolving any differences of opinion between conflicting interests about such issues as the location of a new sewage treatment plant or an adult bookstore.

Referring to local urban governments is rather misleading because it disguises the fact that many different types of local government exist. Indeed, this level of government might be described as an ongoing search for the best type. Seven major types of metropolitan government are outlined in Table 12.1.

Several difficulties and problems inherent in local governance can be seen from this table. Among these are the following:

- Local councils may fail to provide adequate services for all city residents.
- Voluntary co-operation may not work when it is most needed.

Table 12.1	Types of Metropolitan Government			
Type of Government	Examples	Roles	Advantages	Disadvantages
Small independent councils	Common in US	Local planning, representation, regulation. Purchase of community services.	'Public choice' for residents. Payment limited to selected services.	Affluent communities and industrial areas opt out of metropolitan service funding. Exclusionary zoning.
Voluntary co-ordination of local councils	London (UK) Planning Advisory Committee	Exchanging information. Regional advocacy. Preparing regional studies and advisory plans.	Voluntarism, co-operation, and consensus-building. Development of regional consciousness.	Lack of powers of implementation, control, or co-ordination. Record of ineffectiveness.
National, state, provincial administration and service provision	Australian states; many less developed countries	Preparation of metropolitan planning strategies. Supply of power, water, transport services.	Potentially strong implementation powers and finances. Capacity for wide contextual view.	Indirect accountability. Poor political control. Poor recognition of metropolitan-scale issues.
Special boards and districts performing specific functions and services	Canadian provinces; US states (New York has more than 2,000)	Schools, fire, highways, health, water, sewerage, drainage, power, light, gas, housing, and transit.	Easy to establish as new needs evident. Elected boards can maximize local participation.	Lack of co-ordination and planning capacity. Lack of accountability of appointed boards.
Metropolitan governments nominated from local councils (indirectly elected)	Regional districts in British Columbia, e.g., Greater Vancouver	Strategic, environmental, transport, and functional planning. Water supply, waste disposal, transit.	Regional co-ordination of planning and services. Links well with other levels of government.	Lack of direct electoral mandate, low public profile. Reliance on powers from above and political support from below.
Metropolitan governments (directly elected)	Metro Toronto	Vary. Very wide range of powers and services in Metro Toronto.	High public profile and role. Powers to exact compliance from junior jurisdictions. Capacity to take on new functions and extend effectiveness.	Conflicts of interest with other levels of government. Difficulties in adjusting boundaries to match expanding metropolitan extent.
Regional governments	Sweden (1976), New Zealand (1989–92)	Vary from environmental management and resource policy (NZ) to wide-ranging powers and services (Sweden).	Reflects changed scale of social interaction. Intermediary between local authorities and national governments.	Involves radical changes to existing administrative systems. Hard to introduce in federal systems because states/provinces reluctant to cede power.

SOURCE: Adapted from P. Heywood, 'The Emerging Social Metropolis: Successful Planning Initiatives in Five New World Metropolitan Regions', *Progress in Planning* 47 (1997): 221–2. Copyright © 1996 Elsevier Science Ltd.

- Service provision by state/provincial governments or by special boards suffers from lack of co-ordination with other bodies.
- Indirectly elected metropolitan governments lack a direct electoral mandate to justify their actions.
- Directly elected metropolitan governments often conflict with other levels of government.

- Regional governments are difficult to introduce because they are likely to compete with provincial/state governments.

Perhaps most notably, any fragmented political system can involve significant financial disparities between different governments and hence can lead to very different levels of

Box 12.3 Governing the Canadian City

Canada is a federal state with government structured into three, sometimes four, levels—federal, provincial or territorial, and municipal, which is sometimes divided into regional and local.

Although the federal government has no legislative authority over municipalities, federal economic policies have many impacts on urban areas. Further, the federal government owns much land in urban areas and has authority over ports, airports, and railways. In general, provincial and territorial governments are financially responsible for education, health care, and social services, and have many impacts on specific aspects of urban land use. Also, provincial and territorial governments are authorized to create municipalities. There are hundreds of municipalities in each province and territory and many different specific forms. Some municipalities are single-tier; others are part of a two-tier system. Local municipalities are called cities, towns, or townships. If they are part of a two-tier local government they are referred to as the lower tier, with the upper tier being a region, county, or district.

The roles played by municipal governments in Canada are highly variable and subject to changes imposed by provincial or territorial governments. Regardless of specific political and legal arrangements, for most municipal governments throughout Canada the primary concern is with regulating, servicing, and taxing the urban environment. Major services include waste collection and disposal, street maintenance, water supply, public transit, libraries, emergency services, and animal control. In this context they listen to and work with interest groups, including developers, businesses, and citizen groups. As such, the working of these governments needs to be understood in the light of urban political theory.

Some of the duties that in other countries are the responsibility of urban governments are assigned to special-purpose bodies in Canadian municipalities—the best-known example is that of school boards. In recent years the authority of some special-purpose bodies has diminished, with power being assumed by other local or provincial governments. For both governments and special-purpose bodies, income to fund services is usually derived from three sources: grants from provincial governments, property taxes, and user fees.

Until the 1950s major Canadian municipalities were not politically linked to surrounding rural areas, at least partly because of a general assumption that urban and rural areas were quite different. However, as large cities expanded outward some linkage became inevitable. A first type of metropolitan government was established in Montreal in 1920, but in Toronto in 1953 and Winnipeg in 1960 metropolitan governments were first established with some significant authority over areas of expansion. By the 1960s all three of these governments functioned in a similar way, with the central city and the surrounding municipalities each transferring some of their responsibilities to the new level of government. The 1960s and 1970s witnessed further significant changes in the organization of municipal governments in many parts of Canada, with Ontario and British Columbia, for example, adding an additional upper-tier level, while in 1972 Winnipeg merged the core city and 11 suburban areas into a single structure called Unicity.

Two-tier systems are often criticized for being too complex, and recent moves to a single-tier system have occurred in Ontario, Quebec, and Nova Scotia. Most notably, Toronto moved to a single level of government in 1998 when seven municipalities were merged into the City of Toronto, thus resulting in the creation of the fifth largest municipal government in North America. This major restructuring was initiated by the provincial Conservative government and was ambitious or tyrannical—most voters opposed the change—depending on political viewpoint. The restructuring can be seen as the culmination of a process of change initiated in 1953 with the creation of the initial metropolitan region. Interestingly, this strategy is essentially the same as that applied in Winnipeg when Unicity was created.

A recent report commissioned by the Federation of Canadian Municipalities highlighted the deteriorating financial situation of municipalities across Canada, the result of increased responsibilities and reduced revenues (Mirza, 2007). As described above, municipal governments have many responsibilities but receive only about 8 per cent of all tax revenues. What this means is that many needed infrastructural improvements—in water supply, transportation, transit services, recreational facilities, and waste management facilities—are not taking place. Mirza (2007: 21) concludes that 'much of our municipal infrastructure is past its service life and near collapse', and recommends the development of a national plan to eliminate the municipal infrastructure deficit and prepare for the future.

service provision in different parts of the city. In many cities this takes the form of a deprived core and affluent suburbs. The final three types in Table 12.1 are attempts to tackle some of the problems created by political fragmentation. A two-tier metropolitan government was established in 1953 for Toronto, while Winnipeg established a single-tier government in 1972. Recent years have witnessed much municipal reorganization in Canada; Box 12.3 considers some of these governance changes in the context of federal and provincial or territorial authority.

HOW ARE CITIES REALLY GOVERNED?

In principle, three key groups are involved in urban politics. (1) Professional planners—local-level bureaucrats, consultants, and academics—collect and analyze data, prepare reports, and make policy recommendations. (2) Politicians

are responsible for deciding goals and priorities and for making decisions. (3) The people who live in the city elect local politicians and are directly affected by the decisions made by those officials. One way to conceive of urban politics is described in Figure 12.8.

But governing the city is complicated by five challenges. *First*, local governments often have an uneasy relationship with other levels of government because of different interests and perhaps a different ideology. *Second*, a fundamental constraint is the available tax base, which is dependent on the overall economic health of the city. Booming cities have funds available to provide a relatively high level of services, whereas economically depressed cities, which as a rule are in greater need of improved services (from social welfare to parks and policing), lack needed funds to improve services. *Third*, many cities are growing outward, and consequently local governments

FIGURE 12.8 **Professional, political, and public participation in urban policy-making**

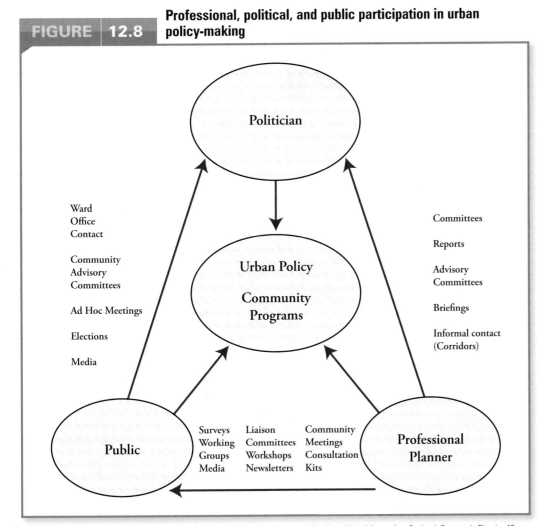

SOURCE: P. Heywood, 'The Emerging Social Metropolis: Successful Planning Initiatives in Five New World Metropolitan Regions', *Progress in Planning* 47 (1997): 214. Copyright ©1997 Elsevier Science Ltd.

An underpass in Montreal in need of repair.

Ximena Griscti/ Alamy/GetStock

may seek to expand the territory over which they have some authority and thus come into conflict with other local governments. *Fourth*, cities are socially and economically fragmented such that there are often very different public interests and concerns in different parts of the city—contested place identities complicate both planning and governance. *Fifth*, some influential special interest groups, such as developers or single-issue citizen groups (i.e., NIMBY or 'not in my back yard' groups), may be able to direct urban policy decisions in their favour and therefore to the disadvantage of others. In some cases, particularly in regard to development, zoning, and polluting industries, this informal influence results from patronage or corruption. The fourth and fifth challenges in particular have prompted much debate about precisely how urban politics unfolds. Five political theories help us to understand the political dynamics of the city.

Pluralism

A theoretical stance conceived in the 1950s and influential for at least the next 30 years, pluralism views urban politics as proceeding in

a relatively orderly and equitable manner, with politicians making decisions following representation by interested groups such as trade unions, the Chamber of Commerce, and other formal organizations. Neo-pluralist logic contends that the number of groups with different perspectives is now much increased, which means that urban politics is a much less orderly endeavour.

Structural logic

Most urban political theories dispute the core pluralist idea, arguing instead that power is concentrated in one or a small number of groups. Structural Marxist logic claims that one particular interest group—business—is able to wield too much influence. The basic logic is that in a capitalist economy the accumulation of capital takes precedence over social goals and, hence, business elites with a narrow understanding of the city based on self-interest are in a commanding position to influence urban politics, perhaps through the city budget.

Regulation concepts

This is the idea that urban politics changes in response to changes in the capitalist mode of production, especially the change, noted in Chapter 7 and more fully described in the preceding chapter, from organized capitalism to disorganized capitalism. In the 1980s, deindustrialization and related economic restructuring prompted a shift from planning solutions to a more market-based approach to urban government. More recently, some evidence indicates a shift away from reactive policies to a more active urban entrepreneurialism, as implied in the next two versions of urban political theory.

Urban regimes

Urban politics is often dominated by one particular regime—an informal but stable coalition—that works with some particular goal in mind. Although regime types often overlap, studies identify several principal types of regime. One regime that is most often found in relatively homogeneous communities favours maintaining the status quo as this is seen as a desirable state of affairs. Another type of regime advocates development, often in the context of a specific large urban project such as downtown redevelopment. Some other regimes strive for social gains for city dwellers rather than development, perhaps focusing on urban conditions for disadvantaged groups. Finally, another regime type focuses on promoting the image of a city, perhaps through revitalization or conservation policies.

Identity politics

The hugely influential idea that a politics of difference is replacing an earlier politics of equality was noted in Chapter 7. In general, this is the idea that our identities are increasingly unsettled, with many of the traditional bases for identity, such as religion, becoming less relevant. Emerging instead are numerous fluid identities related especially to place, local community, multiculturalism, sexuality, gender, opposition to dominant groups, and popular culture. Many of these new ways of thinking about ourselves are related to the identity challenges posed by globalization processes that can be interpreted as contributing to a destabilizing of local identity. Figure 12.9 shows the presence of neighbourhood organizations in one part of Chicago. In some cases, these new identities have a grassroots origin and only come into being for a specific purpose—such as preserving an urban landmark, resisting rent increases, or asserting right of access to some place—and then fade away when no longer needed.

Planning the City

WHY PLAN?

Cities are unplanned—in the sense that, every day, business people are making decisions about where to invest capital, for example, about where to locate a new activity such as a bowling alley or a video store. In addition, residents make decisions daily about where to spend money or where to play, for example, about where to shop or which park to frequent. Although necessarily constrained in various ways, these decisions are not planned. They are, however, important in shaping the social and economic geography of the city: while each decision has only a minimal impact, the cumulative effect of all these decisions is to contribute significantly to that geography. But cities are not left to develop solely through these operations of the free market. City form does not just follow city function.

Cities are also planned. As already described, they are planned as one part of larger urban policies designed to manage the city, for example, to ensure the provision of essential

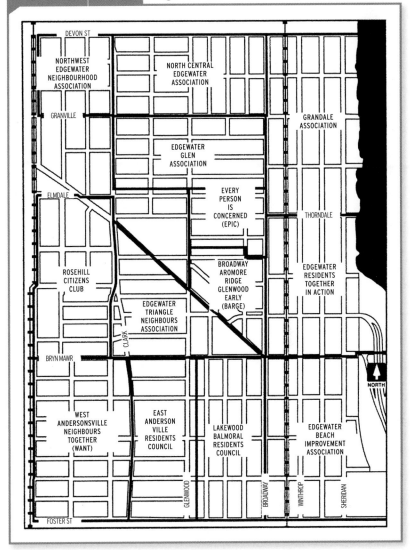

FIGURE 12.9 Neighbourhood organizations in the Edgewater area on Chicago's Northside

SOURCE: B. Badcock, *Making Sense of Cities: A Geographical Survey* (London: Edward Arnold, 2002), 233. © 2002 Blair Badcock. Reproduced by permission of Edward Arnold (Publishers) Ltd.

services such as a sewage system, a water supply, and transportation networks. Further, cities are planned in that many of the decisions that business people and residents make are subject to constraints, such as zoning regulations that affect land use. Such regulations are one component of a larger urban planning exercise intended to influence what is located where inside the city and to influence the spatial spread and temporal growth of the city. It is inconceivable to think of a city today that could develop and effectively function without some planning activity. In other words, city function follows city form.

We might say that urban planning provides for the broad outlines of the city, serving as a template that is moulded into a particular and ever-changing urban landscape by the every-day actions of those who participate in the life of the city.

The politics of planning

Today, the need for urban planning is widely acknowledged by people regardless of political ideology. Of course, this does not mean that there is always unanimous agreement about a particular planning strategy because planning is one of the ways that particular values are imposed on landscapes. Indeed, there is often sharp disagreement between people with different political perspectives, especially concerning the relative roles of government interference on the one hand and the operation of the free market on the other. Planning may be criticized by radical thinkers who interpret it as simply being a tool used by government to benefit private enterprise and to reinforce the unequal geography of the city. Planning may also be criticized by conservative thinkers who sometimes interpret it as causing urban decline because of the rigid enforcement of zoning and other policies.

In short, urban planning is essential, ubiquitous, but often controversial. The fundamental principle underlying all planning activity is that cities cannot be allowed to develop without some controls being imposed by the appropriate level(s) of government. As planning implies a loss of individual rights and a search for the common good, perhaps the key question to be asked about any planning exercise is—planning for whom?

ORIGINS

The first permanent settlements likely developed without any overseeing by some authority.

But some of the earliest large cities did not just emerge; rather, they were at least partially created, as evidenced especially by the use of a rectangular grid street layout, often including a central square. Such a scheme, imposing some order and regularity, required a strong central authority that desired to assert itself on and through the urban landscape. An early use of the grid system was in about 2500 BCE in Mohenjo Daro, located on the Indus River in what is now Pakistan. The grid layout scheme was popularized by Hippodamus of Miletus, a fifth-century BCE Greek architect who planned several settlements, including the harbour town of Peiraeus at Athens and the new city of Rhodes, and it was also widely used throughout the Roman Empire and medieval Europe. Indeed, it has been, and continues to be, so frequently employed that it might be regarded as a normal feature of most planned cities.

Many later cities were also planned, at least in part. Prior to the rapid growth of cities during the Industrial Revolution, most European cities show evidence of the expression of authority on the urban landscape, notably in the orientation of streets and the location of public places such as squares and market areas. These ideas were taken overseas by Europeans and are evident in Canada in the layout of such cities as Vancouver, Quebec City, and Charlottetown. Figure 12.10 shows the 1768 street plan for colonial Charlottetown.

It was not unusual for the central areas of large cities to be designed in a monumental style, with wide and long avenues and intentional focal points at intersections where statues or other imposing structures were located. As with some of the early grid pattern cities, the motivation for this type of planning was primarily to express authority in landscape, not to improve the quality of life for all residents of the city. A famous example is Versailles, close to Paris in France, which was the basis for the design of Washington, DC.

Given these planned origins of many cities, it is not surprising that many urban landscapes, in common with other human landscapes, can be interpreted symbolically—that is, as physical representations of cultural identities. In a series of iconographic studies, Cosgrove (for example, 1984) analyzed the cities of Renaissance Italy in the context of humanistic culture, which was expressed in an urban landscape that displayed wealth through monuments and patronage of the arts. More generally, it is evident that many

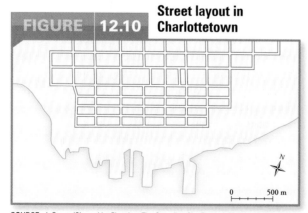

FIGURE 12.10

Street layout in Charlottetown

SOURCE: J. Grant, 'Shaped by Planning: The Canadian City Through Time', in T. Bunting and P. Filion, eds, *Canadian Cities in Transition: Local Through Global Perspectives*, 3rd edn (Toronto: Oxford University Press, 2006), 321.

planned urban landscapes are constructed to represent and legitimate the power that the elite are able to exercise (Duncan, 1990).

IMPROVING THE INDUSTRIAL CITY

While some cities included at least some degree of planning, land use in many other cities was unregulated and uncontrolled and, as a result, these unplanned cities often grew haphazardly in response to specific local circumstances. This situation changed in the first half of the nineteenth century with the onset of rapid urbanization associated with industrialization. As urbanization proceeded it was clear that many cities in Europe and North America had major sanitation problems that could only be solved through the planning and building of citywide sewage systems. Early industrial cities were dirty and overcrowded, inviting diseases such as typhoid, typhus, and cholera. What is often described as modern urban planning developed from the need for essential improvements. While the initial focus of this planning was on improving health, planners soon focused attention as well on related concerns, such as including open parkland areas inside the city and building better housing and neighbourhoods.

These interrelated concerns were evident especially in the work of Frederick Law Olmstead, the American planner acknowledged to be the founder of American urban design and landscape architecture. One of the best-known planned urban open spaces, designed by Olmstead and his partner Calvert Vaux in 1857, is Central Park in New York.

Model industrial villages

Urban planning is not separate from larger political and social concerns. Indeed, the nineteenth-century origins of the modern planning movement show close links with the political philosophy of anarchism, through such notable figures as Patrick Geddes and the geographers Peter Kropotkin and Elisée Réclus, and also with Utopian socialism, for example through the Scottish philanthropist Robert Owen. These links highlight that the urge for modern urban planning was at least partly motivated by a concern for the welfare of city residents. Interestingly, their concern contrasts markedly with that of many city leaders and capitalists, who, as in previous periods, saw in cities opportunities for ostentatious display. Railway stations, banks, and even factories were often constructed in monumental proportions so that today the core of many British cities comprises clear expressions of the Victorian municipal spirit.

Monumental buildings were expensive to build and the pioneering planners envisaged better ways to use money. In some cases, paternalistic employers constructed planned towns for their workers; three of the best-known British examples of these planned working-class communities are Saltaire, Port Sunlight, and Bournville. Each of these new towns included not only residences, but also a range of community facilities and services. Saltaire was built in the 1850s to house people working in Titus Salt's new woollen factory near Bradford in northern England; Port Sunlight was built in the 1880s to house people working in the Lever Brothers soap factory on the south side of the Mersey River in northwest England, and Bournville was built in the 1870s to house people working in the confectionery factory set up by George Cadbury near Birmingham in central England. Figure 12.11 shows the layout of Bournville with the factory, homes, schools, recreational facilities, and other land uses all contained within the community.

PLANNING CITIES AND SUBURBS

As the reference to model industrial villages suggests, by the late nineteenth century there was an interest not just in making improvements to existing urban landscapes, but also in the more exciting prospect of creating new urban places. The model villages were one rather idiosyncratic aspect of this interest.

Beautiful cities

The City Beautiful movement was one of two major planning directions proposed in reaction to the pollution and crowding of industrial cities. Inspired by a prototype city on display at the 1893 Chicago Exposition, this movement advocated constructing aesthetically pleasing cities with a focus on those parts of the city landscape over which governments had some control, such as the layout of streets, public buildings, and public open spaces. The most substantive outcome of this movement was the famous comprehensive integrated plan for Chicago that proposed a series of ring and radial roads, street widening, an integrated public transportation system, a uniform height for public buildings, and extensive open spaces. The plan was not fully implemented but many

FIGURE 12.11   Plan of the model industrial community of Bournville

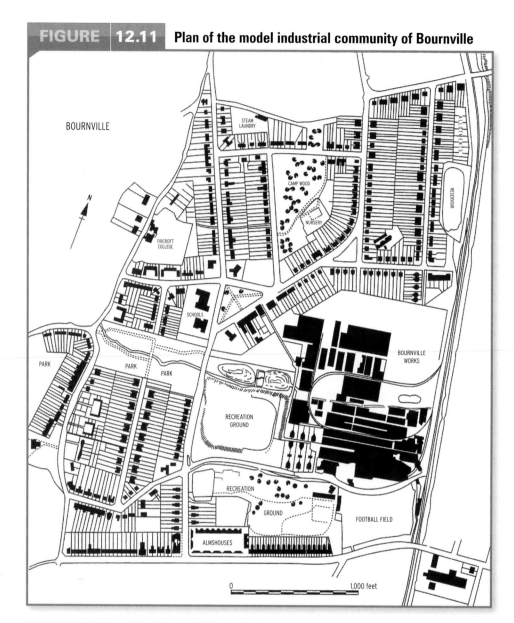

SOURCE: S. Ward, *Planning and Urban Change*, 2nd edn (London: Paul Chapman, 2004), 20. Copyright © 1992 Sage Publications Ltd. Reprinted by permission.

green belt
A planned area of open, partially rural, land surrounding an urban area.

garden city
A planned settlement designed to combine the advantages of urban and rural living; an urban centre emphasizing spaciousness and quality of life.

specific components were approved and put in place.

The movement aroused much interest in other countries, including Canada where city beautification groups arose around the country. Despite the interest generated, however, it appears that, in retrospect, the City Beautiful movement is best seen as a brief effort to impose a social order on urban areas, and possibly also as a modernistic attempt to control the often chaotic growth of large cities.

Garden cities

The second principal direction was initiated in 1898 when Ebenezer Howard introduced the idea of the **garden city**, intended to be a blend of city and nature. Howard was influenced in his thinking by some of the model industrial towns in England and was also reacting against the glorification of industrial capital in the form of the large and expensive buildings so characteristic of the industrial town. Each new city was to be built according to a master plan and to provide a spacious and high-quality environment for working and living. There was to be a concentric pattern of land use, wide streets, low-density housing, public open spaces, and a **green belt**. About 32,000 people were to live in an area of 2,429 ha (6,000 acres). Figure 12.12 shows the planned garden city and its agricultural belt. The seemingly limitless growth of industrializing Britain was to be replaced by intentionally structured urban settlements.

> Howard's singular contribution, influenced by the welter of ideas circulating in later Victorian Britain, was an urban model in the form of an ideal town, built as a satellite. Located at a distance from the parent city, surrounded by an agricultural belt, and developed on land held in common by the community, it was brilliantly conceived. A constellation of such satellites, forming a 'Social City', was in direct answer to many of the problems of the day, both urban and rural. His vision spoke the language of social progress and the perfectability of man. (Cherry, 1988: 65)

Several towns, most notably Welwyn and Letchworth, both near London, were built following a modified version of the garden city plan, and the basic concepts have been highly influential in urban planning generally. The manager of the Letchworth project, Thomas Adams, moved to Canada and was the leading government figure concerned with urban planning for several years after 1914, writing

'The Civic Centre of Calgary as It May Appear Many Years Hence': this frontispiece of a plan by Thomas Mawson shows the influence of the City Beautiful movement, which favoured impressive malls with sweeping views.

"Frontispiece—The Civic Centre as it may appear many years hence" from *The City of Calgary: Past, Present and Future* (1914) by Thomas H. Mawson & Sons, The City of Calgary, Corporate Records, Archives, Town Planning Commission, File 6.

planning legislation and promoting both City Beautiful and garden city principles.

Inspired by the success of Howard's garden city concept, two American architects, Charles

---

FIGURE 12.12 Ebenezer Howard's garden city and its agricultural belt

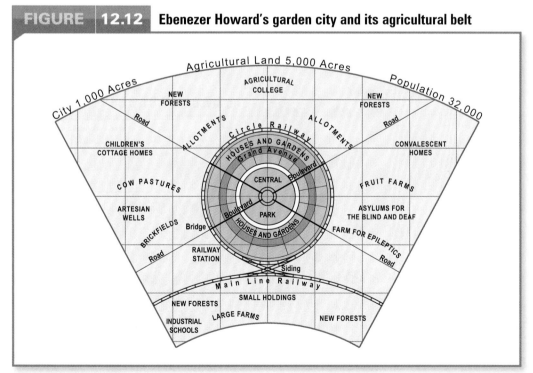

This diagram shows the entire 6,000 acres. Note that radial routes divide the city into sectors and that a rail line encircles the city. Howard also produced plans showing the layout of land uses inside the city and plans showing the pattern of several of these cities.

SOURCE: G. Cherry, *Cities and Plans: The Shaping of Urban Britain in the Nineteenth and Twentieth Centuries* (New York: Arnold, 1988), 66–7. Reproduced by permission of Edward Arnold (Publishers) Ltd.

Stein and Henry Wright, created the garden suburb of Radburn, New Jersey, in the 1920s. This innovative plan was designed to be safe for children, with pedestrian and vehicle traffic kept apart by means of a road network for vehicles only and a network of internal pedestrian pathways. The suburb comprises superblocks sharing an area of common parkland that can be thought of as the heart of the community. Houses are turned around in that back doors face the roadway and front doors face the parkland and pathways. Although the scale of the original plan was never fulfilled, with only 667 homes built, Radburn today is a highly desirable place to live as judged by sales. The Radburn plan has not been widely copied, but one successful Canadian suburb designed in this way is described in Box 12.4.

Box 12.4   Wildwood Park Community in Winnipeg

Two communities in Canada—Kitimat, BC, and Wildwood Park, a residential community in south Winnipeg—are at least partly patterned on the Radburn model. Hubert Bird, a local developer and builder, established Wildwood in the late 1940s after having seen Radburn from the air while flying between New York and Europe. The neighbourhood comprises 286 homes organized around 10 horseshoe-shaped bays (Figure 12.13). The reversed-design concept has the front of some houses, those on the outside of each looped lane, looking onto a community park with pedestrian walkways, and the back looking onto a vehicular lane. Other houses, those on the inside of each looped lane, have the front looking onto a pedestrian walkway that provides direct access to the park and the back looking onto a vehicular lane. Like Radburn, a major motivation is that of separating people and vehicles and also of eliminating through traffic. All homes—bungalow, 1.5-storey, and 2-storey—are single family.

Wildwood Park is enclosed in a bend of the Red River with a golf course and a private school as neighbours. Only one side links the neighbourhood with other Winnipeg residences and with businesses. In common with residents of similar developments, Wildwood residents have a strong sense of community, and property values indicate that homes in the community are considered desirable.

But the reversed-design concept is not without problems, critics, and controversy. Reversing the traditionally public and private faces of homes has consequences for both front and back landscapes. One problem is that living in a reversed house results in social contacts with neighbours being centred on back doors rather than front doors, a circumstance that is different to that experienced in other suburbs. Also, visitors to the community encounter difficulty when first trying to locate a specific home. Among the critics of this type of community, and of garden cities more generally, was the renowned urban commentator, Jane Jacobs. Perhaps the most important criticism is that reversed-design communities have not really worked out quite as the planners intended. In Wildwood, the communal parkland at the front of homes designed to be the place for movement and social interaction is not well used. Rather, the back lanes fulfill both of these roles. Indeed, the back lanes are so well used that they are often very congested. Increased car ownership adds to this congestion and has changed the landscape markedly, with additional garage construction often creating a barrier between home and lane. It is also clear that the original logic behind the design is being questioned by some current residents and unanticipated changes are taking place. For example, a controversy erupted in 2008 when one resident received planning permission to build a 300-square-foot deck on the front of a house despite the local councillor and some residents expressing strenuous objections to this loss of shared green space. The failure of the reversed-design community to function as planners intended is one of the reasons why many planners today advocate streetscapes as the principal public space, a preference that is part of the New Urbanism approach to planning (Gillmor, 2005; Martin, 2001).

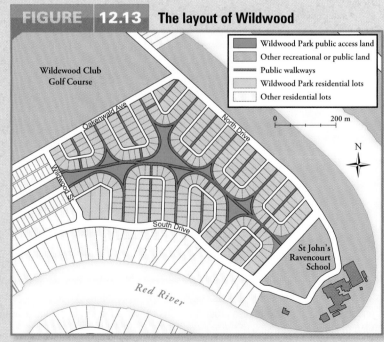

FIGURE | 12.13   The layout of Wildwood

Wildewood Club Golf Course

Oakenwald Ave

North Drive

Wildwood St

South Drive

Red River

St John's Ravencourt School

- Wildwood Park public access land
- Other recreational or public land
- Public walkways
- Wildwood Park residential lots
- Other residential lots

0   200 m

N

SOURCE: D. Gillmor, 'Wildwood Childhood', *Canadian Geographic* (July–Aug. 2005): 57.

Cities of towers

The Swiss architect Le Corbusier (1887–1965) designed totally new modernist cities characterized by public ownership, but saw very few of his designs converted to reality. While other planners advocated garden suburbs and low-rise dwellings, Le Corbusier was planning large, dramatic-looking urban centres with massive multi-storey buildings. Four of the adjectives often applied to Le Corbusier are innovative, influential, admired, and maligned.

Le Corbusier advocated increasing both open space and population density through use of high-rise buildings separated by parkland. Clusters of buildings were to be linked by road or rail. In the 1920s, these were revolutionary ideas. The new planned cities were to have a population of about 3 million and an organized spatial structure with residences located according to social class. In some cases the plan was for people to live and work in the same building. Only two cities have been built on these lines: Chandigarh, the new capital of the Punjab, India, and Brasilia, the new capital of Brazil. Today, both cities have experienced an influx of poor people leading to squatter settlements (discussed in Chapter 13) on the city margins, significantly affecting the form of the city as originally envisaged. Parts of Brasilia today are plagued with crime, violence, traffic congestion, and pollution. It is also possible to see Le Corbusier's influence in the high-rise buildings that were built close to many British city centres during the 1950s and 1960s.

Like Le Corbusier, the urban designs of Frank Lloyd Wright reflected the modernist taste for purity and simplicity in design but were never really fulfilled. Nevertheless, his impact on architectural style has been substantial. Indeed, Wright's vision was not too different from the modern suburb.

New Urbanism

The idea of building ideal communities, evident with the model industrial villages, the Radburn design, and some other plans, was resurrected in the 1980s in the form of what has been called New Urbanism. This was developed in response to the predominant residential trend of twentieth-century urbanism, which, as described later in this chapter, is the spread of suburbs on the outskirts of a city. For many planners this is seen as a regrettable trend because it consumes large areas and creates residential areas that may lack focus and a sense of community. New Urbanism was also based on the experiences of some European urban centres that have maintained a concentrated population without excessive suburbanization, including places such as Barcelona, Florence, and Capri.

Described by Veninga (2004: 461) as 'an architectural and planning ideology that seeks to change the feel, look, and role of urban landscapes', New Urbanism has focused on building communities based on pedestrian movement and mass transit rather than private vehicles. Businesses are readily accessible by walking or by use of public transit. Streets are intentionally narrow in an effort to reduce traffic speed and make the communities friendlier and safer for all residents. Homes have front porches close to the street to encourage social interaction. Each neighbourhood is to have an evident centre. There are a wide variety of dwelling types, including detached single-family homes, condominiums, and apartments designed to be attractive to a range of income groups. Rather like the Radburn concept, parking lots and garages are not located at the front of dwellings but at the rear.

Examples of these planned communities include Seaside, on the Florida panhandle (see photo, p. 271), the first such community and the film location of the 1998 hit, *The Truman Show*; Celebration, Florida, built by the Disney Corporation; and Kentlands, Maryland. There are also Canadian examples—e.g., Mackenzie Towne in Calgary and Cornell in Markham, Ontario—and New Urbanist-inspired communities have been built in other countries such as Jamaica, Turkey, and the Philippines. The intent of the New Urbanism is not simply to plan the city, but also to advocate and sell a particular image of what urban life ought to be. Urban places are being mythologized, and urban dwellers are actively encouraged to behave in particular ways. Indeed, this is an essential theme of *The Truman Show*, in which actor Jim Carrey portrays Truman Burbank, who from birth has been the focus of microscopic media attention and behavioural modification, just like the closed, planned community in which he lives.

THE IMAGE OF THE CITY

Planning does not just involve the imposition of the ideas of planners on landscape. Planners often try to respond to perceived needs of urban dwellers. Consider the following questions that planners might ask:

- What images do residents have of their cities?
- What are residents' mental maps and how do these maps vary from the actual geography of the city?
- Why do residents understand the geography of some areas of a city well, but not other areas?
- How and in what ways do residents invest meaning in the city landscape?
- Are residents able to understand the connections between different parts of the city?

These questions are important both practically (for getting around) and emotionally (for making sense of where we live). Ideally, cities need to be both imageable, meaning that the city landscape evokes strong emotions in the observer, and also legible, meaning that the different parts of the city are easily connected by the observer. Together, imageability and legibility make up the visual quality of a city. A seminal work by Lynch (1960) analyzed the visual quality of urban areas with specific reference to Boston, Jersey City, and Los Angeles.

To determine how residents of different areas perceived their city, people were asked to identify features of the urban landscape. Five kinds of features that give character and coherence to a city were identified:

1. Paths are various routes through the city, from which many parts of the city can be seen.
2. Edges are boundaries between perceived districts.
3. Districts are parts of the city that are in some way perceived as different from surrounding areas; they are equivalent to geographic regions.
4. Nodes are key points in the city such as major intersections.
5. Finally, landmarks are distinctive physical objects such as buildings, statues, or hills.

It is claimed that cities with many and strong features are inherently more interesting places to live and work. Such research is intriguing and appears to offer a valuable approach to urban planning. Indeed, this work has proven influential, especially among those planners predisposed towards New Urbanist ideas.

However, a difficulty with this approach to planning is that, notwithstanding the valuable generalizations that Lynch provides, the image of the city necessarily varies from person to person. Whether or not it really is possible to focus on specific ways to improve the visual quality of urban environments remains debatable. In this context it is not insignificant to note that Lynch's findings for Los Angeles derived from only a limited number of respondents (15).

PLANNING IN PRACTICE

Much of the preceding account of modern urban planning has focused on individual planners and their often utopian aspirations. The reality of planning today in most countries is quite different. Planning is just one more aspect of government intervention, mostly local government, into the lives of individuals and the operation of the free market. Essential? Doubtless yes. Utopian? Not usually. Planning today is a pragmatic activity intended to influence the form of the city and, in most cases, to make the city a more livable environment.

Since World War II, planning in many countries has centred on physical design, urban renewal, and suburban expansion. Planning is typically undertaken in the context of legislation from a senior government that sets parameters for what can and cannot be done, and usually involves public participation. Motivations are usually economic and social. For example, beginning in the 1960s, many Asian cities established urban planning authorities that had as their major concern the economic development of the city. In Canada, planning is well established with plans prepared and implemented by local governments. A notable planned community was the Don Mills suburb of Toronto. Inspired by the garden city concept, Don Mills is characterized by low-density development and separation of land uses. This modernist approach to suburban expansion was promoted at the federal level and has been imitated across Canada. Two generally acknowledged drawbacks of this type of expansion are the related decline

Since the end of World War II, urban planning in many countries has focused on the physical design of new suburbs. Two of the examples noted in the text section 'Planning in Practice', the Don Mills suburb of Toronto and the British new town of Milton Keynes (see Figure 12.14), are shown here.

Construction of Canada's first planned suburb, Don Mills (upper photograph), began in 1953. This development incorporates a system of curved streets and pedestrian walkways. Most of the homes are ranch-style detached houses on large lots, but there are also semi-detached homes and three-storey apartment buildings. In total, over 200 different home designs were built. Some of the homes are built into gently rolling hillsides and many of the mature trees in the area were left untouched when construction took place. This 1955 photograph shows the curved streets, cul-de-sacs, and large front and back yards.

Milton Keynes (lower photograph) was designated as a new town in 1967. The form and style of the neighbourhood is not dissimilar to that of Don Mills. This 2007 photograph shows one of the many different neighbourhoods, adjacent to one of several linear parks that are based on river valleys. There are curved suburban streets, detached homes that are uncrowded (certainly by European standards), and much green space. Each of the neighbourhoods is characterized by a distinct assortment of trees and shrubs.

Planned communities such as Don Mills and Milton Keynes are not without their critics, who bemoan especially the dependence on automobiles if residents wish to move outside of their local neighbourhood setting.

[top] Toronto Star Archives/GetStock.com
[bottom] Jason Hawkes/Corbis

FIGURE | 12.14   British new towns

Most of these towns were established between 1947 and 1950. In the south of England, most were located so as to relieve pressure on London; others were new industrial towns. The governmental motivation was a paternalistic socialism, with the towns intended to be free of class distinctions and displays of personal wealth. Inspired by the garden city concept, most of these towns have proved quite successful but are generally acknowledged to lack any real sense of place. This lack of place identity is hardly surprising given their planned origins, but would doubtless disappoint Lewis Silkin, Minister of Town and Country Planning in the early 1950s, who prophesied that these new towns would be visited by tourists from all over the world.

of city centres and the inevitable dependence on expressways.

In Britain three key acts—the Distribution of Industry Act (1945), the Town and Country Planning Act (1947), and the New Towns Act (1958)—set the scene for future planning. One principal outcome was the creation of 28 new towns (Figure 12.14). In some cases the aim of new town creation was to disperse population from overcrowded cities; in others, the purpose was to support specific regional industrial policies. Probably the most significant consequence of Britain's planning legislation to date is that the growth of large urban areas has been limited.

In 2008 the British government announced plans for a new set of towns, again inspired by the garden city concept, but this time with an emphasis on environment. Fifteen environmentally friendly eco-towns were proposed with each to have its own distinctive identity, to include about 40 per cent social housing, to be organized such that cars are not necessary, and to be zero-carbon settlements. The announcement was greeted with much protest from a variety of groups, especially those concerned about urban encroachment into rural areas, and it seems likely that, at most, only two of the 15 might be built.

In some other countries the worry that large cities are becoming too large—one of the concerns that motivated the British new town agenda—is prompting governments to plan new cities. Concern about the growth of Jeddah and Riyadh is prompting new city development in Saudi Arabia, while Egypt is attempting to dissuade people from moving into Cairo by building 20 new towns with another 45 more in the planning stage.

The United States has seen much less government intervention and lacks a coherent national framework, at least partly because it is a federal state. Local policies prevail and are variable from place to place. Typically, planning at the national scale has been limited to public housing and slum clearance. Zoning regulations are used to ensure that urban land is used for what are judged to be appropriate purposes.

Zoning

Zoning is one of the major tools of urban planning as it functions to regulate land use; it has a major impact on the social and economic geography of the city. The first zoning regulations came into force in the 1920s, and zoning was the principal tool used by urban planners through to the 1950s; in Canada, Kitchener, in 1924, was the first city to introduce a zoning bylaw.

In most cities today, urban plans include a classification of the urban area into distinct zones, for example, residential (usually divided into several classes) and industrial (also usually divided into several classes). The fundamental goal of zoning is to ensure that land uses judged to be incompatible are not located adjacent to each other. Appeals of zoning regulations are possible and variances may be granted.

The positive intent of zoning regulations is clear, but a criticism is that they do lead to inflexibility and perhaps, for some critics, to a boring urban landscape. More important is that, in some American cities, one consequence of zoning is that areas of the city may become accessible only to specific income groups. The New Urbanism is one response to the possible limitations of zoning regulations.

The inner city as cultural capital

Many cities, first in Europe and North America and now throughout the world, have initiated a process of planned redevelopment of city centres focusing on cultural capital such as historic preservation and promotion of arts and crafts activities. This use of culture as an engine of economic growth, one part of the larger transformation of cities from centres of production to centres of consumption, often involves plans to build museums, cinemas, bookshops, and cafés. In Europe, the European Union initiated a competition in 1985 to select an annual City of Culture and this has encouraged cities to plan along these lines. But this type of planning is not limited to more developed countries. African cities such as Zanzibar and Nairobi and the Indian city of Chennai are just three places in the less developed world that are actively pursuing cultural heritage preservation.

Moving Around the City

As noted in Chapter 9, location and interaction are opposite sides of the same coin. Cities occupy locations separated by distance yet need to interact with each other, while, inside a city, geographic facts such as homes, workplaces, and sites of consumption similarly are located apart yet need to interact with each other. Both interurban and intra-urban interaction require transportation and communication services. The importance of transportation and communication as these relate to patterns of urban centres was noted in the previous chapter, while the close relationship between urban transportation modes and networks and the internal structure of cities was discussed earlier in this chapter. The concern now is with how transportation and communication facilitate movement inside cities.

Cities are concentrations of people and activities, and to function effectively they must allow people, goods, and information to move around without significant constraints. Even a cursory consideration of urban centres makes clear the important role of transportation. Roads, but also rail, sometimes air, and, increasingly, telecommunications services are therefore key to understanding how cities function. Most people have to travel to earn money and have to travel again to spend money.

People in cities move around as part of their everyday routine. They commute between home and workplace several days each week; they travel elsewhere in the city to shopping centres, places of entertainment, and to visit friends and relatives. Also, people are in frequent electronic contact with other people, and this movement of ideas and information using telephones, fax machines, computers, and cellphones is increasingly important both economically and socially. Goods also move into, out of, and around the city as part of everyday economic transactions.

If you are in any doubt concerning the importance of movement, consider what happens if a city is without transportation services, perhaps because of blizzard conditions. The likely scenario is that people cannot go to work or purchase food, and emergency services have difficulty responding to urgent medical and other needs. Even in situations where only some portion of a city transport network is unavailable—perhaps because of localized flooding or a damaged bridge—or where some mode of transportation is unavailable—perhaps because of industrial action—lives and activities can be seriously disrupted.

ACCESSIBILITY AND MOBILITY

Two closely related concepts—accessibility and **mobility**—aid understanding of the role played by urban transportation. Accessibility refers to the number of activity sites or opportunities within a certain distance of a particular location. For example, a home in the suburbs may be further from many workplaces and from specialist financial, legal, and medical services than is a home in the city centre, meaning that these places and services are less accessible from the suburban locations. Of course, calculations of accessibility need to

mobility
The ability to move from one location to another.

use an appropriate distance concept—physical distance, time distance, or cost distance, for example.

Mobility refers to the ability to move from one location to another. In general, mobility has increased with technological improvements in transportation and with rising incomes, which has meant that physical distance has become less of a constraining factor or, speaking more geographically, we say that the friction of distance is reduced. In many cities today widespread automobile ownership increases personal mobility. But, of course, mobility is closely related to economic circumstances, with low-income earners more dependent on public transit systems and generally less able to move from place to place than high-income earners. Transportation issues cannot be considered apart from larger questions of social and economic well-being in the city.

ROLE OF INFORMATION TECHNOLOGIES

Information technologies (ITs) are altering communications and transportation services in cities and also prompting changes in personal and business location decisions. Most notably, accessibility can no longer be thought of only in terms of some measure of distance because IT allows people to have virtual accessibility to more and more activities—consider, for example, the rising popularity of on-line shopping. Certainly, fundamental changes are occurring in the conduct of both personal and business communications, and these changes are affecting movement inside the city. It also seems fair to note that we are still in the early stages of an IT revolution.

> Not since the introduction of the automobile age has transportation been faced with technological impacts of this scale. Telecommunications are just beginning to permeate almost every aspect of our lives. As we become busier, we will increasingly rely on it to avoid unnecessary travel. As populations increase, particularly populations within metropolitan areas, we will also increasingly rely on it to avoid congestion on transportation networks and at activity sites. Also, as we spend new time engaged in telecommunications, there will simply be less time available for other activities, including travel. Small effects by

a very large number of persons will aggregate up to large effects on a system-wide basis. (Golob and Regan, 2001: 114)

Changing information technologies

Improvements are taking place in four key components that combine to make up what is usually called information technology. First, computer technologies are improving with more efficient microchips and increased use of wireless communication. Both of these are leading to faster, less expensive, and more portable computers. Second, the Internet is becoming more and more sophisticated. Satellite-based distribution services will contribute to a faster and more efficient system; technologies such as Voice over Internet Protocol enhance multimedia capabilities; the language used is becoming more sophisticated and adaptable, and programs are being developed that will be able to act independently of users. As noted in Chapter 11, everyday use of the Internet has seen an exponential increase in the number of hosts in recent years, and also a substantial increase in the number of users (see Figure 11.15). Third, sophisticated devices such as mobile phones, hand-held Internet devices, and satellite-based positioning systems are becoming commonplace. Fourth, new software packages are designed to take advantage of changing technologies and to serve an ever-expanding list of communications needs and wants.

Impacts

For many people and businesses, these IT advances mean that more and more of their daily routine is being conducted from a home or business base without the need to move around the city. Electronic commerce (e-commerce) is mostly business to business, but also person to business in the form of banking, contacting government agencies, gathering information, telemedicine, and on-line shopping for such items as electronics, clothing, books, music, and travel needs. Most commentators acknowledge the rise of person-to-business e-commerce but do not go so far as to suggest that more traditional shopping behaviour involving travel will decline notably. This is partly because shopping is about much more than making purchases; it is also a social, indeed often a recreational, activity.

IT advances enable more people to work or study from home—to telecommute—a situation that removes the daily need to commute from home to elsewhere in the city. Also, the fact that businesses are tending to subcontract more work means that working from a home base becomes more feasible for some workers. A closely related development is that people may have mobile offices through use of cellphones and portable e-mail and fax devices. IT advances also permit some business activities, including **back-office activities**, to locate away from established office location areas. This decentralization reduces movement into and out of the city centre. However, in the case of **front-office activities**, IT intensifies existing concentration in established office areas, and this concentration usually increases movement into and out of the city centre.

An additional impact of e-commerce is a trend towards more frequent and smaller commercial shipments as IT encourages the rise of services that provide information about such matters as pricing, routing, and possible delays at intermodal facilities.

Overall, it is not easy to determine specific consequences of these various changes to the lives of people and to the ways businesses function; on balance, however, one likely outcome is a reduction in the number of travel trips. Transport planning needs to anticipate and accommodate these ongoing changes.

URBAN TRANSPORTATION PLANNING

Several different levels of government are usually responsible for planning and financing transportation systems, a process that is much complicated by the fact that all these systems are dynamic as the need for new routes and new modes becomes apparent and as older routes and modes require upgrading or perhaps become redundant. Many cities confront planning and financial challenges because previous capital investment in roads, subways, and public transit vehicles becomes less valuable as cities grow and land uses change (see Box 12.3).

Most roads inside the city are local streets in housing areas that do not experience heavy traffic volume, but these streets are critical for personal mobility and for services such as garbage collection and postal delivery. Local governments are usually responsible for these streets, although they are often initially put in place by developers and are also supported through taxes on residences and businesses.

In many cities there are ongoing debates about the respective roles of public transit and private automobile use; indeed, the question of private automobile use—specifically, limiting such use—seems likely to dominate discussions about urban transportation in many cities around the world. In general, public transit systems are well developed in European and some

back-office activities
Repetitive office operations, usually performed using telecommunications, that can be located anywhere in the city, or indeed outside the city, including in relatively low-rent areas.

front-office activities
Skilled occupations requiring an educated, well-paid workforce. Image is important and, although telecommunications play a key role, so is face-to-face contact with others. Hence, these activities favour prestige locations in major office buildings in city centres.

A multi-storey bike parking lot in Amsterdam.
Wendy Kay / Alamy

Asian cities, a fact that at least partly reflects the limited space available for urban expansion, the subsequent congested character of these cities, and also pollution concerns. Singapore limits and charges for private automobile access to the city centre region while some other cities are also implementing charges for access to the city centre. The city centres of Copenhagen and Amsterdam are home to public transit and to bicycles rather than to cars. In contrast, most North American cities are designed with auto-mobile travel in mind.

A related debate concerns the merits of adding new or improving existing highways in a city rather than investing in public transit improvements. Opponents of such expansions and improvements contend that this is bad public policy because more and better high-ways do not relieve congestion and improve movement but rather serve to increase demand. This is simple supply-and-demand logic—if a new highway enhances accessibility to a work-place or mall, then people are likely to use that highway. Also at issue here are costs. Many public transit systems, especially rail systems, are expensive to implement, whereas high-

way projects are more able to fund themselves through gasoline and vehicle taxes.

The question of who pays for transporta-tion services is regularly debated. The principle of public provision of services and transporta-tion infrastructure, funded by governments, is widely accepted but not universally applied. In many countries some components have been privatized: consider Highway 407, a pri-vately built and operated toll road north of Toronto. Also, since 2003, the London under-ground rail system has operated as a partner-ship, with all the infrastructure maintained by private companies, but the underground con-tinues to be owned and operated by a public company.

Questions of social and environmental justice

It costs money for people to move around the city. Most high-income people accept this expense as a necessary and valued expenditure, but for low-income people the cost of move-ment may be excessive. In most cities, transpor-tation planning is a technical activity, usually based on quantitative data concerned with matters such as engineering problems, traffic flows, and predictions of use. As such, there is little concern with questions of social justice. Debates about private versus public transporta-tion and about financing do have social justice implications, but these implications are rarely front and centre.

Transportation systems are intended to improve accessibility, not only for subur-ban commuters travelling between home and downtown, but also for inner-city residents who need to travel to work, shops, hospitals, and recreational sites. But highway improve-ments clearly benefit suburban developers and high-income suburban residents. Figure 12.15 suggests some of the linkages between highway construction and inner-city living.

In most cities, low-income people find that accessibility to work and other important loca-tions is limited because of their low level of personal mobility. In the case of the journey to work, the current trend towards decentraliza-tion penalizes inner-city residents, tending to lead to the creation of employment opportuni-ties outside the city centre and contributing to high unemployment rates in the city centre. If work is available nearby for an inner-city resi-dent, the use of a single fare structure on pub-lic transport means that people moving just a

FIGURE   12.15   **A conceptual description of the effect of freeway construction on cities**

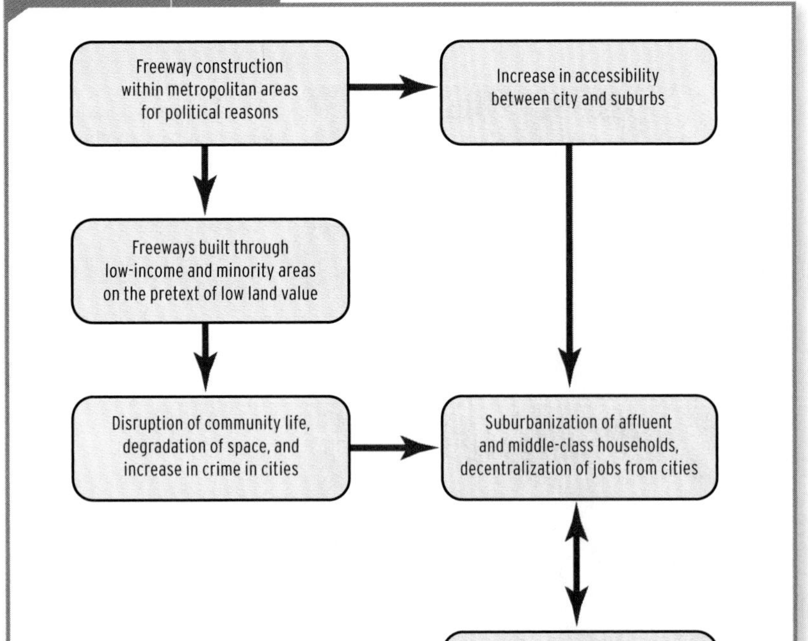

SOURCE: D. Deka, 'Social and Environmental Justice Issues in Urban Transportation', in S. Hanson and G. Giuliano, eds, *The Geography of Urban Transportation*, 3rd edn (New York: Guilford, 2004), 348. Reprinted by permission of the publisher.

short distance pay as much as those travelling across the city.

In addition to promoting urban sprawl, highways and private automobile ownership contribute to congestion in city centres, to pollution problems, and to the incidence of accidents. Not surprisingly, then, highway development and private automobile use are viewed unfavourably by many people concerned with the creation and maintenance of sustainable urban environments. The dispersal that highways encourage, and the widespread neglect of pedestrian needs, may mean that many people lack needed access to hospitals and other essential services.

URBAN TRANSPORTATION AND LAND USE

Transportation is both a major user of urban land—consider the amount of space used especially by roads and parking lots in a typical urban centre—and also affects and responds to land use. Figure 12.16 shows the mutual dependence of transport and land use in a simplified form. The transportation system, including roads and the availability of public transit, affects accessibility. In turn, accessibility is one factor considered when businesses and other land uses make location decisions that affect the pattern of land use. The pattern of land use, along with the transportation system, impacts on the activity patterns of people and businesses contributing to the city's overall travel pattern.

Much of the theoretical work on the relationship between transportation and land use has turned to classical economics, specifically the link between distance and land use that is central to von Thünen's theory of agricultural location. Land use following this logic is explained in terms of land rent, the cost of movement, and the cost of other relevant goods and services, with a decline in transport cost allowing decentralization.

Much of the empirical work on the relationship between transportation and land use has looked at the impact of highway construction, showing that it leads to increases in land values close to the highway. This is because highways enhance accessibility. The specific impacts relate to local circumstances. If accessibility improves in an established area, then few land-use changes occur; if accessibility improves in an undeveloped area, then land-use change is likely. Empirical work on rail transit systems and land use suggests that rail has relatively little impact on accessibility and hence on land use, and that any impacts tend to be highly localized. This is because rail transit necessarily moves fewer people from a specific origin and destination location than do highways that link to local road systems without changing the mode of transport. But the question remains: is it possible for cities built around the private automobile to organize themselves more effectively today (Box 12.5)?

Spreading Outward

Suburbs are as old as cities themselves (Bourne, 1996: 166). Historically, urban areas were concentrated around a central area—perhaps a bridging point across a river, a confluence of two rivers, or a market—and gradually expanded outward in the form of residential suburbs as population increased. Outside the urban area, most land was in agricultural use. As noted earlier, the favoured residences were most often those close to the city centre, with poorer people relegated to the more distant and therefore less accessible homes.

In the more developed world especially, this long-standing spatial organization has been transformed in two general ways. First, beginning in the late nineteenth century increasing numbers of people opted for homes outside of the established urban area, a location decision made possible by improved transportation and, as the twentieth century progressed, increased private automobile ownership. Second, in recent decades the globalization processes that

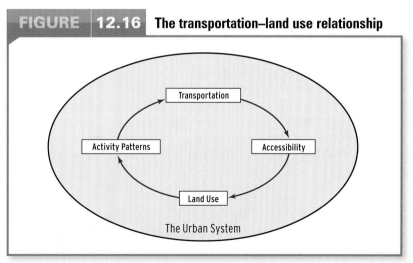

FIGURE | 12.16 The transportation–land use relationship

SOURCE: G. Giuliano, 'Land Use Impacts of Transportation Investments: Highway and Transit', in S. Hanson and G. Giuliano, eds, *The Geography of Urban Transportation*, 3rd edn (New York: Guilford, 2004), 239. Reprinted by permission of the publisher.

have contributed to the world city phenomenon have also worked to transform suburbs and to initiate the growth of new cities beyond established urban areas. These two transformations are described in this section.

SUBURBANIZATION

Suburban growth was mostly in response to changes taking place inside the city, including the establishment of factories and rising population densities. The favoured locations outside the city were directionally biased along transportation routes, and new streetcar suburbs appeared around many large cities in Europe and North America during the second half of the nineteenth century. This was dramatically affected by the introduction of the automobile, which allowed access to the large areas of land that lay between the streetcar routes, meaning that distance from the centre of the urban area became the key variable rather than direction. Most of these new suburbs comprised housing and basic services designed for the working class and/or middle class and represented a compromise between the need to live close to employment in the central city or industrial areas and the desire to be close to open spaces and rural areas.

Suburbanization is a form of decentralization and proceeded apace during the twentieth century. It is especially prevalent where land is readily available, planning regulations are weak, populations are wealthy and can afford large homes, and levels of physical mobility are high. In many countries suburbanization has become the characteristic urban experience, producing

Box 12.5   The Denver Solution

In common with many other large cities, Denver, Colorado, is home to much traffic congestion. The solution being implemented is a mix of relatively conventional freeway improvements and ambitious light-rail development. Freeway improvements include the new E-470 tollway that aims to relieve congestion in the city centre and also added lanes to the I-25 and I-225 freeways (Figure 12.17).

But the light-rail developments are the more anticipated improvements, both because of the hoped-for reduction of highway congestion and because of the planned impacts to land use and city shape. First, a new 31-km (19-mile) double-track light-rail line is linking the city centre with the southeast. The freeway improvements and this transit link are part of the Transportation Expansion (or T-Rex) plan. Also planned are at least six additional rail lines, and by 2016 a further 190 km (119 miles) of light and commuter rail and 29 km (18 miles) of rapid bus transit will spread out from the city centre. These routes are part of the FasTracks plan. Each of the light-rail stations to be built is intended to act as an incentive for development in the immediate vicinity of new residential and commercial land uses. The larger goal is, then, to encourage development inside the city and thus limit expansion of the city.

Will these transportation changes have the desired effects, both reducing congestion and promoting development inside the city? Perhaps the answer is suggested by a September 2005 item in the *Rocky Mountain News* (Steers, 2005).

The light-rail line along Interstate 25 is spurring a development frenzy around the rail stations that will open next year, marking what could be a back-to-the-future change in the way the metro area grows.

The rail line, part of the T-REX project between Broadway and Lincoln Avenue (south of C-470), will have 13 new stations, most of them on the west side of I-25. Plans are under way for new residential and commercial development that will create homes and workplaces for thousands of people within walking distance of the stations.

Ironically, this new growth pattern is a throwback to the way Denver neighborhoods were shaped a century ago, when homes and businesses sprang up along streetcar lines.

The T-REX rail line is next to a freeway, however, which some say is a major detriment to creating pedestrian-oriented zones around the stations. But that hasn't stopped developers from backing new projects along the line.

But critics of these initiatives worry about parking problems at the light-rail stations and question whether they will generate the intended growth. Perhaps the biggest concern is that people really do value the convenience and comfort of car travel and that any reduction of congestion resulting from increased use of rapid transit will only encourage other people to use their cars.

It is useful to think about transportation in Denver in the context of the politics of mobility, that is, 'the political struggle over what type of transportation mode—be it automobile, transit, or walking—is developed in a city, and how urban space is configured to make various modes functional' (Henderson, 2004: 193). This is a spatial struggle about how cities are to grow in the future and about who will benefit from growth.

Continued

city, leading to reductions in the city's tax base, urban unemployment, underutilization of urban infrastructure, and lower property values inside the traditional city.

Box 12.7 outlines two useful attempts to generalize about recent and ongoing urban landscapes changes. The most notable current circumstance is a process of spread outside the traditional city area, leading not merely to the formation of suburbs but, increasingly, as implied by the fifth stage of the Hartshorn and Muller model, to the formation of new and different urban centres (as suggested by the account of exurbanization in the previous chapter).

Box 12.7 Modelling Suburbanization and Sprawl

Geographers have developed several models intended to generalize about the processes of suburbanization and sprawl. The metropolitan evolution model described in the previous chapter is the seminal contribution, relating technological change, transportation developments, and the spread and growth of cities. The fourth and final stage of this model highlights the links between automobile use and suburban growth. Two other useful models are by Erickson (1983) and by Hartshorn and Muller (1989).

Erickson's model: For the period from 1920 to 1980, the evolution of suburbs is divided into three stages, each closely tied to changes in the transportation network that improved accessibility.

- 1920–40: The *spillover and specialization* stage involved some manufacturing activities and related residential developments spilling out beyond the city, usually along railway or water transportation routes.
- 1940–60: The *dispersal and diversification* stage involved much suburban growth as road systems improved and automobile ownership proliferated. Many of these new suburbs were marketed to relatively high-income groups, but there were also low-income suburbs located close to new industrial employment opportunities outside the city.
- 1960–80: The *infilling and multinucleation* stage continued the suburbanization process as areas of land between existing suburbs were filled in. In many city areas, a clear distinction arose between the earlier inner and later outer suburbs. Also, as this residential spread continued, commercial nuclei developed to serve suburban residents.

Hartshorn and Muller model: This model describes suburbanization from about 1950 to the 1990s. Five stages are identified, each associated with about a 10-year period.

- 1950s: Described as *sprawl*, this is a period of residential expansion and dispersal, linked to the road network and based on automobile use, leading to the creation of bedroom communities. This concept of bedroom communities highlights the fact that these were places to sleep, but not to work—employment was still centred in the traditional city centre. Few retailing and other commercial activities were included in the new developments.
- 1960s: This stage, called *independence*, involved the emergence of suburbs that were no longer dependent on the city centre. The creation of more and better road links allowed the growth of suburban shopping malls and industrial parks that began to weaken the link between suburbs and the city centre. This increasing suburbanization involved residential diversification.
- 1970s: The weakening of ties between suburbs and city centre continues as suburbs become *magnets* for commercial and industrial growth, weakening downtown areas in the process. In many cities perimeter highways that circled the city allowed for improved links between suburbs.
- 1980s: Suburban growth continues unabated as many new employment opportunities are attracted to these areas, especially *high-technology* industries. In turn, these activities encourage higher-density residential development in the form of high-rise apartment complexes and condominiums.
- 1990s: By this time, suburbs have *matured* into essentially self-contained urban regions that include most of the economic, social, political, and other services associated with a city.

Both models provide helpful outlines of the suburbanization process. The link between suburbs and transportation is explicit, as is the growing independence of suburbs as they incorporate most if not all of the features of a mature urban landscape—namely, homes marketable to a range of income groups, varied employment opportunities, sites of consumption such as shopping malls, and recreational facilities. As suburbs become more and more developed, the separation from the parent city becomes almost complete, with perhaps just a few essential government functions and some major cultural establishments such as museums being frequented by suburban residents. Increasingly, then, the parent city is barely needed by suburban residents. Indeed, as suggested by Hartshorn and Muller's fifth stage, suburban areas can grow to resemble city centres.

POST-SUBURBIA

Recall the reference at the beginning of this chapter to the many new concepts that have gained currency in urban geography in recent years. Nowhere is this proliferation of new concepts more evident than in accounts of the continuing spread of urbanization. Many commentators see both ongoing suburbanization and a new process of post-suburbanization involving both movement of people and activities to locations beyond the suburbs and changes in existing suburbs. But what does this mean in terms of how people live and work?

Edge cities

Perhaps the first identification of this transformation was Joel Garreau's (1988) use of the term **edge city** to describe the emergence of completely new urban centres, often centred on a highway interchange in a rural area. These new and rapidly growing centres are post-industrial in character, having limited connections with industry; instead of offering manufacturing jobs, they offer work in the professions, management, administration, and skilled labour. Physically, such cities include office buildings and public institutions but few factories. Also, in recent years, as suburban areas have matured, many of them have acquired features formerly associated with city centres to become post-suburban locations. Indeed, edge city development is not always easy to distinguish from changes in suburban areas. As early as the 1980s, Garreau (1988: 4) identified about 200 edge cities, describing their formation as a third wave of urbanization.

> Edge Cities represent the third wave of our lives pushing into new frontiers in this half century. First, we moved our homes out past the traditional idea of what constituted a city. This was the suburbanization of America, especially after World War II.
>
> Then we wearied of returning downtown for the necessities of life, so we moved our marketplaces out to where we lived. This was the malling of America, especially in the 1960s and 1970s.
>
> Today, we have moved our means of creating wealth, the essence of urbanism—our jobs—out to where most of us have lived and shopped for two generations. That has led to the rise of Edge City.

edge city
A centre of office and retail activities located on the edge of a large urban centre.

Edge cities and other post-suburban developments are one component of the increasing decentralization of offices and factories, the decline of manufacturing employment, and the processes of economic restructuring associated with larger globalization processes. They involve a continuing exodus of financial and producer services from older downtowns. An excellent and often cited example of the edge city is Tyson's Corner, Virginia, just outside of Washington, DC: what was a rural corner in the mid-1960s now includes the seventh largest concentration of retail space in the United States and is only one of 17 edge cities close to Washington (Bourne, 1996: 173).

But edge cities are all over the United States landscape today, often dominating the urban landscape. Many of the fastest-growing urban regions—such as those centred on Las Vegas, Denver, Phoenix, Dallas, Austin, Houston, and Atlanta—include numerous edge cities. Edge city growth is sometimes accompanied by a corresponding weakening central city area, or it may occur around a continuing viable downtown that serves as one command centre in the global economy.

These new edge cities are less prominent in some other more developed countries. In many European countries and in Japan, growth outside cities is severely constrained by space limitations, so more dramatic solutions to problems of congestion are being sought (Box 12.8). In some smaller more developed countries, such as Canada and Australia, most city centres continue as viable economic and social locations; also, the national economy may not be sufficient to generate substantial edge city growth. However, many of the largest cities in less developed countries are experiencing urban fragmentation. The particular circumstances of urban expansion in the less developed world is considered in Chapter 13.

Edge cities in context

'Edge city' is not the only term used by geographers to describe these new urban landscapes. Some choose to talk about peri-metropolitan growth, polynucleation, spillover cities, suburban activity centres, boomburb, superburbia, technoburbs, galactic cities, dispersed cities, exploding cities, and more. This proliferation of labels is suggestive of the complexity of the unfolding urban landscape and of how difficult it is to describe. In the case of the United

Box 12.8 Spreading in all Directions, and into the Sea

Consider Tokyo. It is a crowded city with a constant demand for more space. Initially, this demand expressed itself in urban sprawl, but more recently it has involved internal reorganization. The population of about 36 million in the larger urban region is now making ingenious use of the urban area by building upward, downward, and into the sea. Building up is a relatively recent alternative for a city prone to earth tremors. In addition to skyscrapers, there are elevated highways and railways and even multi-storey golf driving ranges. Building down is also popular; some buildings use at least four below-ground levels for retail and office purposes. Perhaps most dramatically, Tokyo is reclaiming land from the sea to permit more expansion. It has even been suggested that the city might extend out onto a floating platform in Tokyo Bay. Tokyo is one example of an ever-changing urban area with some unusual innovations.

Consider Dubai. The largest city in the United Arab Emirates is building a series of islands in the Persian Gulf. There are three massive palm-shaped islands and a cluster of 300 islets that are designed to appear from the air as a map of the world. The intent is that these islands, when complete, will house about 1 million people. These islands are aimed at wealthy people, and are not a response to issues of congestion in Dubai itself. The first residents occupied homes in 2007, but whether or not these ambitious projects will be completed as planned is in some doubt. Dubai is also home to the world's tallest building, Burj Khalifa or the Dubai Tower; construction began in 2004 and the structure was officially opened in January 2010. Also in Dubai, plans were announced in 2008 for the building of the world's first 'moving' skyscraper, an 80-storey tower with each level designed to move independently such that the building will constantly change shape.

Consider our possible future. Norman Nixon, a Florida-based engineer, outlined a floating city project in the late 1990s. Known as the Freedom Ship, the initial proposal envisaged a 1,400-metre-long mobile city capable of housing 60,000 people and with a full array of services including banks, hospitals, and schools. Other similar projects have been envisaged by a Belgian architect, Vincent Callebaut,

with Project Lilypad functioning as a floating ecopolis, and by a Danish architect, Julien de Smedt, who conceived of Mermaid, a 'floating wellness' resort. Some commentators suggest that projects such as these are needed in light of rising sea levels resulting from global warming, but it seems fair to say that, at the present time, these ideas are dreams, rather than feasible future developments.

Burj Khalifa.
dblight/iStock

States, this new landscape of distinct and often separate urban realms has appeared in conjunction with a major shift in population and employment from the Northeast and Midwest to the South and Southwest.

Figure 12.18 generalizes the form of these new urban landscapes in a manner similar to that proposed by the Hartshorn and Muller model described in Box 12.7. The presence of gated communities reflects the need for skilled labour as edge cities are home to many high-tech employment opportunities and to financial and producer services. Many social commentators view the rise of gated communities as a regrettable trend as they are explicitly landscapes of exclusion. In many cases, edge cities exclude the financially and socially disadvantaged population.

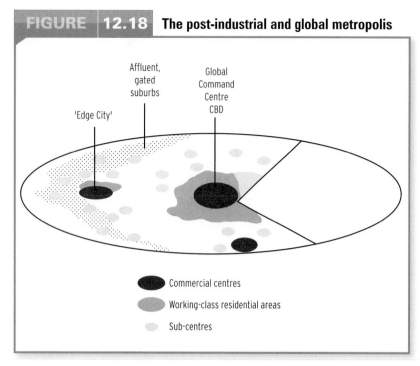

FIGURE **12.18**   **The post-industrial and global metropolis**

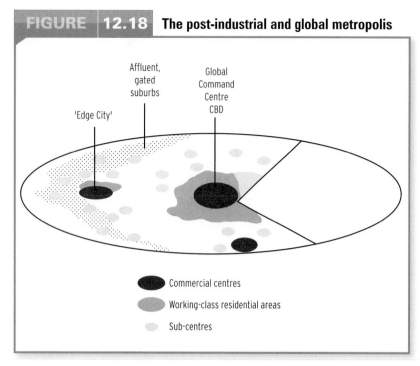

Affluent, gated suburbs

Global Command Centre CBD

'Edge City'

● Commercial centres

● Working-class residential areas

● Sub-centres

SOURCE: T. Hall, *Urban Geography*, 2nd edn (New York: Routledge, 2001), 11. Copyright © 2001 Routledge. Reproduced by permission of Taylor & Francis Books, UK.

In other cases, edge cities are hugely variable places: they are landscapes of excess and of want, of prestigious designer homes and of trailer parks, home not only to the rich and powerful but also to marginalized social and ethnic groups. They are organized in some places and disorganized in others. In many gateway cities, such as Vancouver and Sydney, new immigrants, mostly from Asia, are attracted to outlying suburban areas perhaps because of job opportunities and, in the Canadian case, because of the availability of subsidized housing.

This account of edge cities is an appropriate way to conclude this chapter, highlighting as it does the emergence of postmodern landscapes, the impacts of global-scale processes, the declining importance of distance, and the advantages new urban places often have over older urban places. But perhaps most significant, the emerging landscapes of many edge cities graphically display the tensions between rich and poor people, a theme that is central to many of the discussions in the following chapter.

CHAPTER 12 SUMMARY

CHANGING URBAN THEORY

Major twentieth-century conceptual formulations of the internal structure and identity of cities reflect a wide range of perspectives including those associated with physical science, ecology, positivism, political ecology, urban managerialism, humanism, and postmodernism.

THE INTERNAL STRUCTURE OF THE CITY

Three classic models describe concentric zones, sectors, and multiple nuclei. Cities in different regions of the world typically display internal structures similar in basic principles but often very different in detail. The essentially positivist motivation for these formulations prompted a powerful critique from a political economy perspective.

FROM MODERN TO POSTMODERN

Most recently, postmodernists have questioned all theoretical attempts at explaining urban internal structure on the grounds that any theory privileges one explanation above all others. For much of the twentieth century,

industrial Chicago was regarded as the paradigmatic city, but post-industrial and postmodern Los Angeles has been seen more recently in this light. If such a transition in the urban landscape is occurring, it may be one component of a larger shift from modernism to postmodernism.

URBAN GOVERNANCE

Governments, especially local municipal governments, play a major role in the form and functioning of cities today. There are numerous specific forms of municipal government and, as it is not readily apparent that one is better than all the others, reorganization is not unusual. Governing urban centres is always challenging, not least because they are socially and economically fragmented locations and special interest groups often play important roles.

URBAN POLITICS

The three key groups involved in urban politics are professional planners (who collect and analyze data, prepare reports, and make policy recommendations), politicians (who determine goals and priorities and make decisions), and city residents (who elect politicians and are affected by the decisions made by elected officials). There are five principal bodies of thought concerning how urban politics unfolds—pluralism, structural logic, regulation concepts, urban regimes, and identity politics.

URBAN PLANNING

Cities are both planned and unplanned. Planning imposes values on landscape and is often controversial; it implies a loss of individual rights and a search for the common good. The earliest evidence of town planning is the use of a rectangular grid street layout followed by the building of monumental structures. In the nineteenth century the need for planning to correct serious urban health and overcrowding problems initiated the building of sewage systems and other improvements. Some paternalistic employers built self-contained communities for their workers, while other planners proposed ambitious new urban centres as in the City Beautiful and garden city movements. Notable modernist planners who saw few of their plans come to fruition included Le Corbusier and Frank Lloyd Wright. More recently the New Urbanism movement has proved a popular approach. One of the most widely implemented tools of planning today is zoning.

URBAN IMAGES

What image do residents have of the cities they inhabit? Lynch's work suggests that the typical resident recognizes five key features: paths, edges, districts, nodes, and landmarks.

URBAN TRANSPORTATION AND COMMUNICATIONS

Cities are concentrations of people and activities. To function effectively, people, goods, and information need to move between cities and within cities without significant constraints. Movement is a part of everyday routine. Key concepts are those of accessibility and mobility and these are always changing. The role played by information technologies is of particular importance today, and changing accessibility and mobility are affecting many personal and business location decisions. There are often debates concerning the respective roles of public transit and private automobile use and concerning the implications for larger issues of social and environmental justice.

SUBURBANIZATION

Suburbs are as old as cities but, beginning in the late nineteenth century, suburbanization dominated the urban areas of the more developed world as increasing numbers of people opted for homes outside of the established urban area, a location decision made possible by improved transportation and increased automobile ownership. Suburbanization is a form of decentralization and has involved a reorientation of the city with people and businesses locating in the suburbs and often a corresponding decline of traditional city centres.

POST-SUBURBANIZATION

In recent decades the globalization processes that have contributed to the world city phenomenon have also worked to transform suburbs and to initiate the growth of new cities beyond established urban areas. The result, especially in the United States but also elsewhere in the world, is the formation of edge cities outside the earlier urban landscape. These new cities are post-industrial in character, often including high-technology industries and financial and producer services. They often display great social differences, and in some instances are home to both trailer parks and gated communities. This decentralization of urban activities and residences is such a feature of many countries today that it is known by a wide variety of terms such as peri-metropolitan growth, polynucleation, spillover cities, suburban activity centres, boomburb, superburbia, technoburbs, galactic cities, dispersed cities, exploding cities, and more.

QUESTIONS FOR CRITICAL THOUGHT

1. Despite sharing certain similarities in terms of their form, cities do differ from one part of the world to another. Why do these differences exist? What role does culture play in explaining these differences?

2. The three models of the internal structure of cities (concentric zone model, sector model, and multiple nuclei model) are based on American cities in the early to mid-twentieth century. Of what value are these models to urban geographers in the early twenty-first century? How can we use these models to better understand our cities of today?

3. Compare and contrast the different perspectives described in the section 'Rethinking the City'. Why do we have multiple perspectives or views on cities? Why isn't there just one single point of view?

4. Norton identifies three key groups involved in urban politics—professional planners, politicians, and citizens (the public). How can/should these three groups work together for the good of the city? In other words, how can we better plan our cities?

5. To what extent should the planning of our cities be based on 'ideal' models such as the City Beautiful or garden city?

6. Do you think that the ideals of New Urbanism can be successfully implemented in our cities today?

7. What role do you think transportation and communication technologies will play in the cities of the future. What will these cities look like? Some have argued that we will no longer require the face-to-face contact provided by cities in the future. Is it possible that cities will no longer exist?

8. Why is it so difficult to control urban sprawl? What is the single most important factor underlying sprawl? How can/should urban sprawl be managed or controlled?

FURTHER EXPLORATIONS

(Note: For additional references, see this section in Chapter 11.)

Badcock, B. 2002. *Making Sense of Cities: A Geographical Survey*. New York: Oxford University Press.

Good urban geography textbook covering all the key conceptual and empirical topics and with an informed international focus.

Fyfe, N.R., and J.T. Kenny, eds. 2005. *The Urban Geography Reader*. New York: Routledge.

A valuable collection of classic and recent urban geography writing organized into seven sections: foundations; globalization; restructuring; politics, governance, and inequality; difference; form and symbolism; and technologies.

Hall, P. 1996. *Cities of Tomorrow: An Intellectual History of Urban Planning and Design in the Twentieth Century*, 2nd edn. Oxford: Blackwell.

A scholarly and entertaining history.

Hanson, S., and G. Giuliano, eds. 2004. *The Geography of Urban Transportation*, 3rd edn. New York: Guilford.

A comprehensive collection that covers a wide range of urban transportation topics, but with a strong emphasis on the United States.

Harris, R. 2004. *Creeping Conformity: How Canada Became Suburban, 1900–1960*. Toronto: University of Toronto Press.

An informed and readable account of how the built suburban landscape and the suburban way of life became such important parts of the Canadian urban experience.

Hunt, T. 2004. *Building Jerusalem: The Rise and Fall of the Victorian City*. London: Weidenfeld.

Hunt is a historian and the focus in this book is on the history and architecture of the Victorian city. There are excellent accounts of how city leaders and industrial capitalists tackled or, more often, failed to tackle the problems of rapidly growing and unhealthy cities.

Landry, C. 2006. *The Art of City Making*. Vancouver: University of British Columbia Press.

An exploration of cities as they reassess their potential, strengthen their identity, and adapt to a globalizing world.

Leslie, T.F., and B. Ó hUallacháin. 2006. 'Polycentric Phoenix', *Economic Geography* 82: 167–92.

The seemingly random spread of the Phoenix urban region, although resulting from several conflicting processes, contains a substantial spatial logic related to accessibility, externality, and regulation.

Levy, J.M. 1997. *Contemporary Urban Planning*, 4th edn. Upper Saddle River, NJ: Prentice-Hall.

Standard text on urban planning; includes a good overview of the history of planning along with discussions of current planning objectives and strategies. Focuses on US examples.

Lucy, W.H., and D.L. Philips. 2006. *Tomorrow's Cities, Tomorrow's Suburbs*. Chicago: Planners Press.

A detailed exploration of city decline and renewal.

Michel, D.P., and D. Scott. 2005. 'The La Lucia-Umhlanga Ridge as an Emerging "Edge City"', *South African Geographical Journal* 87: 104–14.

> Discussion of the city of Durban in South Africa in terms of sprawl, decentralization of the inner city, suburban spread, and the rise of an edge city to the north.

Sharpe, C. 2005. 'Just Beyond the Fringe: Churchill Park Garden Suburb in St. John's, Newfoundland', *Canadian Geographer* 49: 400–10.

> An informative account of the successful building of a new suburb intended, among other things, to promote the clearance of downtown slums and to facilitate the orderly expansion of the city.

Ward, S.V. 2004. *Planning and Urban Change*. Thousand Oaks, Calif.: Sage.

> Good overview of urban planning with emphasis on British examples.

ON THE WEB

URBAN PLANNING

Canadian Urban Institute

www.canurb.com/

This home page of the Canadian Urban Institute includes general information on the Canadian urban situation.

Le Corbusier

www.greatbuildings.com/architects/Le_Corbusier.html

Information, photographs, and links relating to some of the architectural achievements of Le Corbusier.

New Urbanism

www.newurbanism.org/

Informative website advocating the creation of compact, walkable urban centres. Suggests we need to revive the lost art of place-making.

URBAN TRANSPORTATION

Transport Canada

www.tc.gc.ca/programs/Environment/UTSP/menu.htm

This site describes a number of innovative sustainable transport projects in the context of Canadian cities.

European Commission

ec.europa.eu/transport/urban/index_en.htm

Information on clean transport projects in the European context.

URBAN SPRAWL

National Geographic

ngm.nationalgeographic.com/ngm/data/2001/07/01/html/ft_20010701.3.html

Very useful discussion of the causes, consequences, and alternatives to urban sprawl.

Sierra Club

www.sierraclub.org/sprawl/

Much useful information on sprawl, with focus on development projects that offer transportation options, help revitalize neighbourhoods, and preserve local values.

David Suzuki Foundation

www.davidsuzuki.org/Climate_Change/Sprawl.asp

Activist website designed to encourage people to become involved in helping stop sprawl.

EDGE CITIES

The Garreau Group

www.garreau.com/main.cfm?action=book&id=1

Joel Garreau, an influential writer on urban topics, first identified edge cities, and The Garreau Group is concerned with the larger goal of understanding world culture and values. This website has useful information and links.

LIVING AND WORKING IN CITIES

As evident in the preceding two chapters, both the location and the spatial organization of cities reflect and often reinforce economic, political, social, and cultural circumstances. Today, globalization and economic restructuring processes play particularly important roles, but urban life cannot be understood without reference also to social relations and social conflict. Most research into urban life has focused on cities in more developed countries, although studies in less developed countries are increasing in number and sophistication. This chapter considers key topics in the geography of urban life with emphasis on more developed countries and looks at some of the features of urban life in less developed world cities. The chapter addresses the following questions:

- Should housing be provided by the state as a social service, or is housing a commodity to be bought and sold?

- Why do people move residences inside the city, and how do they conduct the search process for a new home?

- What are neighbourhoods, and why are some neighbourhoods gentrifying?

- Why are some neighbourhoods segregated on the basis of ethnic identity?

- What are some of the social inequalities evident in cities, and how might they be modified?

- Where do manufacturing and retailing activities locate in the city and is there a transition from cities as places of production to places of consumption?

- Is urban life in the less developed world distinctly different from that in the more developed world?

- What are some of the quality-of-life and environmental challenges faced by cities and city dwellers in Latin America, Africa, India, and China?

Rapid gentrification in the Friedrichshain neighbourhood of Berlin.
John MacDougal/AFP/Getty Images

One way of understanding the urban environment is to consider the many and complex ways in which social structures and individual actions relate to one another. As human geographers we understand group identity to be the mainspring of human society. Members of a single group have a shared understanding of place, and as a result there is little conflict over land use. In any urban area, however, there are many groups, each of which may have a different understanding of a particular place; locations used by many groups, such as roads and parks, can take on a multitude of meanings and may become areas of conflict. Thus, any urban area has two types of space: locations occupied by groups whose members share a common understanding, and locations used by diverse groups and belonging to none. The architect and planner Greenbie (1981) distinguished between proxemic spaces, such as neighbourhoods, and distemic spaces, such as roads. Proxemic spaces reflect group identity, while distemic spaces have various meanings, depending on the social background of the individual. Traditionally, the most successful and stable distemic space is the market.

Social theory focusing on individuals as members of groups encourages us to acknowledge the meanings that different people attach to different locations. Such theory helps us to understand the values that people ascribe to a place and the conflicts that can emerge over land use. Following this general logic, much geographic research on urban life is informed by interpretations of culture, ethnicity, class, and gender. Indeed, a new urban cultural geography analyzes cultural identities and issues, including popular culture, corporate culture, retail culture, and racist attitudes and behaviours. Much of the content of this chapter reflects these identities and issues.

One key message is that cities are simultaneously successes and failures. Ambitious planning projects such as those that inspired Washington, DC, and Brasilia envisaged capitals designed to proclaim power and wealth, with grand avenues, large squares, and impressive public buildings and monuments—all symbols of mastery. But such goals have not prevented these cities from developing massive problems of poverty, homelessness, and crime. Reality often falls short of grand plans. Like the rest of the human world of which they are a part, cities seem to have brought out the best and the worst in humans. Many cities that are centres of civilization for some of their inhabitants are nightmare landscapes for others.

An example of an urban area that is at one and the same time a centre of civilization and a centre of social and environmental chaos is Los Angeles. Socially, Los Angeles has been described by Davis (1992) as 'fortress LA', while a 2008 study observed that it resembles a less developed world city with an elite group but with the majority—many of whom are recent immigrants with limited education—on poverty-level wages. Environmentally, it depends on water imported several hundreds of miles via aqueducts—but it also has an expensive system of channels and dams to handle flooding caused by the annual snowmelt. In addition, it experiences serious ozone pollution, produced when hydrocarbons and nitrous oxides (emitted by automobile exhaust) react in oxygen. Health warnings are issued when the ozone level exceeds a certain figure: in 1987 warnings were issued on 36 days; at Glendora, about 30 km (19 miles) downwind from LA, warnings were issued on 135 days.

Housing

Shelter is a basic human need and housing is the key feature of the urban social and economic landscape. For most of us, our understanding of a city includes the recognition of distinct housing areas or neighbourhoods, with some relatively privileged areas in terms of housing quality and related lifestyle and others that are more deprived. In many respects, the residents of a large detached suburban home on a substantial landscaped lot with multiple garages and the residents of a small dilapidated inner-city home in a crowded, impoverished, crime-ridden area live in two separate worlds. Housing and the areas in which housing is located mirror and enhance larger social inequalities. One useful way to introduce the idea of the city as social space is to identify four overarching social trends that have the potential to affect the social geography of the city (Box 13.1).

HOUSING MARKETS

How is the need for shelter satisfied? In more developed countries there are two ways of thinking about housing. Primarily, it is viewed as a commodity, a consumer good, and for most people housing is their most substantial expenditure and often also a major capital investment.

Box 13.1   Social Trends and the Social Geography of the City

Any focus on the social geography of the city needs to take into account a number of overarching processes and, following Murdie and Teixeira (2006), four ongoing processes that apply in the Canadian context are of note.

Economic restructuring, as it involves a relative decline in manufacturing employment and an increase in service employment, is characteristic of many cities in post-industrial societies; in Canada, these trends have been accelerated through trade liberalization agreements such as the North American Free Trade Agreement (NAFTA). The significance of this restructuring for the social geography of the city is that many of the new service employment opportunities are low-paid, part-time, and temporary, and that low-paid, service-sector employment has relocated from the city centre to the suburbs. Together, these changes lead to increased social inequalities inside the city and also to a spatial mismatch between where many low-income people live (in the city centre) and where suitable jobs are available (increasingly, in the suburbs).

Changes in age structure and family and household formation, such as a decrease in the percentage of younger age groups and an increase in the percentage of the elderly population, are also typical of cities in post-industrial societies. In Canada age structures are spatially variable because of internal migration: for example, Victoria receives many elderly immigrants whereas Calgary receives job-seekers more likely to have young families. Other important changes

in household formation include decreasing household size as fertility declines and more diversified family structures: both contribute to changes in the social geography of the city, especially those involving neighbourhood change. Indeed, some older industrial cities are experiencing significant demographic change as deaths outnumber births, a circumstance with implications especially for health care and educational facilities. A specific example is Pittsburgh, where obstetrics wards have been converted to acute care while public school enrolment has fallen from 70,000 in the late 1980s to 30,000 in 2008.

Increased internationalization, with growing proportions of immigrants coming from Asia, Africa, the Middle East, the Caribbean, and Latin America and decreasing proportions from traditional European source areas, is another trend in many countries. Two major social geographic impacts on Canadian cities are those of diversifying the cultural mosaic and sharpening differences within many immigrant communities, with some newcomers financially well-off and others possessing limited resources. Indeed, in many cities there may be few contacts between these two groups within any given ethnic identity.

Retrenchment of the welfare state, especially as it involves reductions in subsidized housing construction and in welfare payments, has occurred in several provinces during periods of government by relatively right-wing political parties, significantly affecting opportunities for low-income Canadians, including many recent immigrants.

This view encourages the creation of socially variable housing areas. Secondarily, it is viewed as an entitlement, a universal right, regardless of ability to pay. There is a fundamental ideological distinction between these two ways of thinking about housing. In the simplest terms, those with a socialist perspective believe that housing ought to be provided by the state as a social service like education or health care, while those with a capitalist perspective believe that housing is a commodity to be bought and sold without any state intervention. In most countries, attitudes towards housing lie somewhere between those two scenarios.

In the United States there is minimal state intervention, which equates to little public housing and relatively few people receiving financial support to help meet the cost of purchase or rental. In most other more developed countries, including Canada and European countries, many more people are helped in this

way as the state works to ensure a basic level of housing for all people. In Singapore and Hong Kong the state is the principal provider of housing and public housing dominates. Of course, political perspectives are not unchanging and the details of residential location, and therefore of social areas inside cities, are closely related to changing political attitudes over time. As an obvious example, the former Communist countries of Eastern Europe aimed, in principle, for total state intervention and a complete lack of private ownership, but housing policies in these countries are now more in line with the rest of Europe.

Housing as a commodity—something that is bought, sold, and rented—is appropriately analyzed in the context of a market, a locus where buyers and sellers interact to exchange services or goods. But housing is different from most other commodities. Most notably, with just a few exceptions, housing cannot be moved from one place to another and, further, the place where it is located is a critical determinant of value. The value of housing is affected by its spatial relationship with other urban land uses, such as industry, and with the social characteristics of the area within which it is located. Of course, the value of housing is affected also by many specific attributes of the house itself, including lot size, whether or not it is detached, number of levels, amount of living space, and state of repair both outside and inside.

Our ability to purchase many goods is determined essentially by personal financial circumstances, availability of the good, and awareness of options. But the housing market is much more complex than this. Like any other market, the housing market responds to supply-and-demand considerations, but this simple economic logic is disturbed by the fact that the market comprises many different participants who often display different political perspectives. Important constraints on individual residential location decisions are imposed especially by planners (as described in the previous chapter), by speculators who hold land in anticipation of making a profit, by developers and builders who often combine to influence neighbourhood character and housing type, and by real estate agents, lawyers, and financiers (sometimes called 'gatekeepers') who often work, separately or collaboratively, to encourage sales in some areas of the city and not in others. These components of and interventions into the market have many and varied effects.

For example, when prospective purchasers or renters seek information about available homes some relevant information may intentionally not be forthcoming, the result being that they lack possession of all of the facts needed to make an informed decision. In this way, they can be directed to certain areas of the city and denied access to other areas. Also, when sales support and financial assistance are required, some people may be treated differently from others. Such spatially discriminatory lending behaviours, often based on skin colour, were routine in the past and, although illegal today, continue to occur.

Redlining

The common practice known as **redlining**—because the affected areas were often marked in red on maps used by financial institutions—denies capital (mortgage money for purchases

and loans for home improvement) to certain areas of the city. Such denials are based on the perception that the areas are likely to experience decline in property values over time. Such a practice can be understood as a logical consequence of a capitalist market, where lenders are concerned about property values and the prospects for loan repayment, and also as a response to the fact that real estate agents and lenders tend to be conservative and opposed to spatial social change. But, of course, it is also a practice that discriminates between groups. Attempts by governments to prevent redlining are very difficult to enforce. In recent years it has been common practice to withdraw banking and other financial services from many inner-city areas, effectively a form of redlining as this reduces the possibility of home sales.

Areas most likely to be redlined are characterized by one or more of the following: older housing, a high percentage of rentals, low-income residents, temporary residents such as students, proximity to a low-income area, and either ethnic diversity or the presence of an ethnic minority. Such practices are especially evident in the United States where ethnic criteria have played the most influential role; but they were, and continue to be, evident in many other countries. In 2006 a local council in Loughborough, England, effectively banned students from moving into a new housing estate close to a university by informing developers that planning permission was to be granted only if homes are not sold to students or to owners who will rent to students.

The most obvious effects of these interventions into the housing market are, first, to create and reinforce segregated neighbourhoods and, second, to make some properties in some affected areas difficult to sell and therefore liable to fall into disrepair and to be taken over by slum landlords. A related consequence is to encourage home-building in suburban areas that are likely to develop as socially and ethnically homogeneous.

RESIDENTIAL MOBILITY

The decision to move residences within a city can be simply conceptualized like any other migration decision as being prompted by a mix of the push and pull factors described in Chapter 5 that reflect spatial inequalities. Put simply, if they are able, people move residence in order to improve their quality of life. Push factors include having too little living space

redlining
A spatially discriminatory practice favoured by financial institutions that identified parts of the city regarded as high risk in terms of loans for property purchase and for home improvement. The affected areas were typically outlined in red on maps.

and other perceived inadequacies in the design and quality of the home; often these deficiencies relate to the age of the home. Other push factors relate to neighbourhood characteristics such as the availability of recreational facilities, good-quality schooling, ethnic composition, and likelihood of criminal activity. A third set of push factors relate to accessibility, including distance from employment opportunities and quality of transport services. An additional consideration for many people is the desire to move out of rented accommodation and into a private home. Pull factors can be generally viewed as the reverse of push factors, for example, the prospect of adequate rather than the reality of too little living space.

Of course, these push and pull factors are understood differently by different individuals and families that have varying expectations of what a home in the city can offer them. People may be subject to a host of push factors but be unable financially to relocate. These factors also change through time with residential relocation often occurring at significant moments during the life cycle (Table 13.1).

Conceptualizing residential mobility

Two influential models of the residential relocation decision are those developed by Brown and Moore (1970) and by Robson (1975). Both models reflect a larger body of concepts that highlight the role played in spatial decision-making by stress factors and by the differences among a number of urban spaces. The Brown and Moore model, as modified by Popp (1976), is outlined in Figure 13.1. This model, based on the impact of stress, identifies factors that facilitate or hinder relocation.

The urban environment is seen as a continuous source of stimuli to which residents respond, with some stimuli contributing to a state of household stress if there is a disparity between the needs of specific residents and the urban environment. The greater the stress, the greater the likelihood of residential relocation. The concept of place utility reflects the discrepancy between needs and environmental circumstances, with environment including both the dwelling unit and the neighbourhood.

There are four important spaces. The largest of these is known as *action space*. This is the part of the city with which the decision-maker is familiar and comprises a set of places that are subjectively assessed and ranked. The size

Table 13.1	Selected Life-Cycle Events That Can Cause Residential Relocation
Life-cycle events in chronological sequence	
Completion of secondary education	
Completion of tertiary education	
Completion of occupational training	
Marriage	
Separation or divorce	
Birth of first child	
Birth of last child	
First child reaches secondary-school age	
Last child leaves home	
Retirement	
Death of a spouse	

SOURCE: Adapted from M. Pacione, *Urban Geography: A Global Perspective*, 2nd edn (New York: Routledge, 2005), 204.

of the action space likely reflects the amount of time a person has lived in a city and also the extent of her/his spatial behaviour in the city. *Activity space* is a subset of action space. It is the area within which the decision-maker engages in regular and often routine spatial behaviour. An understanding of the extent of these two spaces provides useful background information about the second two types of space that are critical when residential relocation decisions are being considered. *Awareness space* is the area within which the decision-maker has information about housing. The extent of awareness space is linked to the first two categories but is not necessarily congruent with either one of the two. The final category, *search space*, is a portion of the awareness space and is the area within which the search for housing occurs.

Filtering

A useful related concept is that of **filtering**. This is the idea that, through time, a housing unit is occupied sequentially by people from steadily changing income groups. Most usually the sequence is one of downward filtering with income level declining because housing units typically decline in quality as they age and the neighbourhood in which they are located becomes less desirable. One outcome of this process may be housing abandonment, a major problem in many older industrial cities that may contribute to urban financial problems because of unpaid taxes on abandoned properties. In some cases, the reverse scenario applies and upward filtering occurs. This type

filtering
A term used to refer to the fact that, through time, housing units typically experience a transition from being occupied by members of one income group to members of a different income group. Downward filtering is more usual than upward filtering.

FIGURE 13.1   Flow chart of the residential location decision process

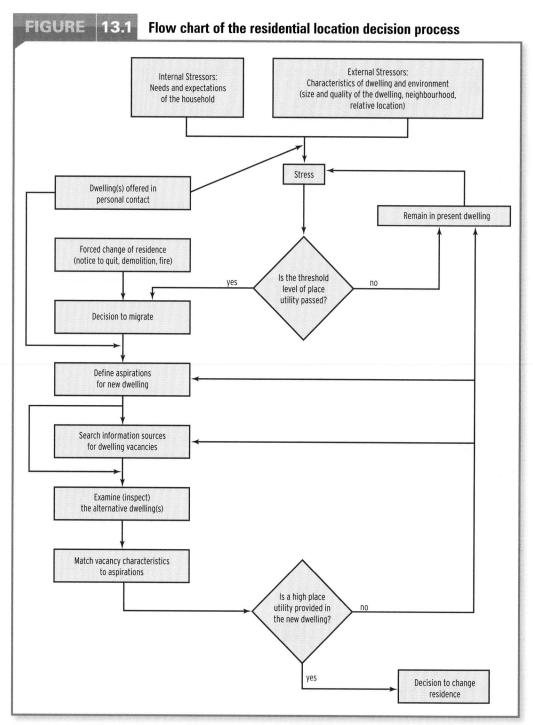

SOURCE: H. Popp, 'The Residential Location Decision-Making Process: Some Theoretical and Empirical Considerations', *Tijdschrift voor Economische en Sociale Geografie* 67 (1976): 303. Published by Blackwell Publishing Ltd.

of change is discussed below in the context of gentrification. The filtering concept can be applied at the level of the individual housing unit and at the neighbourhood level (Box 13.2).

But other factors are at play when people move. Because housing is often the major personal capital investment, people may aim to earn income by buying and then selling at a profit. Even if profit accumulation is not the overriding concern, it is usually a factor. People whose properties have declined in value such that they may have negative equity in their home are much less likely to be able to sell than are homeowners whose properties have increased in value since the time of purchase. For most people such considerations will always play an important role in any decision to relocate and this is likely to complicate significantly the overall decision-making.

Decisions made by individuals and families about where to live in the city reflect existing spatial social variations and then reinforce, or change, those variations. In other words, these decisions contribute to the ever-changing social mosaic of the urban landscape, including the formation of relatively distinct neighbourhoods.

Neighbourhoods

The idea of **neighbourhoods** has been noted on several occasions so far in this chapter, but

what are neighbourhoods? In general terms, they are places inside the city that are both spatially and socially constituted (Martin, 2003: 362). They are part of the visible and material landscape of the city, and also of the symbolic landscape. Most urban residents stereotype neighbourhoods and also the residents of neighbourhoods. But how are these places created, and what do they mean? Two interpretations of neighbourhood are noted.

First, the idea of neighbourhoods as communities where people of similar social and ethnic identity live, work, and worship has roots in nineteenth-century urban change and is central to the Chicago School of urban

> **neighbourhood**
> A formal region inside a city. A part of the city that displays some internal homogeneity regarding type of housing and that may be characterized by a relatively uniform income level and/or ethnic identity and that usually reflects certain shared social values.

Box 13.2   Filtering and Neighbourhood Change

The reciprocal links between residential mobility and neighbourhood quality have prompted urban theorists to develop stage models of neighbourhood change as examples of the filtering process. Early ideas along these lines were presented by Hoyt, the author of the sector model of urban internal structure described in the previous chapter, and his ideas have been further developed. For example, as one part of a substantial empirical analysis of the New York City region, Hoover and Vernon (1962: 192–207) outlined five stages of filtering (technically, four of downward filtering and one of upward filtering) as follows.

Stage 1: New single-family subdivisions are developed, typically on the city outskirts at the time development occurs.

Stage 2: Apartment development. This is a transition stage with new construction and related population growth, but with much of the new construction in the form of multi-family apartment complexes contributing to increasing population density.

Stage 3: This is a downgrading stage during which the existing housing—both single- and multi-family—is converted to allow for higher population densities. Housing quality declines and the neighbourhood ages. This stage may include the spread of ethnically segregated areas of the city.

Stage 4: A thinning-out stage during which population density declines as the average size of household gets smaller and as some units are left vacant, abandoned, or demolished.

Stage 5: In this final stage, a process of renewal commences as new multi-family housing is built, often replacing obsolete units. This is a process of slum renewal that requires substantial public or private

investment and it can be understood as a reversal of the earlier downward filtering process.

As a second example, Downs (1981: 63–7) proposed a five-stage model that describes a downward filtering process (although acknowledging that public planning policies or changes in the housing market might reverse the process and initiate upward filtering). The scheme devised by Hoover and Vernon was based on and applied in the New York City context, whereas this second scheme is intended as a more general conceptualization.

Stage 1: Stable and viable—these neighbourhoods may be either old and stable or new and thriving. There is no evidence of social decay and property values are rising.

Stage 2: Minor decline—these neighbourhoods are usually older and occupied by younger families with limited resources. There are some deficiencies in the housing units and population density is increasing. Property values are relatively stable.

Stage 3: Clear decline—these neighbourhoods are now occupied predominantly by renters rather than owners and many landlords are absentee. Both the physical quality of the dwelling unit and the social quality of the neighbourhood are declining.

Stage 4: Heavily deteriorated—these neighbourhoods have become slums with low-quality housing and a seriously problematic social environment. Property values fall and abandonment is common.

Stage 5: Unhealthy and non-viable—these neighbourhoods are at a terminal point with widespread abandonment. Those who live here are in the lowest economic category.

Continued

Some of the factors that underlie neighbourhood decline or, alternatively, revitalization are listed in Table 13.2. The more revitalization factors are present the higher the demand for property and the greater the likelihood of property improvement; the more decline factors are present the lower the demand for property and the greater the likelihood of property neglect. One consequence of downward filtering evident in some cities, including Regina and Saskatoon, is the donut effect, a term that refers to the situation where population in the central area is growing more slowly than in the surrounding area.

Table 13.2 Factors Underlying Neighbourhood Decline or Revitalization	
Revitalization Factors	**Decline Factors**
High-income households	Low-income households
New buildings with good design or old buildings with good design or historic interest	Old buildings with poor design and no historic interest
Distant from neighbourhoods of very low income	Close to neighbourhoods of very low income or to those shifting to low-income occupancy
In a city gaining (or not losing) population	In a city rapidly losing population
High owner occupancy	Low owner occupancy
Small rental units with owners living on premises	Large rental apartments with absentee owners
Close to strong institutions or desirable amenities, such as a university, a lakefront, or downtown	Far from strong institutions and desirable amenities
Strong, active community organizations	No strong community organization
Low vacancy rates in homes and rental apartments	High vacancy rates in homes and rental apartments
Low turnover and transiency among residents	High turnover and transiency among residents
Little vehicle traffic, especially trucks, on residential streets	Heavy vehicle traffic, especially trucks, on residential streets
Low crime and vandalism	High crime and vandalism

SOURCE: A. Downs, *Neighborhoods and Urban Development* (Washington: Brookings Institution, 1981), 66. Reprinted with permission of the Brookings Institution Press.

geography. This view centres on the association between where people live and the places that they frequent and that serve as centres of social interaction. This understanding is often important to the economic functioning of the housing market, and also has been used to argue for the role played in social life by what are labelled 'neighbourhood effects'—that is, the assigning of behavioural norms to people who live in a neighbourhood and the claim that life choices may either be severely constrained or enhanced because of where one lives. But the neighbourhood–community linkage may be misleading because an important component of neighbourhood is locality—a neighbourhood is a contiguous area—whereas communities may be spatially dispersed rather than clustered.

Second, neighbourhoods can also be viewed as sites of political activity and as contested landscapes, a view that reflects the social changes outlined in Box 13.1. Indeed, Smith (1985) suggested that the best indicator of neighbourhood identity is the presence of neighbourhood activism. Thus, McCann (2003) discussed how concerns about details of urban development in Austin, Texas, affected planning and zoning policies, and Martin (2003: 380) showed that neighbourhoods in Athens, Georgia, 'are produced through daily social life, but especially through strategic activism.' Of course, activism may have sources outside of neighbourhoods, in a working-class community that is not spatially clustered, for example.

These two understandings of neighbourhood are not mutually exclusive but they do highlight two quite different research directions. The community version of neighbourhood is politically innocent, while the activist version centres on political conflict. Both versions are useful to the understanding of the social geography of the city as evident in the text accounts of residential mobility, gentrification, segregation, and homelessness.

GENTRIFYING NEIGHBOURHOODS

The decades since World War II have seen a partial reversal of the tendency for high-status populations to live away from the city centre. Not only is life in the inner city more convenient for those who work in the city centre, but it offers a number of less tangible quality-of-life benefits. The process of people moving in and transforming a formerly derelict or low-quality housing area is known as gentrification (from 'gentry', a term that originally referred to people of high social standing). Today most cities, large and small, in the more developed world are undergoing gentrification (Ley, 1996; Fraser, 2004). Gentrification involves both the upgrading of homes and also changes in neighbourhood character and identity, with new coffee houses, bookstores, and specialty services replacing working-class homes and services such as small grocery stores and small owner-operated neighbourhood restaurants.

According to the Chicago School's ecological model, neighbourhoods decline after reaching maturity, but clearly this is not always the case: gentrification is the redevelopment and revitalization of a declining neighbourhood through rebuilding and other investment. This distinctive feature of some urban areas appears to be another consequence of the economic and social restructuring that has contributed to the rise of the post-industrial city. Two rather different explanations are suggested.

First, Smith (1996) conceived of gentrification in political economy terms as a consequence of capital seeking urban spaces in which a surplus can be accumulated. Inner-city areas that have lacked investment for some time become prime sites for speculative investment. Specifically, the rent gap hypothesis argues that there is a discrepancy between the current value of land and the potential value under a different use, such as gentrification.

Second, other researchers emphasize the role played by social changes, especially the rise of a more socially diverse body of home-buyers, including more childless families, more single people, and more openly gay and lesbian people. Rose (1988), for example, highlights the increasing number of professional

and other women living alone, noting their preference for an environment supportive of their preferred lifestyle. More generally, the rise of an increasingly consumption-oriented society involves middle-class consumers with disposable income seeking homes in areas that provide opportunities to eat out and visit museums and theatres.

Like many changes, gentrification has both positive and negative aspects. While it represents a benefit for middle-class people who want to live close to the city centre, for the poor—especially renters—it represents a threat, since rising house prices may well result in their displacement. The issue of displacement raises an important question: should governments intervene to protect the more vulnerable residents?

Unfortunately, this type of neighbourhood change contributes to social polarization in inner-city areas and may lead to social conflicts. On the other hand, the built environment is upgraded and local governments receive increased revenue from property taxes paid by these 'islands of renewal in a sea of decay' (Berry, 1985). Interestingly, some inner-city areas resist gentrification. The Grandview–Woodlands area of east Vancouver is a case in point: it has retained its distinct character and affordable housing, possibly because residents

Shop fronts on Division Street, the southern border of the Chicago neighbourhood called Wicker Park. Established in the 1870s, the district became home to a series of ethnic communities: first German and Scandinavian, then Polish, then Puerto Rican. By the 1970s it was in serious decline, but artists and musicians helped to spark a revival that eventually began to attract increasing numbers of affluent new residents. Today parts of the district are undergoing gentrification.

Andrew Leyerle

wish to preserve the diversity and identity of the area (Ventimiglia, 2007).

SEGREGATED NEIGHBOURHOODS

In most urban areas there are residential districts that can be distinguished on the basis of income, class, ethnicity, religion, or some other economic or cultural variable. Often, these areas include services as well as residences and, to varying degrees, they are separate worlds—apart from (as opposed to a part of) the larger city.

Geographic analyses of residential areas in cities have built on the early concentric zone, sector, and multiple-nuclei models to examine the spatial segregation of groups according to ethnicity, social status, and ability to pay for desirable locations. Spatial segregation reflects ethnicity, consumer demand, the ability and willingness of suppliers to provide different quality areas, and government intervention. More generally, spatial residential segregation in the Western world reflects a general acceptance of the idea that cities are not communities, but rather institutions serving competing individual interests.

Historically, especially in Europe prior to the Industrial Revolution, the most distinctive urban residential district was the Jewish district, usually labelled the 'ghetto'. During the Industrial Revolution, class divisions became more evident and were expressed in spatial terms. By the beginning of the nineteenth century, British commentators were acutely aware of the distinction between the working-class districts close to the factories and the middle-class districts located elsewhere, especially on the outskirts of the urban area.

Distinct residential areas emerged also in large immigrant-receiving North American cities in the nineteenth century, with distinctions based on ethnicity and class. Once a particular ethnic group was large enough, it settled as a group, usually in an inexpensive area close to employment opportunities. In the United States, black Americans who moved from the South to urban centres in the North and Northeast located in segregated neighbourhoods.

Elsewhere in the world, segregation might be based on religion, as in Belfast, Northern Ireland. Cities in colonial areas usually included separate districts for Europeans, other immigrant groups, and the local population. In South Africa, even before the formal institutionalization of apartheid, cities were clearly divided on explicitly ethnic lines.

In other cases, residential variation may reflect lifestyle preferences or the proximity of major employers, such as universities and hospitals. There is no doubt that society and space are irrevocably entwined and most large cities today include segregated neighbourhoods as part of the larger social spatial mosaic.

Ethnic residential segregation

Explanations for the creation and maintenance of ethnic neighbourhoods have focused on economic and cultural forces. Economic forces play a role in residential segregation generally, with most cities having a number of residential areas distinguished principally by income and access to capital, that is, the ability of residents to pay for homes in the area. But does economic logic apply in the case of ethnic residential segregation?

In many cities, members of ethnic, often recent immigrant, groups have lower incomes than most other residents and are, therefore, more likely to live in lower-quality housing areas. The housing market operates in a capitalist framework and the formation of residential areas can be understood as one logical outcome of the fact that different people have different abilities to pay for housing. But this seemingly innocent economic explanation of segregation is flawed, as most evidence indicates that segregation involving ethnic identity occurs regardless of income and wealth. For many ethnic groups, most notably blacks in American cities, there is no compelling evidence to support the suggestion that segregation is the result of economic processes. Economic forces alone are not an adequate explanation for segregation of ethnic groups and segregation is not to be understood as some natural outcome of economic processes; rather, it is the result of individual and collective human intentions and actions.

Two related cultural forces are the internal cohesion of the group and the desire of non-group members to resist spatial expansion of the group. Internal cohesion is strong when a group self-identifies and sees itself as different and indeed separate from other groups. Together, difference and separation encourage individual members of the group to seek out other members in order to preserve group identity through facilitating interaction with other group members and limiting interaction with members of other groups. In this way, residential segregation performs a critical cultural preservation and heritage function. Further, segregated neighbourhoods, such as the several Chinatowns and also Ghanaian, Portuguese, Italian, and Greek areas in Toronto, serve an important function as socially familiar areas for new immigrants (see Murdie and Teixeira, 2006).

In general, like the groups that occupy them, segregated areas have both inclusive and exclusive characteristics—group members are welcomed but non-group members are likely to be shunned. Segregated neighbourhoods may provide residents with a sense of security, reducing the threat of conflict with members of other groups. They also enable political action, through the formation of neighbourhood associations and perhaps through the election of group members into governmental positions.

Efforts by non-group members to limit the spatial expansion of the group are typically the product of discrimination and prejudice directed at the group. They are most intense when non-group members perceive the group as substantially different. Jewish ghettoization in European cities, black ghettoization in American cities, and the creation of Chinatowns in Canadian and American cities demonstrably are created not by the segregated groups but by the majority urban population and are the product of extreme prejudice. They are best understood as outcomes of institutionalized racism (for the example of Vancouver's Chinatown, see Anderson, 1991).

It is often difficult to determine the relative importance of internal cohesion of the group on the one hand and the desire of non-group members to resist spatial expansion of the group on the other hand. Kaplan and Woodhouse (2004: 583) ask: 'At what point can we say that group location is a result of its own spatial preference or a reaction to the continued spatial avoidance of the majority?' Indeed, it may often be the case that the more internally cohesive a group, the greater the discrimination against the group and vice versa.

Multiculturalism and ethnic residential segregation

Much of the literature reflects circumstances in the US, where ethnic residential segregation

A market outside a housing project in the Paris suburb of Clichy-sous-Bois, with two cars burned during widespread rioting in November 2005. Most of the violence was attributed to the isolation and despair of young men from immigrant (especially North African) backgrounds who, despite their French citizenship, have been unable to find a place in the mainstream society.

AP photo/Jacques Brinon

is an evident phenomenon. However, other parts of the world do not necessarily follow the same pattern.

In Canada, the principle of multiculturalism means that members of immigrant minority ethnic groups are not required to adapt to Canadian social norms while all other Canadians are encouraged to respect minority group differences. It seems clear that ghettoization is not evident in Canadian cities and that most ethnic concentrations are not associated with poverty and deprivation. Australia has similar policies in place while many European countries also favour these outcomes, although usually less formally. If these preferences are implemented, then the urban outcome will be neither complete ethnic segregation nor complete assimilation. In several cities in central England, notably Birmingham and Leicester, about one-third of the urban population is of Caribbean or South Asian origin, and both cities have extensive residential areas dominated by these groups. Also, both cities have recently experienced tensions between members of the two groups as one component of a complex pattern of identity and difference, sometimes based on source area and sometimes based on religious identity and related resentment.

Evidence from Australia, Britain, and from the three major Canadian cities of Vancouver, Toronto, and Montreal indicates that members of visible minorities are not ghettoized (Bauder and Sharpe, 2002). For Vancouver specifically, Hiebert and Ley (2003) distinguish between immigrants from Europe and those from less developed countries and show that, with just a few exceptions, the prevailing circumstance is one of assimilation, although this tends to take longer for groups from less developed countries. Interestingly, there is some association between cultural pluralism and economic marginality but it is not a pervasive circumstance, meaning that residential segregation and other indicators of cultural pluralism are not clearly related to social structural inequality. This tends to be quite unlike the situation in many American cities where the segregation of blacks and Hispanics especially is associated with structural inequalities.

A man holds a sign at a protest against the 2010 Winter Olympic Games on East Hastings Street in Vancouver, blocking traffic.

Ryan Koopmans /Alamy

Poverty and Deprivation

Cities are a collection of unequal places. In many countries, poverty has increasingly become concentrated in cities and, within cities, has increasingly become concentrated in the older inner area, partly because of the disadvantaged people who generally live there.

Cities throughout the more developed world typically experienced a prolonged period of substantial growth, beginning with a process of concentration that created the industrial cities of the nineteenth century followed by a process of decentralization that created the conurbations and urban sprawl of the twentieth century. The rapid expansion of urban industrial areas in the nineteenth century resulted in high-density, poorly serviced housing areas of poor quality. These areas contrast markedly both with the neighbourhoods of pre-industrial cities and with rural villages, where the separation of the poor and the rich was not nearly so marked. In 1845, the slums of Nottingham, England, were described as follows:

> . . . nowhere else shall we find so large a mass of inhabitants crowded into courts, alleys, and lanes as in Nottingham, and those, too, of the worst possible construction. Here they are so clustered upon each other; court within court, yard within yard, and lane within lane, in a manner to defy description. Some parts of Nottingham [are] so very bad as hardly to be surpassed in misery by anything to be found within the entire range of our manufacturing cities. (Hoskins, 1955: 218)

As these nineteenth-century slums were cleared, they were replaced by equally poor forms of low-cost, often high-rise, housing. Both types of slum have been associated not only with poverty and deprivation, but also with high levels of crime, vandalism, and substance abuse. Worsening social circumstances for many residents are evident in the increasing number of homeless people and the rise in criminal activity that is often related to organized gang activity and ethnic tensions. Today, many urban areas are experiencing a number of new problems related to changes in economic base (and associated lack of investment) and decaying infrastructures that exacerbate existing problems. For example, the aging physical infrastructure of most cities in the more developed world today means they lack the financial base to upgrade, and problems such as delays in road repairs and reductions in basic services are often evident, most notably in inner-city areas.

THE URBAN POOR

Although 'ghetto' and 'slum' are not synonyms, deprivation is often associated with distinct ethnic groups and, today, the spatial concentration of poverty often shows a close relationship to ethnic neighbourhoods. In Canada, Aboriginal populations are especially vulnerable, as are black populations in the US. But poverty knows no ethnic boundaries.

Why are there poor people who are deprived of such basic needs as shelter and who are both undernourished and malnourished? Four categories of response to this important question are noted.

1. Some commentators have proposed that a *culture of poverty* typically leads to a **cycle of poverty**. This argument centres on the inability of some people to cope effectively in the larger world such that poverty becomes their normal circumstance. Once in place, this culture of poverty is perpetuated, with poverty, deprivation, and slum living becoming the usual way of life and limiting the prospects for social and spatial integration with the larger urban area. This influential argument has been severely critiqued because of the emphasis placed on people's inability to cope and on deficient child-rearing practices. In short, this is an argument that can be interpreted as blaming the poor for being poor.

2. An extension of this logic focuses on *personal characteristics* that limit the ability of individuals to cope and care for themselves, characteristics such as physical or mental illness, deviant behaviour (criminal activity or substance abuse), personal disaffection with prevailing lifestyles, and lack of education and employment skills.

3. Not surprisingly, as Box 12.2 suggested, Marxism provides one popular explanation with emphasis on *structural class conflict*. According to this logic, the capitalist system necessarily concentrates power and wealth in the hands of a few and also requires a majority who are relatively dispossessed both materially and socially. Extensions of this logic stress the consequences of three factors: economic restructuring, such as the recent deindustrialization that results in both increasing numbers of poorly paid low-level service jobs and increased levels of unemployment; gentrification that removes some homes from the housing market that poor people can access; and the general failure of both public and private sectors to build low-cost housing.

4. Another explanation, as suggested in the account of residential mobility, emphasizes the roles played by those who have the power to control access to urban resources, such as real estate agents and mortgage companies. This is an *institutional* argument centring on the fact that material and knowledge resources are unevenly distributed and on the resulting uneven relations between people in positions of power and authority and the disadvantaged.

The range of theoretical options is stimulating and each idea provides some assistance in understanding the reasons for poverty and in identifying possible solutions. Nevertheless, it cannot be denied that the issues are far from being understood and resolved; most current evidence suggests that urban problems such as poverty, inadequate housing, and homelessness are increasing rather than decreasing.

Indeed, one common response, on the part of rich residents, is to widen the gap between themselves and the poor; the differences between the predatory landscape of a downtown slum and the architecture of a suburban gated community speak directly to the failure of many large cities to provide fair and just communities for all residents.

Recent research suggests that cities, and the countries in which they are located, can suffer serious social and health problems precisely because there are such inequalities. Put simply, compared to places whose populations are relatively equal, places where there is a significant income and lifestyle gap between rich and poor have higher incidences of mental illness, substance abuse, obesity, and criminal activity, have lower educational standards, and have a shorter life expectancy. Significantly, it is not only the poor who suffer in unequal places, but also the rich; this is because inequality increases stress levels for all people, whereas equality

cycle of poverty
The idea that poverty and deprivation are transmitted intergenerationally, reflecting home background and spatial variations in opportunities.

Box 13.3   Eight Mile Road

Consider the notorious example of Eight Mile Road in Detroit, so named because it is located eight miles from the river (Figure 13.2). A familiar name to many—*8-Mile* is the title of a semi-autobiographical film by the rapper, Eminem—this road is the actual and symbolic dividing line between city and suburbs, separating blacks and whites, poor and rich, powerless and powerful, under-privileged and privileged. Eight Mile Road is a major eight-lane highway, frequented by prostitutes and drug dealers and with mixed land uses including pawn shops, topless bars, sex shops, and cheap motels. In 2006, another rapper, Proof, a close friend of Eminem and the best man at his wedding, was murdered outside the Triple C bar on Eight Mile Road.

The urban landscape to the south, the city proper, is more than 80 per cent black and includes landscapes of severe deprivation with burned-out and abandoned buildings. Criminal activity is rampant. Previously a vibrant industrial and working-class residential area, the city has suffered from two circumstances. First, riots in 1967 prompted many white residents to flee to the suburbs, taking their jobs and money with them. Second, industrial restructuring (discussed in Chapter 14), especially the long, drawn-out decline of the automobile industry, has led to huge job losses. Estimates suggest there are about 13,000 homeless people.

The urban landscape to the north, Oakland County, is more than 80 per cent white and is a more typical representation of suburban America, comprising pleasant neighbourhoods of painted homes and parks with names such as Hazel Park, Oak Park, and Pleasant Ridge, as well as coffee shops and attractive office blocks. The difference between city and suburb, north and south of Eight Mile Road, could not be more marked. People live separate lives in these separate communities.

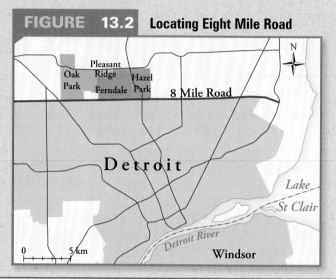

FIGURE   13.2   Locating Eight Mile Road

results in a sense of personal security and a co-operative spirit (Wilkinson and Pickett, 2009).

POVERTY AND URBAN LIFE

In some large American cities it is possible to identify 'welfare neighbourhoods' (DeVerteuil, 2005a), where most people rely on welfare payments and where these payments are the basis for the local economy. These neighbourhoods typically lose health, education, and financial and basic retailing services, including inexpensive supermarkets.

Although members of all social groups are susceptible to poverty, compelling evidence links ethnic minority identity to the formation of slums and also to employment status, level of education, health, age, and gender issues. With reference to employment status and level of education, for most cities the transition to the current post-industrial phase has resulted in major losses of employment opportunities and the deindustrialized areas have not received any new investment. The principal exceptions to this negative trend are the global or world cities discussed in Chapter 11. Elsewhere, the new employment opportunities in the high-technology, service, and tourist industries are rarely open to former members of the industrial labour force, who often lack the needed educational qualifications. Increasing levels of unemployment lead to problems of social service supply and, often, increasing social tension and criminal activity, and also to personal difficulties related to poverty.

Poverty and health problems also go hand in hand as inadequate housing or homelessness increases the risk of hypothermia during cold periods, and of tuberculosis, bronchitis, and skin infections. In countries that lack an effective, free health-care system these illnesses may well go untreated, as they often do in a country such as Canada that does have publicly funded health care. Substance abuse, as one avenue out of despair, is also a major health problem in poor areas of the city.

With reference to age and gender, single mothers and older women living alone are vulnerable, although in many countries the gender-based divisions are being reduced through subsidized daycare and related measures. Most current research on these matters stresses that gender, class, ethnicity, and sexuality are often interwoven and analyses rarely focus on just one of these (Ray and Rose, 2000: 518).

Poverty is also closely related to criminal activity, with poor people over-represented as both perpetrators and victims. There is a close relationship between unemployment and criminal behaviour, but it is not appropriate to say there is a causal relationship. Indeed, it is important to recognize that criminal activity is also related to inequality and social exclusion (Boxes 13.3 and 13.4).

HOMELESSNESS

One of the functions of a city is to provide shelter for those who live there, but not all residents can afford a place of their own and some people rely on institutional settings or public spaces. There are many different forms of accommodation in the typical city and Figure 13.3 itemizes a range of types of accommodation, from the single-family house to public spaces.

Being homeless does not necessarily mean sleeping in the open. Four versions of homelessness are:

- *Rooflessness*. Sleeping 'rough', i.e., in the open air, is the most visible sign of being homeless.
- *Houselessness*. This term applies to people who routinely sleep in shelters.
- *Living in insecure housing*. This circumstance arises when permanent housing is unavailable or when people are obliged to share with others. It may be prompted by domestic violence that results when one partner leaves a family home to live temporarily with friends or family or in a shelter.
- *Living in inadequate accommodation*. Some housing is of such poor quality—overcrowded or otherwise unfit for habitation—that it cannot be considered as a shelter.

Box 13.4 Crime in the City

A major need is that of making cities safer places. Overall, residents of urban areas are much more likely to be victims of crime than are residents of rural areas. Inside the city, adolescents are most likely to be perpetrators and victims. But reliable data are notoriously difficult to obtain as much criminal activity goes unreported. It is, for example, very difficult to collect sound data on violence against women because so much is hidden from view as it takes place behind closed doors and in an intimate social context.

Criminal activity generally is considered newsworthy and newscasts in large cities often highlight inner-city problems, emphasizing crimes such as violent assault and substance abuse and commenting on gang activity and more general social breakdown. Certainly, the occurrence of crime can be understood as a powerful indicator of the failure of urban areas to provide a comfortable environment for all residents. For many, perception of crime affects how they use the urban environment, for example, where (and when) they feel safe and unsafe and also how they understand and relate to other urban residents. A major concern in many downtown areas is street safety.

In many North American cities crime rates have fallen in recent years, a trend that is at least partly explained by changing demographic circumstances with increased numbers of elderly people and decreased numbers of young people.

Other possible explanations may be improved community policing and various policies aimed at preventing crime. Of course, another possibility is that more crimes are not being reported as victims might perceive no advantages to reporting some crimes.

A continuing and highly disturbing feature of urban criminal activity is that younger members of minority groups—for example, in Canada, youth of Caribbean origin and young Aboriginal people—are over-represented both as victims and perpetrators. This circumstance needs to be understood in the context of the systemic racism and discrimination that contributes to social and economic inequality and to exclusion from larger urban society. A Vancouver study showed that areas with high unemployment and a young population were the locations of most criminal activity (Andresen, 2006).

Cities need to be safer places. One typical response to this need is to strengthen police forces and introduce zero-tolerance policies. These responses may lead to more convictions, but most evidence suggests that this is not a sufficient response as it does not typically lead to a reduction in crimes. Indeed, strengthening police forces and introducing zero-tolerance policies may not be an appropriate response to the problem. Rather, there is a need for attitudinal changes, for a focus on the root causes of criminal behaviour, and for a greater focus on prevention.

FIGURE | **13.3**   **Types of accommodation in a post-industrial society**

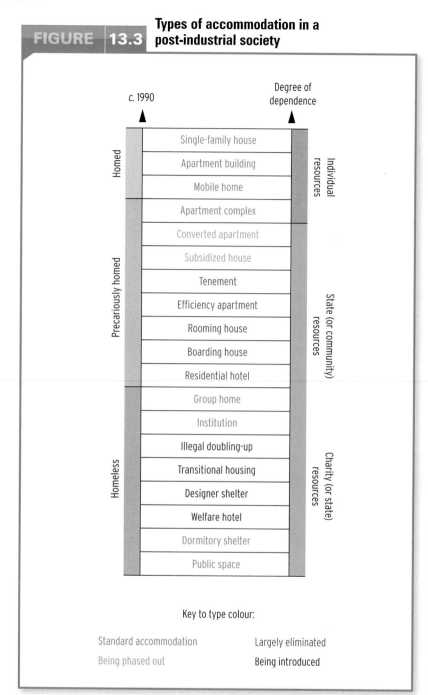

c. 1990   Degree of dependence

Homed
- Single-family house
- Apartment building
- Mobile home

Precariously homed
- Apartment complex
- Converted apartment
- Subsidized house
- Tenement
- Efficiency apartment
- Rooming house
- Boarding house
- Residential hotel

Homeless
- Group home
- Institution
- Illegal doubling-up
- Transitional housing
- Designer shelter
- Welfare hotel
- Dormitory shelter
- Public space

Individual resources

State (or community) resources

Charity (or state) resources

Key to type colour:

Standard accommodation   Largely eliminated
Being phased out   Being introduced

SOURCE: A.R. Veness, 'Home and Homelessness in the United States', *Environment and Planning D* 10(4) (1992): 462. Pion Limited, London.

Consider the following disturbing example of the spatial juxtaposition of privilege and deprivation. New Haven, Connecticut, is the home of Yale University, one of the wealthiest and most prestigious schools in the United States. In 2002 a news report described a tent city adjacent to the university. Established after authorities closed down an overflow homeless shelter, the tent city in turn was quickly cleared by police. The ironies are frightening. Proportionally, Connecticut has more millionaires than any other American state, and yet New Haven has an infant mortality rate comparable to that of Malaysia. It has suffered a devastating number of deaths associated with AIDS, and according to a 2002 report, almost 70 per cent of the city's children are without health insurance. Critics suggest that New Haven is a metaphor for the United States in the early twenty-first century.

Most circumstances of homelessness are not quite so dramatic as the New Haven example, although this serious social problem is experienced by increasing numbers of people. It may be that New Haven's tent city is an instance of what has been called 'the new poverty management' (Wolch and DeVerteuil, 2001). This term refers to the organized efforts undertaken by the state and elite groups to maintain social control over the poor and underprivileged. In recent years, homeless people especially have been subject to involuntary moves between residential units, shelters, prisons, sober-living homes, hospitals, and the street. This cycling approach may have more to do with removing the poor from view than with finding a lasting solution to a social crisis.

In their challenging work *Landscapes of Despair*, Dear and Wolch (1987) describe the social problems prevailing in areas dominated by the mentally ill, many of whom are homeless. Increasingly, these people are moved from one institutionalized setting to another. These settings include residential homes, shelters, prisons, hospitals, rehabilitation centres, hotels, and the street. Hypermobility is then a characteristic feature of this homeless experience and can be understood in two conflicting ways. Hypermobility can be seen as the homeless acting as autonomous agents and coping with their challenging circumstances as well as they can; alternatively, it can be seen as an involuntary response that is indicative of a failure to cope and as causing rifts with earlier social contacts and an increased association with other homeless people.

Homelessness is often precipitated by a specific circumstance such as loss of employment, eviction, or domestic violence. Once homeless, people gradually become labelled as homeless and even self-identify as homeless, a situation that increases the likelihood of remaining

A tent city along the banks of the Saint-Martin canal in Paris in December 2006. Some of the tents were occupied by homeless people, others by their supporters.

AP photo/François Mori

homeless. Box 13.5 summarizes two studies of homeless populations.

How many people are homeless? There is no easy answer to this question, both because there is no precise definition of 'homelessness' and because estimates often reflect vested social and political interests. For the United States in the 1980s, social advocates estimated several million while government officials estimated 250,000–350,000; for Canada, the 1991 census attempted to count the number of homeless people by reference to the use of soup kitchens, but the figure was never released as it was assumed to be inaccurate. A 1987 estimate by the Canadian Council on Social Development suggested 130,000–250,000 homeless people (Peressini and McDonald, 2000: 527–8). One estimate for Britain is 500,000 (Pacione, 2005: 232).

Although few data are available on the financial cost of homelessness to the various levels of government and to society more generally, it is substantial because of the pressures placed on emergency medical facilities, on police services and the justice system, on social service agencies, and on charitable agencies. DeVerteuil (2003: 378) is surely correct to query: 'Might it not be more fiscally sound for governments simply to provide affordable housing?'

Cities as Centres of Production and Consumption

People do not only live in cities; they also work and consume in cities. Historically, people have moved to cities in search of employment and, once employed and settled in the city, their everyday lives are played out in the urban environment. Cities, then, are centres of production and of consumption.

Recall the distinction noted in Chapter 11 between basic and non-basic economic activities. Manufacturing activities are basic: they generate income from outside the city for city residents; retailing and other consumer service activities are non-basic: they circulate income within the city. However, this distinction is disturbed by the rise of many new advanced service activities that are both basic and non-basic. In this context, advanced services include computing, legal, and financial activities that serve both the local urban economy and also the larger global economy.

MANUFACTURING

As noted in the Chapter 11 account of urban growth, urbanization and industrialization go hand in hand. During the Industrial Revolution

Box 13.5   The Homeless Experience

What does it mean to be homeless? In-depth studies that interviewed homeless people in Calgary (Peressini and McDonald, 2000) and in Los Angeles (DeVerteuil, 2003, 2005b) offer some valuable and disturbing insights.

The Calgary study suggests that the typical homeless person is single, male, and poor. Such persons are undereducated and unemployed, but these circumstances are not of their choosing. Rather, they invest much time and energy into seeking employment, contrary to stereotypical perceptions of the homeless as lazy and unwilling to work. Many of these people work on a casual or temporary basis. Also contrary to some claims, they are not choosing to separate themselves from larger society and they do not constitute a deviant subculture. Indeed, many have friends and families and they are homeless despite their best efforts to be and do otherwise. When asked about their personal situations, some respondents understood homelessness as being without a home, but others understood it as being without a source of income or as being unemployed. Reported reasons for being homeless included lack of money, being unemployed, and larger economic circumstances. Many homeless respondents rarely slept on the street, with shelters being their principal means of accommodation.

The Los Angeles study focused on 25 single, homeless women. Interviews uncovered the sequence and variety of residential patterns, the circumstances of movement from one setting to another, and respondents' perceptions of the various settings. Results showed that these single, homeless women lacked employment and other income, that at least some used mobility as a coping mechanism, that sometimes they were not merely passive users of institutions but rather reinterpreted settings in ways that affirmed their identities, and that stability was not necessarily indicative of coping. As an example of hypermobility, one 50-year-old Caucasian woman, Ellie, cycled though 10 settings during a three-year period (Figure 13.4). Much of this was intentional as Ellie often remained at shelters as long as permitted, kept storage units in both Los Angeles and San Francisco, and withheld information about her circumstances from friends.

Other city residents tend to encounter homelessness on a routine but casual basis as it is not uncommon for some homeless people to seek financial support at traffic intersections and downtown traffic arteries and to seek refuge in or adjacent to public buildings. Sometimes, homelessness makes the headlines, as in Vancouver in 2003 when a series of squats were organized by anti-poverty groups to highlight the plight of homeless people and the need for publicly funded housing and welfare reform. But most of the time most people go about their lives with little awareness of the homeless and, certainly, no understanding of how the homeless live their lives.

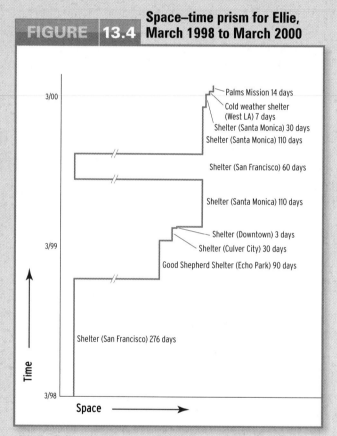

FIGURE 13.4   Space–time prism for Ellie, March 1998 to March 2000

SOURCE: G. DeVerteuil, 'Household Mobility, Institutional Settings, and the New Poverty Management', *Environment and Planning A* 35(2) (2003): 369. Pion Limited, London.

many established urban centres grew rapidly and many new urban centres were established. By the early twentieth century, manufacturing regions came into being and many of the cities in these regions concentrated on just one or a few related manufacturing activities. Inside these cities, it was usual for many industrial activities to be located adjacent to the central business district (a situation explicitly reflected in the concentric zone model) and along railway routes (a situation explicitly reflected in the sector model).

However, these patterns of distribution have changed considerably in recent years. First, most countries have experienced some decentralization of manufacturing activity away from the traditional manufacturing cities, thus weakening the economic base of the old industrial centres and strengthening the new centres. Today, we have many more types of industry, and they favour a much broader range of locations. Inside the city, the central industrial locations have lost manufacturing activities while outer suburban areas have gained. Overall, these various spatial changes are part of larger economic restructuring processes and the decreased friction of distance and are discussed in a larger context in the Chapter 14 account of industry.

Several attempts have been made to categorize different industries and related favoured locations in and near urban areas. Castells and Hall (1994) noted five location types:

1. Industrial complexes of high-technology firms include linked research and development and manufacturing plants. The best-known example is Silicon Valley, south of San Francisco.
2. Science cities focus solely on research and development. A well-known example is Tsukuba, close to Tokyo: 'a national research centre funded totally by central government; its laboratories, engaged on basic research, are government laboratories' (ibid., 67).
3. Technology parks are intended to attract high-tech firms and thus generate industrial growth. These are increasingly common and can be found on the outskirts of most major cities.
4. Technopoles are designed as part of a larger program of regional planning aimed at upgrading depressed areas; they involve a variety of technologies and both public and private funding. Sendai, 300 km (186 miles) north of Tokyo, is one example.
5. Regenerated older areas are often located close to or inside major cities.

These five scenarios highlight the movement of industrial activities outward from inner-city areas. To some degree, their place in the inner city has been taken by service activities, office buildings, and public institutions; as previously noted, one of the principal functions of major cities today is to provide access to providers of advanced business services.

The landscapes of some of these new high-technology industrial areas may leave much to be desired so far as livability is concerned. Silicon Valley, the major centre in the United States for semiconductors and computer software, has been described as a 'generic exopolis,

A classic 'big-box' store, in Scarborough, Ontario.
Dick Hemingway Photographs

homogenized, undifferentiated; a vast inter-nodal matrix of tract housing, apartment complexes, strip malls, and industrial parks' and as 'beige suburbia' (MacCannell, 2005: 92).

RETAILING

Retailing today is a major component of national and urban economies, a significant user of urban land, a key component of the image that residents have of their cities, and also a major leisure activity for urban residents. Retailing is a consumer service, as are entertainment and recreation.

The location of urban retailing

Why do retailing activities locate where they do? One answer identifies eight factors that may influence such decisions (Nelson, 1958):

1. trading area potential—an above-threshold population is needed;
2. access to trading area—retailers must compete for peak land-value locations;
3. growth potential—close to growing populations and rising incomes;
4. business interception—between employment and residential locations;
5. cumulative attraction—adjacent to other retail outlets;
6. compatibility—in an area of compatible uses;
7. minimizing competitive hazard—away from direct competitors; and
8. site economics—ease of local access.

Clearly, not all of these factors will be relevant in any one instance; for example, striving for both cumulative attraction and compatibility is likely to maximize rather than minimize the risks of competition. With several of the eight factors mobility and interaction are important. Low mobility leads to concentration of retailing activities, and high mobility contributes to dispersion. Also important is the size of the urban centre (Simmons and Jones, 2003).

Traditionally, retailing located on streets in the central areas of cities, but a major transformation of retail landscapes that first emerged in the United States was the location of department stores in suburban areas. In 1947 a suburban shopping centre comprising two department stores, several smaller stores, and a large car park opened in Los Angeles. The building of enclosed, climate-controlled shopping malls located in suburban areas soon followed, with the first such centre being Southdale in Minneapolis. There are now about 1,100 enclosed malls in the United States and many others elsewhere in the world.

Since about the 1970s, urban retailing activity has undergone a number of organizational and related locational changes in response to broader post-industrial trends. There are three principal organizational changes. First, the numbers of independent traders have declined sharply as a consequence of competition between large business enterprises. Second, the largest retailers, of both goods and services, are now operating as transnationals, opening stores in more than one country; major examples include Wal-Mart (American) and Marks & Spencer (British). Third, new communications technology has permitted the creation and rapid growth of television shopping networks.

The most evident recent locational change is a further decentralization of shopping centres, many of which are now located in out-of-town areas accessible only by automobile. These new retail areas, often called 'power retail' developments, include many new retailers. Such centres are controversial in many areas because they may compete with small rural retailers, pave over prime arable land, replace attractive green space, bring significant increases in traffic, and require large parking areas. A specific form of this trend is the automobile mall consisting of several dealerships in close proximity. These locational changes challenge the viability of older retailing areas, especially in downtowns, but also the older suburban enclosed shopping malls, many of which were already declining in popularity as suburbs spread further outwards and as demographic circumstances changed. The example of urban retailing in Canada is discussed in Box 13.6.

Box 13.6   Urban Retailing in Canada

In general terms, geographers today often identify three types of retailing area: nucleated centres in the city centre region and the suburbs; ribbon developments along highways; and specialized areas such as an entertainment zone comprising functions that cluster because of mutual interdependence. Typically, the city centre and ribbon areas are unplanned at the outset, whereas the suburban and specialized areas are planned and develop as coherent retailing and service areas.

These three general types of retail area reflect several transformations that have occurred during about the past 60 years, principally because of changes in urban transportation and in the related internal structure of the city. These transformations and the specific spatial consequences can be summarized as follows (Jones, 2000).

1. Before World War II personal mobility was limited and shopping was either local, for basic food items, or downtown, for larger items.

2. A transformation of this pattern began in the 1950s with the establishment of the first shopping centres outside of downtown and proceeded apace for about the next 40 years as personal mobility grew, transportation improved, and suburbs expanded. The earliest of these shopping centres were located close to existing residential areas, but by the 1960s residential and retail growth were occurring at the same time, and by the 1970s it was the shopping centre that was located first and that acted as a spur to residential development. During the 1980s a new type of shopping centre appeared, one that self-identified not as a shopping centre but rather as an entertainment or tourist attraction.

3. During the 1990s a different route was followed by some urban retailers with the establishment of mostly US-based 'big-box' stores (such as Home Depot, Chapters, and Michaels) and power centres in suburban or outlying areas.

4. The 1990s also witnessed an increase in specialized retail clusters, often in gentrified central areas of the city. Five

Continued

types of such clusters are specialty product areas, fashion centres, factory outlets, historic or theme areas, and ethnic strips (Jones and Simmons, 1993).

5. Finally, most central-city areas have reinvented themselves in response to these transformations. Typically, downtowns in wealthy cities have managed to maintain a viable retail environment while those in less wealthy cities struggle to do so.

Figure 13.5 is based on the transformations of the retail environments described above and provides a framework for understanding the complexity of retail environments in the contemporary Canadian city. The classification is based on four criteria: morphology, location, market size, and type. A first distinction is that between centres and strips; a second,

between inner-city and suburban locations; and for inner-city retail centres a third distinction is between unplanned and planned. Within each of the types thus distinguished, the kinds of retail area are ranked according to their position in the intra-urban retail hierarchy. Finally, each of the 28 different classes of retail area identified is classified as either serving spatial or specialized markets.

Table 13.3 indicates another interesting dimension to Canadian urban retailing in the context of service activities more generally. The table identifies six types of business enclave and provides details of regional variations. The percentages indicate the share of each spatial retail type. There are some interesting highlights in the table. For example, retail strips are concentrated in Quebec and Ontario, while retail, office, and hospitality service corridors are located mostly in the West.

FIGURE 13.5 — A typology of the contemporary urban retail system

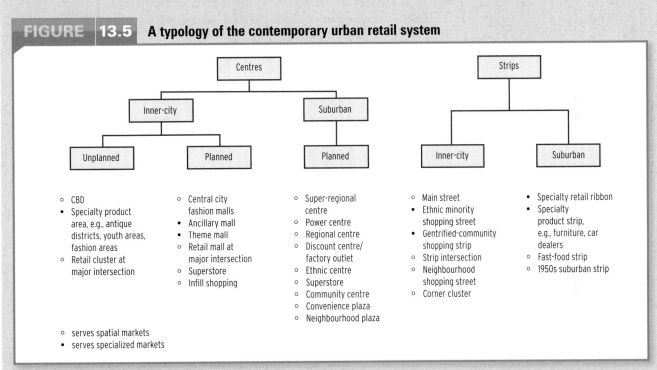

SOURCE: K. Jones and T. Hernandez, 'Dynamics of the Canadian Retail Environment', in T. Bunting and P. Filion, eds, *Canadian Cities in Transition: Local Through Global Perspectives*, 3rd edn (Toronto: Oxford University Press, 2006), 301.

Table 13.3 — Commercial Enclaves in Canada, by Region

Business enclave type by region	Maritimes	Quebec	Ontario	Prairies	BC and Territories	Total
Retail strips	31 (18%)	141 (37%)	141 (28%)	20 (8%)	19 (10%)	352 (23%)
Retail office and hospitality service corridors	9 (5%)	28 (7%)	68 (14%)	43 (18%)	33 (17%)	181 (12%)
Malls, big-box and flagship stores	16 (9%)	29 (8%)	118 (23%)	67 (27%)	61 (31%)	291 (19%)
Entertainment centres and business services	9 (5%)	60 (16%)	61 (12%)	33 (13%)	30 (15%)	193 (13%)
Personal and community service centres	84 (48%)	95 (25%)	89 (18%)	64 (26%)	52 (27%)	384 (25%)
Exurban retail strips	27 (5%)	33 (9%)	26 (5%)	18 (7%)	1 (1%)	105 (7%)

SOURCE: R. Gomez-Insausti, 'The Spatial Structure of the Canadian Business/Commercial Sector: A Study in Supply-side Segmentation', *Progress in Planning* 60:1 (2003): 21. Copyright © 2003 Elsevier Science Ltd.

Shopping, identity, and empowerment

In Canada, retail sales account for about one-third of disposable income, a statistic that emphasizes how important shopping is in contemporary Canadian society. But shopping, of course, is about much more than spending money on such essential items as foodstuffs. Throughout the more developed world, much of our shopping behaviour, both where we shop and what we purchase, is designed not so much to satisfy needs, but rather to say who we are and, sometimes, to enable us to become someone else. As noted in Box 7.13, shopping behaviour reflects identity, reinforces identity, modifies identity, and even creates new identities.

Box 13.6 indicates that developers respond to and initiate consumer trends through store formats. For example, big-box stores targeting the suburban family market locate where space is available and are accessible usually only by car. Entertainment complexes locate in both central and suburban areas to generate and satisfy demand from local residents, tourists, and business travellers. Specialized retail strips sometimes develop a themed format to serve a niche market, perhaps an ethnic market or a gay and lesbian market. Retailers respond to and initiate these consumer trends principally through brand names and marketing and advertising strategies.

City centres

Much research on urban retailing has emphasized the important role played by the city centre—the central business district (CBD) of the three models of internal structure. During the early 1950s, especially, there was an almost exclusive concern with defining and delimiting the CBD. Then Murphy, Vance, and Epstein (1955) changed the terms of the discussion by drawing attention to the internal diversity of CBDs and their constantly changing boundaries. Subsequent studies have also emphasized change in land uses, as well as the fact that not all city centres are the same and that the concept of a CBD often no longer applies.

Industrial and retailing relocation from central to suburban city areas has contributed to a general decline in the city centre area that was previously the CBD, the area where most of the economic life of the city took place, as the advantage that the CBD once held—centrality—becomes less of an advantage. Today, city centres are subject to rapid change, often in response to changing tastes, and they are put to a multitude of uses; moreover, in the largest cities the central areas often contain several spatially distinct land uses, and in these cases the CBD concept has little value.

In many cities processes of urban revitalization have focused on the central city area and, while the details of land use have changed, the city centre remains a viable economic location, albeit not always clearly different in land use from many new suburban areas. Toronto, for example, followed the suburban lead with development of a large indoor shopping complex, the Eaton Centre. Also, many central areas have diversified to take advantage of their historic identity; Byward Market in Ottawa is an interesting example (Tunbridge, 2001).

More generally, many cities are taking advantage of local or regional cultural identity to market themselves to tourist and business consumers. For example, the French government created a new museum of contemporary art—the Centre Georges Pompidou—in the centre of Paris as one way of highlighting the tradition of French artistic creativity, but with the explicit planning intention of upgrading a declining inner-city area. This building is a form of new cultural capital that has emerged as a world-class tourist attraction.

One way to identify the role played by many central areas is to compare daytime and nighttime populations. Washington, DC, grows by 72 per cent during the day, and shrinks correspondingly during the night; for Atlanta the figure is 62 per cent, for Tampa 48 per cent, and for Boston 41 per cent. Perhaps the most remarkable statistic, although admittedly not applying to a traditional city centre, is for Lake Buena Vista, about 20 miles southeast of Orlando, Florida, which increases during the day by 192,238 per cent because of workers commuting to Disney facilities and surrounding businesses. Of course, while Washington, DC, and other cities grow during the day, the reverse is the case for the outlying suburbs that supply the commuters.

Living in the Less Developed World City

Two different surveys in 2005 ranked cities on the basis of quality of life according to such factors as health care, education, security and stability, culture, and infrastructure. Although the specific rankings varied somewhat because of different methodologies, the overall results were very similar. The most desirable cities in which to live are located in several of the European countries, Canada, Australia, and New Zealand. At or near the top of the rankings were Vancouver, Melbourne, Auckland, Vienna, Geneva, and Zurich. The least desirable cities are located in Africa and parts of Asia. At or near the bottom of the rankings were Port Moresby, Algiers, Dhaka, Baghdad, Lagos, Bangui, and Brazzaville.

So, what is life like in a less developed world city? To address this question, we first review the origins, growth, and environments of these cities and then embark on a journey through some of the cities in Latin America, Africa, India, and China. The story told is one of cities that are best described as challenging landscapes presenting a majority of residents with serious problems on a recurring basis.

Notwithstanding these often dire circumstances, residents clearly judge that their best hope for a reasonable quality of life is inside the city, not in a rural area. Most less developed world cities perform the same urban functions as cities in the more developed world, but not usually as effectively. Thus, they provide homes, places to work, and places to consume. The problems of daily life are real and may prove very difficult to resolve, but many residents work extremely hard to cope with these problems and to provide their children with a better life. The problems experienced by the urban poor, especially, are not of their own making but the result of larger economic and social circumstances and, at a root level, these circumstances need to be addressed. One key need is for better education facilities.

COLONIAL ORIGINS AND RECENT GROWTH

As discussed in Chapter 12, many (though by no means all) of the cities in the less developed world have a colonial heritage. Many of the cities in Africa and Latin America, and some in Asia, were essentially creations of European powers developed to serve European needs, a fact that in many cases is reflected in their pre-independence names, their architecture, their locations, and their internal structure. In most instances, the pre-independence name has been changed for powerful symbolic reasons, although colonial architecture has generally remained. Cities were located in places that

suited their role as export centres, but these locations often proved inappropriate for independent countries with radically different economic goals. With regard to internal structure, it was not uncommon for colonial powers to implement some degree of formal spatial segregation on the basis of ethnic identity, a situation that reached its extreme in South Africa before 1994 (see Box 7.9).

It is important to recognize such European-created cities as centres of exploitation: their functions were primarily administrative and military, and they controlled the export of valuable primary products. They failed to generate local growth, and little of the income generated benefited indigenous populations. As a consequence of the importation of indentured labour (see Box 5.3), many of these cities became culturally pluralistic, and as a result some have experienced group frictions. Because the colonial powers failed to initiate local industrial development, the rapidly growing cities of the post-independence period have had no pre-existing industrial base to build on. For many of the cities in the less developed world, a number of the issues discussed here, especially the problems, had their origins in the colonial phase.

As we saw in Chapter 11, rapid city growth in the less developed world is a relatively recent phenomenon. In 1950, only 17 per cent of the population in the less developed world lived in cities; today the percentage is about 44. This growth was the result of large movements of people from rural to urban areas, movements prompted by three general considerations:

1. As we saw in Chapter 4, falling death rates and high birth rates in the less developed world meant high rates of natural increase, and rapid growth of rural populations placed great stress on local rural resource bases.

2. Many workers were prompted to move by economic changes, such as the mechanization and commercialization of agriculture (which reduced agricultural labour requirements) and the importation of manufactured goods (which undermined traditional peasant economies); urban areas offered both greater employment opportunities and higher wage levels.

3. There was a general perception—not always correct—that the quality of life with respect to, for example, provision of basic services was better in cities than in rural areas.

One common feature in rural-to-urban migration has been a tendency to gravitate towards a single well-known city (see Box 5.2); this tendency has been especially strong in less developed countries, where a distinctive feature of urbanization has been the concentration of population in one city, usually the capital. Many less developed countries have a single city that is at least twice the size of the second-largest city; this indicates a high level of primacy (recall the Chapter 11 discussion of the rank size rule). In marked contrast to countries in the more developed world, it is not uncommon for the primate city in a less developed country to have between 20 and 30 per cent of the total population of the country. It can be argued that these exceptionally large cities are too large, given the economic circumstances of the country; hence the large numbers of unemployed and the rise of the **informal sector**.

Urban growth in some parts of the less developed world appears to be slowing down. In both Latin America and much of Asia, declining fertility rates mean that the pool of potential migrants is being reduced; however, the situation in Africa and some parts of South Asia continues as before.

SLUMS

Many cities in the less developed world are economically and socially vibrant places. Tall buildings that are home to major corporations and impressive architecture indicate that these are command centres of national or regional economies, and many of these cities are also increasingly involved in the global economy (Box 13.7).

But, unfortunately, parts of many of these cities suffer from problems of overcrowding, crime, poverty, disease, limited provision of services, traffic congestion, unemployment, damaged environments, and ethnic conflicts. In all these problem situations the poor and underprivileged suffer most. Cities that expand so rapidly that their growth is not controlled may be described as undergoing 'premature urbanization' and always experience problems. The slums of English industrial cities in the mid-nineteenth century were just as bad as those in many cities in the less developed world today, and people moved to them for essentially the same reason: no matter how difficult life there might be, for many people the opportunities outweighed the problems.

With reference to the less developed world, one commentator wondered: 'Does the growth of megacities portend an apocalypse of global epidemics and pollution? Or will the remarkable stirrings of self-reliance that can be found in some of them point the way to their salvation?' (Linden, 1993: 32). The answer is not obvious, although most current evidence suggests that the problems in some cities are many and mostly becoming worse.

The growth of slum areas

Much of the rapid urban growth in the less developed world has involved the emergence and spread of **squatter settlements**: slum areas of uncontrolled expansion on the periphery of a city, made up of low-quality, poorly serviced temporary dwellings. In addition to these new urban areas, many cities include long-standing slum areas, and these two types of urban area are considered together.

Figure 13.6 maps the incidence of urban slums, showing that the greatest problems are in many African, some Latin American, and some Asian countries. Statistics pertaining to slum dwellers indicate the enormity of the problem.

- A 2005 Habitat (the United Nations' human settlements program) estimate suggested that more than one billion people—about one-third of the world's urban

informal sector
A part of a national economy involved in productive labour, but without any formal recognition, control, or remuneration.

squatter settlement
A shanty town; a concentration of temporary dwellings, neither owned nor rented, at the city's edge; related to rural-to-urban migration, especially in less developed countries.

One squatters' community north of Manila, Philippines, is located in a cemetery. Residents live in dwellings built on top of above-ground tombs.
AP photo/Bullit Marquez

FIGURE 13.6 The incidence of urban slums

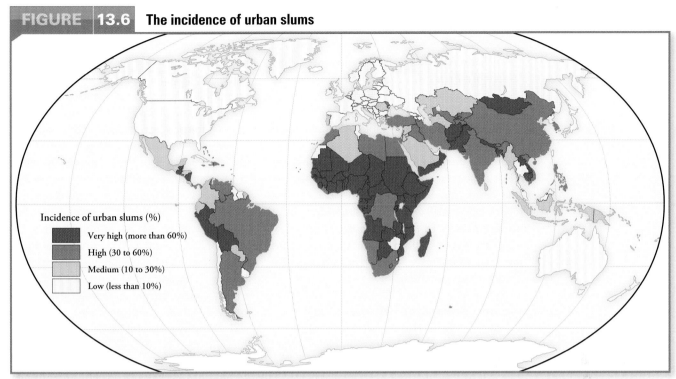

Incidence of urban slums (%)

- Very high (more than 60%)
- High (30 to 60%)
- Medium (10 to 30%)
- Low (less than 10%)

SOURCE: Updated from United Nations Human Settlements Program, *The State of the World's Cities, 2004/2005 Globalization and Urban Culture* (Sterling, Va: Earthscan, 2004), 106.

population—were slum dwellers, with the number increasing rapidly unless significant action is taken, possibly reaching 1.4 billion by 2020 and 3.5 billion by 2050. Annual slum growth rates in Africa are estimated to be 4.5 per cent, in Asia 2.5 per cent, and in Latin America 1.3 per cent.

- In Africa, about 70 per cent of the urban population live in slums. The highest incidence is in Sierra Leone, where slum dwellers comprise 96 per cent of the urban population; the figure for the Central African Republic is 92 per cent, and for Nigeria 79 per cent.

Children scavenging at the municipal dump of La Chureca in Managua, Nicaragua, in November 2006.

AP photo/Cristobal Herrera

- In Asia, the urban slum population for Cambodia is 72 per cent, and for Laos it is 66 per cent.
- In Latin America, 81 per cent of the urban population in Nicaragua and 62 per cent in Guatemala are slum dwellers.

One consequence of rapid and unplanned squatter expansion is often a chaotic, unco-ordinated governance system. In Bangkok, the city core is governed by a single authority, but the larger metropolitan area is divided into about 2,000 small areas, each with its own local government. It is easy to appreciate just how difficult it is to build transportation lines and other components of the urban infrastructure under these circumstances.

Some myths about slums

A 2007 Habitat report listed a number of myths about slums that have limited and con-tinue to limit attempts to improve these areas. Understanding the reality is an important step towards improvement.

Myth—slums serve no purpose. The crucial economic and social roles played by slums are clear, especially the provision of low-cost housing and low-cost services and the cre-ation of social support networks. Slums can be understood as a solution at a particular stage of economic growth involving rapid population growth, industrialization, and urbanization.

Myth—all slum dwellers are poor. Of course, slums and poverty are related—the fact that there are slums is a reflection of poverty. But slums are not home only to poor people, people with little education, people with dys-functional households or no households at all, and people who work as domestic servants, sex workers, manual labourers, and criminals. Rather, slums include people with a wide variety of occupations, including government employees, factory employees, painters, drivers, small entrepreneurs, and office workers. Some educated people with reasonable incomes choose to live in slums.

Myth—slums dwellers are to blame for slums. Slums today result from larger processes of

---

Box 13.7   Globalization and the Less Developed World City

One of the dangers of discussing cities in the less developed world is that it is all too easy to emphasize the many problems at the expense of some of the other changes taking place. One recent and ongoing change concerns globalization processes that can be understood as having both negative and positive impacts.

The often unfortunate social consequences of economic restructuring are not evident only in more developed world cities. Changing job markets combined with reductions in public expenditures and rising prices are evident around the world and frequently contribute to a loss of social stabil-ity. But some cities have responded to these challenges by adopting a strategy similar to that noted for cities in the more developed world, namely, identifying and building on local or regional cultural capital. In Madras, India, for example, a cul-tural complex centred on a crafts museum serves both tour-ist and educational functions, through promoting traditional craft activities such as indigo-dyeing.

But globalization processes have also facilitated the growth of new and vibrant business centres in many cit-ies. A detailed analysis of Accra and Mumbai indicates the links between globalization and the spatial structure of urban corporate geography (Grant and Nijman, 2002). Both cit-ies (and many others, of course) have passed through pre-colonial, colonial, and nationalist phases—each of which restructured urban geographies—and have now entered a

fourth—global—phase of restructuring. This global phase has transformed aspects of the geography of both cities essentially through a significantly increased foreign presence since the early 1980s.

Today, financial and producer services are a rapidly grow-ing sector of the economies of both cities, contributing in important ways to the overall economic health of the city. These activities are highly spatially concentrated, meaning that new centres of corporate control have arisen. Figure 13.7 maps the urban economic geographies during the colo-nial phase, and Figure 13.8 maps these geographies during the global phase. The most notable spatial development is that both cities today have three separate business districts, each tying in with a different level of the larger economy with one global, one national, and one local. Interestingly, the dis-tinctions between these three are not explained in terms of whether companies are under domestic or foreign owner-ship, with both types differentially integrated in the global economy.

The two cities have also undergone similar structural changes. Especially among foreign companies, the fastest-growing aspects of the urban economies are financial and producer services. In Accra, the emphasis in this sector is on communication, real estate, advertising, and consulting. These are important also in Mumbai, as are banking and finance.

Continued

FIGURE 13.7 Economic geographies of Accra and Mumbai during the colonial phase

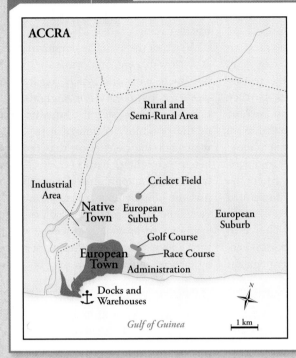

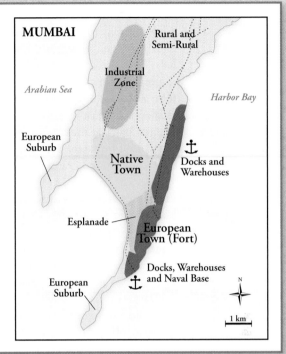

SOURCE: R. Grant and J. Nijman, 'Globalization and the Corporate Geography of Cities in the Less-Developed World', *Annals, Association of American Geographers* 92 (2002): 325, Fig. 1 (MW/RAAG/P1890). Published by Blackwell Publishing Ltd. Used with permission of Taylor & Francis Ltd., http://www.informaworld.com.

FIGURE 13.8 Economic geographies of Accra and Mumbai during the global phase

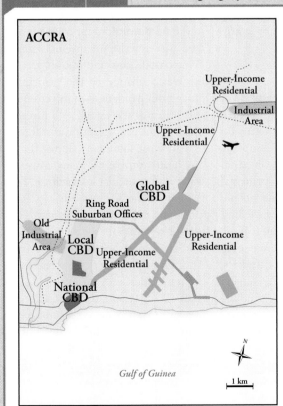

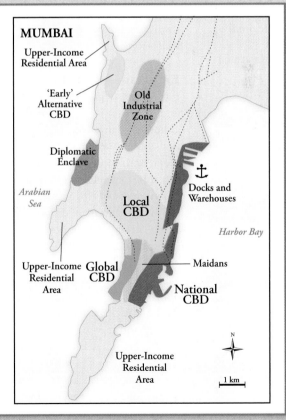

SOURCE: R. Grant and J. Nijman, 'Globalization and the Corporate Geography of Cities in the Less-Developed World', *Annals, Association of American Geographers* 92 (2002): 325, Fig. 11 (MW/RAAG/P1890). Published by Blackwell Publishing Ltd. Used with permission of Taylor & Francis Ltd., http://www.informaworld.com.

population growth and economic change along with specific failings of housing and other national policies. As noted, nineteenth-century slums in industrializing Europe and North America grew for broadly similar reasons.

Myth—slum dwellers are a burden on the economy. In many cities the informal sector is an important part of the urban economy, creating jobs and providing needed services. Further, as they age, squatter settlements may become integral parts of the city, reducing pressure on rental areas inside the city and offering a home base for new migrants. In many instances, squatter settlements have achieved a genuine integration with the larger city as unplanned but necessary growth.

Myth—squatters do not pay rent. The term 'squatting' refers to the legality of landowner-ship and not to the issue of rent payment. An estimated 25 per cent of squatters do pay rent.

Myth—slums are all the same. Some of the older slum areas are based on ethnic, often religious, identity and in some cases these are the current expression of colonial segregation of ethnic groups as discussed earlier. Abidjan and Lagos, for example, include several large ethnic slums. Also, as can be the case elsewhere in the world, some slums are places of cultural resistance with local populations developing survival strategies that further enhance group solidarity.

Employment in slums

For those living in slums the prospect of employment is a principal urban magnet, although many new immigrants to cities do not obtain permanent full-time employment and subsist by becoming involved in the informal sector. There are four principal components of this sector. First, some people are involved in retail distribution, often selling in the street such goods as fresh water, food, newspapers, or jewellery. Second, some people provide services such as car-washing, laundries, gambling, and prostitution. Third, others make items for everyday use and people produce food on small garden areas. Fourth, many people beg and scavenge in order to survive.

The labour market is often segmented on the basis of gender, with women performing the least attractive labour as measured by the quality of the work environment and remuneration. Child labour is also a common circumstance, including working in the home, in the informal sector, and in waged labour.

Children, especially, are valued by some employers because of their willingness to work long hours for little pay in often unsafe and unhealthy environments, such as mining and brick-making.

Much of the available employment is necessarily low-paying, and sometimes very unsafe, but such work is usually regarded by the worker as temporary. For migrants, the attraction to cities is not the squatter settlement or the possibility of informal-sector employment, but the perception that the city offers better schooling, better medical services, a better water supply, and the prospect of permanent employment. As squatter settlements grow, enormous pressure is put on urban governments to provide the needed services, but then, inevitably, extended urban services attract even more migrants. This may be unfortunate for the migrants, because squatter settlements are usually decrepit and lack basic amenities, but it can be viewed in a more positive light: these areas can be seen as reception areas for migrants, providing a transition between rural and urban ways of life.

Health issues and slum environments

Health issues are often discussed in the context of the **epidemiological transition**. This is the idea that development—including improved living conditions and health-care facilities—involves a change in the principal causes of death from pandemics of infection to degenerative diseases associated with old age. In many urban slums this transition is still to take place.

For those living in slums, there are serious health problems, many caused by unsafe water and inadequate sanitation. For example, Bangui in the Central African Republic has a current population of about 700,000 but a sewage system built for a population of 26,000 that has never been extended or improved. Also, Cairo treats less than half of its sewage; the remainder ends up in the Nile or in local lakes. Acute respiratory infections and diarrhea are major causes of death in many urban slums, especially among young children, while malaria, typhoid, and cholera remain major problems in some cities. Air pollution is often at a dangerous level because of the proximity of industrial activities and the often minimal control of emissions.

The issue of environmental conditions in slum areas, and indeed the cities of which

epidemiological transition
A process associated with reductions in fertility and improvements in overall health. As the transition progresses, death and disability from communicable diseases decline in importance relative to problems resulting from non-communicable conditions.

they are a part, has been labelled the 'brown agenda' (Cohen, 1993). This term is an explicit acknowledgement of links between urban and economic growth and environmental deterioration. There are two principal concerns: conventional environmental health problems, such as limited land for housing and lack of services, and problems related to rapid industrialization, such as waste disposal and pollution. The significance of this twofold agenda is that cities in the less developed world are being challenged to address the second set of problems, which remain unsolved even by many cities in the more developed world, at a time when they have not been able to solve the first set of much more fundamental concerns.

Slums in less developed world cities are especially vulnerable to some of the possible consequences of global environmental change discussed in Chapter 4, including both sea-level rise and increased incidence of extreme weather events such as hurricanes. Often located in areas prone to flooding, lacking infrastructure and protection, unplanned and poorly if at all managed, slum areas suffer most in disaster circumstances.

Improving slum areas

In general, cities in the less developed world are experiencing intense pressure to provide additional housing but are incapable of responding adequately. Further, there are clear disparities in the ability of different groups in the city to gain access to improved services—disparities that are a direct reflection of relative power (Lowder, 1994). It is, however, instructive to relate the text comments on counter-urbanization and the discussion on declining fertility in Box 4.2 to this issue. If birth rates were to decline substantially in the less developed world, then at least the problems of squatter settlements might be decreased somewhat.

Given that squatter settlement growth seems inevitable, at least in the short term, how can it best be handled? Three answers are suggested.

1. Most countries have moved away from negative policies such as forced eviction, neglect, and involuntary resettlement. Urban authorities are making efforts to formally integrate squatter settlements into the city proper. For example, beginning in 1994, a plan was developed for Rio de Janeiro that anticipated the integration of the about 600 slums into the city

Accra, Ghana: the Makola market.

Thomas Cockrem/Alamy/GetStock

and that provided assistance for people to improve their housing. Although there have been many setbacks, this plan appears to be improving conditions in many areas. Of course, such ventures are not without critics; some Marxists argue that they serve to rationalize and perpetuate poverty while relieving governments of responsibility for the welfare of people.

2. International aid might be redirected from rural to urban areas. Western governments and international agencies have long focused their attention on rural development, neglecting urban areas. Of course, rural areas need aid, but the neglected reality is that the need for flush toilets is far greater in a congested city than in a relatively sparsely populated rural area.

3. It is not appropriate to simply impose change from outside (recall the Chapter 10 account of the green revolution, for example). Improvements in urban areas need to actively involve local residents as participants. From a Western perspective, it is easy to imagine the urban poor in the less developed world as some anonymous mass, incapable of handling their own affairs, but this is not the case. If the people living in squatter settlements had some security of tenure, they would likely be more willing to work to improve their homes.

LATIN AMERICAN CITIES

Latin America and the Caribbean are characterized by a highly unequal distribution of wealth against a background of rapidly continuing urbanization with rural-to-urban migration affecting the largest cities. Many cities are characterized by marked class segregation and decentralization. Wealthy centres have high-income residents in the central area and in residential enclaves, while many of the poorest people locate on the fringes in squatter settlements and participate in the informal economy (see Figure 12.4). Urban problems abound and appear to be far from resolution. The poor live in an environment rife with violence that is often linked to gang culture and drug-dealing.

São Paulo

The earliest urban centres in Brazil were founded under Portuguese rule and were located on the coast to facilitate the export of sugar and other products. Brazil is typical of the less developed world in that it has experienced explosive urban growth since 1945, especially in recent years. Recent urban growth has accompanied an economic transition from agriculture to industry.

São Paulo, founded in 1554 by Jesuits, remained relatively unimportant until it became the key port for the export of coffee in the late nineteenth century. Today it is the financial and industrial capital of Brazil. Growth has been accompanied by distinct spatial expressions of class. Working-class areas arose close to industrial sites, while elite areas developed on higher ground; in some cases, elite areas are enclosed by security fences and patrolled by guards. Squatter settlements, known as *favelas*, are common on the city's outskirts. About one-third of the *favelas* are on riverbanks and subject to flooding; one-third are on steep slopes and may be subject to landslides and erosion; and about one-tenth are located on waste dumps or landfill sites.

Regardless of wealth, residents encounter challenges moving around the city. Public transportation is woefully inadequate for the needs of poor people while the rich live in fear with armed hijackings a daily occurrence. Indeed, one solution to the traffic problem for the rich is to use helicopters, and the city has one of the largest private helicopter fleets in the world.

Of particular concern in recent years in the slums of São Paulo and Rio de Janeiro is a dramatic rise in criminal gang violence that is exacerbating an already very challenging security situation. Many slum areas are being taken over by gangs, most of which are involved in the drug trade, but in turn the gangs are being challenged by local militias that offer protection to residents but only at a price. The poor are caught in the middle.

Mexico City

Mexico City has grown and continues to grow at an alarming rate. It is engulfed by squatter settlements and the housing services and larger social infrastructure cannot cope. Furthermore, it is located in an unfavourable environment that continually suffers from flooding, subsidence, and shortages of fresh water. The urban explosion began about 1900 as a result of development policies favouring industrialization. The city has doubled in size since 1970 and may have doubled again by about 2020. It is not an exaggeration to describe Mexico City as a social and environmental disaster—a disaster that is constantly worsening because of continuing growth. According to the United Nations Environment Program and the World Health Organization, it is the most polluted of large world cities: levels of sulphur dioxide, suspended particulate matter, carbon monoxide, and ozone exceed health guidelines by a factor of two or more, and levels of lead and nitrogen oxide are almost as bad. The principal source of these pollutants, which affect the health of anyone who breathes the city's air, is the burning of fossil fuels (Middleton, 1994).

But perhaps the greatest challenge facing Mexico City is that it is currently sinking into the ground at an alarming rate, up to 38 cm a year in the worst affected areas. A principal cause of subsidence is the removal of water from subterranean aquifers to quench the thirst of city residents.

AFRICAN CITIES

This region has a very large, poor urban population who often live in life-threatening circumstances. It is experiencing slow and erratic economic growth and, according to some commentators, the gap between rich and poor is ever-widening. As already noted, the incidence of urban slums is high and urbanization continues apace. An especially challenging circumstance is that, in many regions, the push factors away from rural areas—landlessness and famine—are so powerful that, even though the urban pull factors are often weak, rural-to-urban migration still occurs. This means that the life prospects, including housing, health care, and employment, for a new urban immigrant can be particularly problematic.

Lagos

Lagos, until 1991 the capital of Nigeria, combines serious physical and human problems. Because Lagos is located on four islands in a lagoon and with just three bridges linking to the mainland, the people find it very difficult to move around. Much of the site consists of swamp, and the land is unstable. Drainage and sewage problems worsen daily. The population is increasing rapidly because of larger economic issues prompting rural out-migration; indeed, Lagos receives more immigrants than any other city in sub-Saharan Africa. Sometimes described as a nightmare or as an urban jungle, the city has less than 50 per cent sanitation coverage, is disease-ridden, congested, and virtually unlivable. Unlike many other cities in the less developed world, Lagos does not have specific slum areas; rather, almost the entire city might be considered a slum. Despite all this, Lagos has long served as an important economic centre, home to much of the national industrial infrastructure, and also as a national and regional centre that is 'passionately loved by its citizens' (United Nations Human Settlements Program, 2004: 56). Growing rapidly, Lagos may have 17 million people by 2015, making it the ninth largest city in the world (see Table 11.6).

Kinshasa

The Democratic Republic of Congo has large mineral reserves, extensive forests, and some good agricultural land, and has received huge amounts of development aid. The capital, Kinshasa, was a relatively efficient and thriving city until about 1990—but after that date the country and the city alike were plunged into chaos under a corrupt government that was not overthrown until 1997. Much of the foreign aid money was diverted, ethnic tensions developed—partly as a consequence of the huge influx of Rwandan refugees in the early 1990s—and the capital city effectively ceased to function. In Kinshasa rioting began in 1991; both the economy and the urban infrastructure collapsed (the formal economy shrank by 40 per cent, inflation was 8,500 per cent in 1993 and 6,000 per cent in 1994); and diseases such as AIDS, tuberculosis, malaria, and cholera are spreading.

Nairobi

The largest city in East Africa, Nairobi has long had a good reputation as an administrative centre, home to light industry, and gateway to tourist areas. As recently as 1960, the population was about 250,000, but today it exceeds 3 million. As in so many other cities, rapid growth has resulted in huge problems, including the growth of slums Interestingly, unlike most slum areas in other cities, the large, densely populated, and very poor area of Kibera in Nairobi is located in the centre of the city. Ironically, it lies close to the offices of the UN agency, Habitat, and, at least partly because of this proximity, it has been much visited and researched. Notwithstanding this attention, it is, like all other slums, a place of squalor, crime, lack of basic services, and health problems. But like Dharavi slum in Mumbai (discussed below), it is also a place where residents are becoming increasingly organized and many are determined to improve themselves and their place.

INDIAN CITIES

Like other cities elsewhere in the less developed world, Indian cities are characterized by extremes of wealth and poverty. However, even in the poor areas there is often much evidence of individual initiative and related economic activity. Perhaps the greatest need is for effective management of the entire urban region.

Mumbai

Mumbai, the financial capital and richest city in India, made headlines in the summer of 2005 when monsoon rains brought the city to a standstill, a circumstance related to the lack of an effective public information and communication system and the lack of backup power sources. This storm affected most of the city but was nothing new, for about half of Mumbai residents who live in slums are flooded each year during the monsoon season so that streets become rivers of sewage.

One major Mumbai slum made headlines in 2009 as the setting for the movie *Slumdog Millionaire*, which won eight Academy Awards, including for Best Picture. The Dharavi slum, often said to be the largest in Asia, is typical of slum areas—miserable housing, lack of security of tenure, contaminated water supply (for those who have piped water), inadequate sewage services, and so forth. And yet, even here, residents start businesses and small industries

Neighbours look out of their door as media representatives crowd near the shanty of *Slumdog Millionaire* child actor Rubina Ali in Mumbai.

Indranil Mukherjee/AFP/Getty Images

while others work in the informal economy. The people living in this slum produce goods worth about US$500 million each year. There are, for example, potteries, leather manufactures, bakeries, and metal workshops. Many of these businesses are owned and operated on a family basis, while others are owned by an entrepreneur who employs workers. These activities highlight the fact that it is not appropriate to think of the urban poor as some homogeneous group or as people lacking initiative. What is happening today in the Mumbai slums is compelling evidence that many people make real efforts to improve their living conditions, becoming actively involved in the local economy (Shaw, 2003). Overall, Mumbai is seen as a city of opportunity, especially in comparison to Kolkata (Calcutta).

Kolkata

Kolkata, the 'city of joy', is the largest city in India. Physically, it is located in a low-lying area that requires pumping of storm and sewage water. Some 40 per cent of the population either live in slums or are homeless. Most homes lack proper sanitation, and garbage disposal is woefully inadequate. A 1991 survey identified an infant mortality rate of 123 per 1,000 live births (for comparison purposes, the 1991 IMR for the more developed world was

16, for the less developed world 81, and for India 95). The principal reasons for Kolkata's exceptionally high IMR are poor sanitation and overcrowding; specific causes of death include such avoidable problems as diarrhea, diphtheria, tetanus, and measles. Other problems reflect recent increases in industrial activity; it has been estimated that 60 per cent of the population suffer from respiratory tract problems because of poor air quality.

CHINESE CITIES

As home to one-fifth of the world's population, China has many very large cities (Figure 13.9). This figure is suggestive rather than authoritative. As noted in Chapter 11, it is difficult to be precise about the size of many Chinese cities both because data can be flawed and because many are growing exceptionally rapidly. The sheer number of very large cities is remarkable, with about 100 cities greater than one million people. According to some observers the fastest-growing city in the world in recent years is Chongqing, the economic hub of western China located in the middle reaches of the Yangtze River—it has been described by one newspaper correspondent as the 'megalopolis you've never heard of' (Watts, 2006) and also compared to Chicago as it serves as a gateway to largely undeveloped western lands.

A consideration of Chinese cities today needs to appreciate four particular circumstances. First, at the end of the first millennium CE, they were cosmopolitan places, not unlike many of the leading European cities. Second, as China began to limit interactions with other parts of the world, the cities suffered from loss of trade and of ideas. Third, during the second half of the twentieth century, the rapid, uncontrollable growth so evident in most of the less developed world did not occur in Communist China, which proved able to control population movements in a way that is not possible in free-market economies. Fourth, although China continues to limit the numbers of rural-to-urban migrants by offering fewer rights and lower wages to migrants compared to those already living in the cities, estimates suggest perhaps as many as 20 million rural dwellers migrate to urban centres each year.

Aspects of the recent Chinese urban experience differ substantially from other regions of the world, both more and less developed: indeed, 'Chinese cities are different'

(Macmillan, 1996: 20). Most obviously, given that recent Chinese urbanization is taking place in the context of market socialism, not Western capitalism, mainstream theories are not applicable. Rather, urbanization is best understood as a state-controlled process comprising four factors: (1) decentralization of decision–making, leading to more relaxed state control over capital accumulation and land acquisition; (2) creation of development zones on city fringes; (3) rural-to-urban movement related to abolition of the household registration system; and (4) new policies designed to attract foreign investment.

In addition to the above context, three specific differences are especially apparent. First, unlike most countries, China has a relatively dispersed pattern of industrial location and therefore its cities are less specialized. In other words, most are relatively self-sufficient—not dependent, as many cities in the more developed world are, on trade in manufactured items. Basic goods such as cloth, bicycles, and watches are produced in all cities, not just a few. There are no single-industry towns. Second, this industrial dispersion is somewhat countered by investment policies that favour the largest cities. Thus, while most cities have a mixture of industrial activities, industrial activity overall is concentrated in a few relatively large cities. Third, because Chinese cities are growing so rapidly and attracting so much investment, the

| FIGURE | 13.9 | **The largest cities in China** |

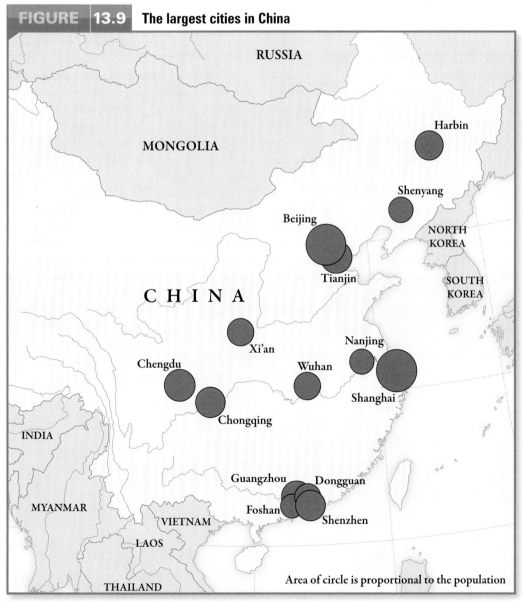

Although mapping Chinese cities is an uncertain exercise, this map does indicate the locations of the largest cities and approximate sizes.

gap between quality of life in the cities and the countryside continues to increase, prompting the idea that China has European cities and an African countryside.

Today, urbanization is not best understood as a by-product of Chinese economic development; rather, it *is* Chinese economic development—the rural areas have a surplus of labour and mostly poorly paid jobs, whereas the mushrooming cities are home to increasing numbers of industrial employment opportunities. Social tensions between rural and urban dwellers are on the rise as lifestyle disparities increase and as rural environments are bulldozed to make way for industrial estates and housing.

Urbanization and the Chinese environment

Chinese cities have many of the same problems that plague other cities in the less developed world, although rarely to the same degree. In Shanghai, for example, only 60 per cent of the population are in residences connected to a sewer. More generally, the UN estimates 400,000 premature deaths in China each year because of air pollution

Also, it is widely acknowledged that Chinese urbanization and industrialization are taking place with minimal regard for the environment, although there are signs that this is changing. It is acknowledged in China that Shenzhen (located north of Hong Kong), which has grown without regard for the environment, is not an example for other cities to follow. The huge expense incurred in preparation for the 2008 Summer Olympics in Beijing entailed some efforts at environmental improvement so that, for example, marathon runners would not be too taxed by poor air quality. However, despite those efforts, air pollution levels were reduced during the period of the Olympics only as a result of factory closures.

As one indication of greater concern for the environment, four Chinese cities—Shanghai, Chongqing, Fatou, and Xi'an—displayed a number of urban innovations at a 2006 architectural show in Venice. Perhaps of even greater significance, in 2006 the city of Shanghai, in partnership with a British engineering group, announced plans to build Dongtan, the world's first eco-metropolis. However, construction has not begun as of 2009 and planning permission has lapsed. Whether or not the city will be built remains uncertain. The ambitious plan involved developing an area of wetland at the mouth of the Yangtze River with the building of a post-industrial sustainable city. The first change was planned to be the construction of housing and services for 50,000 people, a process that would have resulted in displacement of many peasant farmers.

When fully completed, Dongtan was to have half a million people and was to be the first self-sustaining, pollution-free city in the world, with local farmers encouraged to use organic methods in producing all of the city's needs; as well, it was to be a financial and commercial gateway between China and the rest of the world. The city was to be home to light-industrial and high-technology jobs, to a variety of recreational opportunities, and to varied housing types. In what is nominally a Communist state, funding for the project was to come from the global financial system. These ambitious plans are now in disarray and the entire scheme may prove illusory.

The Dongtan proposal and its current status suggest the challenges that our urban world needs to meet, but also the difficulties involved in overcoming these challenges.

CHAPTER 13 SUMMARY

HOUSING MARKETS

In more developed countries there are two different ways of thinking about housing: as a commodity or as a universal right. Usually, the reality lies somewhere between the two. Housing as a commodity—something that is bought, sold, and rented—is appropriately analyzed in the context of a market. This market does not work in the same way for all urban residents. For example, redlining denies capital to certain areas based on the perception that the areas will decline in property values over time.

RESIDENTIAL MOBILITY

Models of residential mobility reflect a larger body of concepts that highlight the role played in spatial decision-making both by stress factors and by a number of different urban spaces. Filtering is the idea that, through time, a housing unit or neighbourhood is occupied sequentially by people from steadily changing income groups.

NEIGHBOURHOODS

These are spatially and socially constituted areas of the city. They can be understood as communities or as sites of political activism.

GENTRIFICATION

The process of people moving in and transforming a formerly derelict or low-quality housing area is known as gentrification. It is sometimes explained in terms of the rent gap hypothesis, which argues there is a discrepancy between the current value of land and the potential value. Alternatively, it is explained in terms of larger social changes. Gentrification often benefits middle-income groups but poses problems for low-income groups who become more spatially marginalized.

SEGREGATION

Most urban areas contain residential districts distinguished on the basis of income, class, ethnicity, religion, or some other economic or cultural variable. Historically, especially in Europe prior to the Industrial Revolution, the most distinctive urban residential district was the Jewish district. Distinct residential areas emerged also in large immigrant-receiving North American cities in the nineteenth century, with spatial differentiation based on ethnicity and class. Cities in colonial areas usually included separate districts for Europeans, other immigrant groups, and the local population. Explanations for the creation and maintenance of ethnic neighbourhoods have focused on economic and cultural forces. In general, two processes are at work: the internal cohesion of the group and the desire of non-group members to resist spatial expansion of the group. Often, these neighbourhoods are appropriately understood in the context of institutionalized racism.

POVERTY IN THE CITY

Cities are unequal places. In many countries, poverty has increasingly become concentrated in cities and, within cities, concentrated in the older inner area, partly because of the disadvantaged people who generally live there. Theoretical explanations for the prevalence of poverty are various. Although members of all social groups are susceptible to poverty, there is compelling evidence of links with ethnic minority identity, level of education, employment status, health, age, and gender. Most current evidence suggests that urban problems such as poverty, inadequate housing, and homelessness are increasing rather than decreasing.

HOMELESSNESS

A principal problem in many large cities is homelessness. Four versions of homelessness are rooflessness, houselessness, insecure housing, and inadequate accommodation. In the United States especially, homeless people have been subject to involuntary moves between residential units, shelters, prisons, detox and rehabilitation centres, hospitals, and the street. This cycling approach may have more to do with removing the poor from view than with finding a lasting solution to a social crisis. Homelessness often is precipitated by a specific circumstance such as loss of employment, eviction, or domestic violence. Once homeless, people may gradually become labelled as homeless and even self-identify as homeless.

MANUFACTURING IN THE CITY

Many countries have experienced some decentralization of manufacturing activity away from traditional manufacturing cities, thus weakening the economic base of the old industrial centres and strengthening the new centres. Much of this decentralization of industrial activity is tied to the fact that there are today many more types of industry and they favour a much broader range of locations. Inside the city, central industrial locations have lost manufacturing activities while outer suburban areas have gained. Overall, these various spatial changes are part of larger economic restructuring processes and the decreased friction of distance.

RETAILING IN THE CITY

Since about the 1970s, urban retailing activity has undergone a number of organizational and related locational changes in response to broader post-industrial trends: the numbers of independent traders have declined sharply as a consequence of competition between large business enterprises; the largest retailers, of both goods and services, are now operating as transnationals; new communications technology has permitted the creation and rapid growth of television shopping networks. The most evident locational change is a decentralization of shopping centres, many of which are now located in out-of-town areas accessible only by automobile.

CHANGING CENTRAL CITY AREAS

Industrial and retailing relocation from central to suburban city areas has contributed to a general decline in the city centre as the advantage of centrality dwindles. In many cities, processes of urban revitalization have focused on the central city area and, while the details of land use have changed, the city centre remains a viable economic location.

CITIES IN THE LESS DEVELOPED WORLD

The colonial heritage of many cities in the less developed world left them ill-prepared to cope with the population explosions that occurred after about 1950. Many of these cities are experiencing social and environmental strains far more severe than those affecting cities in the more developed world. One common feature in rural-to-urban migration has been a tendency to gravitate towards a single well-known city. Regardless of location, many of these cities suffer from overcrowding, crime, poverty, disease, limited provision of services, traffic congestion, unemployment, damaged environments, and ethnic conflicts.

SQUATTER SETTLEMENTS

These are slum areas of uncontrolled expansion on the periphery of a city, made up of low-quality, poorly serviced temporary dwellings. The incidence of slum squatter settlements is highly correlated with the rate of urbanization. As they age, squatter settlements may become integral parts of the city, reducing pressure on rental areas inside the city and offering a base for new migrants, and in many instances squatter settlements have achieved integration with the larger city. One hopeful sign in recent years is that most countries have moved away from negative policies such as forced eviction, neglect, and involuntary resettlement as strategies to solve squatter settlement problems.

LATIN AMERICAN CITIES

Urban problems abound and appear far from resolution. The poor live in an environment rife with violence that is often linked to gang culture and drug-dealing.

AFRICAN CITIES

This region has a very large, poor urban population, who often live in life-threatening circumstances.

INDIAN CITIES

Like other cities elsewhere in the less developed world, Indian cities are characterized by extremes of wealth and poverty. However, even in the poor areas is much evidence of individual initiative and related economic activity.

CHINESE CITIES

The Chinese urban experience is different from other regions of the world, especially because of the role played by the Communist state. One interesting proposal was the planned construction of Dongtan, the world's first eco-metropolis, a post-industrial sustainable city; these plans are currently shelved.

QUESTIONS FOR CRITICAL THOUGHT

1. What factors underlie the social geography of the city (i.e., why has the social geography of (North American) cities taken on the forms we observe today)?

2. What factors induce urban residents to move within the city? To what extent is this intra-urban mobility similar to and different from inter-regional migration discussed in Chapter 5?

3. What are the causes and consequences of gentrification? Explain why you would (or would not) support government incentives that were designed to encourage gentrification.

4. Why does poverty exist? What can/should be done to address the issue of poverty in the city?

5. Suppose that you had the resources needed to instantly create enough units of housing to shelter all those who were homeless. Would the problem of homelessness then be 'solved'?

6. How and why have the retail and industrial landscape of North American cities changed in the post-World War II era?

7. How would you compare living in cities (i.e., the 'urban experience') in more developed nations to that in less developed nations (i.e., how and why are these urban experiences different)?

FURTHER EXPLORATIONS

Boone, C.G., and A. Modarres. 2006. *City and Environment*. Philadelphia: Temple University Press.

A broad-ranging book on the links between urban environments and nature.

Bourne, L.S., and D. Ley, eds. 1993. *The Changing Social Geography of Canadian Cities*. Montreal and Kingston: McGill-Queen's University Press.

An edited collection that addresses a range of social urban topics in the Canadian context; clear emphasis on contemporary social theory.

Brunn, S.D., ed. 2006. *Wal-Mart World: The World's Biggest Corporation in the Global Economy*. New York: Routledge.

International survey by geographers and other academics of this large and complex business, though without discussion of Wal-Mart in Canada.

Davis, M. 2006. *Planet of Slums*. London: Verso.

Powerful and provocative; Davis writes about deprivation and related misery globally as evidenced by the lives of slum cultures and the slum cities where they live.

Drakakis-Smith, D. 2000. *The Third World City*, 2nd edn. New York: Routledge.

An excellent overview of this important topic.

Friedmann, J. 2005. *China's Urban Transition*. Minneapolis: University of Minnesota Press.

A focus on the changing geography of China with emphasis on urbanization.

Gad, G. 1991. 'Toronto's Financial District', *Canadian Geographer* 35: 203–7.

The first in a series on Canadian urban landscapes in this journal.

Hernandez, T., and J. Simmons. 2006. 'Evolving Retail Landscapes: Power Retail in Canada', *Canadian Geographer* 50: 465–86.

Analysis of the power retail landscapes of 'big-box' stores and the consequences for other retailing areas.

Jakle, J.A. 2001. *City Lights: Illuminating the American Night*. Baltimore: Johns Hopkins University Press.

An account of the history and consequences of public lighting in American cities explaining how lighting contributed to the modernization and place-enhancement of cities.

Leitner, H., J. Peck, and E. Sheppard. 2007. *Contesting Neoliberalism: Urban Frontiers*. New York: Guilford.

Discussion of how urban social and political geographies are changing because of the actions of grassroots groups, NGOs, and some progressive city governments.

McGee, T.G., G.C.S. Lin, A.M. Marton, Y.L. Wang, and J. Wu. 2007. *China's Urban Space: Development under Market Socialism*. New York: Routledge.

Authoritative text analyzing the distinctive process of recent rapid urbanization in China, with emphasis on the eastern coastal zone.

Parnell, S., D. Simon, and C. Vogel. 2007. 'Global Environmental Change: Conceptualising the Growing Challenge for Cities in Poor Countries', *Area* 39: 357–69.

Methodological discussion of approaches to the study of possible consequences of global environmental change in less developed world cities.

Walks, R.A., and L.S. Bourne. 2006. 'Ghettos in Canada's Cities? Racial Segregation, Ethnic Enclaves and Poverty Concentration in Canadian Urban Areas', *Canadian Geographer* 50: 273–97.

Detailed analysis using 1991 and 2001 census data to show that ghettos are not a feature of the Canadian urban landscape.

Whitehand, J.W.R., and K. Gu. 2006. 'Research on Chinese Urban Form: Retrospect and Prospect', *Progress in Human Geography* 30: 337–55.

Informative overview of research on Chinese cities.

Williams, S.W. 1997. 'The Brown Agenda', *Geography* 82: 17–37.

Succinct account of the links between poverty, shelter, and environment in the cities of the less developed world, with detailed consideration of Kolkata.

Wilson, D. 2007. *Cities and Race: America's New Black Ghetto*. New York: Routledge.

About the 1990s rise of new black ghettos in Rust Belt American inner-city areas.

ON THE WEB

HOUSING AND NEIGHBOURHOODS

Canada

> www.cmhc-schl.gc.ca/en/

The Canada Housing and Mortgage Corporation website has information about housing and related issues in Canada.

United States

> www.hud.gov/

This Department of Housing and Urban Development website has news and information about housing in the US.

Gentrification

> www.cdc.gov/HEALTHYPLACES/
> healthtopics/gentrification.htm

This Centers for Disease Control and Prevention website is concerned with the health aspects of gentrification.

URBAN PROBLEMS

Improving the Living Environment

> www.bestpractices.org/

This United Nations site contains over 1,100 proven solutions, from more than 120 countries, to the common social, economic, and environmental problems of an urbanizing world.

Urban Institute

> www.urban.org/

This is the site of a 'non-partisan economic and social policy research organization' that includes discussions of urban problems and issues.

Homelessness in Canada

> intraspec.ca/homelessCanada.php

Resources, news, and statistics on homelessness in Canada.

Facts about Homelessness

> www.nationalhomeless.org/
> publications/facts.html

US-based National Coalition for the Homeless website that provides detailed information.

Helping the Homeless

> www.hud.gov/homeless/index.cfm

US Department of Housing and Urban Development website providing information for those who are homeless and those who wish to help the homeless.

CITIES IN THE LESS DEVELOPED WORLD

State of the World's Cities

> www.un-ngls.org/spip.php?article590

Summary of the UN Habitat 2008–9 *Report on Harmonious Cities*. Although one out of every three people living in cities of the developing world lives in a slum, not all slum dwellers suffer the same degree or magnitude of deprivation, nor are all slums homogeneous.

Slum Data

> www.gdrc.org/uem/squatters/
> squatters.html

Information on slum settlements in less developed world cities.

INDUSTRIAL LIVES AND LANDSCAPES

W hy are industries located where they are? In this chapter, the answer once again takes the form of a neo-classical economic theory that displays all the advantages and disadvantages of that approach; the conceptual similarities with both agricultural (Chapter 10) and urban (Chapter 11) location theories are clear.

Following a look at the origins and evolution of twentieth-century industrial landscapes—landscapes that (like so many other phenomena of interest to human geographers) owe their basic character to the many changes associated with the period of the Industrial Revolution—we turn to the world industrial map, emphasizing sources of energy, major world industrial regions, the dramatic rise of Japan and the newly industrializing countries, and the current emergence of China as an industrial country. Next, our discussion of the industrial restructuring that has been apparent since the 1970s (and was introduced in previous chapters) emphasizes the growth and increasing concentration of service activities—a process that requires further consideration of the transition from Fordism to post-Fordism and the deindustrialization evident in the more developed world, as well as a variety of theoretical perspectives, especially forms of Marxism. The chapter concludes with the geography of uneven development and the efforts that states have made to plan industrial growth so that it favours certain regions at the expense of others.

Industrial Buildings near Mt Fuji in Japan.
Adina Tovy/Alamy/GetStock

primary activities
Economic activities concerned directly with the collection and utilization of natural resources.

secondary activities
Economic activities that process, transform, fabricate, or assemble raw materials derived from primary activities; also activities that reassemble, refinish, or package manufactured goods.

tertiary activities
Economic activities involving the sale or exchange of goods and services; includes distributive trades such as wholesaling and retailing, and also personal services.

quaternary activities
Economic activities concerned with handling or processing knowledge and information; typically involve a high level of skill; highly specialized.

In the study of spatial variations and changes in economic activity, human geographers distinguish three levels of economic activity. **Primary activities** such as farming, fishing, forestry, and mining occur where appropriate resources are available. **Secondary activities** convert primary products into other items of greater value. This process is called *manufacturing* ('making by hand') and requires three things: a labour force, an energy supply, and a market. Most manufacturing is done in factories, and the result is a uniform product. **Tertiary activities** are those involved in moving, selling, and trading the goods produced at the first two levels, as well as activities such as professional and financial services. Recently, it has been helpful to identify a fourth level: **quaternary activities** specialize in assembling, transmitting, and processing information and controlling other business enterprises. They include professional and intellectual services, such as those provided by management consultants and educators. Quaternary-sector employment is the favoured growth area in the post-industrial city discussed in Chapter 13. The distinction between tertiary and quaternary is not always made as both involve services; a useful characterization is to note that tertiary activities focus on goods whereas quaternary activities focus on people and information.

Like any classification, this one is not ideal, and the divisions between the four levels are not absolute; but it is still useful. For our purposes, it is convenient to consider all four kinds of economic activity here, with two exceptions: transport-related activities and farming (discussed in Chapters 9 and 10).

The Industrial Location Problem

Locational questions and theories are central issues for geography, and they are just as important for industry as for agricultural and settlement issues. Once again, theories need to be considered in the context of particular social and economic systems, especially when we realize that industries strive to minimize costs, including labour costs, and maximize profits. Many different types of economic organization play roles in the industrial process. Even the household plays a role in production and reproduction, although that role is less important today than it was in the past;

firms, which make up the commercial sector, are a second type of economic organization. Industrial location theory explains why firms locate their factories where they do (Box 14.1). Some firms are owned by one person; others are partnerships, co-operatives, or corporations. Corporations play the most important role in a country's economy.

EARLY LOCATION THEORY

For economists such as Adam Smith, J.S. Mill, and David Ricardo, industrial location was related to the location of agricultural food surpluses that could be used to feed industrial workers. Probably more important was Adam Smith's realization that, under the new doctrine of capitalism, the central consideration in location decisions was purely financial. This incentive remains paramount today. In some form, it has been at the heart of most industrial location theory since the writings of Adam Smith. The only major exception to this generalization is Marx, who saw industrial location as one of many issues best explained in terms of political inequalities. The classic industrial location theory, however, is the least-cost theory of the economist Alfred Weber.

LEAST-COST THEORY

The first attempt to develop a general theory of the location of industry was that of the German economist Alfred Weber (1868–1958), brother of the sociologist Max Weber. His work, published in 1909 and translated into English in 1929 (Friedrich, 1929), is another example of a theory that does not purport to summarize reality. Rather, it aims to prescribe where industrial activities ought to be located; it is a normative model. Weber follows the von Thünen tradition of setting up a number of simplifying assumptions:

1. Some raw materials are ubiquitous, that is, they are found everywhere. Examples suggested by Weber are water, air, and sand.
2. Most raw materials are localized, that is, they are found only in certain locations. Sources of energy fall into this category.
3. Labour is available only in certain locations; it is not mobile.
4. Markets are fixed locations, not continuous areas.
5. The cost of transporting raw material, energy, or the finished product is a direct

Box 14.1 | Factors Related to Industrial Location

Traditionally, a number of factors have been considered in the typical decision process to locate an industry.

Various measures of *distance* have been very important. Distance from sources of raw material has become less important today as a result of improvements in transport and changing industrial circumstances, but it used to be a major consideration because such materials are unevenly distributed and vary in quality, quantity, cost of production, and perishability. Similarly, distance from an energy supply is now declining in importance. During the nineteenth century, it was essential to locate factories near a source of coal, but since then the emergence of new energy sources, such as electricity, has freed industry from this locational constraint. Distance from market is another factor. Like the other distance variables, it tends to decrease in relevance with improvements in transportation.

Availability of *labour* also shows spatial variations in quality, quantity, and cost. Although theoretically perfectly mobile, labour is often limited to specific locations for social reasons or because of political restrictions on movement.

Capital, in the form of tangible assets (such as machinery) or intangible assets (such as money), can be a key locational consideration; capital may be mobile.

Transport was traditionally a key factor, linked to several of the variables already noted.

Each of these factors plays a role in the location decision, although the extent to which each is significant may vary according to the particular production process; in some cases, for example, capital might be substituted for labour. When an industrialist has many options, the number of feasible locations usually is increased.

Also important for understanding industrial location are factors related to the nature of the product, as well as various internal and external economies. With regard to *internal economies*, each manufacturing activity has an optimum size and benefits may be gained from either horizontal integration (enlarging facilities) or vertical integration (involvement in related activities at one location). *External economies* often result when similar industries are located in close proximity.

Finally, industrial location decisions often reflect various human and institutional considerations. The most important human factor is *uncertainty*. Decision-makers do not have adequate information about all the relevant factors. They may wish to minimize their costs, but they are unlikely to be able to do so. We have already distinguished between optimizing and satisficing behaviour, and in the current context this distinction is crucial. Location decisions, then, often reflect uncertainty, 'educated guess' judgements, and chance factors. The central institutional factor is the role played by *government*. Conflicts often arise between economic logic and what governments see as socially desirable.

---

function of weight and distance. Thus, the greater the weight, the greater the cost; similarly, the greater the distance, the greater the cost.

6. Perfect economic competition exists. This means that the industry consists of many buyers and sellers, and no single participant can affect product price.

7. Industrialists are economic operators interested in minimizing costs and maximizing sales.

8. It is assumed that both physical geography (climate and relief) and human geography (cultural and political systems) are uniform.

These assumptions are clearly unrealistic, but they are necessary for Weber to arrive at a least-cost location format. Given these assumptions, industrialists will locate industries at least-cost locations in response to four general factors: transport and labour (interregional factors) and agglomeration and deglomeration (intraregional factors).

Transport costs

Weber's first concern was transport cost. To find the point of least transport cost, two factors must be assessed: distance and the weight being transported. According to Weber, the point of least transport cost is the location where the combined weight movements involved in assembling finished products from their sources and distributing finished products to markets are at the minimum.

Weber also introduced the **material index**: the weight of localized material inputs divided by the finished product weight. If an industry has a material index greater than 1, then the point of least transport cost, and therefore the appropriate location, will be nearer the localized material sources; if an industry has a material index less than 1, then the point of least transport cost, and therefore the appropriate location, will be nearer to the market. The historical importance of orientation to material resources has declined because today fewer industries use heavy, bulky materials and because transportation has improved.

material index
An index devised by Weber and used in industrial location theory to show the extent to which the least-cost location for a particular industrial firm will be either material- or market-oriented.

In general, Weber's predictions are correct. M.J. Webber (1984: 57) gives the example of the soft drink industry where 1 ounce of syrup, a ubiquitous material (water), and a 4-ounce can combine to produce 12 ounces of soft drink and a 4-ounce can. Thus, the material index is $(1 + 4)/(12 + 4) = 0.31$. This material index is less than 1, and indeed the soft drink industry is typically market-oriented. By contrast, industries such as the smelting of primary metals remain predominantly material-oriented, as do many agricultural processing industries, because the cost of materials still forms a substantial share of the total cost.

Locational figures

To tackle such problems, Weber used a graphic approach. A simple version of his figure is shown in Figure 14.1. In this case there are two raw material sources (R_1 and R_2) and a market (M), and the industry locates according to the relative attraction of each of the three. If only ubiquities are used, then the locational figure is reduced to one point, the market. If only one raw material is used and that material loses no weight in the production process, then the industry can locate at the material source, at the market, or anywhere on a straight line between the two. Clearly, the locational figure is needed only in those cases where weight-losing materials are used—but this is the most common type of manufacturing. Thus, the locational figure will usually have more than three sides, generating a complex problem. Geographers usually solve such locational problems algebraically rather than by referring to diagrams.

Labour costs

Although transport costs are the key to Weber's work, he also acknowledged the importance of labour costs. Such costs represent a first distortion of the basic situations already described and can be approached by mapping the spatial pattern of transport costs and then comparing this pattern with that of the relevant labour costs.

To achieve this comparison, Weber introduced the concepts of **isotims** (lines of equal transport costs around material sources and markets) and **isodapanes** (lines of equal additional transport cost drawn around the point of minimum transport cost). Maps of isotims and isodapanes are examples of cost surfaces. Figure 14.2 shows isotims around two material sources and one market, while Figure 14.3 superimposes isodapanes on the isotims. Weber identified a critical isodapane, outside of which industrial location would not occur—industries can be attracted to low-labour-cost locations only if they lie inside the critical isodapane. A low-labour-cost location will attract industrial activity only if the savings in labour cost exceed the additional transportation costs of moving the raw materials to, and the finished products from, the low-cost labour site.

Agglomeration and deglomeration

Transport and labour are interregional considerations. Agglomeration and deglomeration are intraregional considerations that, like labour, can potentially cause deviation from the location with the least transport cost. *Agglomeration* economies result from locating a production facility close to similar industrial plants, allowing plants to share equipment and services, and generating large market areas that aid the circulation of capital, commodities, labour, and information. *Deglomeration* economies are the

isotims
In Weberian least-cost industrial location theory, lines of equal transport costs around material sources and markets.

isodapanes
In Weberian least-cost industrial-location theory, lines of equal additional transport cost drawn around the point of minimum transport cost.

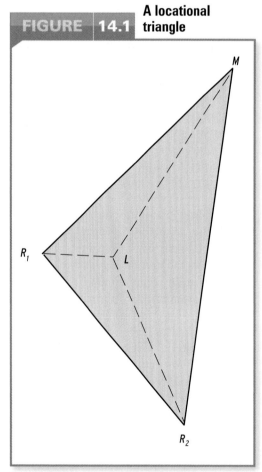

FIGURE 14.1 A locational triangle

The position of the least-cost location (*L*) will depend on the relative attractiveness of each of R_1, R_2, and M.

FIGURE 14.2 A simple isotim map

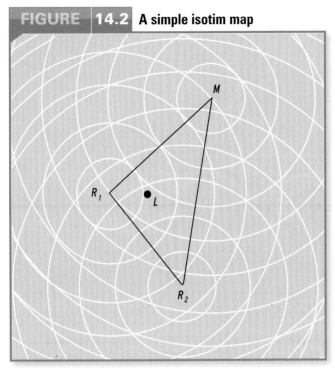

The diagram shows isotims around the two raw-material locations (R_1 and R_2) and the market (M). L is the least-cost location.

FIGURE 14.3 An isodapane map

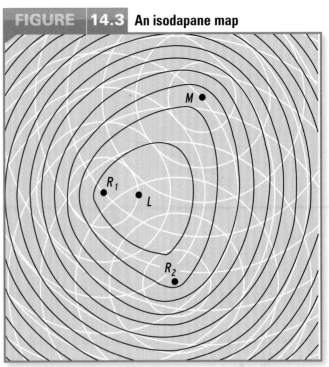

Superimposing isodapanes on an isotim map indicates how far from L an industry may locate to take advantage of a particular low-cost labour location.

Least-cost theory: An evaluation

Weber's theory is normative and does not attempt to describe or summarize reality. The major criticisms of it are directed at its simplifying assumptions. Markets, for example, are not the simple fixed points that Weber chose to assume. Labour markets are characterized by discontinuities associated with age, gender, ethnicity, and skill. Nevertheless, Weber's basic logic is simple and sound (Box 14.2).

Hoover (1948) continued Weber's focus on transport costs and added substantially to it. The Weberian locational figure changes when transport costs are not directly proportional to distance. The cost per unit of distance is less for long hauls than for short hauls (Figure 14.4), and a step structure is often used (Figure 14.5).

MARKET-AREA ANALYSIS

According to proponents of market-area analysis, Weber's fundamental error is his assumption that industrialists seek the lowest-cost location. Rather, for the market-area analyst, industrialists are profit maximizers. Least-cost theory is a form of variable cost analysis (concerned with spatial variations in production costs), while the market-area argument is a form of variable revenue analysis (concerned with spatial variations in revenue).

In market-area analysis, the location that will result in the greatest profit can be determined by identifying production costs at various locations and then taking into account the size of the market area that each location is able to control. The argument is that industries will attempt to monopolize as many consumers as possible—they seek a **spatial monopoly** and, in doing so, exhibit **locational interdependence**.

This argument about seeking a spatial monopoly is relevant, but if least-cost theory tends to de-emphasize demand, then market-area analysis tends to de-emphasize factors other than demand. The most important market-area theorist is Lösch, who (like Christaller) determined that the ideal market area is hexagonal. Like least-cost theory, market-area analysis is a normative approach.

BEHAVIOURAL APPROACHES

Normative approaches allow us to identify rational or best locations and understand aspects of actual locations through comparison. But is such an approach satisfactory? Many

> **spatial monopoly**
> The situation in which a single producer sells the entire output of a particular industrial good or service in a given area.

> **locational interdependence**
> The situation in which competing businesses base their location decisions on those of their rivals.

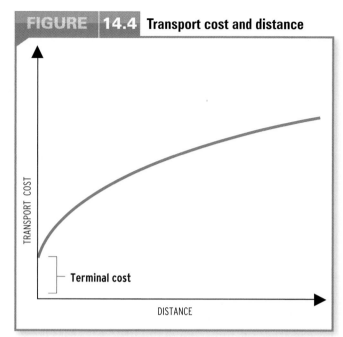

FIGURE 14.4 Transport cost and distance

Note that the cost does not begin at $0, because there is a basic terminal charge.

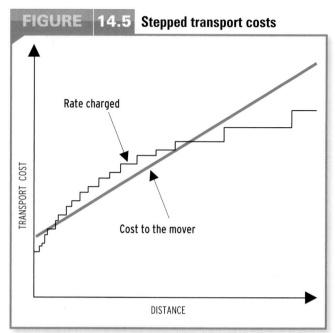

FIGURE 14.5 Stepped transport costs

Here the rate charged does not relate directly to distance because of the rate structures used by the movers.

geographers think not, and prefer to focus on explicit analyses of why things are where they are rather than on abstract analyses of where they ought to be, given certain assumptions. Normative approaches consider why things are where they are only in an indirect fashion. A set of approaches often labelled 'behavioural' attempt to correct this situation.

In brief, behavioural approaches centre on the subjective views of individuals. The concept of the economic operator is rejected and the concept of satisficing behaviour is paramount.

A simple example demonstrates the relevance of this approach. The beginning industrialist does not really debate where to locate; rather, the selected location is the existing location of the industrialist (or the nearest feasible location). Oxford, England, became the home of Morris, a major automobile manufacturer, for this non-least-cost and non-market-area reason. The location selected was the home area of the industrialist William Morris. Admittedly, had that home area been a remote Scottish island, some other location would have been

Box 14.2 | Testing Weberian Theory

Least-cost industrial location theory has been tested many times. A classic example is provided in the application of Weberian concepts to the particular industrial locations of the Mexican steel industry in the 1950s (Kennelly, 1968). The choice of an example dating back half a century is deliberate; least-cost theory was much more relevant then than it is today. The choice of the iron and steel industry is also deliberate, because least-cost theory can still be appropriately applied to this type of heavy industrial activity.

In 1950 the Mexican steel industry consisted of two main steel plants and a number of other plants. Production covered a range of products and accounted for about 50 per cent of Mexican requirements. Location of plants, particularly the major ones, was closely related to the availability of raw material resources, such as iron ore, coke, oil, and market scrap. The location of labour appeared to be of minimal importance, as unskilled labour was quite mobile. Access to market was important, however, with major markets in Monterey and Mexico City. After determining least-cost locations, Kennelly reported that, overall, the 1950 Mexican steel industry was in accord with basic Weberian principles.

Weberian theory is useful in explaining the locations of industries of this type, but its logic weakens with technological change and is less relevant to most contemporary location decisions than to those of the past. Other (especially Marxist) approaches offer insights into industrial location decisions that the traditional theories do not attempt to consider.

selected. Nonetheless, specific locations are often chosen for subjective reasons.

Contemporary industrial firms, especially large corporations, are complex decision-making entities. Location decisions made by small, often single-owner, firms are less likely to be economically rational than those made by large firms, but other considerations compete with least-cost and maximum-profit logic for all firms. It is also clear that decision-makers do not typically have access to all relevant information and, furthermore, that such information is subject to various interpretations. Geographers are well aware of the less than perfect mental maps held by decision-makers. Although behavioural issues are not as easily theorized or quantified as the more directly economic issues, they are always factors to be considered in any analysis of industrial location.

The Industrial Revolution

Before the Industrial Revolution, industrial activity was important but limited. Every major civilization was involved in a few basic industrial activities such as bread-making, brick-making, pottery, and cloth manufacture. These activities were located in the principal urban centres and close to raw materials. In addition, each household was an industrial organization that produced clothing, furniture, and shelter. All industries involved minimal capital and equipment, were structured simply, and were small in size and output. Energy sources (such as wood) and raw materials (such as agricultural products and sand) were relatively ubiquitous, not localized. Given these characteristics, industries could be dispersed, and as a result, there was little evidence of an industrial landscape. Transport systems were limited and markets for products were local rather than regional or national. Everything changed, however, as a result of the developments that we call the Industrial Revolution (Box 14.3).

ORIGINS

Between about the mid-eighteenth and mid-nineteenth centuries, the industrial geography of England was dramatically transformed. In fact, England was the first industrial area. Elsewhere in Western Europe the transformation did not begin until the nineteenth century, but it was accomplished over shorter periods of time.

Why was England first, not China or Japan, and why did this occur in the eighteenth century, not earlier or later? A conventional answer is that the Industrial Revolution was triggered by the rapid onset of new and stable political, legal, and economic institutions throughout much of seventeenth-century Europe—including the rule of law, secure property rights, and social mobility—with England the first country to build on these changes. An alternative elitist answer, proposed by Clark (2007), is that England had a particularly long history of settlement, stability, and security, and that this history created the necessary cultural preconditions for industrialization. These preconditions, Clark contends, are hard work, rationality, frugality, and education. A form of natural selection took place as rich and successful people—who worked hard, were rational, frugal, and well-educated—propagated themselves, while the poor died out. Such an idea, while certainly contentious, might be described as a pattern of differential fertility that favours the rich. According to this argument, then, the greater prevalence of these traits by about 1700 made possible the Industrial Revolution in England.

In general, the term 'Industrial Revolution' is apt, for it was a period of rapid and cumulative change unlike anything that had occurred previously: machines replaced many hands, and inanimate energy sources were efficiently harnessed. However, the term may be misleading to the extent that it suggests that the process of industrialization took the same form in every case. In truth, the process was uneven, and other countries did not simply imitate the English example. Belgium followed the English example most closely, but—unlike England—emphasized the metallurgical industries far more than the textile industries.

The most important component was the rise of large-scale factory production. Developed out of earlier small establishments and domestic household activities, factories were both much larger and more mechanized; they also required more capital. Further, because they relied on new, localized energy sources, they tended to agglomerate. The key energy source was coal, and the first truly industrial landscapes arose in the coalfield areas of England. But these developments were only one part of the Industrial Revolution.

Box 14.3   The Period of the Industrial Revolution

Earlier in this book, we referred to the Industrial Revolution even in contexts without any direct connection to industry. The reason is that the Industrial Revolution involved much more than industrial change. Collectively, these changes, which originated in Europe around the middle of the eighteenth century and continued until roughly the middle of the nineteenth, have contributed much to our contemporary world.

'Industrial Revolution' is the name given to a series of technical changes that involved the large-scale use of new energy sources, especially coal, through inanimate converters. After the first successful steam engine was built in 1712, other new machines were able to use steam to improve industrial output. The introduction of new machines led to the construction of factories close to the necessary energy sources, raw materials, or transport routes. The resulting concentrations of factories promoted urban growth, including the emergence of distinct industrial landscapes and working-class residential areas, since workers needed to live near their places of employment. Industrial cities replaced the earlier pre-industrial forms of settlement. As we saw in Chapter 9, to move raw materials to the factories and finished products to the markets, new transport links were essential; in England they initially took the form of improved roads, then canals and, most critically, railways. Agriculture became increasingly mechanized, and there was a demand for new raw materials, especially wool and cotton, from overseas areas. Overall, the percentage of the total labour force engaged in agriculture declined rapidly.

Technical changes were accompanied by significant changes in demographic characteristics. As described in the demographic transition model (see Chapter 4), the first phase of the Industrial Revolution involved a dramatic reduction in death rates, followed by reductions in birth rates. The interval between these two drops was a period of rapid population growth: the world population was estimated at 500 million in 1650 and at 1.6 billion by 1900. Industrialization and the associated population increases contributed to new areas of high population density, rural-to-urban migration, rapid urban growth, and considerable European movement overseas to the new colonial territories.

A major social and economic change was the collapse of feudal societies and the rise of capitalism. As labour was transformed into a commodity to be sold, the producer became separated from the means of production. Politically, this period witnessed the rise of nationalism and the emergence of the modern nation-state, as well as the expansion of several countries around the globe to create empires. Although these empires were short-lived, they contributed enormously to the contemporary political map—and to the division between more and less developed worlds. The nation-state, as we have already seen, remains one of the most potent forces in the contemporary world. The rise of modernism was linked to the Industrial Revolution through a common emphasis on the practicality and desirability of scientific knowledge. The transition to an industrial way of life was also a transition from tradition to modernity, from *Gemeinschaft* to *Gesellschaft*. Even though new post-industrial forces have emerged since about 1970, the world in which we now live is unmistakably a product of the Industrial Revolution.

There were also important advances in the organization of academic knowledge during this period; indeed, the social sciences are themselves related to these various changes. Geography was one of these disciplines. Major contributions to the advancement of knowledge included the works of Marx and Darwin.

The late eighteenth century witnessed the rise of a new economic system that emphasized individual success and profits: capitalism. The new capitalists seized the opportunities provided by a whole series of technological developments in the metals and clothing industries: a furnace that burned coal to smelt iron (first used in 1709); a spinning jenny for multiple-thread spinning (1767); a steam engine used as an energy source (1769); an energy loom for weaving (1785). These are only a few examples of the new inventions, most of which were developed in England.

Recall our definition and discussion of energy in Chapter 3: energy is the capacity to do work, and the more successfully we can use forms of energy other than our own, the more successfully we can achieve our work goals. Also, recall our definition of technology as the ability to convert energy into forms that are useful to us. The fact that the first agglomerations developed on the coalfields of northern and central England and in the traditional areas of cloth manufacture in northern England reflected the importance of coal for energy and the focus on the metal and clothing industries. These industrial agglomerations caused rapid increases in urban populations as workers moved closer to places of employment.

The basic logic of the Weberian least-cost theory is helpful in understanding these developments. Using localized energy sources meant that industries were located at those sources and, if possible, at the sources of raw materials. The need to reduce costs meant that the Industrial Revolution was paralleled by a transport revolution.

To clarify some of the generalizations above, let us take a closer look at examples of nineteenth-century industrialization, especially the early British iron, steel, and wool industries.

EARLY INDUSTRIAL GEOGRAPHY

Analyses of these early industrial landscapes include traditional descriptive accounts based on Weberian theory and other theoretical accounts that acknowledge the role played by economic power and organizations.

Iron and steel

Prior to the Industrial Revolution, Britain had two principal centres of iron production—one in central England and one in southern England. Both used local ores and relied on wood (charcoal) as their energy source. The iron industry expanded rapidly after coal was substituted for charcoal from 1709 onward and an efficient steam engine was developed. As iron production increased, new uses for it were found: the first cast-iron bridge was built in 1779 in central England, and most new industrial machinery was constructed of iron.

By the early nineteenth century, new iron areas dominated as technological advances continued to favour coalfield locations such as south Wales and north-central England. An important new factor emerged after 1825 when the first steam-powered train was operated in northeast England. The railway boom reached its height in the 1840s and gradually allowed some movement away from coalfields. Railways, together with the discovery of new iron ore sources, allowed a new industrial centre to develop in northeast England. Other major mid-nineteenth-century iron and steel areas were located in the English Midlands, south Wales, and central Scotland.

Textiles

Before the Industrial Revolution, the woollen industry was specialized and spatially concentrated in three areas: southwest England, east-central England, and northern England. The manufacturing process had remained unchanged from the fourteenth century to the late eighteenth century, when various inventions that were part of the larger Industrial Revolution prompted radical changes. Of the two major textile industries, cotton was the first to experience change, since most of the new procedures were more easily applied to cotton than to wool. In both industries, the major changes followed the same pattern: a decline in home production (the 'cottage industry') and a rise in factory production; use of steam engines; rapid expansion in northern England, reflecting the availability of coal; and associated declines in the other two areas. In the case of the textile industry, the principal technological breakthroughs occurred in villages and small towns in northern England, as mechanized factories were constructed near suitable sources of water.

In the case of the cotton industry, raw material came initially from the East Indies, but supplies were irregular and a new source had to be found. The Caribbean and southern United States were suitable source areas, but the task of separating cotton fibres from seeds was so labour-intensive that, even with a slave economy, supplies were limited. The invention of the cotton gin in 1793 allowed American plantations to flourish and English factories to be adequately supplied. The cotton mills of northern England became so dependent on imports from the United States that mill workers were laid off when supplies were reduced during the American Civil War. In the case of wool, the new area of supply was southeastern Australia.

Industrial landscapes

The Industrial Revolution and related rapid population and urban growth combined to create new landscapes. Factory towns arose, especially in northern England, and migrants from the south and rural areas poured into the new employment centres. In environmental and social terms, often the results were disastrous.

Before the onset of industrial change, the landscape was agricultural and predominantly rural. Even before the introduction of steam energy, however, some industrial production was shifting from the household to factories and the landscape was becoming more congested. Probably the first real factory in Britain was a silk mill in north-central England that was completed in 1722; five to six storeys high, it employed 300 people and used water energy. This example shows that coal was not essential for industrialization, but it certainly accelerated the process of change. The first factory was followed by many others as the century

The Tin Fouye Tabankort gas processing plant is the high-tech showpiece of Algeria's natural gas industry.

CP/AP photo/Bruce Stanley

progressed, but they produced neither smoke nor dirt. It was steam energy that resulted in major landscape change. More factories meant more workers crowded into small areas, and the use of coal quickly polluted the landscape. Land was expensive, and the areas of small, terraced houses where workers lived quickly deteriorated into slums. These industrial cities were subject to heavy smoke pollution—a situation that in many locations was not addressed until the 1960s.

Diffusion of industrialization

After about 1825, the technological advances developed in Britain diffused rapidly to mainland Europe (especially Belgium, Germany, and France) and to North America. Britain did not welcome the spread of its industrial innovations and even tried to keep some advances secret, but it did not succeed. In the United States, Pennsylvania and Ohio became the early industrial leaders because of their high-quality coal sources, while in the Russian Empire, Ukraine became the driving force when coal was discovered there. Japan began to industrialize after establishing cultural contact with the United States and rejecting feudalism in 1854.

In Europe the Ruhr region offered not only coal and small, local iron ore deposits, but an excellent location for water transport. As the nineteenth century progressed, the demand for labour became so great that immigrants moved to the Ruhr from elsewhere in Europe,

giving the region an unusually diverse ethnic character.

Sources of Energy

As we saw in Chapter 3, most energy today comes from fossil fuels: oil, natural gas, and coal. Since the beginning of the Industrial Revolution, coal had dominated. In the 1960s, however, oil replaced coal as the most important energy source globally, and in the early 1970s it provided almost 50 per cent of the world's energy. Since then, oil use has declined to about 35 per cent because of high prices, occasionally erratic supplies, more efficient use, and increasing reliance on other energy sources. However, oil continues to be the chief source of industrial energy, and the geographies of oil reserves, production, and consumption are important aspects of contemporary life. As noted in Chapter 3, other principal energy sources are coal (25 per cent of the world's energy), natural gas (21 per cent), biomass and waste (10 per cent), nuclear power (6 per cent), hydroelectricity (2 per cent), and other renewables (1 per cent).

A few countries tend to dominate world production of oil. In 1960 the major sources were the United States, Russia, and Venezuela; only about 15 per cent came from the Middle East. Today, the Middle East dominates oil production, with about 31 per cent of the global total. Similarly, a small number of countries dominate in the production of natural gas: together Russia (21 per cent), the US (19 per cent), and Canada (6 per cent) produce almost 50 per cent of the global total; no other country produces more than about 4 per cent. Five countries are now responsible for about 75 per cent of world coal production: China (41 per cent), the US (19 per cent), India (6 per cent), South Africa (5 per cent), and Russia (5 per cent).

OIL PRODUCTION AND CONSUMPTION

The most important player to be identified in any account of global energy is the Organization of Petroleum Exporting Countries (OPEC), a cartel founded in 1960 by Iran, Iraq, Kuwait, Saudi Arabia, and Venezuela, which were later joined by Algeria, Gabon, Indonesia, Nigeria, Qatar, and the United Arab Emirates. OPEC's principal purpose is to set levels of production in order to determine the price of oil, therefore, OPEC is able to wield enormous influence on the economies of more developed countries—a price increase in 1973, for example, was a leading cause of a widespread recession. In the first years of the twenty-first century, OPEC worked to keep the price of oil around US$22–28 per barrel, but a series of economic uncertainties, political events, and natural disasters combined to increase the price dramatically to more than $150 by 2008, after which prices fell as one part of the larger global recession that began in that year. Box 14.4 offers capsule comments on the 10 countries with the largest known reserves of oil.

As shown in Table 14.1, the United States is the world's leading consumer of oil, with about 24 per cent of global consumption, followed by China and Japan. In general, most of the principal consuming countries are likely to become increasingly dependent on imports. In short, there is a growing gap between the places where oil is produced and where it is consumed. Table 14.1 lists the major oil-consuming countries, while Figure 14.6 shows the major patterns of trade in oil.

Energy supplies in Britain

In Britain the transition from coal to oil began in the 1960s. As we have seen, the Industrial Revolution started in Britain and relied heavily on coal. As local coal sources were depleted,

Table 14.1	Principal Oil-Consuming Countries, 2007
Country	Share of Total (%)
United States	23.9
China (including Hong Kong)	9.3
Japan	5.9
India	3.3
Russian Federation	3.2
Germany	2.9
South Korea	2.7
Canada	2.6
Saudi Arabia	2.5
France	2.3
Brazil	2.4
Mexico	2.3
Italy	2.1
United Kingdom	2.0
Spain	2.0

SOURCE: *BP Statistical Review of World Energy 2008*. Reprinted by permission of the publisher.

imports became necessary, but they proved very expensive. As early as 1945, Britain began to import oil from the Middle East. Cost and politically related supply uncertainties prompted a search for oil and gas in the North Sea. Gas was discovered in 1965 and oil in 1969, and both were soon determined to be available in commercial quantities. By 1980

FIGURE 14.6 Major oil trade movements (million tonnes)

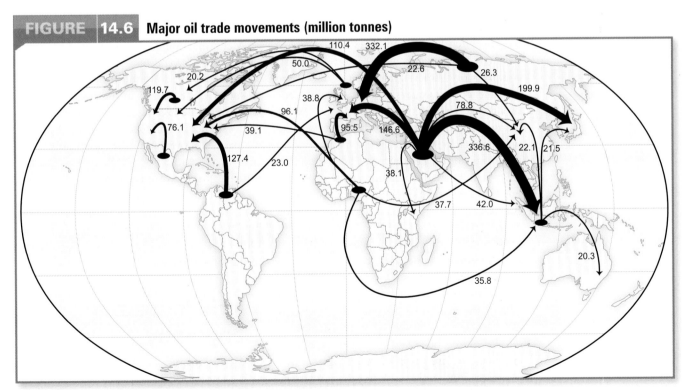

Source: *BP Statistical Review of World Energy 2008*.

The oil rig *Rowan Gorilla III* is towed into Halifax harbour from the Cohasset oil field for a scheduled refit. Nova Scotia's three producing oil fields—Cohasset, Panuke, and Balmoral—are about 250 kilometres southeast of Halifax.

Andrew Vaughan/The Canadian Press

production was equal to domestic demand, with 36 oilfields and 25 gas fields developed. What does the future hold? The current supply areas are being depleted, and the key question is the extent to which declines can be countered by new discoveries.

ALTERNATIVE AND RENEWABLE ENERGY SOURCES

The International Energy Authority (IEA) predicts a rise in global energy demand of 50–60 per cent between 2005 and 2030, so what will happen when the oil runs out? This is one of the most asked questions in early twenty-first century North America and Europe. According to some commentators, some time this century the oil will run out. Nobody knows when because, although demand increases all the time, especially with the rise of China as an industrial power, new reserves are being discovered and new technologies are being developed to make other known reserves, such as the Alberta oil sands, profitable to exploit (some of the environmental impacts of this activity are discussed in Chapter 3). A comprehensive study in 2000 conducted by the United States Geological Survey suggested depletion day was at least two decades away. Despite the uncertainties, the question is pressing. To put energy use in context, consider that human 'progress' has always been tied to increasing energy use, from the use of fire, through agricultural domestication, and then to the Industrial Revolution and the use of fossil fuels.

So far, there are few substitutes for gasoline. Countries such as Brazil and Canada are converting agricultural products to alcohol that can be blended with gasoline; indeed, government regulations in Brazil require that all gasoline sold be mixed with at least 20 per cent ethanol, a by-product of sugar cane. A 2005 projection from the IEA suggested that, at best, ethanol will provide 10 per cent of the world's gasoline by 2025. Other areas, such as Switzerland, are increasingly experimenting with electric cars. Different countries generate electricity in different ways. Nuclear reactors are heavily used in France, Belgium, and South Korea; Denmark is increasingly experimenting with wind; and many countries, such as Norway, India, New Zealand, and Canada, as well as countries in South America and Africa, rely significantly on hydroelectric power.

Perhaps most important is the need to increase the supply of renewable energy sources. As noted in Chapter 3, renewable sources reduce the demand for fossil fuels and can have environmental and other benefits. According to the IEA, renewable energy sources are only 13.3 per cent of the world's total primary energy supply (all sources) and, of that 13.3 per cent, 10.6 per cent is combustible renewables and renewable waste, while hydroelectricity is about 2 per cent. This means that all other sources of renewable energy, including geothermal, tidal, wind, and solar, make up little more than 0.5 per cent of world supply.

Major oil companies are investing heavily in both alternative and renewable energy sources. BP, for example, announced in 2005 its intent to spend US$8 billion in the next decade on the development of alternative and renewable sources. There is, of course, a strong argument for national governments to promote alternatives to oil. Most notably, Sweden has established a goal of weaning itself off oil by about 2020. This is to be accomplished not through constructing any new nuclear power stations but through use of renewable energy sources. Other countries are less ambitious but heading in the same direction.

World Industrial Geography

The global manufacturing system is dominated by three regions: North America, that is, the United States and Canada; Europe, especially

Box 14.4 Oil-Producing Countries

Ranked by quantity of known reserves at the end of 2007, the 10 most important oil-producing countries are as follows (note that the first five countries are all in the Middle East):

- *Saudi Arabia*. Saudi Arabia has by far the largest known reserves (265 billion barrels—about 22 per cent of the global total). It is also the world's largest producer of oil (10.4 million barrels per day). Saudi Arabia is nevertheless economically and politically dependent on the United States.
- *Iran*. Iran has the second-largest proven reserves of 138 billion barrels (about half the Saudi total and 11 per cent of the global total) and is OPEC's second-largest daily producer (4.4 million barrels).
- *Iraq*. Iraq has proven reserves of 115 billion barrels and produces about 2.1 million barrels per day, but production can be erratic. Even before the American-initiated war of 2003 and the resulting political chaos, Iraq was suffering from the war with Iran in 1980–8, the Gulf War in 1991, and more than a decade of sanctions.
- *Kuwait*. Kuwait's known reserves of 102 billion barrels are similar to those of the United Arab Emirates, but its daily production is somewhat lower (2.6 million barrels). Kuwait maintains close ties with the West.
- *United Arab Emirates (UAE)*. There are 98 billion barrels of known reserves and 2.9 million barrels are produced per day. The city of Dubai is the trading and financial centre for much of the larger Middle East region. The UAE is closely linked to the neighbouring state of Qatar, which has known reserves of only 15.2 billion barrels but a relatively high daily output of just over 1 million barrels.
- *Venezuela*. Venezuela has known reserves of 87 billion barrels and produces 2.6 million barrels per day. The largest oil exporter in the Americas, Venezuela is unsympathetic to the United States and to the Western world generally.
- *Russia*. With known reserves of 79 billion barrels, Russia produces 10.0 million barrels per day. Traditionally, Russia has had close links with OPEC, especially Iraq.
- *Kazakhstan*. In addition to Russia, several other states that were part of the former USSR also have oil reserves, notably Kazakhstan with 40 billion barrels of known reserves and a daily production of 1.5 million barrels. These reserves (and those in Azerbaijan) are more accessible since the 2005 completion of a pipeline linking Baku on the Caspian Sea with the Turkish Mediterranean port of Ceyhan. For Europe, this is a strategically significant source of oil outside of the Middle East and Russia.
- *Nigeria*. The largest producer in Africa, Nigeria has 42 billion barrels of known reserves and produces 2.4 million barrels daily.
- *Sudan*. Sudan has 36 billion barrels of proven reserves but produces only 0.5 million barrels daily.

Other countries with known reserves of more than 1 per cent of the global total are Canada, the United States, Algeria, and China. Other countries with daily production of more than 1 per cent of the global total are the United States, Canada, Mexico, Brazil, Algeria, China, and India. It is notable that several countries with relatively low proven reserves are high daily producers. Indeed, the United States is third in daily production after Saudi Arabia and Russia, with 6.9 million barrels per day.

As already noted, the United States is the world's largest importer of oil, and about one-third of its imports come from OPEC countries. Controversial exploratory activity is now underway in the Gulf of Mexico and in Alaska.

A few small countries are anticipating significant wealth from oil in the foreseeable future. The 169,000 people living on the islands of Sao Tome and Principe off the coast of West Africa live in poverty, but the government reached an agreement with Nigeria in 2005 to explore for oil in a jointly owned maritime area. If substantial reserves are discovered, the government is determined not to follow the route of some other African countries where oil revenues have not benefited the majority of people. The recently independent country of East Timor, often described as the world's poorest, is also anticipating income from oil following a 2006 agreement with Australia to share revenues from oil and gas in the Timor Sea. More generally a 2006 agreement between Exxon Mobil and Indonesia's state oil firm, Pertamina, will lead to the development of the large Cepu oil field on Java, and is likely to encourage other big oil companies to invest more in oil and gas throughout Southeast Asia.

It is unfortunate that some countries with large reserves of oil have not benefited significantly from this resource. Nigeria is a prime example, with the coastal producing areas severely damaged both environmentally and socially—instead of marine life the Niger delta is home to warring groups. Oil-producing areas of the Amazon basin in Ecuador are polluted and impoverished. While these are exceptionally negative cases, there is much evidence to suggest that oil states are often dysfunctional, at least partly because a reliance on oil seems to discourage countries from investing in other areas of the economy.

Note that the data in this box cannot give us an indication of when the world will run out of oil. The proven oil reserves in the Middle East and North Africa increased by more than 80 per cent in the period between 1973 and 2007, and other reserves continue to be discovered. However, it is clear that the current dependence on oil cannot be sustained as the ever-increasing demand both in the more developed world and in industrializing countries (especially China) will force prices to rise, as occurred prior to the onset of recession in 2008, such that, at some future date, the supply of oil will be unable to meet demand in an economically feasible way.

Germany, Italy, the United Kingdom, and western Russia; and Pacific Asia, especially Japan, South Korea, and eastern China. These three global regions, identified in more general terms in Chapter 9 (especially Figure 9.16), control the export and import of manufactured goods. Many industries are part of this global manufacturing system through trading activities. Other industries are transnational, owning or controlling production in more than one country. Transnational corporations and other large firms tend to be less committed to specific industrial locations than small firms because of the spatial separation of production from organization (Dicken, 1992).

Referring to individual countries, the United States continues to be the world's leading producer of manufactured goods by far, with China ranked second, Japan third, and Germany fourth. Industrial activity in all of these countries except China is framed within a capitalist social and economic system. Thus, decisions that affect the geography of industry and employment opportunities are made by industrial firms rather than governments. While small firms must cope with larger economic changes, large firms often have the power to influence the economic environment within which they operate. Nevertheless, all firms must take international trends into account. In some cases, firms may find their home markets threatened by imports; in others, firms may rely heavily on export markets.

Industrial firms today are still concerned with costs and profits, and hence the basic logic of least-cost theory and market-area analysis, as previously discussed, continues to be relevant in all countries. While there is no doubt that many location (and other industrial) decisions are not optimal, this may not be crucial in the long term. Thus, firms continue to make location decisions on the basis of the products, production technology, labour costs, sources of raw material and energy, capital availability, markets, and land costs—all considerations implicitly or explicitly contained within the traditional theories. Another key consideration, however, is the political context.

MAJOR INDUSTRIAL REGIONS IN MORE DEVELOPED COUNTRIES

As Figure 14.7 shows, there are five principal centres of industrial activity in the world and a large number of secondary centres. Four of the five principal centres are in the more developed world—eastern North America, Western Europe, western Russia and Ukraine, and Japan. The fifth region is in China, a country of the less developed world. Notably, the Pearl River Delta in southern China (discussed below) is the world's most dynamic industrial

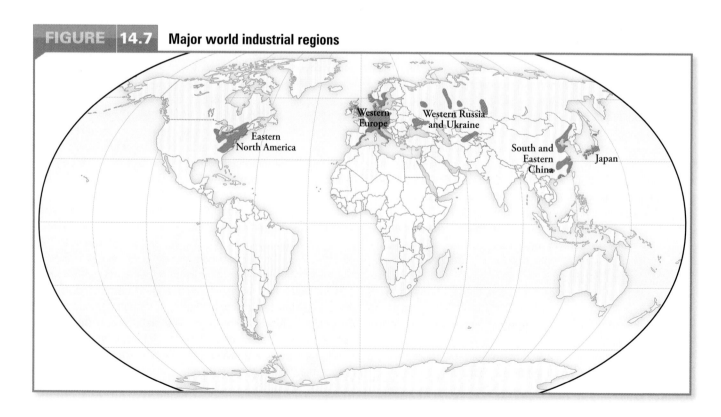

FIGURE 14.7   Major world industrial regions

region today. Together, these five regions account for over 80 per cent of global industrial production.

Eastern North America

This is a highly productive industrial region. Following European settlement, which began in the seventeenth century, it developed an industrial geography based on close ties to Europe, the proximity of raw materials such as coal, iron ore, and limestone, availability of labour, rapid local urbanization and market growth, easy movement of materials and finished goods along the natural waterways, and the building of canals and railways. There are numerous local industrial areas within this region, including southern Ontario and southern Quebec, southern New England, the Mohawk Valley, the southern Lake Erie shore, and the western Great Lakes. Box 14.5 outlines the industrial geography of Canada.

Western Europe

This region is also highly productive and, like the North American region, consists of numerous local industrial areas, the most important of which are central and northern Britain, the Ruhr and mid-Rhine valleys in Germany, and northern Italy. Although Britain had the initial advantage of the earliest industrial start, by the mid-twentieth century it was having difficulty keeping up with more recent industrial developments. Much the same is true of major European coalfield areas. However, the Ruhr Valley in northwestern Germany, which contains the most important coalfield in Europe, developed a large iron and steel industry and has been much more successful in coping with industrial changes. In northern Italy, rapid diversification after World War II transformed a textile region into a textile, engineering, chemical, and iron and steel region using local gas and imported oil.

Box 14.5 | Industry in Canada

Plentiful natural resources mean that Canada has a well-developed primary (resource extraction) industrial sector. Its immense geographic extent and (consequently) dispersed national market pose a challenge to many firms because of high transportation costs. In addition, the primary sector is scattered throughout Canada, while the secondary sector is concentrated in the St Lawrence Lowlands—southern Ontario and southern Quebec. Proximity to the United States has resulted in a manufacturing sector that includes many branch plants of American parent companies.

Following the initial in-movement of Europeans, a series of staple economies developed: fish, fur, timber, wheat, and minerals. By the mid-nineteenth century Canada was a major exporter of primary (resource) products and an importer of secondary (manufactured) products. Manufacturing in Canada at that time was mostly limited to the processing of agricultural and other primary products for both domestic and export markets. Following Confederation in 1867, the Canadian government pursued two policies with direct relevance to the infant industrial geography: subsidizing railway building and introducing tariff protection for manufactured products. The immediate beneficiaries were the established areas of the St Lawrence Lowlands, and a factory system was soon established. Montreal and Toronto in particular became industrial centres.

The capital needed to develop manufacturing industry initially came from Britain, but as British investment dropped off, a branch-plant economy controlled by Americans developed. American firms were thus able to bypass the protective tariffs in place, while Canada benefited by the introduction of financial and other capital and technology. After World War II, the US became less globally competitive (especially compared to Japan), and as Canada became increasingly nationalistic, the branch-plant economy became less and less attractive for industrialists from the US.

Today, the Canadian manufacturing industry is highly regionalized, and many geographers see Canada in terms of heartland and hinterland, or core and periphery. The hinterland produces the resources on which the manufacturing industries of the heartland depend. Toronto and Montreal have fabrication economies, while all other areas of Canada have resource-transforming economies. The two types of economy face different sets of problems. The industrial economies of Toronto and Montreal have suffered because of imports, especially from Japan and other Asian countries. The economy of the rest of Canada, which is largely dependent on raw materials, is subject to fluctuations in world price and demand over which Canadians have no control. At the national level, because Canada has many resources, economies facing problems are usually buffered by the successes of other sectors. At the local level, however, many areas are subject to booms and busts. Single-resource towns, of which there are many, are the most vulnerable (Britton, 1996).

Western Russia and Ukraine

Until the dramatic political upheavals of the late 1980s and early 1990s, industrial location and production in the former USSR and the Eastern European countries were determined by central planning agencies. Recent changes have not, of course, significantly altered the spatial pattern of industrial activities. In the former USSR there are five major industrial areas, four of which are now in Russia. Those industrial centres based in Moscow and Ukraine were established in the nineteenth century. The Moscow area is market-oriented and began its industrial development specializing in textiles; today it is diversified, with a wide variety of metal and chemical industries using oil and gas. In Ukraine industry is centred on a coalfield but also has access to supplies of iron, manganese, salt, and gas; Ukraine is a major iron and steel and chemical industrial area. The other three regions were developed by the USSR government after the 1917 revolution and are located in southern Russia. The Volga area to the east of Ukraine has local oil and gas sources and was a relatively secure location during World War II; its major industries are machinery, chemicals, and food processing. Farther east, the Urals area is both a source of many raw materials, especially minerals, and a major producer of iron and steel and chemicals; industrial growth in the Urals was promoted by the former USSR government because of its great distance from the western frontier. Farthest east is the Kuznetsk area, which is similarly endowed with minerals as well as a major coalfield.

Japan

Japan is relatively small in size; with few industrial raw materials of its own, it is forced to

Box 14.6   Industry in Japan

Japan comprises numerous islands scattered over an area roughly 2,500 km (1,500 miles) from north to south. Although cultural contact with Europe and North America did not begin until after 1854, with the end of feudalism, by 1939 Japan had expanded territorially and developed into an industrial power comparable to those in the West. Industrialization continued after World War II despite the loss of Korea and considerable bomb damage. During the 1950s and 1960s Japan excelled in heavy industry, especially shipbuilding, but by the late 1960s the emphasis had shifted to automobiles and electronic products, and most recently Japan has focused on computers and biotechnology. During each of these three post-World War II phases, Japan has been a world industrial power.

The principal factors behind these remarkable industrial successes are low labour costs, high levels of productivity, an emphasis on technical education, minimal defence expenditures, aid from the United States (motivated by the perception that Japan served as a bulwark against Chinese communism), and a distinctive industrial structure in which small specialized firms are linked to corporate giants such as Nissan and Sony. This structure facilitates rapid acceptance of various new technologies.

Another aspect of this success story is Japanese business's considerable investment overseas. Many Japanese companies operate in North America and Europe especially, and Japanese banks dominate international finance (Table 14.2).

But by the late 1990s, Japan showed signs of a substantial economic downturn with a declining growth rate, increasing number of bankruptcies, corporate indebtedness, and rising unemployment. A major concern is competition for production, with the loss of many manufacturing jobs to China where production costs are lower and competition for high-technology production from South Korea.

Table 14.2	Locations of R&D Facilities and Plants of Nine Japanese Electronics Firms, 1975 and 1991			
Location	R&D Facilities		Production Plants	
	1975	1991	1975	1991
Japan	22	95	207	341
North America	0	13	7	69
EU	0	7	6	50
Asia	0	2	40	123
Other	0	6	20	34

NOTE: The nine electronics firms are Hitachi, Toshiba, Mitsubishi, Matsushita, Sony, Sanyo, Sharp, NEC, and Fujitsu.

SOURCE: Adapted from Y.-M. Yeung and F.-C. Lo, 'Global Restructuring and Emerging Urban Corridors in Pacific Asia', in F.-C. Lo and Y.-M. Yeung, eds, *Emerging World Cities in Pacific Asia* (New York: United Nations University Press, 1996), 37.

import almost everything it needs; and it is a substantial distance from major world markets. But Japan has overcome these obstacles by making use of its large population. Since World War II, Japan has become a major exporter of industrial goods by keeping labour costs low, and by 1970 its highly skilled labour force was increasingly focused on the production of high-quality goods. Japan is now a leading producer of computers and electronic equipment (Box 14.6).

NEWLY INDUSTRIALIZING COUNTRIES

Japan is no longer alone. The success story outlined in Box 14.6 has encouraged many imitators in other Asian countries. During the 1970s, the economies of South Korea, Taiwan, Hong Kong, and Singapore all accelerated rapidly, and those of Malaysia, Thailand, Indonesia, and the Philippines are accelerating today. These eight are the leaders among what have come to be known as the newly industrializing countries, or NICs (others include Brazil, Mexico, Greece, Spain, Portugal, and, most recently, Vietnam). It is notable, however, that several of these economies suffered during the financial crisis of 1997–8, and that all shared in the global recession that began in 2008 and resulted in a shrinking of the world economy in 2009. Industries in the Pacific Rim countries follow the Japanese example of low labour costs and high productivity. All of these countries have high growth rates and are experiencing the same labour shifts from agriculture to industry that much of Europe experienced in the nineteenth century.

The most successful of the NICs is South Korea. Until 1950 South Korea was a poor, less developed country characterized by subsistence rice production; today it is an industrial giant that began with heavy industry, then focused on automobiles, and now specializes in high-technology products. Its experience is the Japanese transformation repeated, and over a period of less than 40 years.

Current evidence suggests that these trends will continue. One motivation for regional free trade agreements in recent years and for continuing European integration has been awareness of the need to respond to the dramatic industrial growth in much of Asia.

To attract transnational corporations, several NICs—including South Korea, Singapore,

Taiwan, Hong Kong, China, the Philippines, and Mexico—and some other less developed countries have set up export-processing zones (EPZs): manufacturing areas that export both raw materials and finished products (Figure 14.8). Today there are more than 3,500 such zones in about 130 countries, employing more than 66 million people, about 40 million of whom are in China. Industries are attracted to these zones for three general reasons: (1) inexpensive land, buildings, energy, water, and transport; (2) a range of financial concessions in such areas as import and export duties; and (3) low workplace health and safety standards and an inexpensive labour force made up largely of young women, who are regarded as less likely than men to be disruptive and more willing to accept low wages and difficult working conditions. There are few advantages other than waged employment for the processing country, and even this advantage varies depending on larger international economic circumstances.

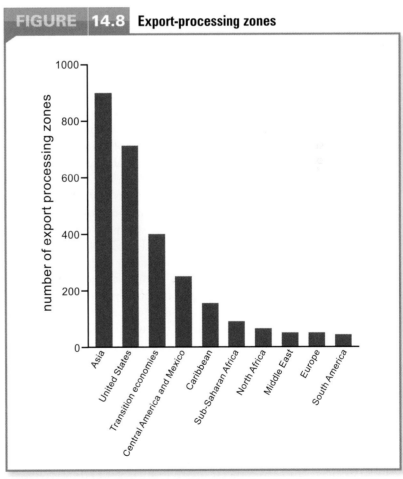

FIGURE 14.8 Export-processing zones

SOURCE: Adapted from J.-P.S. Boyenge, *ILO Database on Export Processing Zones (Revised)* (Geneva: International Labour Office, Sectoral Activities Programme, Working Paper, 2007), 2.

The parallel between these export-processing zones and the agricultural plantations discussed in Chapter 10 is compelling.

There are especially close links between these export-processing zones and high-technology companies; in Mexico EPZs known as maquiladoras are set up within easy reach of the high-tech suppliers in such areas as the Santa Clara Valley in California (Silicon Valley) and the Dallas–Fort Worth area in Texas (Silicon Prairie). The high-tech materials, such as silicon chips, are manufactured in the United States and transported to the maquiladoras for the simple but labour-intensive process of assembly; then the finished product is transported to the US for sale.

INDUSTRY IN CHINA

It is sometimes too easy for people in the Western world to forget that China has historically been a major player in the world economy. It is estimated that about 2,000 years ago it produced about 25 per cent of total world output and that as recently as 1820 it produced about 33 per cent. These numbers highlight the fact that China's poor economic performance for much of the past 200 years, as it failed to follow the European and North American examples, has been atypical. In 1950, for example, China was producing only 5 per cent of total world output. But China is now making up for lost time and growing rapidly in what might be called a second 'industrial revolution'. Indeed, after about 25 years of rapid growth, China is now best described as the workshop of the world. Growth is rapid and ongoing; the port of Qingdao, for example, did not exist in 2004 but is now one of the biggest container ports in the world (see Chapter 9 for a brief account of recent developments in the Chinese transport network). It is widely expected that China will soon become the world's leading manufacturing country, although specific predictions are very difficult in light of larger global economic changes.

As we noted in our discussion of agriculture in Chapter 10, liberalization of China's Communist economy began in 1978 with the opening up of the country to international trade and foreign investment. Since the 1980s, both the agricultural and industrial economies have undergone reform and modernization. These important changes were initiated during the 18-year leadership of Deng Xiaoping (died 1997), who effectively determined that economic growth was more important than continued class struggle.

During the 1950s, industrial development was based on a combination of large technology-intensive, state-funded factories and small labour-intensive, locally organized units (Morrish, 1994, 1997). The large state-funded units made up about 100,000 of the total of about 8 million units and produced 55 per cent of the national industrial output. Despite this high percentage, the strategy proved unsuccessful because the state-owned factories were expensive to operate and highly inefficient. Shortages of consumer goods were commonplace.

A new regional development policy was introduced in 1980 with the establishment of several special economic zones (similar to export-processing zones) that enjoy low taxes and other financial concessions to attract foreign investment. A reform program was initiated in 1984 that involved less state control, decreased subsidies, and increased response to market forces, and that encouraged local collective and private industrial enterprises. One result was that China's industrial production outstripped that of South Korea, increasing at 12 per cent per annum between 1980 and 1990. In 1994 the state relaxed its control even further and, to encourage collective and private enterprise, introduced a new program requiring urban industrial enterprises to respond to market forces directly, without state involvement.

The special economic zones have played a major role in the recent expansion of Chinese industrial activity (Figure 14.9). Four zones were established in 1980 at Shenzhen, Zhuhai, Guangzhou (all located in Guangdong province north of Hong Kong), and Xiamen (in neighbouring Fujian province). These zones have grown dramatically with the introduction of 'labour-intensive factories that manufacture everything from computer keyboards and dishwashers to leather coats and Mighty Morphin Power Rangers' (Edwards, 1997: 12). Most of the foreign investment has come from Hong Kong, attracted by cheap labour and land. Together, the three zones north of Hong Kong, all located in the Pearl River Delta, now form what many consider to be the world's most dynamic industrial region, sufficiently large that it is affecting global trading patterns and investment flows. This delta area attracts about one-fourth of China's foreign direct investment and generates about one-third of its exports. In a sense, it is the early twenty-first-century equivalent of mid-nineteenth-century industrial northern England.

In 1984, 14 locations were designated 'open coastal cities', free to trade outside China, and an additional three larger open zones were created one year later. Another special economic zone, Hainan Island, was established in 1988. By the early 1990s it was clear that the coastal economy was growing too quickly, and the Chinese government began to focus attention on some inland areas. In fact, as discussed in Box 14.7, China's growth has been so dramatic not only in production but also in consumption that it is a cause of some concern to other nations and producers.

INDIA

The example of India highlights many of the general points noted earlier. Following independence in 1947, India was a producer of agricultural products, but today it has a much more diversified industrial structure. Industrial landscapes have been created, and the government has attempted to reduce regional disparities. Since 1951, India has used a series of five-year plans to guide development. Initially, the emphasis was on heavy industry, but later plans have stressed self-reliance and social justice. Given that its industrial transformation did not begin until 1951, India has achieved remarkable success; the main reasons are its substantial market, available resources, adequate labour, and relatively sound government planning. India is also experiencing the benefits of outsourcing as both North American and European countries relocate employment. Of particular note is the movement of call centres for many major businesses, with the United Kingdom losing about 33,000 jobs by the end of 2003.

By 2009, India ranked twelfth in the world in terms of GDP. Indian industrial employment and output will likely continue to grow rapidly in the immediate future; indeed, many observers expect it to emulate China as it also has a very large population. But this must be some years in the future as India's GDP is about half that of China, while its exports of goods and services are much less. One well-known feature of the Indian economy is the vibrant

FIGURE 14.9 Special economic zones in China

1. Pudong District, Shanghai Municipality
2. Xiamen, Fujian Province
3. Shantou, Guangdong Province
4. Shenzhen, Guangdong Province
5. Zhuhai, Guangdong Province
6. Hainan Province

software and information technology industry, especially in Bangalore, yet this sector accounts for only 3 per cent of India's GDP.

Globalization and Industrial Geographies

Globalization processes are playing a key role in contemporary industrial geography, affecting location, organization, and activity. Although these processes have already been discussed, especially in Chapter 9, it is important to review two key issues related to industrial change in both the more and the less developed worlds. Industrial restructuring today amounts to a global shift in industrial investment and activity. This process was well underway by the 1980s, along with the other classic component of globalization—a decline in the friction of distance that is a consequence of new communication technologies. Together these two processes have produced the new and dynamic industrial geography of today. Related to the two globalization processes is the transition from Fordism to post-Fordism.

FORDISM TO POST-FORDISM

To understand the current 'post-Fordist' phase of industrial activity in the more developed

Box 14.7   From Struggling Peasant Economy to Industrial Giant

Napoleon famously stated: 'Let China sleep, for when she awakes, she will shake the world.' China has awoken. Since 1978 its GDP has increased by an average of just under 10 per cent each year, about three times more than the increase for the United States (until the global recession that began in 2008).

China's economy is enormous, with the third largest GDP in the world in 2009, after the United States and Japan. In addition to dominating production, China is increasingly important as a trading nation and as a consumer.

In recent decades, the United States has been the only significant mass market in the world, but that is changing with the rise of China and the emergence of a new and huge mass market. China is now more than just a supplier of cheap products for the rich countries of the world; it is also a rival purchaser of resources and products. Until recently, China was self-sufficient in most primary products but it is now the most important player in global commodity markets. It is the world's largest consumer of grain, meat, coal, and steel. China is even importing rice because so much land is being taken over for housing, factories, and shopping malls. There are several reasons for rising oil prices, but Chinese demand is certainly a significant part of the explanation. The EU, the United States, and Japan are China's three most important trading partners.

The world's biggest net recipient of foreign investment, China now manufactures 60 per cent of the world's bicycles, 40 per cent of the world's socks, and 50 per cent of the world's shoes. It is also by far the largest garment exporter with almost 50 per cent of the global total today. Wages are extremely low: garment workers are paid about US40 cents an hour—less than a third of what their counterparts in Mexico receive. Today, China is the favoured location for multinational firms to contract out manufacturing employment. China is also increasingly a major global consumer: with about one-fifth of the world's population it consumes half of its cement, a third of its steel, and a quarter of its aluminum.

This rise of a new consuming China creates a problem for the United States, not only because many jobs are lost to this growing economy, but also because, as Chinese trade with other countries increases, there is a resultant weakening of American influence. It seems clear that Chinese influence around the world generally will increase, not least in the context of trade, investment, and other economic decisions. China is now heavily involved in other countries that are able to supply needed raw materials, especially Australia, DR of Congo, Angola, Sudan, and Myanmar. Consider, for example, that Chinese firms are currently building or renewing transport infrastructure in DR of Congo at an estimated cost of about US$12 billion in exchange for rights to copper mining. Also, Chinese investment in the oil industry in Sudan has allowed Sudan to ignore Western concerns and sanctions about the conflict in the Darfur region. Today about 10 per cent of China's oil imports are from Sudan. The rise of China also has impacts on the Canadian economy, principally because of the related commodities boom and resulting surge in investment, but also because China is increasing trade with Canada.

Chinese industrial growth has major consequences for the environment. Most of the power plants burn coal, which causes acid rain and pollution with impacts on agriculture and health. In many parts of China there is already insufficient water for both agriculture and industry, and with the melting of Himalayan glaciers that feed China's largest rivers the problem is worsening.

China is indeed a major global producer and consumer, and is playing an increasingly important role in the global economy with each passing year, and all of this is occurring in a country that continues to deny basic democratic rights to its citizens, a fact that must not be forgotten.

world, it is important to understand the Fordism that preceded it. The term 'Fordism' refers to the methods of mass production first introduced by Henry Ford in 1920s America, particularly fragmentation of production—best illustrated by the use of assembly lines that reduced labour time and related costs. In sharp contrast to most nineteenth-century industrial activity, the Fordist system gave workers the necessary income and leisure time to become consumers of the many new mass-produced goods. Economic policies based on the theories of John Maynard Keynes, a British economist, were widely implemented, resulting in rising living standards throughout the 1950s and 1960s. These two decades of sustained economic growth witnessed the rise of the first transnationals, among them Ford, based in the United States (automobiles), Nestlé, based in Switzerland (foodstuffs), and Imperial Chemicals, based in the United Kingdom (chemicals). These early transnationals were limited in their ability to invest outside their home countries, but they were able to take advantage of technological advances in transportation.

This economic boom slowed down in the early 1970s for two principal reasons. First was the termination of the provision in the Bretton Woods Agreement (see Chapter 9) that allowed the United States to convert overseas holdings of US dollars to gold at a fixed rate. When the US ended this practice because of the expenses incurred in the Cold War and the resulting budget deficit, other currencies fluctuated, resulting in price fluctuations and some business losses. The second reason was the 1973 OPEC decision to raise the price of oil—a commodity essential to industrial production.

One important outcome of the recession of the early 1970s was industrial restructuring in the more developed world. Basically this involved deindustrialization, especially in such traditional industrial activities as textiles and shipbuilding and in automobile manufacturing, and corresponding reindustrialization as firms established branch plants overseas, either in NICs or in less developed countries (Box 14.8).

INDUSTRIAL RESTRUCTURING

As Box 14.8 suggests, the transition to post-Fordism entails major changes in industrial geographies, resulting essentially from technological advances and globalization processes. Three technological changes are especially significant:

1. Production technologies, such as electronically controlled assembly lines and automated tools, are increasing the separability and flexibility of the production process.
2. Transaction technologies, such as computer-based, just-in-time inventory control systems, also increase locational and organizational flexibility.
3. Circulation technologies, such as satellites and fibre optic networks, facilitate the exchange of information and increase market size.

Together, these three changes represent a transition to **flexible accumulation**, making it easier for companies to take advantage of spatial variations in land and labour costs and to serve larger markets. The subsequent industrial restructuring takes three principal forms:

> **flexible accumulation** Industrial technologies, labour practices, relations between firms, and consumption patterns that are increasingly flexible.

> **reindustrialization** The development of new industrial activity in a region that has earlier experienced substantial loss of traditional industrial activity.

Box 14.8 Deindustrialization and Reindustrialization

Deindustrialization and reindustrialization are spatial trends caused by economic restructuring: specifically, shifts in investment between the manufacturing and service sectors—or, to put it another way, shifts between production and consumption. Both can be seen as components of the transition to a post-industrial society. Deindustrialization is a reduction in manufacturing in more developed countries that is usually most easily measured by reference to employment data. It is most evident in the older industrial areas and in such industries as iron and steel, textiles, and shipbuilding. It is not so common in high-tech industries such as pharmaceuticals and electronics. The major social consequence is unacceptably high unemployment.

In the United States and the United Kingdom the percentage of workers employed in manufacturing fell from about 40 per cent to about 20 per cent between 1900 and 2000. There are two general causes for this deindustrialization. First, as a result of globalization processes, including the rise of transnationals and improvements in communication technologies, much of the manufacturing in the more developed world is being transferred to less developed countries—especially China. (This decline of manufacturing in the more developed world mirrors the earlier decline of agriculture.) In its place, service-sector activities have become increasingly important.

Deindustrialization is usually seen as a negative development. Because the decline takes place in areas where industry used to be highly concentrated, it initiates a larger economic and social decline. The fact that it tends to occur rapidly, often during periods of economic recession, makes regional and local adjustment all the more difficult.

Reindustrialization at least partially counters industrial decline. This process may take various forms. First, there is an increasing tendency for small and/or new firms to be more competitive. This is most likely to occur outside the traditional industrial areas and reflects the information exchange made possible by electronic means. Second, high-tech industrial activities, especially microelectronics, are expanding rapidly in output, if not also in employment. Such industries are locating in environmentally attractive areas where skilled workers choose to live. A third aspect of reindustrialization is the expanding service industry. Rising incomes and changing lifestyles are bringing significant growth in tourism and recreation industries, with accompanying impacts on environmentally attractive areas. Other service industries, especially banking and information services, are also expanding in major urban centres.

In the more developed world the new industrial landscape is quite different from the nineteenth-century version. Not only is the landscape itself visually different with the decline of smoke-producing heavy industry, but both the work experience and the location of the landscape have changed. The need for coherent regional policies is obvious.

1. The relationship between corporate capital and labour is changing as machines replace people, manufacturing industry declines, and transnationals seek locations with low labour costs (as noted in the account of export-processing zones above).

2. Both the state and the public sector are playing new roles with the shift away from **collective consumption** in areas such as education and health care to joint public–private projects and deregulation.

3. There is a new division of labour at various spatial scales as the new technologies allow corporations to respond rapidly to variations in labour costs (as in the account of export-processing zones, as well as discussions in Chapter 13).

collective consumption The use of services produced and managed on a collective basis.

INFORMATION TECHNOLOGIES AND LOCATION

As the preceding discussion implies, decision-making by industrial firms today, especially concerning location, often differs markedly from the model proposed by Weber. Increasing emphasis on technology and decreasing emphasis on materials and localized energy sources mean that, for many industries, transport cost is no longer the main criterion. Rather, many contemporary firms are trading off between two types of cost: the labour and land costs considered by Weber and a whole set of new costs associated with exchanging information between firms.

A major debate today in studies of industrial activity, especially location decisions, concerns the implications of labour (and, to a lesser extent, land) costs on the one hand and the costs of information exchange on the other hand. For many industrial activities, information can be rapidly exchanged at low cost by electronic means, but there is a danger in such impersonal exchanges if the information is lacking in clarity. In principle, firms that make location decisions on the assumption that they will be able to exchange information successfully are able to seek out locations with low labour costs. The result is a decentralized (deglomerated) industrial pattern. A prime example is the successful expansion of many high-tech Japanese firms to other countries (see Table 14.2). In situations where firms lack confidence in their ability to exchange clear and unambiguous information, they will prefer to locate in close proximity (to agglomerate) in order to facilitate personal, face-to-face exchange of information.

Acknowledging the contemporary relevance of information exchange introduces two possible patterns of industrial location. Decentralization occurs if long-distance electronic information exchange is feasible, but centralization occurs if the information is lacking in clarity and requires personal contact to be effective. Decentralization is typically associated with firms (often transnationals) that mass-produce a standard product and thus do not need to worry about ambiguities in the exchange of information, while centralization occurs when functionally related firms need to exchange information on a person-to-person basis because their products are not identical. Much evidence suggests that firms concentrating spatially achieve a competitive advantage. In the case of the automotive tool, die, and mould industry in southwestern Ontario, a collection of small, specialized firms co-operate closely and have access to a pool of skilled labour, thriving on social networks and on flows of knowledge related to local labour mobility (Holmes et al., 2005). Another example is the mining supply and service industry concentrated in Sudbury, which similarly depends on a pool of skilled labour (Robinson, 2005).

SERVICE INDUSTRIES

It is generally assumed that as an economy progresses, it undergoes a transition from primary (or extractive) to secondary (or manufacturing) to tertiary (or service) activities. Figure 14.10

FIGURE 14.10 Economic growth and employment distribution

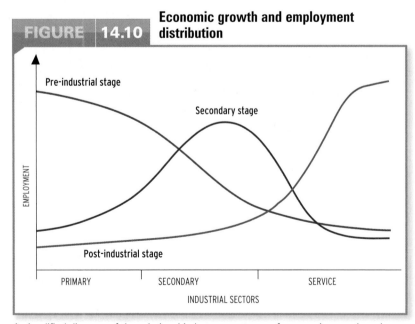

A simplified diagram of the relationship between stages of economic growth and distribution of employment in the three principal economic sectors.

New customer service agents learn US geography at the 24/7 Customer.com training centre in Bangalore, southern India. Their training included coaching in American or British accents, depending on the clients they were to represent, as well as appropriate slang terms and cultural references from sports and television.

CP/AP photo/Namas Bhojani

provides a simple illustration of the relative importance, in terms of employment, of the different sectors of industrial activity. Indeed, some geographers contend that, especially in the more developed world, the emerging post-industrial society is increasingly service-oriented. This may be so, but service industries have existed for a very long time.

According to Daniels (1985: 1), a service 'is probably most easily expressed as the exchange of a commodity, which may either be marketable or provided by public agencies, and which often does not have a tangible form.' Service industries, including transportation, utilities, insurance, real estate, education, health, and government, are crucial components of any economy. The service industry grew along with manufacturing during the Industrial Revolution, but it has achieved its most rapid expansion since World War II. Globally, since 1960, employment in services has increased from about 20 per cent to over 40 per cent (Figure 14.11). In the less developed world, retailing and distribution are among the dominant service activities, while in the more developed world more specialized services such as banking and advertising are also important.

The global variations in employment in industry and in services are indicated in Figures 14.12 and 14.13. It is useful to compare the two figures with Figure 10.15, which maps the percentage of the labour force employed in agriculture. The figures provide some compelling additional evidence of the differences between the more and less developed worlds. Some summary data for selected transition economies are presented in Table 14.3. This table shows the growing role of services in these economies, a result of increasing local demand and also of offshore outsourcing from other countries.

Geographic attempts to explain the location of service industries have focused on the

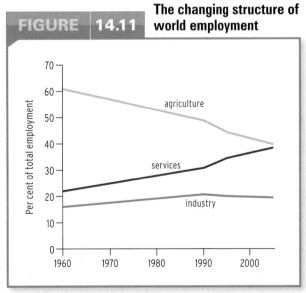

FIGURE 14.11

The changing structure of world employment

SOURCES: International Monetary Fund, *Finance and Development: A Quarterly Magazine* 43, 1 (Mar. 2006); United Nations Development Program, *Human Development Report* (New York: Oxford University Press, 1996), Table 16.

FIGURE | 14.12   Percentage of labour force in industry by country, 2007

Legend:
< 25%
25 - 49%
no data

Figure 10.15, mapping percentages of the labour force in agriculture, showed not only enormous variations between countries but a basic distinction between the more developed and less developed worlds. Although this map of percentages in the industrial labour force shows less variety, it also indicates a distinction between the more developed and less developed worlds. In general, the percentages are between 25 and 49 in the more developed world and under 25 in the less developed world. The world average is 20 per cent. There are significant gender differences: in Canada; 11 per cent of employed women and 33 per cent of employed men work in industry. In Bangladesh the corresponding figures are 9 for women and 11 for men.

SOURCE: Updated from United Nations Development Program, *Human Development Report* (New York: Oxford University Press, 1996), Tables 16 and 32. By permission of Oxford University Press, Inc.

central place model described in Chapter 11. This work has made it clear that service locations can be determined by considering such standard causal variables as transport costs, market location, and economies of scale. In addition, services have an especially strong tendency towards agglomeration. Their locations depend on population, and they tend to cluster. Furthermore, location decisions in the service industry appear to be especially vulnerable to behavioural variables such as the availability of information and the interpretation and use of information. A consideration of behavioural variables helps to explain the clustering tendencies. Information diffusion takes place most rapidly and effectively in local networks. Indeed, some argue that recent changes in the technology of information diffusion have led not only to the clustering of certain services but also to spatial changes in the roles played by urban centres. Others, however, contend that new information technologies are leading to increased decentralization, possibly even (in some cases) home-based service work. Regardless of where service industries locate, there is little doubt that they are able to contribute significantly to increases in productivity and living standards. In Canada, for example, service-sector growth is a crucial component of larger economic prosperity (Grubel and Walker, 1989).

A geography of service industries must also consider the availability of those services that are central to human well-being, among them

Table 14.3	Employment by Sector, 1990 and 2001, Selected Transition Economies					
	1990			**2001**		
Country	Agriculture	Industry	Services	Agriculture	Industry	Services
Bulgaria	18.5	44.2	37.3	26.3	27.6	46.0
Estonia	21.0	36.8	41.8	6.9	33.0	60.1
Poland	25.2	37.0	35.8	19.1	30.5	50.4
Romania	29.1	43.5	27.4	42.3	26.2	31.5
Russia	13.9	40.2	45.6	11.8	29.4	58.8

SOURCE: Adapted from International Labour Organization, *Global Employment Trends* (Geneva: ILO, 2004), 34.

FIGURE 14.13 Percentage of labour force in services by country, 2007

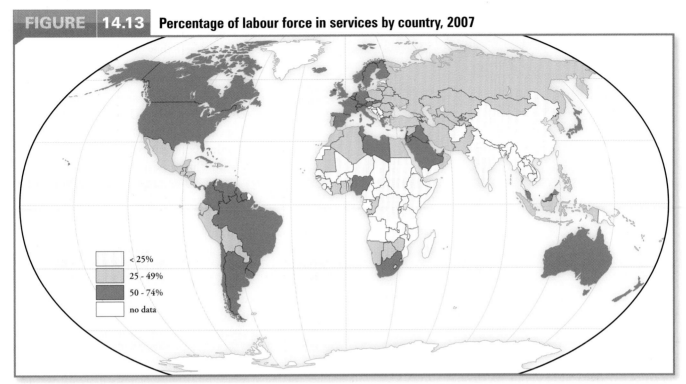

< 25%

25 – 49%

50 – 74%

no data

This map shows the highest values (between 50 and 74 per cent) occurring in the more developed world and a few other countries (especially in Latin America). In general, the less developed world has lower proportions of service workers, often below 25 per cent. The world average is 31 per cent. There are significant gender differences: in Canada, 87 per cent of employed women and 64 per cent of employed men work in services; in Bangladesh the corresponding figures are 12 for women and 30 for men.

SOURCE: Updated from United Nations Development Program, *Human Development Report* (New York: Oxford University Press, 1996), Tables 16 and 32. By permission of Oxford University Press, Inc.

health care and educational facilities. In fact, any study of service industries is a virtual microcosm of contemporary geography, because services are linked to so many other geographic topics and because they are so important in contemporary economies and societies.

OUTSOURCING

Outsourcing refers to the situation where a company hands work that was previously completed in-house to other firms. If this involves work being outsourced to other countries in order to take advantage of inexpensive labour it is a type of **offshoring**, adding to the internationalization of employment. Offshore outsourcing was initially viewed by many in North America and Europe as one indicator of the flexibility of a globalizing economy, but is now viewed more typically in terms of the negative impact of job losses, as many firms have an increasing percentage of employees in countries such as India, China, and Mexico, where inexpensive labour is plentiful.

Service activities generally are increasingly outsourced offshore as companies are able to hand over work to specialist service suppliers located elsewhere. This is occurring largely because of the improved communications associated with the globalizing economy. Some observers estimate that, by 2015, the United States will have lost about 75,000 jobs in legal services (such as case research) to poorer countries, while Europe will have lost about 120,000 computer professional positions. It seems likely that Russia, some Eastern European countries, China, and especially India will gain more and more business employment with the services provided being consumed elsewhere in more developed world countries. The discussion of industry in India noted the importance of offshore outsourcing to industrial growth in that country, and Table 14.4 shows recent and projected increases in information technology and business processing outsourcing (BPO) employment. BPO is the outsourcing of back-office and front-office functions typically performed by white-collar and clerical workers. Examples include accounting, payroll and human resources, and medical coding and transcription.

outsourcing
Paying an outside firm to handle functions previously handled inside the company (or government) with the intent to save money or improve quality; a term often used interchangeably (and incorrectly) with 'offshoring'. Much outsourcing does involve offshoring jobs from more developed countries such that they become much lower-paid jobs in less developed countries.

offshoring
The transferring by a company of production or service provision to another country.

Table 14.4	Information Technology and Business Process Outsourcing Employment in India (thousands)				
	2002	2003	2006	2009	2012
Information Technology	106	160	379	1,004	2,717
Business Process Outsourcing	170	205	285	479	972

NOTE: Data for 2006, 2009, and 2012 are projections.

SOURCE: 'Survey: A World of Work', *The Economist*, 13 Nov. 2004, 10. Reprinted by permission of the publisher.

It is helpful to view outsourcing and off-shoring of manufacturing employment in a historical context. For much of the period since the onset of the Industrial Revolution, companies manufactured products relatively close to where they would be consumed. As many firms became international, they maintained this pattern by increasing their foreign direct investment and producing overseas for their customers overseas. However, it has become increasingly economical for companies to produce parts or products offshore that are then used or sold in the domestic market. Consider the example of the Barbie doll: raw materials are from Taiwan and Japan, assembly is in the Philippines, Indonesia, and China, and the design and final coat of paint come from the United States.

Overall, India is the favoured location for offshoring service employment while China is favoured for offshoring manufacturing employment, although China seems likely to gain more service employment. An important debate centres on the question of who benefits from offshore outsourcing. It is correct to note that employment opportunities are created in countries where they are needed, but this necessarily involves job losses in the more developed countries.

INDUSTRY AND SOCIETY

The industrial geography discussed so far has said little about either humans or the environment. Our concerns have been essentially spatial and economic—a fair reflection of the dominant geographic traditions. Increasingly, however, as we have seen, human geographers are turning their attention to issues related to political and social circumstances. Since about 1970, many researchers have realized that much industrial geography can be understood best by reference to social theory. They argue that industrial activity does not take place in a politically and socially neutral setting, according to the principles of neo-classical economics. Rather, to gain real insights into the geography of industrial capitalism, it is argued, we must turn to contemporary social theory. Unfortunately for beginners, the range of that theory is considerable, but most of it consists of variations on the Marxist themes discussed in Chapter 10. Massey (1984), for instance, produced a Marxist theoretical analysis of the changing industrial geography of the United Kingdom, while Graham et al. (1988) outlined a framework based on Marxism for investigating structural change in industries. Such industrial geographies represent significant departures from the usual Weberian economic theory; instead they give priority to relating spatial and social issues by reference to such topics as class and gender as they have evolved in a capitalist framework.

For example, the fact that transnationals actively seek low-wage locations for manufacturing activities has attracted charges of labour exploitation. High-profile companies such as Nike have been especially subject to criticism. There is also evidence that some transnationals locate plants in countries where the costs of production are kept low not only because the wages are low but also because there are fewer requirements concerning matters such as pollution control.

It is now commonplace to argue that no hypothesis about industrial (or other) locations can be complete unless it is placed in some appropriate social, political, and institutional context. Massey (1984) argued that places vary

Hundreds of members of an advocacy group called 'Global Exchange' demonstrated outside a new 'NikeTown' store in San Francisco in February 1997. Protesting against 'sweatshop labour', they claimed that the company was paying its workers in Indonesia a mere 29 cents per hour.

CP/AP photo/Ben Margot

in many respects, from workplace social relations and the division of labour to local political circumstances, and each of these must be considered in any analysis of industrial location and the changing distribution of employment. Similarly, a focus on workplace social relations often requires consideration of gender differences.

Changing local labour markets

As we saw in Box 14.1, industrial geography traditionally treated labour as either a location factor or a commodity. Reacting against these interpretations, Marxist theory focused on the spatial division of labour—the idea that regional development and employment distribution are spatial expressions of the labour process, subject to control and manipulation. More recently, industrial geographers have focused on the geographies of labour markets, especially as these are affected by the transition from Fordism to post-Fordism. Table 14.5 summarizes the differences between these two phases.

Gendered employment

In most of the more developed world today, a majority of women of working age are in the labour force. This marks a significant change: in Canada, for example, female participation rates in the labour force more than doubled between 1951 and 1981. Yet for the most part the type of work done by females has not changed: a majority still are employed in clerical, service, and low-skill jobs. Why is this so? One clear reason is discrimination by employers. In an account of the steel industry in Hamilton, Ontario, Pollard (1989) cited evidence of sexist hiring practices by the principal steel company, Stelco. Between 1961 and 1978, Stelco received between 10,000 and 30,000 applications from women for production jobs—none of whom was hired. In the same time period, about 33,000 men were hired for production jobs. Following considerable publicity and intervention by the union and the Human Rights Commission, some women were hired. But the evidence suggests that the underlying social problem is much larger. While on the

Table 14.5	Labour Markets: From Fordism to Post-Fordism	
Factor	**Fordist Labour Markets**	**Post-Fordist Labour Markets**
Labour process	Mass production involving large workforces within firms. Productivity achieved by division of labour into detailed standardized tasks.	Flexible, specialized mass production. Productivity achieved by division of labour within firms into core and peripheral categories, with functional flexibility of core workers and numerical flexibility of peripheral workers.
Employment	Substantial proportion in manufacturing. Predominantly male. Most jobs full-time. High degree of job security. Generally full employment.	Small proportion in manufacturing, most in services. Increased female participation in labour force. Many jobs part-time and temporary. Increasing job insecurity. High unemployment.
Wages	Most set by collective bargaining (by industry-wide or nation-wide unions). Income inequalities stable or falling.	De-collectivization and localization of wage determination. Increasing polarization between high-paid full-time workers and low-paid part-time workers.
Labour relations	Highly formalized and confrontational. Labour organized into strong unions. Collective strike action by workers common and protected in law.	Much more individualistic and co-operative. Unions are in decline and solidaristic labour cultures in retreat.
Labour market regulation	Macroeconomic demand management to maintain full employment. Welfare-oriented. Benefits unconditional and funded by progressive taxation system. Extensive employment protection and workplace regulation.	Abandonment of full employment policies. Shift from welfare to workfare and training. Benefits increasingly conditional. Deregulation of employment and workplaces to promote labour flexibility.
Spatial features	Relatively self-contained local labour markets. Distinct local employment structures and labour cultures. Typified by centres of mass production industry. Local disparities in unemployment are minimal.	Local labour markets are still important, but segmentation is greater within than between them. Typified by new local centres of flexible production and service activity. Substantial local disparities in unemployment and growth of localized concentrations of social exclusion.

SOURCE: Adapted from R.L. Martin, 'Local Labour Markets: Their Nature, Performance, and Regulation', in G.L. Clark, M.P. Feldman, and M.S. Gertler, eds, *Oxford Handbook of Economic Geography* (New York: Oxford University Press, 2000), Table 23.1, 459.

job, women were made to feel uncomfortable not only by men in the workplace but by members of the community, especially wives of male employees. Clearly, larger social traditions are at stake in any attempt to change the gendered division of labour.

Industry and environment

Industrial impacts on the environment were discussed in some detail in Chapter 3, and the case of China is noted earlier in this chapter. Here we note briefly the locational implications of the environmental consequences of industrial activity. In many countries, especially democratic countries, industries are facing new rules designed to control pollution by regulating the quantities of waste produced and the disposal of waste products. In principle, environmental considerations add a new cost factor that may affect both location and production decisions. Interestingly, in a study of the effects of the National Environmental Policy Act in the United States, Stafford (1985) concluded that the policy would not lead to major locational changes. As regulations become increasingly more rigorous, however, locational effects will likely become apparent.

The Geography of Uneven Development

Much of our discussion has hinted at the relationship between industry, larger issues of economic change, regional disparities, and the possible value of regional planning. It is immediately evident that spatial inequalities exist at a variety of scales with respect to many factors that affect quality of life. How do we explain these differences and deal with them?

EXPLAINING VARIATIONS IN REGIONAL ECONOMIC GROWTH

Explaining spatial variations in economic development and quality of life is not easy. Geographers are aware that disparities are common at all spatial scales, from the basic global division into two worlds all the way to local differences between, for example, residential areas in a city. Following are six of the many possible explanations for these disparities.

Physical geography

Knowing that human geographers no longer apply environmental determinism simplistically,

you may be surprised by our first proposed explanation. Nevertheless, as we have seen on several occasions in this book, there is often a correlation between physical geography and a particular human pattern, such as the pattern of population distribution and density. On the global scale, there are obviously relationships between economic development and two aspects of physical geography: climate and proximity to a coastline. This is not to suggest that physical geography causes development, but clearly it can play an enabling or constraining role.

A GIS database of information on aspects of physical and human geography for 152 countries (accounting for 99.7 per cent of the world's population) shows that climate and proximity to coast are two essential considerations in a discussion of economic development (Mellinger et al., 2000). Temperate areas within 100 km of sea-navigable waterways are responsible for more than 50 per cent of global economic output, even though they account for only 8 per cent of the inhabited area of the globe. Findings such as these suggest that other areas are unsuited to long-term economic development: interior regions, for example, have high transportation costs, while tropical regions are more subject than others to infectious diseases and the productive quality of their soils is low.

Developmental stages

Societies have often been seen as travelling along a sequential path of development. Marx, for example, envisaged society passing from primitive culture to feudalism, to capitalism, to communism. Rostow (1960) saw a transition through five stages, from traditional society to mass-consumption society (Box 14.9).

Others have seen society as experiencing shifts in the dominant occupational category, from primary to secondary to tertiary. Geographers have considered these various generalizations and contributed to them. In general, the sequence is as follows:

1. In the early stages of development, economy and society are fragmented, trade is limited, and primary activities dominate.
2. Subsequently, as transport improves and regions become more specialized and less subsistence-oriented, manufacturing develops and trading becomes important.
3. Finally, the service sector develops and regions become highly specialized.

Box 14.9 | The Rostow Model of Economic Growth

The Rostow model is a classic example of generalization about the economic development of capitalist states. On the basis of the European experience, Rostow proposed five stages:

1. *Traditional society:* Subsistence agriculture, domestic industry, and a hierarchical social system; a stable population/resource balance.
2. *Preconditions for take-off:* Localized resource development because of colonialism or activities of a multinational corporation; export-based economy; often a dual economy (the Canadian Shield and the Canadian North might be considered to be in this stage).
3. *Take-off to sustained growth:* Exploitation of major resource; possibly radical and rapid political change (the NICs may be in this stage or the next).

4. *Drive to maturity:* Creation of a diverse industrial base and increased trade.
5. *Age of high mass consumption:* Advanced development of an industrial economy; evident in the more developed world when the model was developed in the 1960s.

This model is useful in a human geographic context because it links easily to the demographic transition model (Chapter 4), the mobility transition model (Chapter 5), and the global diffusion of Western cultures (Chapter 7 especially). As with these other models, however, the simplifications in the Rostow model can be risky: what happened in Europe and North America may not happen elsewhere (as Box 4.2 details with regard to fertility). It is also worth remembering that economic growth and development are not necessarily the same thing; recall the discussion of the human development index in Chapter 5.

This idea of developmental stages has also been subject to considerable criticism, however—recall the discussion of world systems theory in Chapter 5.

Staples theory

An alternative view is the staples theory of economic growth. A staple is a primary industrial product that can be extracted at low cost and for which there is a market demand. Originally developed by Canadian historians, the staples theory appears particularly applicable to areas of European overseas expansion. Staple success means economic growth both in the areas of extraction and in the export centres; hence staple production has direct impacts on regional growth. Indeed, other scholars have built models describing this process. Pred (1966), for example, focused on the staple's multiplier effects.

Core and periphery

A fourth general approach to regional differences in economic growth, proposed by Friedmann (1972), is based on the core–periphery concept. A core region is a dominant urban area with potential for further growth. Peripheries include areas of old established settlement characterized by stagnant, perhaps declining, economies, some of which may be former staple production areas; these are called downward transition areas. There are two other

types of peripheral regions: upward transition regions are linked to cores and continue to be important resource areas, and resource frontier regions are peripheral new settlement areas. 'Peripheral regions can be identified by their relations of dependency to a core area' (Friedmann, 1972: 93).

Growth poles

A fifth approach argues that growth does not occur everywhere at the same time; rather, it manifests itself in points of growth. Thus, a new area of resource exploitation will induce economic growth, but not necessarily at the source. Urban centres often serve as growth poles, as Calgary has, for example, for the Alberta oil industry. Although a substantial literature has developed around the idea of growth poles, the concept is still not entirely clear.

New economic geography

Since about 1990, economists, not geographers, have initiated an influential research interest called 'new economic geography'. The impact of this approach is evident in the fact that one of the leading advocates, Paul Krugman, was awarded the 2008 Nobel Prize in economics for his work on trade patterns and location of economic activity. Also, the World Bank employed this approach in the 2009 edition of its flagship annual publication, the *World Development Report*. Subtitled *Reshaping Economic Geography*, the *Report* argues that the best way to promote long-term growth is to acknowledge the inevitability of uneven development and to develop policies that enable geographic concentration of economic activity. Challenging the assumption that economic activities are best spread spatially evenly, the *Report* asserts that attempting to prevent concentration only results in limiting prosperity, and stresses at the same time that development can still be spatially inclusive through government promotion of economic integration. The *Report* explains uneven development in terms of differences in population density, distances between places, and political and other divisions.

Overall, human geographers have not been attracted to this new economic geography, at least partly because it ignores issues of spatial injustice with its failure to recognize that many places are peripheral and undeveloped because of a history of exclusion. As was noted earlier, no hypothesis about industrial (or other) locations can be complete unless it is placed in some appropriate social, political, and institutional context. Certainly, some instances of regional inequality within countries result principally from social and power relations, as evidenced, for example, in the spatial pattern of racism and related landscapes (Rigg et al., 2009). Notwithstanding these criticisms, the new economic geography is able to provide numerous insights into why some areas develop more than others.

Each of these six approaches has been widely applied, but each has deficiencies. As generalizations, they provide insights but not necessarily explanations. Geographers investigating particular spatial and social issues tend to use the most appropriate model for their specific purposes. We now turn to some specific circumstances of uneven development in the context of the less developed world.

DEVELOPMENT ISSUES IN THE LESS DEVELOPED WORLD

The distinction between more and less developed is never straightforward. Because of their recent industrial success, the NICs are frequently referred to as economic miracles, yet they are still generally regarded as belonging to the less developed world. More typical less developed countries have been much less successful industrially, and no miracles are expected of them in the near future.

Certainly, one obvious distinction between the more and the less developed worlds is the degree of industrialization: many countries in the latter group have not yet experienced anything like an industrial revolution. But it is widely hoped that industrialization might lead to employment for the unemployed or underemployed rural poor. In addition, many in the less developed world believe that industrialization would demonstrate economic independence, encourage urbanization, help to build a better economic infrastructure, and reduce dependence on overseas markets for primary products. At the same time, governments and development agencies in most poor countries need to address the issue of significant regional differences within these countries.

Problems

Potentially, industrialization has many advantages. So why is it so difficult for many countries to achieve it? A major obstacle is the legacy of former colonial rule. Historically, European colonial countries saw their colonies both as producers of the raw materials needed for their domestic manufacturing industries and as markets for those industries.

This is a difficult legacy to move beyond. Most countries in the less developed world have neither the necessary infrastructure nor the capital required to develop it. Furthermore, fundamental social problems resulting from limited educational facilities do not encourage the rise of either a skilled labour force or a domestic entrepreneurial class. Finally, the domestic markets in many of these countries lack the spending capacity to make industrial production economically feasible.

Employing the logic of the new economic geography, the lack of development throughout most of sub-Saharan Africa is explained in terms of low population density relative to other major world areas, long distances between the major settlement locations, and deep divisions related to ethnic differences and the political partitioning of Africa—consider, for example, that Africa has more countries relative to the total land mass than any other major world region, and also that each African country has an average of four neighbours (in Latin America the average is 2.3) (World Bank, 2009: 283).

Progress

Despite the many difficulties faced, since World War II considerable industrialization has taken place in many less developed countries. Major successes have been achieved in import substitution—manufacturing goods that were previously imported, often with the help of protective tariffs. Examples include light industries that are not technologically advanced, such as food processing and textiles. Heavy industry has been more difficult to develop, even though many of the raw materials used in the more developed world come from the less developed world. In general, the less developed world contradicts the standard primary–secondary–tertiary sequence. The more typical sequence is primary–basic tertiary–secondary; perhaps a quaternary sector will develop in due course.

Since about 1980 many of these countries have been under sustained pressure from both the IMF and the World Bank to adopt top-down free-market economic policies that will result in structural adjustment. Specifically, governments are encouraged to open their economies to increase international trade, privatize previously state-owned enterprises, and reduce government spending. The logic is that such policies will generate wealth and development that will trickle down through all sectors of the economy to benefit all the people. This approach has been criticized on the grounds that in fact most of the wealth stays with the elite and any trickle-down benefits are minimal.

We saw in Chapter 10 that some failings of the green revolution have been blamed on the preponderance of top-down strategies. An alternative approach favoured by many national development agencies and by most NGOs working in poor countries, such as Oxfam, is to initiate development from the bottom, building on local community strengths and being sensitive to particular local circumstances. Many poor countries are currently employing both approaches to regional economic and social development, and it may be that both are necessary to achieve meaningful growth.

Planning in China

The clearest examples of regional development policies involve the planned economies in socialist and communist states. By definition, these states do not rely on the capitalistic dynamics of private ownership and entrepreneurship. Rather, they depend on the rationalization of industry and the implementation of comprehensive planning. The majority of such states have been in the less developed world. In China, regional policies have included

Women assemble dolls at the Sinomex Toy Factory in Hermosillo, Sonora, Mexico.

Keith Dannemiller/Alamy/GetStock

efforts to industrialize large cities (as during the Great Leap Forward of 1958–9, when perhaps 20 million rural residents moved into cities) and to limit the growth of large cities. The seventh five-year plan in China (1986–90), which focused on uneven development and rapid urbanization, was a direct reflection of the fact that earlier plans had largely failed to address two issues: income differences between urban and rural areas, and the domination of industry by urban areas located on the country's coast. The constants in Chinese planning are state direction, collective ownership of the means of production, and control over population movements—all features rare in capitalist countries.

DEVELOPMENT ISSUES IN THE MORE DEVELOPED WORLD

Spatial variations in economic development and quality of life are normal outcomes of any process of change. Central or core areas will usually experience innovations before outer or peripheral areas, and the uneven distribution of resources prompts spatial variations. Unfortunately, in many parts of the world these natural spatial variations have resulted in differences that society considers inappropriate. Hence, most countries have regional development policies.

The Canadian example

Governments have long intervened in the processes of economic change. In Canada, for example, companies were granted monopolies, the progress of settlement was dictated, and tariff barriers were installed—all before 1900. In the twentieth century, however, such intervention began to play a major role in redressing spatial imbalances. Typically, regional development policies have been designed to encourage growth in depressed, usually peripheral, areas.

A number of regions in Canada qualify as relatively underdeveloped. Indeed, some would contend that almost all of Canada outside the St Lawrence Lowlands qualifies for this description. In the case of the Atlantic provinces—Newfoundland, New Brunswick, Nova Scotia, and Prince Edward Island—economic and related social problems stem from a peripheral location some 1,600 km (1,000 miles) from the major centres, a poor resource base, national tariff policies, and increasingly centralized transport and production systems. Inequality is evidenced by low wages and high unemployment. The Canadian government has attempted to address these issues by stimulating growth through low-interest loans to farmers and others and tax subsidies for industrialists. Some results have been achieved, but the basic problems have not been solved. The principal goals are to foster a self-sustaining entrepreneurial climate, more successful medium and small business enterprises, lasting employment opportunities, and an expanding competitive economy. These are not simple goals in a globalizing world that tends to reinforce the dominance of already successful regions rather than spread investment and opportunity to peripheral locations.

In the eyes of many Canadians, the Canadian Shield and the North—which together make up some 75 per cent of Canada—are frontiers that produce primary products for export that benefit primarily the Canadian core area; in many respects, then, these resource extraction areas are part of the less developed world. Others, notably Native populations, see these areas as a homeland. These very different images are not easy to reconcile (see Bone, 2009).

The European example

The European Union is a second example of a political unit that contains core and peripheral regions and, as a result, regional inequities. As of 2007 there are 27 members, with Romania

and Bulgaria the most recent additions. Recent moves towards enlargement and integration have been taking place within a context of significant regional disparities. The number of peripheral regions with low incomes and high unemployment rates increases with each expansion. Several countries that joined the EU in the 1980s, especially Greece (1981), Spain (1986), and Portugal (1986), brought serious regional development problems, while the subsequent addition of many East European countries has further increased the breadth of regional variations. There are, therefore, very strong arguments for a European regional policy. To date, however, the principal policy has been simply to provide funds to depressed regions. An effective regional policy is clearly needed.

Decaying industrial areas

In core and peripheral regions alike, many areas are depressed as a result of outdated industrial infrastructures (see Box 14.8). A useful way to consider industrial change is to recognize four stages:

1. Infancy: initial primary activities and domestic manufacturing.
2. Growth: the beginnings of a factory system.
3. Maturity: full-scale development of manufacturing and related infrastructure.
4. Old age: decline and inappropriate industrial activity.

Today, the typical depressed region is located on a coalfield, lacks diversification, and is overly reliant on heavy industry. Unemployment is high and out-migration is normal. Depressed regions are similar in principle to peripheral regions that were formerly dependent on a staple, as both are unable to diversify when circumstances change.

The United Kingdom, which has both peripheral and depressed areas, has experienced considerable government intervention since the 1930s. Areas are variously designated as special development areas, development areas, and intermediate areas—designations that change according to political circumstances. Government intervention in the United Kingdom takes the form of development controls, financial incentives, and creation of industrial estates.

We end this chapter with an observation about developmental inequality from Deng Xiaoping, the political leader of China for more than two decades before his death in 1997. Apparently Deng once remarked that it was perfectly in order for some parts of China to become rich before other parts; in fact, this was the way it had to be (Freeberne, 1993: 420). The 2009 *World Development Report* is saying much the same thing.

CHAPTER 14 SUMMARY

LEVELS OF PRODUCTION

Traditionally, three levels are recognized: primary, which is extractive; secondary, which constitutes manufacturing; and tertiary, which is service. A fourth (quaternary) level involves the transfer of information: communications.

LEVELS OF ECONOMIC ORGANIZATION

Two levels are recognized. Households produce and reproduce; as economic organizations they are of decreasing importance. Firms make up the commercial sector; manufacturing firms operate in factories.

FACTORS RELATED TO LOCATION

Industrial location theory explains why factories are located where they are. Traditionally, the relevant factors have included transport; distance from raw material sources, energy supplies, and the market; availability of labour and capital; the nature of the industrial product; internal and external economies; entrepreneurial uncertainty; and governmental considerations.

LEAST-COST THEORY

The typical location decision aims to maximize profits by minimizing costs or maximizing sales (or both). Least-cost theory, developed by Weber and published in 1909, is a normative theory that identifies where industries ought to be located. Following a series of simplifying assumptions, Weber concluded that industries locate at least-cost sites determined by transport costs, labour costs, and agglomeration–deglomeration benefits. Solving locational problems required defining concepts such as material index, locational figure, isodapane, critical isodapane, and isotim. Other researchers, especially Hoover, have improved Weber's formulation of the transport-cost variable.

MARKET-AREA ANALYSIS

Market-area analysis, the second major theoretical approach, centres on profit maximization rather than cost minimization. Lösch is the principal theorist.

BEHAVIOURAL APPROACHES

Behavioural approaches to the industrial location problem focus on the subjective views of the people involved and the common preference for satisficing rather than optimizing behaviour. Behavioural issues are important, albeit more difficult to quantify or theorize than strictly economic considerations. Contemporary industrial firms are, typically, complex decision-making entities.

INDUSTRIAL REVOLUTION

Prior to the Industrial Revolution, industrial activity was domestic, small in scale, and dispersed. A revolution in industrial activity and related landscapes occurred between 1760 and 1860, beginning in Britain. Factories replaced households as production sites, mechanization increased, and localized energy sources were used. All these changes were related to the rise of capitalism, which emphasized individual initiative and profits. New industrial landscapes appeared on the coalfields of Britain and were accompanied by city growth, rural-to-urban migration, and a series of transport innovations. Inner-city slum landscapes developed in many of the new agglomerations on coalfields and in the traditional textile areas of northern England.

ENERGY

Coal was the main fuel during the Industrial Revolution but was replaced by oil in the 1960s. OPEC is a major player in international affairs because of the importance of oil in the contemporary world. Some countries, such as New Zealand, are dependent on hydro-generated electricity supplies, which may be disrupted by droughts.

WORLD INDUSTRY

Industry is responsible for a disproportionate share of manufacturing employment and output. Most of this activity continues to take place in a capitalist economic and social framework. The global manufacturing system is dominated by North America, Western Europe, and the Asian Pacific region. Multinational corporations are increasing in importance.

CANADIAN INDUSTRY

The dominant characteristics of Canada's industrial geography include abundant natural resources, a well-developed extraction industry, a branch-plant economy reflecting the country's proximity to the United States, a dispersed national market, a dispersed primary sector, and highly concentrated secondary and tertiary sectors.

NEWLY INDUSTRIALIZING COUNTRIES

Japan was the first of the NICs and has been followed by other Asian countries (notably South Korea), some Southern European countries, and Brazil and Mexico. Export-processing zones using inexpensive and relatively unskilled labour have been established in some countries (for example, the maquiladoras in Mexico).

INDUSTRY IN THE LESS DEVELOPED WORLD

Industrial growth is highly desired by many less developed countries to bolster weak economies, provide employment, demonstrate economic independence, encourage urbanization, help create a better economic infrastructure, and reduce dependence on overseas markets for their primary products. But industrialization is difficult to achieve, not least because of the legacy of colonialism. Many countries have experienced their greatest industrial successes in import substitution. India has thrived in this regard since 1951, while China opened the doors to international trade and foreign investment and also reduced state control of industries during the 1980s. The most dynamic industrial region in the world today is the Pearl River Delta in south China, just north of Hong Kong.

INDUSTRIAL RESTRUCTURING

Technological changes in production, transaction, and circulation are currently prompting some major changes in the global economy and in national and regional industrial geographies. Offshore outsourcing of both manufacturing and service employment is increasingly important.

INFORMATION EXCHANGE

Industrial location decisions used to be made largely on the basis of transport costs, but today the costs of information exchange and labour are much more important. Efficient electronic transfer of reliable information allows firms to decentralize by seeking out areas of low labour cost. Where information exchange requires personal contact, location patterns tend to be centralized.

SERVICE INDUSTRIES

As part of the process of economic restructuring and the transition from Fordism to post-Fordism, industrial societies are moving to a post-industrial service stage. Because the service industry is so diverse, a wide range of factors may determine locations. Central place theory offers the most important set of explanatory concepts, but behavioural approaches also seem relevant. New information technologies are undoubtedly affecting the location of services.

SOME SOCIAL ISSUES

Industrial geography, like other branches of economic geography, has neglected social issues. Now that industrial geography is becoming more concerned with issues such as the gender division of labour, it is increasingly turning to various social theories.

SPATIAL INEQUALITIES

Correcting spatial inequalities is seen as a government responsibility. Governments in capitalist countries frequently implement policies to improve the economic status of regions that are less developed, either because they are peripheral or because their industrial infrastructure is outdated. The most obvious instances of planning the spatial economy take place in socialist countries. Economic restructuring continues to contribute to uneven development.

QUESTIONS FOR CRITICAL THOUGHT

1. How has the 'industrial location problem' changed from the late nineteenth century/early twentieth century to today? What (if any) locational considerations have remained in place over the past century or so?

2. How relevant/useful is Weber's least-cost theory to understanding industrial location today?

3. What were the effects of the Industrial Revolution on cities and agriculture in the eighteenth and nineteenth centuries?

4. How has the global pattern of industrial production been affected by globalization?

5. To what extent will the expansion of industry in developing nations enhance social and economic well-being in those nations?

6. What are the advantages and disadvantages of promoting tourism as a stimulus for economic growth and development?

7. Why does 'uneven development' exist? Is it possible (and if so, under what circumstances) that all nations could be more or less equally developed?

FURTHER EXPLORATIONS

Bale, J. 1981.*The Location of Manufacturing Activity*. Edinburgh: Oliver and Boyd.

An easy-to-read basic text full of examples of industrial location issues; focuses on UK examples.

Bathelt, H., and A. Hecht. 1990. 'Key Technology Industries in the Waterloo Region: Canada's Technology Triangle (CTT)', *Canadian Geographer* 34: 225–34.

A detailed analysis of 33 high-technology firms showing that the smaller firms are the most research-intensive, that all the firms are strongly linked to local economic activities, and that the strongest locational factors are the availability of skilled labour and local residence or education.

Blackbourn, A., and R.G. Putnam. 1985. *The Industrial Geography of Canada*. London: Croom Helm.

A very readable account that includes discussions of each Canadian region and raises many of the issues that face manufacturing industries in the more developed world.

Britton, J. 1996. *Canada and the Global Economy: The Geography of Structural and Technological Change*. Montreal and Kingston: McGill-Queen's University Press.

Focuses on the importance, for Canada, of openness to external influences, regional variations in resource bases and urbanization, rapid changes in technology and markets, and government strategies.

Cleary, M. 2005. 'Southeast Asian Development: Miracle or Mirage', in M. Philipps, ed., *Contested Worlds: An Introduction to Human Geography*, Burlington, Vt: Ashgate, 229–50.

Focuses on reasons behind the rise of Southeast Asian economies and impacts on environments and peoples.

Crosby, A.W. 2006. 1993. *Children of the Sun: A History of Humanity's Unappeasable Appetite for Energy*. New York: Norton.

A big-picture look at the history of the world's never-ending search for energy sources and our resultant dependence on those sources.

Daniels, P. 1993. *Services in the World Economy*. Oxford: Blackwell.

A thorough account of the service component of industrial activity, with detailed examples.

Edgington, D.W. 1993. 'The New Wave: Patterns of Japanese Direct Foreign Investment in Canada during the 1980's', *Canadian Geographer* 38:28–36.

An analysis of how Japanese investment during the 1980s helped diversify and strengthen the economic importance of southern Ontario.

Harrington, J.W., and B. Warf. 1995. *Industrial Location: Principles and Practices*. New York: Routledge.

A valuable, wide-ranging book that includes both conceptual and empirical analyses.

Hay, A.M. 1976. 'A Simple Location Theory for Mining Activity', *Geography* 61: 65–76.

A useful contribution to a largely neglected area; illustrated by an educational game.

Hoggart, K. 2005. 'Inequalities at the Core: A Discussion of Regionality in the EU and UK', in M. Philipps, ed., *Contested Worlds: An Introduction to Human Geography*. Burlington, Vt: Ashgate, 192–227.

This focus on regional inequalities highlights the multiple dimensions to inequity and the relations between local areas, regions, and states.

Jones, K. 1989. 'Editorial', *The Operational Geographer* 7, 2: 2.

An introduction to a series of useful articles on services and marketing activities in Canada, especially in Ontario.

Krugman, P. 1995. *Development, Geography, and Economic Theory*. Cambridge, Mass.: MIT Press.

A thoughtful book by an important economist that reflects some non-traditional views; regrets the rarity of 'spatial' considerations in economics; useful also for Chapters 10 and 11.

Massey, D. 1995. *Spatial Divisions of Labour*, 2nd edn. New York: Routledge.

Now considered a classic; a valuable introduction to the type of industrial geography that emerged during the 1980s, emphasizing labour and employing a Marxist approach (as opposed to the previous neoclassical focus on raw materials, energy sources, and transport).

Mather, C. 1993. 'Flexible Technology in the Clothing Industry: Some Evidence from Vancouver', *Canadian Geographer* 37: 40–7.

An account of the impact of flexible technology on the Vancouver textile industry; shows that some old technologies are still important and that impacts on labour are largely negative.

Morgan, N., and A. Pritchard. 1998. *Tourist Promotion and Power: Creating Images, Creating Identities*. Toronto: Wiley.

A thoughtful and challenging overview of images in the tourist industry with short case studies; draws on ideas both from the cultural studies and from the human geographic traditions.

Norcliffe, G. 1993. 'Regional Labour Market Adjustments in a Period of Structural Transformation: An Assessment of the Canadian Case', *Canadian Geographer* 38: 2–17.

A detailed account of changes associated with the transformation from mass production to flexible production; demonstrates that the Canadian experience is distinctive because the transformation has been mediated by a specific staple regime.

Piper, L. 2009. *The Industrial Transformation of Subarctic Canada*. Vancouver: University of British Columbia Press.

Insightful study of the rise of twentieth-century industrial economies in the region including Lakes Winnipeg, Athabasca, Great Slave, and Great Bear with emphasis on resource exploitation, ecosystems, and sustainability.

Prestowitz, C. 2005. *Three Billion New Capitalists: The Great Shift of Wealth and Power to the East*. New York: Basic Books.

One of many recently published books focusing on the shift of economic power from West to East, with an emphasis on the rise of China.

Sandberg, L.A. 1989. 'Geographers' Perception of Canada in the World Economic Order', *Progress in Human Geography* 13: 157–75.

An interesting account of recent and current perceptions of Canada in the global economic context, which aids understanding of regional disparities.

Shaxson, N. 2007. *Poisoned Wells: The Dirty Politics of African Oil*. New York: Palgrave Macmillan.

Focusing on six countries in West Africa, Shaxson argues that the increasing importance of African oil in global markets is not good news for the people of the producing countries as it merely allows corrupt political leaders to prosper.

Smith, D.M. 1981. *Industrial Location Theory: An Economic Geographical Analysis*, 2nd edn. New York: Wiley.

A comprehensive account of all aspects of industrial location theory.

Watts, H.D. 1987. *Industrial Geography*. London: Longman.

An excellent survey of the field with a distinctive focus on the human consequences of contemporary industrial change.

Weaver, D.B. 1998. *Ecotourism in the Less Developed World*. New York: CAB International.

Places ecotourism in a larger context, as one component of alternative tourism.

Wolfe, D.A., and M. Lucas, eds. 2005. *Global Networks and Local Linkages: The Paradox of Cluster Development in an Open Economy*. Montreal and Kingston: McGill-Queen's University Press.

A useful collection of articles concerned with clusters of industrial activity across Canada.

Yeung, Y.-M. 2005. 'Emergence of the Pan-Pearl River Delta', *Geografiska Annaler* 87B: 75–9.

Good factual statement detailing industrial growth in this dynamic region.

ON THE WEB

INDUSTRIAL REVOLUTION

Modern History Sourcebook

🔍 www.fordham.edu/halsall/mod/modsbook14.html

Detailed overview of the Industrial Revolution.

ENERGY

BP Facts and Figures

🔍 www.bp.com/productlanding.do?categoryId=6929&contentId=7044622

The annual report produced by BP provides a wealth of data and analysis concerning energy. Covers oil, natural gas, coal, and primary energy. This link is to the 2008 report. To view the most recent report, search for 'BP report' in any search engine.

International Energy Authority

🔍 www.iea.org/

The International Energy Authority, established in November 1974, has gained recognition as one of the world's most authoritative sources for energy statistics. Its massive annual studies of oil, natural gas, coal, and electricity are indispensable tools for scholars as well as energy policy-makers and companies involved in the energy field.

INDUSTRY TODAY

Canada

🔍 info.ic.gc.ca/

Industry Canada publishes this site, which mostly covers news items and governmental matters.

OECD

🔍 www.oecd.org/topicstatsportal/0,3398,en_2825_495649_1_1_1_1_1,00.html

This OECD web page has links to a variety of industrial data.

EconomyWatch

🔍 www.economywatch.com/world-industries/global-industry.html

An overview of global industry with links to specific regional economies.

Embassy of India

🔍 www.indianembassy.org/dydemo/industry.htm

Overview of industrial activity in the 'Dynamic Democracy', India.

China Daily

🔍 www.chinadaily.com.cn/

This site provides regular news articles about all aspects of Chinese life, including industrial change, as do other news sites such as the BBC and CNN.

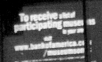

Conclusion

WHERE NEXT?

The Introduction to this book identified three recurring themes: relations between humans and land, regional studies, and spatial analysis. By now you should have a clearer idea of why these themes are so central to human geography.

The Introduction also pointed out that one goal of human geography—and of this book in particular—is to help us understand our changing world as it is today and how it came to be so. Human geography is a discipline of tremendous breadth and diversity, with especially close links to the natural sciences, particularly physical geography, and a healthy academic pedigree. A human geographic perspective—a geographical imagination—is an invaluable asset. The editor of a major world newspaper, *The Times* of London, has written of school geography:

It seemed a discipline that took a child out of the classroom, into the street and the park and said: 'Look what's here! How did it come to be here? What is it made of?' On such empirical enquiry rests all learning. Geography offered such empiricism in the most comprehensible and immediate form. It also offered the basis for argument, for disagreement, for controversy, for the spirit of dialectic that makes for true understanding, not rote learning. (Jenkins, 1992: 193–2)

A giant billboard featuring a carbon counter near Madison Square Gardens in New York City digitally displays the amount of greenhouse gases in the atmosphere.
Spencer Platt/Getty Images

If it is not possible to answer the question that titles this final chapter, we must at least attempt to comment on current changes both in our subject matter itself—human behaviour as it affects the earth's surface—and in our methods of studying it.

Changing Human Geographies

Human geography is an exciting field. If the first chapters of this book suggested to you that human geographers today must look back nostalgically to the golden days when every journey meant new facts to be discovered, catalogued, and understood, by now you probably realize that the end of the exploration/invasion era did not by any means signify an end to the discovery of new facts. Humans are constantly changing landscapes in a myriad of ways—cultural, social, political, and economic. There may not be a wealth of new facts over the horizon, waiting to be discovered, but the facts within the horizon are constantly changing, as is our understanding of those facts. Indeed, the facts are changing at an ever-increasing pace: because there are so many of us to cause change, because technology is always improving, and because we live in a globalizing world that is shrinking as a result of advances in communications and other forms of movement. Six examples of changing human landscapes—the human geography of the future—will clarify these generalizations.

POPULATION

Human geography is about human beings, and probably the single most important change related to human beings is the continuing increase in our numbers. As we saw in Chapter 4, the rate of natural increase in the world population peaked in the late 1980s and is now declining (the fertility transition), but the total number of people continues to increase from the current 6.8 billion. The United Nations estimates 8.0 billion by the year 2025, 9.4 billion by 2050, and a stable population of about 10 billion by 2200. Most of the current growth is taking place in the less developed world; many countries in the more developed world have natural increase rates close to zero (see Table 4.5). The composition of the world population is also changing as the numbers of elderly people increase, and important social transformations are underway as gender roles and family patterns change.

With population growth below replacement levels, population aging, and corresponding declines in the labour force, immigration is likely to become an increasingly important—and divisive—issue in the more developed world. Meanwhile, the shift will continue away from industrial activity and towards 'knowledge' or 'information' as the economic engine of society. Since knowledge travels freely, without regard for physical borders, the prospects for upward social mobility through higher education will increase—but so will the competition that individuals face.

As recently as about 1750, most people lived in rural areas and derived their living from agriculture. Many of the different human group identities apparent today developed in these conditions of relative isolation from others—before the massive changes initiated by the Industrial Revolution. These differences of language, religion, and ethnicity have left a legacy of cultural and social diversity that continues to shape the contemporary world (recall the discussion of regional identities and political aspirations in Box 8.8).

URBANIZATION

The percentage of the world's population living in urbanized areas has reached 50 per cent and continues to increase (see Figure 11.1); the number of very large cities will also continue to increase, especially in the less developed world (see Table 11.6 and Figure 11.6). These ongoing increases will bring further changes in the urban way of life, and if the current problems of housing availability and basic infrastructure evident in urban areas of the less developed world are any indication, many of these changes will be in an undesirable direction. In many more developed countries, cities continue to grow outward and upward, often in response to a new set of postmodern processes; edge cities are now commonplace in areas where sufficient space is available. Such cities are centres of consumption, not production.

HUMAN IMPACTS

More people mean more demands both on resources and on agricultural and industrial outputs. Currently, energy use per capita is much higher in the more developed world than it is in the less developed world. Necessarily, then, energy demands in the less developed world are increasing as expectations rise and industrialization proceeds. The basic conclusion is unavoidable: the impacts on our fragile home, as discussed in Chapter 3, will also increase.

Although it appears that awareness of human impacts is encouraging a shift, at least in the more developed countries, from an anthropocentric to an ecocentric view of the world—an idea discussed in the context of sustainable development in Chapter 3—there is still a long way to go. Indeed, a recent report in the journal *Science*, based on extensive and intensive research on global fisheries, indicated that without major change in current practices there will be little sustainable seafood or fish by mid-century. As one of the authors noted, without real change in how ocean species are managed and harvested, 'We'll be eating sea squid soup and jellyfish pie' because all of the recognizable seafood will be gone (Calamai, 2006). To put the issue in slightly different terms, many human geographers now challenge the basic premise on which industrial societies have been established: the idea that progress depends on conquering nature and increasing production. This idea, centred on what has been called the 'treadmill of production' (Schnaiberg and Gould, 1994: v), can be seen as the cause both of environmental degradation and of inequalities in the social well-being of different groups. Problems of the global environment and social justice are not separate issues.

POLITICAL WORLDS
Every day, the media report on political events of geographic interest. Perhaps most notably, the political map of the world changes: in some cases countries choose to sacrifice some of their national independence as they integrate with other countries, with the most advanced example of this trend being the European Union; in other cases areas within countries assert their independence and strive for nationhood.

This trend towards political disintegration was one factor behind the terrible conflict unleashed by the political collapse of the former Yugoslavia in the early 1990s as different linguistic, religious, and ethnic groups appealed to perceived ethnic identities and historical injustices, and was seen more recently in Sri Lanka when the minority Tamils fought unsuccessfully for autonomy from the majority

Sinhalese. It seems highly probable that there will be further assertions of autonomy by groups who perceive themselves as different from the majority groups in their states.

Certainly, geography aids in any attempt to explain political conflict. For example, to understand the Iraq War that began in 2003 and its aftermath we need to know something about at least five thoroughly geographic subjects: the way the states in the area were created; the spatial distribution of various ethnic groups (Shiite Muslims in the south, Sunni Muslims in the centre, and Kurds in the north); the importance of oil and natural gas to the economies both of producing countries and of their customers in the more developed world and the location of known deposits in Iraq (in the north and south but not in the centre); the relations of dependency that exist between more and less developed worlds; and the geopolitical perception of the United States following the 2001 terrorist attacks.

SAMENESS AND DIFFERENCE
The uniqueness of human beings resides in our biological attributes and in our ability to adapt culturally. We humans have managed to avoid extinction because of our culture—essentially, our ability to develop ideas from experience and to act on the basis of them. Once established, however, culture can be an impediment to change. The fact that most people are most comfortable with familiar patterns goes a long way towards explaining our desire to identify with place in a world where, for many of us, places are changing at a bewildering pace—constantly being created and destroyed. In his book *A Question of Place*, Johnston (1991) argues that geography needs to study what he calls 'milieux': places within which ways of life are constructed and reconstructed—in short, places in their entirety.

An important foundational idea in any discussion of human identity is the fact that humans have the capacity to identify and name different groups, with the most fundamental identification and naming involving 'us' and 'them'. This distinction is typically based on what is loosely called culture, including language, religion, and traditions, and also on place; in other words, on who people are and on where they are.

Recall the social constructionist argument that rejects the essentialist claim that a specific identity or group membership is fixed

Rwandan refugees arriving in Tanzania.
© CIDA photo/Roger LeMoyne

and unchangeable, arguing instead that identities are fluid, contested, and negotiated. From this perspective, self, identity, community, and social reality all are creations of the human mind and should not be regarded as objective entities in some way separate from ourselves. This idea informs numerous geographic studies of difference; one of the great achievements of human geography today is this conceptual broadening and the related gaining of insights into the lives and experiences of people. One of the most notable aspects of the concern with geographies of identity is the realization that dominant societies construct identities of others and also construct the landscapes in which others might be obliged to live. In doing so, they construct landscapes that take certain characteristics—the characteristics of the dominant group—for granted.

Notwithstanding the increasing evidence of cultural difference, based on such considerations as language, religion, ethnicity, gender, sexuality, physical well-being, age, and place, to cite just a few of the bases for identity, it is clear that many countries believe in the existence of and need for a national culture, indeed an essentially unchanging national culture. But is there, or ought there to be, any such thing? Are there any cultural identities, national or otherwise, that have some inalienable right to exist and to remain relatively unchanged? Agreeing with the legitimacy of static, or relatively static,

national identities raises a myriad of questions concerning who has the right to define that identity. Some of the questions that this issue raises were considered especially in Chapters 6, 7, and 8.

A Changing Discipline

Chapters 1 and 2 of this book outlined the evolution of human geography and identified key philosophies, concepts, and techniques of analysis. One purpose of these accounts was to demonstrate how our discipline is always changing, largely in response to the changing needs of society. Human geography, like other academic disciplines, exists to serve society. There is, then, every reason to believe that the discipline described in this book will be a different discipline tomorrow. Attempting to predict the details of such change is hazardous, but we may at least identify six of what seem to be the most pressing societal demands on human geography today.

SPACE AND PLACE

The need for an academic discipline focusing on space and place seems self-evident. Space is important because we need to understand location, distances between locations, and movements between locations if we are to make sense of the world. For this reason the single most important geographic contribution to human life is the map. Maps are the means by which geographers make sense of the world; maps allow us to understand and explain why things are where they are and to express the closely related concepts of space, location, region, and distance. Because maps are so important, geographers have taken advantage of recent technological advances to develop the more sophisticated tool of geographic information systems (GIS).

Place is important because understanding the meaning that people assign to locations is crucial if we are to comprehend their attitudes and behaviours. All people have an innate sense of place, but human geographers are able to refine this innate sense into a knowledge that facilitates understanding of the changing human world. Our recurring discussions of globalization, regional character, and local change have continually reinforced these points.

INTEGRATING HUMAN AND PHYSICAL GEOGRAPHIES

As discussed in Chapter 1, geography occupies 'a very puzzling position within the traditional organization of knowledge . . . . It is neither a purely natural nor a purely social science' (Haggett, 1990: 9).

- *For some*, geography is a single, coherent discipline that integrates physical and human geography. For example, Stoddart (1987: 330) wrote that 'Human geography as an exclusively social science loses its distinctive identity' and 'outside a more general framework physical geography loses its coherence.'
- *For others*, disciplinary coherence is less clear. Johnston (2005: 9) noted that: 'For most outsiders, an encounter with the discipline of geography may suggest that it studies everything, from global environmental change at one extreme to the minutiae of body-space at the other. It spans the physical, environmental, and social sciences, and reaches into the humanities too,' Johnston queried whether geography 'is now coming apart at the seams'.

Perhaps the most helpful way to conceive of geography today is to acknowledge that it is characterized by diversity and fragmentation but that, at the same time, the core ideas—what are called recurring themes in the Introduction to this book—are always evident (Johnston, 2005) (see Box C.1).

It has been claimed that physical geographers are from Mars and human geographers are from Venus (see Viles, 2005). But even if these metaphors contain truth, relationships have always existed between the human and physical worlds, at all spatial scales and at all times since human life first appeared. It is true that much contemporary human geography does not make any explicit reference to physical geography. Yet human beings have always had to take physical geography into account in their shaping of the human world, and human activities have had an ever-increasing impact on the physical world. In order to grasp these fundamental relationships, a basic understanding of the physical world is essential.

To say that human activities are closely related to physical environments is not to validate environmental determinism. Rather, it is to acknowledge a central and pervasive relationship. Humans make decisions, as individuals and as groups, but they make them with a multitude of physical factors in mind. Success in adaptation—that is, coping—means adapting to both the physical environment and relevant human variables.

In a book dealing with geography and social theory, the following statement appears:

> The journey along Mulholland Drive, atop the Hollywood Hills, provides one of the world's great urban vistas. To the south lies the Los Angeles basin, a glittering carpet. To the north, the San Fernando Valley (still part of the City of Los Angeles) unfolds in an equivalent mass of freeways, office towers, and residential subdivisions. There is probably no other place in North America where such an overpowering expression of the human impact on landscape can be witnessed. And yet, the physical landscape cannot be denied. Even in this region of almost 12 million people, the landscape still contains and molds the city. (Dear and Wolch, 1989: 3)

A SYNTHESIZING DISCIPLINE

Notwithstanding debates about what geography may or may not be, the discipline of geography is important to society because it offers the best framework for investigating human impacts on global and local environments. Most ecological research benefits

Box C.1 | Moving Beyond the Introductory Level: Be Prepared to Think Again

As you worked through your human geography course, with this textbook as one guide, you may have noticed that human geography is not quite as neatly structured as you possibly imagined it to be. In this respect, human geography is no different from any other social science. Contrary to first appearances it does not consist of a set of tidy packages, each with content clearly distinguished from all the others. This textbook may be packaged into chapters, each neatly labelled to suggest the contents, but such packaging and labelling necessarily are simplifications of a very complex reality.

Simplification and categorization are essential to enable understanding of complexity. This text has to simplify and categorize. But one danger of working within the framework suggested by this book (or any other book for that matter) is that some very important tensions are lost. In particular, human geography includes several important cases of one particular type of categorization—binary thinking. In the first instance, most of these cases of binary thinking are attempts to handle complexity, but they also have some important consequences for understanding our world and ourselves. This is especially because some binaries go beyond simply dividing into two, as in the case of 'A or B', and also imply a power relationship between the two categories, as in the case of 'A or not-A'.

Perhaps the single most important binary in human geography is that of thinking about people in terms of self and others or, more bluntly put, of 'us' and 'them'. This is of the type 'A or not-A'. This binary is encouraged especially by our membership in a state—we are members of a particular state, but they are not—and refers to one aspect of our identity. In common with other 'A or not-A' ways of thinking about identity, this is not simply about difference. It is also about status. Members of a particular state often see themselves as in some way superior to those who are not members of that state and, hence, this identity binary contributes to conflict circumstances. Similar comments can be made about binaries such as white-skinned/dark-skinned (white or not-white), man/woman (man or not man), and heterosexual/homosexual (heterosexual or not heterosexual).

Another influential human geographic binary concerns the concepts of culture/economy. These concepts are typically assigned to different boxes—recall that Marx packaged economy in the infrastructure and culture in the superstructure. But this binary disguises the evident fact that many places, such as skyscrapers, and many events, such as Christmas, are both cultural and economic, and cannot be understood with reference to just one of the two concepts.

In *Spaces of Geographical Thought* (2005), edited by Cloke and Johnston, additional binaries noted include agency/structure (see structuration theory in Chapter 7), space/place (an evident tension between positivist and humanist positions), nature/culture (an abiding tension in Chapter 3), and local/global (central to the discussion in Chapter 9). We can add others that appear throughout this book: relevant/esoteric; image/reality; control/freedom; inclusion/exclusion; and science/art.

Human geography has many examples of binary thinking. Perhaps the time has come to think instead about building bridges.

---

immeasurably from some understanding of the links between human and physical geography. The nineteenth-century American geographer and Congressman George Perkins Marsh wrote in 1864:

> ... the earth is fast becoming an unfit home for its noblest inhabitant, and another era of equal human crime and human improvidence, and of like duration ... would reduce it to such a condition of impoverished productiveness, of shattered surface, of climatic excess, as to threaten the depravation, barbarism, and perhaps even extinction of the species. (Marsh, 1965: 43)

Focused, like Marsh, on the relationship between humans and land, geographers continue to play an important role in both environmental analysis and environmental education.

HANDLING DATA

Human and physical geography are serving society today through the development and application of remote sensing and electronic data processing. In an increasingly complex world with ever-larger data sets, remote sensing and global positioning systems allow us to acquire data rapidly and repeatedly, while electronic data processing permits the conversion of data into maps and facilitates data analysis. Perhaps the best-known recent advance is the development of geographic information systems—integrated computer systems for the input, storage, analysis, and output of spatial data.

These new ways of handling data are much more than mere extensions of earlier procedures. In an important sense, they are changing the nature of geographic information and the role that such information plays in society. A traditional map is a single product manufactured for users, whereas a GIS can be manipulated by the user: data can be added or removed depending on specific interests, scales can be changed (the user can 'zoom in' on a particular area, for example), and the data can be updated as new information becomes available. Remember, new tools prompt new thoughts.

UNDERSTANDING AND SOLVING PROBLEMS

Human geography facilitates understanding of people and place by explaining how different peoples see and organize space differently. Hence, geographers are acutely aware that there are no simple or universal solutions to problems. Effective changes can be made only following full analysis of spatial variations. Sensitive to spatial variations and the significance of place, human geographers know that solutions to a problem in one area may not be solutions to similar problems in another area. This awareness of the need for sensitivity is evident, albeit in different ways, in all of our major philosophical traditions—empiricist, positivist, humanist, Marxist, and postmodernist. Understanding space and place is important in itself and because it allows us to propose solutions to social and environmental problems. This focus is particularly evident in the vital subdiscipline of cultural geography introduced in Chapters 6 and 7. This progressive subdiscipline is increasingly concerned with highlighting regional and global inequalities and working towards solutions.

VALUES

What underlying values direct human geographic research? The general answer is that human geographic values are socially specific, reflecting the values that legitimate the societies in which we live. As we saw in Chapter 2, whereas the positivist school of geography purported to be value-free, Marxist and humanistic geography contributed the important insight that values are socially constructed and therefore subject to change. This general answer does not, of course, address specific questions about values, such as the relative merits of environmental preservation on the one hand and economic gain on the other

hand (Buttimer, 1974). As discussed on several occasions in this text, the views held on such matters tend to reflect larger ideological preferences.

Being a Human Geographer: Where We Began

By now the advantages of training in human geography should be clear; the advantages of being human geographers follow logically. Training in human geography allows us to understand our world; being human geographers allows us to put that training into practice. Most students of human geography do not formally become human geographers; rather, they find employment in a wide range of professions. Training in human geography encourages understanding of the importance of spatial scale, global and local alike. It emphasizes interactions between humans and land as well as emotional attachments to place; and it develops an appreciation of the major global changes in population numbers, culture, political identity, and land use. This understanding is combined with an enviable set of skills that includes not only traditional quantitative and qualitative research procedures, but also the distinctive tools of cartography, remote sensing, and geographic information systems. It is not surprising that human geographers find employment in such diverse arenas as teaching, business, industry, government, consulting, urban and regional planning, conservation and historical preservation, and libraries and archives. Most important, any training in human geography provides lifelong benefits—benefits that we might summarize under the heading of 'geographical imagination'.

GEOGRAPHICAL IMAGINATION

C. Wright Mills (1959) introduced the important concept of 'the sociological imagination' to describe the value of seeing the connections between one's own experiences and problems and the larger social world. A related and no less valuable perspective has developed over the years in the field of human geography. What is 'the geographical imagination'? Simply put, it is appreciation of the relevance of space and place to all aspects of human endeavour. Peattie (1940: 33) recognized this more than half a century ago when he wrote that geography 'explains ways of living in all their myriad diversity'. More recently, Gregory (1994) offered a detailed commentary on geographical imagination as sensitivity to space and place in the effort to understand ways of life.

Geographical imagination is independent of philosophical preference:

- From an essentially *positivistic* viewpoint, Morrill (1987: 535) observed, 'If there is not a convincing theory of why and how humans create places and imbue them with meaning, then it is time to develop that theory.'
- From an essentially *humanistic* viewpoint, Prince (1961: 231) asserted, 'Good geographical description demands not only respect for truth, but also inspiration and direction from a creative imagination. . . . it is the imagination that gives [the facts] meaning and purpose through the exercise of judgement and insight.' This is similar to what Bunkše (2004: 12) refers to as a 'geographic sensibility'.
- Finally, from a *Marxist* viewpoint, Harvey (1973: 24) observed, 'This imagination enables the individual to recognize the role of space and place in his own biography, to relate to the spaces he sees around him, and to recognize how transactions between individuals and between organizations are affected by the space that separates them.'

Three different geographers, three different theoretical perspectives, one fundamental conclusion.

The concept of geographical imagination is an appropriate note on which to conclude this introductory text. I hope that as students of human geography you have gained an appreciation and a sensitivity to the importance of space and place in understanding the world that is our home.

CONCLUSION SUMMARY

CHANGING HUMAN LANDSCAPES

Although the days of discovering new worlds are over, human geography is by no means fixed or static. As our world is constantly changing, old facts disappear and new ones appear. Among the most important areas where changes are occurring are population numbers, urban growth, and resource use. Political, economic, and cultural circumstances also change at the global scale, the precise consequences of which remain uncertain, although one clear impact is increasing pressure on tribal peoples, who are disappearing at an alarming rate.

CHANGING HUMAN GEOGRAPHY

Human geography is an academic discipline that changes in response to the changing requirements of the society of which it is a part. Space and place are constantly being reinforced as our key concepts; our links with physical geography are a major strength; new technical approaches offer ever-improving ways to handle data, and our increasing understanding of issues is leading to more and better solutions.

BEING A HUMAN GEOGRAPHER

Develop and use your geographical imagination.

FURTHER EXPLORATIONS

All of the following offer excellent, if sometimes idiosyncratic, commentaries on human geography. All are by distinguished geographers and all are well-written and highly entertaining. Read and enjoy!

Gould, P. 1985. *The Geographer at Work*. London: Routledge & Kegan Paul.

Haggett, P. 1990. *The Geographer's Art*. Oxford: Blackwell.

Johnston, R.J. 1985. 'To the Ends of the Earth', in R.J. Johnston, ed., *The Future of Geography*. London: Methuen, 3–26.

Stoddart, D.R. 1986. *On Geography*. Oxford: Blackwell.

Each of the following books provides insights into the value of a geographical imagination when addressing issues and problems evident in the world today.

Abler, R.F., M.G. Marcus, and J.M. Olson, eds. 1992. *Geography's Inner Worlds: Pervasive Themes in Contemporary American Geography*. New Brunswick, NJ: Rutgers University Press.

Brunn, S.D., S.L. Cutter, and J.W. Harrington, eds. 2004. *Geography and Technology*. Boston: Kluwer Academic.

de Blij, H. 2005. *Why Geography Matters: Three Challenges Facing America. Climate Change, the Rise of China, and Global Terrorism*. New York: Oxford University Press.

Janelle, D.G., B. Warf, and K. Hansen. 2004. *World Minds: Geographical Perspectives on 100 Problems Commemorating the 100th Anniversary of the Association of American Geographers 1904–2004: Celebrating Geography—The Next 100 Years*. Boston: Kluwer Academic.

Because there is such a close relationship between geography and many world events—both human and physical—it is essential for geographers to keep abreast of current developments. Two excellent factual sources, both published monthly, are available in major reference libraries: Keesing's *Record of World Events* and *Current History*.

Finally, one of the most useful ways to explore any aspect of geography in greater detail is to refer to the following publications:

Gaile, G.L., and C.J. Willmott, eds. 2003. *Geography in America at the Dawn of the 21st Century*. New York: Oxford University Press.

National Research Council. 1997. *Rediscovering Geography: New Relevance for Science and Society*. Washington: National Academy Press.

ON THE WEB

GEOGRAPHY TOMORROW

Facing the Future

> www.facingthefuture.org/default.aspx

A discussion of some global problems, focusing on scarce resources, poverty, conflict, and the environment.

Also, now is a good time to revisit the sites listed at the end of the Introduction.

GLOBAL PHYSICAL GEOGRAPHY

This Appendix provides an overview of planet earth as a habitable environment for humans and an account of 10 principal global environment regions.

A HABITABLE PLANET

The earth is a habitable planet for humans—the only such planet known to us. By 'habitable', we mean that on this planet both physical processes and physical environments were such that they permitted the emergence and subsequent spread and growth of human life. Indeed, only a series of quite remarkable coincidences have made earth habitable for humans. The planet's distance from the sun, its axial tilt, its 24-hour rotation, and its 'minor' variations in surface relief all combine to spread the energy received from the sun relatively evenly. The evolution of the earth required the presence of water to form oceans, an average surface temperature of 15°C, and the presence of a particular kind of atmosphere. Interestingly, the first life on earth evolved in water, and it was the subsequent movement of life onto land that converted the initial carbon dioxide atmosphere into an oxygen-rich one. This atmospheric transformation included the creation of the ozone layer—ozone is a form of oxygen—that prevents the penetration of ultraviolet rays that would sterilize the land surface. The oxygen-rich atmosphere also allowed the emergence of animal life.

solstice
Occurs twice each year when the sun is vertically overhead at the farthest distance from the equator, once 23.5°N and once 23.5°S of the equator; for example, when at 23.5°N, there is maximum daylight (longest day) in the northern hemisphere and minimum daylight (shortest day) in the southern hemisphere.

equinox
Occurs twice each year, in spring and fall, when the sun is vertically overhead at the equator; on this day the periods of daylight and darkness are both 12 hours long all over the earth.

From our earthbound perspective, the surface of the earth is tremendously varied. The atmosphere is constantly changing; there are mountains as high as 8,854 m (29,048 feet) above sea level (Mount Everest), land areas 408 m (1,338 feet) below sea level (the shores of the Dead Sea), and known ocean depths of 10,927 m (35,849 feet) below sea level (south of the Mariana Islands in the Pacific Ocean). The planet frequently experiences dramatic changes—volcanic eruptions, earthquakes, tidal waves. On another level, all these variations and changes are insignificant. An accurate model of the earth of a size to fit in our hands would feel as smooth as a peach. The earth is approximately spherical in shape, with an equatorial circumference of 40,067 km (24,897 miles) and a polar circumference of 40,000 km (24,855 miles). This shape is the result of gravity, the force that is also responsible for the tendency of the earth's surface area to be layered: a dense substance, rock, is at the bottom, a less dense substance, water, is often above the rock, and the least dense substance, air, is above the other two. Such an arrangement is crucial for human life, which has access to all three essential substances.

The sphere that is the earth moves in two ways. First, it rotates on its axis; one rotation defines a day. Second, it orbits the sun; one revolution defines a year. Rotation imposes a diurnal cycle on much of the life on earth. Revolution, in conjunction with the axis of the earth tilted at 23.5°, produces seasonal climatic changes and variations in length of daylight.

ROTATION

The end points of the axis of rotation are defined as the poles, north and south. All locations on the earth's surface, except the poles, move as the earth rotates, and one circuit of the earth is a line of latitude. Because the earth is a sphere, the longest line of latitude lies midway between the poles: this is the equator. Lines from one pole to the other are lines of longitude. Unlike lines of latitude, these are not parallel; rather, they converge at the poles and are farthest apart at the equator. Lines of latitude are measured in degrees, with the equator defined as 0° and the poles therefore as 90°N and 90°S; lines of longitude are similarly measured in degrees, in this case from an arbitrary 0° line known as the prime meridian that runs through the Greenwich Observatory in London, England. The geographic grid that is commonly used to describe locations is based on latitude and longitude.

REVOLUTION

Revolution around the sun is the earth's second important movement. Figure A1.1 illustrates the four seasonal positions of the earth relative

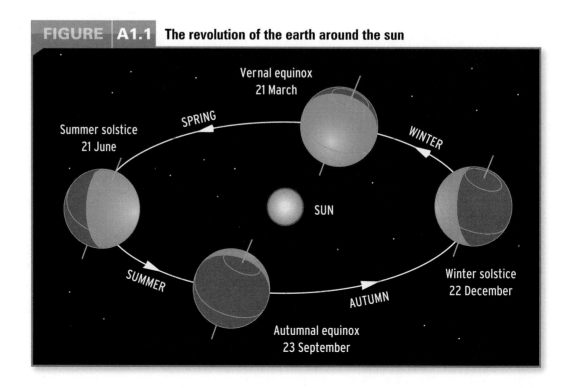

FIGURE A1.1   The revolution of the earth around the sun

Vernal equinox
21 March

Summer solstice
21 June

SPRING

WINTER

SUN

SUMMER

AUTUMN

Winter solstice
22 December

Autumnal equinox
23 September

to the sun and introduces the terms 'equinox' and 'solstice'. The summer **solstice** occurs on 21 or 22 June in the northern hemisphere and on 21 or 22 December in the southern hemisphere. This is the time when the hemisphere reaches maximum inclination towards the sun so that at latitude 23.5°, the sun is directly overhead. The winter solstice occurs on 21 or 22 December in the northern hemisphere and on 21 or 22 June in the southern hemisphere. This is the time when the hemisphere reaches maximum inclination away from the sun. The two **equinoxes**, vernal and autumnal (spring and fall), are the halfway points between the solstices, and on both occasions the axial inclination is at exactly 90° with respect to the sun.

Much human activity is related to these global physical factors. Diurnal, seasonal, and annual cycles are basic to much social and economic organization. The durations of the day and year are defined by earth movements, while the month is defined by lunar movement around the earth. The hour, by contrast, is an arbitrary human division. Another arbitrary definition is the use of the equinoxes and solstices as the beginnings of seasons.

THE EARTH'S CRUST

The crust or outer layer of the earth is composed of rocks—aggregates of minerals—that have continued forming and changing for at least 3,000 million years. It is usual to classify the many rock types as sedimentary (those that were formed at low temperatures on the earth's surface); igneous (those that solidified

In (a), beginning 222 million years ago, there is one land mass, a super-continent called Pangaea. By 180 million years ago (b), Pangaea is breaking up; Laurasia is moving north, and the southern land mass, Gondwanaland, is also breaking away. In (c), 135 million years ago, the process continues, with the beginnings of the North and South Atlantic oceans. In (d), 65 million years ago, Madagascar has separated from Africa and India is moving towards Asia, but Australia remains linked to Antarctica. Map (e) shows the present distribution of land and sea, and (f) shows a likely scenario 50 million years in the future: Australia continues to move north; East Africa is shifting farther east, California (west of the San Andreas fault) has separated from North America, and the Mediterranean Sea is shrinking as Africa moves north.

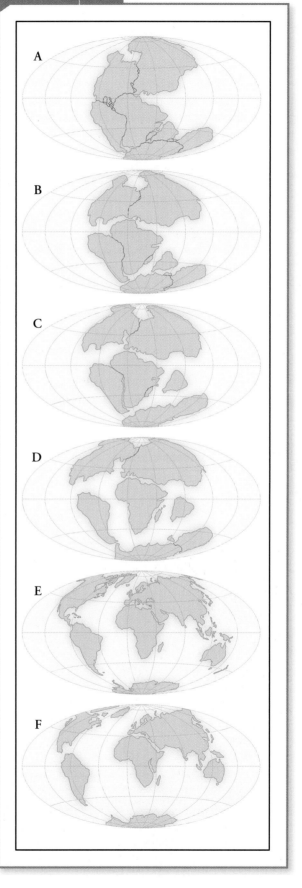

FIGURE A1.2 Moving continents

from a molten state); or metamorphic (those that have changed under high temperature and pressure).

One of the most important features of the earth's surface is that it is part land (29 per cent) and part water (71 per cent). The present distribution of land and water is the result of a long and ongoing process of movement by the **lithosphere**—the earth's crust and the uppermost mantle below the crust. The lithosphere is divided into six major and several minor plates that move relative to one another as a result of movements in the liquid mantle deep inside the earth. The continents quite literally float on this liquid mantle, with some four-fifths of their mass below the surface. The idea that the land surfaces, the continents, also move was suggested as long ago as the seventeenth century when Francis Bacon (1561–1626), an English philosopher, observed that the Atlantic coasts of South America and Africa fitted together like pieces of a jigsaw puzzle. Early scientific evidence was gathered by Wegener (1880–1930), a German meteorologist, but it was not until the 1960s that this idea of moving continents, known as **continental drift**,

achieved scientific credibility. Figure A1.2 illustrates the positions of the continents since outward movements began roughly 225 million years ago (MYA). At the present time, the Atlantic Ocean is widening about 2 cm (0.75 inches) each year; Australia is moving northward, as is Africa, while parts of Africa and western North America may separate from their respective continents.

A second consequence of moving continents is the creation of mountain chains such as the Himalayas. When continents collide—as India and Asia did, perhaps 150 MYA—mountain chains are thrust upwards. The Appalachians are much older, resulting from a collision of North America and Africa perhaps 350 MYA.

Earthquakes are a third consequence of movement; they occur when parts of the crust move over other parts as a result of movements in the liquid mantle. Once stress exceeds rock strength, a sudden movement—an earthquake—occurs. Such movements typically take place along fault lines (such as the San Andreas fault in California), which are lines of crustal weakness. Volcanoes also occur along lines of

lithosphere
The outer layer of rock on earth; includes crust and upper mantle.

continental drift
The idea that the present continents were originally connected as one or two land masses that separated and drifted apart.

FIGURE | A1.3   The major ocean currents

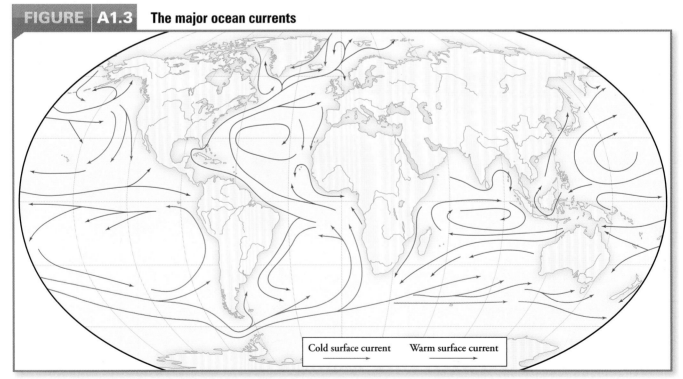

Cold surface current →   Warm surface current →

The prevailing winds and the rotation of the earth combine to cause gyres (large whirlpools). These gyres move warm water away from the equator and cold water towards it. There are five major gyres and 30 major currents.

crustal weakness, often plate boundaries. A volcanic eruption involves liquid from the mantle, magma, being forced upwards and onto the earth's surface.

As already noted, much of the earth's crust is covered by water—approximately two-thirds. Beneath the water is a physical landscape that is constantly changing, while the water itself is also involved in a series of movements. Waves are caused by wind; tides result from the gravitational pull of the moon (and, to a lesser extent, the sun). The most significant movements, however, are ocean currents (Figure A1.3). Warm water moves away from the equator and cold water moves towards the equator.

Not all of the earth's water is in oceans; some is in the form of ice—a leftover from the most recent glacial period. Most of this ice is in two continental ice sheets, the Antarctic and Greenland sheets, which cover about 10 per cent of the land surface and reach thicknesses of several kilometres. As this ice gradually melts, the sea level rises. Some 20,000 years ago, the sea level was 100 m (328 feet) lower than today; the current rise is about 1 cm (0.39 inch) a year.

The physical processes and circumstances identified so far are global and/or long-term, but they are important to us. Understanding past migrations of life forms, for example, requires an appreciation of land movement and climatic change. One example of the possible long-term effect of a physical process—exacerbated by human activity—is the melting of all ice, which would result in a sea-level rise of 60 m (197 feet). Such a rise would flood many of the largest cities in the world.

THE PHYSICAL LANDSCAPE

The physical landscape in which human activities take place (and which is continually modified by those activities) is a product of a series of processes. Crustal movement and the effects already noted are followed by weathering and gradational processes that combine to produce specific landforms. Weathering is the reduction in size of surface rock; gradation is the movement of surface material under the influence of gravity and typically involves water, ice, or wind. In addition to creating landforms, these processes combine with the activities of living organisms to create soil. Plant and animal life

in turn are closely related in a circular fashion to the landscapes created.

Weathering occurs when rock is mechanically broken and/or chemically altered. Temperature change is an important cause of mechanical weathering; water is the principal cause of chemical weathering. Once rocks are weakened in this way, they become especially susceptible to gradational processes. Water, ice, and wind can erode, move, and deposit material. Many of the earth's distinctive physical landscapes are clear evidence of these processes at work.

Water is probably the most important cause of gradation. A continuous transfer of water from sea to air to land to sea is taking place. River erosion creates V-shaped valleys and waterfalls, while river deposition creates flood plains and deltas. Sea water also affects land surfaces, and coastlines undergo constant change; cliffs are the result of erosion, beaches the result of deposition. Other parts of the earth's surface show the effects of ice and snow; during the most recent glacial period, ice covered an area approximately three times greater than that covered today, including Northern Europe and Canada. Areas formerly covered by ice show distinctive evidence both of erosion (e.g., U-shaped valleys) and of deposition (e.g., the low hills called drumlins). Wind erosion results either from wind-blown materials colliding with stable materials (as in sandblasting) or, more often, the wind's removal of loose particles from a surface. Sand dunes are among the best-known products of wind deposition of material.

Each of these processes is affecting parts of the earth's surface at all times. Physical landscapes, then, are subject to continuous modification—almost fine-tuning. In principle, the eventual outcome, without crustal movement or climatic change, would be an almost featureless physical landscape.

Our physical environment is, of course, much more than the distribution of landforms. Soils, vegetation, and climate are additional components of a physical environment. Figures A1.4, A1.5, A1.6, and A1.7 provide general outlines of global landforms, soils, vegetation, and climate respectively.

Figure A1.4 maps the major *cordilleran belts*— the backbones of the continents. The slope and ruggedness of local landforms in particular

FIGURE | A1.4 **Global cordilleran belts**

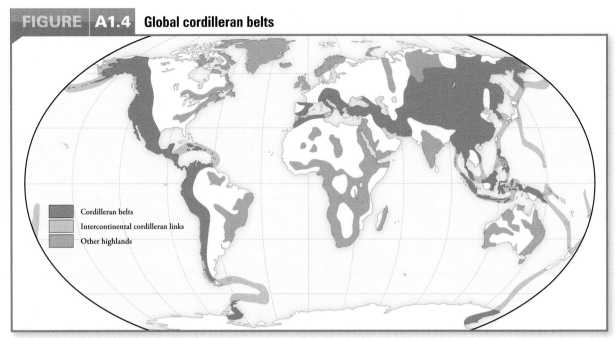

The backbones of the continents, these cordilleran belts also extend through ocean areas.

SOURCE: Adapted from R.C. Scott, *Physical Geography* (St Paul, Minn.: West Publishing, 1989), 295. © 1989 West Publishing Co. Used by permission of Wadsworth Publishing Co.

FIGURE | A1.5 **Global distribution of soil types**

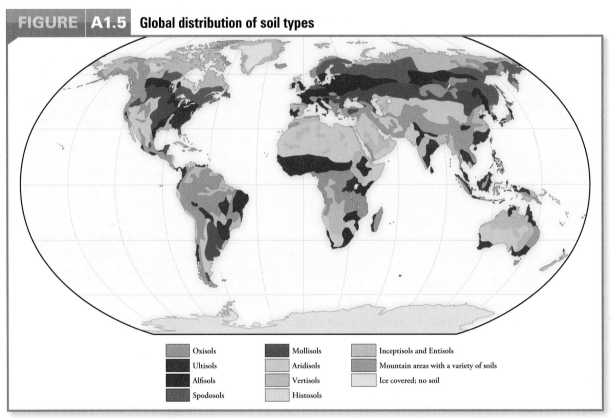

Oxisols are typical of tropical rain forests and are deep soils but not inherently fertile: if the natural vegetation is removed, the soil is soon impoverished. Ultisols, in the humid tropics, are low in fertility. Alfisols, dispersed throughout low- and middle-latitude forest/grassland transition areas, are of moderate to high fertility. Spodosols are associated with the northern forests of North America and Eurasia; low fertility is typical. Mollisols, associated with middle-latitude grasslands, are often very fertile. Aridisols, located in areas of low precipitation, lack water. Vertisols, limited to areas of alternate wet and dry seasons, may be very fertile. Histosols are most common in northern areas of North America and Eurasia; they are composed of partially decayed plant material, usually waterlogged, and typically fertile only for water-tolerant plants. Inceptisols and entisols are immature soils found in environments that are poorly suited to soil development.

SOURCE: Adapted from R.C. Scott, *Physical Geography* (St Paul, Minn.: West Publishing, 1989), 276–7. © 1989 West Publishing Co. Used by permission of Wadsworth Publishing Co.

FIGURE A1.6   Global distribution of natural vegetation

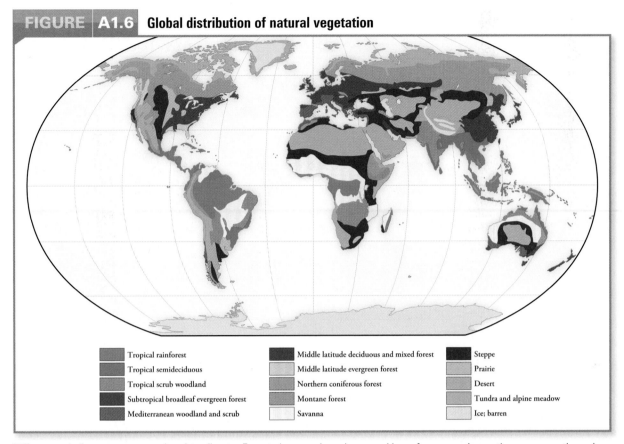

Tropical rainforest	Middle latitude deciduous and mixed forest	Steppe
Tropical semideciduous	Middle latitude evergreen forest	Prairie
Tropical scrub woodland	Northern coniferous forest	Desert
Subtropical broadleaf evergreen forest	Montane forest	Tundra and alpine meadow
Mediterranean woodland and scrub	Savanna	Ice; barren

Differences in forest types are related to climate. Forests become less dense and have fewer species as they move northward or southward from the equator. Deciduous trees dominate the middle latitudes, evergreens the higher latitudes. Savanna, steppe, and prairie are composed chiefly of grasses. Desert vegetation is closely linked with an arid climate. Tundra vegetation is the most cold-resistant.

SOURCE: Adapted from R.C. Scott, *Physical Geography* (St Paul, Minn.: West Publishing, 1989), 232–3. © 1989 West Publishing Co. Used by permission of Wadsworth Publishing Co.

are major considerations in human decisions about the location of economic activities, such as agriculture, industry, and transport.

Soil types are distributed in a systematic fashion (Figure A1.5). There are close general links both to climate and to natural vegetation; these are two of the most important determinants of soil type. Soils are important for their agricultural potential and for their capacity to support buildings. Much of the European migration to the New World was closely tied to the availability of soils suitable for commercial agriculture; in the late nineteenth century, many settlers found suitable grassland soils in Canada, Argentina, and Australia.

Figure A1.6 maps the global distribution of *natural vegetation*. The four principal types are grassland, forest, desert scrub, and tundra. Much of the history of agriculture is intimately tied to their distribution. In circumstances of limited agricultural technology, for example, forested areas can be a significant barrier.

Climate is the characteristic weather of an area over an extended period of time and can be usefully summarized by reference to mean monthly and annual statistics of temperature and precipitation. These means reflect latitude. In Figure A1.7, it is clear that climatic distributions are generally linked to latitude. Three low-latitude, two subtropical, three mid-latitude, and three high-latitude climates are mapped, along with one highland climate. Low-latitude climates are characteristically hot; the distinguishing feature of each of the three is precipitation. The two subtropical climates are less hot, and again the distinguishing feature is precipitation. The three low-latitude climates lack any significant seasonality. The subtropical and mid-latitude climates vary according to the degree of continentality (distance from ocean) and, of course, latitude. All five of these climates have seasonal variations. Proximity to oceans tends to increase precipitation and reduce temperature variations. The three high-latitude climates are cold most of the year

FIGURE | A1.7   Global distribution of climate

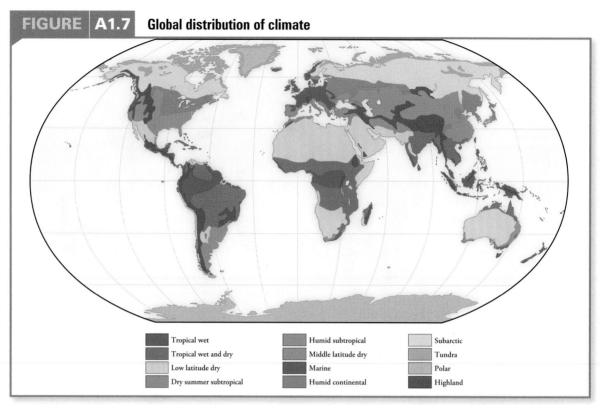

Tropical wet	Humid subtropical	Subarctic
Tropical wet and dry	Middle latitude dry	Tundra
Low latitude dry	Marine	Polar
Dry summer subtropical	Humid continental	Highland

These types of climate, distinguished on the basis of moisture and temperature, are derived from the Köppen classification. Tropical wet, tropical wet and dry, and low-latitude dry climates may be grouped as low-latitude climates; dry summer subtropical, humid subtropical, mid-latitude dry, marine, humid, and continental climates as middle-latitude climates; and subarctic, tundra, and polar climates as high-latitude climates.

SOURCE: Adapted from R.C. Scott, *Physical Geography* (St Paul, Minn.: West Publishing, 1989): 158–9. ©1989 West Publishing Co. Used by permission of Wadsworth Publishing Co.

FIGURE | A1.8   Generalized global environments

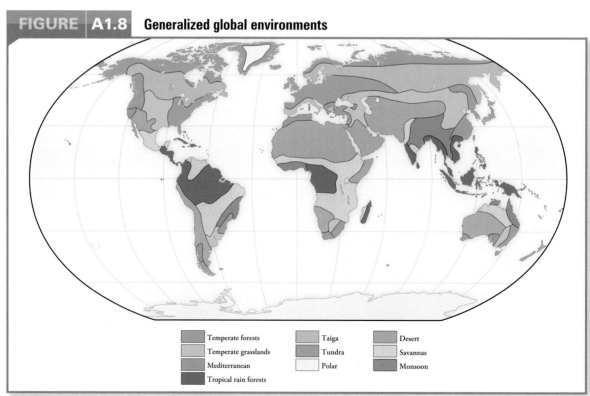

Temperate forests	Taiga	Desert
Temperate grasslands	Tundra	Savannas
Mediterranean	Polar	Monsoon
Tropical rain forests		

Climate is clearly a key consideration in many human decisions. But *weather* is also important. Weather can have a major impact on agricultural yields. Atypical or violent weather can cause floods or droughts, and hurricanes—vast tropical storms—can cause serious damage to populated land areas.

and have low precipitation. Finally, highland climates display many local variations, since temperature and precipitation are affected by altitude in addition to latitude.

Global Environments

The preceding section has highlighted a wide variety of physical processes and environments. Their relationship to human activities is often intimate, but these physical factors do not in any sense determine our human activities. Although we need to be aware of physical factors at all levels, we must not assume any causal relationships. As this book has demonstrated, we humans make the decisions—albeit with many physical factors in mind! This section presents a brief summary of the major global environments (Figure A1.8), integrating much of the material above in a more applied human context. The notion of global environmental regions is useful here. Indeed, those regions are closely correlated with climatic regions in particular. Humans have favoured certain environments and have made changes to all areas they have settled in.

TROPICAL RAIN FORESTS

These areas are located close to the equator and experience high rainfall—typically exceeding 2,000 mm (78 inches) per annum—and high temperatures—24°C to 30°C (75°F to 86°F). They lack seasonal variation. Broadleaf trees dominate, and the vegetation is dense and highly varied. Nevertheless, once the vegetation is removed, the soil is characteristically

Tropical rain forest, Costa Rica.

© Mark Hobson

shallow and easily eroded. Tropical rain forests have not become major agricultural regions. Among the economically valuable products of rain forests are mahogany and rubber, as well as various starchy food plants, such as manioc, yams, and bananas.

MONSOON AREAS

Monsoon areas are characterized by marked seasonality, especially an extended dry season and heavy rainfall. Rainfall may be as high as 2,000 cm (780 inches) in some areas. The

Monsoon area, Bali.

United Nations photo/ P.S. Sudhakaran

Savanna with baobab, Malawi.
V. Last, Geographical Visual Aids

Desert, Death Valley, California.
V. Last, Geographical Visual Aids

typical natural vegetation is deciduous forest, but all monsoon areas have been changed dramatically as a result of human activity. Agriculture has been practised for a long time, with rice as the basic crop in many such areas. Population densities are usually high.

SAVANNAS

Savannas are located to the north and south of the tropical rain forests. They have a distinctive climate: a very wet season and a very dry season. The natural vegetation is a combination of trees and grassland. Africa has the most extensive savanna areas, which are rich in animal species. To date, most savannas are relatively undeveloped economically, but they offer great promise for increases in animal and

Mediterranean area, near St-Tropez, France.
V. Last, Geographical Visual Aids

crop yields. Important grain crops include sorghum in Africa and wheat in Asia.

DESERTS

A desert area has low and infrequent rainfall, perhaps as low as 100 mm (4 inches) per annum and perhaps no rain for 10 years. Areas are typically defined as arid if they receive less than 255 mm (10 inches) of rain per annum and as semi-arid if they receive less than 380 mm (15 inches). There are hot, cold, sandy, and rocky deserts. In all deserts, water is a crucial consideration in human activities. Irrigation and nomadic pastoralism are two human responses to desert environments.

MEDITERRANEAN AREAS

There are only a few Mediterranean environments. They are characterized by long, hot, dry summers and warm, moist winters. The typical location is on the western side of a continent in a temperate latitude. Evergreen forests are the usual natural vegetation, although human activity means that scrub land prevails today. Important crops, especially in the Mediterranean proper, are wheat, olives, and grapes.

TEMPERATE FORESTS

A temperate climate predominates in much of Europe and eastern North America. This climate is characterized by a distinct seasonal cycle, fertile soils, and a natural vegetation composed primarily of mixed deciduous trees. The seasonal cycle gives this area its distinctive character. The many animals and plants have adapted to the variations in heat, moisture,

Temperate grassland, Head-Smashed-In Buffalo Jump, Alberta.

V. Last, Geographical Visual Aids

Boreal forest, Terra Nova National Park, Newfoundland.

V. Last, Geographical Visual Aids

food, and light. Annual loss and growth of leaves is an especially striking adaptation that affects all other forest life forms. Particularly since the beginnings of agriculture, these areas have been attractive to and significantly altered by humans, and they are now dominated by agriculture, industry, and urbanization. These changes began in Europe about 7,000 years ago, in Asia about 6,000 years ago, and in North America about 1,000 years ago. Indeed, the removal of forest cover has been so dramatic that today there is a general recognition of the need to preserve and even expand the remaining forests.

TEMPERATE GRASSLANDS

North American prairies, Russian steppes, South African veld, and Argentine pampas are four of the large temperate grasslands. Located in continental interiors, they have insufficient rainfall for forest cover and tend to show great annual temperature variations. Such grasslands were once occupied by particular large migrating herbivores such as the bison (North America), antelopes, horses, and asses (Eurasia), and pampas deer (South America). After remaining almost devoid of humans until the late seventeenth century, these grasslands became the granaries of the world in the twentieth century. Their soils are typically suited to grains; wheat became dominant in most areas.

BOREAL FORESTS

Boreal forests, often called taiga, are located only in the northern hemisphere—in North America and Eurasia. Their winters are long and cold, their summers brief and warm, and precipitation is limited. The natural vegetation is evergreen forest. Agriculture is very limited, and the major economic importance of taiga consists in forest products and minerals.

TUNDRA

Located north of the boreal forest, this area experiences long, cold winters; temperatures average −5°C (23°F) and there are few frost-free days. Perennially frozen ground—permafrost—is characteristic and presents a severe challenge to human activities. Only a few plants and animals have adapted to this environment because the land has been exposed for only about 8,000 years, since the last retreat of ice caps. There has been little time for soil to form.

Temperate mixed deciduous forest of lower Great Lakes–St Lawrence Valley region.

Don Johnston/All Canada Photos

Tundra, Herschel Island, Yukon.
Philip Dearden

Polar region, Ellesmere Island.
Philip Dearden

POLAR AREAS

Polar conditions prevail in both hemispheres and are best represented by the Greenland and Antarctic ice sheets. Antarctica is a continental land mass under an ice sheet that reaches thicknesses of 4,000 m (more than 13,000 feet). These areas lack vegetation and soil and have proven hostile to human settlement and activity. Indigenous inhabitants of these areas lived in close relationship to the physical environment, subsisting on hunting and fishing products. In recent years, non-indigenous inhabitants have greatly increased in number, particularly because of the discovery of oil.

These brief descriptions of 10 global environments make no attempt to describe all areas. Nor do all areas within a particular environment have uniform characteristics. Regions merge into one another; the tropical rain forest merges into savanna and then into desert, for example. Rarely are there clear, sharp divisions. As a result, many areas are difficult to classify. What this outline has emphasized is that the earth comprises many environments with which humans have to cope. All of these environments have been affected by human activities, and all need to be considered in any attempt to comprehend our human use of the earth.

FURTHER EXPLORATIONS

De Blij, H.J., P.O. Muller, R.S. Williams Jr, C.T. Conrad, and P. Long. 2009. *Physical Geography: The Global Environment*, Second Canadian edn. Toronto: Oxford University Press.

Detailed, comprehensive text on physical geography. Very readable, well illustrated, and student-friendly throughout.

Trenhaile, A.S. 2009. *Geomorphology: A Canadian Perspective*, 4th edn. Toronto: Oxford University Press.

A textbook presenting a systematic explanation of the landforms of Canada.

ON THE WEB

PHYSICAL GEOGRAPHY FOR HUMAN GEOGRAPHERS

Canada

gsc.nrcan.gc.ca/index_e.php

The home page of the Geological Survey of Canada, offering a wide variety of information on physical geography and links to relevant provincial and territorial websites.

United States

www.usgs.gov

US Geological Survey site includes clear overviews of physical geographic landscapes; offers a good account of the application of GIS technologies.

Appendix 2 • • • • •

THE EVOLUTION OF LIFE

LIFE ON EARTH

Air-breathing life cannot exist without oxygen, which was not a part of earth's original atmosphere. However, life in the form of primeval bacteria and algae evolved without oxygen; these life forms consumed carbon dioxide and nitrogen, which were in the original atmosphere, and emitted oxygen as a waste. In addition to adding oxygen to the atmosphere, this process also formed the ozone layer, which filters out harmful ultraviolet radiation from the sun. The first life forms evolved in the seas. As indicated in Table A2.1, there is evidence of life forms as early as 3,500 million years ago (MYA), of an ozone layer 2,500 MYA, and of a breathable oxygenated atmosphere 1,700 MYA. Such early life forms were not affected by the absence of an ozone layer because they lived below the surface of the water. Oxygen-breathing life, initially single- and later multi-celled, appeared following the creation of a suitable atmosphere. Soft-bodied animals, comparable to jellyfish, evolved 650 MYA, and shelled animals about 70 million years later.

About 400 MYA, fishes diversified considerably and gradually some moved into areas of shallow water that were subject to drying up. Amphibious **species** appeared shortly thereafter, moving onto land already occupied by plants and insects, especially close to water. Much of the earth at this time comprised swamp environments, which were well suited to plant growth and provided food for amphibians. One group of amphibians evolved into reptiles: animals with waterproof skins capable of remaining out of the water indefinitely. These reptiles were able to move away from water and soon colonized large areas, dominating much of the earth for about 135 million years. During this period mammals and birds appeared; and when the reptiles vanished, 65 million years ago, the mammals and birds remained. The cause of the extinction of reptiles remains uncertain.

With the absence of reptilian competition, mammals and birds flourished, expanding and diversifying, filling the ecological gaps left by the dinosaurs. Among the new evolutionary lines that began at this time were the primates; today, humans are one of about 200 species of living primates. Apes, a new group of primates, evolved about 25 MYA. Some 10 MYA, mammals reached their greatest diversity. Most available evidence suggests that the human evolutionary line separated from that of the apes about 7 MYA.

Four factors need to be highlighted in this account of the emergence and spread of life on earth. First, the earth was a changing environment in this period. Climatic change, especially a series of glacial periods, was ongoing. Second, the land masses were moving. The most recent movements began with the initial breakup of Pangaea (see Figure A1.2) some 225 MYA. The spread of life has been greatly affected by both of these factors, as land routes around the world have changed. Third, the relationship between the physical earth and the emergence and evolution of life is not fully understood (a topic discussed in Box 3.4). Fourth, even a very basic understanding of life on earth helps us understand that our human species is just one among millions, probably tens of millions, of species.

Human Origins

Although our knowledge continues to increase, there is still much that we do not know about human origins and early history (Table A2.2). Current evidence suggests that the large apes evolved as one primate line in Africa about 25 million years ago and then split into several relatively distinct evolutionary lines. One of these in turn split into two further lines, chimpanzees and humans. The most compelling evidence of a common origin is that humans and chimpanzees differ in only about 1 per cent

> **species**
> A group of organisms able to produce fertile offspring among themselves but not with any other group.

Table A2.1	Basic Chronology of Life on Earth
Time	Event
Physical Origins	
10–20 billion years ago?	Creation of universe
5,000 MYA	Solar system forms
4,500 MYA	Earth forms
Origins of Life	
3,500 MYA	Oldest life forms
3,200 MYA	Oxygen-creating bacteria
2,500 MYA	Ozone shield forms
2,200 MYA	Oxygen in atmosphere
1,700 MYA	Atmosphere breathable
Origins and Evolution of Life Forms	
1,400 MYA	Oxygen-using animals
800 MYA	Multicellular life
650 MYA	Soft-bodied animals
580 MYA	Shelled animals
500 MYA	First fishes
430 MYA	First land plants
360 MYA	First insects
350 MYA	First amphibians
280 MYA	First reptiles
225 MYA	Breakup of Pangaea
220 MYA	First mammals
200 MYA	Reptiles dominant for next 135 million years
190 MYA	First birds
65 MYA	Dinosaurs extinct; birds and mammals diversify; first primates
25 MYA	Grasslands become widespread; first apes
10 MYA	Maximum diversity of mammals
6 MYA	Emergence of earliest human ancestors

NOTE: MYA = million years ago.

FIGURE | A2.1   Probable directions of movement out of Africa by early humans

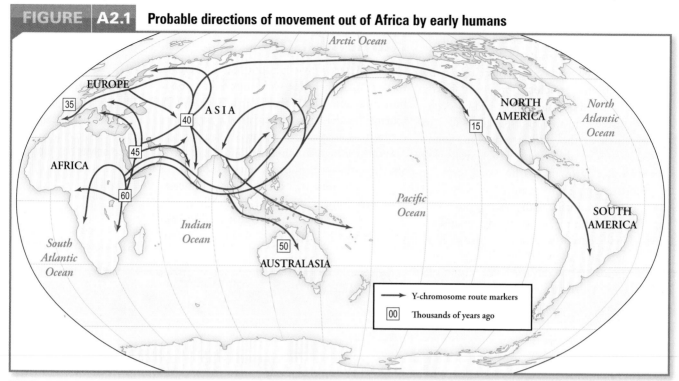

SOURCE: The Genographic Project. At: <www3.nationalgeographic.com/genographic/index.html>.

of their genes; this means that our two species could not have been evolving separately for much more than about 7 million years. But precisely when this split occurred is uncertain. It is generally accepted—using a combination of fossil, geological, and genetic evidence—that this split occurred, that is, human ancestors diverged from the ape line, about 7 million years ago, as confirmed by the 2001 discovery of a skull in Chad, known as *Sahelanthropus*. But some recent genetic research suggests that the split occurred more recently, between 6.3 and 5.4 million years ago, and also that this split was followed by some interbreeding between humans and chimpanzees.

OUR EARLIEST ANCESTORS: AUSTRALOPITHECUS

What was it that distinguished the earliest human ancestors from other primates? Unlike all earlier primates, they were bipedal—they walked upright. This was a critical adaptation that permitted early humans to move over great distances while carrying objects; fossil evidence has been found throughout eastern and southern Africa. The first known **hominid** group is called *Australopithecus*. Currently, most authorities agree that the common ancestor of all hominids was *Australopithecus afarensis*; fossil evidence of this species, discovered in East Africa, has been dated to more than 3.5 MYA. *A. afarensis* had a small brain of about 440 ml (modern human brains average 1,450 ml).

TOWARDS MODERN HUMANS: HOMO HABILIS

The next critical evolutionary event in the human lineage occurred about 3 million years ago when the hominid line split into two types, one of which became extinct and one of which evolved into modern humans. The first representative of the line leading to modern humans appears to be the species *Homo habilis*. Both *Australopithecus* and *Homo habilis* were probably—though not certainly—restricted to Africa, notably the East and South African savanna areas, where they lived by hunting small animals and gathering plants. But *Homo habilis* had a larger brain (about 680 ml), which made possible the development of an important new cultural behaviour: the fabrication of stone tools. Tool-making was a major technological advance. Brain size distinguished *Homo habilis* from other previous hominids and from other hominid lines living at the same time; some evidence even suggests that *Homo habilis* possessed the capacity for speech.

Why did *Australopithecus* evolve, perhaps as long ago as 6 million years? Why did *Australopithecus afarensis*, our common ancestor,

hominids
Bipedal primates; humans and their ancestors.

split into two types about 3 MYA? Why did one of these become extinct about a million years ago? Although we are uncertain of the answers, there is evidence to suggest that global climatic change was a causal factor. Declines in global temperatures correspond with each of these evolutionary events. Such declines prompted drying trends in Africa: drying reduced the extent of tropical forest, increasing that of open woodland and dry savanna. It is in this way that evolutionary changes can be seen as adaptations to changing physical environments.

OUT OF AFRICA: HOMO ERECTUS

Beginning about 1.8 million years ago, *Homo erectus* appeared. This new species was distinguished from *Homo habilis* primarily by a brain size of about 1,000 ml—roughly the lower limit of modern human brain size. Until recently, it was widely accepted that *Homo habilis* evolved into *Homo erectus*, but the 2007 discovery of two hominid fossils in the Turkana Basin area of East Africa provides evidence that *Homo erectus* and *Homo habilis* overlapped in this region for about 500,000 years. *Homo erectus* subsequently spread over much of Africa and into the warm temperate areas of Europe and Asia. Currently this hominid species is thought to have been the first to move out of Africa. Some recent evidence suggests that *Homo erectus* may have reached Asia not long after emerging in Africa. The reasons for this movement are not known, but they probably had to do with climatic change, population increases, and the associated search for food. Again, the basis of subsistence was hunting and gathering.

After the appearance of *Homo erectus*, human biological evolution stabilized for some time. But cultural developments, including the introduction of clothing and shelter, were considerable. Different cultural adaptations were devised for different environments. By about 400,000 years ago, hunting strategies had progressed to include the use of fire and possibly the planned hunting of larger game. An innovation at this time was the introduction of relatively permanent pair bonding—monogamy. Human language use also increased substantially, encouraging the gradual development of human culture. All of these adaptations took place against a background of changing physical environments, as the geological period beginning about 1.5 MYA, the **Pleistocene**, was one of considerable fluctuation in global temperatures. Several extremely cold periods,

Table A2.2	Basic Chronology of Human Evolution
6 MYA	Probable earliest date for evolution of first hominids, the bipedal primate *Australopithecus*, in Africa
3.5 MYA	First fossil evidence of *Australopithecus afarensis* in East Africa
3 MYA	Probable appearance of *Homo habilis* in East Africa; evidence of first tools
1.8 MYA	First appearance of *Homo erectus* in East Africa
1.5 MYA	Evidence of *Homo erectus* in Europe; evidence of chipped-stone tools in Africa; beginning of Pleistocene
700,000 BP	First evidence of chipped-stone tools in Europe
400,000 BP	*Homo erectus* uses fire; develops strategies for hunting large game; first appearance of archaic *Homo sapiens*
195,000–150,000 BP	Appearance of *Homo sapiens sapiens* (in Africa only, according to favoured replacement hypothesis)
80,000 BP	Beginnings of most recent glacial period
50,000 BP	Arrival of humans in Australia
25,000–15,000 BP	Arrival of humans in America
12,000 BP	First permanent settlements
10,000 BP	End of most recent glacial period; beginning of Holocene period

NOTE: MYA = million years ago; BP = before present.

when large areas of the earth were covered in ice, were punctuated by periods with average temperatures comparable to those of today—an alternation of glacial and interglacial phases. These fluctuations continued until as recently as 10,000 years ago and may indeed be continuing; it is possible that we are in an interglacial period at present.

MODERN HUMANS: HOMO SAPIENS SAPIENS

Some 400,000 BP (before present) a new hominid evolved from *Homo erectus*. This species was similar to modern humans, but because there are physical differences between them and modern humans, they are often called archaic *Homo sapiens*. Their mean brain size of 1,220 ml was about 85 per cent that of modern humans. The best-known subset of these archaic humans is the Neanderthals, who first appeared about 130,000 BP and lasted until about 30,000 BP. They lived throughout much of Eurasia during this time, although the population probably never exceeded about 15,000. In the early twentieth century, following the first fossil discoveries, Neanderthals were regarded as subhuman in intellect, but by the 1950s it was not uncommon for them to be regarded as virtually modern in appearance and behaviour. The current interpretation

Pleistocene
Geological time period, from about 1.5 million years ago to 10,000 years ago, characterized by a series of glacial advances and retreats; succeeded by the Holocene.

is a compromise—neither apes nor modern humans. Although Neanderthals coexisted with modern humans, recent genetic research provides no evidence that the two groups ever interbred.

The uncertainty about specific details of human evolution was highlighted by the 2004 discovery on the remote Indonesian island of Flores of what some believe to be a previously unknown *Homo erectus* species, *Homo floresiensis*, which survived until as recently as 13,000 BP. The presumed species was nicknamed the 'Hobbit' because it was no taller than 1 m (3 feet). Subsequent research has divided into two camps: those who claim the discovery was of a human suffering from a genetic illness that caused both body and brain to shrink, and those who find this is a new species.

We have now reached a critical point: the emergence of anatomically modern humans, officially called *Homo sapiens sapiens*. Where and how did this come about? There are two rather different explanations. The multi-regional hypothesis proposes that modern humans evolved independently from archaics, at more or less the same time in a number of different geographic areas.

However, all the currently available paleontological, genetic, and archaeological evidence favours a second hypothesis. This replacement hypothesis proposes a single evolution in a limited geographic area and a subsequent spread to replace archaics elsewhere. Specifically, modern humans, *Homo sapiens sapiens*, evolved from a group of archaics in a single place, in Africa, by 195,000–150,000 BP. This hypothesis received additional confirmation with the 2003 discovery in Ethiopia of anatomically modern human skulls dated to 160,000 BP and the 2005 re-dating of two similar skulls found in 1967 showing that they are about 195,000 years old.

These humans spread throughout the world, replacing existing archaic groups, including the Neanderthals, because of some adaptive advantage. From that time onward, human evolution, insofar as we can determine, has been cultural, not biological. Acceptance of the favoured replacement hypothesis, centred on a single exodus from Africa, has enormous implications for our understanding of ourselves. Simply put, we are all related. Modern humans

settled much of Africa, Asia, and Europe, with movement out of Africa begun about 60,000 BP. The routes and dates shown in Figure A2.1 reflect ongoing genetic research (part of a major study known as the Genographic Project). Of particular note is the evidence that the first humans to leave Africa travelled along the Indian Ocean coast to Arabia, India, and Indonesia, reaching Australia 50,000 BP. This route was probably favoured because of cold temperatures further north and because of the availability of a seafood diet. The next principal moves were into the Middle East, inland Asia, and Europe by about 35,000 BP, and America perhaps by 25,000 BP but certainly by 15,000 BP. These movements of humans were, of course, related to the continuing climatic changes associated with the glacial and interglacial periods. Modern humans began to evolve and spread across the earth during a period of harsh climate. Evidence suggests that favoured locations were at the boundaries of geographic regions, where the first permanent settlements appeared about 12,000 BP. The most recent glacial period ended 10,000 BP, marking the end of the Pleistocene and the beginning of the **Holocene** period.

The fact that early humans effectively settled most areas of the earth is particularly significant. It suggests that humans had some adaptive advantage over other primates, something that allowed them to adjust and survive in harsh new environments where other species could not; this advantage was probably cultural. Brain growth is important, of course, but it does not seem to be the whole story. Several related hypotheses have in common a focus not on competition but rather on social interaction. It seems clear that humans related to one another collaboratively because they trusted one another. Migration to previously unsettled areas enlarged the overall resource base and allowed slow but steady increases in human populations. We can only estimate the number of people in these early stages. *Australopithecus* was restricted to Africa, and population estimates range from 70,000 to 1 million. *Homo erectus* covered much more of the globe and possibly attained a population of 1.7 million. A figure of 4 million seems reasonable for the human population about 12,000 years ago.

Holocene
Literally, 'wholly recent': the post-glacial period that began 10,000 years ago; preceded by the Pleistocene.

FURTHER EXPLORATIONS

Dawkins, R. 2004. *The Ancestor's Tale: A Pilgrimage to the Dawn of Evolution*. Boston: Houghton Mifflin.

A wonderfully entertaining and informative book that looks at the evolutionary tree of life. Loosely based on Chaucer's *Canterbury Tales*.

Feder, K.L., and M.A. Park. 2001. *Human Antiquity: An Introduction to Physical Anthropology and Archaeology*, 4th edn. Toronto: Mayfield.

Introductory physical anthropology textbook with detailed discussion of the origin of humans.

Gore, R. 1997. 'The First Steps', *National Geographic* 191, 2: 72–99.

Well-illustrated, readable articles on human evolution are published regularly in this popular and accessible magazine.

MacDonald, G. 2003. *Biogeography: Space, Time, and Life*. New York: Wiley.

One chapter of this book offers an excellent overview of the relationships between biogeography and human evolution, including an account of the geographic expansion of modern humans.

Passingham, R.E. 1982. *The Human Primate*. Oxford: Freeman.

An excellent book that discusses humans as an animal species.

Powell, J.F. 2005. *The First Americans: Race, Evolution, and the Origin of Native Americans*. New York: Cambridge University Press.

Excellent summary of this complex issue with detailed accounts of several controversies. Uses new archaelogical and geological evidence to question the land bridge hypothesis.

ON THE WEB

HUMAN EVOLUTION AND EARLY MIGRATIONS

The Genographic Project

www3.nationalgeographic.com/genographic/index.html

This remarkable site provides a wealth of information on the Genographic Project.

Glossary

accessibility

A variable quality of a location, expressing the ease with which it may be reached from other locations.

acculturation

The process by which an ethnic individual or group is absorbed into a larger society while retaining aspects of its distinct identity; compare 'assimilation'.

acid rain

The deposition on the earth's surface of sulphuric and nitric acids formed in the atmosphere as a result of fossil fuel and biomass burning; causes significant damage to vegetation, lakes, wildlife, and built environments.

adaptation

The process by which humans adjust to a particular set of circumstances; changes in behaviour that reduce conflict with the environment. See 'cultural adaptation'.

agglomeration

The spatial grouping of humans or human activities to minimize the distances between them.

agricultural revolution

The slow transition in the human way of life, beginning about 12,000 years ago, from foraging to food production through plant and animal domestication.

alienation

The circumstance in which a person is indifferent to or estranged from nature or the means of production.

anarchism

A political philosophy that rejects the state and argues that social order is possible without a state.

animism

A general name for beliefs that attribute a spirit or soul to natural phenomena and inanimate objects.

anthropocentric

Regarding humans as the central fact of the world; stressing the centrality of humans to the detriment of the rest of the world.

apartheid

The South African policy by which four groups of people, as defined by the authorities, were spatially separated between 1948 and 1994.

areal differentiation

From Hartshorne, a synonym for 'regional geography'.

artifacts

The physical objects created by a culture for pleasure (art, toys), work (tools), living (vessels for cooking), or worship (crucifix).

assimilation

The process by which an ethnic group is absorbed into a larger society and loses its own identity; compare 'acculturation'.

authority

The power or right, usually mutually recognized, to require and receive the obedience of others.

back-office activities

Repetitive office operations, usually performed using telecommunications, that can be located anywhere in the city, or indeed outside the city, including in relatively low-rent areas.

biomass

The mass of biological material present in an area, including both living and dead plant material.

capitalism

A social and economic system for the production of goods and services based on private enterprise; under capitalism the producer is separated from the means of production, and the power to exploit resources is limited to relatively few individuals.

carrying capacity

The maximum population that can be supported by a given set of resources.

cartography

The conception, production, dissemination, and study of maps.

caste

A social rank, based solely on birth, to which an individual belongs for life and that limits interaction with members of other castes.

catastrophists

Those who argue that population increases and continuing environmental deterioration are leading to a nightmarish future of food shortages, disease, and conflict.

ceiling rent

The maximum rent that a potential land user can be charged for use of a given piece of land.

census

The periodic collection and compilation of demographic and other data relating to all individuals in a given country at a particular time.

centrifugal forces

In political geography, forces that make it difficult to bind an area together as an effective state; in urban geography, forces that favour the decentralization of urban land uses.

centripetal forces

In political geography, forces that pull an area together as one unit to create a relatively stable state; in urban geography, forces that favour the concentration of urban land uses in a central area.

chain migration

A process of movement from one location to another through time sustained by social links of kinship or friendship; often results in distinct areas of ethnic settlement in rural or urban areas.

chorography

A Greek term revived by nineteenth-century German geographers who used it to refer to regional descriptions of large areas.

chorology

A Greek term revived by nineteenth-century German geographers as a synonym for 'regional geography'.

choropleth map

A thematic map using colour (or shading) to indicate density of a particular phenomenon in a given area.

civilization

A culture with agriculture and cities, food and labour surpluses, labour specialization, social stratification, and state organization.

class

A large group of individuals of similar social status, income, and culture.

Cold War

The period of confrontation without direct military conflict between Western (led by the US) and Communist (led by the USSR) powers that began shortly after the end of World War II and lasted until the early 1990s.

collective consumption

The use of services produced and managed on a collective basis.

colonialism

The policy of a state or people seeking to establish and maintain authority over another state or people.

commercial agriculture

An agricultural system in which the production is primarily for sale.

competitive capitalism

The first of three phases of capitalism, beginning in the early eighteenth century (although some have identified signs of capitalism as early as the late sixteenth century); characterized by free-market competition and laissez-faire economic development.

conservation

A general term referring to any form of environmental protection, including preservation.

constructionism

The school of thought according to which all our conceptual underpinnings (for example, ideas about identity) are socially constructed and therefore contingent and dynamic, not given or absolute (compare 'essentialism').

contextualism

Broadly, the idea that it is necessary to take into account the specific context within which any research is conducted.

continental drift

The idea that the present continents were originally connected as one or two land masses that separated and drifted apart.

conurbation

A continuously built-up area formed by the coalescing of several expanding cities that were originally separate.

core/periphery

The concept that states are often unequally divided between powerful cores and dependent peripheries.

cornucopians

Those who argue that advances in science and technology will continue to create resources sufficient to support the growing world population.

cosmography

The science that maps and describes the entire universe, both heavens and earth; this Greek term is rarely used today.

counter-urbanization

A process of population decentralization that may be prompted by several factors, including the high cost of living in cities, improvements in personal spatial mobility, industrial deconcentration, and advances in information technologies.

creole

A pidgin language that assumes the status of a mother tongue for a group.

critical geography

A collection of ideas and practices concerned with challenging inequalities as these are evident in landscape; focus is critical of capitalism, emphasizes political change and social justice.

cultural adaptation

Changes in technology, organization, and ideology that permit sound relationships to develop between humans and their physical environment.

cultural regions

Areas in which there is a degree of homogeneity in cultural characteristics; areas with similar landscapes.

culture

The way of life of the members of a society.

cybercity

A postmodern circumstance. A city with parts that are so closely connected that location is less important than in the past.

cycle of poverty
The idea that poverty and deprivation are transmitted intergenerationally, reflecting home background and spatial variations in opportunities.

deconstruction
A method of critical interpretation applied to texts, including landscapes, that aims to show how the multiple positioning (in terms of class, gender, and so on) of an author or a reader affects the creation or reading of the text.

deglomeration
The spatial separation of humans or human activities so as to maximize the distances between them.

deindustrialization
Loss of manufacturing activity and related employment in a traditional manufacturing region in the more developed world.

democracy
A form of government involving free and fair elections, openness and accountability, civil and political rights, and the rule of law.

demographic transition
The historical shift of birth and death rates from high to low levels in a population. Mortality declines before fertility resulting in substantial population increase during the transition phase.

demography
The study of human populations.

density
A measure of the number of geographic facts (for example, people) per unit area.

dependence, dependency
In political contexts, a relationship in which one state or people is dependent on, and therefore dominated by, another state or people.

dependency theory
Centres on the relationship between dependence and under-development.

desertification
The process by which an area of land becomes a desert; typically involves the impoverishment of an ecosystem because of climatic change and/or human impact.

development
A term that should be handled with caution because it has often been used in an ethnocentric fashion; typically understood to refer to a process of becoming larger, more mature, and better organized; often measured by economic criteria.

developmentalism
Analysis of cultural and economic change that treats each country or region of the world separately in an evolutionary manner; assumes that all areas are autonomous and proceed through the same series of stages.

devolution
A process of transferring power from central to regional or local levels of government.

dictatorship
An oppressive, anti-democratic form of government in which the leader is often backed by the military.

diffusion
The spread of any phenomenon over space and its growth through time.

discourse
A system of ideas or knowledge that serves as the context through which new facts and ideas are understood. Discourses are used to legitimate the exercise of power.

disorganized capitalism
The most recent form of capitalism, characterized by a process of disorganization and industrial restructuring.

distance
The spatial dimension of separation; a fundamental concept in spatial analysis.

distance decay
The declining intensity of any pattern or process with increasing distance from a given location.

distribution
The pattern of geographic facts (for example, people) within an area.

domestication
The process of making plants and/or animals more useful to humans through selective breeding.

donut effect
A term that refers to a pronounced difference in the growth rates between a core city (slow growth or no growth) and its surrounding areas (faster growth).

doubling time
The number of years required for the population of an area to double its present size, given the current rate of population growth.

dual city
A postmodern circumstance; a city with areas occupied by the rich and powerful and areas occupied by the poor and powerless. Divisions may reflect ethnic, religious, and gender differences.

ecocentric
Emphasizing the value of all parts of an ecosystem rather than, for example, placing humans at the centre, as in an anthropocentric emphasis.

ecology
The study of relationships between organisms and their environments.

economic base theory
A theory that tries to explain the growth or decline of particular regions or cities in terms of 'basic' and 'non-basic' economic activities: 'basic' goods and services are those produced for sale outside the city or region.

economic operator
A model of human behaviour in which each individual is assumed to be completely rational; economic operators maximize returns and minimize costs.

economic rent
The surplus income that accrues to a unit of land above the minimum income needed to bring a unit of new land into production at the margins of production.

ecosystem
An ecological system; comprises a set of inter-acting and interdependent organisms and their physical, chemical, and biological environment.

ecotourism
Tourism that is environmentally friendly and allows participants to experience a distinctive ecosystem.

edge city
A centre of office and retail activities located on the edge of a large urban centre.

empiricism
A philosophy of science based on the belief that all knowledge results from experience and, therefore, gives priority to factual observations over theoretical statements.

energy
The capacity of a physical system for doing work.

entropy
A measure of the disorder or disorganization in a system.

environmental determinism
The view that human activities are controlled by the physical environment.

epidemiological transition
A process associated with reductions in fertil-ity and improvements in overall health. As the transition progresses, death and disability from communicable diseases decline in importance relative to problems resulting from non-com-municable conditions.

equinox
Occurs twice each year, in spring and fall, when the sun is vertically overhead at the equator; on this day the periods of daylight and darkness are both 12 hours long all over the earth.

essentialism
Belief in the existence of fixed unchanging properties; attribution of 'essential' characteris-tics to groups; compare 'constructionism'.

ethnic group
A group whose members perceive themselves as different from others because of a common ancestry and shared culture.

ethnocentrism
A form of prejudice or stereotyping that pre-sumes that one's own culture is normal and nat-ural and that all other cultures are inferior.

ethnography
The study and description of social groups based on researcher involvement and first-hand obser-vation in the field; a qualitative rather than quan-titative approach.

existentialism
A philosophy according to which humans are considered to be responsible for making their own natures; stresses personal freedom, decision-making, and commitment in a world without absolute values outside of individuals' personal preferences.

exonym
A name given to people (or a place) by a group other than the people to which the name refers (or who are not native to the territory within which the place is situated).

exurbanization
The movement of households from urban areas to locations outside the urban area but within the commuting field.

fecundity
A biological term; the ability of a woman or man to produce a live child; refers to potential rather than actual number of live births.

federalism
A form of government in which power and authority are divided between central and regional governments.

feminism
The movement for and advocacy of equal rights for women and men, and commitment to improve the position of women in society.

fertility
Generally, all aspects of human reproduction that lead to live births; also used specifically to refer to the actual number of live births produced by a woman.

feudalism
A social and economic system prevalent in Europe, prior to the Industrial Revolution, in which land was owned by the monarch, con-trolled by lords, and worked by peasants who were bound to the land and subject to the lords' authority.

fieldwork
A means of data collection; includes both quali-tative (for example, observation) and quantitative (for example, questionnaire) methods.

filtering
A term used to refer to the fact that, through time, housing units typically experience a transi-tion from being occupied by members of one income group to members of a different income group. Downward filtering is more usual than upward filtering.

first effective settlement
A concept based on the likely importance of the initial occupancy of an area in determining later landscapes.

flexible accumulation
Industrial technologies, labour practices, rela-tions between firms, and consumption patterns that are increasingly flexible.

forces of production
A Marxist phrase that refers to the raw materials, tools, and workers that actually produce goods.

Fordism
A group of industrial and broader social prac-tices introduced by Henry Ford, including use of the mass-production assembly line and

development of a new relationship with workers characterized by higher wages and shorter working hours; dominant until recently in most industrial countries.

foreign direct investment

Direct investment by a government or multinational corporation in another country, often in the form of a manufacturing plant.

formal region

A region identified as such because of the presence of some particular characteristic(s).

friction of distance

A measure of the restraining effect of distance on human movement.

front-office activities

Skilled occupations requiring an educated, well-paid workforce. Image is important and, although telecommunications play a key role, so is face-to-face contact with others. Hence, these activities favour prestige locations in major office buildings in city centres.

functional region

A region that comprises a series of linked locations.

garden city

A planned settlement designed to combine the advantages of urban and rural living; an urban centre emphasizing spaciousness and quality of life.

gated community

A high-quality residential subdivision or community with access limited to residents and other authorized people such as domestic workers, tradespeople, and visitors. Often surrounded by a perimeter wall, fence, or a buffer zone such as a golf course.

gateway city

A city located at a key point of entry to a major geographic region or country, often a port or major rail centre, through which goods and people pass and in which several different cultural traditions are absorbed and assimilated.

Gemeinschaft

A term introduced by Tönnies; a form of human association based on loyalty, informality, and personal contact; assumed to be characteristic of traditional village communities.

gender

The social aspect of the relations between the sexes.

genocide

An organized, systematic effort to destroy a group defined in ethnic terms; usually the targeted group is seen as living in the 'wrong place'.

gentrification

A process of inner-city urban neighbourhood social change resulting from the in-movement of higher-income groups.

geographic information system (GIS)

A computer-based tool that combines the storage, display, analysis, and mapping of spatially referenced data.

géographie Vidalienne

French school of geography initiated by Paul Vidal de la Blache at the end of the nineteenth century and still influential today, focusing on the study of human-made (cultural) landscapes.

geopolitics

The study of the importance of space in understanding international relations.

Geopolitik

The study of states as organisms that choose to expand in territory in order to fulfill their 'destinies' as nation-states.

gerrymandering

The realignment of electoral boundaries to benefit a particular political party.

Gesellschaft

A term introduced by Tönnies; a form of human association based on rationality and depersonalization; assumed to be characteristic of urban dwellers.

ghetto

A residential district in an urban area with a concentration of a particular ethnic group.

globalization

A complex combination of economic, political, and cultural changes that have long been evident but that have accelerated markedly since about 1980, bringing about a seemingly ever-increasing connectedness of both people and places.

green belt

A planned area of open, partially rural, land surrounding an urban area.

gross domestic product (GDP)

A monetary measure of the value at market prices of goods and services produced by a country over a given time period (usually one year); provides a better indication of domestic production than GNP.

gross national product (GNP) or gross national income (GNI)

A monetary measure of the value at market prices of goods and services produced by a country, plus net income from abroad, over a given period (usually one year).

guild

An association of people who share the same trade or job skill. Historically, guilds were formed to protect shared interests and to ensure some uniformity of practice.

heartland theory

A geopolitical theory of world power based on the assumption that the land-based state controlling the Eurasian heartland held the key to world domination.

hegemony

A social condition in which members of a society interpret their interests in terms of the world view of a dominant group. Thus, dominant groups are dominant at least partly because they have the, perhaps unintended, support of other groups.

historical materialism

An approach associated with Marxism centred on the material basis of society that attempts to understand social change by reference to historical changes in social relations.

Holocene

Literally, 'wholly recent': the post-glacial period that began 10,000 years ago; preceded by the Pleistocene.

homeland

A cultural region especially closely associated with a particular cultural group; the term usually suggests a strong emotional attachment to place.

hominids

Bipedal primates; humans and their ancestors.

humanism

A philosophy centred on such aspects of human life as value, quality, meaning, significance, and spirituality.

hybrid city

A postmodern circumstance. A city lacking conventional communities, containing instead new cultural categories including cultural hybrids.

hypothesis

In positivist philosophy, a general statement deduced from theory but not yet verified.

iconography

The description and interpretation of visual images, including landscape, in order to uncover their symbolic meanings; the identity of a region as expressed through symbols.

idealism

A humanistic philosophy according to which human actions can be understood only by reference to the thought behind them.

ideal type

A term introduced by the sociologist Max Weber to refer to a hypothetical norm used to deepen understanding of the real world through comparison.

idiographic

Concerned with the unique and particular.

image

The perception of reality held by an individual or group.

imperialism

A relationship between states in which one is dominant over the other.

Industrial Revolution

The process that converted a fundamentally rural society into an industrial society beginning in England *c.* 1750; primarily a technological revolution associated with new energy sources.

informal sector

A part of a national economy involved in productive labour, but without any formal recognition, control, or remuneration.

infrastructure (base)

A Marxist concept that refers to the economic structure of a society, especially as it gives rise to political, legal, and social systems.

innovation

The introduction of a new invention or idea, especially one that leads to change in human behaviour or production processes.

interaction

The relationship or linkage between locations.

interaction theory

A body of theories explaining movements of goods and people between locations.

international division of labour

A term referring to the current tendency for high-wage and high-skill employment opportunities, often in the service sector, to be located in the more developed world, whereas low-wage and low-skill employment opportunities, often in the industrial sector, are located in the less developed world.

irredentism

The view held by one country that a minority living in an adjacent country rightfully belongs to the first country.

isochrones

Lines on a map of equal travel time from a given starting point. One example of an isoline. Isolines generally allow map readers to infer change with distance and to estimate specific values at any location on the map.

isodapanes

In Weberian, least-cost, industrial-location theory, lines of equal additional transport cost drawn around the point of minimum transport cost.

isopleth map

A map using lines to connect locations of equal data value.

isotims

In Weberian least-cost industrial location theory, lines of equal transport costs around material sources and markets.

landscape

A major concern of geographic study; the characteristics of a particular area especially as created through human activity.

landscape school

American school of geography initiated by Carl Sauer in the 1920s and still influential today; an alternative to environmental determinism, focusing on human-made (cultural) landscapes.

Landschaftskunde

A German term, introduced in the late nineteenth century, best translated as 'landscape science'; refers to geography as the study of the landscapes of particular regions.

latitude

Angular distance on the surface of the earth, measured in degrees, minutes, and seconds, north and south of the equator (which is the line of 0° latitude); lines of constant latitude are called parallels.

law

In positivist philosophy, a hypothesis that has been proven correct and is taken to be universally true; once formulated, laws can be used to construct theories.

less developed world

All countries not classified as 'more developed' (see 'more developed world').

life cycle

The process of change experienced by individuals over their lifespans; often divided into stages (such as childhood, adolescence, adulthood, old age), each of which is associated with particular forms of behaviour.

limits to growth

The argument that both world population and world economy may collapse because available world resources are inadequate.

lingua franca

An existing language used as a common means of communication between different language groups; compare 'pidgin'.

lithosphere

The outer layer of rock on earth; includes crust and upper mantle.

locale

The setting or context for social interaction; a term introduced in structuration theory that has become popular in human geography as an alternative to 'place'.

location

A term that refers to a specific part of the earth's surface; an area where something is situated.

locational interdependence

The situation in which competing businesses base their location decisions on those of their rivals.

location theory

A body of theories explaining the distribution of economic activities.

longitude

Angular distance on the surface of the earth, measured in degrees, minutes, and seconds, east and west of the prime meridian (the line of 0° longitude that runs through Greenwich, England); lines of constant longitude are called meridians.

malapportionment

A form of gerrymandering, involving the creation of electoral districts of varying population sizes so that one party will benefit.

malnutrition

A condition caused by a diet lacking some food necessary for health.

Maoism

The revolutionary thought and practice of Mao Zedong (1893–1976), based on protracted revolution to achieve power and socialist policies after power is achieved.

Marxism

The body of social and political theory developed by Karl Marx (1818–83), in which mode of production is seen as the key to understanding society. Marxism emphasizes the importance of labour; argues that throughout history the state

has enabled a small dominant class to exploit the masses; and sees class struggle as the key to historical change.

material index

An index devised by Weber and used in industrial location theory to show the extent to which the least-cost location for a particular industrial firm will be either material- or market-oriented.

mental map

The individual psychological representation of space.

mentifacts

Those mental or non-physical elements of culture; the values held by members of a group.

mercantilism

A school of economic thought dominant in Europe in the seventeenth and early eighteenth centuries that argued for the involvement of the state in economic life so as to increase national wealth and power.

minority language

A language spoken by a minority group in a state in which the majority of the population speaks some other language; may or may not be an official language.

mobility

The ability to move from one location to another.

model

An idealized and structured representation of the real world.

mode of production

A Marxist term that refers to the organized social relations through which a human society organizes productive activity; human societies are seen as passing through a series of such modes.

modernism

A view that assumes the existence of a reality characterized by structure, order, pattern, and causality.

monarchy

The institution of rule over a state by the hereditary head of a family; monarchists are those who favour this system.

more developed world

(According to a United Nations classification) Europe, North America, Australia, Japan, New Zealand, and the former USSR.

mortality

Deaths as a component of population change.

multiculturalism

A policy that endorses the right of ethnic groups to remain distinct rather than to be assimilated into a dominant society.

multilingual state

A state in which the population includes at least one linguistic minority.

nation

A group of people sharing a common culture and an attachment to some territory; a term difficult to define objectively.

nationalism
The political expression of nationhood or aspiring nationhood; reflects a consciousness of belonging to a nation.

nation-state
A political unit that contains one principal national group that gives it its identity and defines its territory.

nativism
Intense opposition to an internal minority on the grounds that the minority is foreign.

neighbourhood
A formal region inside a city. A part of the city that displays some internal homogeneity regarding type of housing and that may be characterized by a relatively uniform income level and/or ethnic identity and that usually reflects certain shared social values.

neo-colonialism
Economic relationships of dominance and subordination between countries without equivalent political relationships; often develops after political colonialism ends and the former colony achieves independence, but may also occur without prior political colonialism.

nomothetic
Concerned with the universal and the general.

normative
Focusing on what ought to be rather than what actually is; in normative theory, the aim is to seek what is rational or optimal according to some given criteria.

nuptiality
The extent to which a population marries.

offshoring
The transferring by a company of production or service provision to another country.

oligarchy
Rule by an elite group of people, typically the wealthy.

organized capitalism
The second phase of capitalism, beginning after World War II; characterized by increased growth of major corporations and increased involvement by the state in the economy.

Orientalism
A term used to refer to Western views of the Orient, implying a view of the periphery from the centre; although the term is closely associated with post-colonial theory, especially the work of Edward Said, the idea of 'the Orient' as a foil for the West dates back to classical Greece.

other
Philosophically, an ambiguous term, derived from Hegel and widely used in feminist and post-colonial theory, referring to subordinate groups as they are seen by and contrasted to dominant groups; implies both difference and inferiority.

outsourcing
Paying an outside firm to handle functions previously handled inside the company (or government) with the intent to save money or improve quality; a term often used interchangeably (and incorrectly) with 'offshoring'. Much outsourcing does involve offshoring jobs from more developed countries such that they become much lower-paid jobs in less developed countries.

ozone layer
Layer in the atmosphere 16–40 km (10–25 miles) above the earth that absorbs dangerous ultraviolet solar radiation; ozone is a gas composed of molecules consisting of three atoms of oxygen (O_3).

pandemic
A term used to designate diseases with very wide distribution (a whole country, or even the world); 'epidemic' diseases have more limited distribution.

participant observation
A qualitative method in which the researcher is directly involved with the subjects in question.

patriarchy
A social system in which men dominate, oppress, and exploit women.

perception
The process by which humans acquire information about physical and social environments.

phenomenology
A humanistic philosophy based on the ways in which humans experience everyday life and imbue activities with meaning.

phenotype
Any physical or chemical trait that can be observed or measured.

physiological density
Population per unit of cultivable land.

pidgin
A new language designed to serve the purposes of commerce between different language groups; typically has a limited vocabulary; compare 'lingua franca'.

place
Location; in humanistic geography, 'place' has acquired a particular meaning as a context for human action that is rich in human significance and meaning.

placelessness
Homogeneous and standardized landscapes that lack local variety and character.

place utility
A measure of the satisfaction an individual derives from a location relative to his or her goals.

Pleistocene
Geological time period, from about 1.5 million years ago to 10,000 years ago, characterized by a series of glacial advances and retreats; succeeded by the Holocene.

pollution
The release into the environment of substances that degrade land, air, or water.

population aging

A process in which the proportion of elderly people in a population increases and the proportion of younger people decreases, resulting in increased median age of the population.

population momentum

The tendency for population growth to continue beyond the time that replacement-level fertility has been reached because of the relatively high number of people in the child-bearing years.

population pyramid

A diagrammatic representation of the age and sex composition of a population. By convention, the younger ages are at the bottom, males are on the left, and females on the right.

positionality

The ideological preference and the identity (for example, ethnic background, language, religion, gender, sexual preference, and age) of the researcher as this relates to the subjects of the research.

positivism

A philosophy that contends that science is able to deal only with empirical questions (those with factual content), that scientific observations are repeatable, and that science progresses through the construction of theories and derivation of laws.

possibilism

The view that the environment does not determine either human history or present conditions; rather, humans pursue a course of action that they select from among a number of possibilities.

post-Fordism

A group of industrial and broader social practices evident in industrial countries since about 1970; involves more flexible production methods than those associated with Fordism.

postmodernism

A movement in philosophy, social science, and the arts based on the idea that reality cannot be studied objectively and that multiple interpretations are possible; compare 'positivism'.

power

The capacity to affect outcomes; more specifically, to dominate others by means of violence, force, manipulation, or authority.

pragmatism

A humanistic philosophy that focuses on the construction of meaning through the practical activities of humans.

primary activities

Economic activities concerned directly with the collection and utilization of natural resources.

primate city

The largest city in a country, usually the capital city, which dominates its political, economic, and social life.

principle of least effort

Considered to be a guiding principle in human activities; for human geographers, refers to minimizing distances and related movements.

producer services

Activities that offer a wide range of services to multinational and other companies that need to respond quickly to changing circumstances, including banking, insurance, marketing, accountancy, advertising, legal matters, consultancy, and innovation services; in recent years, the fastest-growing sector of national economies in most of the more developed countries.

projection

Any procedure employed to represent positions of all or a part of the earth's spherical (three-dimensional) surface onto a flat (two-dimensional) surface.

proxemics

The study of personal space, the invisible boundaries that protect and compartmentalize each individual.

public goods

Goods that are freely available to all or that are provided (equally or unequally) to citizens by the state.

qualitative methods

A set of tools used to collect and analyze data in order to subjectively understand the phenomena being studied; the methods include passive observation, participation, and active intervention.

quantitative methods

A set of tools used to collect and analyze data to achieve a statistical description and scientific explanation of the phenomena being studied; the methods include sampling, models, and statistical testing.

quaternary activities

Economic activities concerned with handling or processing knowledge and information; typically involve a high level of skill; highly specialized.

queer theory

Ideas developed in gay and lesbian studies and concerned with oppressed sexualities in terms of both social rights and cultural politics.

questionnaire

A structured and ordered set of questions designed to collect unambiguous and unbiased data.

race

A subspecies; a physically distinguishable population within a species.

racism

A particular form of prejudice. Attributing characteristics of superiority or inferiority to a group of people who share some physically inherited characteristics.

raster

A method used in GIS to represent spatial data; divides the area into numerous small cells and pixels, and describes the content of each cell.

rational choice theory

The theory that social life can be explained by models of rational individual action; an extension of the economic operator concept to other areas of human life.

recycling
The reuse of material and energy resources.

redlining
A spatially discriminatory practice favoured by financial institutions that identified parts of the city regarded as high risk in terms of loans for property purchase and for home improvement. The affected areas were typically outlined in red on maps.

region
A part of the earth's surface that displays internal homogeneity and is relatively distinct from surrounding areas according to some criterion or criteria; regions are intellectual creations.

regionalization
A special kind of classification in which locations on the earth's surface are assigned to various regions, which must be contiguous spatial units.

reindustrialization
The development of new industrial activity in a region that has earlier experienced substantial loss of traditional industrial activity; see 'deindustrialization'.

relations of production
A Marxist term referring to the ways in which the production process is organized, specifically the relationships of ownership and control.

remote sensing
A variety of techniques used for acquiring and recording data from points that are not in contact with the phenomena of interest.

renewable resources
Resources that regenerate naturally to provide a new supply within a human lifespan.

replacement-level fertility
The level of fertility at which a couple has only enough children to replace themselves.

representation
A depiction of the world, acknowledging impossibility to be exact as all such depictions are affected by the researcher's identity.

restructuring
In a capitalist economy, changes in or between the various components of an economic system resulting from economic change.

rimland theory
A geopolitical theory of world power based on the assumption that the state controlling the area surrounding the Eurasian heartland held the key to world domination.

sacred space
A landscape particularly esteemed by an individual or a group, usually (but not necessarily) for religious reasons.

sampling
The selection of a subset from a defined population of individuals to acquire data representative of that larger population.

satisficing behaviour
A model of human behaviour that rejects the rationality assumptions of the economic operator model, assuming instead that the objective is to reach a level of acceptable satisfaction.

scale
The resolution levels used in any human geographic research; most characteristically refers to the size of the area studied, but also to the time period covered and the number of people investigated.

scientific method
The various steps taken in a science to obtain knowledge; a phrase most commonly associated with a positivist philosophy.

secondary activities
Economic activities that process, transform, fabricate, or assemble raw materials derived from primary activities; also activities that reassemble, refinish, or package manufactured goods.

sense of place
The deep attachments that humans have to specific locations such as home and also to particularly distinctive locations.

sexism
Attitudes or beliefs that serve to justify sexual inequalities by incorrectly attributing or denying certain capacities either to women or to men.

sex ratio
The number of males per 100 females in a population.

sexuality
In some feminist and psychoanalytic theory, interpreted as a cultural construct rather than as a biological given. Aligned with power and control.

simulation
Representation of a real-world process in an abstract form for purposes of experimentation.

site
The location of a geographic fact with reference to the immediate local environment.

situatedness
An idea that rejects notions of researcher authority and impartiality—knowledge is not neutral and cannot be acquired in some detached and disembodied manner. Rather, all knowledge is partial and located somewhere.

situation
The location of a geographic fact with reference to the broad spatial system of which it is a part.

slavery
Labour that is controlled through compulsion and is not remunerated; in Marxist terminology, one particular mode of production.

socialism
A social and economic system that involves common ownership of the means of production and distribution.

social physics
An approach to aggregate human movement and interaction based on the assumption that such phenomena are analogous to processes in the physical sciences.

society
Refers to the interrelationships that connect individuals as members of a culture.

sociofacts
Those elements of culture most directly concerned with interpersonal relations; the norms that people are expected to observe.

solstice
Occurs twice each year when the sun is vertically overhead at the farthest distance from the equator, once 23.5°N and once 23.5°S of the equator; for example, when at 23.5°N, there is maximum daylight (longest day) in the northern hemisphere and minimum daylight (shortest day) in the southern hemisphere.

sovereignty
Supreme authority over the territory and population of a state, vested in its government; the most basic right of a state understood as a political community.

space
Areal extent; a term used in both absolute (objective) and relative (perceptual) forms.

spatial monopoly
The situation in which a single producer sells the entire output of a particular industrial good or service in a given area.

spatial preferences
Individual (sometimes group) evaluation of the relative attractiveness of different locations.

spatial separatism (or spatial fetishism)
A phrase critical of spatial analysis for treating space or distance as a cause without reference to humans.

species
A group of organisms able to produce fertile offspring among themselves but not with any other group.

spectacle
A term referring to places and events that are carefully constructed for the purposes of mass leisure and consumption.

squatter settlement
A shanty town; a concentration of temporary dwellings, neither owned nor rented, at the city's edge; related to rural-to-urban migration, especially in less developed countries.

state
An area with defined and internationally acknowledged boundaries; a political unit.

state apparatus
The institutions and organizations through which the state exercises its power.

stock resources
Minerals and land that take a long time to form and hence, from a human perspective, are fixed in supply.

subsistence agriculture
An agricultural system in which the production is not primarily for sale but is consumed by the farmer's household.

suburb
An outer commuting zone of an urban area; associated with social homogeneity and a lifestyle suited to family needs.

superorganic
An interpretation of culture that sees it as above both nature and individuals and therefore as the principal cause of the human world; a form of cultural determinism.

superstructure
A Marxist concept that refers to the political, legal, and social systems of a society.

sustainable development
A term popularized by the 1987 report of the World Commission on Environment and Development, referring to economic development that sustains the natural environment for future generations.

symbolic interactionism
A group of social theories that see the social world as a social product with meanings resulting from interaction.

system
A set of interrelated components or objects linked together to form a unified whole.

tariff
A tax or customs duty on imports from other countries.

technology
The ability to convert energy into forms useful to humans.

telecommuters
People who work at home using a computer to complete their tasks and to communicate with others.

teleology
The doctrine that everything in the world has been designed by God; also refers to the study of purposiveness in the world and to a recurring theme in history, such as progress or class conflict.

tertiary activities
Economic activities involving the sale or exchange of goods and services; includes distributive trades such as wholesaling and retailing, and also personal services.

text
A term that originally referred to the written or printed page, but that has broadened to include such products of culture as maps and landscape; postmodernists recognize that there may be any number of realities, depending on how a text is read.

theory
In positivist philosophy, an interconnected set of statements, often called assumptions or axioms, that deductively generates testable hypotheses.

time–space convergence
A decrease in the friction of distance between locations as a result of improvements in transportation and communication technologies.

topography
A Greek term, revived by nineteenth-century German geographers to refer to regional descriptions of local areas.

topological map
A diagram that represents a network as a simplified series of straight lines.

toponym
Place name; evidence provided by place names can be crucial in a historical study of movement and settlement if other sources of information are unavailable.

topophilia
The affective ties that humans have with particular places; literally, love of place.

topophobia
The feelings of dislike, anxiety, fear, or suffering associated with a particular landscape.

transnational
A large business organization that operates in two or more countries.

undernutrition
Diet inadequate to sustain normal activity.

urbanism
The urban way of life; associated with a declining sense of community and increasingly complex social and economic organization as a result of increasing size, density, and heterogeneity.

urbanization
The spread and growth of cities.

urban sprawl
The largely unplanned expansion of an urban area; typically discontinuous, leaving rural enclaves.

vector
A method used in GIS to represent spatial data; describes the data as a collection of points, lines, and areas, and describes the location of each of these.

vernacular region
A region identified on the basis of the perceptions held by people inside and outside the region.

verstehen
A research method, associated primarily with phenomenology, in which the researcher adopts the perspective of the individual or group under investigation; a German term, best translated as 'sympathetic' or 'empathetic understanding'.

welfare geography
An approach to human geography that maps and explains social and spatial variations.

well-being
The degree to which the needs and wants of a society are satisfied.

world systems theory
A body of ideas that suggests a division of the world into a core, semi-periphery, and periphery, stressing that the periphery is dependent on the core; has numerous implications for an understanding of the less developed world.

References

INTRODUCTION

Bone, R.M. 1992. *The Geography of the Canadian North*. Toronto: Oxford University Press.

Fairgrieve, J. 1926. *Geography in School*. London: University of London Press.

Gritzner, C.F. 2002. 'What Is Where, Why There, and Why Care?', *Journal of Geography* 101: 38–40.

Harding, K. 2007. '"It's Just Too Late in Nunavut"', *Globe and Mail*, 12 Jan.

Lewis, P.F. 2002. *Careers in Geography*. Washington: Association of American Geographers.

CHAPTER 1

Chiasson, P. 2006. *The Island of Seven Cities: Where the Chinese Settled When They Discovered North America*. Toronto: Random House Canada.

Hart, J.F. 1982. 'The Highest Form of the Geographer's Art', *Annals, Association of American Geographers* 72: 1–29.

Hartshorne, R. 1939. *The Nature of Geography: A Critical Survey of Current Thought in the Light of the Past*. Lancaster, Penn.: Association of American Geographers.

Heffernan, M. 2003. 'Histories of Geography', in S.L. Holloway, S.P. Rice, and G. Valentine, eds, *Key Concepts in Geography*. London: Sage, 3–22.

Johnson, D. 2005. 'From a Lost World: A New Breed of Romantic Hero—Explorer, Writer, Polymath', *Times Literary Supplement*, 22 July, 3–4.

May, J.A. 1970. *Kant's Concept of Geography and Its Relation to Recent Geographical Thought*. Toronto: University of Toronto Press.

Menzies, G. 2002. *1421: The Year China Discovered the World*. London: Bantam Press.

Nisbet, R. 1980. *History of the Idea of Progress*. New York: Basic Books.

Ward, R.G. 1960. 'Captain Alexander Maconochie, R.N., K.H., 1787–1860', *Geographical Journal* 126: 459–68.

Wright, J.K. 1926. 'A Plea for the History of Geography', *Isis* 8: 477–91.

———. 1947. 'Terra Incognitae: The Place of Imagination in Geography', *Annals, Association of American Geographers* 37: 1–15.

CHAPTER 2

Bunge, W. 1968. 'Fred K. Schaefer and the Science of Geography', *Harvard Papers in Theoretical Geography, Special Papers Series*, Paper A (mimeographed).

———. 1979. 'Fred K. Schaefer and the Science of Geography', *Annals, Association of American Geographers* 69: 128–32 (condensed version of Bunge, 1968).

Dohrs, F.E., and L.M. Sommers, eds. 1967. *Introduction to Geography: Selected Readings*. New York: Crowell.

Febvre, L. 1925. *A Geographical Introduction to History*. London: Routledge & Kegan Paul.

Hartshorne, R. 1955. '"Exceptionalism in Geography" Re-examined', *Annals, Association of American Geographers* 45: 205–44.

———. 1959. *Perspective on the Nature of Geography*. Chicago: Rand McNally.

Holland, P., et al. 1991. 'Qualitative Resources in Geography', *New Zealand Journal of Geography* 92: 1–28.

Huntington, E. 1927. *The Human Habitat*. New York: Van Nostrand.

Jackson, R.H., and R. Henrie. 1983. 'Perceptions of Sacred Space', *Journal of Cultural Geography* 3, 2: 94–107.

Lewthwaite, G.R. 1966. 'Environmentalism and Determinism: A Search for Clarification', *Annals, Association of American Geographers* 56: 1–23.

Martin, G.J. 1989. 'The Nature of Geography and the Schaefer-Hartshorne Debate', in J.N. Entrikin and S.D. Brunn, eds, *Reflections on Richard Hartshorne's The Nature of Geography*. Washington: Association of American Geographers, Occasional Publication, 69–88.

Minshull, R. 1967. *Regional Geography: Theory and Practice*. London: Hutchinson.

Monmonier, M. 1991. *How to Lie with Maps*. Chicago: University of Chicago Press.

Peet, R. 1989. 'World Capitalism and the Destruction of Regional Cultures', in R.J. Johnston and P.J. Taylor, eds, *A World in Crisis: Geographical Perspectives*, 2nd edn. Oxford: Blackwell, 175–99.

Raper, J.F., and N.P.A. Green. 1989. 'The Development of a Tutor for Geographic Information Systems', *British Journal of Educational Technology* 20: 164–72.

Relph, E. 1976. *Place and Placelessness*. London: Pion.

Saarinen, T.F. 1974. 'Environmental Perception', in I.R. Manners and M.W. Mikesell, eds, *Perspectives on Environment*. Washington: Association of American Geographers, Commission on College Geography Publication 13, 252–89.

Schaefer, F. 1953. 'Exceptionalism in Geography: A Methodological Examination', *Annals, Association of American Geographers* 43: 226–49.

Semple, E. 1911. *Influences of Geographic Environment*. New York: Henry Holt.

Spate, O.H.K. 1952. 'Toynbee and Huntington: A Study in Determinism', *Geographical Journal* 118: 406–28.

Tobler, W. 1970. 'A Computer Movie', *Economic Geography* 46: 234–40.

Tuan, Yi-Fu. 1979. *Landscapes of Fear*. Oxford: Blackwell.

———. 1982. *Segmented Worlds and Self: Group Life and Individual Consciousness*. Minneapolis: University of Minnesota Press.

———. 1983. 'Geographical Theory: Queries from a Cultural Geographer', *Geographical Analysis* 15: 69–72.

———. 1984. *Dominance and Affection*. New Haven: Yale University Press.

Whittlesey, D. 1954. 'The Regional Concept and the Regional Method', in P.E. James and C.F. James, eds, *American Geography: Inventory and Prospect*. Syracuse, NY: Syracuse University Press, 19–68.

Wood, D. 1993. 'The Power of Maps', *Scientific American* 268, 5: 88–93.

Yeates, M.H. 1968. *An Introduction to Quantitative Analysis in Economic Geography*. Toronto: McGraw-Hill.

CHAPTER 3

Agnew, C. 1990. 'Green Belt around the Sahara', *Geographical Magazine* 62, 4: 26–30.

Anderson, D. 1999. 'Abrupt Climatic Change', *Geography Review* 13, 1: 2–6.

Bahn, P., and J. Flenley. 1992. *Easter Island: Earth Island*. New York: Thames and Hudson.

Blacksmith Institute. 2006. *The World's Most Polluted Places: The Top Ten*. New York: Blacksmith Institute.

Carson, R. 1962. *Silent Spring*. Boston: Houghton Mifflin.

Diamond, J. 2005. *Collapse: How Societies Choose to Fail or Succeed*. New York: Viking Penguin.

Dregne, H.E. 1977. 'Desertification of Arid Lands', *Economic Geography* 53: 322–31.

Environment Canada. 1991. *The State of Canada's Environment*. Ottawa: Supply and Services Canada.

Friedman, T.L. 2008. *Hot, Flat, and Crowded: Why We Need a Green Revolution—And How It Can Renew America*. New York: Farrar, Straus and Giroux.

Gale, R.J.P. 1992. 'Environment and Economy: The Policy Models of Development', *Environment and Behavior* 24: 723–37.

Glance, N.S., and B.A. Huberman. 1994. 'The Dynamics of Social Dilemmas', *Scientific American* 270, 3: 76–81.

Goodall, C. 2008. *Ten Technologies to Save the Planet*. London: Profile Books.

Goudie, A. 1981. *The Human Impact: Man's Role in Environmental Change*. Oxford: Blackwell.

Grove, R.H. 1992. 'Origins of Western Environmentalism', *Scientific American* 267, 1: 42–7.

Hardin, G. 1968. 'The Tragedy of the Commons', *Science* 162: 1243–8.

Johnston, R.J. 1992. 'Laws, States and Superstates: International Law and the Environment', *Applied Geography* 12: 211–28.

——— and P.J. Taylor. 1986. 'Introduction: A World in Crisis?', in R.J. Johnston and P.J. Taylor, eds, *A World in Crisis: Geographical Perspectives*. Oxford: Blackwell, 1–11.

Katz, E., A. Light, and D. Rothenberg, eds. 2000. *Beneath the Surface: Critical Essays in Deep Ecology*. Cambridge, Mass.: MIT Press.

Kaplan, R.D. 1994. 'The Coming Anarchy', *Atlantic Monthly* 273, 2: 44–76.

———. 1996. *The Ends of the Earth: A Journey at the Dawn of the 21st Century*. New York: Random House.

Kelly, K. 1974. 'The Changing Attitudes of Farmers to Forest in Nineteenth Century Ontario', *Ontario Geography* 8: 67–77.

Kirchner, J.W. 1989. 'The Gaia Hypothesis: Can It Be Tested?', *Reviews of Geophysics* 27, 2: 223–35.

Lomborg, B. 2001. *The Skeptical Environmentalist: Measuring the Real State of the World*. New York: Cambridge University Press.

———. 2007. *Cool It: The Skeptical Environmentalist's Guide to Global Warming*. New York: Knopf.

Lovelock, J. 1979. *Gaia: A New Look at Life on Earth*. Toronto: Oxford University Press.

———. 1988. *The Ages of Gaia*. New York: W.W. Norton.

———. 2006. *The Revenge of Gaia: Why the Earth Is Fighting Back and How We Can Still Save Humanity*. London: Penguin.

McKibben, B. 2006. 'The Coming Meltdown', *New York Review of Books* 53, 1 (12 Jan.): 16–18.

Marsh, G.P. 1965 [1864]. *Man and Nature, or Physical Geography as Modified by Human Action*, ed. D. Lowenthal. Cambridge, Mass.: Harvard University Press.

Ponting, C. 1991. *A Green History of the World*. New York: St Martin's Press.

Powell, J.M. 1976. *Environmental Management in Australia, 1788–1814*. New York: Oxford University Press.

Repetto, R. 1990. 'Deforestation in the Tropics', *Scientific American* 260: 36–42.

Ruckelshaus, W.N. 1989. 'Toward a Sustainable World', *Scientific American* 261, 3: 166–75.

Ruddiman, W.F. 2005. *Plows, Plagues, and Petroleum: How Humans Took Control of Climate*. Princeton, NJ: Princeton University Press.

Simon, J., and H. Kahn, eds. 1984. *The Resourceful Earth*. Oxford: Blackwell.

Smil, V. 1987. *Energy, Food and Environment*. New York: Oxford University Press.

———. 1989. 'Our Changing Environment', *Current History* 88: 9–12, 46–8.

———. 1993. *Global Ecology: Environmental Changes and Social Flexibility*. New York: Routledge.

———. 2005. 'The Next 50 Years: Fatal Discontinuities', *Population and Development Review* 31, 2: 201–36.

Thomas, W.L., et al., eds. 1956. *Man's Role in Changing the Face of the Earth*. Chicago: University of Chicago Press.

Tuan, Y.F. 1971. *Man and Nature*. Washington: Association of American Geographers, Commission on College Geography, Resource Paper No. 10.

Wilkinson, H.R. 1963. *Man and the Natural Environment*. Hull, UK: University of Hull, Department of Geography, Occasional Papers in Geography No. 1.

Wilson, E.O. 1989. 'Threats to Biodiversity', *Scientific American* 261, 3: 108–16.

World Commission on Environment and Development. 1987. *Our Common Future*. Oxford: Oxford University Press.

CHAPTER 4

Boserup, E. 1965. *The Conditions of Agricultural Change*. London: Allen and Unwin.

Daniel, M.L. 2000. 'The Demographic Impact of HIV/AIDS in Sub-Saharan Africa', *Geography* 85: 46–55.

Demeny, P. 1974. 'The Populations of the Underdeveloped Countries', *Scientific American* 231: 148–59.

Doenges, C.E., and J.L. Newman. 1989. 'Impaired Fertility in Tropical Africa', *Geographical Review* 79: 101–11.

Dwyer, D.J. 1987. 'New Population Policies in Malaysia and Singapore', *Geography* 72: 248–50.

Eberstadt, N. 2002. 'The Future of AIDS', *Foreign Affairs* 81, 6: 22–45.

Economist, The. 1993. 'Eastern Germany: Living and Dying in a Barren Land', 331, 23 (Apr.): 54.

———. 2002. 'The Next Wave' (19 Oct.): 75–6.

Ehrlich, P. 1968. *The Population Bomb*. New York: Ballantine Books.

Hall, R. 1993. 'Europe's Changing Population', *Geography* 78: 3–15.

Jowett, J. 1993. 'China's Population: 1,133,709,738 and Still Counting', *Geography* 78: 401–19.

King, R. 1993. 'Italy Reaches Zero Population Growth', *Geography* 78: 63–9.

Meadows, D.H., et al. 1972. *The Limits to Growth*. New York: Universe Books.

Nelson, F. 2006. 'Where Have All the Babies Gone?', *The Spectator*, 4 Mar., 24.

Peterson, P.G. 1999. 'Gray Dawn: The Global Aging Crisis', *Foreign Affairs* 78: 42–55.

Ridker, R.G., and E.W. Cecelski. 1979. 'Resources, Environment and Population: The Nature of Future Limits', *Population Bulletin* 34, 3: 3–4.

Robey, B., S.O. Rutstein, and L. Morris. 1993. 'The Fertility Decline in Developing Countries', *Scientific American* 269, 6: 60–7.

Toolis, K. 2000. 'While the World Looks Away', *The Guardian Weekend*, 2 Dec.

United Nations. 2002. *World Population Aging: 1950–2050*. New York: United Nations, Department of Economic and Social Affairs, Population Division.

Vanderpost, C. 1992. 'Regional Patterns of Fertility Transition in Botswana', *Geography* 77: 109–22.

World Health Organization. 2003. *The World Health Report, 2003*. Geneva: World Health Organization.

CHAPTER 5

Barberis, M. 1994. 'Haiti', *Population Today* 22, 1: 7.

Bongaarts, J. 1994. 'Can the Growing Human Population Feed Itself?', *Scientific American* 271, 3: 36–42.

Brandt, W. 1980. *North-South: A Programme for Survival*. London: Pan.

Chew, S.C., and R.A. Denemark, eds. 1996. *The Underdevelopment of Development: Essays in Honor of Andre Gunder Frank*. Thousand Oaks, Calif.: Sage.

Collier, P. 2007. *The Bottom Billion: Why the Poorest Countries Are Failing and What Can Be Done About It*. New York: Oxford University Press.

Conway, G. 2008. 'Presidential Address: The Food Crisis', *Geographical Journal* 174: 269–73.

Davis, M. 2001. *Late Victorian Holocausts: El Niño, Famines, and the Making of the Third World*. New York: Verso.

Degg, M. 1992. 'Natural Disasters: Recent Trends and Future Prospects', *Geography* 77: 198–209.

Dickenson, J., et al. 1996. *A Geography of the Third World*, 2nd edn. New York: Routledge.

Dowden, R. 2008. *Africa: Altered States, Ordinary Miracles*. London: Portobello Books.

Gee, M. 1994. 'Apocalypse Deferred', *Globe and Mail*, 9 Apr., D1, D3.

Gould, P., and R. White. 1986. *Mental Maps*, 2nd edn. Boston: Allen and Unwin.

Grigg, D. 1977. 'E.G. Ravenstein and the "Laws" of Migration', *Journal of Historical Geography* 3: 41–54.

Halfacree, K., and P.J. Boyle. 1993. 'The Challenge Facing Migration Research: The Case for a Biographical Approach', *Progress in Human Geography* 17: 333–48.

Kaufmann, D., A. Kraay, and M. Mastruzzi. 2006. *Governance Matters V: Aggregate and Individual Governance Indicators for 1996–2005*. Washington: World Bank.

Ignatieff, M. 1996. 'Review of *The Ends of the Earth*, by R.D. Kaplan', *New York Times Book Review*, 31 Mar.

Kaplan, R.D. 1994. 'The Coming Anarchy', *Atlantic Monthly* 273, 2: 44–76.

———. 1996. *The Ends of the Earth: A Journey at the Dawn of the 21st Century*. New York: Random House.

Lee, E.S. 1966. 'A Theory of Migration', *Demography* 3, 1: 47–57.

Mahmud, A. 1989. 'Grameen Bank Bangladesh: A Workable Solution', *Geographical Magazine* 61, 10: 14–16.

Moon, B. 1995. 'Paradigms in Migration Research: Exploring "Moorings" as a Schema', *Progress in Human Geography* 19: 504–24.

Petersen, W. 1958. 'A General Typology of Migration', *American Sociological Review* 23: 256–65.

Porter, G. 1992. 'The Nigerian Census Surprise', *Geography* 77: 371–4.

Ravenstein, E.G. 1876. 'Census of the British Isles, 1871; Birthplaces and Migration', *Geographical Magazine* 3: 173–7, 201–6, 229–33.

———. 1885. 'The Laws of Migration', *Journal of the Statistical Society* 48: 167–227.

———. 1889. 'The Laws of Migration', *Journal of the Statistical Society* 52: 214–301.

Smil, V. 1987. *Energy, Food, Environment: Realities, Myths and Options*. Oxford: Clarendon.

———. 2000. *Feeding the World: A Challenge for the Twenty-First Century*. Cambridge, Mass.: MIT Press.

Sowden, C. 1993. 'Debt Swaps—For or Against Development', *Geographical Magazine* 25, 12: 56–9.

Taylor, P.J. 1989. 'The Error of Developmentalism in Human Geography', in D. Gregory and R. Walford, eds, *Horizons in Human Geography*. London: Macmillan, 303–19.

Tinker, H. 1974. *A New System of Slavery: The Export of Indian Labour Overseas 1830–1920*. London: Oxford University Press.

Todd, H. 1996. *Women at the Center: Grameen Bank Borrowers after One Decade*. Boulder, Colo.: Westview Press.

Toronto Star. 2006. 'Banking on the Poor', 16 Oct., A18.

United Nations Development Program. 1996. *Human Development Report, 1996*. New York: Oxford University Press.

———. 2002. *Human Development Report 2002: Deepening Democracy in a Fragmented World*. New York: Oxford University Press.

Wallerstein, I. 1979. *The Capitalist World Economy*. Cambridge: Cambridge University Press.

Wolpert, J. 1965. 'Behavioural Aspects of the Decision to Migrate', *Papers of the Regional Science Association* 15: 159–69.

World Bank. 1998. *Global Development Finance, 1998: Analysis and Summary Tables*. Washington: World Bank.

Young, L. 1996. 'World Hunger: A Framework for Analysis', *Geography* 81: 97–110.

Zelinsky, W. 1971. 'The Hypothesis of the Mobility Transition', *Geographical Review* 61: 219–49.

CHAPTER 6

Allen, C. 2006. *God's Terrorists: The Wahhabi Cult and the Hidden Roots of Modern Jihad*. New York: Little, Brown.

Biswas, L. 1984. 'Evolution of Hindu Temples in Calcutta', *Journal of Cultural Geography* 4: 73–84.

Carneiro, R. 1970. 'A Theory of the Origin of the State', *Science* 169: 733–8.

Cartwright, D. 1988. 'Linguistic Territorialization: Is Canada Approaching the Belgian Model?', *Journal of Cultural Geography* 8: 115–34.

Childe, V.G. 1951. *Man Makes Himself*. New York: Mentor Books.

Diamond, J. 1997. *Guns, Germs and Steel: A Short History of Everybody for the Last 13,000 Years*. London: Random House.

Fagan, B. 2004. *The Long Summer: How Climate Changed Civilization*. New York: Basic Books.

Francaviglia, R.V. 1978. *The Mormon Landscape*. New York: AMS Press.

Friedl, J., and J. Pfeiffer. 1977. *Anthropology: The Study of People*. New York: Harper and Row.

Gale, D.T., and P.M. Koroscil. 1977. 'Doukhobor Settlements: Experiments in Idealism', *Canadian Ethnic Studies* 9: 53–71.

Giddens, A. 1991. *Introduction to Sociology*. New York: Norton.

Harvey, D.W. 1990. 'Between Space and Time: Reflections on the Geographical Imagination', *Annals, Association of American Geographers* 80: 418–34.

Hoernig, H., and M. Walton-Roberts. 2006. 'Immigration and Urban Change: National, Regional, and Local Perspectives', in T. Bunting and P. Filion, eds, *Canadian Cities in Transition: Local Through Global Perspectives*, 3rd edn. Toronto: Oxford University Press, 408–18.

Huntington, E. 1924. *Civilization and Climate*. New Haven: Yale University Press.

Huxley, J.S. 1966. 'Evolution, Cultural and Biological', *Current Anthropology* 7: 16–20.

James, P.E. 1964. *One World Divided*, 2nd edn. Toronto: Xerox College Publishing.

Jordan, T.G. 1988. *The European Culture Area: A Systematic Geography*, 2nd edn. New York: Harper and Row.

——— and M. Kaups. 1989. *The American Backwoods Frontier: An Ethnic and Ecological Interpretation*. Baltimore: Johns Hopkins University Press.

Kearns, K.C. 1974. 'Resuscitation of the Irish Gaeltacht', *Geographical Review* 64: 82–110.

Kroeber, A.L., and T. Parsons. 1958. 'The Concepts of Culture and of Social System', *American Sociological Review* 23: 582–3.

Lewis, B. 2003. *The Crisis of Islam: Holy War and Unholy Terror*. New York: Weidenfeld and Nicholson.

McCrum, R., W. Cran, and R. MacNeil, eds. 1986. *The Story of English*. London: BBC.

Mackay, J.R. 1958. 'The Interactance Hypothesis and Boundaries in Canada: A Preliminary Study', *Canadian Geographer* 3, 11: 1–8.

McWhorter, R. 2002. *The Power of Babel: A Natural History of Language*. New York: W.H. Freeman.

Meinig, D.W. 1965. 'The Mormon Culture Region: Strategies and Patterns in the Geography of the American West, 1847–1964', *Annals, Association of American Geographers* 55: 191–220.

Park, C. 1994. *Sacred Worlds: An Introduction to Geography and Religion*. New York: Routledge.

Russell, R.J., and F.B. Kniffen. 1951. *Culture Worlds*. New York: Macmillan.

Sauer, C.O. 1925. 'The Morphology of Landscape', *University of California Publications in Geography* 2: 19–53.

Simpson-Housley, P. 1978. 'Hutterian Religious Ideology, Environmental Perception, and Attitudes towards Agriculture', *Journal of Geography* 77: 145–8.

Smil, V. 1987. *Energy, Food and Environment*. New York: Oxford University Press.

Sommers, B.J. 2008. *The Geography of Wine: How Landscapes, Cultures, Terroir, and the Weather Make a Good Drop*. New York: Plume.

Spencer, J.E., and R.J. Horvath. 1963. 'How Does an Agricultural Region Originate?', *Annals, Association of American Geographers* 53: 74–92.

Toynbee, A.J. 1935–61. *A Study of History*, 12 vols. New York: Oxford University Press.

Urry, J. 2000. *Sociology beyond Societies: Mobilities for the Twenty-First Century*. New York: Routledge.

Wittfogel, K. 1957. *Oriental Despotism: A Comparative Study of Total Power*. New Haven: Yale University Press.

Zelinsky, W. 1973. *The Cultural Geography of the United States*. Englewood Cliffs, NJ: Prentice-Hall.

CHAPTER 7

Abler, R.F., M.G. Marcus, and J.M. Olson, eds. 1992. *Geography's Inner Worlds: Pervasive Themes in Contemporary American Geography*. New Brunswick, NJ: Rutgers University Press.

Alvarez, A. 2001. *Governments, Citizens and Genocide: A Comparative and Interdisciplinary Approach*. Bloomington: Indiana University Press.

American Anthropological Association. 1998. *Statement on Race*. At: <www.aaanet.org/stmts/racepp.htm> (4 Apr. 2006).

Berry, B.J.L. 1992. 'Review of Geography's Inner Worlds: Pervasive Themes in Contemporary American Geography', *Urban Geography* 13: 490–4.

Blaut, J.M. 1980. 'A Radical Critique of Cultural Geography', *Antipode* 12: 25–9.

Chalk, F., and K. Jonassohn. 1990. *The History and Sociology of Genocide*. New Haven: Yale University Press.

Chouinard, V. 1997. 'Guest Editorial. Making Space for Disabling Differences: Challenging Able-ist Geographies', *Environment and Planning D: Society and Space* 15: 379–87.

Dear, M.J. 1988. 'The Postmodern Challenge: Reconstructing Human Geography', *Transactions, Institute of British Geographers* (new series) 13: 262–74.

Doyal, L., and I. Gough. 1991. *A Theory of Human Need*. London: Macmillan.

Entrikin, N. 1991. *The Betweenness of Place*. Baltimore: Johns Hopkins University Press.

Evans, D. 2001. 'Spatial Analyses of Crime', *Geography* 86: 211–23.

——— and D.T. Herbert. 1989. *The Geography of Crime*. London: Routledge.

Evans, R. 1989. 'Consigned to the Shadows', *Geographical Magazine* 61, 12: 23–5.

Eyles, J., and W. Peace. 1990. 'Signs and Symbols in Hamilton: An Iconology of Steeltown', *Geografiska Annaler* 72B: 73–88.

Fennell, D.A. 1999. *Ecotourism*. London: Routledge.

Francaviglia, R.V. 1978. *The Mormon Landscape*. New York: AMS Press.

Garreau, J. 1981. *The Nine Nations of North America*. Boston: Houghton Mifflin.

Giddens, A. 1984. *The Constitution of Society*. Cambridge: Polity Press.

Gould, S.J. 1981. *The Mismeasure of Man*. New York: Norton.

———. 1985. 'Human Equality Is a Contingent Fact of History', in S.J. Gould, ed., *The Flamingo's Smile*. New York: Norton, 185–98.

———. 1987. 'Bushes All the Way Down', *Natural History* 96, 6: 12–19.

Hale, R.F. 1984. 'Vernacular Regions of America', *Journal of Cultural Geography* 5: 131–40.

Hamnett, C. 2003. 'Editorial. Contemporary Human Geography: Fiddling While Rome Burns?', *Geoforum* 34: 1–3.

Harvey, D.W. 1989. *The Condition of Postmodernity*. Oxford: Blackwell.

Herbert, D.T., and D.M. Smith, eds. 1989. *Social Problems and the City: New Perspectives*. Oxford: Oxford University Press.

Hopkins, J.S.P. 1990. 'West Edmonton Mall: Landscape of Myths and Elsewhereness', *Canadian Geographer* 34: 2–17.

Isajiw, W. 1974. 'Definitions of Ethnicity', *Ethnicity* 1: 111–24.

Jackson, E.L., and D.B. Johnson. 1991. 'Geographic Implications of Mega-Malls, with Special Reference to West Edmonton Mall', *Canadian Geographer* 35: 226–32.

——— and ———, eds. 1991. 'The West Edmonton Mall and Mega Malls', *Canadian Geographer* 35, 3: 226–305.

Jackson, P. 1989. *Maps of Meaning: An Introduction to Cultural Geography*. London: Unwin Hyman.

——— and S.J. Smith. 1984. *Exploring Social Geography*. London: Allen and Unwin.

Jakle, J.A. 1985. *The Tourist: Travel in Twentieth-Century North America*. Lincoln: University of Nebraska Press.

James, P.E. 1974. *One World Divided*, 2nd edn. Toronto: Xerox College Publishing.

Jayne, M. 2006. 'Cultural Geography, Consumption and the City', *Geography* 91: 34–42.

Johnston, L. 1997. 'Queen(s') Street or Ponsonby Poofters? Embodied Hero Parade Sites', *New Zealand Geographer* 53, 2: 29–33.

Johnston, R.J. 1986. 'Individual Freedom and the World Economy', in R.J. Johnston and P.J. Taylor, eds, *A World in Crisis: Geographical Perspectives*. Oxford: Blackwell, 173–95.

———. 1991. *A Question of Place: Exploring the Practice of Human Geography*. Oxford: Blackwell.

Jones, K., and G. Moon. 1987. *Health, Disease and Society: An Introduction to Medical Geography*. London: Routledge.

Kennedy, K.A.R. 1976. *Human Variation in Space and Time*. Dubuque, Iowa: Brown.

Keynes, J.M. 1926. *The End of Laissez-faire*. London: Hogarth Press.

Kobayashi, A. 1993. 'Multiculturalism: Representing a Canadian Institution', in J. Duncan and D. Ley, eds, *Place/Culture/Representation*. London: Routledge, 205–31.

Ley, D., and K. Olds. 1988. 'Landscape as Spectacle: World's Fairs and the Culture of Heroic Consumption', *Environment and Planning D: Society and Space* 6: 191–212.

McKittrick, K., and L. Peake. 2005. 'What Difference Does Difference Make to Geography?', in N. Castree, A. Rogers, and D. Sherman, eds, *Questioning Geography*. New York: Blackwell, 39–54.

McQuillan, A. 1993. 'Historical Geography and Ethnic Communities in North America', *Progress in Human Geography* 17: 355–66.

Mann, M. 2004. *The Dark Side of Democracy: Explaining Ethnic Cleansing*. Cambridge: Cambridge University Press.

Mead, G.H. 1934. *Mind, Self and Society*. Chicago: University of Chicago Press.

Monk, J. 1992. 'Gender in the Landscape: Expressions of Power and Meaning', in K. Anderson and F. Gale, eds, *Inventing Places: Studies in Cultural Geography*. Melbourne: Longman Cheshire, 123–38.

Montague, A., ed. 1964. *The Concept of Race*. New York: Collier.

Nolen, S. 2008. 'Understanding the Turmoil in South Africa', *Globe and Mail*, 24 May, F4–5.

Nostrand, R.L., and L.E. Estaville Jr. 1993. 'Introduction: The Homeland Concept', *Journal of Cultural Geography* 13, 2: 1–4.

O'Hare, G., and H. Barrett. 1993. 'The Fall and Rise of the Sri Lankan Tourist Industry', *Geography* 78: 438–42.

Osborne, B.S. 1988. 'The Iconography of Nationhood in Canadian Art', in D. Cosgrove and S. Daniels, eds, *The Iconography of Landscape: Essays on the Symbolic Representation, Design and Use of Past Environments*. New York: Cambridge University Press, 162–78.

Pain, R. 1992. 'Space, Sexual Violence and Social Control: Integrating Geographical and Feminist Analyses of Women's Fear of Crime', *Progress in Human Geography* 15: 415–31.

Pawson, E., and G. Banks. 1993. 'Rape and Fear in a New Zealand City', *Area* 25: 55–63.

Raitz, K.B. 1979. 'Themes in the Cultural Geography of European Ethnic Groups in the United States', *Geographical Review* 69: 79–94.

Robinson, M. 1999. 'Cultural Conflicts in Tourism: Inevitability and Inequality', in M. Robinson and P. Boniface, eds, *Tourism and Cultural Conflicts*. New York: CABI, 1–32.

Rooney, J.F., Jr. 1974. *A Geography of American Sport*. Reading, Mass.: Addison-Wesley.

Shields, R. 1989. 'Social Spatialisation and the Built Environment: The Example of West Edmonton Mall', *Environment and Planning D: Society and Space* 7: 147–64.

Smith, D.M. 1973. *A Geography of Social Well-Being in the United States*. New York: McGraw-Hill.

Smith, S.J. 1987. 'Fear of Crime: Beyond a Geography of Deviance', *Progress in Human Geography* 11: 1–23.

Stein, H.F., and G.L. Thompson. 1992. 'The Sense of Oklahomaness: Contributions of Psychogeography to the Study of American Culture', *Journal of Cultural Geography* 11, 2: 63–91.

Thomas, B., and D. Dorling. 2007. *Identity in Britain: A Cradle-to-Grave Atlas*. Bristol, UK: Policy Press.

Trépanier, C. 1991. 'The Cajunization of French Louisiana: Forging a Regional Identity', *Geographical Journal* 157: 161–71.

United Nations Development Program. 1996. *Human Development Report, 1996*. New York: Oxford University Press.

Wagner, P.L. 1975. 'The Themes of Cultural Geography Rethought', *Yearbook: Association of Pacific Coast Geographers* 37: 7–14.

Wheat, S. 1994. 'Taming Tourism', *Geographical Magazine* 67: 16–19.

Zelinsky, W. 1980. 'North America's Vernacular Regions', *Annals, Association of American Geographers* 70: 1–16.

CHAPTER 8

Anderson. B. 1983. *Imagined Communities: Reflections on the Origins and Spread of Nationalism*. London: Verso.

Bunge, W. 1988. *Nuclear War Atlas*. Oxford: Blackwell.

Carroll, W.K. 1992. *Organizing Dissent: Contemporary Social Movements in Theory and Practice, Studies in the Politics of Counter-Hegemony*. Toronto: Garamond Press.

Clark, G.L., and M.J. Dear. 1984. *State Apparatus: Structures and Language of Legitimacy*. Boston: Allen and Unwin.

Cohen, S.B. 1991. 'Global Geopolitical Change in the Post Cold War Era', *Annals, Association of American Geographers* 81: 551–80.

———. 2003. *Geopolitics of the World System*. Lanham, Md: Rowman & Littlefield.

de Blij, H. 1992. 'Political Geography of the Post Cold War', *Professional Geographer* 44: 16–19.

East, W.G., and J.R.V. Prescott. 1975. *Our Fragmented World: Introduction to Political Geography*. London: Macmillan.

Elsom, D. 1985. 'Climatological Effects of a Nuclear Exchange: A Review', in D. Pepper and A. Jenkins, eds, *The Geography of Peace and War*. Oxford: Blackwell, 126–47.

Evans, R. 1991. 'Legacy of Woe', *Geographical Magazine* 63, 6: 34–8.

Fukuyama, F. 1992. *The End of History and the Last Man*. New York: Free Press.

Hartshorne, R. 1950. 'The Functional Approach in Political Geography', *Annals, Association of American Geographers* 40: 95–130.

Hirsch, P. 1993. 'The Socialist Developing World in the 1990s', *Geography Review* 7, 2: 35–7.

Huntington, S.P. 1993. 'The Clash of Civilizations', *Foreign Affairs* 72, 3: 22–49.

———. 1996. *The Clash of Civilizations and the Remaking of World Order*. New York: Simon & Schuster.

Johnston, R.J. 1982. *Geography and the State: An Essay in Political Geography*. New York: St Martin's Press.

———. 1985. *The Geography of English Politics*. London: Croom Helm.

———. 1993. 'Tackling Global Environmental Problems', *Geography Review* 6, 2: 27–30.

Jones, S.B. 1954. 'A Unified Field Theory of Political Geography', *Annals, Association of American Geographers* 44: 111–23.

Kasperson, R.E., and J.V. Minghi, eds. 1969. *The Structure of Political Geography*. Chicago: Aldine.

Klare, M.T. 2001. 'The New Geography of Conflict', *Foreign Affairs* 80: 49–61.

Kohr, L. 1957. *The Breakdown of Nations*. Swansea: Christopher Davies.

Krebheil, E. 1916. 'Geographic Influences in British Elections', *Geographical Review* 2: 419–32.

Lemon, A. 1996. 'Lesotho and the New South Africa: The Question of Incorporation', *Geographical Journal* 162: 263–72.

Lorenz, K. 1967. *On Aggression*. London: Methuen.

Mackinder, H.J. 1919. *Democratic Ideals and Reality*. New York: Henry Holt.

Messick, D.M., and D.M. Mackie. 1989. 'Intergroup Relations', *Annual Review of Psychology* 40: 45–81.

Montague, A. 1976. *The Nature of Human Aggression*. New York: Oxford University Press.

Morris, D. 1967. *The Naked Ape*. New York: McGraw-Hill.

Ohmae, K. 1993. 'The Rise of the Region State', *Foreign Affairs* 76, 2: 78–87.

O'Loughlin, J. 1992. 'Ten Scenarios for a "New World Order"', *Professional Geographer* 44: 22–8.

Openshaw, S., P. Steadman, and O. Greene. 1983. *Doomsday: Britain after Nuclear Attack*. Oxford: Blackwell.

Paddison, R. 1983. *The Fragmented State: The Political Geography of Power*. New York: St Martin's Press.

Rokkan, S. 1980. 'Territories, Centres and Peripheries', in J. Gottman, ed., *Centre and Periphery*. London: Sage, 163–204.

Royle, S. 1991. 'St Helena: A Geographical Summary', *Geography* 76: 266–8.

Taylor, P.J. 1989. *Political Geography: World Economy, Nation State and Locality*, 2nd edn. London: Longman.

The Times (London). 2006. 'Comment: Till the Votes Do Them Part', 20 May.

Vining, D.R., Jr. 2005. 'On Empires', *Geography* 90: 177–9.

Wilkinson, P. 1971. *Social Movement*. London: Macmillan.

Wood, W.B. 2001. 'Geographic Aspects of Genocide: A Comparison of Bosnia and Rwanda', *Transactions, Institute of British Geographers* (new series) 26: 57–75.

Zurick, D. 1999. 'Lands of Conflict in South Asia', *Focus* 45, 3: 33–7.

CHAPTER 9

Appleton, J.H. 1962. *The Geography of Communications in Great Britain*. Toronto: Oxford University Press.

Blainey, G. 1968. *The Tyranny of Distance*. London: Macmillan.

Carrothers, G.A.P. 1956. 'An Historical Review of the Gravity and Potential Concepts of Human Interaction', *Journal of the American Institute of Planners* 22: 94–102.

Cleary, M., and R. Bedford. 1993. 'Globalisation and the New Regionalism: Some Implications for New Zealand Trade', *New Zealand Journal of Geography* 20: 19–22.

Cliff, A.D., P. Haggett, and J.K. Ord. 1986. *Spatial Aspects of Influenza Epidemics*. London: Pion.

——— and M.R. Smallman-Raynor. 1992. 'The AIDS Pandemic: Global Geographical Patterns and Local Spatial Processes', *Geographical Journal* 158: 182–98.

Conkling, E.C., and M.H. Yeates. 1976. *Man's Economic Environment*. Toronto: McGraw-Hill.

Dicken, P. 1992. *Global Shift: The Internationalization of Economic Activity*. London: Paul Chapman.

Freeman, D.B. 1985. 'The Importance of Being First: Preemption by Early Adopters of Farming Innovations in Kenya', *Annals, Association of American Geographers* 75: 17–28.

Friedman, T.L. 2005. *The World Is Flat: A Brief History of the Twenty-First Century*. New York: Farrar, Straus, and Giroux.

Garrison, W.L. 1960. 'Connectivity of the Interstate Highway System', *Regional Science Association, Papers and Proceedings* 6: 121–37.

Gould, P. 1993. *The Slow Plague: A Geography of the AIDS Pandemic*. Oxford: Blackwell.

Hagerstrand, T. 1951. 'Migration and the Growth of Culture Regions', *Lund Studies in Geography B*, 3.

———. 1967. *Innovation Diffusion as a Spatial Process*, trans. A. Pred. Chicago: University of Chicago Press.

Hall, E.T. 1966. *The Hidden Dimension*. Garden City, NY: Doubleday.

Harvey, D. 1969. *Explanation in Geography*. London: Arnold.

Hilling, D. 1977. 'The Evolution of a Port System: The Case of Ghana', *Geography* 62: 97–105.

Hoad, D. 2002. 'The World Trade Organisation, Corporate Interests, and Global Opposition: Seattle and After', *Geography* 87: 148–54.

Hoyle, B.S., and R.D. Knowles. 1992. *Modern Transport Geography*. London: Belhaven Press.

Isard, W. 1956. *Location and Space Economy*. New York: Wiley.

——— et al. 1960. *Methods of Regional Analysis: An Introduction to Regional Science*. Cambridge, Mass.: MIT Press.

Janelle, D.G. 1968. 'Central Place Development in a Time-Space Framework', *Professional Geographer* 20: 5–10.

———. 1969. 'Spatial Reorganization: A Model and Concept', *Annals, Association of American Geographers* 59: 348–64.

Kansky, K. 1963. *The Structure of Transportation Networks*. Chicago: University of Chicago, Department of Geography, Research Paper No. 84.

Kniffen, F.B. 1951. 'The American Covered Bridge', *Geographical Review* 41: 114–23.

Levinson, M. 2006. *The Box: How the Shipping Container Made the World Smaller and the World Economy Bigger*. Princeton, NJ: Princeton University Press.

Lowe, J.C., and S. Moryadas. 1975. *The Geography of Movement*. Boston: Houghton Mifflin.

Massey, D. 2002. 'Globalization: What Does It Mean for Geography?', *Geography* 87: 293–6.

Norberg, J. 2001. *In Defence of Global Capitalism*. Stockholm: Timbro.

Pyle, G. 1969. 'The Diffusion of Cholera in the United States in the Nineteenth Century', *Geographical Analysis* 1: 59–75.

Raitz, K.B. 1973. 'Ethnicity and the Diffusion and Distribution of Cigar Tobacco Production in Wisconsin and Ohio', *Tijdschrifte voor Economische en Sociale Geografie* 64: 293–306.

Reilly, W.J. 1931. *The Law of Retail Gravitation*. New York: Free Press.

Rigg, J. 2001. 'Is Globalization Good?', *Geography Review* 14, 4: 36–7.

Rogers, E.M. 1962. *Diffusion of Innovations*. New York: Free Press.

Saul, J.R. 2005. *The Collapse of Globalism and the Reinvention of World*. New York: Overlook Press.

Smallman-Raynor, M.R., A.D. Cliff, and P. Haggett. 1992. *Atlas of AIDS*. Oxford: Blackwell.

Taafe, E.J., R.L. Morrill, and P. Gould. 1963. 'Transport Expansion in Underdeveloped Countries: A Comparative Analysis', *Geographical Review* 53: 503–29.

Ullman, E.L. 1956. 'The Role of Transportation and the Bases for Interaction', in W.L. Thomas Jr, ed., *Man's Role in Changing the Face of the Earth*. Chicago: University of Chicago Press, 862–80.

Warntz, W. 1957. 'Transportation, Social Physics and the Law of Refraction', *Professional Geographer* 9: 2–7.

Watson, J.W. 1955. 'Geography: A Discipline in Distance', *Scottish Geographical Magazine* 71: 1–13.

Wilson, A.G., and R.J. Bennett, eds. 1986. *Mathematical Methods in Human Geography and Planning*. Chichester: Wiley.

Zipf, G.K. 1949. *Human Behavior and the Principle of Least Effort*. Cambridge, Mass.: Addison-Wesley.

CHAPTER 10

Arkell, T. 1991. 'The Decline of Pastoral Nomadism in the Western Sahara', *Geography* 76: 162–6.

Atkins, P., and I. Bowler. 2001. *Food in Society: Economy, Culture, Geography*. London: Arnold.

Barnes, T.J. 2001. 'Retheorizing Economic Geography: From the Quantitative Revolution to the "Turn"', *Annals, Association of American Geographers* 91: 546–65.

Barrett, H., and A. Browne. 1991. 'Environmental and Economic Sustainability: Woman's Horticultural Production in the Gambia', *Geography* 76: 241–8.

Bassett, T.J. 1988. 'The Political Ecology of Peasant-Herder Conflicts in the Northern Ivory Coast', *Annals, Association of American Geographers* 78: 453–72.

Binford, L. 1968. 'Post-Pleistocene Adaptations', in L. Binford and S. Binford, eds, *New Perspectives in Archaeology*. Chicago: Aldine, 313–41.

Boserup, E. 1965. *The Conditions of Agricultural Change*. London: Allen and Unwin.

Carlyle, W.J. 1994. 'Rural Population in the Canadian Prairies', *Great Plains Research* 4: 65–87.

Chisholm, M. 1962. *Rural Settlement and Land Use: An Essay in Location*. London: Hutchinson.

Conzen, M.P. 1971. *Frontier Farming in an Urban Shadow*. Madison: State Historical Society of Wisconsin.

Fernandez-Armesto. F. 2001. *Food: A History*. London: Macmillan.

Friedmann, H. 1991. 'Changes in the International Division of Labour: Agri-Food Complexes and Export Agriculture', in W. Friedland et al., eds, *Towards a New Political Economy of Agriculture*. Boulder, Colo.: Westview Press, 65–93.

Goodall, C. 2008. *Ten Technologies to Save the Planet*. London: Profile Books.

Griffin, E. 1973. 'Testing the Von Thünen Theory in Uruguay', *Geographical Review* 63: 500–16.

Grigg, D.B. 1974. *The Agricultural Systems of the World: An Evolutionary Approach*. New York: Cambridge University Press.

———. 1992. 'Agriculture in the World Economy: An Historical Geography of Decline', *Geography* 77: 210–22.

Grossman, L.S. 1993. 'The Political Ecology of Banana Exports and Local Food Production in St Vincent, Eastern Caribbean', *Annals, Association of American Geographers* 83: 347–67.

Hall, P.G., ed. 1966. *Von Thünen's Isolated State*. Oxford: Pergamon.

Harman, O. 2004. 'The Essential Element', *New Republic* (30 Aug.): 31–7.

Harris, D.R. 1996. 'Preface', in D.R. Harris, ed., *The Origins and Spread of Agriculture and Pastoralism in Eurasia*. Washington: Smithsonian Institution Press, ix–xi.

Harvey, D.W. 1973. *Social Justice and the City*. London: Arnold.

Horvath, R.J. 1969. 'Von Thünen's Isolated State and the Area around Addis Ababa, Ethiopia', *Annals, Association of American Geographers* 59: 308–23.

Huggins, D.R., and J.P. Reganold. 2008. 'No-Till: The Quiet Revolution', *Scientific American* 299, 5: 70–7.

Isard, W. 1956. *Location and Space Economy*. New York: Wiley.

Johnson, H.B. 1962. 'A Note on Thünen's Circles', *Annals, Association of American Geographers* 52: 213–20.

Jonasson, O. 1925. 'Agricultural Regions of Europe', *Economic Geography* 1: 277–314.

King, L.J. 1969. *Statistical Analysis in Geography*. Englewood Cliffs, NJ: Prentice-Hall.

Leaman, J.H., and E.C. Conkling. 1975. 'Transport Change and Agricultural Specialization', *Annals, Association of American Geographers* 65: 425–37.

McCallum, J. 1980. *Unequal Beginnings: Agriculture and Economic Development in Quebec and Ontario until 1870*. Toronto: University of Toronto Press.

McDowell, L. 2000. 'Feminists Rethink the Economic: The Economics of Gender/The Gender of Economics', in G.L. Clark, M.P. Feldman, and M.S. Gertler, eds, *The Oxford Handbook of Economic Geography*. Toronto: Oxford University Press, 497–517.

Marsden, T.K. 1997. 'Creating Space for Food: The Distinctiveness of Recent Agrarian Development', in D. Goodman and M.J. Watts, eds, *Globalising Food: Agrarian Questions and Global Restructuring*. New York: Routledge, 169–91.

Morrish, M. 1994. 'China Takes the Road to Market', *Geographical Magazine* 67, 4: 43–5.

Muller, P.O. 1973. 'Trend Surfaces of American Agricultural Patterns: A Macro-Thünen Analysis', *Economic Geography* 49: 228–42.

Norton, W., and E.C. Conkling. 1974. 'Land Use Theory and the Pioneering Economy', *Geografiska Annaler* 56B: 44–56.

Overton, M. 1996. *Agricultural Revolution in England: The Transformation of the Agrarian Economy, 1500–1850*. Cambridge: Cambridge University Press.

Peet, J.R. 1969. 'The Spatial Expansion of Commercial Agriculture in the Nineteenth Century: A Von Thünen Interpretation', *Economic Geography* 45: 283–301.

Raitz, K.B. 1973. 'Ethnicity and the Diffusion and Distribution of Cigar Tobacco Production in Wisconsin and Ohio', *Tijdschrifte voor Economische en Sociale Geografie* 64: 293–306.

Rogers, A., and F. Chew. 2005. 'Ducks for Bugs: Rice Farming in Malaysia', *Geography Review* 18, 5: 34–7.

Salamon, S. 1985. 'Ethnic Communities and the Structure of Agriculture', *Rural Sociology* 50: 323–40.

Sauer, C.O. 1952. *Agricultural Origins and Dispersals*. New York: American Geographical Society.

Schlebecker, J.T. 1960. 'The World Metropolis and the History of American Agriculture', *Journal of Economic History* 20: 187–208.

Seaborne, A. 2001. 'Crop Diversification in Canada's Breadbasket: Land Use Changes in Saskatchewan's Agriculture', *Geography* 86: 151–8.

Sinclair, R. 1967. 'Von Thünen and Urban Sprawl', *Annals, Association of American Geographers* 57: 72–87.

Slatford, R., and I. Fishpool. 2002. 'India and the Green Revolution', *Geography Review* 15, 2: 2–5.

Smil, V. 2001. *Enriching the Earth: Fritz Haber, Carl Bosch and the Transformation of World Food Production*. Cambridge, Mass.: MIT Press.

Smith, J.R. 1925. *North America*. New York: Harcourt Brace and Co.

Smith, W. 1984. 'The "Vortex" Model and the Changing Agricultural Landscape of Quebec', *Canadian Geographer* 28: 358–72.

Swyngedouw, E. 2000. 'The Marxian Alternative: Historical-Geographical Materialism and the Political Economy of Capitalism', in E. Sheppard and T.J. Barnes, eds, *A Companion to Economic Geography*. Malden, Mass.: Blackwell, 41–59.

Wolpert, J. 1964. 'The Decision Process in Spatial Context', *Annals, Association of American Geographers* 54: 537–58.

Young, L.J. 1991. 'Agriculture Changes in Bhutan: Some Environmental Questions', *Geographical Journal* 157: 172–8.

Zimmerer, K.S. 1991. 'Wetland Production and Smallholder Persistence: Agriculture Change in a Highland Peruvian Region', *Annals, Association of American Geographers* 81: 443–63.

CHAPTER 11

Berry, B.J.L. 1967. *Geography of Market Centres and Retail Distribution*. Englewood Cliffs, NJ: Prentice-Hall.

Borchert, J.R. 1967. 'American Metropolitan Evolution', *Geographical Review* 57: 301–32.

Brush, J.E. 1953. 'The Hierarchy of Central Places in Southwest Wisconsin', *Geographical Review* 43: 350–402.

Campaign to Protect Rural England. 2005. *Your Countryside, Your Choice*. London: Campaign to Protect Rural England.

Cater, J., and T. Jones. 1989. *Social Geography: An Introduction to Contemporary Issues*. New York: Arnold.

Christaller, W. 1966. *Central Places in Southern Germany*, trans. C.W. Baskin. Englewood Cliffs, NJ: Prentice-Hall.

Church, R.L., and T.L. Bell. 1988. 'An Analysis of Ancient Egyptian Settlement Patterns Using Location-Allocation Covering Models', *Annals, Association of American Geographers* 78: 701–14.

Davies, S., and M. Yeates. 1991. 'Exurbanization as a Component of Migration: A Case Study in Oxford County, Ontario', *Canadian Geographer* 35: 177–86.

Earle, C.V. 1977. 'The First English Towns of North America', *Geographical Review* 67: 34–50.

English, P.W., and R.C. Mayfield, eds. 1972. *Man, Space and Environment*. New York: Oxford University Press, 601–10.

Friedmann, J. 1986. 'The World City Hypothesis', *Development & Change* 17: 69–83.

———. 2002. 'Intercity Networks in a Globalizing Era', in A. Scott, ed., *Global City-Regions*. New York: Oxford University Press, 119–36.

Ghose, R. 2004. 'Big Sky or Big Sprawl: Rural Gentrification and the Changing Cultural Landscape of Missoula, Montana', *Urban Geography* 25: 528–49.

Hall, P. 1966. *The World Cities*. London: Heinemann.

——— et al. 1973. *The Containment of Urban England*. London: Allen and Unwin.

Iversen, L., et al. 2005. 'Is Living in a Rural Area Good for Your Respiratory Health? Results from a Cross-sectional Study in Scotland', *Chest* 128, 4: 2059–67.

Keddie, P.D., and A.E. Joseph. 1991. 'The Turnaround of the Turnaround? Rural Population Change in Canada, 1976 to 1986', *Canadian Geographer* 35: 367–79.

Lösch, A. 1954. *The Economics of Location*, trans. W.H. Woglom. New Haven: Yale University Press.

Meyer, D.R. 1980. 'A Dynamic Model of the Integration of Frontier Urban Places into the United States System of Cities', *Economic Geography* 56: 120–40.

Muller, E.K. 1976. 'Selective Urban Growth in the Middle Ohio Valley, 1800–1860', *Geographical Review* 66: 178–99.

Olds, K. 2001. *Globalization and Urban Change: Capital, Culture, and Pacific Rim Mega-Projects*. New York: Oxford University Press.

Pahl, R.E. 1966. 'The Rural/Urban Continuum', *Sociologia Ruralis* 6: 299–327.

Robinson, J. 2005. 'Urban Geography: World Cities, or a World of Cities', *Progress in Human Geography* 29: 757–65.

Rozman, G. 1978. 'Urban Networks and Historical Stages', *Journal of Interdisciplinary History* 9: 65–92.

Sargent, C.G. 1975. 'Towns of the Salt River Valley, 1870–1930', *Historical Geography Newsletter* 5, 2: 1–9.

Sassen, S. 1991. *The Global City: New York, London, Tokyo*. Princeton, NJ: Princeton University Press.

———, ed. 2002. *Global Networks—Linked Cities*. New York: Routledge.

Short, J. 1991. *Imagined Country: Environment, Culture and Society*. New York: Routledge.

Sjoberg, G. 1960. *The Pre-Industrial City: Past and Present*. Glencoe, Ill.: Free Press.

Smailes, A.E. 1957. *The Geography of Towns*. London: Hutchinson.

Statistics Canada. 1999. *1996 Census Dictionary* (Catalogue no.92-351-UIE). Ottawa.

Taylor, P. 2004. *World City Network: A Global Urban Analysis*. New York: Routledge.

——— and R.E. Lang. 'Commentary: The Shock of the New; 100 Concepts Describing Recent Urban Change', *Environment and Planning A* 36: 951–8.

Vance, J.E. 1970. *The Merchant's World: The Geography of Wholesaling*. Englewood Cliffs, NJ: Prentice-Hall.

———. 1971. 'Land Assignment in Pre-Capitalist, Capitalist and Post-Capitalist Societies', *Economic Geography* 47: 101–20.

———. 1990. *The Continuing City: Urban Morphology in Western Civilization*. Baltimore: Johns Hopkins University Press.

Wheatley, P. 1971. *The Pivot of the Four Quarters: A Preliminary Enquiry into the Origins and Character of the Ancient Chinese City*. Chicago: Aldine.

Wirth, C. 1938. 'Urbanism as a Way of Life', *American Journal of Sociology* 44: 46–63.

Yarwood, R. 2005. 'Beyond the Rural Idyll: Images, Countryside Change and Geography', *Geography* 90: 19–31.

Zook, M.A. 2005. *The Geography of the Internet Industry*. New York: Blackwell.

CHAPTER 12

Badcock, B. 2002. *Making Sense of Cities: A Geographical Survey*. New York: Oxford University Press.

Bassett, K., and J. Short. 1989. 'Development and Diversity in Urban Geography', in D. Gregory and R. Walford, eds, *Horizons in Human Geography*. London: Macmillan, 175–93.

Bourne, L.S. 1996. 'Reinventing the Suburbs: Old Myths and New Realities', *Progress in Planning* 46: 163–84.

Brunn, S., and J. Williams. 1983. *Cities of the World: World Regional Urban Development*, 2nd edn. New York: Harper and Row.

Cherry, G. 1988. *Cities and Plans: The Shaping of Urban Britain in the Nineteenth and Twentieth Centuries*. New York: Arnold.

Cooke, P. 1990. 'Modern Urban Theory in Question', *Transactions, Institute of British Geographers* 15: 331–43.

Cosgrove, D. 1984. *Social Formation and Symbolic Landscape*. London: Croom Helm.

Dear, M.J., and S. Flusty. 1998. 'Postmodern Urbanism', *Annals, Association of American Geographers* 88: 50–72.

Deka, D. 2004. 'Social and Environmental Justice Issues in Urban Transportation', in S. Hanson and G. Giuliano, eds, *The Geography of Urban Transportation*, 3rd edn. New York: Guilford, 332–55.

Duncan, J.S. 1990. *The City as Text: The Politics of Landscape Interpretation in the Kandyman Kingdom*. New York: Cambridge University Press.

Erickson, R.A. 1983. 'The Evaluation of the Suburban Space Economy', *Urban Geography* 4: 95–111.

Firey, W. 1947. *Land Use in Central Boston*. Cambridge, Mass.: Harvard University Press.

Ford, L. 1996. 'A New and Improved Model of Latin American City Structure', *Geographical Review* 86: 437–40.

Garreau, J. 1988. *Edge City: Life on the New Frontier*. Toronto: Doubleday.

Gillmor, D. 2005. 'Wildwood Childhood', *Canadian Geographic* (July–Aug.): 54-61.

Giuliano, G. 2004. 'Land Use Impacts of Transportation Investments: Highway and Transit', in S. Hanson and G. Giuliano, eds, *The Geography of Urban Transportation*, 3rd edn. New York: Guilford, 237–73.

Golab, T.F., and A.C. Regan. 2001. 'Impacts of Information Technology on Personal Travel and Commercial Vehicle Operations: Research Challenges and Opportunities', *Transport Research, Part C* 9: 87–121.

Grant, J. 2006. 'Shaped by Planning: The Canadian City Through Time', in T. Bunting and P. Filion, eds, *Canadian Cities in Transition: Local Through Global Perspectives*, 3rd edn. Toronto: Oxford University Press, 320–37.

Hall, T. 2001. *Urban Geography*, 2nd edn. New York: Routledge.

Harris, C.D., and E.L. Ullman. 1945. 'The Nature of Cities', *Annals, American Academy of Political and Social Science* 37: 7–17.

Hartshorn, T.A., and P.O. Muller. 1989. 'Suburban Downtowns and the Transformation of Metropolitan Atlanta's Business Landscape', *Urban Geography* 10: 373–95.

Harvey, D.W. 1969. *Explanation in Geography*. London: Arnold.

———. 1973. *Social Justice and the City*. London: Arnold.

———. 1982. *The Limits to Capital*. Oxford: Blackwell.

———. 1989. *The Urban Experience*. Oxford: Blackwell.

———. 1996. *Justice, Nature and the Geography of Difference*. Oxford: Blackwell.

———. 2000. *Spaces of Hope*. Berkeley: University of California Press.

———. 2003. *The New Imperialism*. Clarendon Lectures in Geography and Environmental Studies. Oxford: Oxford University Press.

———. 2005. *A Brief History of Neoliberalism*. Oxford: Oxford University Press.

Henderson, J. 2004. 'The Politics of Mobility and Business Elites in Atlanta, Georgia', *Urban Geography* 25: 193–216.

Heywood, P. 1997. 'The Emerging Social Metropolis: Successful Planning Initiatives in Five New World Metropolitan Regions', *Progress in Planning* 47, 3: 159–250.

Hoyt, H. 1939. *The Structure and Growth of Residential Neighbourhoods in American Cities*. Washington: Federal Housing Administration.

Kaplan, D.H., J.O. Wheeler, and S.R. Holloway. 2004. *Urban Geography*. New York: Wiley.

Lynch, K. 1960. *The Image of the City*. Cambridge, Mass.: MIT Press.

McGee, T. 1967. *The Southeastern Asian City*. New York: Praeger.

Mann, P. 1965. *An Approach to Urban Sociology*. New York: Routledge.

Martin, M.D. 2001. 'The Landscapes of Winnipeg's Wildwood Park', *Urban History Review/Revue d'histoire urbaine* 30, 2: 156–75.

Mirza, S. 2007. *Danger Ahead: The Coming Collapse of Canada's Municipal Infrastructure*. Ottawa: Federation of Canadian Municipalities.

Nijman, J. 2000. 'The Paradigmatic City', *Annals, Association of American Geographers* 90: 135–45.

O'Connor, A. 1983. *The African City*. London: Hutchinson.

Pacione, M. 2005. *Urban Geography: A Global Perspective*, 2nd edn. New York: Routledge.

Park, R.E., and E.W. Burgess. 1921. *Introduction to the Science of Sociology*. Chicago: University of Chicago Press.

Steers, S. 2005. 'Planners Betting on Light-rail Boom', *Rocky Mountain News*. At: <www.rockymountainnews.com/drmn/local/article/0,1299,DRMN_15_4091710,00.html> (16 Nov. 2005).

Taylor, P.J., and R.E. Lang. 2004. 'Commentary: The Shock of the New; 100 Concepts Describing Recent Urban Change', *Environment and Planning A* 36: 951–8.

Veninga, C. 2004. 'Spatial Prescriptions and Social Realities: New Urbanism and the Production of Northwest Landing', *Urban Geography* 25: 458–82.

Ward, S. 2004. *Planning and Urban Change*, 2nd edn. London: Paul Chapman,

White, M. 1987. *American Neighborhoods and Residential Differentiation*. New York: Russell Sage Foundation.

Yeoh, B.S.A. 1996. *Contesting Space: Power Relations and the Urban Built Environment in Colonial Singapore*. Kuala Lumpur: Oxford University Press.

CHAPTER 13

Anderson, K.J. 1991. *Vancouver's Chinatown: Racial Discourse in Canada, 1875–1980*. Montreal and Kingston: McGill-Queen's University Press.

Andresen, M.A. 2006. 'A Spatial Analysis of Crime in Vancouver, British Columbia: A Synthesis of Social Disorganization and Routine Activity Theory', *Canadian Geographer* 50: 487–502.

Bauder, H., and B. Sharpe. 2002. 'Residential Segregation of Visible Minorities in Canada's Gateway Cities', *Canadian Geographer* 40: 204–22.

Berry, B.J.L. 1985. 'Islands of Renewal in Seas of Decay', in P. Peterson, ed., *The New Urban Reality*. Washington: Brookings Institution, 69–96.

Brown, L.A., and E.G. Moore. 1970. 'The Intra-Urban Migration Process: A Perspective', *Geografiska Annaler* 52B: 1–13.

Castells, M., and P. Hall. 1994. *Technopoles of the World: The Making of the Twenty-first Century Industrial Complexes*. New York: Routledge.

Cohen, M. 1993. 'Megacities and the Environment', *Finance and Development* 30, 2: 44–7.

Davis, M. 1992. *City of Quartz: Excavating the Future in Los Angeles*. London: Vintage Books.

Dear, M.J., and J.R. Wolch. 1987. *Landscapes of Despair*. Princeton, NJ: Princeton University Press.

DeVerteuil, G. 2003. 'Household Mobility, Institutional Settings, and the New Poverty Management', *Environment and Planning A* 35: 3661–79.

——. 2005a. 'Welfare Neighborhoods: Anatomy of a Concept', *Journal of Poverty* 9: 23–41.

——. 2005b. 'The Relationship between Government Assistance and Housing Outcomes among Extremely Low-income Individuals: A Qualitative Inquiry in Los Angeles', *Housing Studies* 20: 383–99.

Downs, A. 1981. *Neighborhoods and Urban Development*. Washington: Brookings Institution.

Fraser, J.C. 2004. 'Beyond Gentrification: Mobilizing Communities and Claiming Space', *Urban Geography* 25: 437–57.

Gomez-Insausti, R. 2003. 'The Spatial Structure of the Canadian Business/Commercial Sector: A Study in Supply-side Segmentation', *Progress in Planning* 60: 13–34.

Grant, R., and J. Nijman. 2002. 'Globalization and the Corporate Geography of Cities in the Less-Developed World', *Annals, Association of American Geographers* 92: 320–40.

Greenbie, B.B. 1981. *Spaces: Dimensions of the Human Landscape*. New Haven: Yale University Press.

Hiebert, D., and D. Ley. 2003. 'Assimilation, Cultural Pluralism, and Social Exclusion among Ethnocultural Groups in Vancouver', *Urban Geography* 24: 16–44.

Hoover, E.M., and R. Vernon. 1962. *Anatomy of a Metropolis: The Changing Distribution of People and Jobs within the New York Metropolitan Region*. Cambridge, Mass.: Harvard University Press.

Hoskins, W.G. 1955. *The Making of the English Landscape*. London: Hodder and Stoughton.

Jones, K., and T. Hernandez. 2006. 'Dynamics of the Canadian Retail Environment', in T. Bunting and P. Filion, eds, *Canadian Cities in Transition: Local Through Global Perspectives*, 3rd edn. Toronto: Oxford University Press, 287–305.

—— and J. Simmons. 1993. *Location, Location, Location: Analyzing the Retail Environment*, 2nd edn. Toronto: Nelson Canada.

Kaplan, D.H., and K. Woodhouse. 2004. 'Research in Ethnic Segregation I: Causal Factors', *Urban Geography* 25: 579–85.

Ley, D. 1996. *The New Middle Class and the Remaking of the Central City*. Oxford: Oxford University Press.

Linden, E. 1993. 'Megacities', *Time*, 11 Jan., 30–40.

Lowder, S. 1994. 'Urban Development in Ecuador', *Geography Review* 7, 4: 2–6.

McCann, E.J. 2003. 'Framing Space and Time in the City: Urban Politics, Urban Policy, and the Politics of Scale', *Journal of Urban Affairs* 25: 159–78.

MacCannell, D. 2005. 'Silicon Values: Miniaturization, Speed, and Money', in C. Cartier and A.A. Lew, eds, *Seductions of Place: Geographical Perspectives on Globalization and Touristed Landscapes*. New York: Routledge, 91–102.

MacMillan, B. 1996 'Chinese Cities', *Geography Review* 9, 5: 2–6.

Martin, D.G. 2003. 'Enacting Neighborhood', *Urban Geography* 24: 361–85.

Middleton, N. 1994. 'Environmental Problems in Mexico City', *Geography Review* 7, 4: 16–18.

Murdie, R.A., and C. Teixeira. 2006. 'Urban Social Space', in T. Bunting and P. Filion, eds, *Canadian Cities in Transition: Local Through Global Perspectives*, 3rd edn. Toronto: Oxford University Press, 154–70.

Murphy, R.E., J.E. Vance, and B.J. Epstein. 1955. 'Internal Structure of the CBD', *Economic Geography* 31: 21–46.

Nelson, R.L. 1958. *The Selection of Retail Locations*. New York: Dodge.

Pacione, M. 2005. *Urban Geography: A Global Perspective*, 2nd edn. New York: Routledge.

Popp, H. 1976. 'The Residential Location Decision Making Process: Some Theoretical and Empirical Considerations', *Tijdschrifte voor Economische en Sociale Geografie* 67: 300–6.

Peressini, T., and L. McDonald. 2000. 'Urban Homelessness in Canada', in T. Bunting and P. Filion, eds, *Canadian Cities in Transition: The Twenty-First Century*, 2nd edn. Toronto: Oxford University Press, 525–43.

Ray, B., and D. Rose. 2000. 'Cities of the Everyday: Socio-spatial Perspectives on Gender, Difference, and Diversity', in T. Bunting and P. Filion, eds, *Canadian Cities in Transition:*

The Twenty-First Century, 2nd edn. Toronto: Oxford University Press, 502–24.

Robson, B.T. 1975. *Urban Social Areas*. New York: Oxford University Press.

Rose, D. 1988. 'A Feminist Perspective on Employment Restructuring and Gentrification: The Case of Montreal', in J. Wolch and M. Dear, eds, *The Power of Geography: How Territory Shapes Social Life*. Winchester, Mass.: Unwin Hyman, 118–38.

Rowland, D. 1982. 'Living Arrangements and the Later Family Life Cycle in Australia', *Australian Journal of Ageing* 1: 3–6.

Shaw, A. 2003. 'Planning and Local Economies in Navi Mumbai: Processes of Growth and Governance', *Urban Geography* 24: 2–15.

Simmons, J., and K. Jones. 2003. 'Growth and Change in the Location of Commercial Activities in Canada: With Special Attention to Smaller Places', *Progress in Planning* 60: 55–74.

Smith, N. 1996. *The New Urban Frontier: Gentrification and the Revanchist City*. New York: Routledge.

Smith, R.L. 1985. 'Activism and Social Status as Determinants of Neighbourhood Identity', *Professional Geographer* 37: 421–32.

Tunbridge, J.E. 2001. 'Ottawa's Byward Market: A Festive Bone of Contention?', *Canadian Geographer* 45: 356–70.

United Nations Human Settlements Program. 2004. *The State of the World's Cities, 2004/2005: Globalization and Urban Culture*. Sterling, Va: Earthscan.

Veness, A.R. 1992. 'Home and Homelessness in the United States', *Environment and Planning D* 10: 445–68.

Ventimiglia, A. 2007. 'The Inside Story: Curbing Gentrification', *Canadian Geographic* 127, 3: 9–12.

Watts, J. 2006. 'Invisible City', *The Guardian*, 15 Mar. At: <www.guardian.co.uk/world/2006/mar/15/china.china>.

Wilkinson, R., and K. Pickett. 2009. *The Spirit Level: Why More Equal Societies Almost Always Do Better*. London: Allen Lane.

Wolch, J., and G. DeVerteuil. 2001. 'Landscapes of the New Poverty Management', in J. May and N.J. Thrift, eds, *TimeSpace*. New York: Routledge, 149–68.

CHAPTER 14

Bone, R.M. 2009. *The Canadian North: Issues and Challenges*, 3rd edn. Toronto: Oxford University Press.

Britton, J. 1996. *Canada and the Global Economy: The Geography of Structural and Technological Change*. Montreal and Kingston: McGill-Queen's University Press.

Clark, G. 2007. *A Farewell to Alms: A Brief Economic History of the World*. Princeton, NJ: Princeton University Press.

Daniels, P.W. 1985. *Service Industries: A Geographical Appraisal*. New York: Methuen.

Dicken, P. 1992. *Global Shift: The Internationalization of Productive Activity*. London: Chapman.

Edwards, M. 1997. 'China's Gold Coast', *National Geographic* 191, 3: 2–31.

Freeberne, M. 1993. 'The Northeast Asia Regional Development Area: Land of Metal, Wood, Water, Fire and Earth', *Geography* 78: 420–32.

Friedmann, J. 1972. 'A General Theory of Polarized Development', in N.M. Hansen, ed., *Growth Centers in Regional Economic Development*. New York: Free Press, 82–102.

Friedrich, C., trans. 1929. *Alfred Weber's Theory of the Location of Industries*. Cambridge, Mass.: Harvard University Press.

Graham, J., et al. 1988. 'Restructuring in U.S. Manufacturing: The Decline of Monopoly Capitalism', *Annals, Association of American Geographers* 78: 473–90.

Grubel, H.G., and K. Walker. 1989. *Service Industry Growth: Causes and Effects*. Vancouver: Fraser Institute.

Holmes, J., T. Rutherford, and S. Fitzgibbon. 2005. 'Innovation in the Automotive Tool, Die and Mould Industry: A Case Study of the Windsor-Essex Region', in D.A. Wolfe and M. Lucas, eds, *Global Networks and Local Linkages: The Paradox of Cluster Development in an Open Economy*. Montreal and Kingston: McGill-Queen's University Press, 119–53.

Hoover, E.M. 1948. *The Location of Economic Activity*. Toronto: McGraw-Hill.

Kennelly, R.A. 1968. 'The Location of the Mexican Steel Industry', in R.H.T. Smith, E.J. Taafe, and L.J. King, eds, *Readings in Economic Geography: The Location of Economic Activity*. Chicago: Rand McNally, 126–57.

Massey, D. 1984. *Spatial Divisions of Labour: Social Structures and the Geography of Production*. New York: Methuen.

Mellinger, A.D., J.D. Sachs, and J.L. Gallup. 2000. 'Climate, Coastal Proximity, and Development', in G.L. Clark, M.P. Feldman, and M.S. Gertler, eds, *The Oxford Handbook of Economic Geography*. Toronto: Oxford University Press, 169–94.

Morrish, M. 1994. 'China Takes the Road to Market', *Geographical Magazine* 67, 4: 43–5.

———. 1997. 'The Living Geography of China', *Geography* 82: 3–16.

Pollard, J.S. 1989. 'Gender and Manufacturing Employment: The Case of Hamilton', *Area* 21: 377–84.

Pred, A. 1966. *The Spatial Dynamics of U.S. Urban-Industrial Growth, 1800–1914*. Cambridge, Mass.: MIT Press.

Rigg, J., et al. 2009. 'The World Development Report 2009 "Reshapes Economic Geography": Geographical Reflections', *Area* 34: 128–36.

Robinson, D. 2005. 'Sudbury's Mining Supply and Service Industry: From a Cluster "In Itself" to a Cluster "For Itself"', in D.A. Wolfe and M. Lucas, eds, *Global Networks and Local Linkages: The Paradox of Cluster Development in an Open Economy*. Montreal and Kingston: McGill-Queen's University Press, 155–76.

Rostow, W.W. 1960. *The Stages of Economic Growth*. Cambridge: Cambridge University Press.

Stafford, H.A. 1985. 'Environmental Protection and Industrial Location', *Annals, Association of American Geographers* 75: 227–40.

Webber, M.J. 1984. *Industrial Location*. Beverly Hills, Calif.: Sage.

World Bank. 2009. *World Development Report: Reshaping Economic Geography*. Washington: World Bank.

Yeung, Y.-M., and F.-C. Lo. 1996. 'Global Restructuring and Emerging Urban Corridors in Pacific Asia', in F.-C. Lo and Y.-M. Yeung, eds, *Emerging World Cities in Pacific Asia*. New York: United Nations University Press, 17–47.

CONCLUSION

Bunkše, E.V. 2004. *Geography and the Art of Life*. Baltimore: Johns Hopkins University Press.

Buttimer, A. 1974. *Values in Geography*. Washington: Association of American Geographers, Commission on College Geography, Resource Paper No. 24.

Calamai, P. 2006. 'Seafood Species Face Extinction', *Toronto Star*, 3 Nov., A3.

Cloke, P., and R.J. Johnston, eds. 2005. *Spaces of Geographical Thought*. Thousand Oaks, Calif.: Sage.

Dear, M., and J. Wolch. 1989. 'How Territory Shapes Social Life', in J. Wolch and M. Dear, eds, *The Power of Geography: How Territory Shapes Social Life*. Boston: Unwin Hyman, 3–18.

Gregory, D. 1994. *Geographical Imaginations*. Oxford: Blackwell.

Haggett, P. 1990. *The Geographer's Art*. Oxford: Blackwell.

Harvey, D.W. 1973. *Social Justice and the City*. London: Arnold.

Jenkins, S. 1992. 'Four Cheers for Geography', *Geography* 77: 193–7.

Johnston, R.J. 1991. *A Question of Place: Exploring the Practice of Human Geography*. Oxford: Blackwell.

———. 2005. 'Geography—Coming Apart at the Seams?', in N. Castree, A. Rogers, and D. Sherman, eds, *Questioning Geography*. New York: Blackwell, 9–25.

Marsh, G.P. 1965 [1864]. *Man and Nature; or, Physical Geography as Modified by Human Action*. Cambridge, Mass.: Belknap Press.

Mills, C. Wright. 1959. *The Sociological Imagination*. New York: Oxford University Press.

Morrill, R.L. 1987. 'A Theoretical Imperative', *Annals, Association of American Geographers* 77: 535–41.

Peattie, R. 1940. *Geography as Human Destiny*. Port Washington, NY: Kennikat Press.

Prince, H.C. 1961. 'The Geographical Imagination', *Landscape* 11, 2: 22–5.

Schnaiberg, A., and K.A. Gould. 1994. *Environment and Society: The Enduring Conflict*. New York: St Martin's Press.

Stoddart, D.R. 1987. 'To Claim the High Ground: Geography for the End of the Century', *Transactions, Institute of British Geographers* (new series) 12: 327–36.

Viles, H. 2005. 'A Divided Discipline?', in N. Castree, A. Rogers, and D. Sherman, eds, *Questioning Geography*. New York: Blackwell, 26–38.

Index

● ● ● ● ●